Brain Anatomy and Neurosurgical Approaches

Eberval Gadelha Figueiredo
Nícollas Nunes Rabelo
Leonardo Christiaan Welling
Editors

Brain Anatomy and Neurosurgical Approaches

A Practical, Illustrated, Easy-to-Use Guide

Springer

Editors
Eberval Gadelha Figueiredo
Division of Neurological Surgery
University of Sao Paulo
São Paulo, São Paulo, Brazil

Nícollas Nunes Rabelo
Neurosurgery Department
University of Sao Paulo
São Paulo, Brazil

Leonardo Christiaan Welling
Department of Neurosurgery
State University of Ponta Grossa
Ponta Grossa, Paraná, Brazil

ISBN 978-3-031-14819-4 ISBN 978-3-031-14820-0 (eBook)
https://doi.org/10.1007/978-3-031-14820-0

This Springer imprint is published by the registered company Springer Nature Switzerland AG
The registered company address is: Gewerbestrasse 11, 6330 Cham, Switzerland

Foreword

When Eberval Gadelha Figueiredo asked me to contribute a chapter on the far lateral approach to *Brain Anatomy and Neurosurgical Approaches*, I was pleased to see this effort dedicated to detailing the anatomical knowledge that underlies our essential surgical approaches. This knowledge is crucial when dealing with challenging lesions and situations in the operating room.

Brain Anatomy and Neurosurgical Approaches is a beautiful compilation of 29 illustrated chapters that takes the reader on a journey from the superficial anatomy (scalp, skull, cortical gyri, and sulci) to the cisternal and ventricular anatomy to the central core and brainstem. Dr. Figueiredo and his coeditors, Nícollas Nunes Rabelo and Leonardo Christiaan Welling, do not shy away from the most complex anatomical areas, such as the skull base, temporal bone, foramen magnum, orbit, and cavernous sinus. The book covers the full range of approaches in the armamentarium of modern tumor and vascular neurosurgeons. Furthermore, it is written by authoritative experts in the field. The contributions of Dr. Figueiredo's Brazilian colleagues further advance their traditions of excellence in open neurosurgery and exquisite cadaveric dissection.

I congratulate the authors and editors of this textbook on producing a valuable resource for residents and practicing neurosurgeons to broaden their repertoire and increase their confidence based on the anatomical foundations of neurosurgical approaches in the tradition of the great neuroanatomist Albert L. Rhoton Jr. This book captures the key surgical anatomy needed to advance one's microsurgical skills. I expect we will see this book in the operating rooms, laboratories, and libraries of neurosurgeons around the world.

Michael T. Lawton

Department of Neurosurgery
Barrow Neurological Institute
St Joseph's Hospital and Medical Center
Phoenix, AZ, USA

Preface

Neurosurgery is an extremely attractive surgical specialty, and it is constantly evolving, whether with new surgical instruments, microscopes, exoscopes, neuronavigation systems, neuroendoscopy, or intraoperative imaging. It is undoubtful that all auxiliary methods play an important role in the final result. But as the name itself refers, they are "auxiliary," and the basis of all surgical knowledge is neuroanatomy. Neurosurgery is an art that mixes with science, and the constant acquisition of knowledge and operative skills keeps us fresh and inspired. While there is a difference in natural skills among surgeons, it should be borne in mind that practice combined with study is what guarantees expertise.

When we initially conceived the ***Brain Anatomy and Neurosurgical Approaches: A Practical, Illustrated, Easy-to-Use Guide*** project, we were faced with the initial difficulty of how to make this book more attractive, how to integrate anatomy with neurosurgical practice, and which projections and views of operative field that each approach provides. For this, we have the collaboration of renowned experts who were carefully chosen, professionals with extensive anatomical and surgical knowledge and who may have assumed the current positions in the neurosurgical community after much training and study and that, in a practical way, enrich this work. We designed this book for neurosurgical residents, young neurosurgeons, and professionals with extensive experience. We hope that this book may provide the tools for improving tachnical and anatomical with knowledge and that it may impact the management and outcomes of our patients, who are the main motivators to continue in this career with so many personal and surgical challenges.

São Paulo, Brazil — Eberval Gadelha Figueiredo
São Paulo, Brazil — Nícollas Nunes Rabelo
Ponta Grossa, Paraná, Brazil — Leonardo Christiaan Welling

Acknowledgments

And in the end, it's not the years in your life that count. It's the life in your years—Abraham Lincoln

This book, entitled *Brain Anatomy and Neurosurgical Approaches—A Practical, Illustrated, Easy-to-Use Guide*, aims to offer fundamental practical knowledge for neurosurgeon, emphasizing surgical approaches and its respective surgical anatomy.

No project like this comes about without the help of several expert colleagues with outstanding microsurgical and anatomic expertise. Their efforts have the merits for all value that this book may have. We are indeed grateful to all authors and collaborators for their valuable work. In addition, we would like to thank the editorial and production teams at *Springer Nature*, as always, an extremely talented and most pleasant group to work with. Notwithstanding, special thanks to our family, parents, wives, sons, and friends who supported us in harsh moments and always believed in this project.

Finally, we would like to thank our former teachers and mentors for their guidance, encouragement, and friendship.

Eberval Gadelha Figueiredo, MD, PhD
Nícollas Nunes Rabelo, MD, PhD
Leonardo Christiaan Welling, MD, PhD

Contents

About the Editors

Eberval Gadelha Figueiredo, MD, PhD completed a fellowship in microsurgery at the Barrow Neurological Institute (USA). He holds a PhD in sciences and is an associate professor of neurosurgery and lecturer in the School of Medicine at the University of São Paulo (USP), as well as head of the Vascular Neurosurgery Group, Division of Neurological Surgery (HCFMUSP). He has been postgraduate coordinator at the University of São Paulo's Department of Neurology since 2015 and chief editor of Brazilian Neurosurgery, the official Brazilian neurosurgery journal, since 2011. He is the president of Brazilian Neurosurgical Society (2021–2022) and associate member of Harvard Medical School and Harvard University Alumni since 2015.

Nícollas Nunes Rabelo, MD, PhD is a neurosurgeon and member of the Brazilian Society of Neurosurgery and SBN Academic League Director (2021–2022). He has a postgraduate qualification in Neurocritical Care Medicine from the Sirio Libanês Hospital and fellowship in vascular and skull base neurosurgery at the University of São Paulo (HCFMUSP); PhD and Post PhD student from School of Medicine at the University of São Paulo (FMUSP) and is an adjunct researcher in the same University; MBA in health management from Fundação Getúlio Vargas and Sinpain pos graduated. He is professor at Atenas Medical School and is a postgraduate from Neurocritical Care Medicine of Sirio Libanês—IEP. He is member of Brazilian Neurosurgical Society board as SBN academic leagues (2021–2022).

Leonardo Christiaan Welling, MD, PhD is a neurosurgeon and member of the Brazilian Society of Neurosurgery. He obtained a postgraduate qualification in neurointensive care from Sirio Libanês Hospital (São Paulo) and completed a fellowship in microneuroanatomy at the Portuguese Beneficence Hospital (São Paulo). He holds a doctoral degree in medical sciences from the School of Medicine at the University of São Paulo (FM–USP) and completed the same university's postdoctoral neurology program. He is a member of the Congress of Neurological Surgeons (USA) and a Professor of Neurosurgery at the State University of Ponta Grossa (UEPG).

Basic Concepts in Microsurgery

Gustavo Badino Krahembühl, Nicollas Nunes Rabelo, Leonardo Christiaan Welling, and Eberval Gadelha Figueiredo

1 Introduction

The term "expert" is practically all the dictionaries of all languages. It refers to an individual with special skills or knowledge who master a certain human knowledge or practice. However, in a popular language, it takes on different meanings. The expert is also not the "know-it-all" who knows a little bit of everything; he has knowledge and information about various subjects, but has not mastered any of them. This is a generalist; he only stays on the surface of each area and does not develop the ability to solve anything complex or understand anything in depth. While he is a "know-it-all," he is a "dominate-nothing."

The expert digs deeper into an area of knowledge and dominates it, to the point of fully competently exercising the knowledge he has developed.

According to Confucius, ***"The essence of knowledge is to apply it, once possessed."*** This is different from the memorizer, which has all the information but does not know how to apply any. Note that the expert does not have only theoretical knowledge; in addition to knowledge, he has competence; moreover, to have action competence and expertise, not just memorize or understand. In this scenario, in addition to knowledge, practice and laboratory training make the "neurosurgery expert" or "expert neurosurgeon" differ from professionals in other areas of expertise, whether in medicine or beyond.

G. B. Krahembühl (✉) · E. G. Figueiredo
Neurosurgery Department, University of São Paulo, São Paulo, PR, Brazil

N. N. Rabelo
Neurosurgery Department, Athens University Center, Passos, MG, Brazil

L. C. Welling
Neurosurgery Department, State University of Ponta Grossa, Ponta Grossa, PR, Brazil

E. G. Figueiredo et al. (eds.), *Brain Anatomy and Neurosurgical Approaches*,
https://doi.org/10.1007/978-3-031-14820-0_1

Microneurosurgery is not exclusively and directly linked to the surgical microscope, but is a conceptual way of drafting and executing all phases of surgery using delicate techniques for manipulating the central nervous system tissues. The surgery starts outside the operating room with careful preoperative evaluation, meticulously reviewing each moment of the procedure that is going to take place. Moreover, essential strategies include mental preparation, analysis of previous experiences, good knowledge of microanatomy, high-quality neuroanesthesia, collaboration, and operating room staff well-trained.

There are three sentences uttered by neurosurgery icons that inspire those who aspire to be an expert in microsurgery, and it could not go unmentioned at the beginning of this book.

> "*Simple, clean, while preserving normal anatomy. Being clean is quick and effective. Surgery is art - you should be one of those artists.*" Juha Hernesniemi

> "*Despite this, such complex, labyrinthine access through the skull and brain requires accurate preoperative planning and preparation of a prospective surgical concept (including anticipated variants) based on a firm knowledge of anatomy microtechniques and surgical experience. These elements constitute the art of microneurosurgery.*" Gazi Yaşargil

> "*Know and respect the brain, the greatest of divine creations, touch the specimens, know how to use its natural ways, learn to enter, solve the problem and leave without causing damage.*" Evandro de Oliveira

2 Become an Expert Cost Time

As in most surgical specialties, neurosurgery is characterized by evolutions and achievements as remarkable as the history of humankind. The relationship between technological evolutions and scientific advancement allows a better execution of contemporary neurosurgeries.

According to the quote from T.S. Eliot, "*Time present and time past are both perhaps present in time future and time future contained in time past …*"; thus, denoting the importance of "time" in both the literary and figurative sense of the word. It is evident that for the practice of neurosurgery, especially microneurosurgery in state of the art, "TIME" is undoubtedly the main factor for excellence. Thanks to the dedicated time to laboratory training, the microneurosurgeon can improve and evolve, dedicate time to studying microanatomy, dedicate time to patients, and abdicated time from personal issues. All of this "time" allows excellence and constant evolution in the noble art of neurosurgery. The time is too expensive, and time is priceless. The vast experience is acquired after a long time of "flying hours" in the microscope and the operating room, treating more distinct pathologies that affect the nervous system, whether central or peripheral. Most of the surgeons know the concept "more I know, more I see and more I see, more I know, "; more primordial to mention in this timeline the importance of the pioneers of microsurgery and microneurosurgery, people who were ahead of their time:

Peardon Donaghy, MD. Professor of Neurosurgery University of Vermont, Burlington.
Theodor Kurze, MD. Professor of Neurosurgery University of Southern California, Los Angeles.
Harry J. Buncke, Jr. MD. Professor of Plastic and Reconstructive Surgery University of Stanford, Palo Alto, California.
Julius H. Jacobson, II, MD. Professor of Vascular Surgery Mount Sinai Hospital, New York, NY.
William F. House, MD. Professor of Otolaryngology University of Southern California, Los Angeles [1].

Due to their contributions to the understanding and better understanding of the most diverse pathologies, these professionals were fundamental in implementing new technologies, such as the microscope, as a tool in treating diseases that needed greater visualization and precision in the surgical approach.

Historically, the first record of the use of the scope was by Lougheed and Tom of Toronto. In 1961, they published a paper method of introducing blood into the subarachnoid space in the region of the circle of Willis in dogs", but Theodore Kurze in Los Angeles performed the first clinical application into the nervous system. This surgeon started the microsurgical approach to the petrous bone in the middle fossa with Dr. House. In Burlington, more exactly in 1960, Jacobson and Donaghy performed an endarterectomy on the middle cerebral artery. A few years later, in Zürich, the work of Krayenbühl and his student Yasargil who became known as the " Father of the Modern Microneurosurgery" demonstrated worldwide how to operate intracranial vessels, especially intracranial aneurysms and arteriovenous malformations. Since then, much time has passed; the technical art of neurosurgery has evolved significantly from when the discipline was founded [1, 2].

3 Watching and Learning

It is possible to achieve progress and implement the technique by finding self-reflection and absorbing the excellence of other surgeons, respect for constructive feedback from colleagues, and the ability to learn from mistakes and experiences of one's own and other professionals. These factors are critical for each surgeon's growth and personal development. During the residency program, some problems cannot be completely resolved. For example, manual dexterity and temperamental skills are needed to improve and pursuit with passion during every phase of the neurosurgeon career. Each surgery must be meticulously observed by watching older and more experienced colleagues operate, critically reviewing their surgical videos, and trying alternative and potentially innovative methods to enhance their surgical delicacy and precision to understand the pathologies and related anatomy better [3–5].

4 Decision-Making Process

The execution of operative maneuvers results from a decision-making process that involves a continuum, ranging from intuitive (subconscious) to analytical (conscious), that will produce a specific result [3, 4]. The quality of that result will be affected by the ability to switch between analytical and intuitive processes more effectively [3], as it is shown by professional athletes, as higher the performance and experience, more intuitive and subconscious are their actions (for example, IndyCar Series Drivers can repeat the same movements performed during the driving experience (real), as with their eyes closed during simulator). The technical maturation of the microneurosurgeon involves some stages. Initially, the operative procedure is interpreted as a series of actions that must be performed for a good outcome to be achieved. In the second phase, the mental formation of an image with pattern recognition for the operative situation itself is important. Finally, whether subconsciously or automatically, making a surgical decision implies a deeper understanding of the immediate consequences of the intraoperative maneuvers to be performed. These three stages are developed during the residency period [3–5].

The learning process may never end by itself, but it is markedly accelerated during residency, providing a reasonable time for studying intraoperative decision-making to deal with the transition from analytical to the subconscious decision-making process. Moreover, this change is needed because the conscious analytical process consumes a significant portion of a surgeon's mental capacity during the surgery. The continuous conscious analytical process, if not well-managed, may decrease the operative efficiency [3, 4], improving fatigue and increasing surgeon stress which can lead to irreversible results. To achieve a greater result, the surgeon must look for inconsistencies and remains flexible in adjusting the operative plan, when necessary, look for consistent actions dealing with unexpected situation aiming surgical rearrangement strategy to achieve a positive outcome [3].

An expert surgeon uses initially during the procedure a subconscious operative mode. When an operative variation or high-risk situation appears, at that moment, the surgeon's level of *self-awareness* will change to conscious operative mode to adapt the operative strategy. This dynamic cycle based on intraoperative findings and adaptive maneuvers will show the neurosurgeon's expertise. The main focus during the years of learning and training is improving that ability [3, 4].

Technical fluency and efficiency are the ground base to become an expert, and the time invested in development technical learning plays an essential role in increasing the surgeon's efficiency. One of the ways to gain more experience and achieve expertise is when always it is possible to review the own surgical videos critically, aiming to do better and quicker every step of the procedure [3]. Performing new strategies abandon time-consuming, unnecessary approaches and maneuvers, handling the pathology appropriately with minimal disruption of the patient's normal/healthy tissues [3–5].

5 Why to Training?

Nowadays, the expectations on neurosurgeons are extremely high based on excellent results, even with minimal complication rates. These events are highly dependent on individual technical skills based on the efficiency of the previous training, which will provide the surgeon to deliver an outstanding result. On the other hand, there are some training issues, especially with animal models. The training is getting harder because of an increasing moral and ethical pressure to reduce its use. These resources were in the past fundamentally relevant to training, and it has changed the practice of microsurgical, helping a generation of microsurgeons to improve and develop their technique. Some animal models were introduced and widely used during the mid-twentieth century. Since then, high numbers of ex vivo training concepts and quality control measures have been proposed, aiming to reduce the number of animals without compromising the quality and outcome of the training [6]. In the last decade, surgical training is being revolutionized by two concepts: objective assessments of surgical skills and development of surgical skills in a simulation laboratory setting. Acquiring surgical skills in the laboratory setting can help move the microsurgical learning curve from the patient to the laboratory, and this will, in turn, improve patient safety substantially [7]. The first is the objective assessments of surgical skills. The second is the development of surgical skills in a simulation laboratory setting [7] using a unique set of instruments and, therefore, a unique set of surgical principles that differ from standard surgical practices. Since the introduction by Prof. Yaşargil in the 1960s, microneurosurgical techniques have been the essentials of neurosurgical practice, and it is well-known that the relation of these principles to work developed in the laboratory [8] (Fig. 1a, b). It also involves very refined movements that predominantly employ the intrinsic muscles of the hands as opposed to the wrist movements applied in general surgery. It is widely acknowledged that microsurgery is technically more demanding than general surgery [9, 10]. The combination of stereoscopic vision, visuospatial skills, precision and, dexterity is imperative. These factors are the keystone that must be aimed at during the learning curve [7, 10, 11].

It is very important to emphasize that training greatly influences both the practical/technical and the posture of the neurosurgeon. It is important to note that the posture of an expert surgeon precedes him even before starting the procedure; it is through a proper posture that the most stressful situations, for example, rupture of an aneurysm, will shape the outcome of the situation, whether positive or negative; a neurosurgeon should be seen as a captain of a ship; if the captain of the ship does not demonstrate a correct posture (Disruptive behavior), most likely all your subordinates, whether they are anesthetists, assistants, residents, surgical team, and nursing staff, will succumb and will not be able to do their jobs and help effectively. The operating room is an intimidating setting for most neurosurgery residents, particularly during their early years of residency. Training is essential to be prepared for these occurrences, and effectively communicating can prevent or minimize the risks

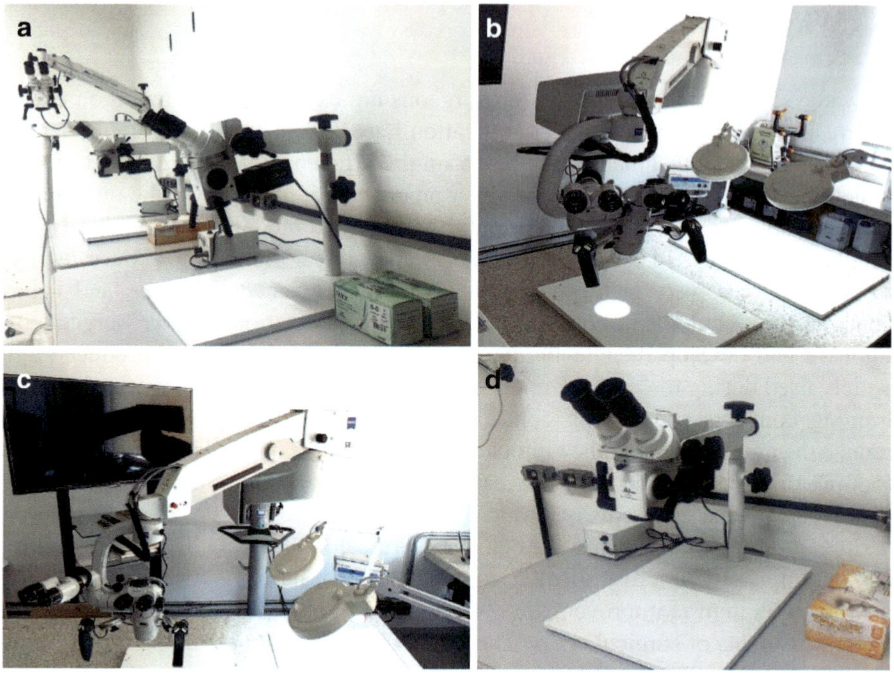

Fig. 1 (**a**) Microsurgery laboratory residence program of CHS – Sorocaba –SP. (**b**) Microsurgery laboratory University of São Paulo. (**c**) Surgical microscope. (**d**) Table training microscope

associated with these events to maximize patient safety; the appropriate behaviors for a surgeon always with respect, self-awareness, and kind speech will define operating room climate and appropriate etiquette [12, 13].

6 Materials and Tools

It is fundamental concerning the microsurgical instruments, the importance of quality, standardization, and material maintenance. When possible, high-quality instruments should always be used from the very beginning of training. These instruments are essential for performing microsurgery, and it is very common for beginner trainees to start with damaged or low-quality "training" instruments for practicing. Nevertheless, inadequate equipment can force candidates to compensate the deficit within the inferior instruments with unnecessary, awkward movements, and it may promote the acquisition of incorrect handling techniques leading to addictive behavior. Once learned, incorrect techniques are very difficult to "reprogram" in a later phase of training and clinical practice. Normally, using a standardized set of

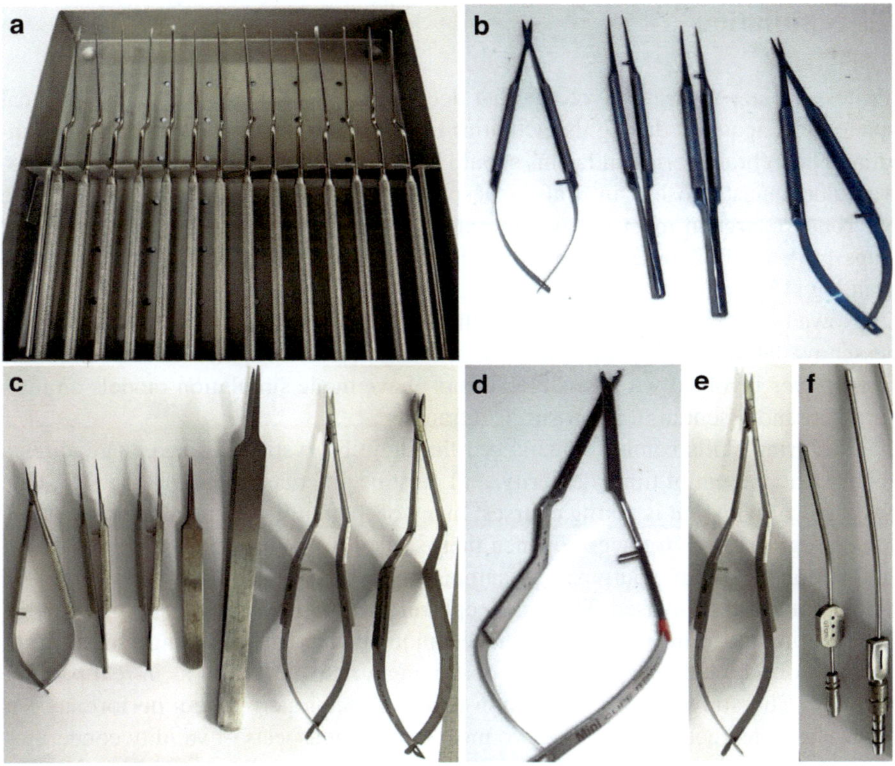

Fig. 2 (**a**) 5 Rhoton's Micro dissectors. (**b**) Curved microscissors, Straight microforceps +, Straight microscissors. (**c**) Needle holder, Straight microforceps. (**d**) Microvessel clip applicator. (**e**) Yasargil microscissors. (**f**) Microsurgical suction tubes

instruments (Fig. 2b) for training will respond to all the demands of many centers to operate with similar "workstations" with standard microscopes (Fig. 1d) and instrument setups (Fig. 2b).

An important part of the development of microsurgical expertise is the instruments care during the training activities. Learning the proper care to handle and clean these delicate instruments must also be part of the pretraining phase to reduce damages risks and costs [6, 14]. Some materials must be let in counter to the beginning of the training (Surgical microscope – Fig. 1c: Straight microforceps, Fig. 2c: Curved microforceps, Micro dissectors, Fig. 2a: Needle holder, Fig. 2c: Straight microscissors, Curved microscissors, Fig. 2e: Microsurgical suture materials like flat body needles, 8-0 to 12-0 sutures nylon or polypropylene (monofilament)) [6]. In vivo training (Microsurgical suction tubes—Fig. 2f, Vessel-dilating fórceps, Irrigation tips, Microvessel clamps, and Microvessel clip applicator [6]—Fig. 2d) is essential to develop a good training methodology.

7 Simulation

Professor Gazi Yasargil advocates that deep knowledge of the microneurosurgical anatomy is acquired during the activities in cadaveric laboratories, and gentle handling of cerebral arteries and veins should be practiced first when performing microvascular anastomoses in rats. Despite basic theoretical knowledge, the microneurosurgeon must always be trained to act in small, deeper, and profound gaps [15]. In this context, simulation has been an integral part of microsurgical training [15].

Nowadays, two situations present themselves more frequently; the first is the lesser availability of time to dedicate to training in the workplace. Furthermore, the legal issues involved with possible "errors" have made simulation models an integral part and essential of microsurgical training.

Mastering skills cannot be achieved through observation alone as it requires a significant amount of time, dexterity, and must be practiced regularly [15, 16]. One way to deal with that is taking courses; many of them use various models in a stepwise progression as trainees advance their skills [16]. When discussing the use of simulation for training purposes, it is important to consider both the validity and the fidelity of the method used. These concepts refer to the appropriateness of the model in preparing the trainee to perform the technique in the clinical setting and the accurateness of the method employed in portraying a realistic model as in real practice [15–17]. The ability of the simulation model to accurately assess performance is known as construct validity. The two most important aspects of validity concerning microsurgical simulation are predictive and concurrent validity, that is, the ability of the simulation model to predict future performance at the task and the comparison with models that assess a similar outcome. However, these are perhaps the most difficult to assess [15].

Some studies have shown that construct validity is reliably seen in many models used for microsurgical simulation training [18–20]. As mentioned previously, predictive validity is challenging to measure. In a systematic review by Dumestre et al. [21], the only model successfully establishing this was the rat femoral artery that is widely recognized as the benchmark in microsurgical simulation [15, 16, 21].

Fidelity is defined as the ability of the practice scenario to represent and recreate a real-life situation. It will be used as a simulation model (low-fidelity models tend to use nonliving material such as silicone tubing), (medium-fidelity models use cadaveric animal or human tissue), and (high-fidelity models are performed on live animal tissue such as a rat femoral artery) [15].

Basic models are usually low-fidelity simulations of simple tasks, such as simple microsurgical suturing, allowing the learner to become familiarized with the microscope and microsurgical instruments and how to tie microsurgical knots. These simulations have low cost, are easy to construct [22, 23], and reduce the number of live animals required to teach the more complex microsurgical techniques. Because of the low fidelity of models, there is a limit to the level and quality of skill that models can reach. Helping in the beginning, as it showed in many microsurgical simulation courses, uses simple interrupted sutures on a straight incision made on a rubber glove as the initial microsurgical task to allow the learner to familiarize with the microsurgical instruments and learn how to perform a simple microsurgical suture [15, 23, 24].

Southern et al. [25] and Yenidunya et al. [26] describe microsurgical simulation models where the trainee passes a microsurgical suture through fenestrations in beads fixed at different heights and angles. This task has the advantage of requiring the trainee to learn how to adjust and deal with the microscope focus and handle sutures and instruments in different angles and planes [15].

In another simulation model, Hosnuter et al. [27] describe the use of polyethylene stretch film membrane to simulate the task of adventitial stripping. A slightly more advanced model, which is considerably more expensive though, is the practice [28, 29]. This model allows the practice of nerve repair and arterial and venous anastomosis using polyethylene tubes combined with a similar artificial adventitia with the addition of luer-lock circulation allows assessment of the quality of anastomosis [15]. Another inanimate material that has been described for microsurgical simulation training is the Japanese noodle ("shirataki konnyaku"), which according to Prunieres [30], is structurally similar to cadaveric animal tissue. Malik et al. [31] have used various inanimate models to practice basic microsurgical skills in the laboratory and at home using three types of magnification: tabletop microscope, jeweler's microscope, and iPad tablet. They found that there was no statistically significant difference in the development of basic microsurgical skills between the types of magnification [15].

Intermediate models are high-fidelity simulations that allow the learner to dissect small vessels and perform a microanastomosis in an environment that closely resembles microsurgical practice in live humans. These models allow the trainee to learn correct tissue handling and practice vessel anastomosis as well as form an understanding of the physiology of blood flow [15].

8 Cadaveric Animal Tissues

Similar to inanimate materials, animal cadaveric tissue reduces the requirement for alive animals to teach microsurgical skills. Animal cadaveric tissue offers a more realistic representation of real clinical practice regarding tissue consistency than synthetic materials. Turkey or chicken cadaveric tissue is a very popular model used by many authors [32–36]. These tissues offer the opportunity for trainees to practice arterial and venous anastomosis on vessels ranging from 1.25 to 1.69 mm diameter in the wing model, 2.04 mm artery, and 1.45 mm vein in the thigh model, and 4.5 mm in the chicken aorta [34, 35, 37]. The chicken wing has the added advantage of the opportunity to perform a nerve repair [38]. The materials for this model are cheap, easily accessible and, closely resemble clinical practice concerning tissue quality and vessel size [15, 35, 36].

Cadaveric animal tissue can also be obtained from animals used in other laboratory experiments, such as cryopreserved rat vessels [39, 40] and spleens obtained following splenectomy on pigs [41]. Another form of cadaveric animal tissue that can be used to practice microsurgery is that of the earthworm, an easily accessible tissue in various sizes that closely resembles the texture of the arterial wall [42]. Bovine placental veins can also be used and have higher anatomical relevance and more similar consistency and texture to real human vessels than other cadaveric animal tissue, without the requisite for written maternal consent, which is needed for practice on human placental tissue [15, 18].

The addition of a circulation system, such as cannulation of the vessels and connection to a membrane pump that creates pulsatile flow, can increase the fidelity of the simulation and allow tests of patency [43, 44]. Simple patency tests can also be performed by injecting fluid into the cadaveric vessels following the anastomosis [15].

8.1 Cadaveric Human Tissues

Because of the correct anatomy, cadavers are being used in many courses worldwide and adopted by many centers in Europe and the United States. The human cadaveric advantage compared to animals is the correct anatomy, consistency, and tissue texture [45, 46].

This, therefore, gives high face, content, and construct validity. Perfused fresh human cadavers, achieved with the cannulation of large vessels and perfusion of the circulation with indocyanine green, can create a high-fidelity simulation that closely resembles alive human tissue [45]. Another alternative that offers tissue of similar qualities to clinical practice is the human umbilical cord or placenta [15].

8.2 Live Animal Models

The main purpose of animal models is to reproduce disease processes, test biomedical devices or compounds, and help provide answers concerning disease pathogenesis, prevention, and treatment in human and veterinary medicine. However, animal models are also fundamental in the training of biomedical scientists and even more important for the preclinical training of future surgeons. Live animal models are still considered the gold standard for microsurgical simulation because of their close resemblance to clinical microsurgical practice. These models allow practicing the dissection of vessels and flaps and performing an anastomosis on living tissue with similar physiology to real practice on humans (perfusion and clotting.) There are, however, several disadvantages to the animal model that have led to the development of alternative simulation models. Lately, several ethical issues regarding the use of animals for research and training need to be considered, and several rules and regulations need to be followed, which makes the process more complicated and expensive [15, 16, 23]. Live animal models allow practicing advanced microsurgical skills by raising flaps based on the rat epigastric vessels that have a diameter of approximately 0.2 to 0.3 mm [47–49]. Bodin et al. [50] described a live pig simulation model for breast reconstruction such as the superior epigastric artery perforator, the superior gluteal artery perforator, and the transverse gracilis musculocutaneous flap [15].

While training in large animals is similar to the surgical technique used in humans, experimental microsurgery is substantially different from macrosurgery; it requires specific surgical (micro-)instruments, very fine suture materials, optical magnification, and perfect hand-eye coordination in a very small, indirect field of view [7].

9 Technology Simulating Neurosurgery

Thanks to technological advances in imaging, computation, virtual reality (VR), 3D printing, even robotic, nowadays, these tools are even more present during the initial acquisition and refinement of surgical skills based on simulation models [51].

As the field of neurosurgery continues to advance, the technology simulating tools are in constant evolution too, and it has shown to be a promising field that will help the surgeon achieve mastery and experience; now it has become clearer that the operating room is not the ideal place for the initial acquisition and refinement of surgical skills. The finesse of improving maneuvers in clinical practice can rarely be repeated if the failure occurs. Simulation affords surgeons the possibility and train with higher realistic situations thanks to the advance of technology [51].

Simulation trainers are classified as physical models, virtual reality, and mixed-reality simulators [51–53]. Animals and human cadavers, so-called physical models, have been considered standard methods for surgical training in the past [51, 54, 55]. However, they have several limitations, including biological and practical restrictions, such as biohazard safety, tissue rigidity, length of preservation, ethical issues, and a growing law bureaucracy, which will make it difficult to access fresh specimens, associated with a higher cost.

Simulators are beneficial in training residents and for the continuing education of subspecialized neurosurgeons by learning in a fail-safe scenario [51, 56], reducing costs as time goes by being more affordable unlike classic models as live animals and fresh specimens, VR simulator, the mixed-reality simulator will become even more frequently during the training experience [49, 51], robots are capable of providing virtual data, superior spatial resolution and geometric accuracy, superior dexterity, faster maneuvering, and non-fatigability with a steady motion. Robotic surgery also allows simulation of virtual procedures, which turn out to be of great succor for young apprentice surgeons to practice their surgical skills in a safe environment, as previously mentioned. It also allows senior professionals to rehearse difficult cases before being involved in considerable risky procedures [57].

10 Global Rating Scales

Rating scales are objective assessment tools that elsewhere may apply task-specific rating scales to assess the microsurgical skills of trainees and evolution. They are also used to demonstrate an improvement in microsurgical skills as a measure of success of the microsurgical training course by comparing performance before and after a microsurgical course.

The history of Objective Assessment Tools in Microsurgery dates back in the literature to 1993. Starkes et al. conducted studies in Ontario surgical trainees undergoing 40-h microsurgical training courses recording various dependent measures, including accuracy of suture placement, time to complete the task, and subsequent vessel patency [35]. The researchers reported that completing a task was the most sensitive measure of experience and skill [35]. In 1998, the same group developed a standardized microsurgical test in which participants were assessed on the elementary task of performing two microsurgical sutures to repair a slit on a surgical glove [35]. This pioneering work in objective assessments in microsurgery was succeeded in 2003 by Grober et al., who adapted validated objective assessment tools in general surgery (namely global rating scales and hand motion analysis) to microsurgery.

Since the mid-90, various tools have been developed and validated on multiple microsurgical simulation models and environments. GRS or other systems assess various microsurgical skills. For example, the rating scale developed by Atkins et al. [16] assesses respect for tissue, micro instrument handling, time and motion, anastomosis performance, and vessel patency. Microsurgical performance can be evaluated by expert direct face-to-face observation or reviewing video recordings [8].

The structured assessment of microsurgery skills tool measures dexterity, visuospatial ability, operative flow, and judgment using 12 items scored on a 1 to 5 Likert scale [19, 31, 33]. The addition of an errors list to this tool provides better knowledge of the mistakes made during microsurgical practice [15].

11 Conclusion

Fortunately, the talent is distributed randomly worldwide, and there is no strict relationship with socioeconomic level. It is not possible to determine who or whom will be favored. The talent occurs by chance, which is worldwide well-observed in sports. Unfortunately, this does not apply to the surgical world, especially microneurosurgery. Much time, money, and dedication during the training are needed to achieve mastery. Besides access to appropriate materials, frequent visits to the laboratory are essential. The formation of an expert involves much more than knowing how to handle the microscope, microsurgical instruments, understanding anatomy, pathologies, and the necessary maneuvers. However, it is more related to knowing yourself better, knowing your limits, always wanting to overcome them ethically with passion, and always looking for ways to self-assess, self-criticize, and learn from the mistakes and successes of the senior surgeons. Moreover, a quote by René Leriche (la philosophie de la Chirurgie, 1951) "every surgeon carries within himself a small cemetery, where from time to time he goes to pray, a place of bitterness and regret, where he must look for an explanation for his failures."

Finally, a more comprehensive definition of what an expert is as follows:

> *"An expert is not recognized for what he says; the expert is recognized for the results he has achieved. Expertise is the maximum degree of intelligence when knowledge and competence come together and generate results. This is the definition of expertise consistent with the conceptual era. We should not go to school to memorize information; we study to be experts. The information era is over. So, it is time to study and develop your expertise".*

The authors.
Good studies!

References

1. Yaşargil MG. Microsurgery: applied to neurosurgery. Stuttgart: Georg Thieme Verlag; 1969.
2. Rhoton AL, Surgeons CN. Rhoton's cranial anatomy and surgical approaches operative techniques and instrumentation for neurosurgery. Oxford: Oxford University Press; 2019.
3. Cohen-Gadol AA. The art of microneurosurgery and passion for technical excellence. J Neurosurg. 2018;130:1023–7. https://doi.org/10.3171/2018.9.JNS182475.
4. Francis DMA. Surgical decision-making. ANZ J Surg. 2009;79(12):886–91. https://doi.org/10.1111/j.1445-2197.2009.0.
5. Crebbin W, Beasley SW, Watters DAK. Clinical decision making: how surgeons do it. ANZ J Surg. 2013;83(6):422–8. https://doi.org/10.1111/ans.12180.
6. Tolba RH, Czigány Z, Osorio Lujan S, Oltean M, Axelsson M, Akelina Y, et al. Defining standards in experimental microsurgical training: recommendations of the European Society for Surgical Research (ESSR) and the International Society for Experimental Microsurgery (ISEM). Eur Surg Res. 2017;58(5–6):246–62. https://doi.org/10.1159/000479005.
7. Ramachandran S, Ghanem AM, Myers SR. Assessment of microsurgery competency-where are we now? Microsurgery. 2013;33(5):406–15. https://doi.org/10.1002/micr.22111.
8. Hernesniemi J. Principles of microneurosurgery for safe and fast surgery. In: Sindou M, editor. Practical handbook of neurosurgery. Vienna: Springer; 2009. p. 245–56. https://doi.org/10.1007/978-3-211-84820-3.
9. Studinger R, Bradford M, Jackson I. Microsurgical training: is it adequate for the operating room? Eur J Plast Surg. 2005;28:91–3.
10. Chan WY, Srinivasan JR, Ramakrishnan VV. Microsurgery training today and future. J Plast Reconstr Aesthet Surg. 2010;63:1061–3.
11. Lascar I, Totir D, Cinca A, Cortan S, Stefanescu A, Bratianu R, Udrescu G, Calcaianu N, Zamfirescu DG. Training program and learning curve in experimental microsurgery during the residency in plastic surgery. Microsurgery. 2007;27:263–7.
12. Achor T, Ahn J. Becoming the "captain of the ship" in the OR. J Orthop Trauma. 2014;28:S18–9.
13. Cochran A, Elder WB. Effects of disruptive surgeon behavior in the operating room. Am J Surg. 2015;209:65–70.
14. VanderKam V. Care of microvascular surgical instruments. Plast Surg Nurs. 1999;19:31–4.
15. Evgeniou E, Walker H, Gujral S. The role of simulation in microsurgical training. J Surg Educ. 2018;75(1):171–81. https://doi.org/10.1016/j.jsurg.2017.06.032.

16. Chan WY, Matteucci P, Southern SJ. Validation of microsurgical models in microsurgery training and competence: a review. Microsurgery. 2007;27(5):494–9.
17. Evgeniou E, Loizou P. Simulation-based surgical education. ANZ J Surg. 2013;83(9):619–23.
18. Atkins JL, Kalu PU, Lannon DA, Green CJ, Butler PE. Training in microsurgical skills: does course-based learning deliver? Microsurgery. 2005;25(6):481–5.
19. Chan WY, Figus A, Ekwobi C, Srinivasan JR, Ramakrishnan VV. The "round-the-clock" training model for assessment and warm up of microsurgical skills: a validation study. J Plast Reconstr Aesthet Surg. 2010;63(8):1323–8.
20. Temple CL, Ross DC. A new, validated instrument to evaluate competency in microsurgery: the University of Western Ontario Microsurgical skills acquisition/assessment instrument [outcomes article]. Plast Reconstr Surg. 2011;127(1):215–22.
21. Dumestre D, Yeung JK, Temple-Oberle C. Evidence-based microsurgical skill-acquisition series part 1: validated microsurgical models—a systematic review. J Surg Educ. 2014;71(3):329–38.
22. Soto-Miranda MA, Ver Halen JP. Description and implementation of an ex vivo simulator kit for developing microsurgery skills. Ann Plast Surg. 2014;72(6):S208–12.
23. Fanua SP, Kim J, Shaw Wilgis EF. Alternative model for teaching microsurgery. Microsurgery. 2001;21(8):379–82.
24. Brosious JP, Tsuda ST, Menezes JM, et al. Objective evaluation of skill acquisition in novice microsurgeons. J Reconstr Microsurg. 2012;28(8):539–42.
25. Southern SJ, Ramakrishnan V. Dexter: a device for the assessment of microsurgical instrumentation and instruction of trainees. Microsurgery. 1998;18(7):430–1.
26. Yenidunya MO, Tsukagoshi T, Hosaka Y. Microsurgical training with beads. J Reconstr Microsurg. 1998;14(3):197–8.
27. Hosnuter M, Tosun Z, Savaci N. A nonanimal model for microsurgical training with adventitial stripping. Plast Reconstr Surg. 2000;106(4):958–9.
28. Yen DM, Arroyo R, Berezniak R, Partington MT. New model for microsurgical training and skills maintenance. Microsurgery. 1995;16(11):760–2.
29. Weber D, Moser N, Rosslein R. A synthetic model for microsurgical training: a surgical contribution to reduce the number of animal experiments. Eur J Pediatr Surg. 1997;7(4):204–6.
30. Prunieres GJ, Taleb C, Hendriks S, et al. Use of the Konnyaku Shirataki noodle as a low fidelity simulation training model for microvascular surgery in the operating theatre. Chir Main. 2014;33(2):106–11.
31. Malik MM, Hachach-Haram N, Tahir M, Al-Musabi M, Masud D, Mohanna PN. Acquisition of basic microsurgery skills using home-based simulation training: a randomised control study. J Plast Reconstr Aesthet Surg. 2017;70(4):478–86.
32. Rodriguez JR, Yanez R, Cifuentes I, Varas J, Dagnino B. Microsurgery workout: a novel simulation training curriculum based on nonliving models. Plast Reconstr Surg. 2016;138(4):739e–47e.
33. Masud D, Haram N, Moustaki M, Chow W, Saour S, Mohanna PN. Microsurgery simulation training system and set up: an essential system to complement every training programme. J Plast Reconstr Aesthet Surg. 2017;70(7):893–900.
34. Ramachandran S, Chui CH, Tan BK. The chicken aorta as a simulation-training model for microvascular surgery training. Arch Plast Surg. 2013;40(4):327–9.
35. Bates BJ, Wimalawansa SM, Monson B, Rymer MC, Shapiro R, Johnson RM. A simple cost-effective method of microsurgical simulation training: the Turkey wing model. J Reconstr Microsurg. 2013;29(9):615–8.
36. Couceiro J, Castro R, Tien H, Ozyurekoglu T. Step by step: microsurgical training method combining two nonliving animal models. J Vis Exp. 2015;99:e52625.
37. Jeong HS, Moon MS, Kim HS, Lee HK, Yi SY. Microsurgical training with fresh chicken legs. Ann Plast Surg. 2013;70(1):57–61.
38. Beth Grossman L, Komatsu DE, Badalamente MA, Braunstein AM, Hurst LC. Microsurgical simulation exercise for surgical training. J Surg Educ. 2016;73(1):116–20.

39. Lausada NR, Escudero E, Lamonega R, Dreizzen E, Raimondi JC. Use of cryopreserved rat arteries for microsurgical training. Microsurgery. 2005;25(6):500–1.
40. Kim E, Singh M, Akelina Y, Shurey S, Myers SR, Ghanem AM. Effect of microvascular anastomosis technique on end product outcome in simulated training: a prospective blinded randomized controlled trial. J Reconstr Microsurg. 2016;32(7):556–61.
41. Maluf Junior I, da Silva AB, Groth AK, et al. An alternative experimental model for training in microsurgery. Rev Col Bras Cir. 2014;41(1):72–7.
42. Leclere FM, Trelles M, Lewbart GA, Vogelin E. Is there good simulation basic training for end-to-side vascular microanastomoses? Aesthet Plast Surg. 2013;37(2):454–8.
43. Nam SM, Shin HS, Kim YB, Park ES, Choi CY. Microsurgical training with porcine thigh infusion model. J Reconstr Microsurg. 2013;29(5):303–6.
44. Olabe J, Olabe J. Microsurgical training on an in vitro chicken wing infusion model. Surg Neurol. 2009;72(6):695–9.
45. Carey JN, Rommer E, Sheckter C, et al. Simulation of plastic surgery and microvascular procedures using perfused fresh human cadavers. J Plast Reconstr Aesthet Surg. 2014;67(2):e42–8.
46. Rayan N, O'Donoghue DL, Rayan G. The use of cadaver tissue in advanced microsurgical training. J Okla State Med Assoc. 2010;103(8):365–8.
47. Rodriguez A, Alvarez A, Aguirrezabalaga J, Martelo F. The anteromedial thigh flap as a training model of a perforator flap in rat. J Reconstr Microsurg. 2007;23(5):251–5.
48. Schonauer F, La Rusca I, Gonzales YRE, Romagnuolo M, Molea G. Laboratory model of a microvascular free flap in the rat: the epigastric flap. Minerva Chir. 2002;57(4):537–41.
49. Ozkan O, Koshima I, Gonda K. A supermicrosurgical flap model in the rat: a free true abdominal perforator flap with a short pedicle. Plast Reconstr Surg. 2006;117(2):479–85.
50. Bodin F, Diana M, Koutsomanis A, Robert E, Marescaux J, Bruant-Rodier C. Porcine model for free-flap breast reconstruction training. J Plast Reconstr Aesthet Surg. 2015;68(10):1402–9.
51. Rehder R, Abd-El-Barr M, Hooten K, Weinstock P, Madsen JR, Cohen AR. The role of simulation in neurosurgery. Childs Nerv Syst. 2016;32(1):43–54. https://doi.org/10.1007/s00381-015-2923-z.
52. Bohm PE, Arnold PM. Simulation and resident education in spinal neurosurgery. Surg Neurol Int. 2015;6:33.
53. Bova FJ, Rajon DA, Friedman WA, Murad GJ, Hoh DJ, Jacob RP, Lampotang S, Lizdas DE, Lombard G, Lister JR. Mixedreality simulation for neurosurgical procedures. Neurosurgery. 2013;73(Suppl. 1):138–45.
54. Chan S, Conti F, Salisbury K, Blevins NH. Virtual reality simulation in neurosurgery: technologies and evolution. Neurosurgery. 2013;72(Suppl. 1):154–64.
55. Cohen AR, Lohani S, Manjila S, Natsupakpong S, Brown N, Cavusoglu MC. Virtual reality simulation: basic concepts and use in endoscopic neurosurgery training. Childs Nerv Syst. 2013;29:1235–44.
56. Musacchio M, Smith AP, McNeal C, Munoz L, Rothenberg DM, Von Roenn K, Byrne RW. Neuro-critical care skills training using a human patient simulator. Neurocrit Care. 2010;13:169–75.
57. Ahmed SI, Javed G, Mubeen B, et al. Robotics in neurosurgery: a literature review. J Pak Med Assoc. 2018;68(2):258–63.

Part I
The Surface

Surgical Anatomy of the Scalp

Leonardo Christiaan Welling, Nicollas Nunes Rabelo, and Eberval Gadelha Figueiredo

1 Introduction

When we analyze the harmony of the human body, the craniofacial region is of fundamental importance. In this context, the aesthetic results after neurosurgical interventions should be considered. Advances in surgical techniques to approach the most diverse intracranial pathologies evolve rapidly. Parallel to better functional results, better aesthetic results are also demanded by patients [1, 2].

Among the skills required in neurosurgery, one of the main ones is to centralize the bone flap so that the approach to the intracerebral lesion is not compromised. A poorly positioned head or a poorly planned surgical incision can compromise the entire surgical act. According to the region to be approached and the angles needed for this, the incision in the scalp is planned. Also, in this planning, anatomical knowledge of vascularization, scalp innervation, and the possibility of hiding the scar, or, if this is not possible, minimizing aesthetic defects are essential steps in any cranial approach [3].

Despite all the recommended care, the biggest aesthetic problems, or possible dissatisfaction, are more observed in young women and hairless people. In addition to the aesthetic result, it is important to consider the intracranial pathology or the main suspicion.

L. C. Welling (✉)
Neurological Surgery Department, State University of Ponta Grossa, Ponta Grossa, PR, Brazil

N. N. Rabelo
Neurological Surgery Department, Athens University Center, Passos, MG, Brazil

E. G. Figueiredo
Neurological Surgery Department, University of São Paulo, São Paulo, SP, Brazil

E. G. Figueiredo et al. (eds.), *Brain Anatomy and Neurosurgical Approaches*,
https://doi.org/10.1007/978-3-031-14820-0_2

For example, if therapeutic complementation with radio and chemotherapy is necessary, suture dehiscence and secondary infection may occur. In pathologies with potential for local recurrence, the incision must also be carefully planned. In cases where there is a prediction of reoperation or expansion of the craniotomy margins, the first approach, if performed with a linear cut, will allow better incisional alternatives for subsequent interventions.

2 Historical Aspects

As the neurosurgical specialty advances and patients survive the procedures, the final cosmetic result becomes more relevant. For this, the orientation of the surgical incision will directly affect the healing process and the cosmetic result. When surgeons visualize lines on the skin's surface to guide their incisions, the Langer lines (also known as tension lines) are primarily remembered [4].

Austrian researcher Karl Langer (1819–1887) is universally known for his famous lines. Born in Vienna, where he spent most of his life, he was a professor of anatomy at Joseph's Academy [4]. In 1861, Langer began publishing a series of five articles detailing his work on the physical and mechanical properties of the skin.

Langer's original experiment included making wounds all over the body of a large series of cadavers - of different ages and body constitutions - analyzing the relationship of each small lesion to the adjacent ones. Through these incredibly careful efforts, Langer was able to draw his famous boundaries. Unfortunately, Langer's work was little appreciated and ignored during his lifetime.

As his original publications were written in the classical German of his day, it was not until 1978 that a translation of his work was made available to the English-speaking world.

These days, not all surgical incisions in the body follow Langer's lines, and other references such as lines perpendicular to the direction of muscle contraction may be more useful in certain patients. However, Langer's work has helped improve the practice of surgery and forensic medicine and an understanding of the skin's many fascinating properties [4, 5].

3 Scalp Layers

Through the SCALP mnemonic, we can define the five layers of tissue that cover the skull: Skin, Connective tissue (or subcutaneous tissue), Aponeurosis, Loose areolar tissue, and Periosteum (or pericranium) (Fig. 1). Externally, the skin is the

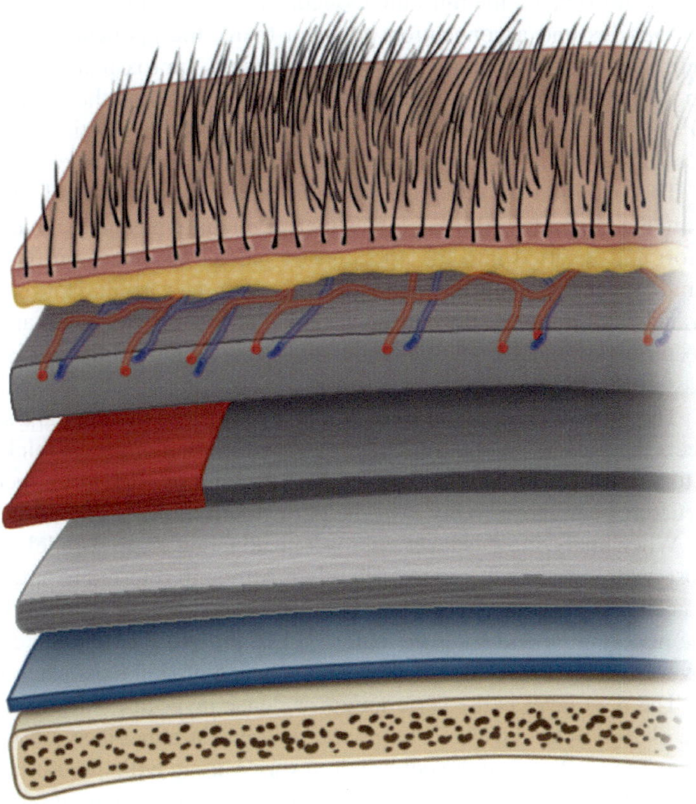

Fig. 1 The five layers of tissue that cover the skull: Skin, Connective tissue (or subcutaneous tissue), Aponeurosis, Loose areolar tissue, and Periosteum (or pericranium) are demonstrated

first layer. Its thinnest part is at the apex of the skullcap, measuring about 3 mm. Its thickest part is in the occipital region, and its thickness can reach 8 mm. With advancing age, its thickness decreases, being more evident in bald people. Subcutaneous connective tissue has firmly adhered. It is observed that below the dermis are located the numerous hair follicles, sebaceous, and sweat glands [6].

The second layer is the subcutaneous connective tissue, with an average thickness of 4–7 mm, and is characterized by fibrous, firm, dense, and vascular that connects the skin to the underlying epicranial aponeurosis [7]. Several fibrous septa divide this layer into small compartments. Also, in this layer, the blood vessels of greater caliber are in this topography, and there is a large subcutaneous vascular plexus [7].

The third layer, the musculoaponeurotic layer, comprises the frontal and occipital muscles and a connection between them, the aponeurotic galea covering the entire calvaria. This thin, inelastic membrane is approximately 1–2 mm thick, made up of connective tissue, and its lateral extension joins the temporoparietal fascia. The galea is firmly adhered to the overlying subcutaneous tissue, but is loosely connected to the underlying pericranium through an areolar layer of connective tissue.

The occipital muscles are thin and quadrilateral, formed by tendinous fibers from the lateral two-thirds of the superior nuchal line and the mastoid. Its fibers are short and inserted into the aponeurotic galea. An extension of the galea fills the small space between the occipital muscles. Its vascularization is performed by branches of the posterior auricular and occipital artery. Innervation is done by the posterior auricular branch of the facial nerve, and its function is to pull the scalp posteriorly [5].

The loose areolar tissue layer below the aponeurotic galea is a structure that connects the musculoaponeurotic layer with the pericranium. Its thickness ranges from 1 to 3 mm. Due to their looseness, the galea and the strongly superimposed surface structures can move freely over the periosteum of the skull bone [8].

When analyzing this layer microscopically, numerous parallel fibrous laminae are observed, loosely adhered to each other. Although referred to as an avascular space, it is largely vascularized by tiny vascular branches, and some of these pass through the pericranium and enter the outer slab of the skull. Due to its characteristics, this layer is easily dissected and allows the traction of skin flaps when necessary [9].

The next layer is the pericranium, a dense layer of connective tissue adhered to the skull's outer surface. The points of greatest adherence are in the supraorbital area. The thickness of the pericranium varies along with the skull, and it is generally thinner than the aponeurotic galea. In cranial sutures, the pericranium becomes continuous with the endosteum. Its blood supply comes from vessels in the scalp that traverse the loose areolar tissue. It becomes continuous with the endosteum (the periosteum on the skull's inner surface) at the suture lines of the skull. The pericranium receives its blood supply mainly from branches of the main scalp vessels that traverse the loose areolar tissue. Another source of blood supply is the perforating vessels in the skull that also reach the pericranium [5].

4 Sensitive Innervation of the Scalp

Sensory innervation of the temporal region and scalp is transmitted via eight nerves originating from the trigeminal nerve or cervical plexus. The first division of the trigeminal (V1) contributes to the supraorbital and supratrochlear nerves. The maxillary division of the trigeminal (V2) carries sensory information from the zygomatic-temporal nerve. The auriculotemporal nerve projects its fibers through the mandibular division (V3) of the trigeminal nerve.

The cervical plexus also contributes to sensory innervation. The third occipital nerve is a dorsal branch of the third cervical nerve (C3). The greater occipital nerve is a dorsal branch of the second cervical nerve (C2). The greater auricular nerve and the lesser occipital nerve are also branches of the cervical plexus.

The terminal branches of the supratrochlear nerve may be responsible for sensitivity up to the anterior, middle third of the scalp. Similarly, the supraorbital nerve divides into a superficial branch and a deep branch. The superficial branch carries fibers related to frontal sensitivity. On the other hand, the deep branch ascends below the frontal muscle, but at the hairline, it becomes superficial and goes to the frontoparietal transition and eventually reaches the vertex [10].

The zygomatic nerve originates from the V2 segment of the trigeminal nerve, and its origin from the maxillary nerve occurs at the level of the pterygopalatine fossa. Upon entering the orbit via the inferior orbital fissure, it follows a path along the lateral wall of the orbit and divides into the zygomaticotemporal and zygomaticofacial nerves. When traversing the inferolateral angle of the orbit, the zygomaticotemporal nerve sends an anastomotic branch to the lacrimal nerve. About 1 cm posterior and inferior to the frontozygomatic suture, this nerve pierces the temporal fascia. As it superficializes towards the subcutaneous tissue, it innervates the scalp over a small area of the anterior part of the temporal region [10, 11].

In a situation parallel and posterior to the superficial temporal artery, we find the auriculotemporal nerve, which, in its preauricular course, passes over the posterior third of the zygomatic arch, above the temporal fascia, and emits its terminal branches that will carry sensory information from the anterior temporal region, external acoustic meatus, and anterior aspect of the ear.

The branches of the cervical nerves to the scalp are the greater auricular nerve, the lesser occipital nerve, the greater occipital nerve, and the branches of the third cervical nerve.

The greater auricular nerve innervates the skin overlying the mastoid and the underside of the pinna. At the level of the lower pole of the parotid gland, this neural structure branches, and its posterior branch crosses over the upper third of the sternocleidomastoid muscle until reaching the skin that covers the mastoid.

It is observed that one of its terminal filaments crosses the pinna and carries sensory information from the earlobe and the shell. The lesser occipital nerves, the auricular branch of the vagus nerve, and the auricular branch of the facial nerve also make connections with the greater auricular nerve.

As a cutaneous branch of the cervical plexus, specifically from C3, the lesser occipital nerve ascends via the posterior margin of the sternocleidomastoid muscle to the occipital region. As it approaches the skull, it crosses the deep fascia and enters the subcutaneous layer providing sensory innervation to the occipital part of the scalp. Topographically, it is located 25 mm lateral to the occipital artery and 70 mm lateral to the external occipital bulge [12–15].

The greater occipital nerve represents the continuation of the dorsal medial branch of C2. In its initial path to the scalp, this neural structure has its path superior to the inferior oblique muscle and below to the semispinatus of the head. As it moves upward, it crosses the semispinatus and trapezius. The moment that it reaches the subcutaneous tissue is around 35 mm lateral to the midline and 9 mm inferior to the external occipital bulge [16]. Its most distal destination is the skin overlying the coronal suture. This path follows the branches of the occipital artery and carries sensory fibers from the medial region of the occipital and parietal region. The anatomical intersection between the greater occipital nerve and the occipital artery, seen in up to 50% of people, may be an additional trigger for migraine [16]. The third occipital nerve and the lesser occipital nerve are also anastomosed with the greater occipital nerve [6, 11, 16].

The dorsal medial branch of the third occipital nerve is represented as the third occipital nerve. When crossing the semispinatus muscles of the head, head splenium, and trapezius, the third occipital nerve, about 5 cm below the inion, becomes superficial, directs about 3 mm from the midline, and carries sensitive information from this region. It is also a structure with connections to the cutaneous branches of the greater and lesser occipital nerves [14].

5 Motor Innervation of the Scalp

The motor innervation of the musculoaponeurotic layer of the scalp is represented by two branches of the facial nerve: the frontal branch and the posterior auricular branch.

The anterior part of the occipitofrontal muscle and the anterior and superior auricular muscles are innervated by the frontal branch of the facial nerve. Since they are in the direction of frontotemporal craniotomies (Fig. 2) applied in most neurosurgical approaches, the accuracy in the knowledge of its path and depth is essential. In the upper margin of the parotid gland, the frontal branch arises from the temporofacial division of the facial nerve. It depicts a deep path to the parotid masseteric fascia and crosses the zygomatic arch in its middle third, within the temporoparietal fascia. Its course through the temporal region occurs along the inferior surface of the temporoparietal fascia. It should be observed that the frontal branch can be subdivided into branches. According to Zani et al., the presence of a single branch is described in about 16–28% of individuals. Two branches seem to be more frequent, being described in 32–52% of individuals. Three or four branches are less frequent, occurring in 4–16% of cases. According to the division, the branches can be divided into anterior, middle, and posterior groups [17].

The posterior branch, which is of no surgical importance since it appears in a few individuals, partially innervates the anterior auricular and temporoparietal muscles. Already of extremely relevant surgical importance, since it is directed to the frontal muscle and part of the orbicularis oculi, is the frontal branch of the facial nerve. This, as already said, can be presented in an isolated way or divided into two branches (anterior and middle). Its path as soon as it leaves the parotid gland follows the parotid-masseteric fascia. It crosses the surface of the zygomatic arch and, on the inferior surface of the temporoparietal fascia, it crosses the temporal region and

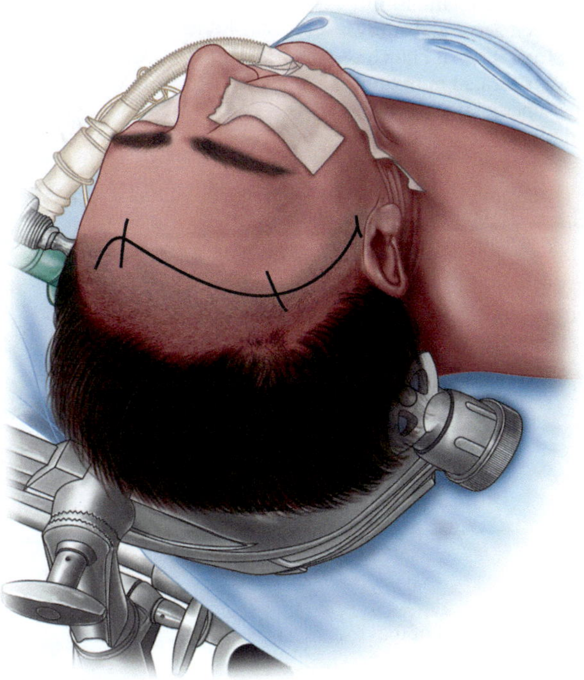

Fig. 2 The skin planning incision for a frontotemporal craniotomy. The anterior part of the occipitofrontal muscle and the anterior and superior auricular muscles are innervated by the frontal branch of the facial nerve, which is adjacent to the neurosurgical trajectory

ascends about 10–20 mm laterally to the epicanthus of the eye [18]. When reaching the lateral border of the frontal muscle, it goes below it and innervates it. The anterior and middle branches often have anastomosed connections to each other and innervate the orbicularis oculi muscle. During its course, the frontal branch is always located in front of the superficial temporal artery (Fig. 3). Only the tiny branch with a posterior direction crosses the superficial temporal artery to project the auricular muscles [2, 18].

Externally, the path of the facial nerve can be projected through a line called the "Pitanguy line" that starts at a point 5 mm below the tragus, and the other point is arranged 10 mm posterolateral to the edge of the eyebrow (Fig. 4). To increase the safety of any procedure in frontotemporal topography, a curved zone that starts from the middle third of the zygomatic arch and ends 2.5 cm above the lateral and superior face of the ipsilateral orbit is established as a safe area. This knowledge, associated with the perception of depth of the fascia around the frontal branch of the facial nerve, reduces the chances of injury to this structure during a surgical procedure [18].

In the context of frontotemporal approaches, several temporal muscle dissection techniques have been developed, but the most used are interfascial and subfascial dissection and myocutaneous flap in a single plane [1–3].

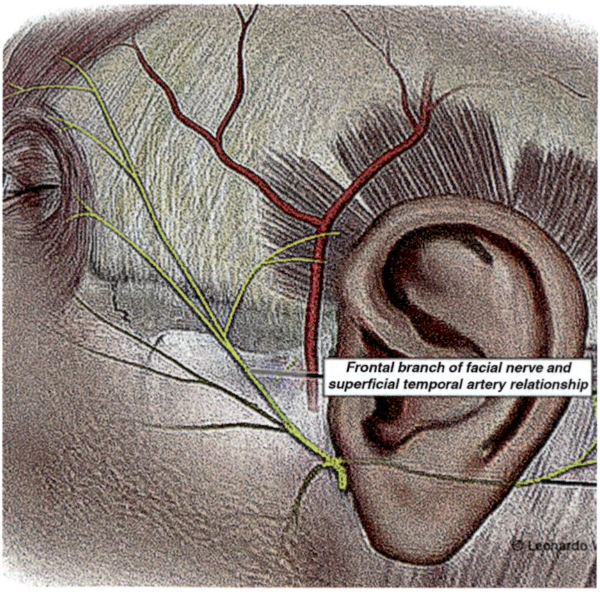

Fig. 3 The relationship between nerve and artery: the frontal facial branch is always located in front of the superficial temporal artery

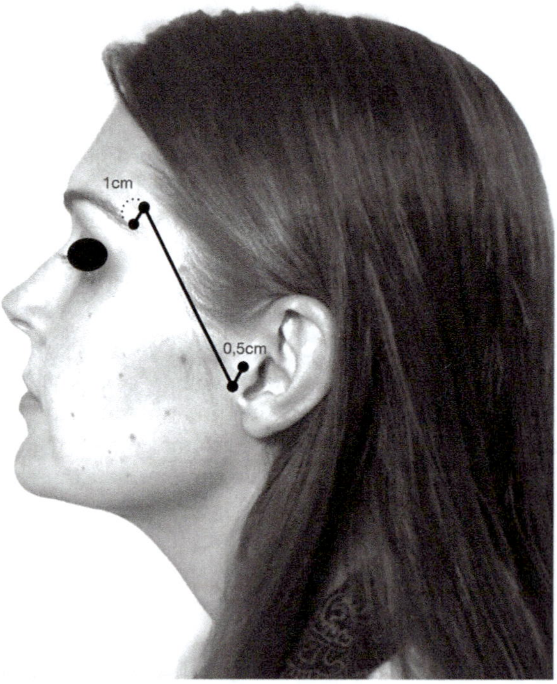

Fig. 4 The path of the facial nerve can be projected through a line called the "Pitanguy line" that starts at a point 5 mm below the tragus, and the other point is arranged 10 mm posterolateral to the edge of the eyebrow

The other branch, known as the posterior auricular nerve, arises from the facial nerve after it emerges from the stylomastoid foramen. Its trajectory follows a superior direction between the external acoustic meatus and the mastoid process. It runs posteriorly along the superior nuchal line and inserts into the occipital muscle, giving communicating branches to great auricular and lesser occipital nerves.

6 Vascular Supply

The vascularization of the scalp is formed of the superficial temporal, posterior auricular, and occipital arteries, which are branches of the external carotid artery. For the frontal region, the greatest irrigation comes from the internal carotid system, through the supraorbital and supratrochlear arteries. Perforating meningeal branches (which cut through the bone) are responsible for a small contribution. As it ascends in the frontal portion of the scalp, the supratrochlear artery, with its superficial terminal branch, slightly directs itself towards the midline and is in the subcutaneous tissue. The supraorbital artery with its superficial subbranches ascends laterally to the supratrochlear artery and significantly contributes to the vascular supply of the frontal scalp [6].

6.1 Superficial Temporal Artery (STA)

In addition to the scalp, the superficial temporal artery supplies the upper lateral half of the face, part of the parotid gland, and part of the temporomandibular joint. Its course is directly related to the preauricular region and crosses the posterior third of the zygomatic arch superiorly [19]. At first, its path is within the frontotemporal fascia, but it becomes more superficial and bifurcates into the frontal and posterior parietal branches as it ascends towards the top. In the topography of the bifurcation, its diameter varies from 1.8 to 2.7 mm. In about 70% of individuals, this bifurcation occurs above the zygomatic arch, usually 2–4 cm above it; about 4–26% at the level and 8–11% below the zygomatic arch [19–22]. Although rare, in about 3% of cases, the artery does not bifurcate, and its course occurs with only one frontal branch. It is observed that the superficial temporal artery is the continuation of the external carotid artery after the origin of the maxillary artery. Before dividing into frontal and parietal branches, the superficial temporal artery emits three important branches: the middle temporal artery, the zygomatic-orbital artery, and the superior auricular artery.

The first branch, the middle temporal artery, originates from the superior margin of the zygomatic arch. In some situations, the artery may appear about 1 cm below the zygomatic arch. As it passes through the superficial layer of the temporal fascia, it penetrates the fat and then enters the deeper layer of the temporal fascia. It runs along the inferior surface of the temporal fascia and communicates with the deep temporal arteries (branches of the maxillary artery). The middle temporal artery supplies the temporal muscle.

The second branch, the zygomatic orbital artery, originates just after the middle temporal artery and proceeds to the lateral orbital wall. In this topography, the eyelid and lacrimal arteries, branches of the ophthalmic artery, anastomose with branches of the zygomatic orbital artery. In some patients, its origin comes from the frontal branch of the superficial temporal artery and not from its main trunk. Along its path, it emits cutaneous branches that irrigate the adjacent skin. In individuals in whom the orbital zygomatic artery is absent, which can occur in up to 20% of cases, the supply of the lateral orbital surface is provided by small branches of the superficial temporal artery and the transverse facial artery [6].

The third branch to be described still appears in its path over the temporal region, the superior auricular branch, a small artery that supplies part of the ear [22].

The frontal branch of the superficial temporal artery, one of its terminal branches, has a superior, tortuous course that sends ascending and descending branches to the temporal area. At the bifurcation level, it has a mean diameter between 1.6 and 2.1 mm [22, 23]. Tayfur et al. describe its length as approximately 11 cm. After giving rise to a transverse branch in the forehead topography, its main continuation has a superomedial trajectory. It lies on the frontal muscle, becoming superficial, until the subdermal level, as it approaches the midline. All scalp layers in the frontal region are supplied by the frontal branch of the superficial temporal artery, which makes it important for this region [19]. The other terminal branch, the parietal branch of the superficial temporal artery, runs slightly superiorly and posteriorly. Its course is represented by a vertical zone that originates 1 cm anterior to the tragus and extends posteriorly for 4 cm. Its diameter at origin varies between 1.6 and 1.8 mm [19, 20, 22].

Although in practice, interrupting its flow causes little damage to the scalp (mainly due to a vast network of anastomoses), it is nevertheless important when it comes to making reconstructive flaps or extracranial-intracranial anastomoses. Palpable in the temporal region can be easily mapped in cases where it needs to be incorporated into the myocutaneous flap.

6.2 *Occipital Artery*

Approximately 2 cm distal to the origin of the facial artery (which originates from the anterior surface of the external carotid artery) and on the posterior surface of the external carotid artery is the origin of the occipital artery (OA). Its average diameter in this region is about 3 mm. In its initial course, it is directed superiorly and posteriorly in the depth of the posterior belly of the digastric muscle. It traverses anteriorly the internal carotid artery, internal jugular vein, vagus, accessory, and hypoglossal nerves. On the floor of the temporal bone, it crosses the medial surface of the mastoid when it then reaches the occipital area. This region is at a depth of insertion of the sternocleidomastoid muscle, splenius capitis, longissimus capitis. In its final course, when directed medially, it is located over the semispinalis muscle. It then projects superiorly, about 4 cm from the midline, when it emerges through the deep cervical fascia, trapezius, and sternocleidomastoid muscle and divides into its terminal occipital branches.

Along its course, the occipital artery emits meningeal branches that enter the skull through the jugular foramen; the stylomastoid artery, which supplies the facial nerve as it enters the stylomastoid foramen; a branch toward the antrum of the mastoid; as well as vessels that supply the digastric, stylohyoid, longissimus capitis, trapezius, and splenium muscles. It is observed that the skin in the region is also vascularized through perforating musculocutaneous branches. Soon after appearing in the scalp's posterior part, the OA ascends tortuously over the occipitalis muscle and the galea within the subcutaneous cellular tissue and divides into its terminal occipital branches. The terminal occipital branches distribute to the scalp, the main arterial system of the posterior scalp, and pericranium.

From a surgical point of view, many of the terminal arterial branches are in direct company with the greater occipital nerves, eventually intertwining [5, 6].

6.3 Posterior Auricular Artery

On the posterior surface of the external carotid artery, above the digastric and stylohyoid muscles, the artery called the posterior auricular artery appears. Before reaching the mastoid process, it ascends between the parotid gland and the styloid process. Also, in the neck, at the level of the external acoustic meatus, its small branches perforate the deep fascia and are divided into two: the auricular branch and the occipital branch. The auricular branch ascends under the posterior auricular muscle, and its branches vascularize the skin on the posterior surface of the pinna. The superior auricular artery, which is a branch of the superficial temporal artery, anastomoses with the auricular branch of the posterior auricular artery. In contrast, the occipital branch runs laterally through the mastoid process, over the insertion of the sternocleidomastoid muscle, and ends over the occipital muscle and supplies the posterolateral region of the scalp [6].

7 Venous Drainage

As in other body segments, the veins accompany the arteries; however, on the scalp, they are not necessarily close, and in some cases, the distance can reach 3 cm between the artery and its corresponding veins. The anastomotic network is wide and connects to the diploic veins. The larger venous trunks, represented mainly by the supratrochlear, supraorbital, superficial temporal, occipital, and posterior auricular veins, are formed by the convergence of small veins.

The supratrochlear and supraorbital veins drain the anterior region of the scalp. The veins that drain the lateral aspect of the scalp are formed by the frontal vein and the parietal veins. These unite above the zygomatic arch and form the superficial temporal veins. They are more superficial and run parallel to the superficial temporal artery, and they are located about 3 cm posterior to the artery still in the temporoparietal fascia. The superficial temporal vein descends anterior to the external acoustic meatus and receives the face's middle and transversal temporal veins. In the topography, the parotid gland joins the maxillary vein and forms the retromandibular vein.

The posterior auricular vein drains the posterolateral region of the scalp. When the posterior auricular vein joins with the posterior division of the retromandibular veins, the external jugular vein is formed.

The occipital veins that drain the posterior scalp are formed from small veins that converge with each other. As it follows its corresponding artery, it traverses the insertions of the trapezius and sternocleidomastoid muscles in the skull and joins the deep cervical and vertebral veins. There is variability to where the occipital vein flows. In some people, it drains into the internal jugular vein and in others into the posterior auricular vein and subsequently into the external jugular vein [6].

8 Healing Process

Knowledge of the wound healing process also applies to the cranial region. There are three phases: inflammatory, proliferative, and maturation phase. The first phase, also known as the inflammatory phase, lasts about 3 days and is characterized by the migration of polymorphonuclear leukocytes at first, followed by the migration of macrophages and mononuclear leukocytes. Tissue damage is the initial event that triggers the entire restoration process. Immediately, the body begins hemostasis with the contraction of small vessels nearby, platelet aggregation, activation of the coagulation cascade, and formation of a fibrin matrix. This fibrin network acts as a barrier to prevent wound contamination and as a basis for the healing process, supporting cell migration and stimulating growth factors.

The second phase, or proliferative phase, includes reepithelialization, collagen matrix synthesis, and neovascularization. Such events start around the third day after the injury and last for a few weeks. The third phase, or maturation phase, is characterized by organized collagen deposition. It is of the greatest clinical importance in the healing process, as it is when the wound receives greater support. This final phase begins 21 days after the injury and can last for up to a year. Regarding the tensile strength, it is important to note that the closed wound recovers about 20% of its original tensile strength after 3 weeks, and about 60–70% in 6 weeks, and 80% in 1 year.

Technically, it is always important to cut the skin perpendicular to its surface and avoid oblique incisions from a neurosurgical point of view. Likewise, in closing, the suture needle must follow a path perpendicular to the epidermis, and the distance and depth of the suture must be comparable on both incision margins. Edges that are uneven decrease the chance of proper healing.

In addition, careful approximation of the wound edges is also important for reepithelialization. The tension to bring the edges of the incision together should be light to avoid excessive inversion. If greater tension is applied, strangulation of the wound edges and subsequent tissue ischemia can occur, impairing complete wound healing [24].

Scalp incision and postoperative pain.

The physical processes of skin incision, traction, and hemostasis used for any kind of craniotomy stimulate nerve fibers and specific nociceptors of the scalp, resulting in postoperative pain. During the first 24 h after craniotomy, 87% of patients have pain. In some cases, the pain persists after craniotomy in 3% for every year of life.

With the local intradermal infiltration of the scalp with bupivacaine and lidocaine combined with the postoperative use of dipyrone, patients reported low pain levels in the first 12 h after surgery. This may contribute to the well-being of patients and their hemodynamic stability, which could permit their early discharge. Nevertheless, in some cases, it is needed to use drugs treatment such as codeine, morphine, tramadol, anti-inflammatory nonsteroid such as cyclooxygenase-2 inhibitors, and gabapentin [25].

9 Conclusions

Good aesthetic and often functional results start with planning the skin incision. The ideal incision should provide ample vascular support and not compromise microsurgical visualization at the angles needed to access intracranial pathology.

If the surgical approach and functional outcomes are not compromised, there is no reason to ignore or compromise the cosmetic outcome. Thus, the neurosurgeon must consider the aesthetic aspect an obligation for all patients, even those with unfavorable pathologies.

References

1. Figueiredo EG, Welling LC, Preul MC, Sakaya GR, Neville I, Spetzler RF, et al. Surgical experience of minipterional craniotomy with 102 ruptured and unruptured anterior circulation aneurysms. J Clin Neurosci. 2016;27:34–9.
2. Welling LC, Figueiredo EG, Wen HT, Gomes MQT, Bor-Seng-Shu E, Casarolli C, et al. Prospective randomized study comparing clinical, functional, and aesthetic results of minipterional and classic pterional craniotomies. J Neurosurg. 2015;122(5):1012–9.

3. Welling LC, Figueiredo EG. Temporal muscle atrophy: not only the approach but the way you get it. J Plast Reconstr Aesthet Surg. 2018;71(3):445–6.
4. Carmichael SW. The tangled web of Langer's lines. Clin Anat. 2014;27(2):162–8.
5. Thomaidis VK. Scalp and temple. In: Cutaneous flaps in head and neck reconstruction. Berlin: Springer; 2014. p. 13–76. https://doi.org/10.1007/978-3-642-41254-7_2.
6. Standring S. Gray's anatomy: the anatomical basis of clinical practice. 40th ed. Edinburg: Churchill Livingstone/Elsevier; 2008.
7. Hayman AL, Shukla V, Ly C, Taber KH. Clinical and imaging anatomy of the scalp. J Comput Assist Tomogr. 2003;27(3):454–9.
8. Tremolada C, Candiani P, Signorini M, Vigano M, Donati L. The surgical anatomy of the subcutaneous fascial system of the scalp. Ann Plast Surg. 1994;32(1):8–14.
9. Casanova R, Cavalcante D, Grotting JC, Vasconez LO, Psillakis JM. Anatomic basis for vascularized outer-table calvarial bone flaps. Plast Reconstr Surg. 1986;78(3):300–8.
10. Jeong SM, Park KJ, Kang SH, Shin HW, Kim H, Lee HK, et al. Anatomical consideration of the anterior and lateral cutaneous nerves in the scalp. J Korean Med Sci. 2010;25(4):517.
11. Hwang K, Suh MS, Lee SI, Chung IH. Zygomaticotemporal nerve passage in the orbit and temporal area. J Craniofac Surg. 2004;15(2):209–14.
12. Cohen-Gadol A, Kemp W III, Tubbs RS. The innervation of the scalp: a comprehensive review including anatomy, pathology, and neurosurgical correlates. Surg Neurol Int. 2011;2(1):178.
13. Kemp WJ, Cohen-Gadol AA. A review of skin incisions and scalp flaps for the retromastoid approach and description of an alternative technique. Surg Neurol Int. 2011;2:143.
14. Tubbs RS, Fries FN, Kulwin C, Mortazavi MM, Loukas M, Cohen-Gadol AA. Modified skin incision for avoiding the lesser occipital nerve and occipital artery during retrosigmoid craniotomy: potential applications for enhancing operative working distance and angles while minimizing the risk of postoperative neuralgias and intraoperative hemorrhage. J Clin Neurosci. 2016;32:83–7.
15. Tubbs RS, Mortazavi MM, Loukas M, D'Antoni AV, Shoja MM, Chern JJ, et al. Anatomical study of the third occipital nerve and its potential role in occipital headache/neck pain following midline dissections of the craniocervical junction: laboratory investigation. J Neurosurg Spine. 2011;15(1):71–5.
16. Janis JE, Hatef DA, Ducic I, Reece EM, Hamawy AH, Becker S, et al. The anatomy of the greater occipital nerve: part II. Compression point topography. Plast Reconstr Surg. 2010;126(5):1563–72.
17. Zani R, Fadul R, Dias Da Rocha MA, Santos RA, Alves MCA, Ferreira LM. Facial nerve in rhytidoplasty: anatomic study of its trajectory in the overlying skin and the most common sites of injury. Ann Plast Surg. 2003;51(3):236–42.
18. Pitanguy I, Ramos AS. The frontal branch of the facial nerve: the importance of its variations in face lifting. Plast Reconstr Surg. 1966;38(4):352–6.
19. Tayfur V, Edizer M, Magden O. Anatomic bases of superficial temporal artery and temporal branch of facial nerve. J Craniofac Surg. 2010;21(6):1945–7.
20. Chen TH, Chen CH, Shyu JF, Wu CW, Lui WY, Liu JC. Distribution of the superficial temporal artery in the Chinese adult. Plast Reconstr Surg. 1999;104(5):1276–9.
21. Marano SR, Fischer DW, Gaines C, Sonntag VKH. Anatomical study of the superficial temporal artery. Neurosurgery. 1985;16(6):786–90.
22. Pinar YA, Govsa F. Anatomy of the superficial temporal artery and its branches: its importance for surgery. Surg Radiol Anat. 2006;28(3):248–53.
23. Edizer M, Beden U, Icten N. Morphological parameters of the Periorbital arterial arcades and potential clinical significance based on anatomical identification. J Craniofac Surg. 2009;20(1):209–14.
24. Gantwerker EA, Hom DB. Skin: histology and physiology of wound healing. Facial Plast Surg Clin N Am. 2011;19(3):441–53.
25. Santos CMT, Pereira CU, Chaves PHS, de Lima Tôrres PTR, da Paixão Oliveira DM, Rabelo NN. Options to manage postcraniotomy acute pain in neurosurgery: no protocol available. Br J Neurosurg. 2021;35(1):84–91.

Surgical Anatomy of the Skull

Renan Salomão Rodrigues, Marcelo Medeiros Felipe, and Ricardo Castro de Oliveira Filho

1 Ethmoid Bone

The ethmoid is an unpaired bone that articulates 13 cranial and facial bones, part of the nasal cavities, orbit, and anterior cranial base. It is composed of three main structures: Cribriform plate, perpendicular plate, and ethmoid labyrinth.

The cribiform plate is a horizontal plate, which articulates with the ethmoidal notch in the frontal bone, composing the foramen cecum. From the cribriform plate arises a vertical bony proeminence—the crista galli—where the faux cerebri will insert itself [1]. The cribriform plate also has a series of foramina—the cribriform foramina—which give it its name and where the olfactory nerves transverse.

The perpendicular plate is a vertical, thin lamina that projects inferiorly from the cribriform plate, forming the nasal septum's posterior part. Septal nasal cartilage will attach to the perpendicular plate and superior and middle nasal concha project laterally from the perpendicular plate into the nasal cavity.

The ethmoidal labyrinth is formed by the ethmoidal cells—or ethmoidal sinuses—which vary in number, size, and disposition and form the lateral mass of the ethmoid bone. They are disposed between the perpendicular plate and the orbital plate, arranged in anterior, middle, and posterior groups. Lateral to the ethmoidal labyrinth is the orbital plate, or papyracea plate, an exceedingly thin osseous wall dividing the nasal cavity from the orbit [2]. The arterial irrigation of the nasal

R. S. Rodrigues (✉)
Department of Neurosurgery, University of São Paulo, São Paulo, SP, Brazil

M. M. Felipe
Department of Neurosurgery, Hospital Municipal Dr. José de Carvalho Florence, São Jose dos Campos, SP, Brazil

R. C. de Oliveira Filho
Department of Neurosurgery, Federal University of Rio de Janeiro, Rio de Janeiro, RJ, Brazil

E. G. Figueiredo et al. (eds.), *Brain Anatomy and Neurosurgical Approaches*,
https://doi.org/10.1007/978-3-031-14820-0_3

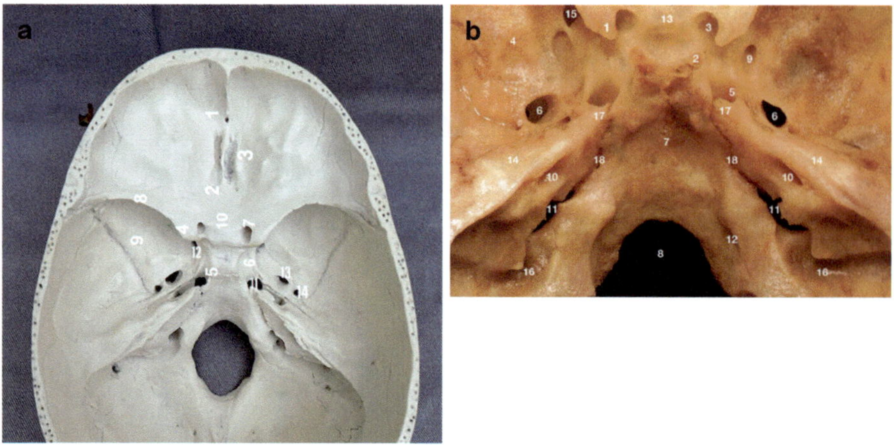

Fig. 1 (**a**) Superior view of the skull base highlighting the ethmoid and sphenoid bones. (*1*) Crista galli. (*2*) Ethmoid bone. (*3*) Cribiform plate. (*4*) Anterior clinoid process. (*5*) Posterior clinoid process. (*6*) Middle clinoid process. (*7*) Optic canal. (*8*) Lesser sphenoid wing. (*9*) Greater sphenoid wing. (*10*) Planum sphenoidale. (*11*) Foramen lacerum. (*13*) Foramen ovale. (*14*) Foramen spinosum. (**b**) Other representation of superior view of the skull base highlighting the ethmoid and sphenoid bones. (*1*) Anterior clinoid process. (*2*) Posterior clinoid process. (*3*) Optic canal. (*4*) Greater sphenoid wing. (*5*) Foramen lacerum. (*6*) Foramen ovale. (*7*) Clivus. (*8*) Foramen magnum. (*9*) Foramen rotundum. (*10*) Internal auditory canal. (*11*) Jugular foramen. (*12*) Hypoglossal canal. (*13*) Tuberculum sella. (*14*) Petrous temporal bone. (*15*) Superior orbital fissure. (*16*) – Sigmoid sulcus. (*17*) Apex. (*18*) Petroclival fissure

septum and the conchae is mainly provided by the sphenopalatine artery and the anterior and posterior ethmoidal arteries. Meanwhile, the innervation is due to the anterior and posterior ethmoidal branches from the nasociliary nerve [1] (Fig. 1a, b).

2 Sphenoid Bone

An unpaired bone localized in the center of the skull base, in intimate contact with the nasal cavity inferiorly and the sellar region superiorly, making it an important approach for pituitary surgery [2] (Fig. 1a, b).

3 Osseous Relationships

It articulates mainly with the squamous part of the temporal bone laterally, the clival part of the occipital bone posteriorly, the ethmoid and frontal bones anteriorly, and

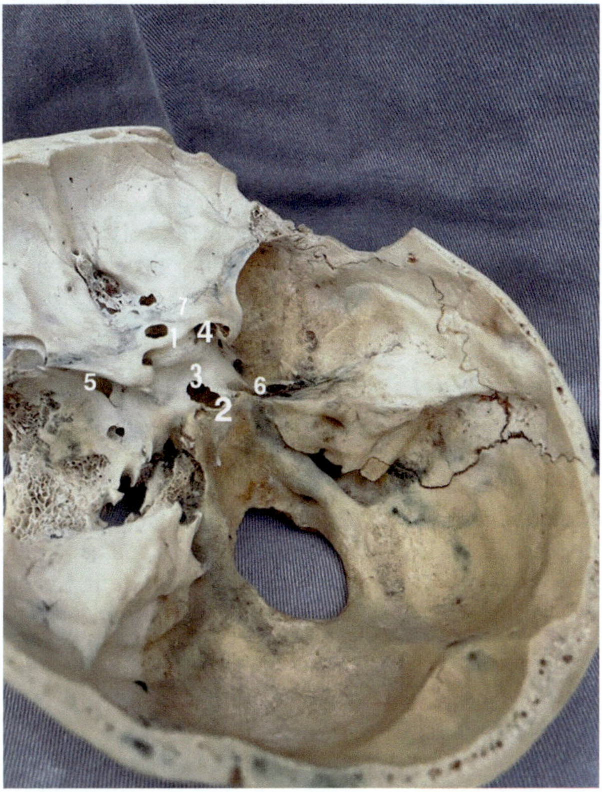

Fig. 2 Oblique view of the skull base highlighting the sphenoid and ethmoid bones. (*1*) Tubercullum sellae anterior to the chiasmatic groove. (*2*) Dorsum sellae. (*3*) Sella turcica. (*4*) Optic strut. (*5*) Superior orbital fissure. (*6*) Ossified Gruber ligament. (*7*) Planum sphenoidale

articulates with the parietal zygomatic and palatine bones [2]. It consists of a central body—containing the hypophyseal fossa, the sphenoidal sinus, and the planum sphenoidale—two greater and two lesser wings [3] (Fig. 2).

The greater wings form the anterior middle fossa and have a series of foramina; from anterior to posterior: foramen rotundum, foramen ovale, and the foramen spinosum.

The lesser wing constitutes the posterior half of the anterior fossa, and its sphenoidal ridge ends medially in the anterior clinoid. Between the greater and lesser wings is located the superior orbital fissure [2]. The optic canal is located between the lesser wing and the body of the sphenoid bone and is separated from the superior orbital fissure by the optic strut, which ends superiorly at the anterior clinoid [3].

Contained in the body is the pituitary fossa. It is anteriorly limited by the tuberculum sellae with its two bony prominences—the middle clinoid process—and posteriorly limited by the dorsum sellae—with the posterior clinoid process. The tuberculum sellae is divided from the planum sphenoidale by the ocular groove—where the optic chiasm sits. Lateral to the body of the sphenoid is the carotid sulcus, where the internal carotid artery transverses during its intracavernous course. Projecting inferiorly, one finds the medial and lateral pterygoid processes—disposed perpendicularly—representing the insertion site of the pterygoid muscles. This V-shaped space is called the pterygoid fossa [3].

Lateral to the body is another osseous prominence—the sphenoidal lingula—site of insertion of the petrolingual ligament, between the lingula and the petrous apex. This ligament roofs Dorello's canal and the abducens nerve [3] (Fig. 2).

4 Vascular Relationships

The cavernous segment of the internal carotid artery will course above the carotid sulcus lateral to sphenoidal body, the clinoid segment is located below the anterior clinoid process, and Supra clinoid segment will course lateral to the pituitary fossa. In contrast, middle cerebral artery has its course parallel to sphenoid ridge. Basilar artery rests partially over the clinoidal part of the sphenoid bone [3]. The cavernous sinus is located laterally to the sphenoid bone.

5 Neural Relationships

The olfactory tracts and the basal surface of the frontal lobe rest over the planum sphenoidale and lesser wing, while the temporal lobes rest above the greater wing.

Inside the superior orbital fissure passes the cranial nerves III, IV, VI, and V1 on its way to the orbit. V2 enters through foramen rotundum, V3 through foramen ovale, and the middle meningeal artery through the foramen spinosum [3].

6 Sphenoidal Sinus

The sphenoidal sinus is one of the paranasal sinuses, perhaps the most important to the neurosurgeon, given the importance of the transsphenoidal endoscopic approaches to the skull base and sellar region. It is located inside the sphenoidal

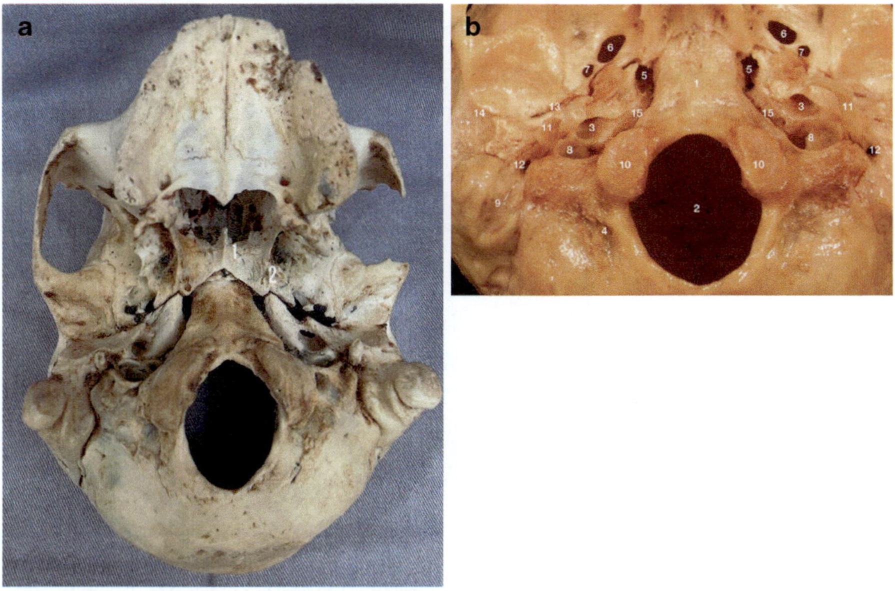

Fig. 3 (**a**) Inferior view of the skull base: (*1*) Vomer. (*2*) Pterigoyd process—perpendicular and horizontal plates. (**b**) Other representation of inferior view of the skull base. (*1*) Clivus. (*2*) Foramen magnum. (*3*) Carotid canal. (*4*) Condylar canal. (*5*) Foramen lacerum. (*6*) Foramen ovale. (*7*) Foramen spinosum. (*8*) Jugular foramen. (*9*) Mastoid notch. (*10*) Occital condyle. (*11*) Styloid process. (*12*) Stylomastoid foramen. (*13*) Petrotympanic fissure. (*14*) Squamotympanic fissure. (15) Petroclival fissure

body and opens to the nasal cavity by the sphenoid Ostia below the superior conchae. It has variable size and pneumatization, being classified in three categories:

Conchal—non-pneumatized—more common in children

Presellar—pneumatization up to the anterior part of the pituitary fossa

Sellar—most common, pneumatized up to the clivus

It possesses a series of septae variables in size, shape, orientation, implantation, and width, creating asymmetric cavities [3].

Its posterior wall has a series of structures of utmost importance to the neurosurgeon: At the center of the posterior wall, one can find the impression of the pituitary fossa, medial to the carotid prominence, and superolateral the osseous impression of the optic nerves. Among these structures is the Opticocarotid recess [3] (Figs. 3a, b and 4).

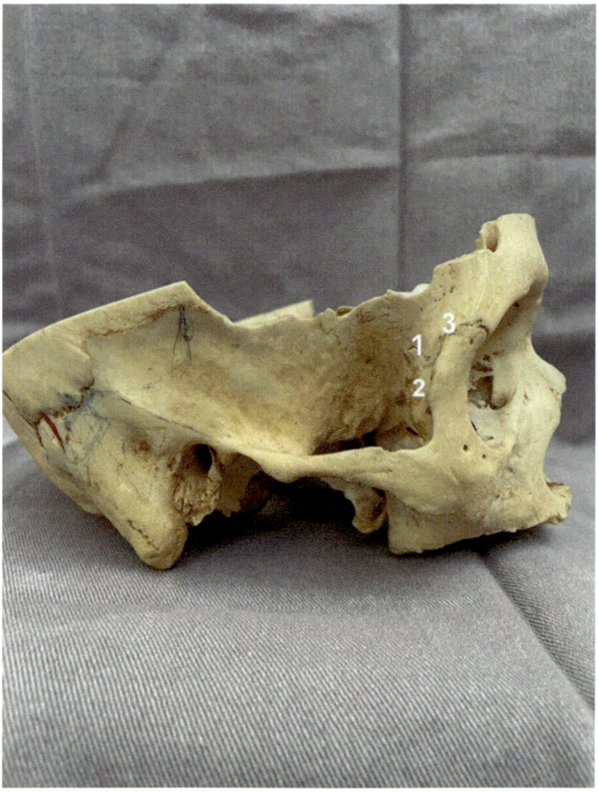

Fig. 4 Lateral view of the skull highlighting the sphenoid bone. (*1*) Pterion. (*2*) Sphenoid bone. (*3*) Keyhole

7 Temporal Bone

The temporal bone is a complex part located inferolateral on the skull and is divided into the squamosal, mastoid, tympanic, styloid, and petrous segments. It is quite important to understand the anatomy of the temporal bone due to several numbers of skull base approaches. The squamosal part is located laterally and encloses the temporal lobe. The mastoid part is pneumatized and located posterior to the external acoustic meatus (Fig. 5). The petrous part is medial and contains important neurovascular structures like the components of the internal ear, facial nerve, and carotid canal. The tympanic part forms part of the wall of the tympanic cavity and the external acoustic meatus. The styloid part is a thin projection that goes down and attaches several muscles [4] (Figs. 5, 6, and 7).

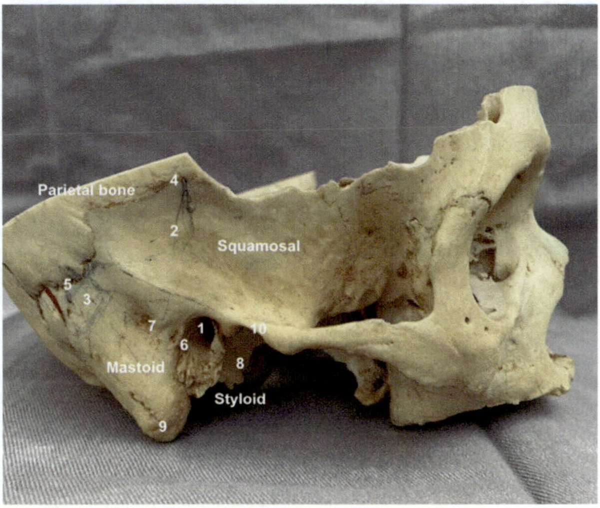

Fig. 5 Lateral view of the skull exposing the squamosal, mastoid and styloid segments of the temporal bone. (*1*) External acoustic meatus. (*2*) Superior temporal line. (*3*) Supramastoid crest. (*4*) Squamosal suture. (*5*) Parietomastoid suture. (*6*) Suprameatal spine of Henle. (*7*) Suprameatal triangle. (*8*) Mandibular fossa. (*9*) Mastoid process. (*10*) Posterior part of the zigomatic process

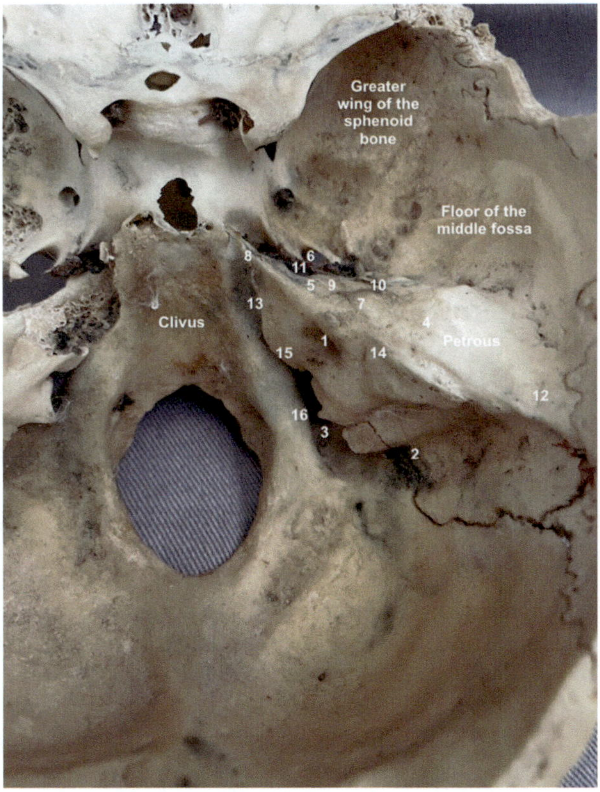

Fig. 6 Superior view of the skull base presenting the middle and posterior fossa. (*1*) Internal acoustic meatus. (*2*) Sigmoid sulcus. (*3*) Jugular foramen. (*4*) Tegmen. (*5*) Trigeminal impression. (*6*) Foramen ovale. (*7*) Arcuate eminence. (*8*) Apex. (*9*) Roof of the carotid canal. (*10*), Impression of the great petrosal nerve. (*11*) Foramen lacerum. (*12*) Base of the petrous part. (*13*) Petroclival fissure. (*14*) Superior border of the petrosal segment. (*15*) Inferior border of the petrosal segment. (*16*) Jugular process

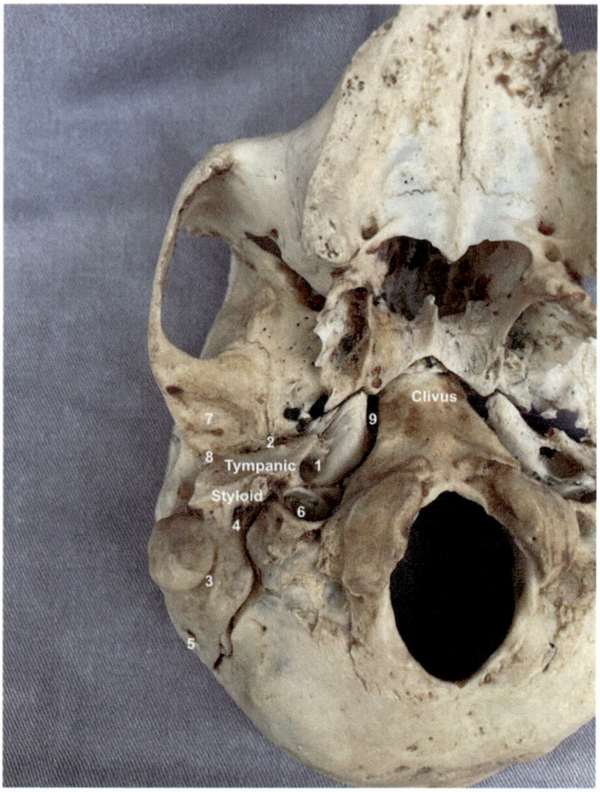

Fig. 7 Inferior view of the skull base. (*1*) Carotid canal. (*2*) Petrotympanic fissure. (*3*) Mastoid notch. (*4*) Stylomastoid foramen. (*5*) Emissary foramen. (*6*) Jugular foramen. (*7*) Mandicular fossa. (*8*) Squamotympanic fissure. (*9*) Petroclival fissure

8 Lateral Surface

The superior temporal line (n.2, Fig. 5) continues posteroinferior as the supramastoid crest (n.3, Fig. 5) and joins into the upper edge of the zygomatic arch. The supramastoid crest represents the floor of the middle fossa when seen on the skull's external surface. The supramastoid crest joins with the squamous suture (n.4, Fig. 5) at the level of the lateral end of the petrous ridge [5]. The parietomastoid suture (n.5, Fig. 5) separates the parietal bone and the mastoid bone, and its junction with the squamous suture is located on the level of the anterior edge of the sigmoid and transverse sinuses junction. The asterion is the junction between the occiptomastoid, parietomastoid, and lambdoid sutures and is over the junction of the lower part of the transverse and sigmoid sinuses [5, 6].

The suprameatal spine of Henle (n.6, Fig. 5) located at the posterosuperior edge of the external meatus represents the level of the semicircular canal and tympanic segment of the facial nerve. The suprameatal triangle (n.7, Fig. 5) is a depressed area behind the spine of Henley that represents the mastoid antrum located deeply [4].

9 The Tympanic Part

The tympanic part is a concave curved plate anterior to the mastoid process and forms the anterior wall, floor, and part of the posterior wall of the external acoustic meatus. It also forms the posterior wall of the mandibular fossa (n.8, Fig. 5). Between the tympanic and petrous parts, there is the petrotympanic fissure [4] (n.2, Fig. 7).

10 The Mastoid Part

Located at the posterior part of the temporal bone, it has a downward projection that forms the mastoid process (n.9, Fig. 5), where the sternocleidomastoid, splenius capitis, longissimus captis, and the posterior belly of the digastric muscles attach to it. The mastoid bone contains the air mastoid cells and the rigid part nominated mastoid antrum [4, 7]; medial, there is a groove called mastoid notch (n.3, Fig. 7). Beside and medial to the notch, there is the occipital groove, where lies the occipital artery. The facial nerve exits through the stylomastoid foramen (n.4, Fig. 7). In the posterior border of the mastoid process, an emissary vein to the sigmoid sinus and a dural branch from the occipital artery enter into an emissary foramen (n.5, Fig. 7). The sigmoid sinus represents the posterior limit of the mastoid cavity, and by a posterior sight, we can find the sigmoid sulcus (n.2, Fig. 6) descending along its posterior surface [7].

By the approaches through the mastoid and exposing its contents by mastoidectomy, we access an important landmark to expose the posterior fossa anterolateral – sinodural angle. This posterior fossa dura mater area is surrounded by the sigmoid sinus posteriorly, the superior petrosal sinus superiorly, the otic capsule anteriorly, and the jugular bulb inferiorly. The distance between the apex of the jugular bulb and the superior petrosal sinus is also an important landmark of the size of the exposure that can be achieved by opening Trautman's triangle. This distance is reduced if there is a high jugular bulb [4, 5]. After its angle, the sigmoid sinus curves medially and forward to the jugular foramen (n.3, Fig. 6/n.6, Fig. 7). The inferior limit of the mastoid cavity is the superior aspect of the jugular foramen located in the jugular bulb [7]. The medial limit of the mastoid cavity is a solid bone containing the otic capsule and the bony labyrinth. The tegmen (n.4, Fig. 6) in the floor of the middle cranial fossa is the roof of the mastoid segment.

The tympanic cavity is a narrow air-filled space between the tympanic membrane laterally and the promontory containing the auditory and vestibular labyrinth medially and communicates posteriorly with the mastoid antrum and anteriorly through the Eustachian tube with the nasopharynx. The middle fossa is the roof of the tympanic cavity, where there is a thin plate called tegmen tympani [8].

11 The Squamous Part

The squamous bone forms some parts of the floor and lateral wall of the middle cranial fossa, besides the posterior part of the zygomatic arch and the upper part of the mandibular fossa. The convex surface of the squamosal part is located laterally, giving attachment to the temporalis muscle on the outer surface. Its cerebral surface is concave, enclosing the temporal lobe and connecting to the greater wing of the sphenoid bone anteriorly [8, 9] (Fig. 6). Its inferior surface has the site of the mandibular fossa where the condyle joins. The posterior part of the zygomatic process (n.10, Fig. 5) joins to the squama through its anterior and posterior roots. The latter blends posteriorly into the suprameatal crest. Between and at the lower margin of these two roots, the mandibular fossa (n.7, Fig. 7) is delimited in front by the articular tubercle and posteriorly by the postglenoid tubercle [7, 10]. The squamotympanic fissure (n.8, Fig. 7) is located between the medial part of the squamosal and the tympanic part below the mandibular fossa. The squamous part forms the upper posterior wall and roof of the external acoustic meatus.

12 The Petrous Part

The petrous part of the temporal bone is located between the sphenoid and occipital bones. The main structures found inside are the vestibular and acoustic labyrinth, facial nerve, internal acoustic meatus, and carotid canal [4]. In the upper surface of the petrous part are the trigeminal impression (n.5, Fig. 6) behind the foramen ovale (n.6, Fig. 6) and arcuate eminence (n.7, Fig. 6). At the inferior part, there is the jugular fossa. The anterior part of the petrous bone between the greater wing of the sphenoid and the occipital bone is the apex (n.8, Fig. 6). This anterior portion has a foramen called the carotid canal (n.1, Fig. 7), where the carotid artery enters medially in the skull base and becomes the petrous segment of the carotid artery. The lateral genu of the petrous carotid at the junction of the vertical and horizontal segments is below and medial to the cochlea [4, 11].

Its anterior surface represents the floor of the middle cranial fossa, and its irregular feature is grooved by the trigeminal impression medially and the roof of the carotid canal anterolateral (n.9, Fig. 6). The dura mater that covers the trigeminal ganglion is nominated Meckel's cave. A shallow depression lateral to the trigeminal impression roofs the internal acoustic meatus, and more lateral to this, there is the arcuate eminence overlying the superior semicircular canal. The area behind the arcuate eminence overlies the posterior and lateral semicircular canals. Laterally, the tegmen roofs the mastoid antrum and tympanic cavities and the canal for the tensor tympani [4, 5, 8]. Drilling the tegmen from above exposes the incus, head of the malleus, the tympanic segment of the facial nerve, and the superior and lateral semicircular canals. Anterior to the tegmen passes the greater petrosal nerve (n.10, Fig. 6) in front of the arcuate imminence crossing the floor of the middle fossa until joins with the geniculate ganglion toward the foramen lacerum (n.11, Fig. 6). The greater petrosal nerve runs beneath the dura of the middle floor in the sphenopetrosal groove, despite in 16% of the specimens there is not any bone over the geniculate ganglion, thus exposing the facial nerve and geniculate ganglion to injury during subtemporal middle fossa approach through peeling of the dura mater [5, 8].

The posterior surface of the petrosal part faces the posterior cranial fossa and is continuous with the mastoid part. In between the base (n.12, Fig. 6) and the apex of the petrous bone is situated the internal auditory meatus. It is divided into superior and inferior halves by the transverse crest. In the area below the transverse crest enters the cochlear nerve anteriorly and the inferior vestibule nerve posteriorly. Above the transverse crest, a vertical crest called Bill's bar separates into an anterior part where the facial nerve enters and into a posterior part where the superior vestibule nerve goes into [7, 9].

If we observe the inferior surface of the petrous bone, we can notice a very irregular surface. Its apex is connected medially to the clivus by the petroclival fissure (n.13, Fig. 6/n.9, Fig. 7). Behind there is the carotid canal and the jugular fossa that contains the jugular bulb.

The superior border along the petrous ridge lies the superior petrosal sinus, and the tentorium cerebelli attaches to it, except where the posterior trigeminal root crosses. The inferior petrosal sinus resides along the inferior border and connects with the cavernous sinus anteriorly and the jugular bulb posteriorly (n.14 and 15, Fig. 6). The jugular fossa on the posterior surface of the petrous part joins with the jugular notch on the jugular process (n.16, Fig. 6) of the occipital bone to form the margins of the jugular foramen. The sigmoid sinus descends along the posterior surface of the mastoid and turns forward on the occipital bone to pass through the petrosal part of the jugular foramen. The lower end of the petro-occipital fissure resides the jugular foramen.

The jugular foramen is divided into a lateral opening – the sigmoid part – that receives de sigmoid sinus and the medial part – the petrosal part – that receives the inferior petrosal sinus. Between these two parts of the jugular foramen, the intrajugular part is where the glossopharyngeal, vagus, and accessory nerves enter [7]. The petrosquamosal suture articulates the temporal squama and the petrous part laterally. The bone labyrinth is formed by the vestibule, the semicircular canals, and the cochlea. The vestibule is medial to the tympanic cavity, posterior to the cochlea, and superior to the jugular bulb. Its aqueduct connects with the endolymphatic sac in the posterior surface inferolateral to the internal acoustic meatus [4, 7, 8]. The semicircular canals are located posterosuperior to the vestibule. Since the superior semicircular canal is in close relation to the middle fossa floor, it is the most susceptible to damage during the middle fossa approach to the internal acoustic meatus [4, 8].

The facial nerve is divided into cisternal, labyrinthine, tympanic, mastoid, and styloid segments. Three of these five segments are located inside the bone. The first is the labyrinthine segment located in the petrous part from the internal acoustic meatus to the geniculate ganglion. Then, the facial nerve turns laterally and posteriorly along the medial face of the tympanic cavity below the lateral semicircular canal being the tympanic segment. At the midpoint of the lateral semicircular canal, the facial nerve turns vertically downward adjacent to the mastoid part of the temporal bone and is called the mastoid segment [12, 13].

13 The Styloid Part

In this small portion of the temporal bone, there is a spicule called the styloid process, which projects from the inferior border of the tympanic bone into the infratemporal fossa (Figs. 5 and 7). The muscles styloglossus, stylopharyngeus, and stylohyoid are attached to the styloid process. Posteriorly, there is the stylomastoid foramen where the facial nerve emerges [4].

14 Frontal, Parietal, and Occipital Bone

The skull is made up of 28 different bones, most connected by fibrous sutures. It is divided into three parts: calvaria, skull base, and facial skeleton. The calvaria is formed by the frontal, parietal, occipital, sphenoid, and occipital bones. These are divided into the main cranial sutures: coronal, sagittal, and lambdoid sutures. The bones that make up the skull base connect and form a network of canals and foramina that allow the passage of neurovascular structures that pass from the endocranial (inner surface) to the exocranial (outer surface) compartment [14].

15 Frontal Bone

The frontal bone is a unique and wide bone formed through the junction of the metopic suture, constituting a single structure. The frontal bone articulates with the parietal, zygomatic, sphenoid, ethmoidal, lacrimal, maxillary, and nasal bones. The frontal bone consists of three parts; the squamous part, the orbital part, and the nasal part. The squamous part is the widest and smoothest part (Fig. 8).

On the outside of the frontal bone, around the midline, there are two elevations known as the frontal eminences in the paramedian region. Among these structures are the superciliary arches that connect in the midline with the name glabella. Laterally, the supraorbital margin forms the orbital rim that contains the orbital notch (which can sometimes appear as a foramen) (n.2, Fig. 8) through which the supraorbital nerves and veins pass [2, 14, 15].

Inferior to the glabela (n.3, Fig. 8) is the nasal notch and nasal spine, which articulate with the ethmoid's nasal bones and perpendicular plate. The cranial surface of the squamous portion of the frontal bone contains the sagittal groove, in which the superior sagittal sinus resides. The sulcus boundaries extend inferiorly,

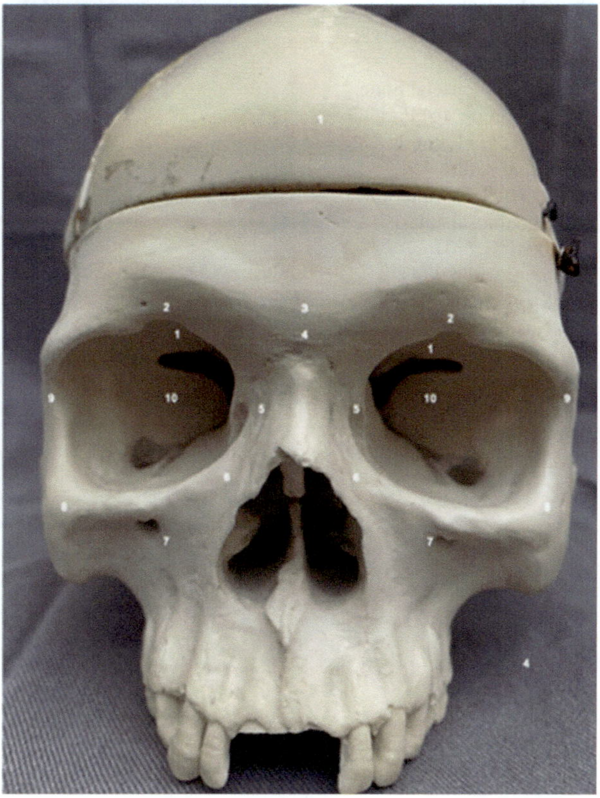

Fig. 8 Frontal view of the skull. (*1*) Frontal bone. (*2*) Supraorbital foramen or notch. (*3*) Glabella. (*4*) Nasion. (*5*) Lacrimal bone. (*6*) Maxilla. (*7*) Infraorbital foramen. (*8*) Zygomatic bone. (*9*) Fronto-zygomatic suture. (*10*) Sphenoid bone

forming the frontal ridge to which the cerebral falx connects. The orbital portion of the frontal bone is formed by the two orbital plates connected by the ethmoidal notch, which is filled by the cribiform plate of the ethmoidal. The lower surface of each orbital plate contains a small depression under the zygomatic process called the lacrimal fossa. The orbital portion of the frontal bone contains the frontal sinus and the frontonasal duct [2, 15].

Among the main craniometric points (reference points on the skull, correlation of anatomical accidents of the cerebral cortex with points on the cranial surface) of the frontal bone and surroundings, there are Nasion (n.4, Fig. 8): located at the frontonasal angle; it corresponds internally to the ethmoidal crest (or crista galli), in the midline of the anterior or frontal fossa. Glabella (n.3, Fig. 8): mid-frontal protuberance, located between the ciliary arches, above the nose, related to the superior sagittal sinus (anterior part) and interhemispheric fissure. Pterion (n.12, Fig. 9): defined by the H formed by the junction of coronal, squamous, sphenoparietal, sphenofrontal, and sphenotemporal sutures. Stephanion (n.5, Fig. 9): stitch at the junction of the coronal suture with the superior temporal line. It corresponds on the cerebral surface to the intersection of the precentral and inferior frontal sulci [16–19].

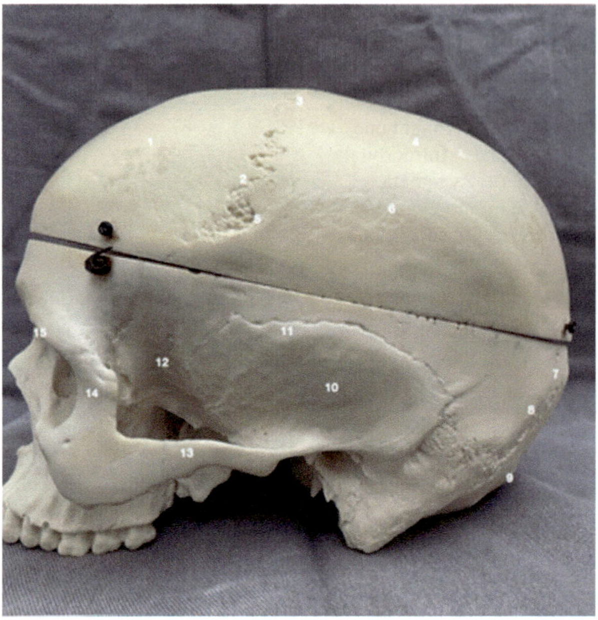

Fig. 9 Lateral view of the skull. (*1*) Frontal bone. (*2*) Coronal suture. (*3*) Bregma. (*4*) Parietal bone. (*5*) Stephanion. (*6*) Parietal eminence. (*7*) Lambda. (*8*) Lambdoid suture. (*9*) Inion. (*10*) Temporal bone. (*11*) Squamosal suture. (12) Pterion. (13) Zygomatic arch. (14) Frontozygomatic suture. (15) Nasio

16 Parietal Bone

The parietal bones, two in number, articulate in the midline, through the sagittal suture, to form the sides and roof of the skull. In addition to relating to each other, in the midline through the sagittal suture, they articulate in the anterior part with the frontal bones (n.1, Fig. 10), through the coronal suture (n.2, Fig. 10), and articulate inferolateral with the temporal (n.10, Fig. 9) and sphenoid bones through the squamous sutures (n.11, Fig. 9), and also with the occipital bones through lambdoid sutures (n.8, Fig. 9). A prominence marks the outer surface of the parietal bone near the central region of the bone, called the parietal eminence (n.6, Fig. 10) [2, 15, 17, 18].

On the inner surface of the parietal bone, there is sulcus for superior sagittal sinus and sulcus for middle meningeal artery inferiorly [2, 15]. The main craniometric point of the parietal bone is the Eurion, located at the end of the largest transverse diameter of the head, at the most prominent point of the parietal tuberosity. It corresponds on the cortical surface to the supramarginal gyrus [19, 20].

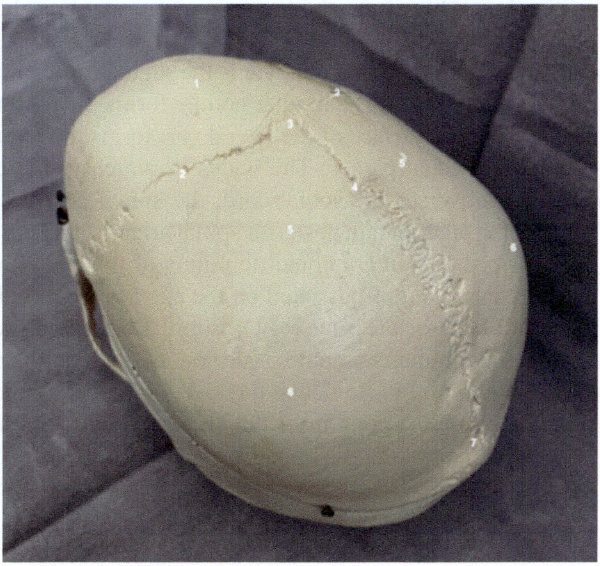

Fig. 10 Posterior-superior view of the skull. (*1*) Frontal bone. (*2*) Coronal suture. (*3*) Bregma. (*4*) Sagital suture. (*5*) Parietal bone. (*6*) Parietal eminence. (*7*) Lambda

17 Occipital Bone

The occipital bone comprises the largest portion of the posterior part of the skull. It has three parts; squamous part, basilar part, and lateral part. The occipital bone articulates with the parietal bone through the lambdoid suture (n.8, Fig. 9), with the temporal and sphenoid bone through the occipitomastoid suture. On the outside, the most prominent part of the occipital bone is the external occipital bulge, specifically, the inion (n.9, Fig. 9), which marks the location of the confluence of the sinuses (or Torcula of Herophilus), where the nuchal ligament and the trapezius muscle are inserted. The occipital planum is the thinnest portion of bone superiorly. Inferior to occipital planum are a series of nuchal lines. The upper and lower nuchal lines are oriented transversely. The superior nuchal line connects medially to the external occipital bulge. The midline nuchal extends from the external occipital bulge to the foramen magnum. The inner surface of the scaly part contains the inner occipital bulge (where the confluence of the sinuses is located) [19–22].

The basilar part of the occipital bone extends from the foramen magnum to the upper part of the clivus, which articulates with the dorsum sella and the sphenoid bone. The outer surface of the basilar part contains the pharyngeal tubercle. The lateral parts of the occipital bone form the walls of the foramen magnum; on its inferior surface, the occipital condyles are located. Anterior to the occipital condyle is the condylar fossa and the condylar canal, through which an emissary's vein passes. The hypoglossal canal is a tunnel inside the condyle, which crosses the hypoglossal nerve (XII) and the meningeal branch of the ascending pharyngeal artery [5, 7, 10].

The hypoglossal canal is an important anatomic landmark for extreme-lateral (far-lateral) approaches to the ventral part of the brainstem. On the external surface, extending laterally to the condyle is the jugular process, in which the jugular nodule is located anteriorly. The jugular nodule forms the posterior wall of the jugular foramen. The upper surface of the lateral part forms the jugular tubercle, over which the hypoglossal canal passes. The widest foramen of the occipital bone, the foramen magnum, traverses the spinal cord, the spinal accessory nerve (XI), the vertebral arteries, the anterior and posterior spinal arteries, and the alar ligaments [5, 7, 15, 16].

Among the main craniometric points of the occipital bone, the following stand out: Inion (n.9, Fig. 9): located on the external occipital bulge, internally related to the confluence of the sinuses; Opistion: Located at the midpoint of the posterior edge of the foramen magnum; Asterion: at the junction of lambdoid, parietomastoidea, and occipitomastoidea sutures. It lies on the junction of the transverse and sigmoid sinuses. A point immediately above the asterium corresponds to the preoccipital notch, marking the boundary between the temporal and occipital lobes at the inferolateral edge of the cerebral hemisphere. Opistocranion: corresponds to the most prominent cranial point [17–20].

18 Conclusion

The following bones and their relationships are of the utmost importance for the neurosurgeon when studying the cranial skull: Frontal, parietal, occipital, sphenoid, ethmoid, and temporal bones. Also relevant is the relationship between the bones and the neural and vascular structures found during surgical approaches. The study of them provides strategies to manneurgery approach to skull base.

References

1. Yu M, Wang SM. Anatomy, head and neck, ethmoid bone. In: StatPearls. Treasure Island, FL: StatPearls Publishing; 2020.
2. Hendricks BK, Patel AJ, Hartman J, Seifert MF, Cohen-Gadol A. Operative anatomy of the human skull: a virtual reality expedition. Oper Neurosurg. 2018;15(4):368–77. https://doi.org/10.1093/ons/opy166.
3. Rhoton AL. Cranial anatomy and surgical approaches, The Congress of Neurological Surgeons. Philadelphia: Lippincott Williams and Wilkins; 2003.
4. Rhoton AL Jr. Overview of temporal bone. Neurosurgery. 2007;61(Suppl_4):7–60.
5. Rhoton AL Jr. Osseous relationships. Neurosurgery. 2007;61:S4–S84.
6. Sampson HW, Montgomery JL, Henryson GL. Atlas of the human skull. 2nd ed. College Station: Texas A&M University Press; 1991.
7. Rhoton AL Jr. Jugular foramen. Neurosurgery. 2000;47(Suppl_3):S267–85.
8. Rhoton AL Jr. The anterior and middle cranial base. Neurosurgery. 2002;51(4 Suppl):S273–302.

9. Stredney D, Wiet GJ, Bryan J, et al. Temporal bone dissection simulation–an update. Stud Health Technol Inform. 2002;85:507–13.
10. Rhoton AL Jr. The foramen magnum. Neurosurgery. 2000;47(Suppl):S155–93.
11. Funaki T, Matsushima T, Peris-Celda M, Valentine RJ, Joo W, Rhoton AL Jr. Focal transnasal approach to the upper, middle, and lower clivus. Neurosurgery. 2013;73(2 Suppl Operative):ONS155–90.
12. Brackmann DE. Translabyrinthine/transcochlear approaches. In: Sekhar LN, Janecka IP, editors. Surgery of cranial base tumors. New York: Raven Press; 1993. p. 351–65.
13. House WF. Middle cranial fossa approach to the petrous pyramid. Arch Otolaryngol. 1963;78:460–9.
14. Osborn AG. Brain: imaging, pathology and anatomy. Philadelphia: Lippincott Williams and Wilkins; 2012.
15. Cohen-Gadol A. The neurosurgical atlas—skull anatomy. https://www.neurosurgicalatlas.com/volumes/operative-neuroanatomy/supratentorial-operative-anatomy/skullanatomy.
16. Gusmão S, Silveira RL, Arantes A. Pontos referencias nos acessos cranianos. Arq Neuropsiquiatr. 2003;61:305–8.
17. Gusmão S, Silveira SL, Cabral G, Arantes A. Topografia cranioencefálica: aplicações neurocirúrgicas. Arq Bras Neurocir. 1998;17(2):59–71.
18. Gusmão S, Reis C, Silveira RL, Cabral G. Relationships between the coronal suture and the sulci of the lateral convexity of the frontal lobe: neurosurgical applications. Arq Neuropsiquiatr. 2001;59(3-A):570–6.
19. Ribas GC. Applied cranieal-cerebral anatomy: brain architecture and anatomically oriented microneurosurgery. Cambridge: Cambridge University Press; 2018.
20. Ribas GC, Yasuda A, Ribas EC, Nishikuni K, Rodrigues AJ Jr. Surgical anatomy of microsurgical sulcal key points. Neurosurgery. 2006;59(4 Suppl 2):NOS177–210.
21. Tubbs RS, Salter G, Oakes WJ. Superficial surgical landmarks for the transverse sinus and torcular herophili. J Neurosurg. 2000;93(2):279–81.
22. Reis CV, Arantes A, Nicolato A, Gusmão S. Delimitação dos acessos cranianos. Arq Bras Neurocirur. 2012;31(3):135–45.

Surgical Anatomy of the Temporal Bone and Transtemporal Approaches

Gustavo Rassier Isolan, Jander Moreira Monteiro, Marcelo Moro da Rocha, and Joel Lavinsky

1 Introduction

The temporal bone is divided differently according to age. In the fetus, it is split into three portions: squamous (encompasses the anterior and superior aspect, constituted by a thin irregular and circular plate), petrous (posterior and internal to the squamous portion, rigid and often described as a pyramid, with its base forming the skull external surface and the apex situated anteromedially), and the tympanic part (situated inferiorly to the squamous and externally to the petrous portion, round-shaped).

As the fetus grows, the squamous and petrous portions merge, forming the petrosquamous suture and developing posteroinferiorly, molding the mastoid part of the temporal bone. The tympanic bone grows medially, forming an incomplete canal corresponding to the inferior, anterior, and posterior walls of the external auditory meatus. The squamous portion composes the upper wall of this canal. The styloid process is the last to develop, only after birth, from the tympanic bone. A fully developed temporal bone is didactically divided into five portions: mastoid, petrous, squamous, tympanic, and styloid (Fig. 1).

G. R. Isolan (✉) · J. M. Monteiro
The Center for Advanced Neurology and Neurosurgery (CEANNE), Porto Alegre, RS, Brazil

Mackenzie Evangelical Faculty of Paraná (FEMPAR), Curitiba, PR, Brazil

M. M. da Rocha
Department of Otorhinolaryngology, Santa Casa de Porto Alegre, Porto Alegre, RS, Brazil

J. Lavinsky
Department of Otorhinolaryngology, Santa Casa de Porto Alegre, Porto Alegre, RS, Brazil

Federal University of Rio Grande do Sul, Porto Alegre, RS, Brazil

E. G. Figueiredo et al. (eds.), *Brain Anatomy and Neurosurgical Approaches*,
https://doi.org/10.1007/978-3-031-14820-0_4

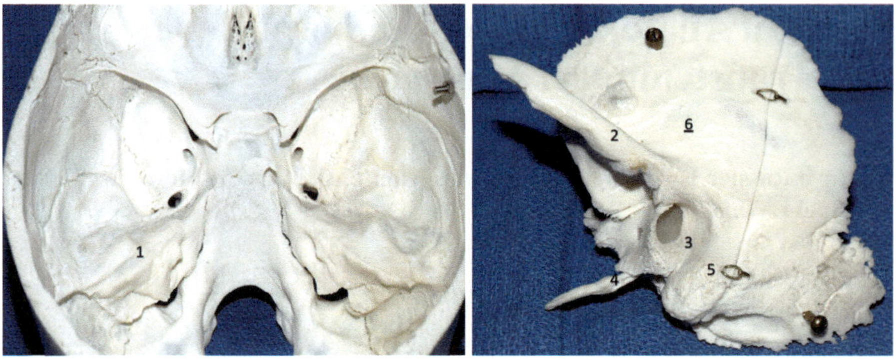

Fig. 1 The parts in the temporal bone are: (*1*) petrous; (*2*) zygomatic; (*3*) tympanic; (*4*) styloid; (*5*) mastoid; and (*6*) squamous

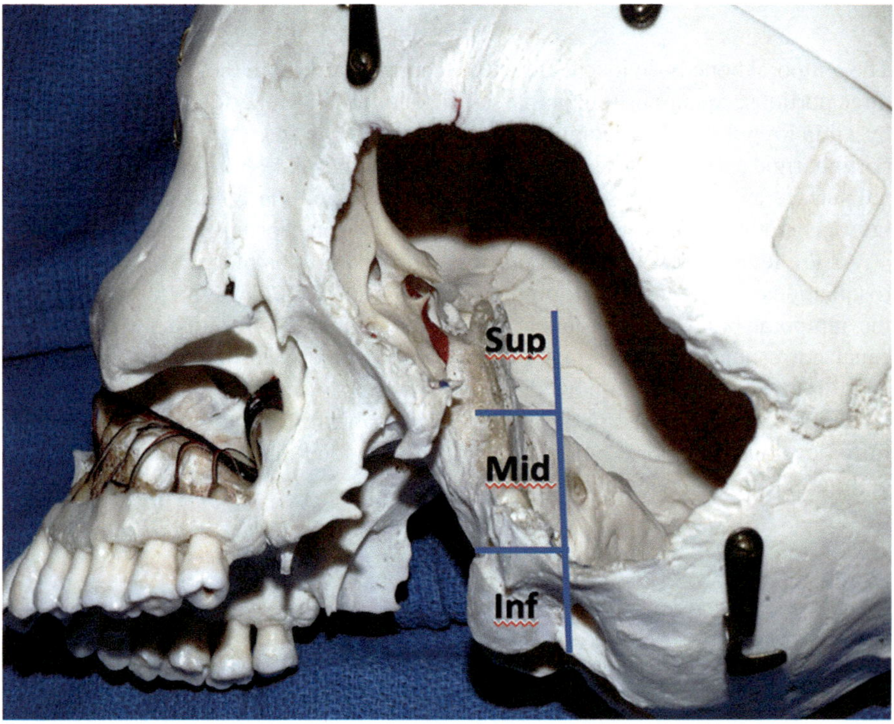

Fig. 2 The temporal bone was removed. Note that the transtemporal approach can reach the superior (Sup) and middle (Mid) thirds of the clivus, but not the inferior (Inf) third, which corresponds to the anterior border of the foramen magnum

This chapter presents the microsurgical anatomy of the temporal bone and its approaches and illustrative cases to correlate this anatomy with intraoperative findings (Fig. 2). Intraoperative images are from patients that authors operated on. The authors performed the dissections.

2 Temporal Bone Anatomy

2.1 Squamous Portion

This thin and shell-shaped portion constitutes most of the temporal bone's lateral surface and is the bigger of the temporal bones. Its plate has a medial cerebral and a lateral temporal side. The anteroinferior part of this bone originates the zygomatic process, contributing to forming the zygomatic arch along with the zygomatic bone.

The squamous bone is divided into three portions:

- A vertical part, with a flat and thin plate forming the lateral wall of the middle cranial fossa, corresponding to the projection of cerebral sulci and gyri (one of these corresponds to the middle meningeal artery) and covered by the temporal muscle.
- The horizontal or inferior part prolongs anteriorly, forming the zygomatic process, the bony roof of the glenoid fossa, and the roof of the external meatus and mastoid cavity.
- The retromeatal portion corresponds to the posterior wall of the external meatus and the anterior wall of the mastoid cavity. In this region, posterosuperiorly to the external meatus can be found the Henle's Spine (or suprameatal spine), an important anatomical landmark corresponding to the attic medially. The cribiform area is posterior to the spine, where projections of mastoid cells appear over the temporal surface.

An important surgical landmark is the temporal line, or supramastoid crest, where the temporal muscle attaches inferiorly. It extends from the zygomatic process posteriorly and usually corresponds to the level of the middle cranial fossa dura [1, 2].

2.2 Mastoid Portion

It is a conic projection of the petrous portion, adhered to the squamous and tympanic portions, and located at the posteroinferior border of the lateral surface of the temporal bone. The mastoid is flat at birth, and as it pneumatizes and its lateral portion grows, the stylomastoid foramen is medially pushed. The inferior part of the mastoid apophysis serves as an attachment to the sternocleidomastoid muscle laterally and to the posterior belly of the digastric muscle medially, which is immediately lateral to the stylomastoid foramen.

The mastoid part of the temporal bone shelters the lengthier segment of the facial nerve (known as the mastoid segment) until its exit at the stylomastoid foramen. Also, from the tympanomastoid suture emerges the auricular branch of the vagus nerve (Arnold's nerve) at the temporal bone, innervating the posterior aspect of the external auditory canal.

The sigmoid sinus, an intradural structure, forms a sulcus in the posterior aspect of the mastoid bone as it passes from a lateral to medial direction. The mastoid foramen is evident near the posterior limit of the mastoid cortex, transmitting a single emissary vein from the sigmoid sinus (Figs. 3, 4, and 5) [3–6].

2.3 Petrous Bone

This portion is a pyramid-shaped structure, with an apex projected anteromedially and a base composing part of the lateral surface of the skull, articulated with the squamous, tympanic, and mastoid portions of the temporal bone. It is divided into three portions: superior (molding part of the middle fossa, with an anterosuperior direction), inferior (related to the great vessels and nerves of the neck), and posterior (related to the posterior fossa) (Figs. 6 and 7).

Superior surface: Forms a major portion of the middle fossa floor. In its lateral region lies the petrosquamous suture, where the petrous bone merges with the temporal squama (and forms the Koerner's septum). Its anterior margin, known as the anterior angle, is free, and two small openings are seen, forming the musculotubal canal. The superior opening, narrower, forms the canal for the tympanic tensor

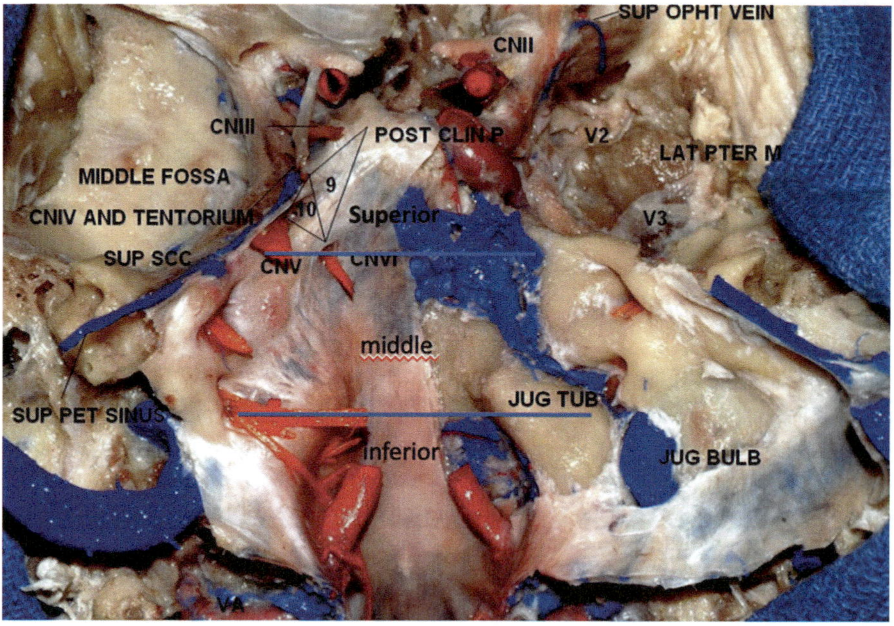

Fig. 3 Posterior view of the inferomedial paraclival (*9*) and inferolateral paraclival (*10*) triangles. The dura in the right clivus was peeled away. Jug. Tub, jugular tubercle; CN, cranial nerve; Post. Clin. P., posterior clinoid process; Lat. Pter. M., lateral pterygoid muscle; Sup. Opht. Vein, superior ophthalmic vein; Sup. Pet. Sinus, superior petrosal sinus; Sup. SCC., superior semicircular canal. The blue lines delimit the upper, middle, and lower clivus

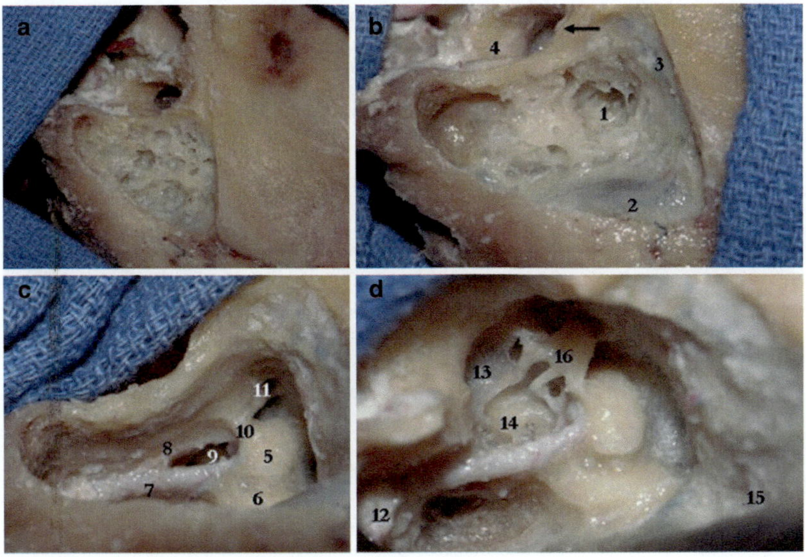

Fig. 4 (**a**) Anatomical knowledge of the mastoid portion of the temporal bone is essential to perform petrosectomies. (**b**) (*1*) Mastoid antrum; (*2*) sigmoid sinus; (*3*) mastoid tegmen; (*4*) external acoustic meatus. (**c**) (*5*) lateral semicircular canal; (*6*) posterior semicircular canal; (*7*) facial nerve; (*8*) chorda tympani nerve; (*9*) facial recess; (*10*) bony septum; (*11*) epitympanum. (**d**) (*12*) tendon of digastric muscle's posterior belly; (*13*) tympanic membrane; (*14*) promontory; (*15*) sinodural (Citelli's) angle; (*16*) incus

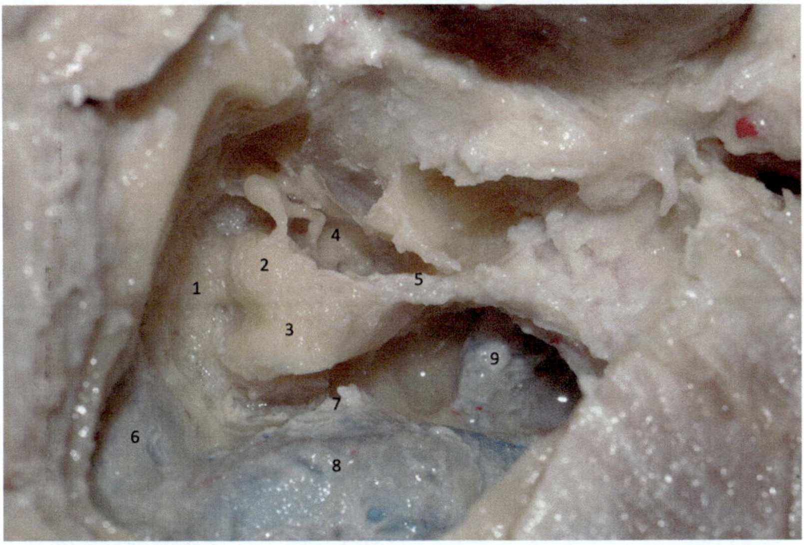

Fig. 5 Important landmarks are exposed at the end of a mastoidectomy. Superior (*1*), lateral (*2*), and posterior (*3*) semicircular canals. Promontory (*4*), facial nerve (mastoid portion) (*5*), temporal lobe dura (tegmen) (*6*), presigmoid dura (*7*), sigmoid sinus (*8*), and jugular bulb (*9*)

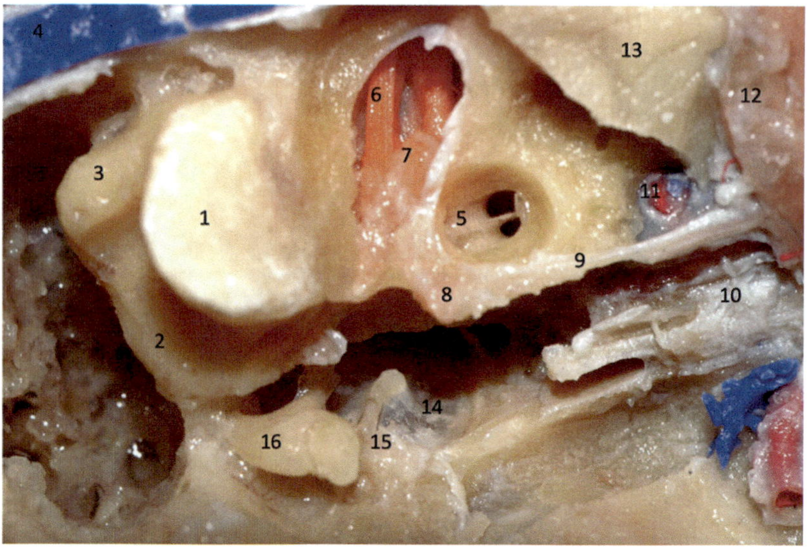

Fig. 6 Middle fossa dissection to expose the main landmarks for the middle fossa approach. (*1*) Superior semicircular canal; (*2*) posterior semicircular canal; (*3*) lateral semicircular canal; (*4*) superior petrosal sinus; (*5*) cochlea; (*6*) facial nerve; (*7*) superior vestibular nerve; (*8*) geniculate ganglion; (*9*) greater superficial petrosal nerve; (*10*) tensor tympani muscle; (*11*) intra-petrosal internal carotid artery; (*12*) V3; (*13*) petrous apex; (*14*) tympanic membrane; (*15*) malleus; (*16*) incus

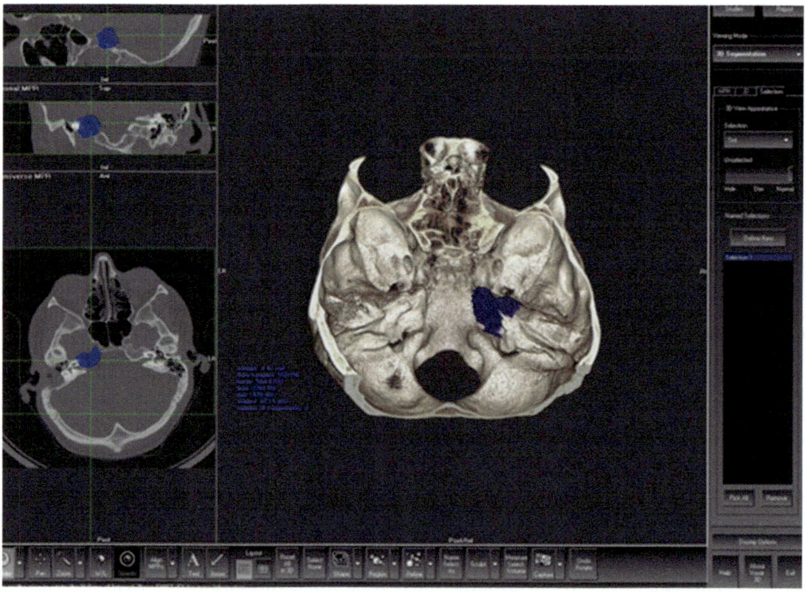

Fig. 7 Volumetry of the petrous apex showing the corridor to the superior and middle clivus

muscle, while the lower one, wider, forms the canal for the osseous portion of the Eustachian tube. Next to the central region of the anterior surface lies the arcuate eminence (an important landmark corresponding to a projection of the anterior semicircular canal at the middle fossa). Anterolateral to the arcuate eminence, the tegmen tympani forms the roof of the tympanic cavity.

Anterosuperiorly, toward the pyramid apex, lies the facial hiatus (where the superior petrosal artery, a branch of the middle meningeal artery, enters the middle ear and nourishes the geniculate ganglion and the tympanomastoid segment of the facial nerve) and the groove of the greater superficial petrosal nerve (which runs from the geniculate ganglion). The superior opening of the tympanic canaliculi is more lateral, parts the small superficial petrosal nerve and the superior tympanic artery. The tegmen tympani is a horizontal thin bone plate, posterolateral to the hiatus, and forms the roof of the tympanic cavity, Eustachian tube, and mastoid antrum, keeping it apart from the middle fossa.

The posterior surface is a vertical plate, forming part of the posterior cranial fossa anterior limit. The sulcus represents its superior limit for the superior petrosal venous sinus. Inferiorly, at the level of the posterior angle of the pyramid, the petrous bone funds with the occipital bone, exactly at the topography of the inferior petrosal sinus. On this surface lies midway, between the base and apex of the pyramid, the fundus of the internal acoustic meatus, with the foramina for the facial and the vestibulocochlear nerves (VII-VIII CNs). A little more lateral in this surface is the endolymphatic sac. A vertical bone plate separates it from the vestibule and contains many perforations destined to pass nervous filaments (cribiform area) and seals the lateral limit of the internal acoustic meatus. The internal acoustic meatus is divided transversally into two compartments (superior and inferior) by a bony crest known as the falciform crest. A vertical bone plate called Bill's Bar divides the superior segment in the anterior (containing the facial nerve) and posterior space (where the superior branch of the vestibular nerve). There is the passage of the cochlear branch of the CN VIII anteriorly and the inferior branch of the vestibular nerve posteriorly in the inferior compartment. Inferiorly to the fundus of the internal acoustic meatus lies the cochlear aqueduct, also known as the perilymphatic duct.

We can find the subarcuate fossa between the meatus opening and the superior petrous sulcus, a shallow depression channel to the subarcuate artery and veins. In a caudal and posterior position, between the acoustic meatus and the sigmoid sulcus, is a small bony fissure indicating the opening of the vestibular aqueduct, containing an extension of the membranous labyrinth, the endolymphatic duct, connected to the endolymphatic sac, coated by a thin bone sheet. The superior and inferior petrous sinus end up, respectively, in the sigmoid sinus and the jugular bulb. The sigmoid sinus, a continuation of the transverse sinus, drives inferomedial and debouch at the jugular bulb.

Inferior Surface: It is situated in a horizontal plane. This surface lies anterior to the styloid process and posterior to the carotid canal, the jugular fossa. Its articulation with the occipital bone forms the jugular foramen. This foramen is divided into two parts: pars vascularis, a posterolateral vascular compartment (where passes the internal jugular vein, CN X, Arnold branch of the CN X, CN XI, and posterior meningeal artery), and the pars nervosa, anteromedially (where passes the CN IX, Jacobson branch of the CN IX, and the inferior petrosal sinus).

The jugular bulb, dilation of the internal jugular vein, lies in the jugular fossa, an oval hollowed area ahead of the jugular foramen lateral compartment, medial to the facial nerve mastoid segment and inferior to the labyrinth. Inferior to the anterior compartment of the jugular foramen and lateral to the foramen magnum, the hypoglossal nerve exits the skull through the hypoglossal canal.

Anterior to the jugular fossa is the carotid canal, formed by the junction of the tympanic and petrous bones. It is an entrance to the skull for the internal carotid artery, along with its venous and sympathetic nerve plexus. Small apertures next to the external foramina, known as the caroticotympanic canaliculi, serve as pathways for arteries and nerves into the middle ear. Therefore, the internal carotid artery penetrates the skull cavity through the carotid canal, describing an arch inside the temporal bone, directed to the petrous apex, being that its route passes inferior to the cochlea and medial to the tympanic ostium of the Eustachian tube, leaving the skull through the foramen lacerum. Inferiorly, at the junction of the carotid canal with the jugular fossa, the glossopharyngeal nerve petrous ganglion is situated, from where leaves its tympanic nerve (Jacobson's nerve), besides the tympanic branch of the ascendant pharyngeal artery. Posteriorly to the jugular fossa, the petrous bone merges the jugular process of the occipital bone [6].

2.4 Tympanic Bone

The smaller the temporal bone portions, the inferior the posterior and anterior walls of the external acoustic meatus. The tympanic bone articulates medially with the petrous bone through the petrotympanic suture and superiorly with the squamous bone, forming the notch of Rivinus, and is located right ahead the mastoid, where they merge, forming the tympanomastoid suture, posteroinferiorly at the external meatus. In its medial end, the tympanic membrane annulus is inserted into the tympanic (or annular) sulcus. The anterior limit of the tympanic bone composes the posterior wall of the glenoid fossa as well. A persistent foramen of Huschke may remain patent at the inferior wall, potentially leading to a canal cholesteatoma. It usually closes around the age of 5 years old.

2.5 Styloid Process

A long, pointed process, measuring from 20 to 25 mm and anteroinferiorly oriented, originated from the tympanic bone, located just lateral to the petrous bone. Posterolateral to the styloid process is the stylomastoid foramen. As the mastoid process is not developed yet in the newborn, the facial nerve, exiting through the stylomastoid foramen, is more exposed in the neck. The auricular branch of the X CN may occasionally pass through this foramen. The muscles attached to this process are the stylopharyngeus, stylohyoid, and styloglossus. Also, the stylohyoid ligament (related to Eagle's Syndrome – a cause of odynophagia) extends from the tip to the hyoid bone, and the stylomandibular ligament ends on the mandibular angle.

2.6 Eustachian Tube

A slender osteocartilaginous tube with a triangular lumen connects the anterior wall of the tympanic cavity (protympanum) to the nasopharynx's lateral wall and serves primarily to equalize air pressure inside and outside of the eardrum, but also committed to protection and clearance of the middle ear. It is orientated anteromedially, measuring around 37 mm around the age of seven, when it reaches adult size. Its posterolateral third, tympanic, is completely involved by the petrous bone, runs along the semicanal for the tympanic tensor muscle, which is enveloped by a thin bone sheet, and very close to the cochlea posteriorly and the carotid canal anteriorly. Its anterior two-thirds are involved by cartilage, down to the pharynx end, and opens intermittently. The narrowest part of the auditory tube is called the isthmus, at the junction of the bony and cartilaginous portions.

3 Inner Ear

Situated in the petrous portion of the temporal bone, in the apex, and the most medial anatomical structure compounding the ear, it is constituted by the cochlea, semicircular canals, saccule, and utricle, making up the membranous labyrinth, filled by endolymph. These organs are held within the bony labyrinth, called the "otic capsule," a compact bone filled by perilymph. The inner ear receives terminations of the vestibulocochlear nerves through its medial side and is crossed by the facial nerve [7].

The endolymph has a very distinct composition from the perilymph, as it is similar to intracellular fluid with high potassium concentrations. Perilymph has a composition similar to other extracellular fluids like plasma or cerebrospinal fluid, richer in sodium ions [6, 7].

3.1 Membranous Labyrinth

This system, sheltered by the bony labyrinth, is composed of four ducts (the three semicircular canals and the cochlear duct) and two sacs (saccule and utricle) and is fluid-filled by endolymph, a liquid originated from the cerebrospinal fluid and secreted by the stria vascularis, which is a network of capillaries located in the spiral ligament of the cochlea. It follows the shape of the bony labyrinth.

The utricle and saccule are connected by a straight duct called the utriculosaccular duct. Next to it, a branch known as the endolymphatic duct connects to the endolymphatic sac, a structure located in the subdural space at the posterior face of the temporal bone. These structures aid in the regulation of fluid pressure inside the inner ear.

3.2 Utricle

Larger of the two sacs, situated within the posterosuperior portion of the vestibule, just anterior to the ampulla of the three semicircular canals. The macula, a small thickening lying at the floor of the utricle, is responsible for sensing movements in the centrifugal and vertical plane and receives innervation from the utricular branch of the superior vestibular nerve. It is one of the vestibular organs accountable for balancing.

3.3 Sacule

The smaller of the two sacs of the membranous labyrinth lies in an anteroinferiorly situation inside the vestibule. The macula sacculus, also represented by a thickening on its wall, is innervated by the saccular branch of the inferior vestibular nerve and senses movements in the horizontal plane. The sacule is connected to the utricle by the utriculossacular duct and to the cochlear duct through the ductus reuniens, a narrow opening at the floor of the sacule, therefore connecting the vestibular and cochlear apparatus. Next to the utriculossacular duct is the endolymphatic duct, extending through the vestibular aqueduct to reach the endolymphatic sac at the posterior cranial fossa.

3.4 Vestibular Aqueduct, Endolymphatic Sac and Duct

From the utriculossacular duct (between utricle and sacule) branches the endolymphatic duct, entering the vestibular aqueduct, a bony aperture in the posterior ridge of the temporal bone, and emerges at the posterior surface of the petrous bone to the posterior cranial fossa. In this region is seen dilatation of the system called endolymphatic sac, an extradural pouch related to the endolymph resorption, sitting at the endolymphatic sac fossa. Obstruction of this system is the basis for developing hydrops in experimental animals, as the duct and the sac may act as a valve to regulate endolymph pressure.

3.5 Cochlear Aqueduct

Analog to the vestibular aqueduct, the cochlear aqueduct contains the perilymphatic duct (or periotic duct), part of the membranous labyrinth that flows perilymph and connects the scala tympani on the basal turn of the cochlea (next to the round window) to the subarachnoid space of the posterior cranial fossa, through the inferior portions of the petrous bone.

3.6 Semicircular Canals

Composed of three membranous ducts containing endolymph and immersed in perilymph, these structures are connected to the utricle vestibule through five openings. They are responsible for detecting and encoding the angular motion of the head. Each of the three canals is nearly orthogonal with one another. Each canal includes an enlarged ampulla on one of its ends, where sensory hair bundles project from the epithelium. Angular motion of the head displaces these bundles, transmitting movement information to the brain through terminal nervous fibers. The superior vestibular nerve innervates the lateral and superior canals.

On the other hand, the posterior canal is innervated by the inferior vestibular nerve. The lateral canal is parallel to its contralateral correspondent, but the superior canal is parallel to the contralateral posterior semicircular canal. The horizontal semicircular canal projects to the mastoid cavity, just behind the antrum. The superior canal extends upwards as the arcuate eminence in the middle fossa [6].

3.7 Cochlea

Right ahead of the vestibule, this bony structure with two and one-half to two and three-quarter turns measuring around 3 centimeters twist around a central bony axis, the modiolus (or spiral lamina). It is the organ responsible for hearing and participates in hearing transduction.

The spiral shape of the cochlea allows for differing frequencies, stimulating specific areas of the organ. The base is positioned through the internal acoustic meatus, receiving cochlear branches as the apex is directed anteromedially. Inside the osseous cochlea runs an extension of the membranous labyrinth called the cochlear duct. The cochlear duct is attached externally to the walls of the cochlea, creating two canals: the scala tympani and the scala vestibule, both filled with perilymph. These two spaces are continuous with each other, joining only at the apex of the cochlea (the helicotrema), but separated through the turns of the cochlea by the scala media, filled with endolymph. The Reissner membrane separates the scala media from the scala vestibule, and the basilar membrane separates it from the scala tympani. This system exhibits three openings: one to the vestibule, another to the middle ear, represented by the round and oval windows, and a third, near the round window, where parts the cochlear aqueduct, connecting the scala tympani with the subarachnoid space. The scala media contains the Organ of Corti, where a single row of inner and three or four rows of outer hair cells are disposed, supported by the basilar membrane, responsible for the mechanotransduction of the sound.

4 Petroclival Region

The petroclival region is one of the skull base regions with the most difficult surgical access. Transtemporal approaches are an important corridor for the petroclival region. The clivus can be divided into three parts. A lower portion corresponds to the anterior border of the foramen magnum, a middle portion related to the pons and the prepontine and cerebellopontine angle cisterns, and an upper portion related to the interpeduncular cistern and the sellar and parasellar regions. The division of the clivus in a sagittal perspective of the magnetic resonance examination is important in selecting the best surgical approach.

5 Infratemporal Fossa

The zygomatic process of the temporal bone divides the temporal from the infratemporal fossa (ITF). There is no anatomical floor located in the ITF. The roof is the greater wing of the sphenoid bone, and the medial wall is the lateral pterygoid plate, the lateral wall is the ramus of the mandible, and the posterior wall is the articular tubercle of the temporal bone and the spine of the sphenoid bone. This anatomical region has communication through the pterygomaxillary fissure with the pterygopalatine fossa. The medial pterygoid fascial layer and the medial pterygoid muscle are the boundaries between the ITF and the parapharyngeal space (Fig. 8) [1–3, 8].

5.1 *Muscular Relationships*

To understand the surgical approaches to the ITF, it is necessary to know the pterygoid muscles and the masseter and temporalis muscles. The tendon of the temporalis muscle is sitting in this space [9].

The masseter muscle is covered by the masseteric fascia and is formed by three muscular layers (superficial, middle, and deep). The superficial layer has a tendinous septum between muscle fibers and is a continuation of an aponeurosis that arises from the zygomatic bone at the maxillary process and the anterior part of the inferior border of the zygomatic arch. This layer then inserts into the lateral surface of the mandibular ramus and angle. The middle layer arises from the medial portion of the anterior part of the zygomatic arch and from the inferior border of the posterior part and then inserts into the central part of the mandibular ramus. The deep layer originates from the zygomatic arch more deeply and inserts into the coronoid process and upper part of the mandibular ramus. The masseter muscle is supplied by the masseteric branch that has its origin in the second part of the maxillary artery, which is deep to the muscle, and by two other arteries that run superficial to the muscle: the facial and transverse facial arteries. The masseter is innervated by the masseteric branch of the mandibular nerve.

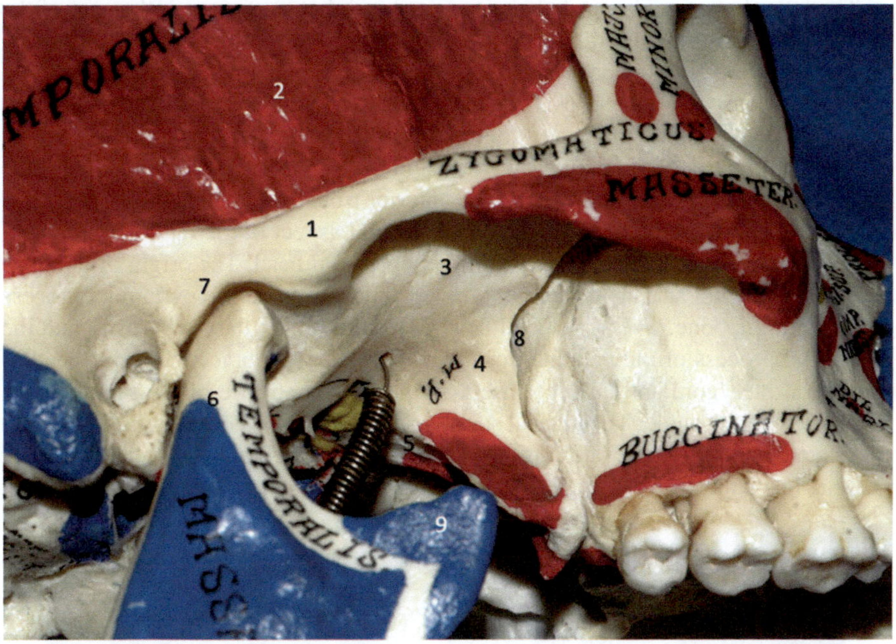

Fig. 8 Limits of the ITF and bone relationships. Lateral view of the ITF. Note that the depression of the mandible (open mouth) gives more access to the ITF laterally. (*1*) The zygomatic process of the temporal bone; (*2*) temporal fossa; (*3*) greater wing of the sphenoid; (*4*) lateral pterygoid plate; (*5*) medial pterygoid plate; (*6*) mandibular ramus; (*7*) articular tubercle of the temporal bone; (*8*) pterygomaxillary fissure; (*9*) coronoid process

The temporal muscle is a fan-shaped muscle that lies in the temporal fossa. This muscle has a main portion, along with the anteromedial, anterolateral, and middle lateral bundles. It also arises in the temporal fossa, and its fibers insert as two separate tendinous heads (superficial and deep) that run deep to the zygomatic arch and insert in the coronoid process and the anterior border of the ramus of the mandible. Although it is difficult to individualize the tendons separately at their insertion, the superficial part is more related to the lateral surface of the coronoid process. One important consideration for the surgical approaches is the anatomical study of the fascia related to the temporal muscle. We identified the temporoparietal fascia (superficial temporal fascia) that blends superiorly in the galea aponeurotica and the temporal fascia (deep temporal fascia). Between this was fat tissue and loose areolar tissue (temporoparietal fat pad). There was fat tissue always between the deep temporal fascia and the temporal muscle fibers and fat tissue superficial to the superficial temporal fascia, although with a variable amount in the different specimens. In our dissections, the temporal branch of the facial nerve was found inside the temporoparietal fat pad but superficially, meaning that it was related to the internal surface of the superficial temporal fascia. The fascial nerve fibers were also identified inside the fat tissue on the superficial temporal fascia. The deep temporal fascia split into superficial and deep laminae attached to the lateral and medial surfaces of the zygomatic bone [3–5].

The lateral pterygoid muscle has an upper head that originates in the roof of the ITF and in the infratemporal crest (the inferior portion of the greater wing of the temporal bone) and a lower head that originates in the lateral surface of the lateral pterygoid plate. The insertion is the same for both parts in the pterygoid fovea, which is a convex space in the neck of the mandible. The vascular supply is from the pterygoid branches of the maxillary artery and the ascending palatine branch of the facial artery. The innervation is from the anterior trunk of the mandibular nerve.

The medial pterygoid muscle lies in the medial aspect of the ramus of the mandible. It has a large deep head and a minor superficial head. The origin of the deep head is the medial surface of the lateral pterygoid plate and the pyramidal process of the palatine bone. The superficial head is from the maxillary tuberosity and the pyramidal process of the palatine bone. The two heads embrace the lower head of the lateral pterygoid and unite to insert on the medial surface of the ramus of the mandible near its angle. The arterial supply is from the pterygoid ramus of the maxillary artery and the innervation by the medial pterygoid branch of the mandibular nerve (Figs. 9 and 10).

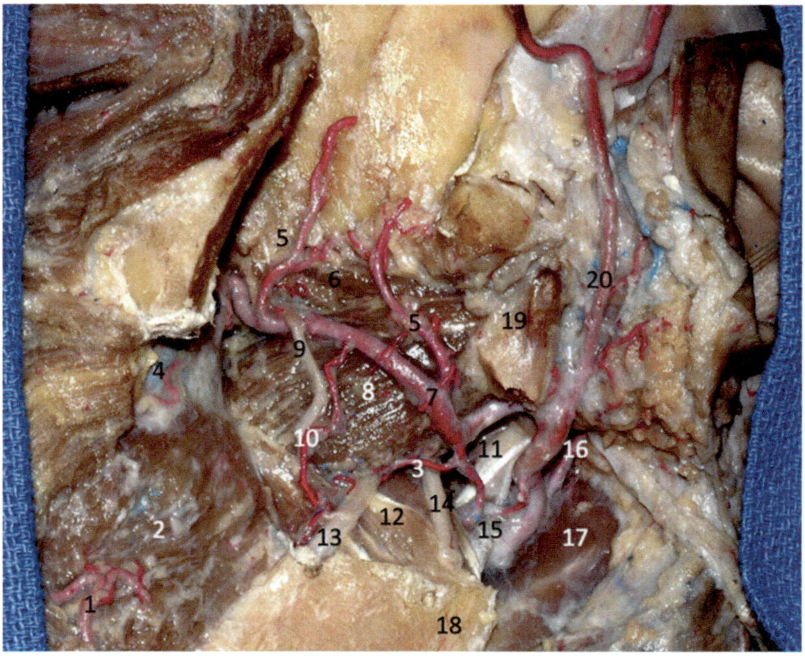

Fig. 9 Lateral view of the ITF. The zygoma and part of the mandible were removed. The condylar process was left to show the insertion of the lateral pterygoid muscle in the pterygoid fovea in the neck of the mandible. (*1*) Facial artery; (*2*) buccinator muscle; (*3*) posterior superior alveolar artery; (*4*) sphenopalatine artery; (*5*) anterior and posterior deep temporal arteries; (*6*) lateral pterygoid muscle (upper head); (*7*) maxillary artery; (*8*) lateral pterygoid muscle (lower head); (*9*) buccal nerve; (*10*) buccal artery; (*11*) styloid process; (*12*) medial pterygoid muscle; (*13*) lingual nerve; (*14*) inferior alveolar nerve; (*15*) external carotid artery; (*16*) posterior auricular artery; (*17*) digastric muscle (posterior belly); (*18*) the angle of the mandible; (*19*) condylar process; (*20*) superficial temporal artery

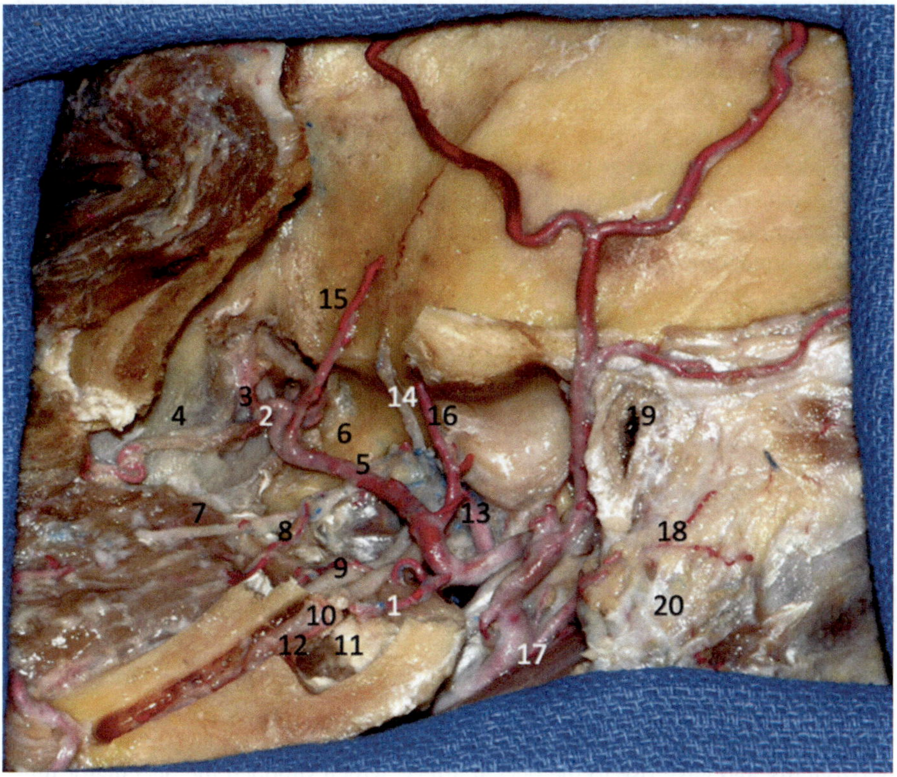

Fig. 10 Lateral view of the ITF. The lateral pterygoid process and the condylar process were removed. (*1*) posterior superior alveolar artery; (*2*) infraorbital artery; (*3*) sphenopalatine artery; (*4*) descending palatine artery (not injected); (*5*) maxillary artery; (*6*) lateral pterygoid plate; (*7*) buccal nerve; (*8*) buccal artery; (*9*) lingual nerve; (*10*) inferior alveolar nerve; (*11*) medial pterygoid muscle; (*12*) inferior alveolar artery; (*13*) middle meningeal artery; (*14*) deep temporal nerve; (*15*) anterior deep temporal artery; (*16*) posterior deep temporal artery; (*17*) digastric muscle (posterior belly); (*18*) posterior auricular artery; (*19*) external acoustic meatus; (*20*) mastoid

6 Pterygopalatine Fossa

The pterygopalatine fossa (PPF) has the shape of an inverted cone, having as apex the greater palatine canal. Its lateral boundary is the pterygomaxillary fissure (PMF), which communicates the PPF with the infratemporal fossa. The other boundaries of the PPF are the following: the maxilla, anteriorly; the medial plate of the pterygoid process and greater wing of the sphenoid process, posteriorly; the palatine bone, medially; and the body of the sphenoid process, superiorly. The maxillary artery (MA), arising medially to the neck of the mandible and bending in an anterior, medial, and slightly superior direction, enters the PPF through the PMF, giving several branches before entering the sphenopalatine foramen as the sphenopalatine artery. The MA can be divided into three parts: the mandibular, pterygoid, and pterygopalatine portions [9].

The contents of the PPF are the following: (1) maxillary division of the trigeminal nerve and its branches; (2) pterygopalatine ganglion; (3) pterygopalatine portion of MA and its branches; (4) venous network surrounding the MA. The maxillary nerve innervates the lateral aspect of the cheek, the temple, and the maxillary teeth. Its branches are the zygomatic nerve, one or two posterior superior alveolar nerves, infraorbital nerve, and roots to the sphenopalatine ganglion. The maxillary artery is a significant source of blood supply to the face's deep structures and gives rise to the middle meningeal artery. The pterygopalatine ganglion gives rise to postganglionic fibers to the lacrimal gland and glands in the nasal and nasopharyngeal mucosa. It is the largest parasympathetic ganglion in the body.

The PPF is a neurovascular hub in the middle face. It communicates with the foramen lacerum, infratemporal fossa, middle cranial fossa, nasal cavity, orbit, and pharynx via foramina and fissures [1, 2, 4, 5].

6.1 *Vascular Relationships*

External carotid artery: The external carotid artery extends from the level of the upper edge of the thyroid cartilage plate to a point behind the neck of the mandible, between the tip of the mastoid process and the angle of the mandible. It divides within the parotid gland into the superficial temporal and maxillary arteries, also called the internal maxillary artery.

Maxillary Artery: The MA is found deep into the mandibular ramus. It passed horizontally and gave rise to the buccal artery, which supplied the buccinator muscle. Then, it turned medially and crossed the PMF to arrive in the PPF, thus becoming the pterygopalatine portion of the MA.

Branches arising from this portion of the MA were located at approximately one third of the height of the maxillary sinus' posterolateral wall. As previously described, the artery entered the PMF in an anterior, medial, and superior direction. While in the PMF, the MA originated two branches, namely, the infraorbital artery (IOA) and the posterosuperior alveolar artery (PSAA). Both arteries were located in the posterior wall of the maxilla. The PSAA entered the posterosuperior alveolar foramen, while the IOA entered the infraorbital fissure.

After giving off these branches, the MA continued to the PPF. It then split into three arteries: the descending palatine artery (DPA), which supplied the palate; the artery of the pterygoid canal (or Vidian artery (VA)); and the sphenopalatine artery (SPA). The PSAA and IOA branched from the MA with two different patterns. A short common trunk arose from the MA in the first one, subsequently bifurcating to form both arteries. In the second pattern, the PSAA and IOA branched separately.

Vertebral artery: The vertebral artery courses in the cervical region within the transverse foramina. After leaving the transverse process of the first cervical vertebra, the vertebral artery turns medially and enters the posterior fossa through the atlanto-occipital membrane. In its initial intracranial course, this artery passes through the lateral cerebellar-medullary cistern and reaches the bulbopontine sulcus, where it joins its counterpart on the opposite side to form the basilar artery. Generally, this union occurs at the level of the lower third of the clivus. The vertebral artery branches include several small branches to the anterolateral, lateral, and posterolateral portions of the cord, the anterior spinal artery, and, in more than 70% of cases, the posteroinferior cerebellar artery.

7 Surgical Approaches

7.1 Middle Fossa Approach

The microsurgical removal of vestibular schwannomas via a middle fossa craniotomy (MFC) is indicated in cases of intracanalicular tumors to preserve functional hearing. This approach offers the opportunity for total tumor resection with acceptable surgical risks. Results of MFC compare favorably with other modalities, with good hearing preservation in more than 60% of cases and good postoperative facial nerve function in more than 95%. The morbidity is acceptable, and the most common complications include facial weakness and CSF leak in 5% of cases.

The middle cranial fossa approach is indicated for vestibular schwannomas in patients with:

1. Intracanalicular tumors extend less than 1.5 cm into the cerebellopontine angle (without brainstem contact).
2. Serviceable hearing (>50 dB pure-tone average and 50%-word recognition score).
3. Evidence of tumor growth (MRI) and/or functional deterioration (vertigo or hearing loss).
4. No contraindications for a supratentorial craniotomy (age > 70 years and ASA > 2–3).

Anteriorly, the limit of the dissection is the middle meningeal artery, which is lateral to the greater superficial petrosal nerve. The arcuate eminence marks the position of the superior semicircular canal and may be readily apparent in some patients. Medially, the superior petrosal sinus runs along the petrous ridge. The labyrinthine portion of the facial nerve lies immediately posterior to the basal turn of the cochlea. Bill's bar separates the facial and superior vestibular nerves. Slightly posterior and lateral to this area is the vestibule and ampullated end of the superior semicircular canal. The geniculate ganglion can be identified by tracing the greater superficial petrosal nerve posteriorly to it. If the tegmem is unroofed, the geniculate is slightly anterior to the head of the malleus. The internal auditory canal (IAC) lies approximately on the same axis as the external auditory canal; this relationship is useful in orienting the surgical field. There are several techniques described to reach the internal auditory canal by the middle fossa approach (Figs. 11, 12, and 13) [1, 2, 8, 9]:

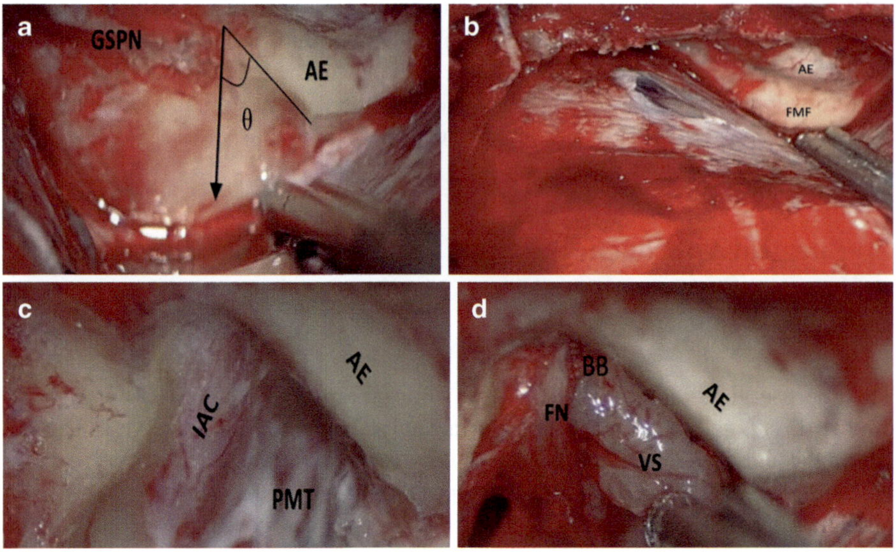

Fig. 11 Illustrative case of an intracanalicular vestibular schwannoma resected through middle fossa approach. Surgical anatomy of the temporal bone as viewed from the middle fossa approach. (**a**) The IAC lies along a line located at a 60° angle from the axis of the superior semicircular canal. The internal auditory canal is skeletonized through the entire length. (**b, c**) Bone is removed around the porus acusticus, uncovering the dura of the posterior fossa. (**d**) Tumor is dissected away from the facial and cochlear nerves, rather than the nerves being retracted away from the tumor. GSPN, greater superficial petrosal nerve; AE, arcuate eminence; θ, 45 to 60° angle; PMT, post-meatal triangle; IAC, internal auditory canal; CN, cochlear nerve; IVN, inferior vestibular nerve; VS, vestibular schwannoma; FN, facial nerve; FMF, floor of the middle fossa

- **House**: Following the greater superior petrosal nerve to the geniculate ganglion and then to the labyrinthine segment of the facial nerve.
- **Fisch**: Blue-lining the superior semicircular canal by drilling the arcuate eminence's bone. The arcuate eminence is used as the landmark to locate the superior semicircular canal, but the profile of the arcuate eminence is highly variable. The axis of the IAC lies along a line located at a 60° angle from the axis of the superior semicircular canal (Fig. 11).
- **Garcia-Ibañez**: The greater superior petrosal nerve and the axis of the superior semicircular canal form a 120° angle; the internal auditory canal axis bisects this angle.

7.2 Anterior Petrosectomy

Anterior petrosectomy is an excellent approach for extradural tumors located in the petrous apex (cholesteatomas, chondrosarcomas, cholesterol granuloma) or as an access route for tumors in the upper third of the clivus and extending into the middle

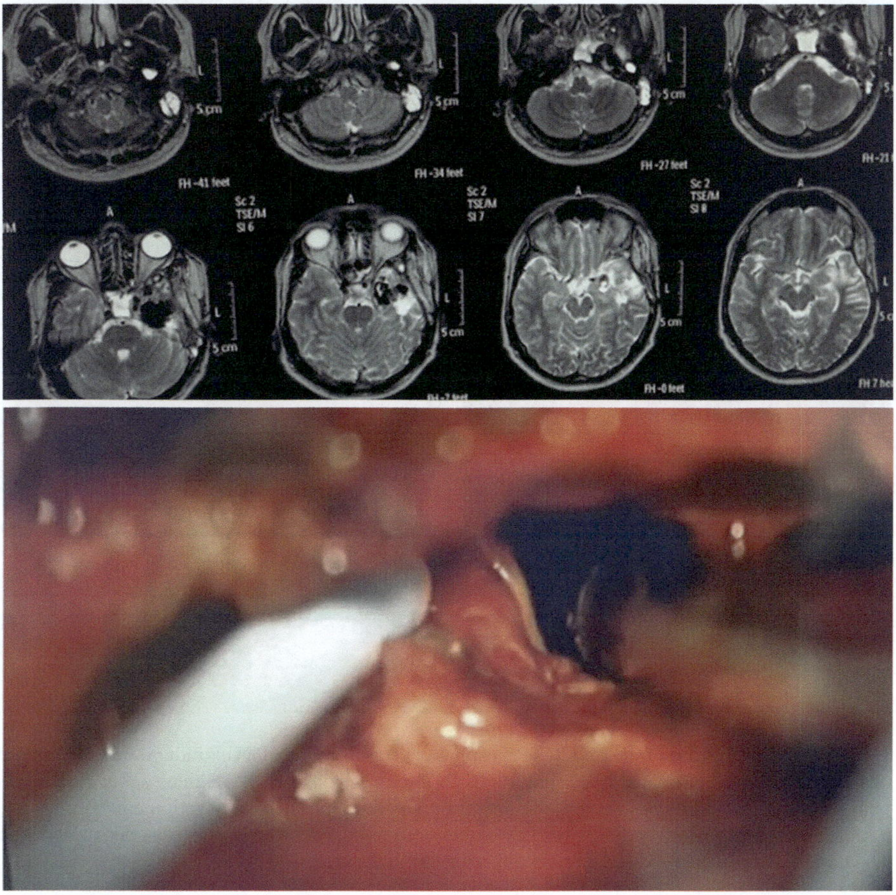

Fig. 12 In this case of aneurysmatic bone cyst of the middle fossa, it is crucial to use Doppler intraoperatively to precisely locate the position of the petrous part of the internal carotid artery, which in this case was inside the cyst, without any correlation with the classical anatomy of the middle fossa

fossa (sphenopetroclival meningeoma). The approach is classically performed through temporal, frontotemporal, or cranio-orbito-zygomatic craniotomy, followed by middle fossa peeling (MFP) and extradural exposure of the petrous apex. The peeling can be performed in an anterior-posterior direction (especially if there is disease inside the cavernous sinus) or in a posterior-anterior direction. The lateral limit is the intrapetrous carotid artery, the medial limit is the superior petrosal sinus, the posterior limit is the internal acoustic meatus and the cochlea, and the anterior limit is the trigeminal nerve. It is recommended that this approach be thoroughly trained in a microsurgical laboratory before performing surgery on a patient. In China, some groups choose to perform an intradural anterior petrosectomy through a subtemporal approach. In these cases, the dura over the petrous apex is incised

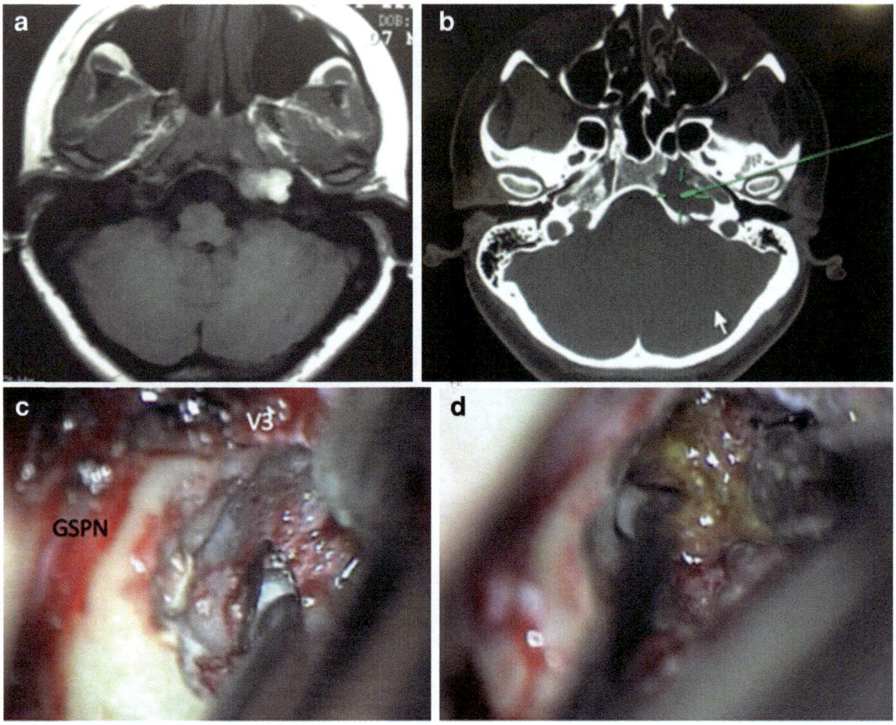

Fig. 13 Illustrative case of the middle fossa approach for resection of an extradural dermoid tumor of the petrous apex. (**a**) Axial MRI shows a hyperintense signal on T1-weighted sequences at the petrous apex. (**b**) Although knowing the anatomy is paramount, neuronavigation can confirm the middle fossa anatomical findings. (**c**) Left subtemporal extradural approach was performed. (**d**) The tumor was resected. GSPN, greater superior petrosal nerve

intraduraly after the temporal lobe is retracted superiorly. Lumbar drains are especially useful in anterior extradural petrosectomy, reducing the pressure on the temporal lobe. Some anatomic variations of the petrous apex should be studied preoperatively. Petrous apex pneumatization was found in 14.2% of patients, while unilateral pneumatization was found in 2.5%. In our previous study. The prevalence reported in the literature ranges from 5% to 10%. Knowledge of petrous apex pneumatization prior to the procedure is essential because it influences skull base reconstruction when it is necessary to be careful to avoid liquoric fistula [9–12].

7.3 Translabyrinthine Approach

The objective of the translabyrinthine approach is to achieve the most direct lateral approach to the cerebellopontine cistern. This approach allows early and consistent facial nerve identification from the labyrinthine portion up to the brainstem. The advantage of the translabyrinthine approach is the wide exposure of the entire length of the internal auditory canal and the cerebellopontine angle without cerebellar retraction. Since most of the approach is extradural, there is a lower risk of damage to neurovascular structures. If there is damage to the facial nerve during manipulation, this route allows great mobilization of the facial nerve for a direct anastomosis in the same surgical time. However, its main disadvantage is hearing loss (Figs. 14, 15, 16, 17, 18, and 19).

The main indication for the translabyrinthine approach is vestibular schwannoma surgery. However, it is also well indicated to treat lesions in patients with associated hearing loss (meningiomas, facial nerve tumors, petrous apex

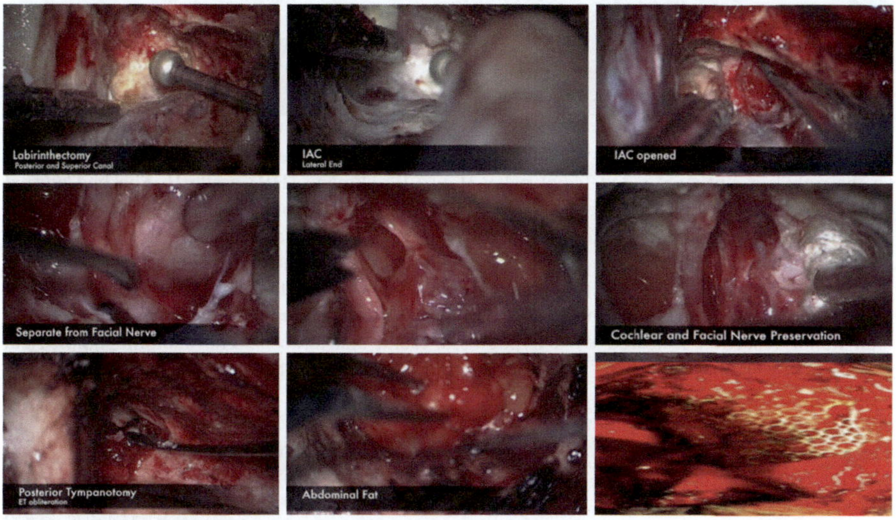

Fig. 14 Illustrative case of a small vestibular schwannoma resected via translabyrinthine approach

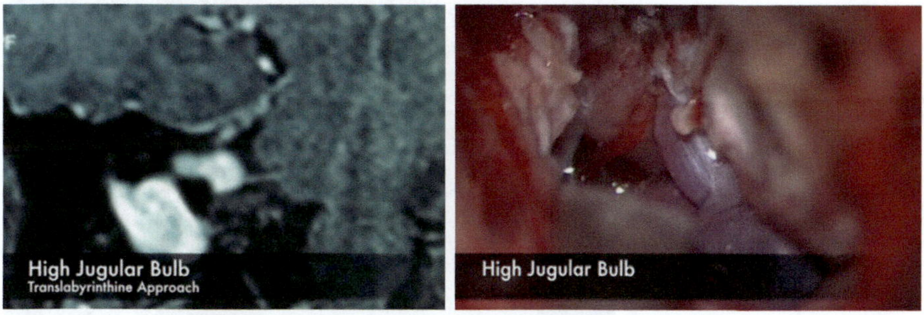

Fig. 15 Understanding the correlation between radiological images and anatomy (and its variations) is the first step to avoid bad surgical approaches and surgical catastrophes. This case illustrates a patient with a high jugular bulb close to the internal acoustic meatus. A translabyrinthine approach was chosen in order to avoid jugular bulb lesions. The patient had no adequate hearing before surgery

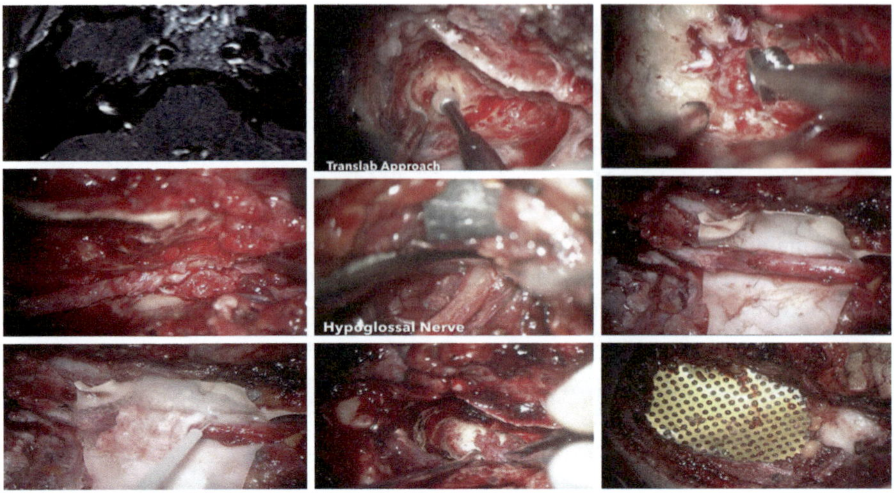

Fig. 16 Illustrative case of a 30-year-old woman referred to our service with severe hearing loss on the right ear and right-sided facial palsy House-Brackmann (HB) grade II. She had normal otoscopy and no vestibular caloric reflex. MRI axial T1-weighted contrast-enhanced image showed a 5 mm lesion in the lateral portion of the right internal auditory canal, extending to the adjacent portion of the facial and cochlear nerve canals, with contrast-enhancement to gadolinium, suggestive of schwannoma. It has opted for conservative treatment. In the following 24 months, the facial palsy progressed to a HB IV. In the same surgery, we performed a translabyrinthine approach with total tumor resection followed by a hypoglossal-facial anastomosis. The anatomopathological study was compatible with facial nerve hemangioma. No hearing changes

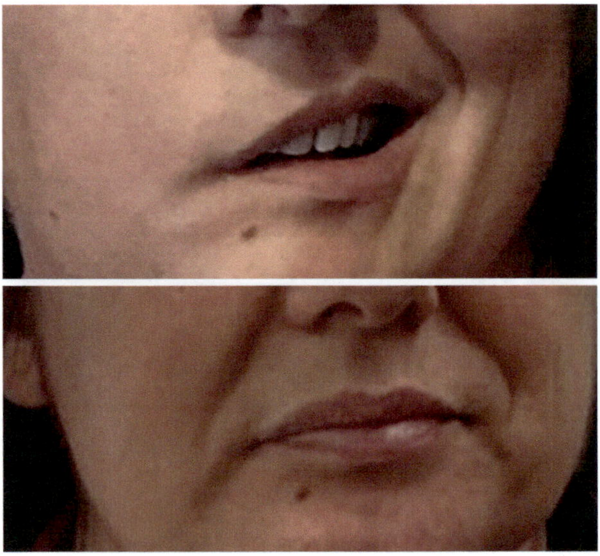

Fig. 17 Facial nerve function before surgery (above) and 4 months after hypoglossal-facial anastomosis (below)

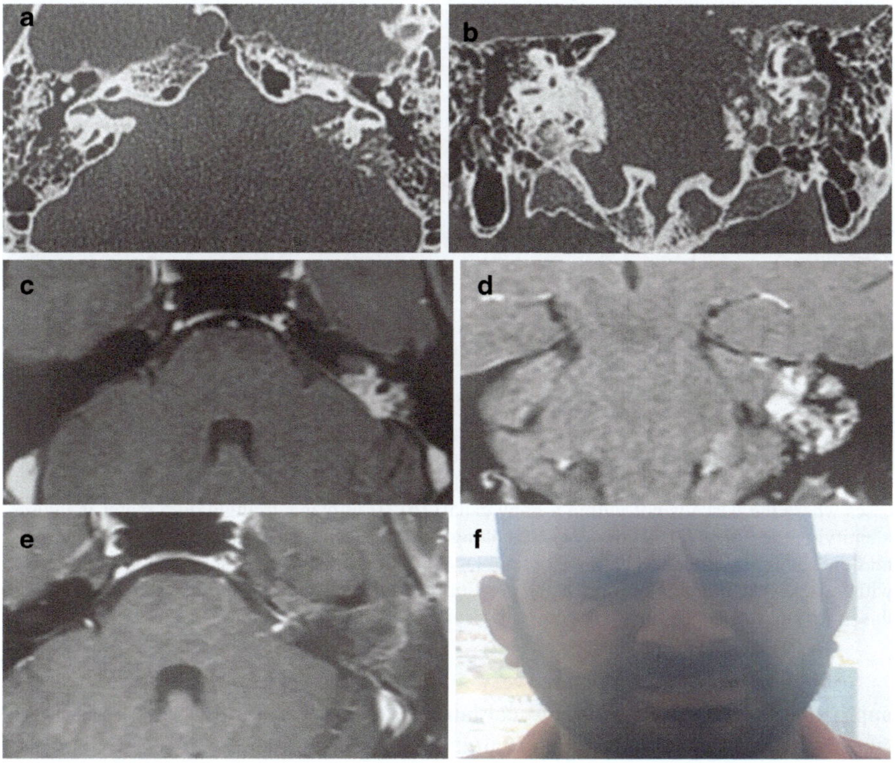

Fig. 18 Temporal bone axial (**a**) and coronal (**b**) CT Scan shows bone destruction of the petrous portion of the temporal bone. Axial (**c**) and coronal (**d**) T1-weighted MRI with gadolinium showing a tumor within the mastoid portion of the temporal bone with involvement of Trautman's triangle dura mater. T1-weighted gadolinium-enhanced MRI 3 months after surgery showing total resection of the lesion (**e**) with preservation of the facial nerve (**f**)

cholesteatomas, posterior fossa, cholesterol granuloma, and glomus). In addition, this access route can be used in vestibular neurectomy due to disabling vertigo, in facial nerve repair/decompression due to temporal bone fractures, and in auditory brainstem implants.

Translabyrinthine access can be performed for tumors of any size. The indication of the translabyrinthine approach depends on the preoperative level of hearing. The candidate patient must have a tonal mean worse than 50 dB and/or a speech discrimination score worse than 50%. However, it is also indicated in tumors larger than 2 cm, regardless of hearing level, as the chance of hearing preservation is minimal.

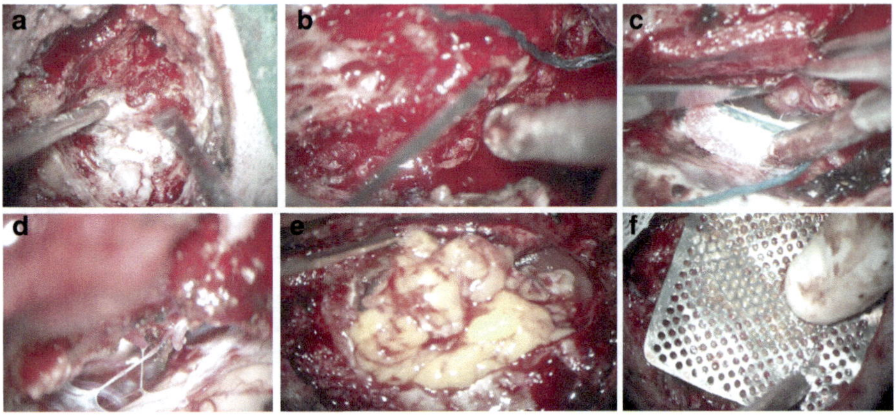

Fig. 19 Intraoperative view of Fig. 18. Mastoidectomy is performed (**a**), the mastoid portion of the facial nerve is identified through anatomical parameters, but also by electrical stimulation (in cases where the nerve is not within a bony canal) (**b**). Trautman's triangle dura mater is incised, identifying the tumor (**c**). Tumor resection along with the Trautman's triangle dura mater (**d**). The mastoid cavity is obliterated with abdominal fat (**e**). Before that, the Eustachian tube is obliterated with muscle. The titanium mesh must be positioned firmly, pressing the fat to avoid CSF leak. For this, it is important to fix the plate with screws on the edges of the temporal bone (**f**)

Some anatomical points are very important to perform a translabyrinthine approach. There are some anatomical landmarks the surgeon must respect for safe surgery. Microsurgical laboratory training is crucial to acquire full knowledge of this anatomy [11, 12].

The facial nerve's second knee (bend) is at the same level of the lateral semicircular canal concavity and immediately anterior. The jugular bulb lies medial to the mastoid segment of the facial nerve. The superior petrous venous sinus represents the tentorial margin and is located at the angle between the dura mater of the middle and posterior cranial fossa. The superior semicircular canal lies completely superior to the lateral semicircular canal. The common crus is the connection between them. The subarcuate artery can be identified within the superior semicircular canal and represents the superior limit of the IAC. The junction between the ampule and the vestibule is a landmark for the superior limit of the IAC. The ampulla of the posterior semicircular canal is the inferior limit of the IAC. The medial wall of the vestibule is the lateral wall of the IAC. The falciform crest is perpendicular to and originates below the vestibule. It presents itself in front of the surgeon, and the transverse crest points to the tegmen. The transverse crest separates the superior and inferior vestibular nerves. The lateral end of the IAC is the Bill's bar. This is the most reliable landmark to identify the facial nerve. It separates the facial from the superior vestibular nerve. This is the posterior limit of Falopio's canal. The facial nerve can also be identified, in the lateral end of the IAC, medial and inferior to the ampulla of the superior semicircular canal.

7.4 *Combined Supra/Infratentorial Presigmoid Approach ("Petrosal Approach")*

The surgery is performed with the patient in the supine position, with the head turned to the opposite side. The skin incision is made in a semicircular shape, starting from the temporal region, 4 cm above the zygomatic arch, passing 3 cm behind the ear, extending two centimeters behind the mastoid tip. To avoid postoperative CSF leak, the skull base is reconstructed using temporal muscle fascia, which is dissected with the mastoid periosteum, craniocervical fascia, and sternocleidomastoid muscle separated from its insertion, forming a large vascularized flap. It is then rotated backward at the end of the surgery to cover the entire surgical field. The mastoid cortex is drilled, identifying the labyrinth and the facial nerve canal. These channels are not open. Two trepanations above and two below the sigmoid venous sinus are performed, and, with a high-speed drill, a craniotomy is performed, exposing the middle and posterior fossa (retrosigmoid). The superior petrosal, sigmoid and transverse venous sinuses are exposed. Retrofacial mastoid cells are removed up to the jugular bulb. The anterior dura is exposed to the sigmoid sinus. The zygomatic and supralabyrinthine cells are removed, keeping the semicircular canals and middle ear intact. The superior petrosal venous sinus is sectioned anterior to its entry into the sigmoid sinus. Prior to this maneuver, it must be ligated or microclipped. Next, the tentorium is incised, initially perpendicular to the superior petrosal venous sinus for two to 3 cm and then medially, parallel to the transverse sinus for another 3 cm. This maneuver allows for ample exposure of the cerebellum, separating it from the posterior aspect of the temporal lobe in an "open book." Care must be taken to preserve the vein of Labbé, which has variable anatomy and usually enters the transverse sinus 10 mm before its junction with the sigmoid sinus.

Preoperative assessment of venous anatomy is critical to planning this approach. The tentorium incision is continued up to the tentorial notch, where the fourth cranial nerve is exposed and preserved. A few small basal bridging veins at the anterior base of the temporal lobe are coagulated and sectioned, allowing ample subtemporal exposure. The placement of retractors with Leyla spatula support on the temporal lobe and cerebellum should be avoided. In general, light and dynamic spatulation performed by the assistant in a noncontinuous manner is sufficient to expose the entire petroclival region and the III to VIII cranial nerves. The trigeminal nerve can usually be seen displaced posteriorly and superiorly. The tumor is devascularized by bipolar coagulation of its dural insertion. After the tumor's intracapsular resection (cytoreduction) is performed, with an ultrasonic aspirator, the dissection of the tumor capsule from the nerves, basilar artery, and superior cerebellar and posterior cerebral arteries is made possible. The abducens nerve is very thin and fragile. Dorello's canal is located medially to cranial nerves VII and VIII, and this region should be approached only after extensive tumor resection. Tumor extensions to the cavernous venous sinus' posterior part are resected following the trigeminal nerve. All petrous bone and clivus infiltration are removed with a diamond bur. After total removal, the dura mater is closed watertightly with fascia graft and fibrin glue.

The translabyrinthine presigmoid approach is similar, but with additional labyrinthine removal by drilling the semicircular canals, which will add approximately 1.5 cm of surgical exposure, allowing for better visualization of midline structures. The opening of the semicircular canals causes deafness and is only indicated for patients with no previous viable hearing. In total petrosectomy, the surgical procedure is initially the same as described for the presigmoid approach (or anterior petrosectomy), complete with additional removal of the semicircular canals and cochlea (posterior petrosectomy). The petrous internal carotid artery is exposed throughout its course within the temporal bone until it enters the cavernous sinus. This approach is especially useful for very large lesions that cross the midline in patients who are already deaf. Although facial nerve transposition will cause postoperative facial paralysis, which improves within 3 months after surgery to House-Brackmann grade 1–2, this maneuver should be avoided. The Eustachian tube is closed with muscle to prevent postoperative CSF leak. If there is not a short space between the sigmoid sinus and the labyrinth, in patients with preserved hearing and a non-dominant sigmoid sinus, a trans-sigmoid approach has been reported [13]. However, the concept of a non-dominant sigmoid sinus is controversial, and there is no way to predict that venous infarction will not occur. One possible solution is to make a retrosigmoid approach to the posterior fossa component in one time and a pterional approach in a second surgical time (Figs. 20, 21, 22, and 23).

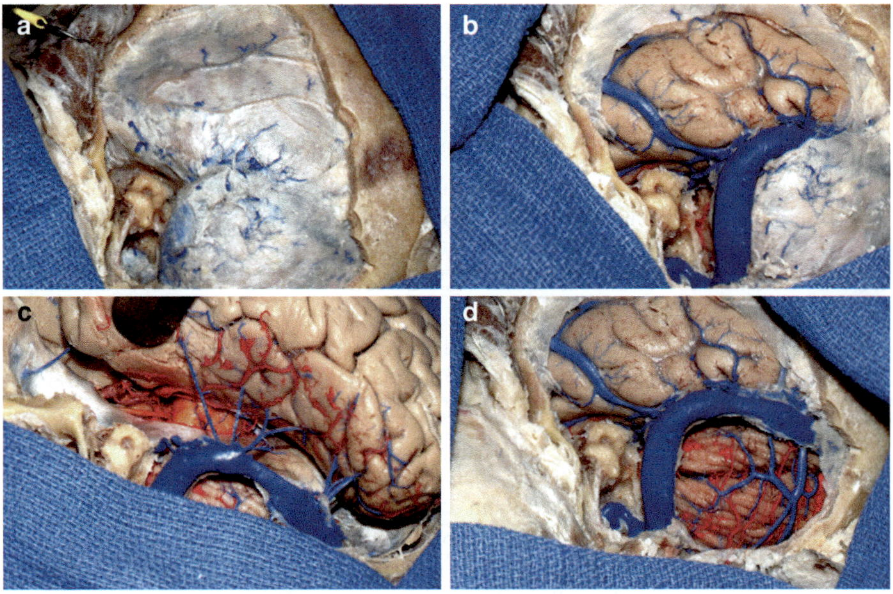

Fig. 20 Posterior petrosal approach. (**a**) The mastoid part of the temporal bone is drilled to expose the presigmoid dura mater. It is unnecessary to individualize the facial nerve and semicircular canals as performed in this dissection. (**b**) The presigmoid dura mater is opened, and the superior petrosal sinus must be exposed. (**c**) The temporal lobe is gently retracted upwards. All care must be taken to preserve the Labbé Vein at this stage. (**d**) For large tumors, the retrosigmoid dura can be opened to drain cerebrospinal fluid at the beginning of the surgery

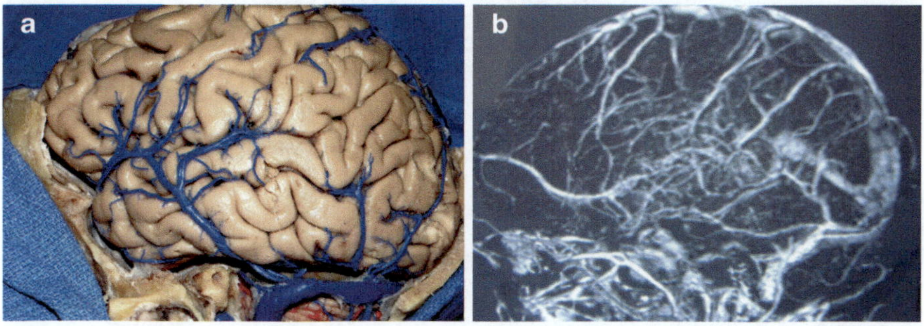

Fig. 21 The study of the position of the drainage of the inferior anastomotic vein (Labbé) is performed to assess when the posterior petrosal approach is intended. Labbé's vein commonly drains at the junction of the transverse and sigmoid sinuses. (**a**) Anatomical piece. (**b**) Venous MRI

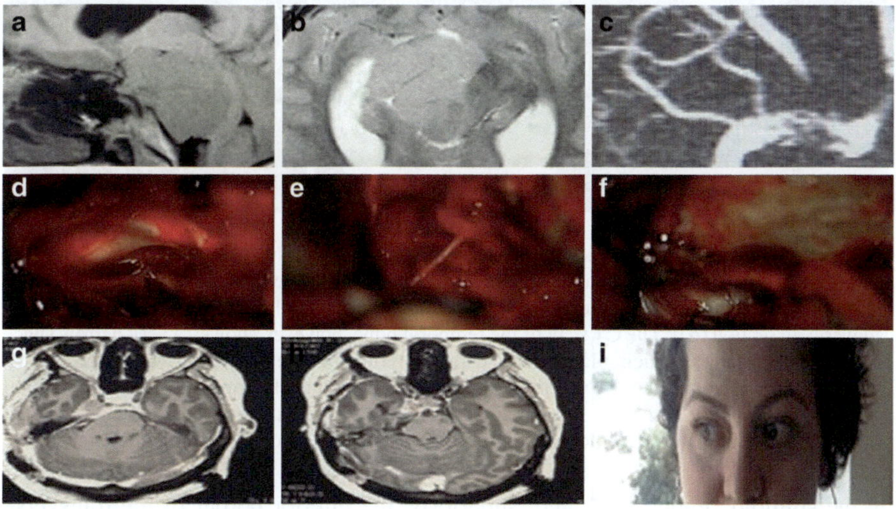

Fig. 22 The posterior petrosal approach is indicated for petroclival meningiomas with insertion into the middle and superior clivus (**a**) and extension into the middle fossa (**b**). Magnetic resonance venography shows drainage of the vein of Labbé at the junction of the transverse and sigmoid venous sinuses (**c**). Intraoperative image of the oculomotor (**d**) and trochlear nerves (**e**) after tumor resection. The facial nerve is within its canal and does not need to be exposed (**f**). Postoperative MRI with gadolinium (**g, h**). Patient without postoperative paresis of the cranial nerves (**i**)

7.5 Retrosigmoid Approach

The retrosigmoid approach is a modification of the suboccipital approach. While the suboccipital approach is positioned close to the midline, the retrosigmoid approach projects more anteriorly and laterally. It allows wide exposure of structures from the tentorium to the foramen magnum. It has the anterior limit of the craniotomy defined

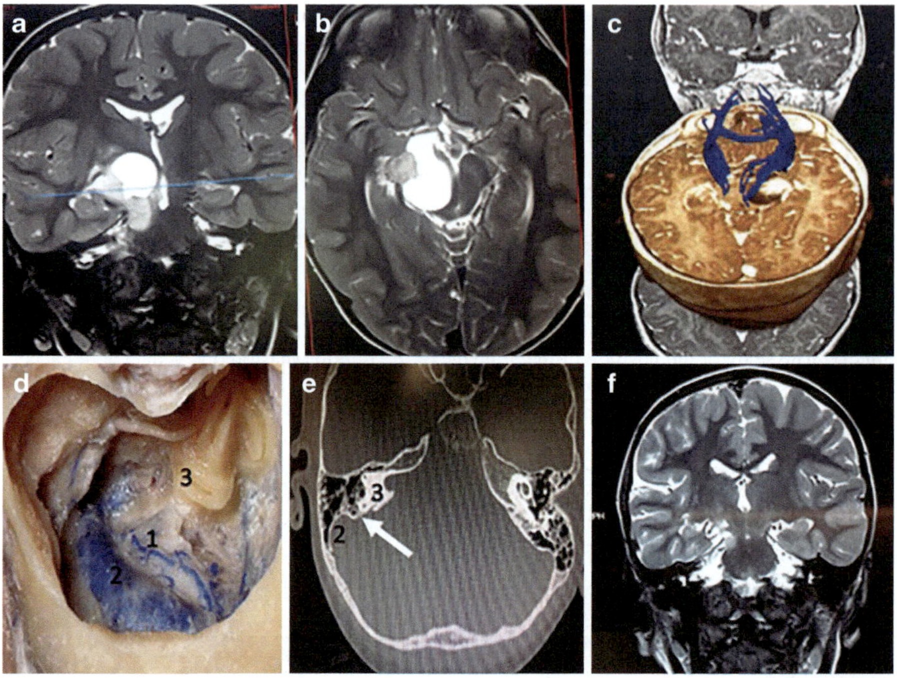

Fig. 23 Illustrative case of a 9-year-old boy with progressive left-sided hemiparesis. The cause of the symptoms was a massive thalamo-mesencephalic glioma (**a–c**). Due to involvement of the middle and posterior fossae, a posterior petrosal approach was planned at first (as shown in a cadaveric dissection (**d**)). The CT scan of the temporal bone, however, showed a very narrow space between the posterior semicircular canal and the sigmoid sinus (**e**, white arrow). Based on this preoperative imaging finding, we opted for a transtemporal approach through the middle temporal gyrus. Subtotal tumor resection was obtained (**f**) and the patient improved from neurological symptoms. (*1*) Trautman's triangle; (*2*) sigmoid venous sinus; (*3*) posterior semicircular canal

by the sigmoid venous sinus and, superiorly, the inferior border of the transverse venous sinus. This reduces the need for cerebellar retraction when compared with the suboccipital approach. The main advantage of this access route is the possibility of hearing preservation and adequate exposure of the lower structures of the cerebellopontine angle.

There are many indications through the retrosigmoid route, especially for tumors with little intracanalicular extension and predominantly extracanalicular. The main indication is vestibular schwannoma surgery, especially when there is a possibility of hearing preservation and if the tumor extends less than 1 cm into the internal auditory canal. The indication for retrosigmoid access depends on the preoperative level of hearing. The patient must have a tonal average better than 50 dB and/or a speech discrimination score better than 50%. The retrosigmoid approach can be performed for tumors of any size, especially in extracanalicular tumors. Other indications include posterior fossa meningiomas, epidermoid cyst, chondroma, chondrosarcoma, chondroblastoma, chordoma, trigeminal schwannoma, basilar artery aneurysms, microvascular decompression (V, VII, IX, and X), and neurectomies (trigeminal, vestibular, and glossopharyngeal). The retrosigmoid approach provides wide exposure of the lower third of the cerebellopontine angle, making it possible to visualize tumors that extend to the foramen magnum. The translabyrinthine approach, on the other hand, has its lower limit at the height of the jugular bulb. It can also be useful in revision surgeries in which other routes (translabyrinthine or middle fossa) were used in previous surgeries (Figs. 24, 25, 26, and 27).

Like the translabyrinthine approach, for the retrosigmoid approach, there are also some anatomical landmarks for safe surgery. The transverse venous sinus lies in a horizontal line that passes through the zygomatic arch and the external auditory canal. The sigmoid venous sinus sits in a vertical line that passes 4–5 cm posterior to the retro-auricular sulcus and level with the mastoid. The anterior inferior cerebellar artery (AICA) originates inferior or medial to the basilar artery, and its course involves the V, VII, and VIII pairs. It may form a loop between VII and VIII cranial nerves. The superior petrous vein (Dandy's vein) usually originates at the level of the trigeminal nerve root in the pons. It drains into the superior petrosal venous sinus and is responsible for venous drainage of the brainstem and cerebellar hemispheric. Through a posterior fossa approach, the superior and inferior vestibular nerves are proximal to the surgeon within the internal auditory canal.

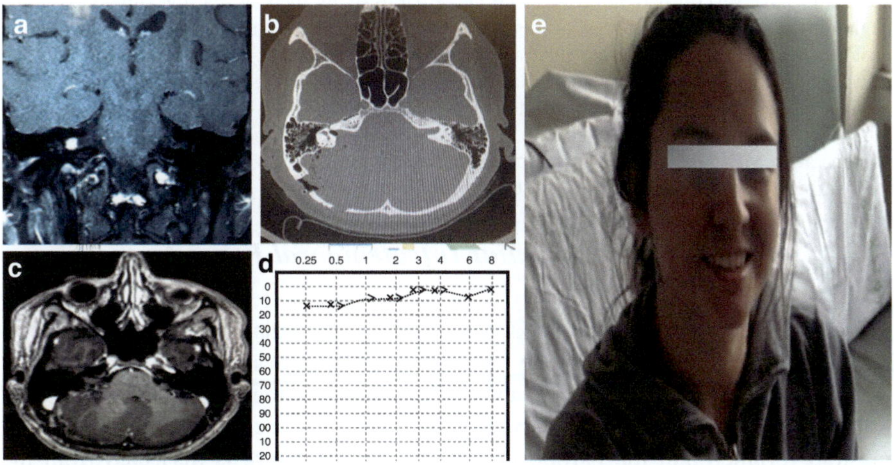

Fig. 24 Facial and hearing preservation can be achieved even in the initial part of the learning process. (**a**) A retrosigmoid approach was performed to resect this small right vestibular schwannoma. (**b**) The postoperative CT scan shows the posterior wall of the internal acoustic meatus drilled out. (**c**) Postoperative gadolinium-enhanced T1-weighted magnetic resonance imaging sequence shows total tumor resection. (**d**) Post-operative audiogram showing hearing preservation on the right side; (**e**) Preservation of the facial nerve function

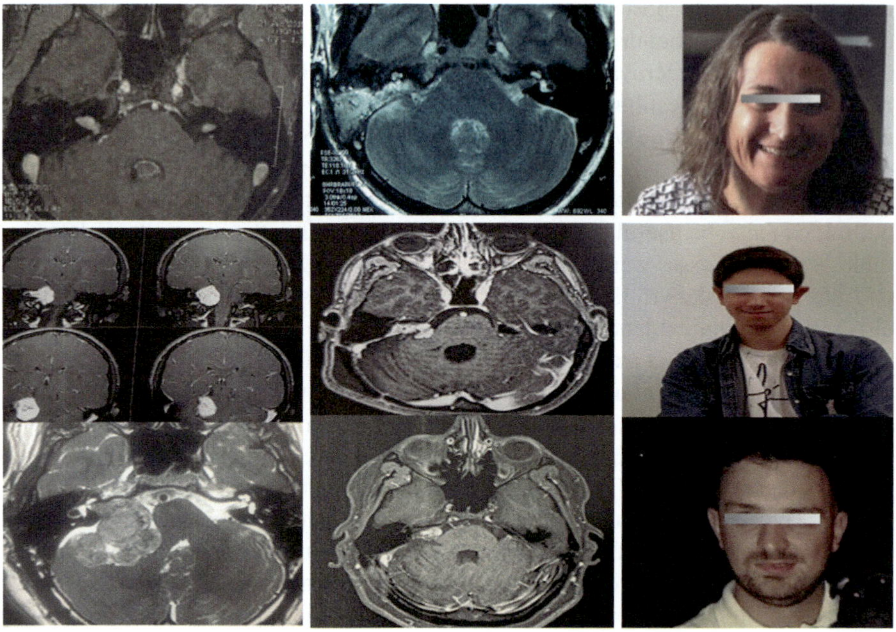

Fig. 25 At the beginning of the neurosurgical learning process, achieving good outcomes is paramount and will reflect the future of the skull base surgeon career. In our opinion, there is no justification to have several patients with facial nerve paralysis by the fact that the surgeon is at the beginning of his learning process. In this case, it is better to achieve a partial resection without a postoperative neurological deficit than a total resection with a post-operative deficit. Although resection of small vestibular schwannomas is less risky, we still can preserve facial nerve function in bigger tumors, even if it is necessary to perform a subtotal or near-total resection. The left column shows a preoperative MRI of other vestibular schwannomas. The middle column shows the postoperative MRI, and the right column the clinical status 3 months after surgery. There was no fair hearing on the right side in none of the patients

7.6 Subtemporal Preauricular Infratemporal Fossa Approach

This approach begins with neck extension and 45°-contralateral head rotation. The beginning of the incision is made on the frontal skin, descending in front of the external auditory canal and with anterior extension in the neck. The skin flap is reflected anteriorly. The facial nerve is identified and dissected between the stylomastoid foramen and the parotid gland. This last structure is dissected from the masseteric fascia to facilitate the anterior displacement of the mandible without traction of the facial nerve. The temporal muscle is displaced down, and the zygomatic arc is resected. The attachment of the stylomandibular and sphenomandibular ligaments is cut to displace anteroinferiorly the mandibular condyle. To gain more space, sometimes, the condyle of the mandible must be resected. The next step is identifying and dissection of the neurovascular structures in the neck. The digastric

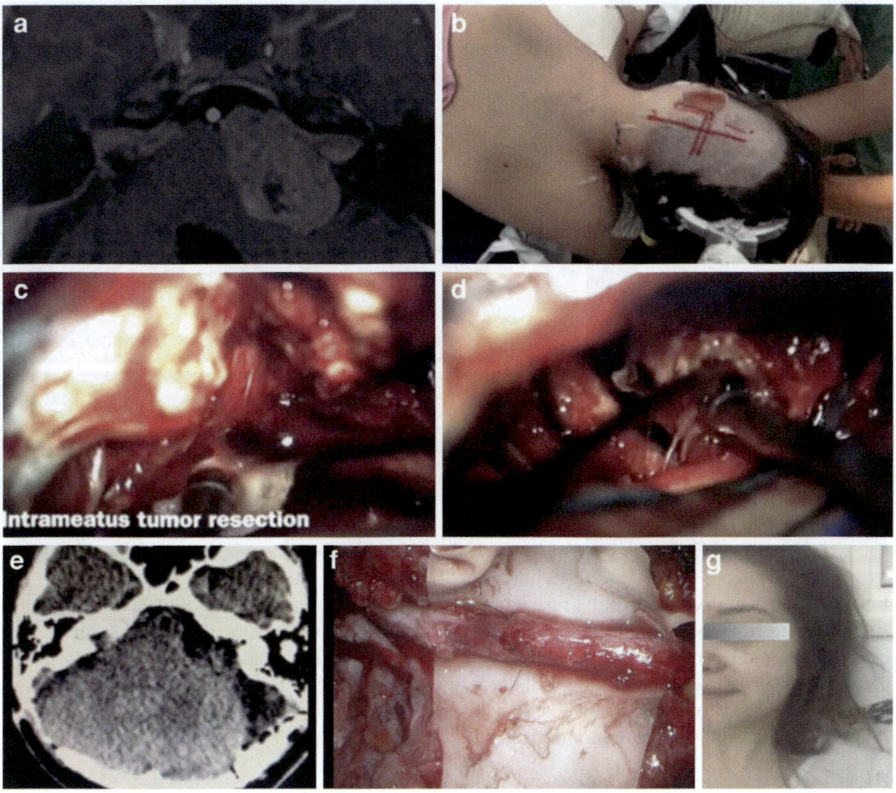

Fig. 26 19-year-old female with NF2. The right vestibular schwannoma was resected in another hospital a year before. The MRI showed a large vestibular schwannoma on the left side (**a**). A retrosigmoid approach was performed (**b**), and the intracanicular (**c**) and cisternal (**d**) parts of the tumor were resected. Although the facial nerve was anatomically preserved, the facial nerves potentials were lost at the end of the surgery. Day one postoperative CT scan (**e**). Since there was no recovery of the facial function in the next 2 months, a hypoglossal-facial anastomosis was performed (**f**). Four months after the procedure, the patient recovered her facial function (**g**)

muscle is divided, and the styloid process is resected. A pterional craniotomy is performed with the posteroinferior limit just above the glenoid fossa. The next step is peeling the middle fossa with identification of the arcuate eminence, middle meningeal artery, greater and lesser superficial petrosal nerves, V3, V2, and V1 entering into the superior orbital fissure. The part of the greater wing of the sphenoid bone in the floor of the middle fossa is drilled around the foramen rotundum and foramen ovale to expose, respectively, the V2 entering into the pterygopalatine fossa and the V3 entering into the ITF. The eustachian tube and the tensor tympani muscle are resected. The intrapetrous portion of the ICA is identified. The cervical ICA is displaced forward. An anterior petrosectomy (by drilling the Kawase triangle) can be done to expose the clivus (Fig. 28).

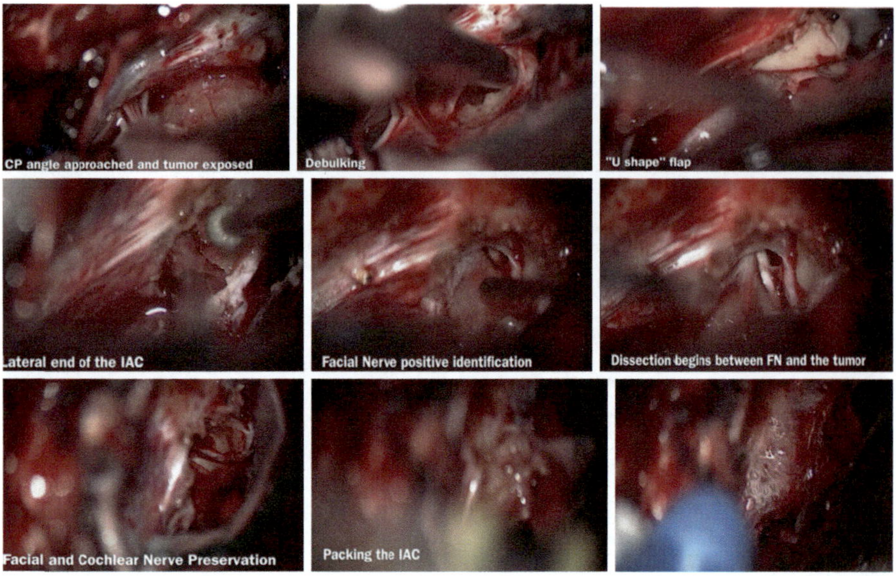

Fig. 27 Intraoperative view of the resection of a vestibular schwannoma, step by step. The microsurgical anatomy laboratory training should be the initial step for every skull base surgeon. The next step should be to watch surgeries of a very experienced skull base surgeon. Only one should perform the surgery. For the initial vestibular schwannoma surgeries, even in developing countries, all armamentarium must be at disposal in the operative room, like ultrasonic aspirator, surgical microscope, neurophysiologic monitoring, and fibrin glue

7.7 Postauricular Transtemporal Approach

The head is positioned in the Mayfield headrest and turned contralaterally. A C-shaped incision is made from behind the ear to the neck. According to Fisch's description, the external ear is transected, everted, and closed as a blind sac. The sternocleidomastoid muscle is detached from the mastoid. The following structures are identified and dissected: common carotid artery; internal carotid artery (ICA); external carotid artery; internal jugular vein; cranial nerves IX, X, XI, and XII; ascending pharyngeal artery; and posterior auricular and occipital arteries, which can be ligated during the surgery. The mastoid is drilled to skeletonize the mastoid portion of the facial nerve with its subsequent anterior transposition. The next step is a posterior fossa craniotomy and exposition of the sigmoid sinus. The posterior belly of the digastric muscle, the stylohyoid muscle, and the styloid process are dissected and removed.

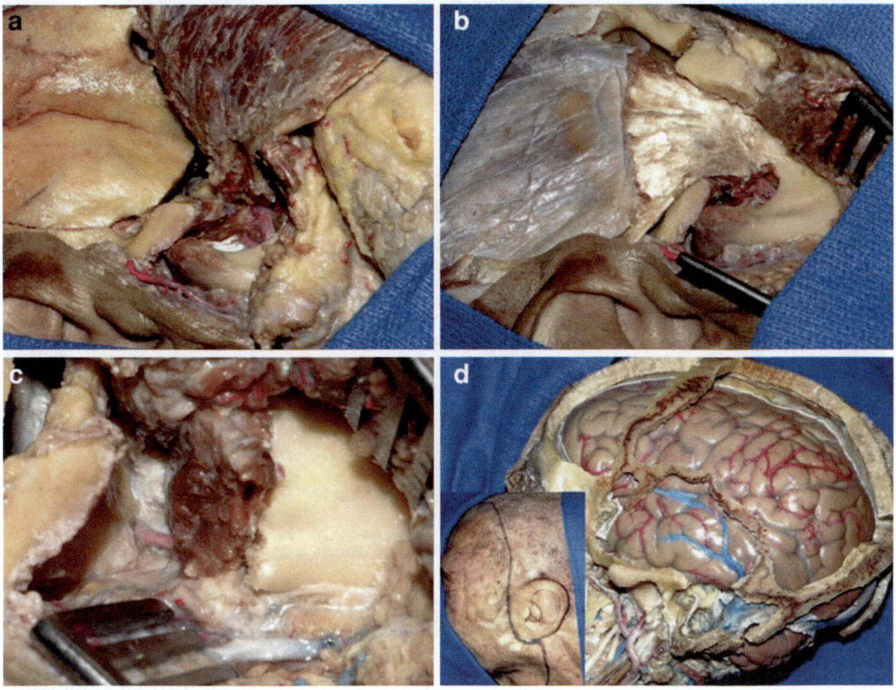

Fig. 28 Subtemporal preauricular ITF approach. (**a**) The preauricular incision on the right side was performed, the flap was displaced anteriorly, and the zygomatic process was removed. (**b**) The temporal muscle is displaced down. (**c**) Exposition of the middle meningeal artery and V3 in the ITF. (**d**) Relations among ITF, mastoid, and temporal lobe; the inset shows the incision. (*1*) Superficial temporal fascia; (*2*) tendon of the temporal muscle; (*3*) mandibular incisura and maxillary artery; (*4*) superficial temporal artery; (*5*) mandibular condyle; (*6*) middle meningeal artery entering in the foramen spinosum; (*7*) V3; (*8*) sigmoid sinus exposed after mastoidectomy; (*9*) mastoid portion of the facial nerve

During the surgery, the sigmoid sinus is ligated between the mastoid emissary vein (there is an anterior and posterior, but the posterior is bigger), proximally, and the sigmoid venous sinus, distal to the tumoral extension. The skin of the external ear canal and the tympanic membrane are removed. The intrapetrous segment of the ICA is exposed by drilling the carotid canal. The eustachian tube is identified and, during the surgery, will be obliterated with wax and fascia. The surgeon must identify the lower cranial nerves leaving the jugular foramen by opening the dura of the posterior fossa and finding the fifth and twelfth nerves and the structures of the cerebellopontine angle, as well as the posterior inferior cerebellar artery (PICA) and the superior cerebellar artery (Figs. 29, 30, and 31).

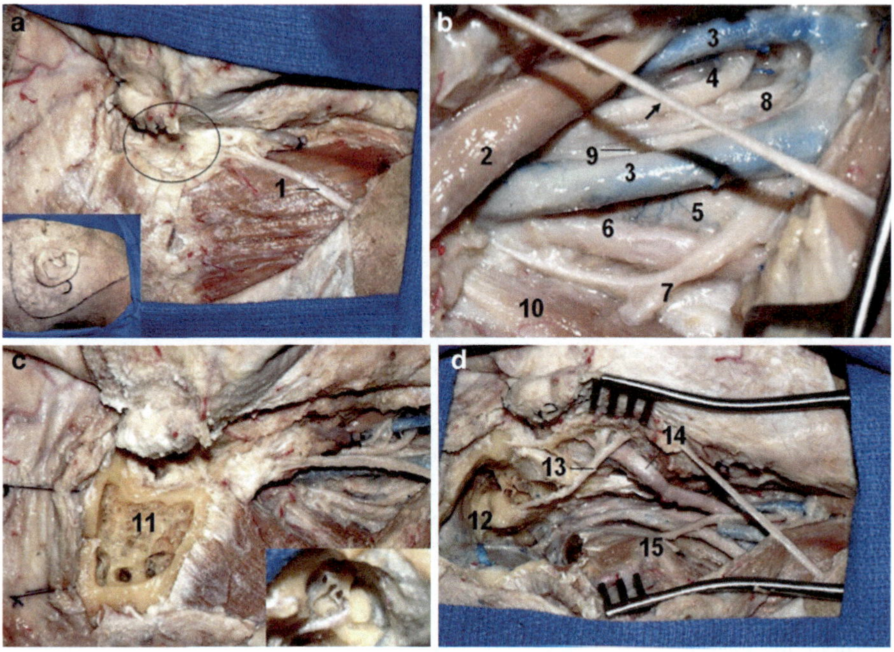

Fig. 29 Combined infratemporal and posterior fossa approach. (**a**) After the incision (inset), the flap is reflected anteriorly, and the external meatus is sectioned and closed in a blind sac, as Ugo Fisch (oval) described. (**b**) The neurovascular structures are exposed in the neck to proximal control. The arrow indicates the greater auricular nerve. (**c**) The petrosectomy is performed. The inset shows the facial nerve, the ossicles, and semicircular canals. (**d**) Final anatomical view of the neurovascular structures in the neck and the presigmoid and temporal fossa dura after petrosectomy. The semicircular canals are preserved, and the facial nerve is transposed anteriorly. (*1*) Great auricular nerve; (*2*) digastric muscle (posterior belly); (*3*) duplicated internal jugular vein; (*4*) glossopharyngeal nerve; (*5*) internal carotid artery; (*6*) sympathetic trunk; (*7*) accessory nerve; (*8*) hypoglossal nerve; (*9*) vagus nerve; (*10*) inferior oblique nerve; (*11*) mastoid drilled; (*12*) facial nerve; (*13*) semicircular canals; (*14*) superficial temporal artery; (*15*) the transverse process of C1

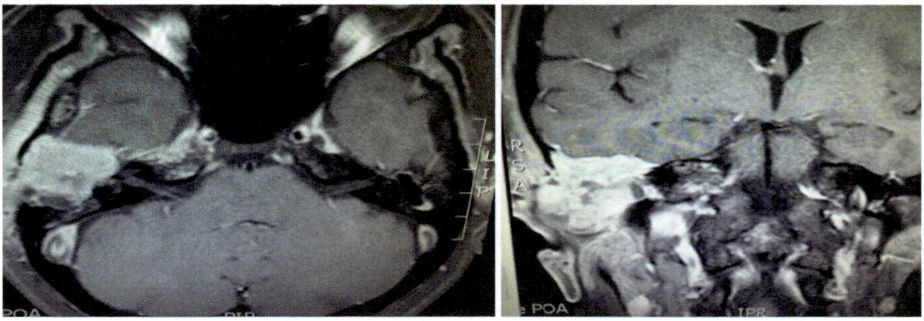

Fig. 30 Temporal bone carcinoma (illustrative case)

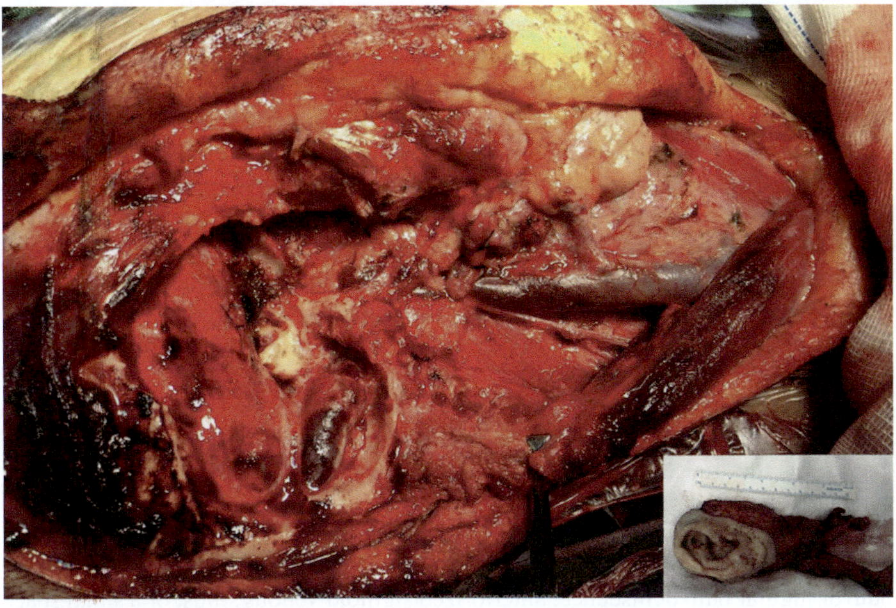

Fig. 31 Temporal bone carcinoma should be resected aggressively and "en bloc"

8 Discussion

The temporal bone has a complex and intricate anatomy. Its internal structures and its adjacencies must be understood in depth and under a three-dimensional vision. This type of knowledge is acquired only in the microsurgical laboratory. Recently, combining a laboratory study of the temporal bone and its correlation with imaging exams has provided new insights for neurosurgeons, otologists, and head and neck surgeons. Transtemporal approaches are a versatile weapon to target diseases in the petroclival region, jugular foramen, and cerebellopontine cistern. They have two main advantages: decreasing the distance between the surgeon and the lesion in these regions and minimizing brain retraction [11, 12].

Although these concepts of skull base approaches became popular in the 90s by forming multidisciplinary teams that included otologists and neurosurgeons, the evolution of microsurgical techniques, the accumulated experience, and the results obtained over the years relegated a more limited role for transtemporal approaches. These approaches still have clear indications and should be part of every skull base surgeon's approach arsenal, but they should not be part of an ideological apology that preaches that every skull base tumor should be resected with an aggressive skull base approach.

Some examples: extradural tumors within the temporal bone should be resected with transtemporal approaches. Petrous apex chondrosarcomas, paragangliomas, cholesterol granuloma, epidermoid tumor inside the temporal bone, aneurysmatic bone cyst, cholesteatoma, and endolymphatic sac tumors are some examples. On the other hand, not all intradural petroclival tumors need to be resected with transtemporal approaches. In vestibular schwannomas, we reserve the translabyrinthine approach for small tumors that reach the lateral end of the internal acoustic meatus in patients with impaired hearing. We observed that in these cases, the facial nerve is found at the lateral end of the meatus at the beginning of the surgery (with intraoperative neurophysiological monitoring), and the tumor is safely resected. This type of approach avoids cerebellar retraction. However, for larger tumors and/or those in patients with preserved hearing, the retrosigmoid approach is versatile and effective. The same is true for petroclival meningiomas, in which transtemporal approaches are reserved for selected cases, such as tumors with supratentorial extension and those whose lower part is above the inferior clivus). For all other cases, the retrosigmoid approach may be sufficient.

Although we have used transtemporal approaches in our routine to treat various types of skull base tumors, preoperative imaging studies to verify contraindication factors for the transtemporal approach are crucial in all cases. These studies are summarized by temporal bone CT scan and venous MRI.

In temporal bone CT, we must pay attention to the following anatomical variations that may compromise the approach through the temporal bone: high jugular bulb or jugular bulb dehiscence, anteriorized sigmoid venous sinus, or a small space between the posterior semicircular canal and the sigmoid venous sinus (common in children).

In venous MRI imaging (or even in the conventional brain angiography, in the venous phase, although this is an invasive procedure), the dominance of the transverse sinus can be studied. This is not so accurate because this is an anatomical study and does not measure flow. Nevertheless, it can give an idea of the soon-to-be skeletonized sigmoid venous sinus drainage. This assessment is crucial to study the drainage pattern of the inferior anastomotic vein (vein of Labbe). Although rare, the vein of Labbe may drain or more anteriorly into the junction of the transverse and sigmoid venous sinus, or even in the superior petrosal sinus. This finding indicates that any superior retraction of the temporal lobe should be avoided, a relative contraindication for the transtemporal approach.

In cases where the vein of Labbé drains at the junction of the transverse and sigmoid venous sinus (usual position), there will likely be no problem of superior temporal lobe retraction. However, the care with the vein of Labbé should be maximized whenever the surgeon needs to pull the temporal lobe to access the crural and ambient cisterns that are medial to the middle incisural space. Retractors with Leyla spatula support should be avoided. Even when the surgeon is focused in a deep field of surgery, he should always be attentive to the position of the vein of Labbé in the outer part of the surgical field. The approaches with the potential to damage the vein of Labbé are the intradural subtemporal approaches and posterior petrosectomy (supra/infratentorial presigmoid).

The study of the superior petrosal vein (Dandy vein) is no less important. This vein (as well as the vein of Labbé) must always be preserved. Although authors postulate that it can be safely ligated, we find this conduct fearful, as there are reports of deadly complications after its ligation.

9 Conclusion

The temporal bone may be an important route to reach deep intracranial tumors. Although it may not be the first choice in most cases, it is a safe option when the more traditional approaches may not be available or may not provide wide exposure. The skull base surgeon must master the anatomical knowledge of the temporal bone so that these approaches are part of its armamentarium. Microanatomical laboratory training is the gold standard method to prepare a skull base surgeon for these approaches.

References

1. da Silva SA, Yamaki VN, Solla DJF, Andrade AF, Teixeira MJ, Spetzler RF, Preul MC, Figueiredo EG. Pterional, Pretemporal, and Orbitozygomatic approaches: anatomic and comparative study. World Neurosurg. 2019;121:e398–403. https://doi.org/10.1016/j.wneu.2018.09.120.
2. Cavalcanti DD, Morais BA, Figueiredo EG, Spetzler RF, Preul MC. Accessing the anterior Mesencephalic zone: orbitozygomatic versus subtemporal approach. World Neurosurg. 2018;119:e818–24. https://doi.org/10.1016/j.wneu.2018.07.272.
3. Isolan GR, Braga FLS, Campero A, Landeiro JA, Araújo RML, Ajler P, Sakaya GR, Rabelo NN, Brito JS, Teixeira MJ, Figueiredo EG. Microsurgical and endoscopic anatomy of the cavernous sinus. Braz Neurosurg. 2020;39(2):83–94. https://doi.org/10.1055/s-0040-1709740.
4. Krayenbühl N, Isolan GR, Hafez A, Yaşargil MG. The relationship of the fronto-temporal branches of the facial nerve to the fascias of the temporal region: a literature review applied to practical anatomical dissection. Neurosurg Rev. 2007;30(1):8–15.; discussion 15. https://doi.org/10.1007/s10143-006-0053-5.
5. Krayenbühl N, Isolan GR, Al-Mefty O. The foramen spinosum: a landmark in middle fossa surgery. Neurosurg Rev. 2008;31(4):397–401.; discussion 401–2. https://doi.org/10.1007/s10143-008-0152-6.
6. Bernardo A, Evins AI, Visca A, Stieg PE. The intracranial facial nerve as seen through different surgical windows: an extensive anatomosurgical study. Neurosurgery. 2013;72(2 Suppl Operative):ONS194–207. ; discussion ONS207. https://doi.org/10.1227/NEU.0b013e31827e5844.
7. Santos FP, Longo MG, May GG, Isolan GR. Computed tomography evaluation of the correspondence between the arcuate eminence and the superior semicircular canal. World Neurosurg. 2018;111:e261–6. https://doi.org/10.1016/j.wneu.2017.12.030.
8. Isolan GR, Rowe R, Al-Mefty O. Microanatomy and surgical approaches to the infratemporal fossa: an anaglyphic three-dimensional stereoscopic printing study. Skull Base. 2007;17(5):285–302. https://doi.org/10.1055/s-2007-985193.
9. Isolan GR, Wayhs SY, Lepski GA, Dini LI, Lavinsky J. Petroclival meningiomas: factors determining the choice of approach. J Neurol Surg B Skull Base. 2018;79(4):367–78. https://doi.org/10.1055/s-0037-1608654.
10. Adams Pérez J, Rassier Isolan G, Pires de Aguiar PH, Antunes AM. Volumetry and analysis of anatomical variants of the anterior portion of the petrous apex outlined by the kawase triangle using computed tomography. J Neurol Surg B Skull Base. 2014;75(3):147–51. https://doi.org/10.1055/s-0033-1356491.
11. Traynelis VC. The geometry of education: patterns for growth. Clin Neurosurg. 2005;52:1–5.

12. Oliveira LM, Figueiredo EG. Simulation training methods in neurological surgery. Asian J Neurosurg. 2019;14(2):364–70. https://doi.org/10.4103/ajns.AJNS_269_18.
13. Kinoshita Y, Zomorodi AR, Friedman AH, Sato H, Carter JH, Bawornvaraporn U, Nakamura H, Fukushima T. Retrolabyrinthine transsigmoid approach to complex parabrainstem tumors in the posterior fossa. J Neurosurg. 2021;8:1–6. https://doi.org/10.3171/2021.5.JNS204130.

Surgical Anatomy of the Sulci and Gyrus of the Brain

Eduardo Carvalhal Ribas and Guilherme Carvalhal Ribas

1 Introduction

The encephalon is the part of the central nervous system that lies within the intracranial space. It is divided by the tentorium into an infratentorial compartment, which contains the brainstem and cerebellum, and a supratentorial compartment housing the brain, which is formed by the diencephalon and the telencephalon constituted by two cerebral hemispheres separated by the falx.

The phylogenetic and embryological development of the brain surface is based on its invagination processes, the fissures and sulci being natural extensions of the subarachnoid space, with depths ranging from 1 to 3 cm approximately and harboring small gyri within its sulcal space, known generically as transverse gyri [1]. Also, the brain undergoes a process of circular curvature, placing each thalamus as the morphological center of each cerebral hemisphere.

E. C. Ribas
Neurosurgeon at Hospital Israelita Albert Einstein, São Paulo, SP, Brazil

Neurosurgeon at Hospital das Clinicas University of São Paulo Medical School, São Paulo, SP, Brazil

G. C. Ribas (✉)
Neurosurgeon at Hospital Israelita Albert Einstein, São Paulo, SP, Brazil

University of São Paulo Medical School, São Paulo, SP, Brazil

Albert Einstein Medical School, São Paulo, SP, Brazil

Department of Neurosurgery, University of Virginia, Charlottesville, VA, USA
e-mail: guilherme@neuroribas.com.br

E. G. Figueiredo et al. (eds.), *Brain Anatomy and Neurosurgical Approaches*,
https://doi.org/10.1007/978-3-031-14820-0_5

These processes led to a typical superficial organization of sulci and gyri. Each mammalian species has a highly characteristic cerebral surface, with humans having a higher complexity and almost two thirds of the neocortex hidden away in the depth of the sulci [2].

The cerebral gyri constitute a continuum of the cerebral surface, and their appearance as distinct gyri is only superficial since they are continuous along with the entire sulcal depth and along its extremities. Therefore, each gyrus must be understood as a region and not as a well-defined and distinct structure [3].

The sulci may be continuous or interrupted, once or at several points, and some sulci may be doubled over a certain part of their trajectory [4]. The sulcal pattern has a considerable intersubject variability if individual sulci are seen in detail, and there is also a considerable variation between the two hemispheres of the same individual.

2 Brain Subdivision into Lobes

In order to favor its better understanding, organize its nomenclature, and apply this knowledge in clinical practice, the cerebral hemispheres were arbitrarily divided into lobes, regions, and compartments. As these divisions correspond to different concepts and have been created according to different criteria, they partly overlap, complement each other, and are particularly useful for a better understanding of brain architecture [5].

Louis Pierre Gratiolet (1815–1865), a prominent French anatomist, was the first to propose a brain subdivision into lobes [6], giving their names in correspondence to the skull bones overlying each one. The insula, first described by Johann Christian Reil (1759–1813) in 1809, was also considered a distinct cerebral lobe [7]. The concept of a limbic lobe was introduced by Paul Broca (1824–1880) in 1877, initially including the cingulate and parahippocampal gyri [8] and later modified by others to include "medial portions of the frontal, parietal and temporal lobes," region septal, peririnal cortex, entorhinal cortex, and the various parts of the hippocampal formation [9]; became officially recognized as a distinct lobe only in the 1998 edition of Terminologia Anatomica [10]. The conception of a paralimbic belt surrounding the limbic lobe also includes the caudal orbitofrontal cortex, the anterior portion of the insula, and the temporopolar cortex.

Currently, the International Anatomical Terminology describes each cerebral hemisphere as being divided into six lobes: frontal, parietal, occipital, temporal, insular, and limbic (Fig. 1a) [11].

Initially, the sutures that delimit the different skull bones were also used to delimit the cerebral lobes, but were replaced by superficial cerebral landmarks as the understanding of the superficial organization of the brain developed.

The lateral sulcus of the brain, also called the Sylvian fissure, is promptly and easily identifiable. It has a stem, which extends from the carotid cistern to the anterior Sylvian point on the superolateral surface of the brain, and gives rise to three rami: anterior horizontal, anterior ascending, and posterior [12]. The anterior

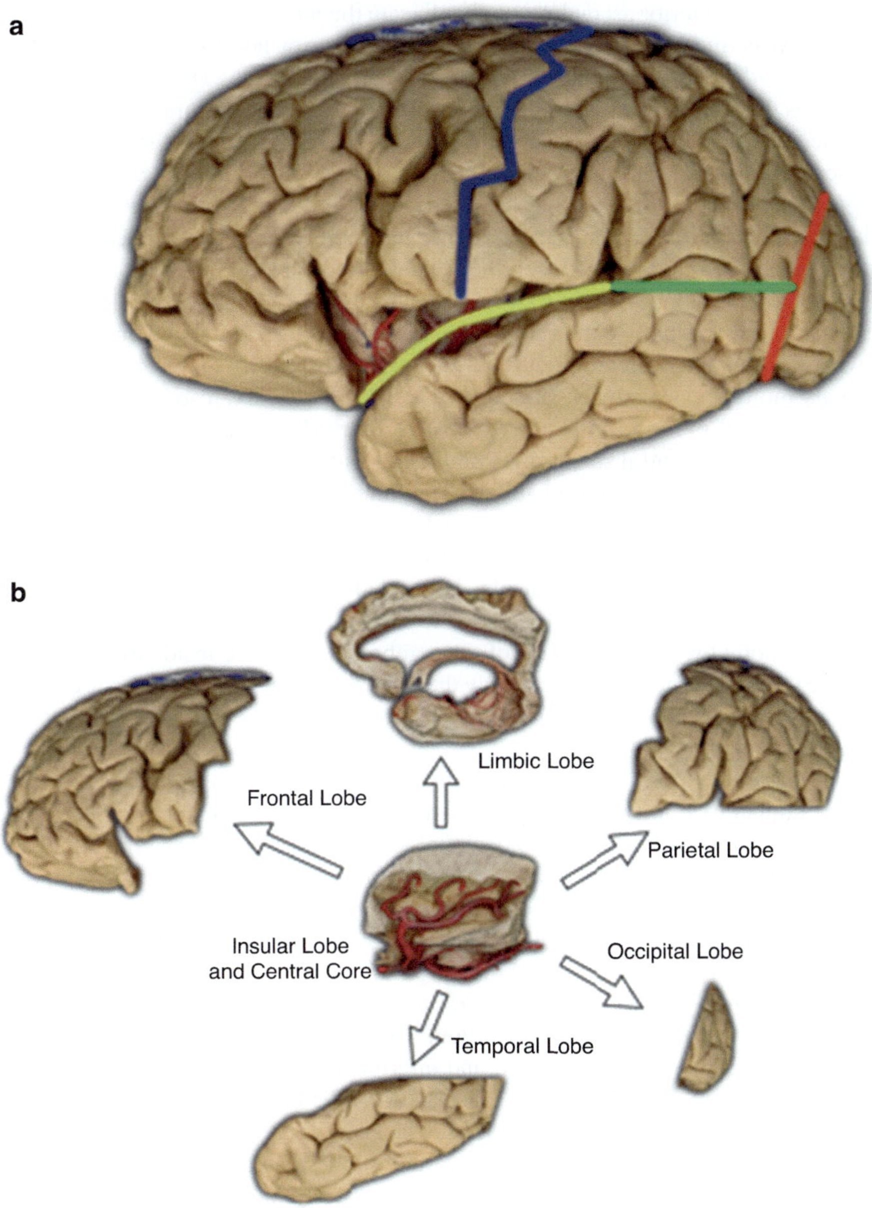

Fig. 1 (**a**) Each cerebral hemisphere is divided into six lobes (frontal, parietal, occipital, temporal, insular, and limbic) by using the lateral sulcus of the brain, also called the Sylvian fissure (yellow line), the central sulcus (blue line), the parietotemporal line (red line), and a line connecting the posterior end of the Sylvian fissure to the parietotemporal line midpoint (green line). (**b**) The disposition and arrangement of cerebral lobes can be understood in relation to the central core of the brain: frontal lobe is anterior and superior, parietal lobe is posterior and superior, occipital lobe is posterior, temporal lobe is inferior, insular lobe is its lateral aspect, and limbic lobe encircles the central core

horizontal and anterior ascending rami delineate the *pars triangularis* in the inferior frontal gyrus, the *pars orbitalis* anterior to the anterior horizontal ramus, and the *pars opercularis* posterior to the anterior ascending ramus. The posterior ramus separates the frontal and parietal lobes superiorly from the temporal lobe inferiorly.

The central sulcus was occasionally referred to as the *fissure of Rolando* due to its accurate description by Luigi Rolando (1773–1831) in his text *Della Strutura degli Emisferi Cerebrali* in 1829 [13], is oblique and can also be regularly recognized at the superolateral surface of the brain. Its most superior aspect is found a little behind the midpoint between the frontal and occipital poles [14] and extends downward and forward with a trajectory that resembles a lengthened italic S [15].

This sulcus is usually continuous [4] and separates the frontal and parietal lobes, having the pre- and postcentral gyri anteriorly and posteriorly to this sulcus with the primary motor and somatosensory cortical areas of the cortex, respectively. The pre- and postcentral gyri are both superiorly and inferiorly connected, encircling both extremities of the central sulcus and comprising altogether a lengthened and oblique ellipse excavated by the central sulcus. The inferior connection corresponds to the subcentral gyrus, delineated anteriorly and posteriorly by the anterior and posterior subcentral rami of the Sylvian fissure, respectively; it is usually found at the superolateral surface of the brain, but can also be located already inside the Sylvian fissure. The superior connection corresponds to the paracentral lobule of Ecker [16] already disposed along the medial surface of the cerebral hemisphere inside the interhemispheric fissure, delineated anteriorly by the paracentral sulcus and posteriorly by the ascending and distal part of the cingulated sulcus, that is, the marginal ramus of the cingulated sulcus.

The parietotemporal line is an imaginary line drawn from the parietooccipital sulcus's upper extremity on the brain's lateral surface to the preoccipital notch and represents the anterior limit of the occipital lobe. Another imaginary line, drawn as a posterior continuation of the Sylvian fissure to the parietotemporal line midpoint, separates the parietal lobe superiorly from the temporal lobe inferiorly.

The frontal, parietal, occipital, and temporal lobes are exposed prima facie at the superolateral surface of the brain, being promptly recognized. The insula is located at a depth of the Sylvian cistern, covered superiorly by the frontoparietal operculum and inferiorly by the temporal operculum. Its surface is separated from the other cerebral lobes, its anterior, superior, and inferior limiting insular sulci. The limbic lobe is present at the medial and basal surfaces of the brain, bounded by the cingulate sulcus superiorly and the collateral sulcus laterally, respectively [1, 17].

The disposition and arrangement of cerebral lobes can also be understood with the brain's central core [5]. The concept of a central core of the brain is based on the clear anatomical delimitation of a central region within each hemisphere between the brainstem, the ventricular cavities, and the cerebral lobes, which has important clinical and surgical implications. It is composed of the insular surface, basal ganglia, and thalamus and is connected with the rest of the supratentorial compartment by a cerebral isthmus (represented by the continuation of the internal, external, and extreme capsules toward the cerebral lobes). It also houses part of the anterior commissure, amygdalofugal pathways, and the region of the innominate substance. This block is at the morphological center of each cerebral hemisphere and plays an important role in integrating information, motor and sensory functions, emotion, and cognition. From an anatomical and architectural point of view, the central core is covered laterally by the insula, medially by the supratentorial ventricular surfaces, and embraced by the limbic lobe, which describes a "C"-shaped ring around the thalamus. The frontal, parietal, occipital, and temporal lobes are placed anteriorly, superiorly, posteriorly, and inferiorly with the central core.

3 Cerebral Surfaces, Sulci and Gyri

Each cerebral hemisphere has three surfaces (superolateral, medial, and basal), three margins (superior, inferior, and medial), and three poles (frontal, temporal, and occipital).

The superolateral surface of the brain, also referred to as cerebral convexity, faces the concave inner surface of the skull (Fig. 2a, b). At this surface, three frontal horizontal gyri (superior, middle, and inferior frontal gyri) converge anteriorly to form the frontal pole, three temporal horizontal gyri (upper, middle, and inferior temporal gyri) converge anteriorly to form the temporal pole, and three occipital gyri (superior, middle, and inferior occipital gyri) converge posteriorly to form the occipital pole. Two oblique, ascending gyri are clearly seen at the center of this surface: precentral and postcentral gyri. Immediately posteriorly, a quadrangular region called superior parietal lobule can be noted superiorly, and the inferior parietal lobule inferiorly, formed by two curved gyri: the supramarginal gyrus, which can be seen as the posterior continuation of the superior temporal gyrus and whose trajectory curves over the posterior margin of the Sylvian fissure; and the angular gyrus, which corresponds to a posterior and curvilinear extension of the middle temporal gyrus forming a horse-shoe "C" shape [1].

Although a considerable intersubject variability exists, some sulcal uniformity can be noted at the superolateral surface of the brain. The superior frontal and superior temporal sulci are usually continuous, whereas the inferior frontal and inferior temporal sulci are usually interrupted. The pre- and postcentral sulci are always interrupted: the first having interruptions due to the frontal gyri connections to the precentral gyrus and the second having interruptions due to the postcentral gyrus connections with the superior parietal lobule. The intraparietal sulcus separates the superior and inferior parietal lobules and is usually clearly identifiable with an anterior-posterior disposition. The intermediate sulcus of Jensen is located between the supramarginal and angular gyri and can arise from the intraparietal sulcus, superior temporal sulcus, or both. The occipital gyri have higher intersubject variability.

The medial surface of the brain lies at a depth of the interhemispheric fissure, where both hemispheres face each other and are anatomically more constant between individuals. The corpus callosum is the main human commissural fiber bundle and is found at the center of this surface, having first the callosal sulcus and then the cingulate gyrus describing an arc around it. Inferior to the rostrum of the corpus callosum, the cingulate gyrus curves and joins the rectus gyrus forming the cingulate pole. The rectus gyrus harbors the inferior rostral sulcus and is separated from the cingulate gyrus by the superior rostral sulcus. Three vertical gyri can be identified posterior to the cingulate pole and form the septal region: anterior paraolfactory gyrus, posterior paraolfactory gyrus, and paraterminal gyrus. Posterior to the splenium of the corpus callosum, the cingulate gyrus presents a narrowing, called the isthmus, and continues inferiorly with the parahippocampal gyrus.

The superior frontal gyrus curves along the superior cerebral margin and is also seen at the medial surface of the brain. Similarly, the precentral and postcentral gyri also curve along the superior cerebral margin and at the medial surface of the brain form together with the paracentral lobule. The cingulate sulcus separates the superior frontal gyrus and the paracentral lobule from the cingulate gyrus and posteriorly curves superiorly to constitute the ascending ramus of the cingulate sulcus separating the paracentral lobule from the precuneus. The paracentral sulcus originates more anteriorly from the cingulate sulcus and separates the superior frontal gyrus from the paracentral lobule. The precuneus has a quadrangular shape, is incompletely separated from the cingulate gyrus by the subparietal sulcus rami, and is deeply separated anteriorly from the paracentral lobule, the ascending ramus of the cingulate sulcus, and posteriorly from the cuneus by the parieto-occipital fissure. The cuneus has a triangular shape, with its vertex pointing anteriorly, and is separated by the calcarine fissure from the lingual gyrus (also known as the medial occipitotemporal gyrus). The calcarine fissure is divided into two parts by the origin of the parieto-occipital sulcus: its proximal part is directly inferior to the cingulate isthmus and caudal base of the precuneus, and its distal part lies between the cuneus and the lingual gyrus.

Fig. 2 Cortical anatomy of the brain, represented in drawings and anatomical dissections. **(a, b)** Superolateral surface; **(c, d)** medial surface; **(e, f)** basal surface; **(g, h)** insular surface. *AcG* Accessory Gyrus, *ALS* Anterior Limiting Sulcus, *AngG* Angular Gyrus, *CalcF* Calcarine Fissure, *CaS* Callosal Sulcus, *CC* Corpus Callosum, *CerHem* Cerebellar Vermis, *Ch* Optic Chiasm, *CiG* Cingulate Gyrus, *CiIst* Cingulate Isthmus, *CiMR* Cingulate Marginal Ramus, *CiS* Cingulate Sulcus, *ColS* Collateral Sulcus, *CS* Central Sulcus, *Cu* Cuneus, *For* Fornix, *FusG* Fusiform Gyrus, *IFG* Inferior Frontal Gyrus, *IFS* Inferior Frontal Sulcus, *III* Third Ventricle, *ILS*, Inferior Limiting Sulcus, InsCS Insular Central Sulcus, *IntS* Intermediate Sulcus of Jensen, *IOG* Inferior Occipital Gyrus, *IOS* Inferior Occipital Sulcus, *IPS* Intraparietal Sulcus, *IRS* Inferior Rostral Sulcus, *ITG* Inferior Occipital Gyrus, *ITG* Inferior Temporal Gyrus, *ITS* Inferior Temporal Sulcus, *IV* Fourth Ventricle, *LatVen* Lateral Ventricle, *LiG* Lingual Gyrus, *LInsG* Long Insular Gyri, *LInsG* Long Insular Gyrus, *Med* Medulla Oblongata, *MFG* Middle Frontal Gyrus, *Mid* Midbrain, *MOG* Middle Occipital Gyrus, *MTG* Middle Temporal Gyrus, *OffS* Olfactory Sulcus, *OlfTr* Olfactory Tract, *OrbG* Orbital Gyri, *OrbS* Orbital Sulcus, *OTS* Occipito-temporal Sulcus, *ParaCL* Paracentral Lobule, *ParaCS* Paracentral Sulcus, *ParaHipG* Parahippocampal Gyrus, *POF* Parieto-occipital Fissure, *PostCG* Postcentral Gyrus, *PostCS* Postcentral Sulcus, *PreCG* Precentral Gyrus, *PreCS* Precentral Sulcus, *PreCu* Precuneus, *RG* Rectus Gyrus, *RinS* Rinal Sulcus, *SFG* Superior Frontal Gyrus, *SFS* Superior Frontal Sulcus, *SInsG* Short Insular Gyri, *SLS* Superior Limiting Sulcus, *SOG* Superior Occipital Gyrus, *SOS* Superior Occipital Sulcus, *SPL* Superior Parietal Lobule, *SRS* Superior Rostral Sulcus, *STG* Superior Temporal Gyrus, *STS* Superior Temporal Sulcus, *SupraMG* Supramarginal Gyrus, *SyFi* Sylvian Fissure, *TransG* Transverse Gyrus, *Ver* Cerebellar Vermis

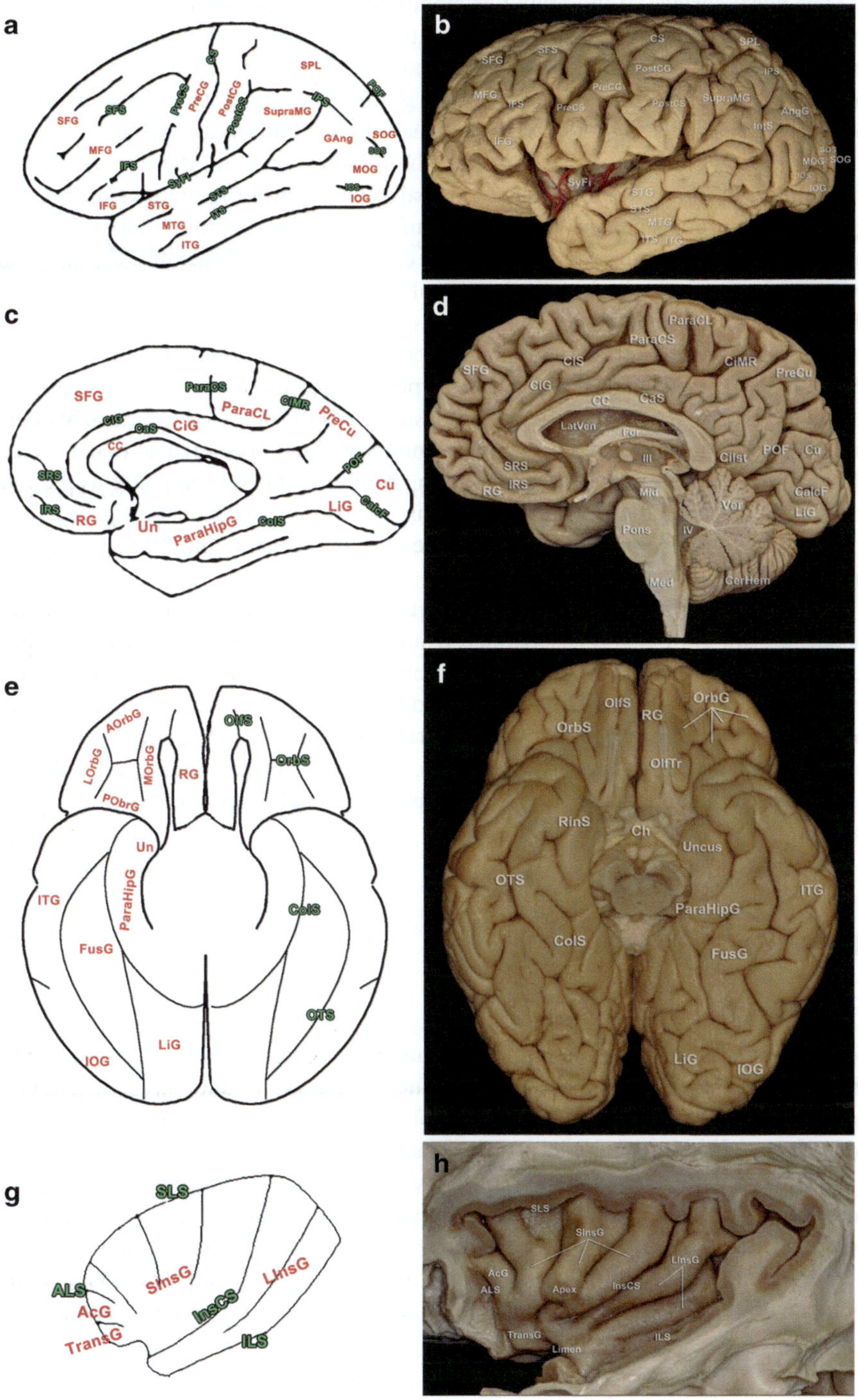
a
SFG
SFS
MFG
IFS
IFG
PreCS
PreCG
CS
PostCG
PostCS
SPL
IPS
SupraMG
POF
GAng
SOG
SOS
MOG
IOS
IOG
SyFi
STG
STS
MTG
ITS
ITG
b
c
SFG
ParaCS
ParaCL
CIMR
PreCu
CiS
CiG
CaS
CC
POF
Cu
SRS
IRS
RG
Un
ParaHipG
ColS
LiG
CalcF
d
e
AOrbG
LOrbG
MOrbG
POrbG
OlfS
OrbS
RG
Un
ParaHipG
ITG
FusG
ColS
OTS
LiG
IOG
f
g
SLS
ALS
AcG
TransG
SInsG
InsCS
LInsG
ILS
h

The basal surface of the brain lies along the base of the skull. The anterior fossa supports the fronto-orbital surface, constituted laterally by the orbital gyri (subdivided by the "H"-shaped orbital sulcus into anterior, posterior, lateral, and medial orbital gyri) and medially by the rectus gyrus. The olfactory sulcus harbors the olfactory tract inside it and is found between the medial orbital and the rectus gyri. The temporo-occipital surface lies over the middle fossa and the tentorium and is constituted by three longitudinal gyral strips: (1) laterally is the inferior temporal gyrus that continues posteriorly with the inferior occipital gyrus; (2) medially is the parahippocampal gyrus, which anteriorly curves over itself forming the uncus and posteriorly continues both with the cingulate and lingual gyri; (3) and in between there is the fusiform gyrus (also known as the lateral occipitotemporal gyrus). The superior and medial surface of the parahippocampal gyrus, called the subiculum, is flat and supports the inferior aspect and pulvinar of the thalamus. This surface is separated laterally by the hippocampal fissure from the dentated gyri, which continues posteriorly as the fasciolar gyri (also known as fasciola cinerea, gyrus of Andreas Retzius, or retrosplenial gyri) circling the splenium, and subsequently as the indusium griseum over the trunk of the corpus callosum. The indusium griseum also extends along the genu and rostrum of the corpus callosum and finishes within the paraterminal gyrus at the septal region.

The surface of the insula has a triangular shape, with the limen insulae at its anteroinferior vertex, and is separated from the other cerebral lobes by its anterior, superior, and inferior insular limiting sulci. The central sulcus of the insula subdivides it into an anterior portion, composed of short gyri, which usually converge to form an apex, and a posterior portion, formed by long insular gyri. The insula has an anterior surface, related to the anterior limiting sulcus, where transverse and accessory gyri can be identified; the first is immediately superior to the limen insulae and the second superior to the first.

4 Sulcal Key Points

Detailed knowledge of the cerebral surface, formed by sulci and turns, is essential for the interpretation of imaging exams and intraoperative guidance. Once identified, the cerebral sulci can be used by the neurosurgeon as landmarks, as microneurosurgical corridors, and also as an aid for cortical mapping techniques in order to identify specific sites related to cortical functions.

Although the sulci and the gyri of the brain are easily identified in standard magnetic resonance images, their accurate visual transoperative recognition is notoriously difficult because of their common anatomic variations and their arachnoid, cerebrospinal fluid, and vessel coverings. Ribas et al. studied sulcal and gyral key points with more constant anatomical cranial-cerebral relationships, mostly formed

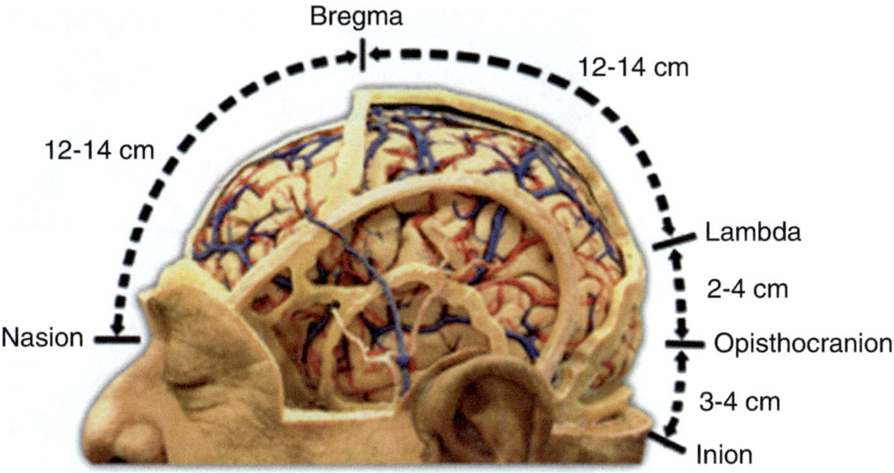

Fig. 3 In normal adults and along the cranial midline surface, the Bregma site is found 12–14 cm posterior from the Nasion, the Lambda site 12–14 cm posterior from the Bregma, the Ophistocranion site 2–4 cm posterior from the Lambda, and the Inion site 3–4 cm posterior from the Ophistocranion

by the main sulci extremities and/or intersections, and by the gyral sites that underlie particularly prominent cranial points [12, 18, 19]. These key points together constitute a neurosurgical anatomic framework that can help in the understanding of other cerebral sulci, lesions, and that can be used to orient the placement of supratentorial craniotomies and to ease the initial transoperative identification of brain sulci and gyri. These findings were based in normal adults, and many of them require the initial identification of the Bregma site (12–14 cm posterior from the Nasion) and of the Lambda site (12–14 cm from the Bregma) (Fig. 3). Importantly, anatomical landmarks can be used as complementary, and not substitutive, to stereotactic and navigation systems when available and transoperative functional or neurophysiological testing is paramount for eloquent cerebral areas identification.

4.1 Frontotemporal Keypoints (Fig. 4)

4.1.1 Anterior Sylvian Point (ASyP) × Anterior Squamous Point (ASqP)

Brain Surface: Corresponds to the enlargement of the Sylvian fissure inferior to the triangular part and anterior/inferior to the opercular part of the inferior frontal gyrus.

Skull Projection: Underneath the most anterior segment of the squamous suture (superior to the sphenosquamous suture and just posterior to the H central bar that characterizes the Pterion).

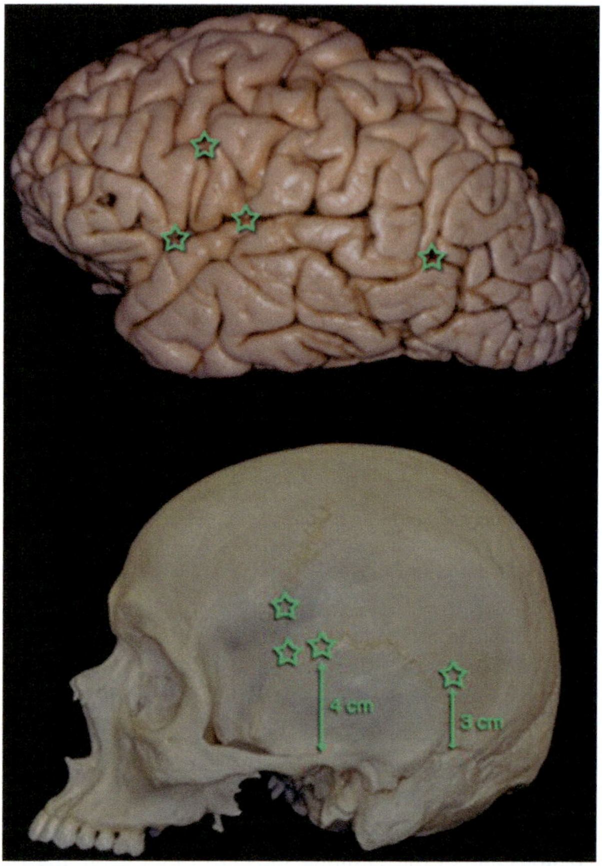

Fig. 4 The Frontotemporal Keypoints, with their brain surface and skull projection correspondence

4.1.2 Inferior Rolandic Point (IRP) × Superior Squamous Point (SSqP)

Brain Surface: Corresponds to the intersection of the central sulcus (or its prolongation) with the sylvian fissure, located at an average distance of 2.3 ± 0.5 cm posterior to the anterior sylvian point, along the Sylvian fissure.

Skull Projection: At the intersection of the squamous suture with a 4 cm vertical line originated at the preauricular depression, which usually corresponds to the most superior segment of the squamous suture.

4.1.3 Inferior Frontal Sulcus/Precentral Sulcus Meeting Point (IFS/PreCS) × Stephanion (St)

Brain Surface: Corresponds to the most posterior aspect of the inferior frontal sulcus, where this sulcus (or its posterior prolongation) meets the precentral sulcus; 2.8 ± 0.6 cm superior to the sylvian fissure.

Skull Projection: Underneath the Stephanion, which corresponds to the craniometric point given by the intersection of the coronal suture with the superior temporal line; average distance from the Bregma along the coronal suture: 7.8 ± 0.7 cm.

4.1.4 Posterior Extremity of the Superior Temporal Sulcus (postSTS) × Temporoparietal Point (TPP)

Brain Surface: It corresponds to its most clearly distal segment identified as a single sulcal trunk before its frequent distal bifurcation and is located 1.4 ± 0.6 cm posteriorly and inferiorly to the posterior sylvian point (end of the sylvian fissure). Along a 30–45° posterior and oblique trajectory, this point is anteriorly related with the *atrium* of the lateral ventricle.

Skull Projection: Underneath a cranial point located 3 cm vertically above the meeting point between the (horizontal) parietomastoid suture and the (oblique) squamous suture (Fig. 5).

4.2 Posterior Frontal Keypoints (Fig. 5)

4.2.1 Superior Frontal Sulcus/Precentral Sulcus Meeting Point (SFS/PreCS) × Posterior Coronal Point (PCoP)

Brain Surface: Corresponds to the most posterior aspect of the superior frontal sulcus, where this sulcus (or its posterior prolongation) meets the precentral sulcus; usually anterior to the "omega" region of the precentral gyrus. Along its perpendicular coronal plane, this point is related with the floor of the body of the lateral ventricle (superior thalamic surface), just posterior to the interventricular foramen of Monro.

Skull Projection: Underneath a cranial point located 2 cm posterior to the coronal suture and 3 cm lateral to the sagittal suture.

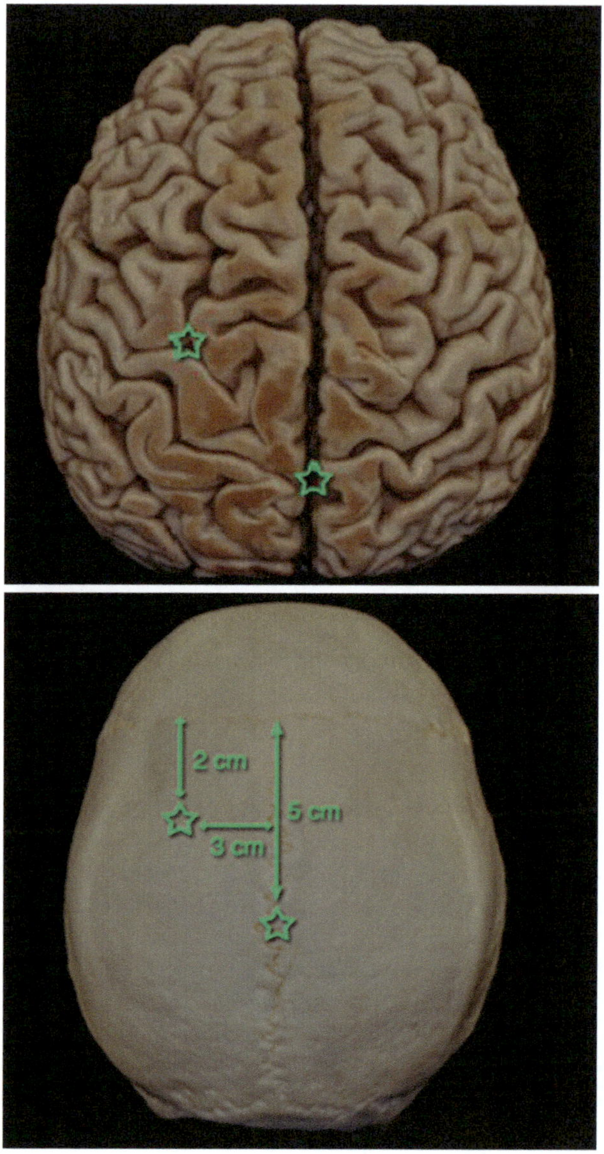

Fig. 5 The Posterior Frontal Keypoints, with their brain surface and skull projection correspondence

4.2.2 Superior Rolandic Point (SRP) × Superior Sagittal Point (SSP)

Brain Surface: Corresponds to the intersection of the central sulcus with the interhemispheric fissure.

Skull Projection: Underneath a cranial point located 5 cm posterior to the *Bregma*, and just lateral to the sagittal suture.

4.3 Parietal Keypoints (Fig. 6)

4.3.1 The Intraparietal Sulcus/Postcentral Sulcus Meeting Point (IPS/PostCS) × Intraparietal Point (IPP)

Brain Surface: Corresponds to the most anterior aspect of the intraparietal sulcus, where this sulcus (or its prolongation) meets (or is continuous) with the inferior segment of the postcentral sulcus. Along a 30–45° posterior and oblique trajectory, this point is anteriorly related with the *atrium* of the lateral ventricle.

Skull Projection: Underneath a cranial point located 6 cm anterior to the *Lambda*, and 5 cm lateral to the sagittal suture.

4.3.2 Supramarginal Gyrus (SMG) × *Euryon* (Eu)

Brain Surface: The anterior and superior aspect of the Supramarginal gyrus is located underneath the *Euryon*.

Skull Projection: The *Euryon* corresponds to the craniometric point that is given by the most prominent aspect of the parietal tuberosity; it can also be identified as the cranial point where a vertical line originated from the most posterior aspect of the mastoid tip reaches the parietal tuberosity, immediately superior to the superior temporal line. The *Euryon* is located anteriorly and inferiorly to the previous cranial point (Intraparietal Point) that lies over the intraparietal sulcus/postcentral sulcus meeting point.

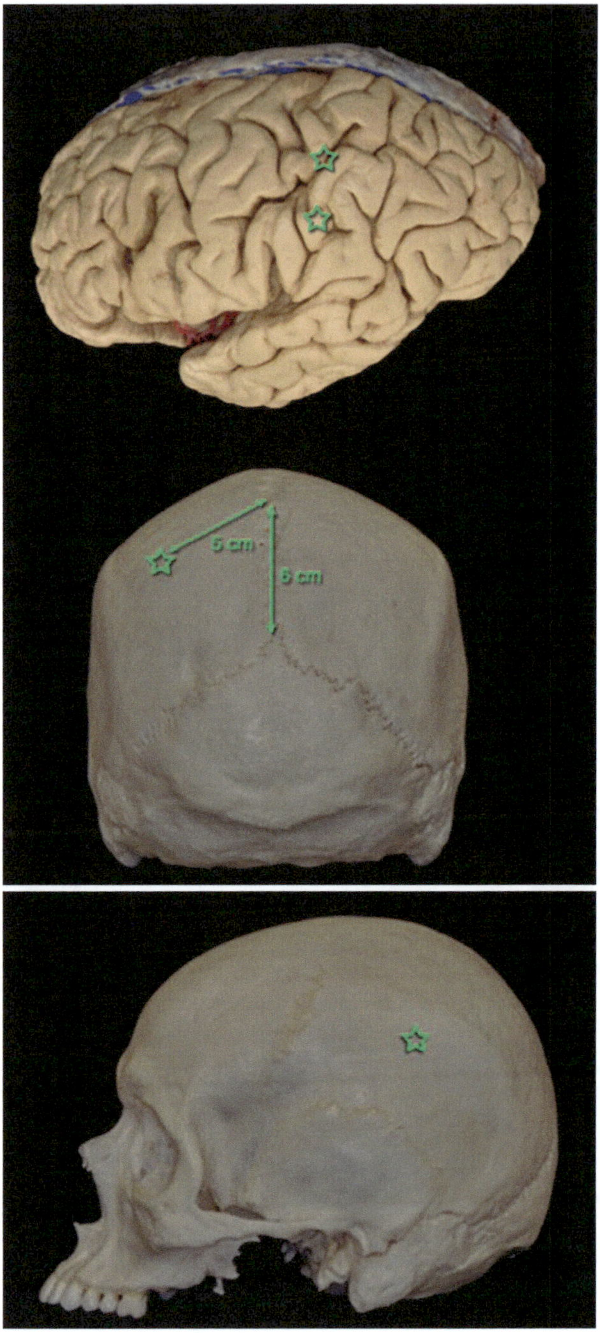

Fig. 6 The Parietal Keypoints, with their brain surface and skull projection correspondence

4.4 *Occipital Keypoints (Fig. 7)*

4.4.1 External Occipital Fissure Depth of the Parieto-Occipital Sulcus (EOF/POS) × Lambdoid/Sagittal Point (La/Sa)

Brain Surface: The External Occipital Fissure corresponds to the most superior aspect and to the depth of the parieto-occipital sulcus, always seen on the convexity brain surface, hence indicating the limit between the precuneus and the cuneus along the interhemispheric fissure.

Skull Projection: Located underneath the angle that there is between the lambdoid and the sagittal sutures.

4.4.2 Distal End of the Calcarine Fissure (dCaF) × Opisthocranion (OpCr)

Brain Surface: Corresponds to the emergence of the calcarine fissure on the occipital convexity, at the level of the basal tip of the cuneus.

Skull Projection: Underneath the Opisthocranion, craniometric point that corresponds to the usually evident most prominent point of the occipital bossa, 3.0 ± 0.9 cm inferiorly to the *Lambda*, 1.7 ± 0.5 cm superiorly to the occipital (lingual gyrus) base, and 3–4 cm superiorly to the *Inion* (external occipital protuberance, torcular base) (Fig. 7).

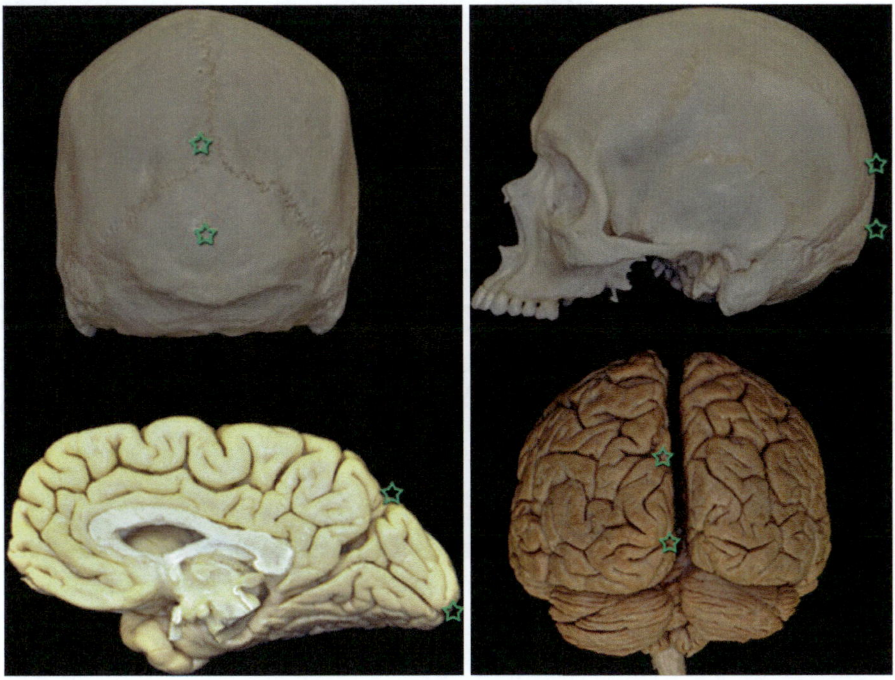

Fig. 7 The Occipital Keypoints, with their brain surface and skull projection correspondence

4.5 Temporo-Occipital Base Keypoints (Fig. 8)

4.5.1 Preauricular Depression (PreAuDepr) and Parietomastoid Suture/Squamous Suture Meeting Point (PaMaSut/SqSut Meet Pt)

The preauricular depression (immediately anterior to the tragus) and the parietomastoid suture/squamous suture meeting point (usually palpated as a slight depression above the mastoid process) define the anterior and the posterior upper limits of the petrous bone, hence standard burrholes just above these points will avoid bone and allow the exposure of the superior petrous surface.

Since the *Asterion* lies over the transverse sinus, a burrhole above it can extend a basal temporo-occipital craniotomy along the tentorium (Fig. 8).

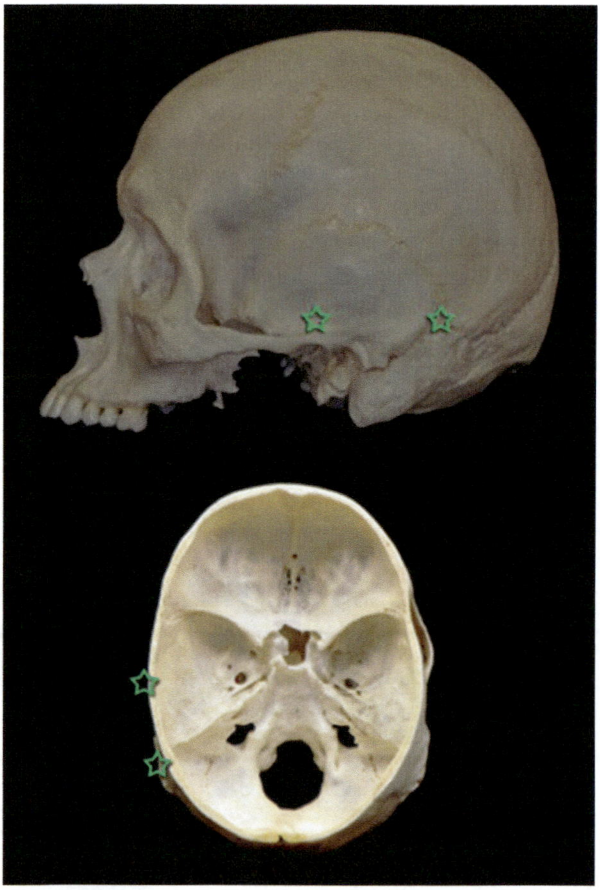

Fig. 8 The Temporo-occipital Base Keypoints

4.6 *Suboccipital Base Keypoints (Fig. 9)*

4.6.1 *Asterion* (Ast) and OccipitoMastoid Suture/Mastoid Notch Point (OccMaSut/MaNaPt)

Since the *Asterion* lies over the transverse sinus, and the occipitomastoid suture at the level of the mastoid notch's posterior aspect lies over the posterior margin of the sigmoid sinus, burrholes done at these two sites are related respectively with the upper and with the lower aspects of the cerebellopontine angle and constitute the natural limits of the suboccipital exposures (Fig. 9).

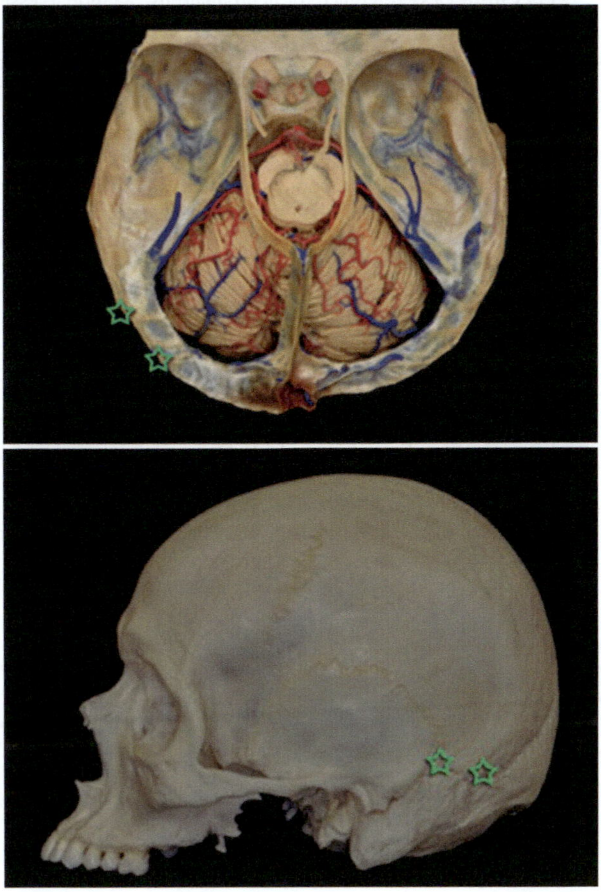

Fig. 9 The Suboccipital Base Keypoints

References

1. Ribas GC. The cerebral sulci and gyri. Neurosurg Focus. 2010;28(2):E2.
2. Braak H. Architectonics of the human telencephalic cortex. Berlin: Springer; 1980. https://doi.org/10.1007/978-3-642-81522-5.
3. Ribas G. Applied cranial-cerebral anatomy: brain architecture and anatomically oriented microneurosurgery. New York: Cambridge University Press; 2018.
4. Ono M, Kubik S, Abernathey CD. Atlas of the cerebral sulci. Stuttgart: G. Thieme Verlag/ Thieme Medical Publishers; 1990.
5. Ribas EC, Yağmurlu K, de Oliveira E, Ribas GC, Rhoton A. Microsurgical anatomy of the central core of the brain. J Neurosurg. 2018;129(3):752–69.
6. Pearce JMS. Louis Pierre Gratiolet (1815–1865): the cerebral lobes and fissures. Eur Neurol. 2006;56(4):262–4.
7. Lockard I. Desk reference for neuroanatomy: a guide to essential terms. New York: Springer; 1977. https://doi.org/10.1007/978-1-4684-0050-2.
8. Broca P. Sur la circonvolution limbique et la scissure limbique. Anthropol. 1877;12:646–57.
9. Nieuwenhuys R, Voogd J, van Huijzen C. The human central nervous system. 4th ed. New York: Springer; 2008. 967 p.
10. Federative Committee on Anatomical Terminology. Terminologia anatomica: international anatomical terminology. Stuttgart: Thieme; 1998.
11. de Anatomia SB, de Terminologia Anatômica C, da Terminologia Anatômica CF. Terminologia anatômica: terminologia anatômica internacional. São Paulo: Manole; 2001.
12. Ribas GC, Ribas EC, Rodrigues CJ. The anterior sylvian point and the suprasylvian operculum. Neurosurg Focus. 2005;18(6B):E2.
13. Türe U, Yaşargil MG, Friedman AH, Al-Mefty O. Fiber dissection technique: lateral aspect of the brain. Neurosurgery. 2000;47(2):417–27.
14. Gray H, Warwick R, Williams PL. Gray's anatomy. 35th ed. London: Longman; 1973.
15. Yousry T. Localization of the motor hand area to a knob on the precentral gyrus. A new landmark. Brain. 1997;120(1):141–57.
16. Déjérine J. Anatomie des centres nerveux. Paris: Rüeff; 1895.
17. Ribas G, Ribas E, Rodrigues AJ Jr. Demonstração estereoscópica dos sulcos e giros cerebrais. Rev Med. 2006;85(3):78–90.
18. Ribas GC, Yasuda A, Ribas EC, Nishikuni K, Rodrigues AJ. Surgical anatomy of microneurosurgical sulcal key points. Neurosurgery. 2006;59(4 Suppl 2):ONS177–210; discussion ONS210–211.
19. Ribas GC, Rodrigues AJJ. The suprapetrosal craniotomy. J Neurosurg. 2007;106(3):449–54.

Surgical Anatomy of the Temporal Lobe

Jander Moreira Monteiro and Gustavo Rassier Isolan

1 Introduction

The temporal lobe is the most heterogeneous segment of the human brain. It is formed by many allocortex and limbic system structures. The mesocortex constitutes the delineation between the temporal isocortex and part of the limbic system. This region is a frequent site of origin for epileptic focus and brain tumors. Approximately 9% of all epilepsy cases are classified as temporal lobe epilepsy—the total prevalence of epilepsy may range from 4 to 57 cases per 1000 persons, being more prominent in developing and tropical countries. Temporal lobe epilepsy is recurrently refractory to antiepileptic drugs and is the most prevalent type of partial epilepsy suitable for surgery approaches [1].

Following the heterogeneity of the area, temporal lobe epilepsy presents in two different ways: neocortical and mesial temporal. The evaluation and surgical treatment for neocortical temporal lobe epilepsy resemble those used for extratemporal epilepsy in general. Meanwhile, the therapeutics for mesial temporal epilepsy are specific, mostly based on its distinguished physiopathological process: mesial temporal sclerosis, usually affecting the hippocampus [2].

Two established surgical interventions for mesial temporal epilepsy are anterior temporal lobectomy and the more conservative amygdalohippocampectomy. Our purpose is to present our findings regarding the microsurgical anatomy of the temporal lobe and the importance of this anatomy in selective amygdalohippocampectomy and relate it to the three main approaches to this procedure: transsylvian approach, Niemeyer's technique, and subtemporal technique. Also, as the title

J. M. Monteiro · G. R. Isolan (✉)
The Center for Advanced Neurology and Neurosurgery (CEANNE), Porto Alegre, RS, Brazil

Mackenzie Evangelical Faculty of Paraná (FEMPAR), Curitiba, PR, Brazil

E. G. Figueiredo et al. (eds.), *Brain Anatomy and Neurosurgical Approaches*, https://doi.org/10.1007/978-3-031-14820-0_6

suggests, we aim to describe and discuss the temporal white fibers and the relation of this subcortical anatomy with cortical and subcortical brain mapping, which is especially useful in glioma surgeries.

2 Surgical Anatomy

2.1 Temporal Lobe Limits

On the lateral surface, the temporal lobe is delineated superiorly by the posterior ramus of the Sylvian fissure and posteriorly by the lateral parietotemporal line. The parietotemporal line is an imaginary line extending from the impression of the parieto-occipital fissure to the preoccipital notch. The temporo-occipital line is another similar line that separates the temporal from the parietal lobe. This line runs from the end of the posterior ramus of the Sylvian fissure to the midpoint of the lateral parietotemporal line. The limit between the temporal and occipital lobes on the basal surface is delineated by a line that connects the preoccipital notch to the inferior end of the parieto-occipital fissure—the basal parietotemporal line. For a didactic purpose, these delineations worked well but are artificial and elusive regarding the biological reality [3] (Figs. 1, 2, 3, 4, 5, 6, 7, and 8).

2.2 Temporal Lobe Sulci

The principal sulci identified in the temporal lobe are the superior, middle, and inferior temporal sulci, angular, hippocampal, uncal, rhinal, and collateral sulci. The superior temporal sulcus extends from the anterior aspect of the lateral surface of the temporal pole to the angular gyrus and is the most consistent sulcus in morphology [4]. The middle temporal sulcus is parallel to the superior temporal sulcus and

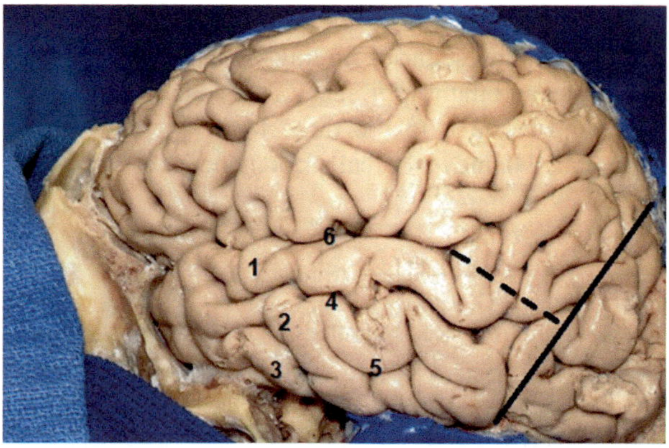

Fig. 1 Lateral view of a left hemisphere with temporal sulci, gyri, and limits: 1—superior temporal gyri; 2—middle temporal gyri; 3—inferior temporal gyri; 4—superior temporal sulcus; 5—middle (inferior) temporal sulcus; 6—Sylvian (lateral) fissure; discontinued line—temporo-occipital line; continued line—parieto-occipital line

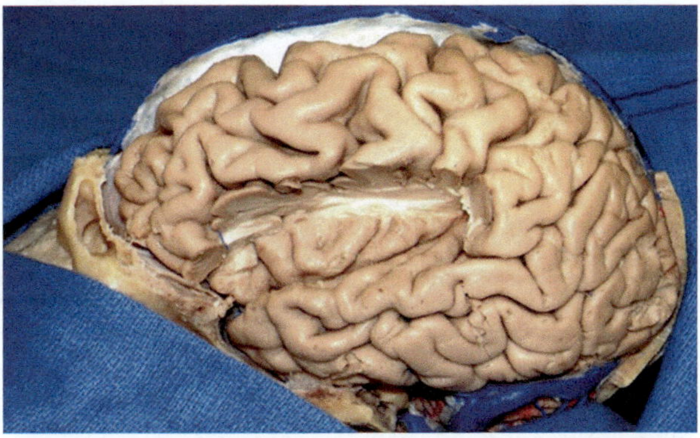

Fig. 2 Frontal operculum resection, exposing insular gyri and transverse temporal gyri

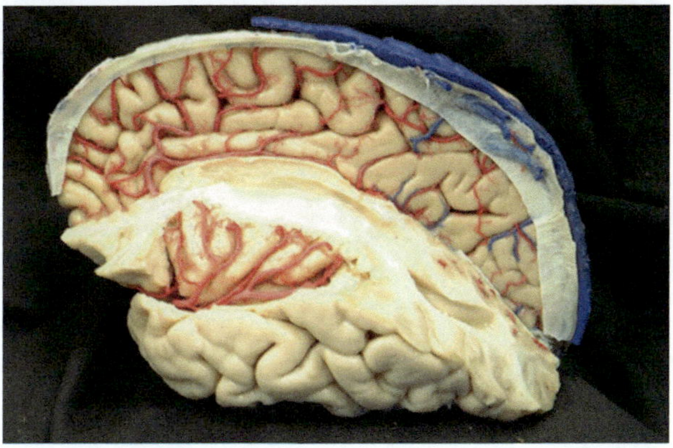

Fig. 3 The M2 branches of the middle cerebral artery travel in the sulci of the insular lobe, and, in the transsylvian approach, surgeons work in the small windows delimited by these arterial branches

extends into the posterior portion of the inferior parietal lobule. The inferior temporal sulcus was not identified in all brains. It is present in about one-third of the individuals [5]. The inferior and middle temporal sulci were highly variable in morphology. The collateral sulcus was uninterrupted in 100% of the brains, while the inferior temporal sulcus was interrupted 100%. The rhinal sulcus is a short sulcus that was identified in all brains. The patterns of the interruptions in the sulcus of the temporal lobe and the terminations and connections of them are detailed in the study performed by Ono [5] (Figs. 1, 2, 3, 4, 5, 6, 7, and 8).

2.3 *Temporal Gyri*

The following principal gyri in the temporal lobe were identified: superior, middle, and inferior temporal gyri, fusiform, and parahippocampal gyri. The superior temporal gyrus (T1) can be divided into anterior, middle, and posterior segments. Three or four small gyri originate from the anterior portion of the superior temporal gyrus that goes into the depth of the anterior Sylvian fissure [6]. A sulcus divides this short

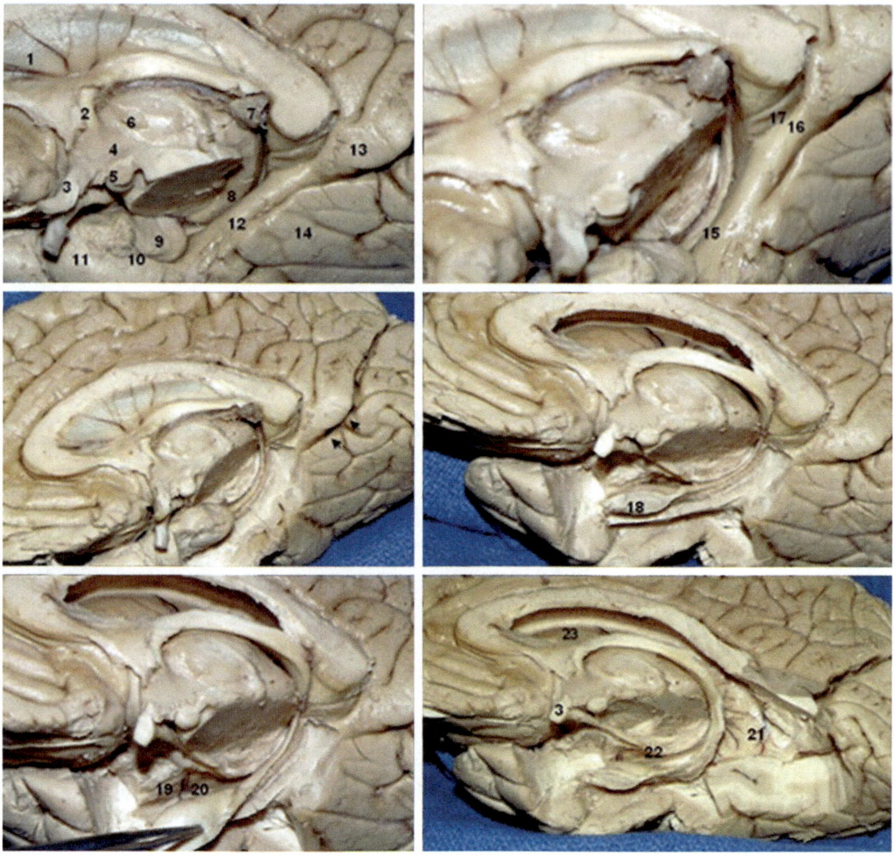

Fig. 4 Medial surface of a right-sided brain hemisphere: 1—septum pellucidum; 2—fornix (anterior column); 3—optic chiasm; 4—hippocampus; 5—mammillary body; 6—interthalamic adhesion; 7—pineal gland; 8—pulvinar nuclei (thalamus); 9—uncus; 10—uncal sulcus; 11—temporal pole; 12—parahippocampal gyrus; 13—cingulate gyrus; 14—lingual gyrus; 15—hippocampal sulcus; 16—isthmus; 17—fasciolar gyrus; 18—hippocampal head; 19—amygdala; 20—inferior choroid point; 21—ventricular atrium; 22—optic tract; 23—head of caudate nucleus; 24—anterior commissure; arrows—anterior calcarine sulcus

anterior gyrus from two bigger parallel gyri that run from the posterior part of the T1 to the ultimate end of the Sylvian fissure. These gyri sitting posteriorly are called transverse temporal gyri or Heschl gyri, containing the primary auditory cortex. The transverse gyrus of Heschl usually has two convolutions outlined by transverse sulci that run obliquely and are usually larger on the left side [4]. Behind the Heschl gyrus lies a triangular space called the *planum temporale*. The *planum temporale* is larger on the left side due to the left hemisphere dominance for speech [4]. This anatomical information is important because the length of the left Sylvian fissure is also larger than the right [4]. The superior surface of the superior temporal gyrus forms the temporal operculum. The middle temporal gyrus (T2) can be divided into

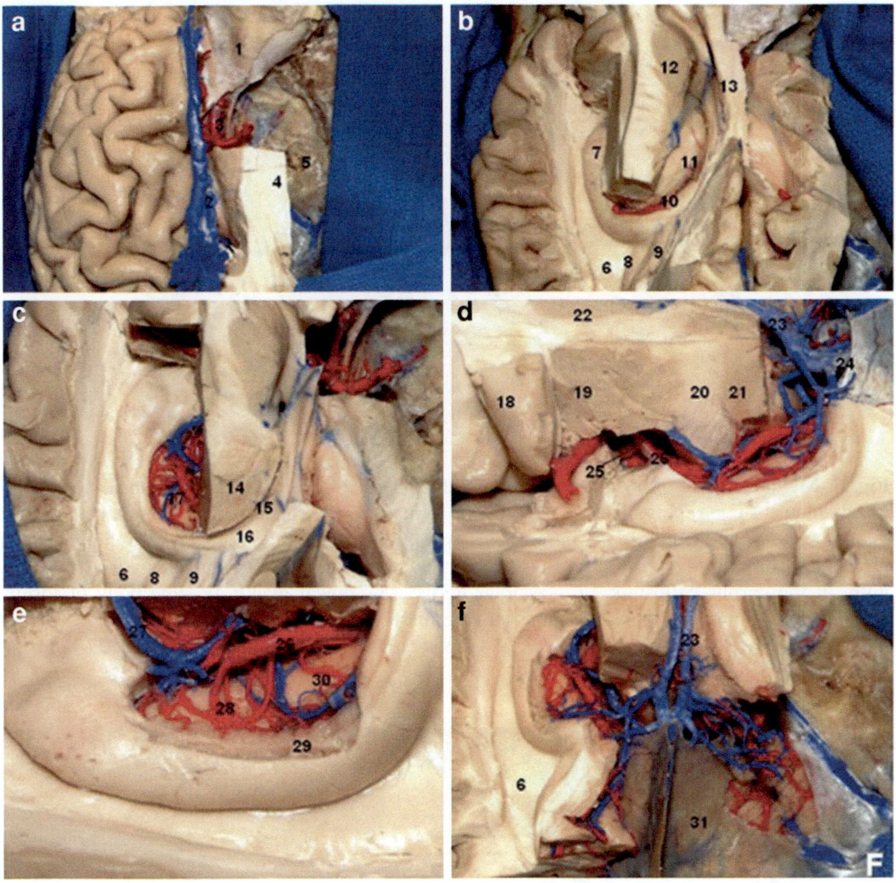

Fig. 5 Superior (**a**–**c**, **f**) and lateral (**d**, **e**) views of the hippocampus and surrounding areas: 1—orbital roof; 2—superior sagittal sinus; 3—internal carotid artery; 4—internal capsule; 5—middle fossa floor; 6—collateral triangle; 7—hippocampus; 8—calcar avis; 9—corpus callosum; 10—choroid fissure; 11—thalamus; 12—caudate nucleus head; 13—fornix; 14—pulvinar nuclei (thalamus); 15—choroid fissure (atrial portion); 16—crura fornices; 17—foramen of Monro; 18—short gyri of insula; 19—globus pallidus; 20—internal capsule; 21—pulvinar nuclei (thalamus); 22—caudate nucleus; 23—internal cerebral vein; 24—a vein of Galen; 25—oculomotor nerve; 26—posterior cerebral artery; 27—basal vein of Rosenthal; 28—middle hippocampal artery; 29—dentate gyrus; 30—posterior hippocampal artery; 31—cerebellar tentorium

anterior, middle, and posterior segments. It is separated from the superior temporal gyrus by the superior temporal sulcus and the inferior temporal gyrus by the middle temporal sulci/inferior temporal sulci. The three segments of the middle temporal sulci represent an interrupted sulcus that separates the middle and inferior temporal gyri. The middle segment of the middle temporal sulcus may be equivalent to the central part of the inferior temporal sulcus [5]. The anterior pole of the middle temporal gyrus is surrounded by the anterior part of the superior and inferior temporal gyri. The inferior temporal gyrus (T3) can be divided into anterior, middle, and

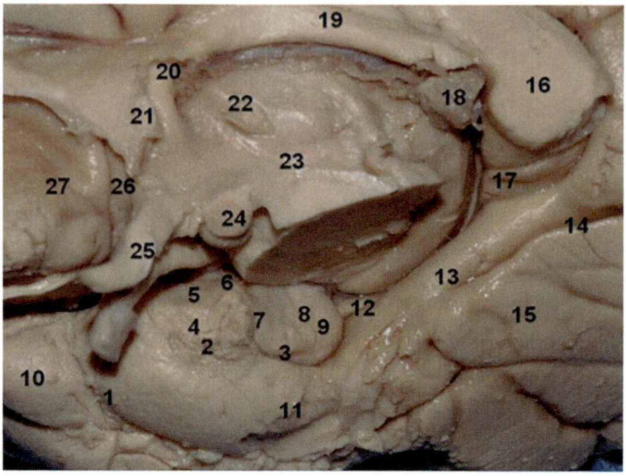

Fig. 6 Medial surface of a right-sided brain hemisphere: 1—rhinal sulcus; 2—uncal notch; 3—uncal sulcus; 4—gyrus ambiens; 5—semilunar gyrus; 6—entorhinal sulcus; 7—uncinate gyrus; 8—the band of Giacomini; 9—intralimbic gyrus; 10—temporal pole; 11—parahippocampal gyrus; 12—fimbria; 13—isthmus; 14—anterior calcarine sulcus; 15—lingual gyrus; 16—splenium of the corpus callosum; 17—fasciolar gyrus; 18—pineal gland; 19—the body of fornix; 20—anterior column of fornix; 21—anterior commissure; 22—interthalamic adhesion; 23—hypothalamic sulcus; 24—mammillary body; 25—optic chiasm; 26—paraterminal gyrus; 27—subcallosal gyrus

Fig. 7 Inferior view of the brain: 1—middle cerebral artery (M1 segment); 2—limen insulae; 4—middle cerebral artery (M2 segment); 5—calcarine artery (P4 segment); 6—oculomotor nerve; 7—pituitary gland; 8—optic chiasm; 9—internal carotid artery; 10—uncus; 11—dentate gyrus

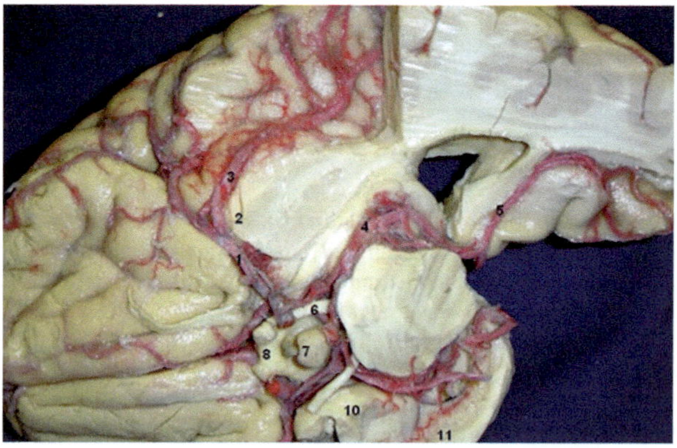

posterior segments. It is separated from the lateral temporo-occipital gyrus by the inferior temporal sulcus. It is small compared to the other temporal gyri. The gyri of the temporal lobe convexity, except for the superior temporal gyrus, are difficult to identify due to their anatomical variability [4]. The fusiform gyrus (lateral temporo-occipital gyrus) (T4) is separated from the parahippocampal and lingual gyri by the collateral sulcus and has a "lazy" form. It has an anterior and posterior relation with the inferior temporal gyrus and the medial temporo-occipital gyrus. The

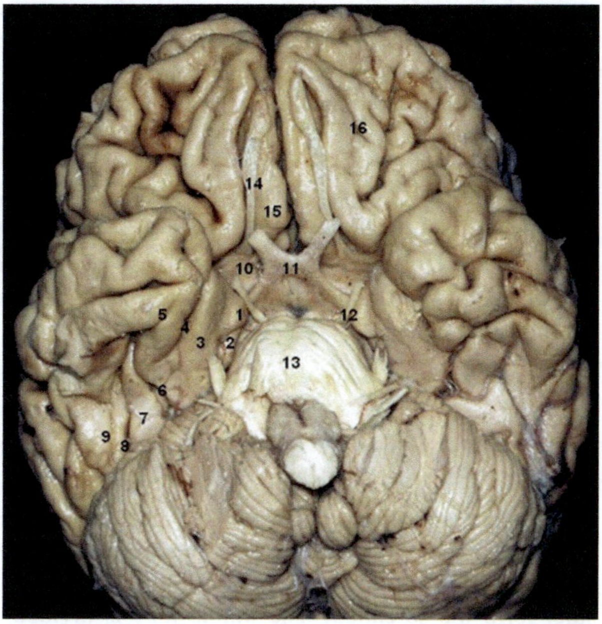

Fig. 8 Inferior view of brain's basal surface: 1—uncus; 2—gyrus ambiens; 3—parahippocampal gyrus; 4—rhinal sulcus; 5—fusiform gyrus; 6—collateral sulcus; 7—fusiform gyrus; 8—occipitotemporal sulcus; 9—inferior temporal gyrus; 10—anterior perforated substance; 11—optic chiasm; 12—oculomotor nerve; 13—pons; 14—olfactory tract; 15—gyrus rectus; 16—orbital gyri

parahippocampal (T5) gyrus sits between the uncus and the isthmus of the cingulate gyrus. It is separated from the fusiform gyrus by the collateral sulcus. The parahippocampal gyrus has a posterior segment called the subiculum. The hippocampal sulcus separates the subiculum from the hippocampus. The parahippocampal gyrus also has an anterior segment called the piriform lobe, formed by the uncus and entorhinal area. The uncus is separated from the parahippocampal gyrus by the uncal sulcus. The uncus has an anterior segment that covers the amygdala. This part is composed of the semilunar and ambient gyri, which are separated by the semilunar sulcus. The posterior segment of the uncus forms part of the hippocampus and the subiculum. The subiculum is divided into prosubiculum, subiculum proper, presubiculum, and parasubiculum. The entorhinal area does not have a precise delineation but is considered to extend into the posterior segment of the parahippocampal gyrus [7] (Figs. 1, 2, 3, 4, 5, 6, 7, and 8).

2.4 Considerations on Temporal Gyri and Sulci

Regarding the division of the brain in lobes, for a didactic purpose, this delineation worked well but is artificial and elusive regarding the biological reality [3]. It has not considered the complex and intricated white fiber tracts and fascicles that communicate with brain lobes, representing more accurately brain functions. Regarding temporal lobe sulci, the superior temporal sulcus is the most consistent in morphology [4]. The inferior temporal sulcus is present in about one-third of the individuals [5]. The patterns of the interruptions analyzed in our dissection regarding the sulci of the temporal lobe and the terminations and connections are similar to the study performed by Ono [5].

The transverse gyrus of Heschl usually has two convolutions outlined by transverse sulci that run obliquely and are usually larger on the left side [4]. The planum temporale is larger on the left side due to the left hemisphere dominance for speech [4]. This anatomical information is important because the length of the left Sylvian fissure is also larger than the right [4]. The gyri of the temporal lobe convexity are, except for the superior temporal gyrus, difficult to identify due to their anatomical variability [4].

2.5 *Hippocampus*

The hippocampus forms an arc with an enlarged anterior extremity. The hippocampus belongs to the limbic lobe, separated from the adjacent cortex by the cingulate, subparietal, anterior calcarine, collateral, and rhinal sulci. Do these discontinuous sulci form the so-called? The limbic lobe is divided into limbic and intralimbic gyrus [8]. The limbic gyrus is formed by the parahippocampal, cingulate, and subcallosal gyri. The intralimbic gyrus has two segments: anterior and superior. The anterior segment is part of the paraterminal gyrus and septal region and is referred to as the prehippocampal rudiment. The superior segment is the indusium griseum, a neuronal lamina on the corpus callosum extending to the hippocampus. The indusium griseum is covered by the medial and lateral longitudinal striae on each side of the midline. The limbic lobe can also be divided structurally in the allocortex (hippocampus, proximal part of the subiculum, and indusium griseum) and periallocortex (cingulate and parahippocampal gyri), which is a transitional cortex between the allocortex and isocortex [7].

Structurally, the hippocampus comprises two laminas that roll up inside the other—the Cornu Ammonis (hippocampus proper) and the gyrus dentatus. Both of them are allocortex. These structures are separated by the hippocampal sulcus, which is divided into vestigial and superficial hippocampal sulci. This last part can be seen on the temporal lobe surface. The Cornu Ammonis, from the intraventricular surface to the vestigial hippocampal sulcus, can be divided into six layers. These layers are alveus, stratum oriens, stratum pyramidale, stratum radiatum, stratum lacunosum, and stratum molecularis. The Cornu Ammonis has a heterogeneous structure in coronal sections and is divided into four fields: CA_1, CA_2, CA_3, and CA_4. CA_1 is the larger portion and continuous with the dentate gyrus, CA_2 has a dense and narrow stratum pyramidal, CA_3 corresponds to the genu of the Cornu Ammonis, and CA_4 is situated within the gyrus dentatus. CA_1 is considered the "vulnerable sector" to hypoxia, and CA_3 is the "resistant sector." The gyrus dentatus, also known as gyrus involutus, is a concave lamina that envelopes CA_4. It is formed by three layers: stratum granulosum, stratum molecular, and polymorphic layer. It is separated from the Cornu Ammonis by the vestigial hippocampal sulcus. The margo denticulatus is a part of the dentate gyrus that has a toothed appearance [9]. It is separated from the subiculum by the superficial hippocampal sulcus and from the fimbria by the fimbriodentate sulcus. The gyrus dentatus and CA_4 are called area dentate [10]. It is important to recognize that the visible part of the gyrus dentatus has different names concerning the hippocampus but is, in fact, the same structure. It is called margo denticulatus in the body, the band of Giacomini in the uncus, and the fasciola cinerea in the tail.

For didactic reasons, we think that the complex anatomy of the hippocampus can be described separately as head, body, and tail. Each of these regions is divided into intraventricular and extraventricular parts, as described by Duvernoy [7] (Fig. 9).

2.6 Hippocampal Head

In the intraventricular part, we identify three or four internal digitations sagittally oriented (digitations hippocampi). A vertical digitation that corresponds to the medial surface of the uncus sometimes can be identified. The digitations hippocampi are covered by part of the gyrus dentatus. The fimbria gives an alveus covering the hippocampal digitations at the junction of the body and head of the hippocampus. Another anatomical particularity is that there is no choroid plexus on the head of the hippocampus. The uncal recess is a prolongation of the temporal horn of the lateral ventricle anterior to the hippocampus that extends into the deeper portion of the uncus [9]. The basal and lateral nuclei of the amygdala "overhang" the hippocampal head and are almost joined together.

The extraventricular or uncal part of the anterior segment of the uncus is described below. The posterior segment of the uncus has an inferior and medial surface. The inferior surface is hidden in the uncal sulcus and can be visualized after resectioning the subjacent parahippocampal gyrus. It is divided into the band of Giacomini,

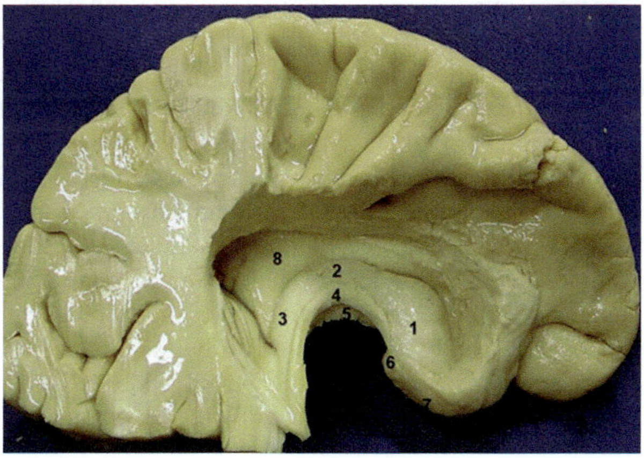

Fig. 9 Anatomical specimens of temporal pole, superior view, with an opened temporal horn to show mesial temporal structures: 1—head of the hippocampus; 2—body of the hippocampus; 3—tail of the hippocampus; 4—fimbria-fornix; 5—dentate gyrus; 6—tail of dentate gyrus; 7—uncus; 8—collateral eminence

external digitations, and inferior surface of the uncal apex. The band of Giacomini is the segment of the gyrus dentatus at the level of the hippocampal head. The external digitations are two or three small lobules anterior to the band of Giacomini that are the "inversion images" of the digitations hippocampi. The inferior surface of the uncal apex is separated from the band of Giacomini posteriorly by a small sulcus. The uncal apex is formed by CA_3 and CA_4 covered by the alveus and is the caudal end of the uncus. The medial surface is divided into the terminal segment of the band of Giacomini, which is localized on the upper lip of the uncal sulcus, the medial surface of the uncal apex, and the uncinate gyrus, which joins with the gyrus ambiens and is anterior to the band of Giacomini (Fig. 9).

2.7 Hippocampal Body

The intraventricular part of the hippocampal body sits at the floor of the temporal horn of the lateral ventricle and is limited laterally by the collateral eminence (corresponds in the basal brain to the collateral sulcus) medially by the fimbria. It is covered by the choroid plexus, which is "attached" to a double layer, formed by ependyma and pia, that constitutes the tela choroidea. The tela choroidea is attached to the taenia of the tela choroidea. The inferior choroidal point, also known as velum terminale of Aeby, is a triangular lamella attached to the superior surface of the uncus where the taenia of the fimbria and stria terminalis unites. The term inferior choroidal point is not precise because there are no identified superior nor posterior choroidal points. The extraventricular part is formed by the gyrus dentatus, fimbria, and superficial hippocampal sulcus (Fig. 9).

2.8 Hippocampal Tail

The intraventricular part is limited medially and laterally by the fimbria and the collateral trigone, respectively. The choroidal plexus on this portion is larger (choroid glomus). Posteriorly, it reaches an intraventricular protrusion, the calcar avis. The hippocampal tail can be divided into initial, middle, and terminal segments in the extraventricular part. The initial segment is present as the margo denticulatus. The gyrus dentatus has many extensions penetrating deeply into the hippocampus. In the middle segment, the margo denticulatus becomes smooth and narrow, forming the fasciola cinerea. Still, in the middle segment, the fimbria ascends to join the crus fornices, and it is possible to identify the gyrus fasciolaris (CA_3 covered by alveus) separated from the fasciola cinerea by the sulcus dentofasciolaris. Contrary to the hippocampal body, CA_1 appears in the tail at the surface of the parahippocampal gyrus, sometimes producing the gyri of Anders Retzius [11]. The terminal segment of the hippocampal tail is formed by the subsplenial gyrus (Fig. 9).

2.9 *Hippocampal Vascularization and Vascular Relationships*

According to their frequencies, the hippocampal arterial vasculature may be divided into six groups [12] (Fig. 10):

- *Group A*: Mixed arterial vasculature originating from the anterior choroidal artery (AChA), posterior cerebral artery (PCA), anterior infratemporal artery (AIA), and splenic artery (SA). Present in 46.3% of hippocampi.
- *Group B*: Main origin at the temporal branches—main inferotemporal trunk (MIT), middle inferotemporal artery (MIA), posterior inferotemporal artery (PIA), AIA, or main branch of PCA. Present in 20.4% of hippocampi.
- *Group C*: AIA is the main branch of the hippocampus. Present in 14.8% of hippocampi.

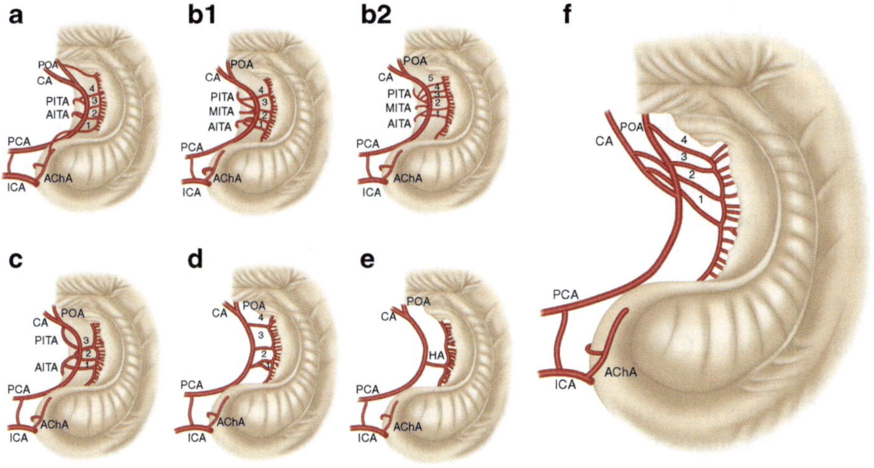

Fig. 10 Intra- and extraventricular view of the hippocampus vascularization groups. Hippocampal arteries present significant anatomical variations that have direct implications to surgical procedures such as amygdalohippocampectomy. According to their origin, the hippocampal vascularization patterns can be categorized into six different groups. (**a**) Group A presents mixed irrigation, with the hippocampal arteries originating from the anterior choroidal, posterior cerebral, anterior inferior temporal, and splenial arteries. (**b1**) Group B presents hippocampal arteries originating from the inferior temporal branches and PCAs' main trunk. (**b2**) Also, Group B, but in a situation where the inferior temporal arteries arise from a common trunk. Arrowhead identifies a common inferior temporal artery. (**c**) Group C presents the anterior inferior temporal artery as the main feeder of the hippocampus. (**d**) Group D hippocampal arteries originate from the posterior cerebral arteries' main trunk. The hippocampal arteries are the largest contributors to the hippocampus blood supply in this group. (**e**) The trunk of the posterior cerebral artery originates from Uchimura's artery—a vessel that irrigates practically the whole hippocampus. (**f**) A new anatomical variation in which the hippocampal vessels arise exclusively from the POA, CA, and SA. *CA* calcarine artery, *ICA* internal carotid artery, *PCA* posterior cerebral artery, *POA* parieto-occipital artery, *AITA* anterior inferior temporal artery, *MITA* medial inferior temporal artery, *PITA* posterior inferior temporal artery, *AChA* anterior choroidal artery. 1, 2, 3, and 4: Hippocampal arteries. (Reprinted from Isolan et al. [12])

- *Group D*: HAs originating from the main branch of PCA. Present in 13% of hippocampi.
- *Group E*: A single hippocampal artery (HA) with the origin at the main branch of PCA. This single artery covered all of the structure and is named Uchimura's artery. Present in 3.7% of hippocampi.
- *Group F*: The hippocampal vessels arose exclusively from the parieto-occipital artery (POA), calcarine artery (CA), and splenic artery (SA). Present in 1.8% of hippocampi.

Regarding hippocampal vascularization, in 26.6% of hemispheres, one of the hippocampal vessels originates from the lateral posterior choroidal arteries and 36.6% from the splenial artery [13]. Regarding the variations, the most frequent pattern is characterized by mixed origins of the hippocampal arteries. This corresponds to Group A in the exhaustive studies carried out by Erdem et al. [13]. The hippocampal arterial supply was divided into five groups in this study according to the hippocampal supply origin. There is an average of 4.7 arteries per hemisphere supplying the hippocampus [13].

While the middle and posterior hippocampal arteries supply the hippocampal body and tail, the hippocampal head and uncus are supplied by the anterior hippocampal artery. An anastomosis can be present among these arteries [14]. In few cases, the arteries that supply the hippocampus can originate from a trunk: some important anatomical considerations are that these arteries vascularize the hippocampus by penetrating the dentate gyrus, the fimbrodentate sulcus, and the hippocampal sulcus. This, however, can be identified only if the fimbria is removed [13].

2.9.1 Anterior Choroidal Artery

The AChA has its origin in the posterior wall of the ICA. The AChA gives branches to the hippocampus in 31.5% of brain hemispheres, although some evidence suggests that it can be up to 50%. The HAs from the AChA have a similar conformation to the HAs from the PCA and follow the hippocampal sulcus in the uncus region. However, the arteries from the AChA supply mainly the head of the hippocampus [12].

2.9.2 Posterior Cerebral Artery

The PCA branches are divided into three groups: central (I), ventricular and choroid plexus (II), and cerebral branches (III). The ones related to hippocampal vascularization are the second and third. In the ventricular/choroidal plexus group, the posterolateral and posteromedial choroidal arteries supply the superior tela choroidea, choroid plexus of the lateral and third ventricle, thalamus, and internal region of the caudate nucleus. The cerebral group includes HAs, AIA, MIA, PIA, POA, CA, and SA. They supply the hemispheres' external occipital and ventral faces, comprehending the inferotemporal, lateral occipitotemporal, medial occipitotemporal, and parahippocampal gyri. On the medial face, they supply the uncus, hippocampus, superior surface of the parahippocampal gyrus, and posterior areas of parietal and occipital lobes, including the precuneus and cuneus [12].

2.9.3 HA and Its Parent Vessels

The HA is one of PCA's first branches after its first segment (P1), as it arises directly from the PCA's main trunk, in up to 17.6% of the hemispheres. Although rare, it can also arise from the MIT (less than 1%). AIA, MIA, and PIA branch arteries to the hippocampus in, respectively, 57.4%, 0.5%, and 12.4% of the hemispheres.

The POA, which also arises from the PCA (P2 or P3 segments), may branch arteries to the hippocampus in up to 48% of hemispheres. The CA, medially and ventrally located to the POA, can emit hippocampal arteries in 31.5% of the hemispheres. Last, the SA (P3) has HA emerging in half of the hemispheres [12].

2.9.4 Vascular Arcades and Anastomoses

Anastomoses connecting the AChA or its branches to the PCA, POA, and CA (or their branches) can be seen in almost half of the brain hemispheres. These anastomoses connect the uncal branches of the AChA to the HAs of the cited vessels. In most cases, the anastomoses have an arcuate conformation.

At the level of the choroid plexus of the lateral ventricle's temporal horn, anastomoses have been identified connecting the AChA or its branches to posterior choroidal arteries (PChA) descending from the PCA. Both lateral PChA branches and distal AChA branches provide blood supply to important structures, such as the lateral geniculate nucleus, and therefore, it is essential to recognize these anastomoses to avoid visual field deficits [12].

2.10 Amygdala

The amygdala can be divided into "temporal or principal amygdala" and "extratemporal or extended amygdala." The first one can be subdivided into basolateral, corticomedial, and central groups and can be described as located into the uncus. The temporal amygdala blends with the globus pallidus superiorly, forming the anterior part of the roof of the temporal horn. Its inferior portion constitutes the anterior wall of the temporal horn. Therefore, the amygdala is closely related to the temporal horn, limiting this structure by designing its anterior wall and the major part of its anterior roof. Anteriorly, the amygdala merges with the rest of the uncus through the semilunar gyrus and relates to the entorhinal area anteroinferiorly [15].

The amygdala represents a bottom line for the limbic system and is especially implicated in fear and aversive responses. Some evidence also points to at least some involvement with positive emotions, such as a minor role in reward processing [16]. Although these functions might be harmed with the amygdalectomy, it is worth noting that not much impairment is expected from the removal of the amygdala in mesial temporal epilepsy surgeries once this region is presumably already damaged in such situations [17].

2.11 Choroidal Fissure

The choroidal fissure is a C-shaped cleft between the thalamus and the fornix localized from the foramen of Monro to the temporal horn of the lateral ventricle. It gives attachment to the choroid plexus in the lateral ventricle. This plexus, filling the choroidal fissure, is also C-shaped, revolving around the thalamus superiorly, posteriorly, and inferiorly. In the temporal horn, the choroidal fissure is located between the stria terminalis of the thalamus, superomedially, and the fimbria, inferolaterally, forming the inferior choroidal point (its inferior termination) next to the lateral geniculate body and behind the uncus.

The choroidal fissure might be didactically divided into three segments: body, atrial, and temporal. The body is located in the body of the third ventricle, precisely between the body of the fornix and the superior surface of the thalamus. The atrial part is situated amid the crus fornices and the pulvinar region of the thalamus. Finally, the temporal part is located between the fimbria of the fornix and the inferolateral thalamus, medially to the temporal horn. The choroidal fissure is a natural corridor and the best anatomic surgical landmark used for resection en bloc of the mesial temporal structures.

It takes around 8 weeks of embryonic progress for the epithelial roof of the third ventricle to invaginate itself medially and create the choroidal fissure. The fact that no nervous tissue is formed between the ependyma and the pia mater at this event causes an interesting anatomic singularity for the neurosurgical approach. It allows this area to become a proper access path to several structures with diminished risk for injuries and neurologic deficits [18, 19].

2.12 Vascular Relationships

Although there is a great number of arteries and branches that are important concerning the amygdalohippocampectomy, for a didactical purpose, we will briefly discuss the internal carotid artery, posterior communicating artery, anterior choroidal artery, posterior cerebral artery, and middle cerebral artery.

The ICA has a supraclinoid segment that enters the intradural space on the carotid cistern on the medial side of the anterior clinoid process below the optic nerve and runs superiorly dividing into ACA and MCA. The posterior communicating artery emerges from the inferolateral wall of the ICA and runs posteromedially, passing close or even adherent to the dura of the posterior clinoid process to pierce the interpeduncular cistern. Here, it joins to the posterior cerebral artery, between P1 and P2 segments.

The posterior cerebral artery arises as to the terminal branches of the basilar arteries in the interpeduncular cistern. The P1 segment is not related to the uncus. The thalamoperforating arteries are the main branches of P1. The P2 segment running into the crural cistern is related to the posteromedial segment of the uncus. The medial posterior choroidal, the short and long circumflex, and the hippocampal arteries arise from this P2 segment. The lateral posterior choroidal artery arises from the P2 segment that runs into the ambient cistern and enters the temporal horn via choroidal fissure. The P2 segment bifurcates in inferolateral and superomedial trunks. The inferolateral trunk gives rise to the inferotemporal branches (anteroinferior, inferomedial, inferoposterior, and inferotemporo-occipital). After loops into the ambient cistern, the superomedial trunk gives rise to the calcarine and parieto-occipital arteries.

The anterior choroidal artery arises in most cases from the posterolateral wall of the internal carotid artery between the posterior communicating artery (PComA) and the carotid bifurcation. There are branches from the inferomedial wall of the ICA between PComA and AChA that supply the optic tract and posterior perforated substance.

The basal vein of Rosenthal is the main venous channel that drains the mesial temporal region. The basal vein originates below the anterior perforated substance by the union of the deep middle cerebral, inferior striate, olfactory, fronto-orbital, and deep middle cerebral veins. It runs posteriorly under the optic tract and medially and inferiorly to the anterior portion of the crus cerebri. This point corresponds to the most inferior and medial part in its course before its termination in the vein of Galen. The hippocampal veins drain to the inferior ventricular vein and then to the basal vein.

2.13 *Optic Radiation*

The most important relationships for neurosurgeons correlate with the anatomical surface of the temporal lobe and the optic radiations. Axons of the neurons in the lateral geniculate nucleus project to the visual cortex in the occipital lobe via the geniculocalcarine tract (optic radiation). The anterior inferior fibers of the optic radiation curve toward the pole of the temporal lobe before making a sharp curve to return posterolaterally. This curve is called "temporal knee" [20], "temporal loop"

[21], or "Meyer-Archambault loop" [22]. The essential knowledge is related to the distance of the temporal tip and the most anterior fibers of Meyer's loop. According to Yasargil [3], this measurement cannot be precisely addressed with the fiber microdissection technique due to the dense network of fibers, especially in the area of the sagittal stratum, which would destroy one fiber tract in the identification of the other and vice versa. However, this subject is controversial.

In our fiber dissections, the sequential steps followed those addressed by Yasargil [3]. We began taking the cortex out, which exposed the arcuate fibers. The arcuate fibers interconnect adjacent cortex areas and are U-shaped when they have to circumvent the bottom of a sulcus between adjacent gyri. Dissecting these most superficial layers of white matter underlying the cortex of the temporal lobe and occipital lobe and the area around the Sylvian fissure, the superior fasciculus longitudinalis was identified. After this fascicle is peeled away, part of the sagittal stratum is identified, although it is difficult to disentangle from the external sagittal stratum [4]. The sagittal stratum is formed by the occipitofrontal fasciculus and the posterior thalamic peduncle, but it is difficult to individualize the different tracts. The sagittal stratum passes from the temporal lobe toward the occipital lobe, and fibers of the optical radiation are included here. The "occipital-temporal projection system" is a chain of short cortico-cortical fibers running external to the sagittal stratum next to the optic radiation that interconnects the anterior temporal cortex to the occipital pole [23]. The cortex of the insula was removed, and extreme capsule was identified. The extreme capsule is a thin fiber system that connects the frontal and temporal opercula with the insula. After this subsequent external-to-internal dissection, the following structures were identified: claustrum, external capsule, putamen, globus pallidus, and internal capsule. The optic radiation constitutes part of the posterior thalamic peduncle, which is part of the internal capsule. The frontal and temporal lobes are connected more anteriorly by the uncinate fascicle, which overlies the amygdala. Its fibers gather under the cortex of the limen insulae, where they form a very compact bundle. The inferior fronto-occipital fasciculus (IFOF) occupies the extreme and external capsules during part of its course. This fascicle carries fibers between the midportion of the superior temporal gyrus (auditory association cortex) and the dorsolateral and mesial prefrontal cortex, between the prefrontal cortex and midportion of the inferior temporal cortex (visual association cortex) and multimodal cortex of the superior temporal sulcus, and between the dorsolateral prefrontal cortex and the parahippocampal area. It is associated with the uncinate fasciculus at the limen insulae.

Part of the optic radiations from the lateral wall of the temporal horn become thicker more posteriorly due to the inclusion of the fibers of the inferior fronto-occipital fasciculus to become the sagittal stratum. The inferior end of the optic radiation does not go inferior to the inferior temporal sulcus. The optic radiation can be divided into three groups of the same size. One is called the posterior bundle and is part of the sagittal stratum. These fibers do not do any anterior curve. The other is the central bundle that makes a partial anterior curve and courses within the sagittal stratum posteriorly. The other is the anterior bundle, also known as Meyer's loop, which courses around the lateral half of the tip of the temporal horn before entering into the sagittal stratum. In the posterior end of the temporal lobe, these fibers turn and become inferior to the occipital horn to approach the inferior bank of the calcarine fissure.

In our dissections, we noted that the full extension of the lateral and superior wall of the temporal horn was covered by optic radiation, and even the lateral half of the tip of the temporal horn, reaching the anterior edge of uncal recess and extending a few millimeters (range 2 mm) anterior to the tip of the temporal horn [24]. The inferior and medial walls of the temporal horn are free of optic radiation, except at the level of the lateral geniculate body.

At coronal sights, the temporal lobe seems to be connected to the basal forebrain through a peduncle resembling a tree's roots. This structure is widely called the temporal stem. Although the terminology is largely used, the temporal stem is not a well-established concept. It ranges from being only the fibers that pass between the limen insula and the temporal horn to all the fibers that pass under the inferior limiting insular sulcus. At its larger definition, the brain stem comprises the anterior commissure, the uncinate fasciculus, the inferior fronto-occipital fasciculus, Meyer's loop of the optic radiations, and inferior thalamic fibers, measuring up to 8.2 mm [25, 26].

Access to the amygdala and hippocampus through the temporal horn is crucial at the transsylvian approach to the medial temporal lobe. Thereby, the neurosurgical value of knowing the temporal stem anatomy is related to this method, once the gateway through the structures described above is generally directed through the temporal stem [25, 26] (Figs. 11, 12, 13, 14, 15, 16, and 17).

2.14 *Inferior Fronto-Occipital Fasciculus*

The inferior fronto-occipital fasciculus (IFOF) is a large association bundle of fibers (white matter) that, as the name suggests, connects the frontal and occipital lobes. It runs through the insular lower portions within the extreme and external capsules. Although it runs above and posteriorly the uncinate fasciculus, there is no defined limit between them in the white matter. It also has a close relationship with the optic radiation, running parallel and above them [27]. Its fibers connect the frontal gyri; run under the superior, anterior, and inferior insular limiting sulci; and reach temporal, parietal, and occipital regions [28].

Based on the intraoperative neurophysiological subcortical stimulation during surgery in awake patients, it is associated with the IFOF contributions to awareness; elaboration of visual information for motor planning, reading, and attention; as well as a role in semantic components of language. Therefore, transinsular surgical approaches must be made with caution. A deep transinsular approach, mostly at the dominant hemisphere (but maybe even in the non-dominant side), can damage the IFOF and may lead to language deficits [28] (Figs. 11, 12, 13, 14, 15, 16, and 17).

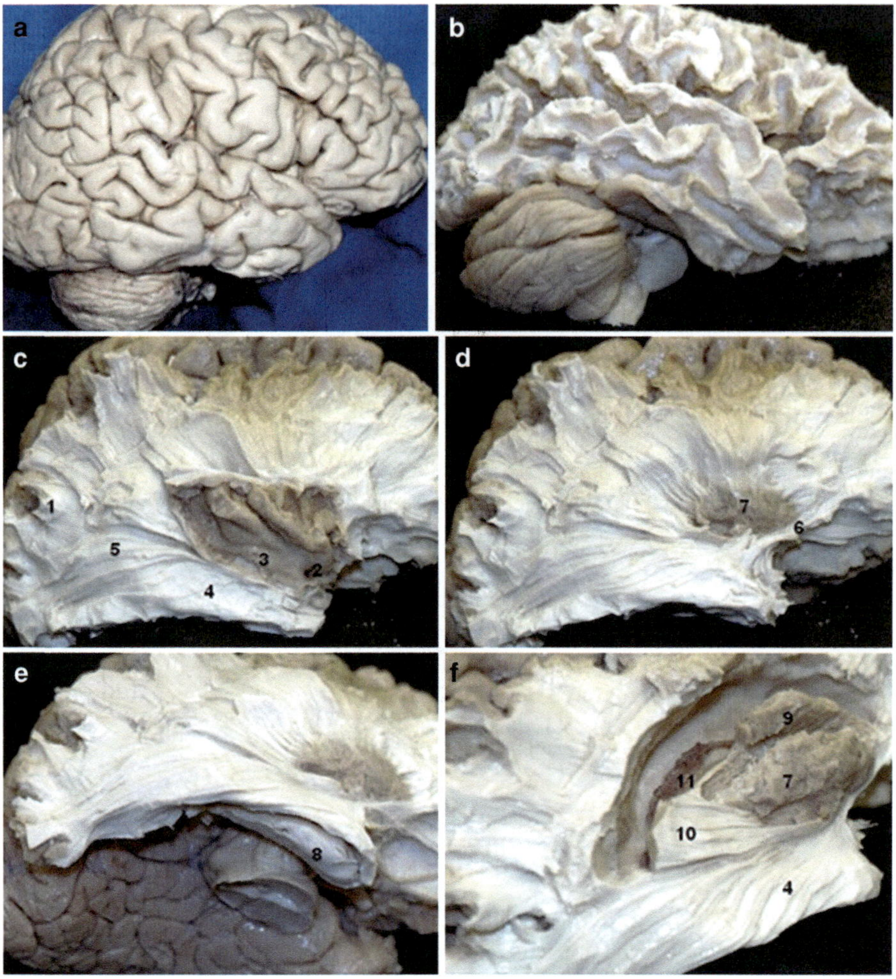

Fig. 11 Lateral view of right-sided brain hemisphere showing step-by-step white fiber dissection. (**a**) Lateral view of the right side of the brain; (**b**) grey matter removed, exposing the U-fibers in the white matter; (**c**) further white matter dissection, exposing the Meyer's loop, sagittal stratum, and insular grey matter; (**d**) grey matter of the insula removed; (**e**) inferolateral view of the dissected right-side brain, showing the inferior view of the hippocampus; after dissection of the central core, exposing the basal nuclei and its relation with the temporal horn of the lateral ventricle; (**f**) superior and lateral view of the Meyer's loop and optic radiation. 1—U-fibers; 2—Sylvian vallecula; 3—insular cortex; 4—Meyer's loop; 5—sagittal stratum; 6—uncinate fasciculus; 7—putamen; 8—hippocampus; 9—caudate nuclei; 10—inferior fronto-occipital fasciculus; 11—choroid plexus

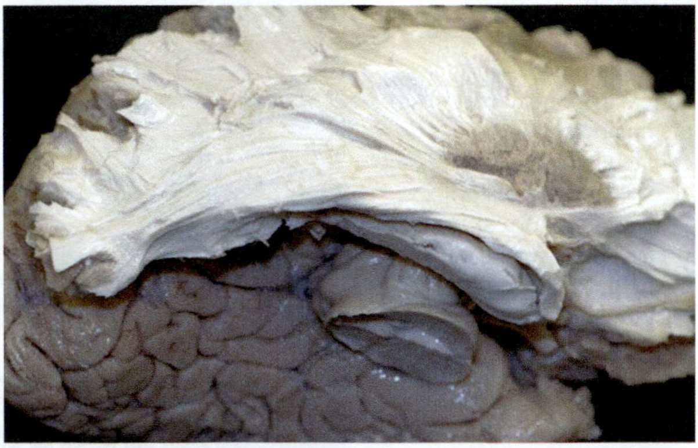

Fig. 12 Closer look to a latero-inferior view of white fiber dissection, showing the hippocampus and its relation to Meyer's loop

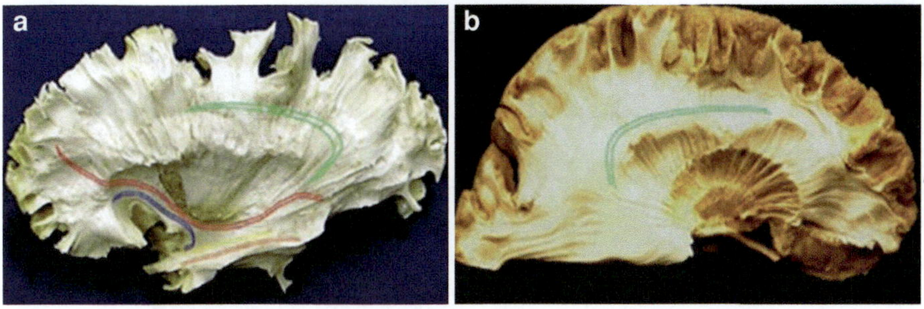

Fig. 13 White fiber dissection and schematics of the fasciculi. (**a**) Long association fiber representation. Arcuate fasciculus (green); inferior fronto-occipital fasciculus (red); middle longitudinal fasciculus (yellow); inferior longitudinal fasciculus (orange). (**b**) Long association fiber representation. Arcuate fasciculus (green)

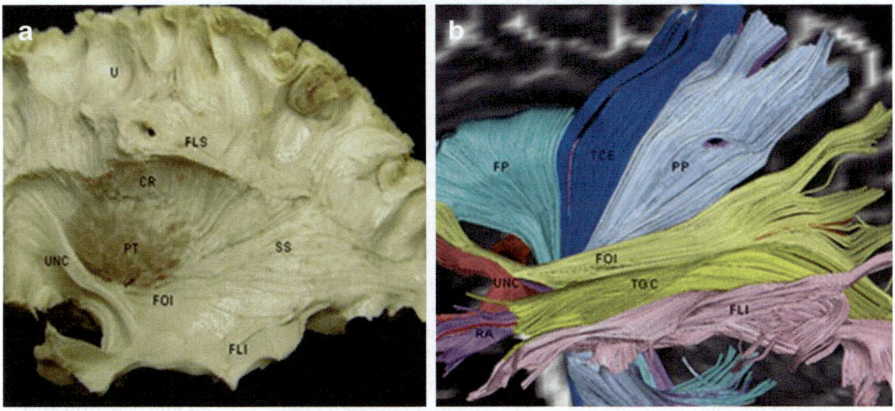

Fig. 14 (**a**) White fiber dissection showing association fibers and (**b**) white fiber schematics based on tractography. *UNC* uncinate fasciculus, *PT* putamen, *FOI* inferior fronto-occipital fasciculus, *FLS* superior longitudinal fasciculus, *SS* sagittal striatum, *CR* corona radiata, *FLI* inferior longitudinal fasciculus, *U* U-fibers, *FP* frontopontine fibers, *TCE* corticospinal tract, *PP* parietopontine fibers

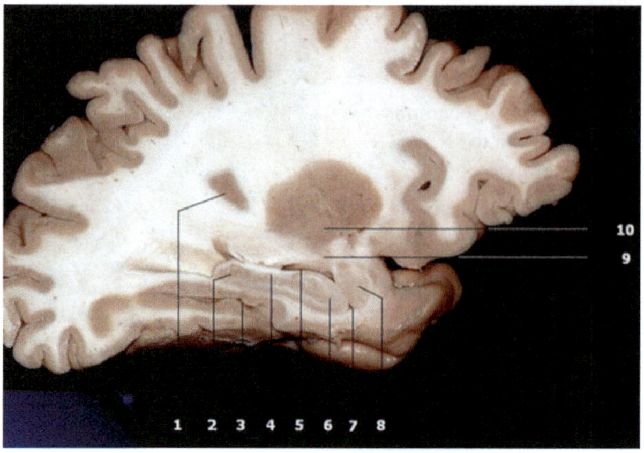

Fig. 15 Parasagittal brain section through the hippocampus: 1—insular cortex; 2—caudal part of the hippocampus; 3—collateral sulcus; 4—alveus and fimbria-fornix; 5—temporal horn of lateral ventricle; 6—rostral part of the hippocampus; 7—temporal horn of lateral ventricle; 8—amygdala; 9—tail of caudate nucleus; 10—putamen

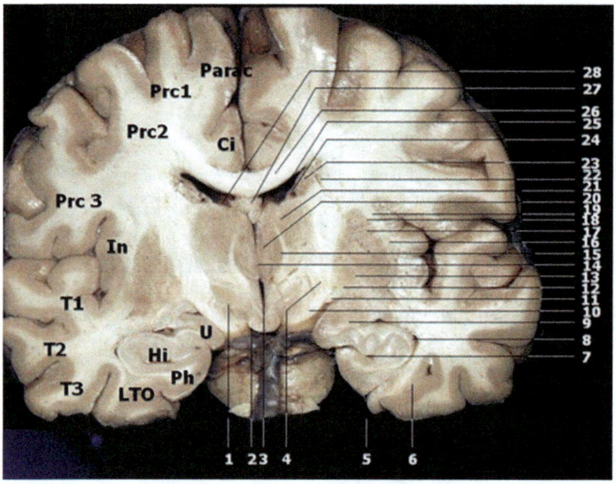

Fig. 16 In a coronal section, we can see that there are no basal ganglia in the superior limiting insular sulcus before the internal capsule. It is easy to damage the corona radiata and disrupt important descending (motor) and ascending (sensitive) fibers. We can also see that the inferior limiting insular sulcus is directly related to the temporal stem. There are several white matter fasciculi in the temporal stem, which could cause different deficits if damaged: 1—lenticular fasciculus; 2—mammillary bodies; 3—third ventricle; 4—posterior limb of internal capsule; 5—collateral sulcus; 6—occipitotemporal sulcus; 7—hippocampal fissure; 8—temporal horn of lateral ventricle; 9—amygdala; 10—cerebral peduncle; 11—optic tract; 12—medial globus pallidus; 13—internal medullary lamina (thalamus); 14—interthalamic adhesion; 15—mammillothalamic tract; 16—claustrum; 17—putamen; 18—external capsule; 19—extreme capsule; 20—anterior nucleus of thalamus; 21—ventral anterior nucleus of thalamus; 22—caudatolenticular transcapsular grey stria; 23—caudate nucleus; 24—stria terminalis; 25—body of lateral ventricle; 26—corpus callosum; 27—fornix; 28—choroid plexus; U—uncus; Hi—hippocampus; Ph—parahippocampus; LTO—lateral occipito-temporal gyrus; T1—superior temporal gyrus; T2—middle temporal gyrus; T3—inferior temporal gyrus; In—insula; Prc 1—superior peduncle of precentral gyrus; Prc2—middle peduncle of precentral gyrus; Prc3—inferior peduncle of precentral gyrus; Parac—paracentral gyrus; Ci—cingulate gyrus

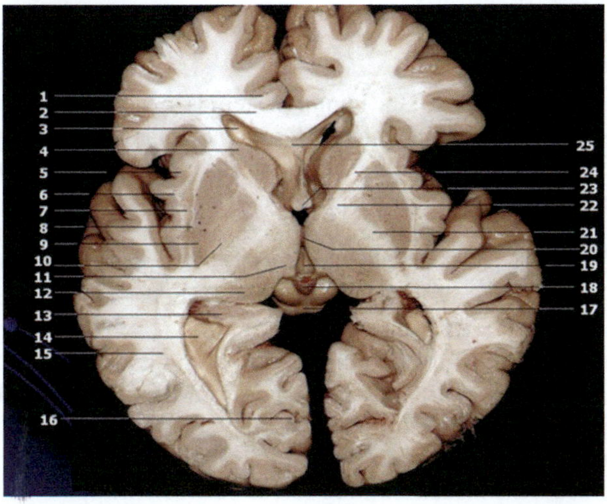

Fig. 17 Axial brain section through the lentiform nuclei, emphasizing all white matter layers from the insular cortex to the internal capsule: 1—cingulate gyrus; 2—genu of the corpus callosum; 3—frontal horn of lateral ventricle; 4—head of caudate nucleus; 5—insular cortex; 6—extreme capsule; 7—claustrum; 8—external capsule; 9—putamen; 10—globus pallidus; 11—thalamus; 12—pulvinar nucleus (thalamus); 13—choroid plexus; 14—occipital horn of lateral ventricle; 15—tapetum (corpus callosum); 16—calcarine sulcus; 17—quadrigeminal plate; 18—pineal gland; 19—third ventricle; 20—interthalamic adhesion; 21—posterior limb of the internal capsule; 22—genu of the internal capsule; 23—anterior column of fornix; 24—anterior limb of the internal capsule; 25—septum pellucidum

2.15 *Uncinate Fasciculus*

The uncinate fasciculus (UF), named after its "hook" shape, is a bundle of association fibers that connects the anterior and medial portions of the temporal lobe to the orbital portion and the pole of the frontal lobe. It lies under the cortex of the limen insulae, anteroinferiorly to the IFOF and the optic radiation (Meyer's loop) [27].

It is hypothesized that the UF is just part of a semantic network that involves the UF, the inferior longitudinal fasciculus (ILF), and the IFOF, with the UF being the link between the occipitotemporal pathway of the ILF and the frontal lobe. As it has a close relation to the IFOF, it is also associated with semantic components of language, although to a lesser extent. Some studies also attribute to the UF some naming functions (objects or actions) and face recognition, although studies show mild to no speech deficits after UF resection (tumor or epilepsy surgery). Another important factor that should be considered is that slow-growing lesions, such as a low-grade glioma, induce neuroplasticity, allowing redistribution of language functions to healthier sites [29] (Figs. 11, 12, 13, 14, 15, 16, and 17).

2.16 Temporal Stem

Although not precisely defined, the temporal stem is a widely used term referring to the fibers passing underneath the inferior limiting insular sulcus. The name "temporal stem" is derived from the appearance, in the coronal plane, that there is a "peduncle" of white fibers, which pass deep through the anterior aspect of the insula, connecting the temporal lobe to the central core, and it may look like the "roots of a tree." Its concept ranges from just the fiber bundles passing between the limen insula and the anterior end of the temporal horn to all the fibers passing beneath the length of the inferior limiting insular sulcus (ILS) [26, 28].

In its larger definition, the temporal stem has been divided into two components, anteromedial and posterolateral, the first being formed by the ventral amygdalofugal fibers, UF, anterior aspects of the IFOF, and anterior commissure and superior extension of the amygdala toward the globus pallidus, and the latter by the fibers that pass underneath the insular limiting sulcus between the tip of the temporal horn and lateral geniculate body-posterior aspects of IFOF and anterior commissure, and optic radiation fibers [26, 28] (Figs. 11, 12, 13, 14, 15, 16, and 17).

2.17 Language in the Temporal Lobe

The advance from a localizationist paradigm to a hodotopical model (network of cortical areas and their connecting pathways) has been largely having taken place with modern techniques that allow scientists to study language function in vivo, such as direct electrical stimulation (DES) in awake craniotomies, diffusion tensor imaging and fiber tractography, functional MRI, magnetoencephalography, and navigated transcranial magnetic stimulation. These techniques provide a more accurate insight into the actual organization and functioning of the brain, while traditional models, built on evidence from patients with cortical lesions, do not explain the multiples aspects of speech function and its connections with other areas [30].

Anatomical and tractographical studies have facilitated the identification of white matter tracts involved in language functions. Intraoperative DES provides a unique opportunity to discover the functional roles of cortical hubs and subcortical pathways [31].

Dysarthria is related to the lateral precentral and the postcentral gyri on both sides of the brain, suggesting that the sensory system modulates the articulatory process. Speech articulation is presumed to be a bilateral process within the face motor cortices [30, 31]. In anomia, speech is partially preserved, but nouns are difficult to retrieve. The epicenter of this function seems to be located in the posterior superior temporal gyrus and the inferior parietal lobule of the dominant hemisphere, corresponding to the "classical" Wernicke's area. However, other areas may also be involved (inferior precentral gyrus, close to the lateral sulcus and middle frontal and the precentral gyri, which is also the junction between the dorsal and ventral premotor cortices) [30, 31].

Semantic and phonological processes are widely distributed throughout the cortex in the dominant hemisphere. Semantics is associated with the junction of the posterior superior temporal gyrus and the supramarginal gyrus, pars triangularis, pars opercularis, and dorsal premotor cortex. Phonology is associated with the middle superior temporal gyrus, the pars opercularis, and the junction of the dorsal and ventral premotor cortices. The localization of semantic and phonological processes in the pars triangularis and the pars opercularis suggests that the classical Broca's area could be involved in higher order tuning of language [31].

The arcuate fascicle (AF), deep portion of the superior longitudinal fascicle that connects the posterosuperior temporal cortex and the inferior parietal cortex to areas in the frontal lobe, is associated with several language dysfunctions such as Broca's aphasia, alexia, agraphia, conduction aphasia, reduced comprehension, anomia, dyslexia, Wernicke's aphasia, and transcortical sensory aphasia. Lesions in the uncinate and inferior longitudinal fasciculi are associated with semantic dementia [31].

2.17.1 Language Input

The first stage in the process of understanding language involves visual, auditory, or somatosensory input. Visual perception is the first stage of a language process elicited by visual input. The visual occipital cortex is connected to the fusiform gyrus (both visual object form area (VOFA) and visual word form area (VWFA)) by the inferior longitudinal fascicle, and stimulation of this pathway triggers visual paraphasia.

Subcortical DES of the occipitotemporal white matter tracts on the dominant hemisphere induced different lower and upper fiber symptoms. In the same patients, stimulation of lower fibers induced alexia, while stimulating the upper fibers induced anomia. Therefore, words and objects seem to be recognized by two distinct and parallel pathways originating from the visual occipital cortex.

Auditory input from the thalamus is processed in multiple locations, such as the posterosuperior temporal areas and the supramarginal, angular, posterior middle temporal, superior, and middle occipital gyri. These locations are interconnected by U-fibers [30, 32].

Somatosensory input information from the thalamus is processed in the superior parietal lobule, the precuneus, the superior cingulate cortex, the SMA, the primary motor cortex, the prefrontal cortex, the middle frontal gyrus, and the orbitofrontal areas, which are also interconnected by U-fibers [30, 32].

2.17.2 Semantic Versus Phonological Pathways

Visual and auditory language processes seem to branch into two main pathways, ventral semantic and dorsal phonological, which interact and work in parallel. In auditory language processing, the ventral stream could be involved in mapping sound to meaning, and the dorsal stream in mapping sound to articulation. This dual model was built for visual language processing based on DES-induced separation between phonemic and semantic processes, demonstrating that these processes do not occur serially but in parallel [30].

2.17.3 Ventral Semantic Stream

The IFOF subserves the direct pathway of the ventral semantic stream. The IFOF connects the posterior occipital lobe and the VOFA to frontal areas such as the dorsolateral prefrontal cortex and the pars orbitalis. DES of the IFOF resulted in semantic paraphasia [26, 30].

In addition to the direct pathway, there is an indirect pathway with a relay in the temporal pole. This indirect ventral stream is subserved by the anterior part of the inferior longitudinal fascicle, which connects the VOFA to the temporal pole. From the temporal pole, the information is relayed to the pars orbitalis by the UF. This sub-network can be bypassed to the direct pathway (IFOF) and functionally compensated, explaining the lack of speech deficits following UF damage.

Another sub-network, the middle longitudinal fasciculus, could be involved in the ventral semantic stream. This pathway connects the angular gyrus to the superior temporal gyrus. However, this fasciculus has not elicited naming disorders, so its specific functional role remains unknown [30].

2.17.4 Dorsal Phonological Stream

The dorsal pathway is subserved by the superior longitudinal fascicle and can be divided into two components. The deeper one is the AF, which runs from the posterior section of the middle and inferior temporal gyri, arches around the insula, and moves toward the ventral premotor cortex and the pars opercularis.

In several studies, phonemic paraphasia was observed after DES of white matter around the superior and posterior parts of the superior insular sulcus. The VOFA, involved in both phonological and semantic processes, is the posterior cortical origin of the dorsal pathway.

Stimulation of the AF induces conduction aphasia without semantic disorders. Conduction aphasia is a phonemic paraphasia combined with repetition disorders. These findings support the role of the AF in phonological processing [30, 32].

The more superficial component of the dorsal stream is subserved by the anterior part of the superior longitudinal fascicle, as anarthria was observed after DES [32]. This bundle connects the posterior part of the superior temporal gyrus and the supramarginal gyrus to the ventral premotor cortex. Therefore, this system integrates somatosensory and auditory information with phonological-phonemic information translated into articulatory motor programs. This loop is also used during word repetition [30, 32].

2.18 Brain Hodotopy

Brain hodotopy is an oppositional concept to the long-established topological and fixed vision of brain organization. Hodotopy sees the nervous system as a dynamic entity with an interconnected and ever-changing network, contrasting with the localizationist view. It has been applied in neuro-oncological surgeries for presurgical planning, aiming to preserve noble regions of the brain, admitting and foreseeing its dynamic nature [33, 34]. Nevertheless, hodotopy is still disregarded in epilepsy surgeries [33]. This is an inquisitive situation considering that the hodotopical concept is especially fit to epilepsy cases, once the disturbance present in such situations seems to be a particularly good source for functional reorganization [35]. In order to avoid potential functional deficits, as mentioned above, we hope that hodotopy will be better explored for epilepsy surgery soon.

2.19 Sylvian Cistern

The lateral sulcus (Sylvian fissure), easily observable from a lateral view, is the largest sulcus of the brain, which makes it a trendy access in neurosurgery. It sits over the insula and is designed to wrap the frontal, parietal, and temporal opercula. The Sylvian fissure transitions between the basal cisterns and the subarachnoid space over the convexities [36]. Concerning the Sylvian fissure, the proximal aspect of the Sylvian cistern is at the level of the origin of the middle cerebral artery from the internal carotid artery. It narrows distally as the frontal and temporal lobes approach each other over a length of 15–20 mm. The relation between the frontal and temporal lobe on the Sylvian fissure is inconstant, difficult for dissection. The arachnoid membranes in the lateral sulcus can be transparent and fragile or thickened and tough. Professor Yasargil divides this relation into four categories, in crescent order of difficulty at dissection. The cistern is large over the insula. The structures sitting in the Sylvian cistern are the superficial and deep Sylvian veins, the middle cerebral artery, and the origin of its branches [36–39].

2.20 Superficial Sylvian Veins

The superficial Sylvian veins are one or more large veins that are generally related to the temporal part of the lateral sulcus. There is much variation in the number and course of the superior and deep Sylvian veins. Generally, they drain into the sphenoparietal sinus and, less frequently, into the cavernous sinus. Although there is a rich collateral system in these veins, the purpose of the neurosurgeon is to preserve all of them, once it is not possible to know what vein sacrifice will lead to venous infarction and neurological deficit [37, 39] (Fig. 18).

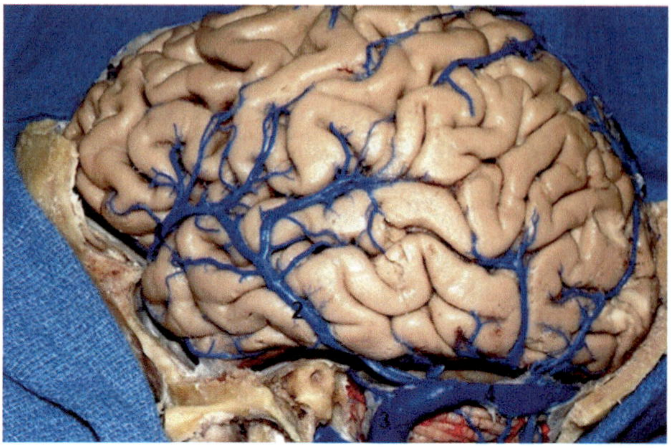

Fig. 18 Lateral view of left-sided brain hemisphere showing superficial temporal drainage system: 1—superficial middle cerebral vein (superficial Sylvian vein); 2—inferior anastomotic vein (vein of Labbe); 3—sigmoid sinus; 4—transverse sinus

2.21 Middle Cerebral Artery

The middle cerebral artery (MCA) originates from the internal carotid artery. Its first segment, M1, ranges from its origin to the bifurcation and gives two groups of branches, the temporal vessels and the perforating vessels. The temporal vessels or superior lateral group are, ordered proximal to distal, the uncal artery (this artery originates from ICA in 70% of the cases), polar temporal artery, and anterior temporal artery. The M1 segment divides in the M2 segment at the limen insulae. The M2 segment is composed of superior and inferior trunks.

There are many variations in the vascularization pattern of the temporal lobe by these arteries, but one very important variation is the false bifurcation of the MCA, which is a large branch from the lateral wall of the proximal M1 segment that gives the impression of a true early bifurcation. The perforating vessels, the inferior medial group, or "lenticulostriate vessels" are 2–15 [36]. The most common pattern is all the lenticulostriate arteries arising from one single large artery. The course of the M1 segment is not only straight but also C- or S-shaped. The true bifurcation of the MCA always occurs at the high point of the limen insulae. The fenestration of the proximal part of the M1 occurs in 0.1–0.3% [40]. An accessory MCA can be present in 0.5% of cases and originates from the proximal or distal A1 segment [40]. When large branches arise from either the superior or the inferior trunks close to the bifurcation, an impression of a "pseudo-trifurcation" or "pseudo-tetrafurcation" may be given [41]. However, true trifurcation, tetrafurcation, or even pentafurcation is present only in few cases [36] (Fig. 19).

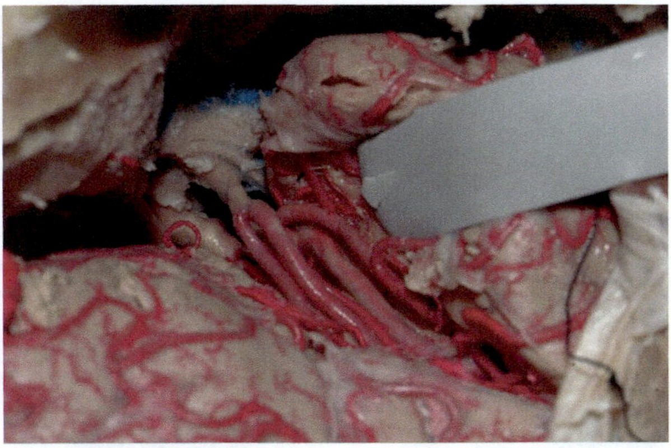

Fig. 19 Pseudo-tetrafurcation of the middle cerebral artery

3 Temporal Surgery for Epilepsy

Resective surgery to treat epilepsy was first described by Horsley in 1886 [42]. Since then, with the advent of electroencephalography and electrocorticography, temporal surgery has been continually improving. In early practice, hippocampus preservation was thought to be mandatory to avoid memory deficits, but it has been proved that seizure control cannot be achieved without mesial temporal structure resection [43]. The goal of treatment is to reach a state of "seizure freedom," as it improves the patient's quality of life (education, employment, social life).

Several techniques have been proposed for resection of temporal structures to be less resective and more efficient, reducing the chance of neurological deficits. However, it is not well established which technique (selective or not) has better seizure control and lower deficit risk (Fig. 20).

The selective amygdalohippocampectomy and the anterior temporal lobectomy are the two techniques used to treat mesial temporal lobe epilepsy. There are two different anatomical routes to perform the selective amygdalohippocampectomy, the transsylvian and the transcortical approaches. Our dissections and results were exposed in the transsylvian way because this route is more anatomical, although more skills are required due to work at the Sylvian fissure.

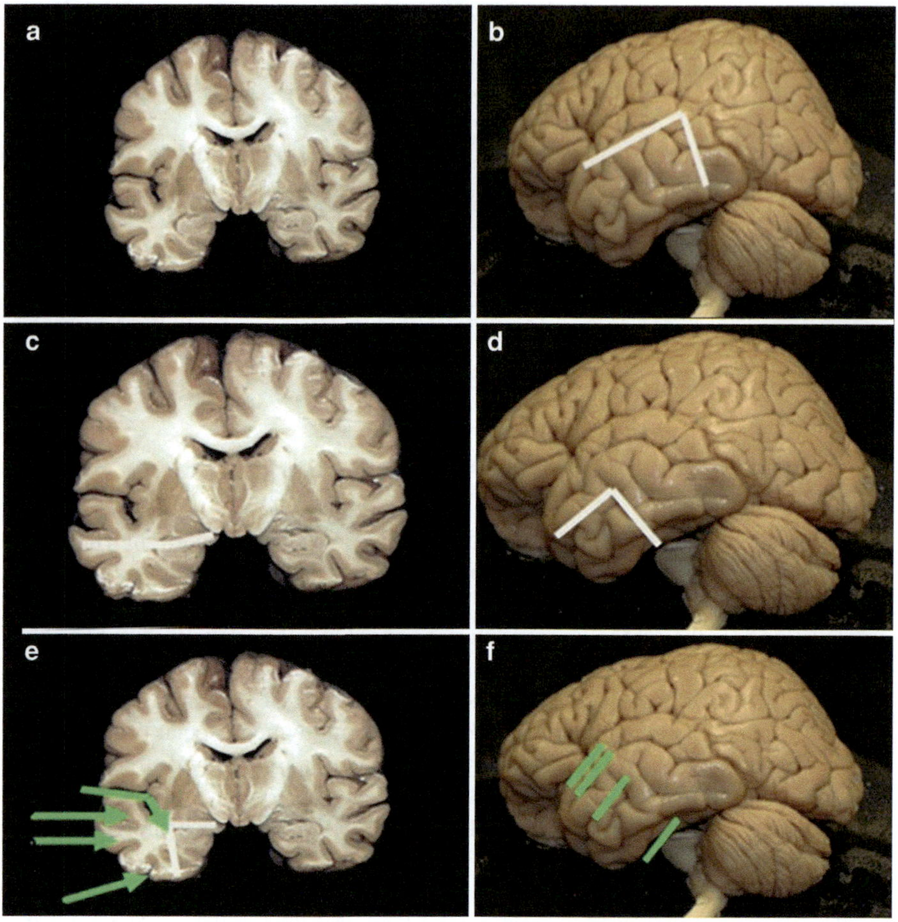

Fig. 20 Sagittal and coronal views of anatomical specimens of the brain, with schematics of temporal approach in epilepsy surgery. (**a**, **b**) show anterior temporal lobectomy. (**c**, **d**) show Spencer's technique. (**e**, **f**) show multiple possibilities of selective amygdalohippocampectomy (transsylvian, transtemporal, or subtemporal)

3.1 Neurological Deficits Following Temporal Epilepsy Surgery

Visual field deficits (VFDs) are a major debate in epilepsy surgery, a controversy especially intense at transsylvian approaches. Around 36% of the patients subjected to the procedure have some VFDs, but this statistic can be shifted to much higher values depending on the study and approach selected. Although a fair amount of these patients is not aware of their deficits, the loss can lead to important problems, such as driver's license acquisition [44, 45].

Some experts argue that damage to the branches that supply the optic radiations and the lateral geniculate body, mostly coming from the anterior choroidal artery, is the main source of VFDs after epilepsy surgeries. However, there is evidence supporting otherwise, mainly experience with surgeries for parkinsonism. In such situations, the obliteration of the very anterior choroidal artery does not seem to lead to VFDs [44].

Presently, the predominant view is aligned with the idea of direct damage to the optic radiations. When approaching the temporal horn through the floor of the Sylvian fissure, the anterior bundle of optic radiations (Meyer's loop) can be impaired [24]. The exact location of this fiber is not consensual. Over time, it has been estimated to be between 20 and 60 mm behind the temporal pole [44]. Through modern techniques, different studies found different but similar values, ranging between 22 and 43 mm. Anyway, Meyer's loop position certainly admits a spectrum of anatomic varieties [45–47].

Choi et al. [24] proposed a triangular area where the temporal horn could be approached with diminished risk for optic radiations. This area is situated between Meyer's loop and the optic tract. The base of the triangle would be near the amygdala, at the level of the limen insula. The anterior edge of the lateral geniculate body would be at the apex of the triangular area. Thereby, a transsylvian incision should be made adjacent to 5 mm of the inferior limiting sulcus. Another proposed option is the use of fiber tracking to predict optic radiation damage in surgeries. Some studies demonstrated the usefulness of diffusion tensor imaging techniques for this purpose. Nevertheless, this is not a real-time procedure and does not consider brain shift [45, 47–49].

Another structure worth mentioning is the uncinate fasciculus, which connects orbitofrontal to temporopolar areas and presumably supports mnemonic associations such as face and object recognition [50]. At transcortical SAH, incisions after the limen insula posterior limits fatally injure this bundle. This might be related to the memory deficits described after the procedure. Also related to the limen insula, the inferior occipitofrontal fasciculus can be impaired in incisions to the inferior limiting sulcus, which might produce paraphrasing episodes [51].

3.2 Standard Temporal Lobectomy

The standard temporal lobectomy (anterior temporal lobectomy) consists of resection of mesial and lateral temporal structures, which can be done en bloc or separately. The resection of lateral structures allows the visualization and en bloc resection of the hippocampus.

The patient lies in a supine position with contralateral head rotation. Pterional craniotomy is followed by a durotomy that exposes the Sylvian fissure and the temporal lobe. The corticectomy starts in the lateral temporal surface, 4–5.5 cm posterior to the temporal tip (if dominant hemisphere, the incision must be closer). A slight anterior deviation in the T1 gyrus may be useful to avoid Heschl's gyrus (primary auditory cortex). Upon reaching the upper temporal border, a subpial dissection is performed through the medial surface to preserve the MCA and its branches. The insula is then exposed. The resection is expanded from the lateral gyri through the fusiform gyrus up to the collateral sulcus, and the temporal horn of the lateral ventricle is opened. From now, the head of the hippocampus can be seen. The temporal stem can now be resected in the inferior insular limiting sulcus. The temporal neocortex (lateral portion) can be removed by sectioning the leptomeninges lateral to the temporal horn. Further mesial resection can be done if an en bloc anterior temporal lobectomy is the goal. Care must be taken during resection of mesial structures, not to damage the posterior cerebral artery, the basal vein of Rosenthal, the third cranial nerve, and the midbrain and not to go medially or superiorly to the globus pallidus. The entorhinal cortex is resected to the anterior portion of the parahippocampal gyrus, which is then dissected. The fimbria is laterally dissected to expose the hippocampal sulcus and the Ammon's horn arteries. The hippocampal feeders can be coagulated, and the hippocampus and parahippocampus can now be resected en bloc. Posterior hippocampal structures may be aspirated [43]. After hemostasis, closure is done to finish the procedure.

3.3 The Selective Amygdalohippocampectomy

3.3.1 Transsylvian Approach

The beginning of the transsylvian approach is by the opening of the Sylvian fissure. The sharp dissection should be performed during the whole opening of the Sylvian fissure, except the thickened arachnoid bands, which should be divided with microscissors or tenotome. A delicate aspirator on moist Cottonoid sponges can be used. The exposure is medial to the Sylvian veins from the carotid bifurcation to the middle cerebral artery bifurcation and beyond to expose the anterior one-third of the insula and 1–2 cm of the M2 segments.

The next step is to identify the lateral branches of the M1 segment. The inferior trunk of the M2 is followed in the insular limiting sulcus pars inferior and gently mobilized to allow coagulation and division of 2–5 perforating branches entering here. At this point, where these small branches were coagulated, a small incision (1–2 cm) was performed anteromedial to the M2 segment. This incision opens the UF. In the majority of the cases, this incision is localized between the temporopolar and anterior temporal arteries. At this level, in the most anterior part of the incision, the amygdala is found few millimeters in-depth by using the surgical aspirator. With the tip of a delicate dissector, the temporal horn of the lateral ventricle was entered a little bit posteriorly. This maneuver is useful to orientation regarding the entire amygdala. The next step is the removal of the amygdala. Special attention must be given to avoid a medial basal extension, which can cause a lesion in the optic tract. This structure should be identified. After that, the parahippocampal gyrus is removed subpially. The "extratemporal amygdala" is not removed. The pial and arachnoid membranes adjacent to the carotid cistern and ambient cistern were identified after subpial resection of the anterior part of the parahippocampal gyrus. These membranes were opened, exposing the uncal and anterior choroidal arteries, entering the choroidal fissure. At this stage, the following structures were identified: optic tract, the basal vein of Rosenthal, cerebral peduncle, third nerve, posterior cerebral artery (P1 and P2 segments), and posterior communicating artery. The next step was the removal of the hippocampus, for which the microscope was turned posteroinferiorly. The temporal horn was opened 2 cm posteriorly from its tip, exposing the trigone and plexus choroid. The choroid plexus is displaced medially to identify the tela choroidea and the choroidal fissure. The tela choroidea was opened to visualize the taenia fimbria of the lateral peduncle. The lateral branches of the anterior choroidal artery were divided, but the main stem and the medial branches must always be preserved. The parahippocampal gyrus was displaced laterally after the opening of the choroidal fissure to visualize the P2 segment, posteromedial choroidal artery, and collicular artery. The Ammon's horn arteries arise from the P2 segment and anterior choroidal artery and enter the sulcus hippocampus just proximal to the origin of the posterolateral choroidal artery. These arteries are coagulated and sectioned. The hippocampus was transected transversally at the level of the posterior rim of the cerebral peduncle, where the P2 bifurcates to form the inferolateral and superomedial trunks. The hippocampal veins and inferior ventricular veins, and the parahippocampal veins into the collateral sulcus, were sectioned. The most posterior part of the optic tract and the lateral geniculate body can be found at the level of P2–P3 bifurcation. This is the local where the fimbria ascends to become the crus of the fornix. The next step was the gentle spreading and opening of a sulcus in the floor of the temporal horn lateral to the hippocampus, which is an extension of the collateral eminence. This sulcus was opened anteroposteriorly in the direction of the peduncle where the rhinal and collateral sulci are found. The vessels originating from the inferolateral trunk (P3) supplying the parahippocampus were sectioned. The other vessels must be preserved. The parahippocampus was elevated and resected en bloc via the subpial plane. If there is no deep mesial herniation, the tentorium can be visualized. The neurovascular structures are protected by the pia and a double-layer arachnoid. The resected specimen has approximately 4 cm length, 1.5 cm breadth, and 2 cm depth [52]. Careful hemostasis is performed, the dura is closed, and the bone flap is replaced.

3.3.2 Transtemporal (Niemeyer's Technique)

Transtemporal selective amygdalohippocampectomy is a transcortical technique to reach the amygdala and the hippocampal formation through the temporal lobe. It was first described by the Brazilian neurosurgeon Paulo Niemeyer, who used T2 (middle temporal gyrus) to perform the cortical incision [43]. Nevertheless, the transtemporal technique encompasses further possibilities for primary incision nowadays, including the anterior part of the superior temporal gyrus and the superior temporal sulcus [53]. To better enlighten the procedure, our description sticks to Niemeyer's technique, the most commonly used.

After a temporal craniotomy and removing the bone flap and inferior reflection of the dura, an incision was performed at the middle temporal gyrus. The skin incision is made anterior to the tragus and above the zygoma. The entry point must be no farther than 3.5 cm from the temporal tip at the dominant hemisphere. The cortical incision, planned to be 2 cm, is where the endopial dissection is made to initiate the approach to the temporal horn. To guide the pathway, the use of neuronavigation can help [54, 55]. The use of neuronavigation optimizes the bone exposure for later cortical opening, indicating the optimal craniotomy point [43].

When the temporal horn was accessed, brain retractors were used to enhance the temporal horn exposition. The resection of mesial temporal structures began after an entry at the parahippocampal gyrus, at its most ventrolateral surface, and advanced to the uncus, medially situated. After removing these structures, the edge of the tentorium and the basal cisterns were viewable, allowing the amygdalectomy to be performed [54, 55].

The next step was focused on the removal of the hippocampus. Choroid plexus and choroidal fissure were identified. The medial and superior limit of the resection was the choroidal fissure, and the posterior limit was at the level of the posterior part of the body of the hippocampus. Starting from the finished amygdalectomy, the resection progressed posteriorly up to the tectal plate. The hippocampal tail was then removed alongside the hippocampal lateral margin, isolating the main structure. Only then can the body of the hippocampus be dissected, usually through suction technique. While small hippocampal vessels must be coagulated, small branches of the choroidal artery and posterior cerebral artery must be preserved, to which the surgeon should pay considerable attention. Finally, after inspection and hemostasis, the dura might be closed, followed by bone flap, temporalis muscle, and scalp [54, 55].

3.3.3 Subtemporal Technique

The subtemporal technique is another transcortical method to perform selective amygdalohippocampectomy. Tomokatsu Hori developed it at the beginning of the 1990s. Since then, several amendments have been proposed, such as the subtemporal transparahippocampal amygdalohippocampectomy (described below), but the general original idea remains the same [43].

A temporal craniotomy was performed. In this procedure, the craniotomy (8 cm × 5 cm flap) is performed so that the sphenoid bone is exposed and removed. After opening the dura, a retractor was placed under the uncus. Afterward, the anterior uncus was elevated to inspect the fusiform, parahippocampal gyri, tentorium, and ambient cistern. This access, for its positional features, requires the cerebrospinal fluid to be drawn [53].

The cortical incision was made at the uncus. The oculomotor nerve was identified at the ambient cistern, offering a reference for this cut (10–15 mm posteriorly). Suction excision of cortex and white matter opened the way to the temporal horn. Once the amygdala shapes the temporal horn anterosuperior wall, it becomes exposed through the former procedure. An incision at the parahippocampal gyrus cleared the way to the anterior hippocampus [53]. The rest of the procedure follows the same technique of the transsylvian approach.

Subpial aspiration split the amygdala from the hippocampus, previously connected through a thin layer of neural tissue. The parahippocampal gyrus around the inferolateral hippocampus must be removed for the later removal of the hippocampus itself. Two centimeters of the anterior hippocampus are then sectioned. Hippocampal arteries, branches from anterior choroidal and posterior cerebral arteries, must be coagulated. Using suction and a bipolar cautery, an additional 1 cm of the hippocampus is removed. Only then is the amygdala resected. For that, an ultrasonic aspirator might be used [53].

The main advantage of the subtemporal technique is the absence of incision at the temporal stem and temporal neocortex [53]. Hori argues that this benefit potentially preserves cognitive functions [54]. However, negative aspects of the procedure have been raised, such as an imminent risk for the vein of Labbé and the limited sight of the aimed structures [53, 54].

3.4 Anterior Temporal Lobectomy (ATL) Versus Selective Amygdalohippocampectomy (SAH)

The two main surgical approaches for mesial temporal epilepsy are still subject to debate once no standardized criteria have shown the superiority of one of the procedures convincingly. Performing a meta-analysis involving 626 patients, Kuang et al. [56] found no difference in seizure control nor verbal memory deficits in 1 year postoperatively. A comparative meta-analysis concluded that ATL improved the chance to accomplish seizure-free conditions but did not measure neuropsychological outcomes [57]. This agrees with the meta-analysis of Hu et al. [58], in which ATL showed better chances to control seizure frequency. In this work, IQ scores were similar between ATL and SAH patients. In an 18-year follow-up study, Hemb et al. [59] evaluated 108 patients and stated that the surgical technique does not interfere with the seizure control rate.

Analyzing a postoperative 5 years, Malikova et al. [60] observed similar seizure control between ATL and SAH, but better results in SAH for memory deficits and visual field defects, besides a lower amygdala and hippocampal volume reduction. This is in concordance with the study of Witt et al., in which they showed that memory preservation is related to hippocampal integrity after surgical treatment for mesial temporal epilepsy [61].

Wendling et al. [62] assessed 60 patients completely seizure-free after 7 years from epilepsy surgery and found that visual encoding, verbal short-term memory, visual short-term memory, and visual working memory were better in SAH patients. The percentage of patients that reached seizure-free state was slightly higher among ATL patients, while the quality of life was somewhat superior in SAH patients, but both were not statistically significant.

Since the mesial temporal lobe processes facial emotional recognition, it is reasonable to wonder what implication mesial temporal surgeries have on this matter. Wendling et al. [63] investigated 60 patients (ATL = 33; SAH = 27) and 30 healthy control subjects with the Ekman 60 faces test. All individuals had similar scores recognizing surprise, happiness, anger, and sadness. SAH patients scored poorer for disgust, while ATL patients had the worse scores for fear recognition.

We still do not have an absolute answer for what surgical approach should be the first choice for mesial temporal epilepsy patients. Although some studies show that ATL has a slightly enhanced chance to free the patient from seizures, others establish better neuropsychological outcomes for SAH. Ecological neuropsychological tests and further outcome evidences are necessary to consider indication criteria. In the meantime, the choice should be based on an individualized assessment of the patient and the surgeon's evaluation of the scenario.

4 Illustrative Cases

Figures 21, 22, 23, 24, and 25 show radiological, surgical, and anatomical images of temporal surgeries as practical examples of anatomical knowledge applied to microneurosurgery.

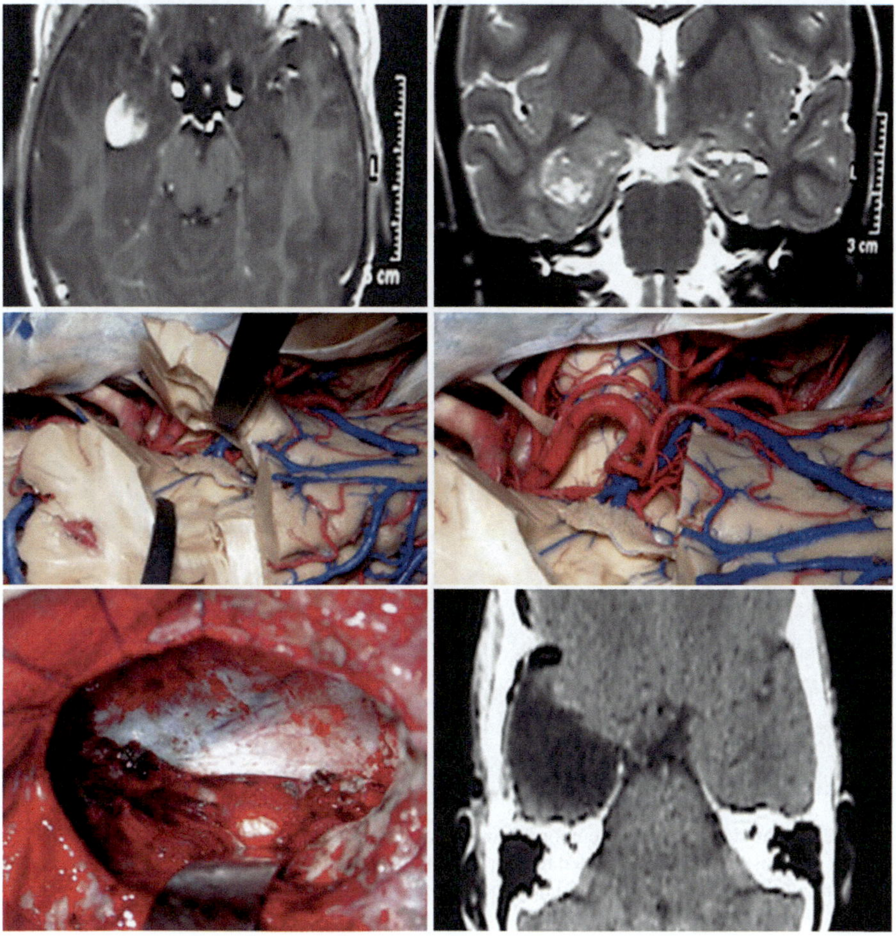

Fig. 21 Right-sided temporal ganglioglioma in a pediatric patient with drug-resistant temporal lobe epilepsy. The patient submitted to standard anterior temporal lobectomy. Excellent epilepsy outcome (Engel 1) in a 10-year follow-up

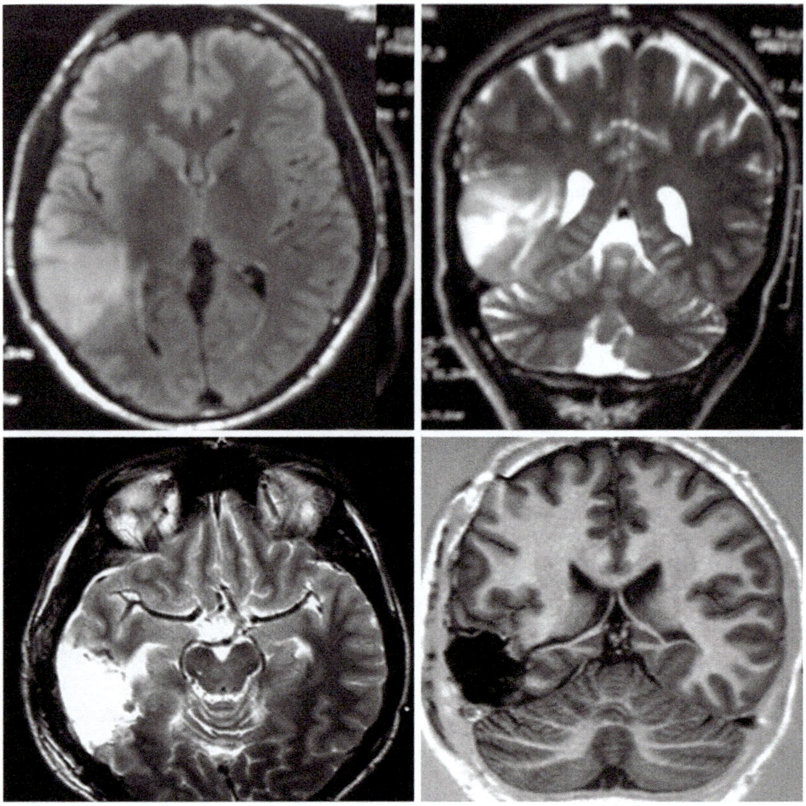

Fig. 22 Right-sided posterior temporal low-grade glioma. Surgery achieved total resection. The patient had postoperative right-sided temporal hemianopia. This case in which intraoperative brain mapping of the optic radiation could have been done with awake surgery

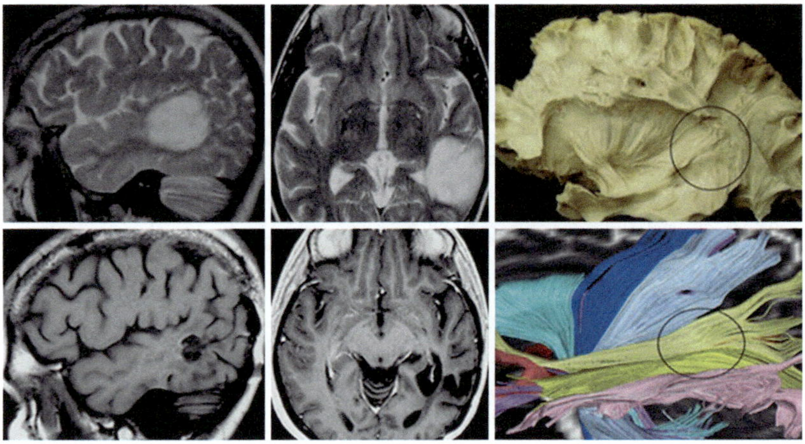

Fig. 23 Left-sided posterior temporal low-grade glioma. Total resection was achieved through the middle temporal gyrus with the preservation of optic radiations

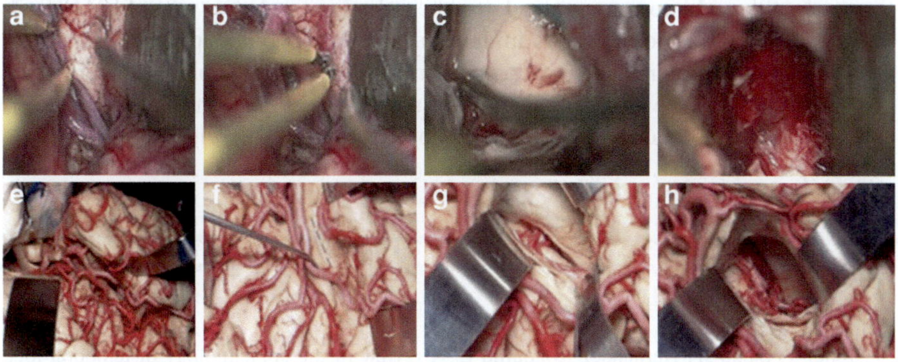

Fig. 24 Selective transsylvian amygdalohippocampectomy. (**a**) Opening of Sylvian fissure; (**b**) T2 corticectomy; (**c**) opening of the temporal horn of lateral ventricle to perform amygdalohippocampectomy; (**d**) surgical site after complete resection of mesial temporal structures; (**e–h**) are the same previously described stages, respectively, performed in a microanatomical laboratory

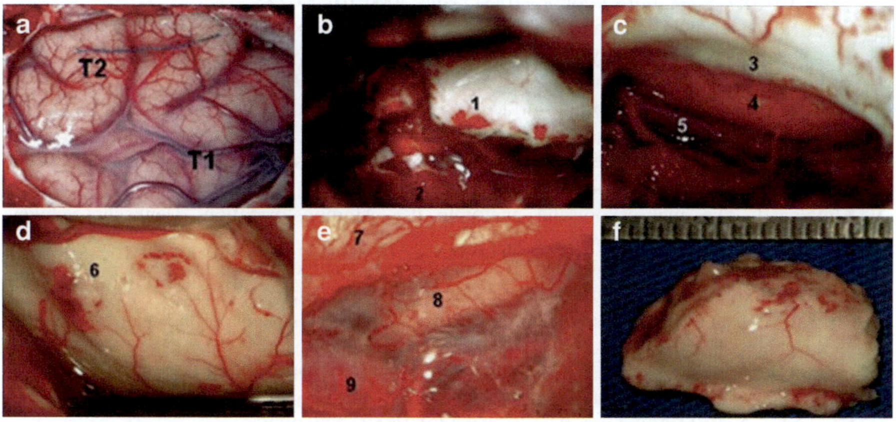

Fig. 25 Surgical stages of right-sided Niemeyer's selective amygdalohippocampectomy (except **a** (left side)). (**a**) Exposition of superior temporal gyrus; (**b**) resection of the amygdala after reaching the temporal horn of lateral ventricle; (**c**) opening of choroid fissure to expose vascular structures of the ambient cistern; (**d**) normal-sized hippocampus with its digitations; (**e**) preserved arachnoid and visualization of ambient and crural cistern structures (see-through); (**f**) en bloc hippocampal resection; 1—head of the hippocampus; 2—choroid plexus; 3—fimbria-fornix; 4—parahippocampus; 5—posterior cerebral artery; 6—hippocampal digitations; 7—cerebellar tentorium; 8—oculomotor nerve; 9—posterior cerebral artery (P2 segment). (Reprinted from Isolan et al. [63])

5 Conclusion

Although the different techniques applied in managing temporal lobe epilepsy have a similar degree of success, some may have more complications than others. Except for transsylvian and subtemporal techniques, all others have great potential of quadrantanopia due to optic radiation damage. Usually, this is, however, an incidental campimetry finding and not referred to by the patient as an issue. The subtemporal approach, although more anatomical, has the risk of injury to the vein of Labbé and, as a result, temporal lobe venous infarction. The transsylvian approach, also very anatomical, must be precisely performed through the planum polare to avoid damage on the IFOF in the temporal stem or even Meyer's loop. Overall, the best surgical approach for temporal lobe diseases is often the one with which the surgeon is best adapted.

References

1. Téllez-Zenteno JF, Hernández-Ronquillo L. A review of the epidemiology of temporal lobe epilepsy. Epilepsy Res Treat. 2012;2012:630853. https://doi.org/10.1155/2012/630853. Epub 2011 Dec 29.
2. Vellutini E, Garzon E, Inuzuka KM, Brock RS, Dal 'Col Lucio JE. Epilepsy surgery. In: Ramina R, Aguiar PH, Tatagiba M, editors. Samii's essentials in neurosurgery. Berlin: Springer-Verlag; 2014. p. 493–506.
3. Yasargil MG, Türe U, Yasargil DC. Impact of temporal lobe surgery. J Neurosurg. 2004;101(5):725–38. https://doi.org/10.3171/jns.2004.101.5.0725.
4. Gloor P. The temporal lobe and limbic system. New York: Oxford University Press; 1997.
5. Ono M, Kubik S, Abernathey CD. Atlas of the cerebral sulci. Stuttgart: Thieme; 1990.
6. Yasargil MG. Microneurosurgery IV A. Stuttgart: Thieme; 1994.
7. Duvernoy HM. The human brain, surface, blood supply, and three-dimensional sectional anatomy. 2nd ed. Vienna: Springer-Verlag; 1999.
8. Broca P. Anatomie comparée des circonvolutions cérébrales: le grand lobe limbique. Annu Rev Anthropol. 1878;1:385–498.
9. Klingler J. Die makroskopische anatomie der Ammonsformation. Denkschr Schweiz Naturforsch Ges. 1948;78:1–80.
10. Blackstad TW. Commissural connections of the hippocampal region in the rat, with special reference to their mode of termination. J Comp Neurol. 1956;105:417–537.
11. Retzius G. Das Menschenhirn—Studien in der makroskopischen morphologie. Stockholm: Kgl. Buchdr; 1896.
12. Isolan GR, Stefani MA, Schneider FL, Claudino HA, Yu YH, Choi GG, Telles JPM, Rabelo NN, Figueiredo EG. Hippocampal vascularization: proposal for a new classification. Surg Neurol Int. 2020;11:378. https://doi.org/10.25259/SNI_708_2020.
13. Erdem A, Yaşargil G, Roth P. Microsurgical anatomy of the hippocampal arteries. J Neurosurg. 1993;79(2):256–65. https://doi.org/10.3171/jns.1993.79.2.0256.
14. Marinković S, Milisavljević M, Puskas L. Microvascular anatomy of the hippocampal formation. Surg Neurol. 1992;37(5):339–49. https://doi.org/10.1016/0090-3019(92)90001-4.
15. Wen HT, Rhoton AL Jr, de Oliveira E, Cardoso AC, Tedeschi H, Baccanelli M, Marino R Jr. Microsurgical anatomy of the temporal lobe: part 1: mesial temporal lobe anatomy and its vascular relationships as applied to amygdalohippocampectomy. Neurosurgery. 1999;45(3):549–91; discussion 591–2. https://doi.org/10.1097/00006123-199909000-00028.
16. LeDoux JE, Damasio AR. Emotions and feelings. In: Kandel ER, Schwartz JH, Jessell TM, Siegelbaum SA, Hudspeth AJ, editors. Principles of neural science. 5th ed. New York: McGraw-Hill; 2013. p. 1079–94.
17. Bernasconi N, Bernasconi A, Caramanos Z, Antel SB, Andermann F, Arnold DL. Mesial temporal damage in temporal lobe epilepsy: a volumetric MRI study of the hippocampus, amygdala and parahippocampal region. Brain. 2003;126(Pt 2):462–9. https://doi.org/10.1093/brain/awg034.
18. Nagata S, Rhoton AL Jr, Barry M. Microsurgical anatomy of the choroidal fissure. Surg Neurol. 1988;30(1):3–59. https://doi.org/10.1016/0090-3019(88)90180-2.

19. Isolan GR, de Oliveira E, Recalde R. Estudo microanatômico da fissura coroidéia na abordagem dos ventrículos e cisternas cerebrais [Microanatomical study of the choroidal fissure in ventricular and cisternal approaches]. Arq Neuropsiquiatr. 2005;63(3B):801–6. Portuguese. https://doi.org/10.1590/s0004-282x2005000500015. Epub 2005 Oct 18.
20. Flechsig P. Anatomie des menschlichen Gehirns und Rückenmarks auf Myelogenetischer Grundlage. Leipzig: Thieme; 1896.
21. Meyer A. The connections of the occipital lobes and the present status of the cerebral visual affections. Trans Assoc Am Phys. 1907;22:7–1.
22. Cushing H. The field defects produced by temporal lobe lesions. Brain. 1922;44:341–96.
23. Tusa RJ, Ungerleider LG. The inferior longitudinal fasciculus: a reexamination in humans and monkeys. Ann Neurol. 1985;18(5):583–91. https://doi.org/10.1002/ana.410180512.
24. Choi C, Rubino PA, Fernandez-Miranda JC, Abe H, Rhoton AL Jr. Meyer's loop and the optic radiations in the transsylvian approach to the mediobasal temporal lobe. Neurosurgery. 2006;59(4 Suppl 2):ONS228–35; discussion ONS235–6. https://doi.org/10.1227/01.NEU.0000223374.69144.81.
25. Sincoff EH, Tan Y, Abdulrauf SI. White matter fiber dissection of the optic radiations of the temporal lobe and implications for surgical approaches to the temporal horn. J Neurosurg. 2004;101(5):739–46. https://doi.org/10.3171/jns.2004.101.5.0739.
26. Ribas EC, Yagmurlu K, Wen HT, Rhoton AL Jr. Microsurgical anatomy of the inferior limiting insular sulcus and the temporal stem. J Neurosurg. 2015;122(6):1263–73. https://doi.org/10.3171/2014.10.JNS141194. Epub 2015 Apr 10.
27. Párraga RG, Ribas GC, Welling LC, Alves RV, de Oliveira E. Microsurgical anatomy of the optic radiation and related fibers in 3-dimensional images. Neurosurgery. 2012;71(1 Suppl Operative):160–71; discussion 171–2. https://doi.org/10.1227/NEU.0b013e3182556fde.
28. Ribas EC, Yağmurlu K, de Oliveira E, Ribas GC, Rhoton A. Microsurgical anatomy of the central core of the brain. J Neurosurg. 2018;129(3):752–69. https://doi.org/10.3171/2017.5.JNS162897. Epub 2017 Dec 22.
29. Duffau H, Gatignol P, Moritz-Gasser S, Mandonnet E. Is the left uncinate fasciculus essential for language? A cerebral stimulation study. J Neurol. 2009;256(3):382–9. https://doi.org/10.1007/s00415-009-0053-9. Epub 2009 Mar 6.
30. Vanacôr CN, Isolan GR, Yu YH, Telles JPM, Oberman DZ, Rabelo NN, Figueiredo EG. Microsurgical anatomy of language. Clin Anat. 2021;34(1):154–68. https://doi.org/10.1002/ca.23681. Epub 2020 Sep 23.
31. Tate MC, Herbet G, Moritz-Gasser S, Tate JE, Duffau H. Probabilistic map of critical functional regions of the human cerebral cortex: Broca's area revisited. Brain. 2014;137(Pt 10):2773–82. https://doi.org/10.1093/brain/awu168. Epub 2014 Jun 25.
32. Duffau H. Stimulation mapping of white matter tracts to study brain functional connectivity. Nat Rev Neurol. 2015;11(5):255–65. https://doi.org/10.1038/nrneurol.2015.51. Epub 2015 Apr 7.
33. Duffau H. Toward the application of the hodotopical concept to epilepsy surgery. World Neurosurg. 2011;75(3-4):431–3. https://doi.org/10.1016/j.wneu.2010.12.012.
34. De Benedictis A, Duffau H. Brain hodotopy: from esoteric concept to practical surgical applications. Neurosurgery. 2011;68(6):1709–23; discussion 1723. https://doi.org/10.1227/NEU.0b013e3182124690.
35. Baciu M, Perrone-Bertolotti M. What do patients with epilepsy tell us about language dynamics? A review of fMRI studies. Rev Neurosci. 2015;26(3):323–41. https://doi.org/10.1515/revneuro-2014-0074.
36. Yasargil MG. Microneurosurgery I. Stuttgart: Thieme; 1984.
37. Wen HT, Rhoton AL Jr, de Oliveira E, Castro LH, Figueiredo EG, Teixeira MJ. Microsurgical anatomy of the temporal lobe: part 2—sylvian fissure region and its clinical application. Neurosurgery. 2009;65(6 Suppl):1–35; discussion 36. https://doi.org/10.1227/01.NEU.0000336314.20759.85.
38. Yasargil MG. Microneurosurgery IV B. Stuttgart: Thieme; 1996.

39. Yasargil MG, Wieser HG. Selective microsurgical resection. In: Wieser HG, Elger CE, editors. Presurgical evaluation of epileptics. New York: Springer-Verlag; 1987. p. 352–60.
40. Yasargil MG, Krisht AF, Türe U. Microsurgery of insular gliomas: part II: opening of the sylvian fissure. More Contemp Neurosurg. 2002;24(12):1–5.
41. Isolan GR, Aguiar PHP, Aires R, Meister C, Stefani MA. Middle cerebral artery pseudotetrafurcation: anatomic report and review of middle cerebral artery variations. Neurosurg Q. 2010;20(4):284–7.
42. Horsley V. Brain surgery. Br Med J. 1886;2(1345):670–7. https://doi.org/10.1136/bmj.2.1345.670.
43. Al-Otaibi F, Baeesa SS, Parrent AG, Girvin JP, Steven D. Surgical techniques for the treatment of temporal lobe epilepsy. Epilepsy Res Treat. 2012;2012:374848. https://doi.org/10.1155/2012/374848. Epub 2012 Mar 22.
44. Yeni SN, Tanriover N, Uyanik O, Ulu MO, Ozkara C, Karaağaç N, Ozyurt E, Uzan M. Visual field defects in selective amygdalohippocampectomy for hippocampal sclerosis: the fate of Meyer's loop during the transsylvian approach to the temporal horn. Neurosurgery. 2008;63(3):507–13; discussion 513–5. https://doi.org/10.1227/01.NEU.0000324895.19708.68.
45. Van Gompel JJ, Welker KM. Visual field mapping to prevent visual field deficits in epilepsy surgery: seeing the problem. Neurology. 2014;83(7):578–9. https://doi.org/10.1212/WNL.0000000000000701. Epub 2014 Jul 11.
46. Rubino PA, Rhoton AL Jr, Tong X, de Oliveira E. Three-dimensional relationships of the optic radiation. Neurosurgery. 2005;57(4 Suppl):219–27; discussion 219–27. https://doi.org/10.1227/01.neu.0000176415.83417.16.
47. Yogarajah M, Focke NK, Bonelli S, Cercignani M, Acheson J, Parker GJ, Alexander DC, McEvoy AW, Symms MR, Koepp MJ, Duncan JS. Defining Meyer's loop-temporal lobe resections, visual field deficits and diffusion tensor tractography. Brain. 2009;132(Pt 6):1656–68. https://doi.org/10.1093/brain/awp114. Epub 2009 May 21.
48. Chen X, Weigel D, Ganslandt O, Buchfelder M, Nimsky C. Prediction of visual field deficits by diffusion tensor imaging in temporal lobe epilepsy surgery. NeuroImage. 2009;45(2):286–97. https://doi.org/10.1016/j.neuroimage.2008.11.038. Epub 2008 Dec 16.
49. Borius PY, Roux FE, Valton L, Sol JC, Lotterie JA, Berry I. Can DTI fiber tracking of the optic radiations predict visual deficit after surgery? Clin Neurol Neurosurg. 2014;122:87–91. https://doi.org/10.1016/j.clineuro.2014.04.017. Epub 2014 May 5.
50. Von Der Heide RJ, Skipper LM, Klobusicky E, Olson IR. Dissecting the uncinate fasciculus: disorders, controversies and a hypothesis. Brain. 2013;136(Pt 6):1692–707. https://doi.org/10.1093/brain/awt094. Epub 2013 May 6.
51. Kucukyuruk B, Richardson RM, Wen HT, Fernandez-Miranda JC, Rhoton AL Jr. Microsurgical anatomy of the temporal lobe and its implications on temporal lobe epilepsy surgery. Epilepsy Res Treat. 2012;2012:769825. https://doi.org/10.1155/2012/769825. Epub 2012 May 21.
52. Gonçalves-Ferreira A, Miguéns J, Farias JP, Melancia JL, Andrade M. Selective amygdalohippocampectomy: which route is the best? An experimental study in 80 human cerebral hemispheres. Stereotact Funct Neurosurg. 1994;63(1-4):182–91. https://doi.org/10.1159/000100313.
53. Park TS, Bourgeois BF, Silbergeld DL, Dodson WE. Subtemporal transparahippocampal amygdalohippocampectomy for surgical treatment of mesial temporal lobe epilepsy. Technical note J Neurosurg. 1996;85(6):1172–6. https://doi.org/10.3171/jns.1996.85.6.1172.
54. Burchiel KJ, Zacest AC, Spencer D. Selective amygdalohippocampectomy. In: Winn HR, editor. Youmans neurological surgery. 6th ed. Philadelphia: Elsevier; 2011.
55. Isolan GR, Azambuja N, Paglioli Neto E, Paglioli E. Anatomia microcirúrgica do hipocampo na Amígdalo-hipocampectomia seletiva sob a perspectiva da técnica de Niemeyer e método pré-operatório para maximizar a corticotomia [Hippocampal microsurgical anatomy regarding the selective amygdalohippocampectomy in the Niemeyer's technique perspective and pre-operative method to maximize the corticotomy]. Arq Neuropsiquiatr. 2007;65(4A):1062–9. Portuguese. https://doi.org/10.1590/s0004-282x2007000600031.

56. Kuang Y, Yang T, Gu J, Kong B, Cheng L. Comparison of therapeutic effects between selective amygdalohippocampectomy and anterior temporal lobectomy for the treatment of temporal lobe epilepsy: a meta-analysis. Br J Neurosurg. 2014;28(3):374–7. https://doi.org/10.3109/02688697.2013.841854. Epub 2013 Oct 7.
57. Josephson CB, Dykeman J, Fiest KM, Liu X, Sadler RM, Jette N, Wiebe S. Systematic review and meta-analysis of standard vs selective temporal lobe epilepsy surgery. Neurology. 2013;80(18):1669–76. https://doi.org/10.1212/WNL.0b013e3182904f82. Epub 2013 Apr 3.
58. Hu WH, Zhang C, Zhang K, Meng FG, Chen N, Zhang JG. Selective amygdalohippocampectomy versus anterior temporal lobectomy in the management of mesial temporal lobe epilepsy: a meta-analysis of comparative studies. J Neurosurg. 2013;119(5):1089–97. https://doi.org/10.3171/2013.8.JNS121854. Epub 2013 Sep 13.
59. Hemb M, Palmini A, Paglioli E, Paglioli EB, Costa da Costa J, Azambuja N, Portuguez M, Viuniski V, Booij L, Nunes ML. An 18-year follow-up of seizure outcome after surgery for temporal lobe epilepsy and hippocampal sclerosis. J Neurol Neurosurg Psychiatry. 2013;84(7):800–5. https://doi.org/10.1136/jnnp-2012-304038. Epub 2013 Feb 13.
60. Malikova H, Kramska L, Vojtech Z, Liscak R, Sroubek J, Lukavsky J, Druga R. Different surgical approaches for mesial temporal epilepsy: resection extent, seizure, and neuropsychological outcomes. Stereotact Funct Neurosurg. 2014;92(6):372–80. https://doi.org/10.1159/000366003. Epub 2014 Oct 28.
61. Witt JA, Coras R, Schramm J, Becker AJ, Elger CE, Blümcke I, Helmstaedter C. Relevance of hippocampal integrity for memory outcome after surgical treatment of mesial temporal lobe epilepsy. J Neurol. 2015;262(10):2214–24. https://doi.org/10.1007/s00415-015-7831-3. Epub 2015 Jul 3.
62. Wendling AS, Hirsch E, Wisniewski I, Davanture C, Ofer I, Zentner J, Bilic S, Scholly J, Staack AM, Valenti MP, Schulze-Bonhage A, Kehrli P, Steinhoff BJ. Selective amygdalohippocampectomy versus standard temporal lobectomy in patients with mesial temporal lobe epilepsy and unilateral hippocampal sclerosis. Epilepsy Res. 2013;104(1–2):94–104. https://doi.org/10.1016/j.eplepsyres.2012.09.007. Epub 2012 Sep 28.
63. Wendling AS, Steinhoff BJ, Bodin F, Staack AM, Zentner J, Scholly J, Valenti MP, Schulze-Bonhage A, Hirsch E. Selective amygdalohippocampectomy versus standard temporal lobectomy in patients with mesiotemporal lobe epilepsy and unilateral hippocampal sclerosis: post-operative facial emotion recognition abilities. Epilepsy Res. 2015;111:26–32. https://doi.org/10.1016/j.eplepsyres.2015.01.002. Epub 2015 Jan 16.

Surgical Anatomy of the Insula

Carlos Perez-Vega, Ricardo A. Domingo, Erik H. Middlebrooks, and Alfredo Quiñones-Hinojosa

1 Historic

The human brain's structural and functional anatomy is studied based on multiple morphological divisions, including lobes, regions, and compartments. A particular area of interest for neuroanatomists across history is the anatomy of the insular lobe. First described by Johann Christian Reil in 1809, it is a portion of the brain covered by the frontal, parietal, and temporal opercula in the depth of the Sylvian fissure [1]. The surgical approach for insular lesions and the complete anatomical exposure of the insula have been subject of discussion across the literature as current techniques for Sylvian fissure opening, the relationship with eloquent white matter tracts deep to the insula, and vascular structures make surgery in this area a high-risk procedure. This chapter describes the anatomy of the Sylvian fissure and insular lobe, with a special scope on surgical approaches and technical nuances.

1.1 Sylvian Fissure

The Sylvian fissure is the most important landmark on the lateral surface of the brain; it is located between the frontal, parietal, and temporal lobes [2]. The superficial portion of the fissure comprises three main sulci: the anterior horizontal,

C. Perez-Vega · R. A. Domingo · A. Quiñones-Hinojosa (✉)
Department of Neurologic Surgery, Mayo Clinic, Jacksonville, FL, USA
e-mail: Quinones-Hinojosa.Alfredo@mayo.edu

E. H. Middlebrooks
Department of Neurologic Surgery, Mayo Clinic, Jacksonville, FL, USA

Department of Radiology, Mayo Clinic, Jacksonville, FL, USA
e-mail: Middlebrooks.Erik@mayo.edu

E. G. Figueiredo et al. (eds.), *Brain Anatomy and Neurosurgical Approaches*, https://doi.org/10.1007/978-3-031-14820-0_7

anterior ascending, and posterior sulci, all beginning at the anterior Sylvian point [3]. Opening the anterior horizontal and ascending sulci provides access to the upper anterior portion of the insula; the remaining portion of the insular cortex can be accessed through the posterior sulcus [2]. The deep portion of the fissure is comprised of two compartments: sphenoidal and operculoinsular [2, 3]. The sphenoidal compartment can be found behind the sphenoidal ridge, and it is formed medially by the medial and proximal portions of the superior temporal gyrus and laterally by the posterior and lateral orbital gyri of the temporal lobe; this compartment contains the M1 segment of the middle cerebral artery (MCA), lenticulostriate arteries, and deep Sylvian veins [2–4]. The operculoinsular compartment is located deep to the superficial rami of the Sylvian fissure, and it can be further divided into the opercular and insular cleft; the opercular cleft is located between the frontal-parietal-temporal operculum, and the insular cleft can be found between the medial surface of the opercula and the insula [2, 3, 5].

1.2 The Insula

The insular cortex can be found deep within the Sylvian fissure and constitutes the medial wall of the operculoinsular compartment [2, 3, 6]. As previously mentioned, it is covered by portions of the frontal, parietal, and temporal lobes and separated from these by the circular sulcus (or "limiting sulcus") (Fig. 1a) [2, 7]. The anterior part of the lateral surface of the insula is covered by the frontal and parietal opercula, while the temporal operculum covers the inferior surface [2, 4, 7].

The insula is considered one of the least understood regions of the brain and has been described as a limbic or paralimbic integration cortex [8]. It plays an important

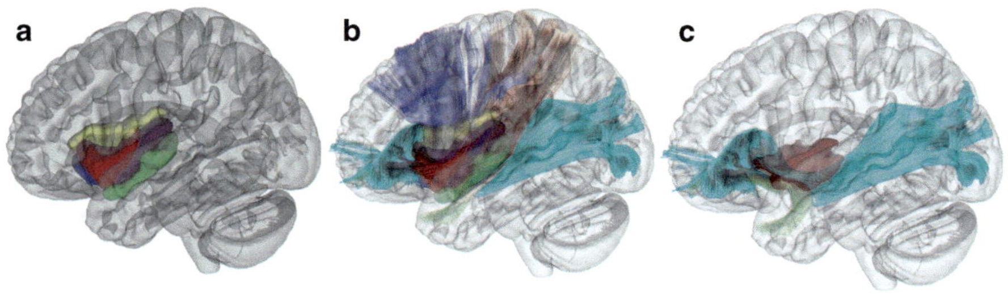

Fig. 1 3-Dimensional brain reconstruction of the insular cortex and related white matter tracts. (**a**) The left insula is showed deep to the Sylvian fissure; the anterior red portion represents the short gyri, while the long gyri are indicated in purple, which are divided by the central sulcus. The circular limiting sulcus surrounds the insula, consisting of the anterior limiting sulcus (blue), superior limiting sulcus (yellow), and inferior limiting sulcus (green). (**b**) White matter tracts deep to the insular cortex include the claustro-cortical fibers (blue), extreme capsule (orange), inferior fronto-occipital fasciculus (cyan), and uncinate fasciculus (green). (**c**) Anatomical representation of the claustrum (red) deep to the insular cortex. The inferior fronto-occipital fasciculus (cyan) and uncinate fasciculus (green) traverse through the ventral claustrum

role in cognitive processes, including language and speech, processing of auditory stimuli, autonomic cardiovascular regulation, and other visceromotor/viscerosensitive responses [8, 9]. A meta-analysis published by Kurth et al. evaluating 1800 functional brain images concluded the presence of four functionally distinct regions within the insula: a sensorimotor region in the mid-posterior insula, a central olfato-gustatory region, a socioemotional region located anterior-ventral, and a cognitive anterior-dorsal region [10].

The insula has a pyramid shape, with the tip pointing towards the base of the brain (Fig. 1a) [3, 11]. Its deepest sulcus is the central sulcus, which divides the insular surface into anterior and posterior portions. The anterior portion consists of the anterior, middle, and posterior short gyri, divided by the anterior and precentral sulci, respectively. The posterior portion consists of the anterior and posterior long gyri separated by the postcentral sulcus. The central sulcus typically aligns in the same anteroposterior plan as the Rolandic central sulcus [2, 3, 12]. Two other important landmarks of the insula include the pole and the apex. The pole is located at the converge of the three short gyri, at the anteroinferior edge of the insula; the apex is defined as the most prominent lateral part of the convexity, usually located in the middle short gyrus [2, 5].

The extreme capsule, claustrum, external capsule, and striatum are located deep to the central portion of the insular cortex (Fig. 1b, c). Most of the blood supply of the insula comes from perforators of the M2 and M3 segments of the MCA. The M1 segment usually bifurcates at the limen insulae, and the trunks of the M2 segment run along the insular cortex. The deep middle cerebral vein (MCV) is the predominant vein draining the insular cortex, while the superficial Sylvian vein (SSV) is the largest vein within the posterior ramus of the Sylvain fissure [13, 14].

1.3 Associated White Matter Tracts

In recent years, neurosurgeons have applied functional MRI and MRI-tractography to surgical decision-making in preoperative and intraoperative settings. Identifying white matter tracts and functional areas involved by brain lesions significantly improves the extent of resection and reduces potential postoperative complications or deficits [15, 16]. As shown in Fig. 1, the insula is intimately involved with the extreme capsule underneath it and the claustrum and external capsule. The inferior part of the extreme capsule is formed by the inferior fronto-occipital fasciculus (IFOF), which connects the frontal lobe with the temporal and occipital lobes; the posterior part of the capsule, the claustrum-cortical fibers, represents the connection between the dorsal part of the claustrum and the brain cortex. The uncinate fasciculus forms the anteroinferior part of the extreme and external capsule as it connects the fronto-opercular area to the temporal pole [17–19].

1.4 Insular Cortex and Associated Lesions

A variety of lesions can be found closely associated with the insula. These include low- and high-grade gliomas, arteriovenous malformation, Sylvian bifurcation aneurysm, cavernoma, cortical dysplasia, and cortical tubers (tuberous sclerosis) [20–22]. According to prior studies, 25% of low-grade gliomas and 10% of high-grade gliomas are located in this region [23]. Patients with perisylvian lesions usually present with seizure episodes demonstrating certain ictal features of the adjacent *operculae*: motor and somatosensory signs of the face, language alterations, laryngeal/pharyngeal signs, neurovegetative signs, among others [24].

Historically, two different surgical approaches to the insula have been identified: the transylvian and the transcortical approaches [25]. Two different pathological entities, a cerebral cavernous malformation and a glioma, are described below with specific technical nuances and surgical anatomy for lesion resection through these two surgical approaches.

2 Clinical Condition: Cerebral Cavernous Malformation

Cavernous malformation (CM) is defined as a cluster of dilated sinusoidal channels with no smooth muscle or elastic layer, and it is considered the most common vascular malformation in the brain. Prevalence across different studies ranges between 0.4% and 0.6%, and the annual risk of hemorrhage is estimated to range between 0.7% and 1.1% per lesion. Patients most likely present with new-onset seizures, and headaches are not uncommon. Surgical resection of CM of the insula consists of complete removal of the lesion, including the surrounding gliosis.

A case of a 34-year-old male is presented. The patient suffered from an episode of disorganized thoughts and gasping breaths that led to a generalized seizure; he was transferred to a medical facility where a thorough neurological examination was conducted. His past medical history was relevant for attention deficit hyperactivity disorder. He was started on levetiracetam, with some side effects characterized by short-term memory problems.

3 Physical Examination

The patient was alert and oriented to person, place, and time. Neurological examination was unremarkable with no evidence of cranial nerve deficits, speech alterations, or motor and sensitive changes.

4 Anatomy Characteristics

Brain MRI revealed a well-circumscribed hyperintense T2 lesion located anterior to the precentral sulcus of the left insula, with associated perilesional edema (Fig. 2). According to functional MRI (fMRI), spoken and non-spoken sentence completion

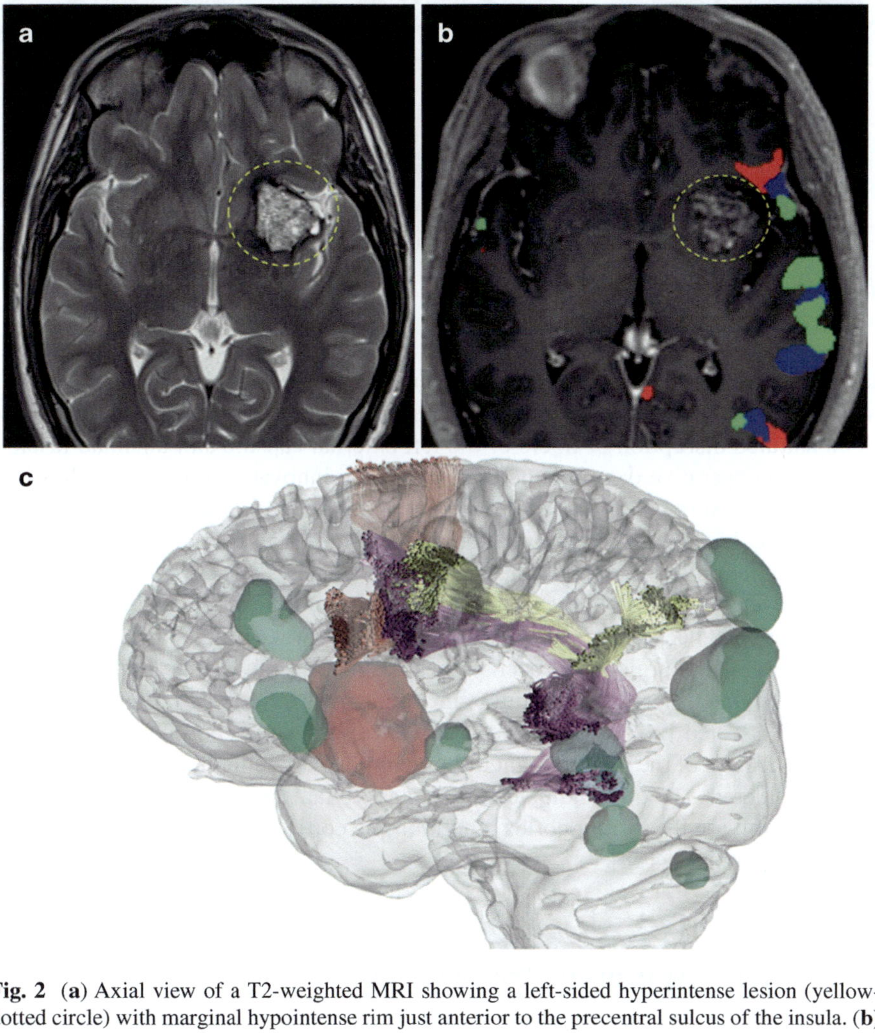

Fig. 2 (**a**) Axial view of a T2-weighted MRI showing a left-sided hyperintense lesion (yellow-dotted circle) with marginal hypointense rim just anterior to the precentral sulcus of the insula. (**b**) Axial view of a contrast-enhancing T1-weighted functional MRI showing the close proximity of this lesion (yellow-dotted circle) to opercular language areas in the dominant left frontal operculum and superior temporal gyrus. Colors blue and green indicate non-spoken and spoken sentence completion, respectively; color red demonstrates activation with semantic decision task. (**c**) 3-Dimensional reconstruction of the functional MRI showing the cavernous malformation (red) surrounded by relevant functional areas (green) and the underlying eloquent language tracts, including frontal aslant tract (red fibers), arcuate fasciculus (purple), and superior longitudinal fasciculus 3 (SLF3; yellow). (**d, e**) Axial view of a T2-weighted and T1 post-contrast MRI showing the extent of resection of the left insular cavernous malformation

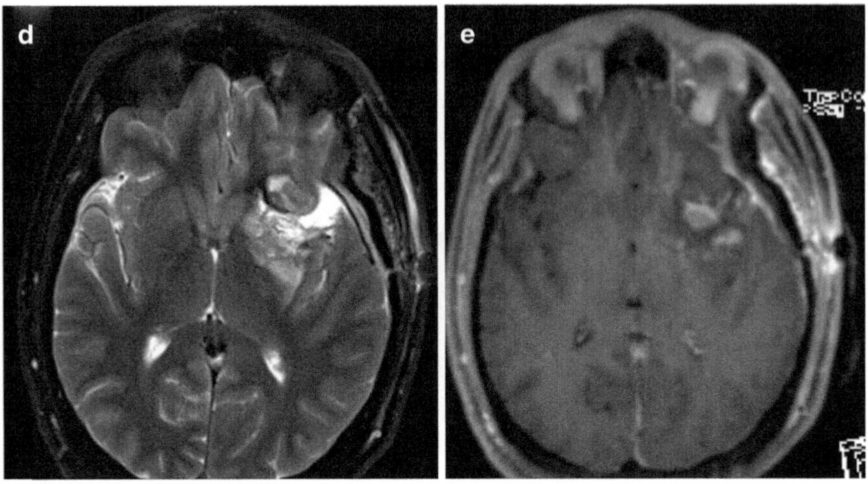

Fig. 2 (continued)

tasks show activation at the left frontal operculum and left superior temporal gyrus, sparing the precentral gyrus at the left subinsular area. Diffusion MRI tractography demonstrated disruption of fibers of the left external capsule and uncinate fasciculus. The preoperative data obtained through these innovative imaging studies help us to plan the surgical approach appropriately by determining language-dominant hemisphere, language-specific functional areas, and establishing safe surgical corridors for the transcortical and transsylvian approaches.

As proven on the fMRI, this lesion surrounded multiple eloquent areas in the left hemisphere that deemed this case highly complex. A wake brain mapping is considered the preferred surgical approach in these patients as the lesions are located deep into the Sylvian fissure, and multiple functional areas have to be preserved as brain tissue is dissected [26, 27]. The ideal surgical approach is chosen considering the specific location of the predominant tumor component within the insula. The patient was considered a good candidate for a transsylvian approach as the lesion was closely involving the insular cortex, and multiple cortical eloquent areas were identified preoperatively [28].

5 Anatomy Approach: Transsylvian

The patient underwent an awake, left-sided, pterional craniotomy for lesion resection; the pterional approach is preferred as it directly exposes the Sylvian fissure. The patient remained under conscious sedation during the first portion of the surgery. The bone flap is shaped and turned to expose the Sylvian fissure (Fig. 3a), and the dura is opened in a semicircular fashion (Fig. 3b). At this moment, the patient

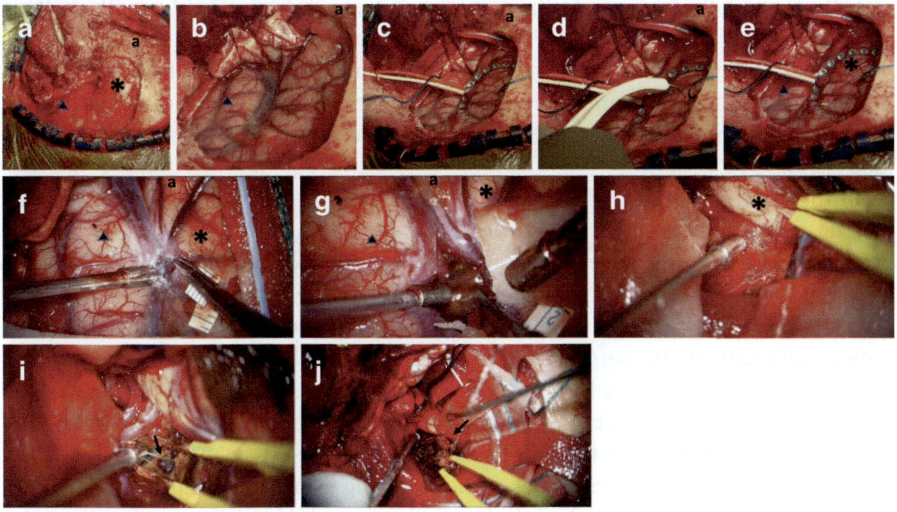

Fig. 3 Left-sided pterional craniotomy for insular cavernous malformation resection. (**a**) After the bone flap is shaped and turned, a portion of the dura overlying the frontal (asterisk) and temporal (arrowhead) lobes is exposed. (**b**) The dura is opened in a semicircular fashion in order to expose the frontal and temporal (arrowhead) lobes surrounding the Sylvian fissure. (**c, d**) 360° electrocorticography for intracranial monitoring of cortical electrical activity is used, alongside the Ojemann stimulator for brain mapping; as this area showed high functional activity on the preoperative fMRI, the role of awake brain surgery and mapping is highly important for function preservation and maximize extent of lesion resection. (**e**) Two positive functional areas for language were identified in the left frontal lobe. (**f**) Opening and dissection of the Sylvian fissure. Under the microscope, the Sylvian fissure is opened by releasing arachnoid adhesions between the frontal (asterisk) and temporal (arrowhead) lobes, and carefully dissecting vascular structures. (**g**) The Sylvian fissure is opened following branches of the middle cerebral arteries into deeper portions of the fissure. (**h**) The insular cortex had a yellowish discoloration due to the vascular nature of this lesion. (**i, j**) The insular cortex is mapped and opened, and the cavernous malformation is identified (arrow). Dissection is carried with the bipolar cautery, preserving a layer of hemosiderin ring around the lesion to ensure complete resection. a, indicates the anterior portion of the craniotomy

is awakened and asked to perform multiple tasks, including language and motor exercises, according to preoperative neurocognitive testing. The brain was mapped with the Ojemann stimulator from 2 up to 6 mA to identify eloquent cortex specific to speech function as indicated by the preoperative fMRI; this has to be carefully performed until the functional areas are fully characterized (Fig. 3c–e). A small window negative for a function was found anteriorly surrounding the Sylvian fissure. Two large veins are found to run parallel to the fissure and are mobilized superiorly/inferiorly to the temporal and frontal side of the fissure using hydrodissection to enlarge the arachnoid space and open the Sylvian fissure (Fig. 3f). Veins of the fissure should be preserved, but sometimes further dissection requires some sacrifice [28].

Arachnoid adhesions are released using sharp scissors, and at this moment, an artery is usually identified. In this case, an arterial branch of the M3 was found.

These arteries help to separate the frontal and temporal lobes and usually help to define the dissection plane [28]. Progressive splitting of arachnoid adhesions is carried down following this artery into the Sylvian fissure until anterior adhesions between the frontal and temporal lobes are released (Fig. 3g). This dissection plane becomes easier in deeper portions of the fissure as larger arteries separate the two lobes [29].

In this particular case, after a generous opening of the Sylvian fissure is completed, it was noted that the frontal lobe was overriding the temporal lobe; this is usually one of the most adherent areas of the fissure, and spreading dissection from inside-out is the best technique for opening these tissue planes [28]. In the Sylvian cistern, dissection is carried deep to the M2 branches to the M1 segment; this is a critical segment of the surgery as arteries and veins have to be unscrambled to approach the insular cortex [28, 30]. Dissection of these segments is followed deep beneath the operculum; alternating between deep and superficial artery dissection helps open the operculum from both sides. When these two planes of dissection meet, the insular cortex is revealed. In general, the transsylvian approach involves opening the superficial and deep Sylvian cisterns while protecting the opercular arteries and perforators and preserving superficial Sylvian veins [28, 30, 31].

The insular cortex had a yellowish discoloration due to the vascular nature of the lesion and its proximity to the surface (Fig. 3h). Dissection of the cavernoma is carried with the bipolar cautery first superficially and then followed by the anterior surface (Fig. 3i, j). Subcortical motor and language mapping is continued to avoid injuring adjacent eloquent areas and allowing the surgeon to push the extent of resection to the limit. A layer of hemosiderin ring is preserved around the cavernoma to ensure complete lesion resection and preservation of surrounding brain tissue. Long perforators coming off the MCA posteriorly and supplying the corona radiata must be preserved to avoid potential complications. Proximally, lateral lenticulostriate arteries supplying the internal capsule and basal ganglia can also supply tumors around this area. These arteries are considered the most medial limit of lesion resection [28]. Hemostasis is achieved once the lesion is resected, the dura is closed watertight, and the bone flap is placed.

6 Clinical Condition: Glioma

Gliomas are the most common primary intra-axial brain tumors in adults. The incidence ranges between 4.7 and 5.7 per 100,000 persons [32]. Neurosurgical advances in the last few years, including brain mapping, confocal microscopy, 5-Aminolevulinic acid, and optical coherence tomography (OCT), have improved the extent of resection in patients harboring intra-axial tumors, further impacting the quality of life and overall survival [33].

A case of a 47-year-old female is presented. She first experienced migraines, odd smells, and staring spells that prompted a thorough neurological evaluation; an initial MRI scan showed a right-sided insular lesion, and the patient underwent surgical intervention for biopsy at an outside institution. The lesion was consistent with a WHO Grade 2 astrocytoma. She presents for further surgical planning after 1 month of the initial biopsy.

7 Physical Examination

The patient was alert and oriented to person, place, and time. Neurological examination was unremarkable with no evidence of cranial nerve deficits, speech alterations, or motor and sensitive changes.

8 Anatomy Characteristics

A well-defined intra-axial lesion is identified at the right insula and adjacent superior temporal gyrus. A central contrast-enhancing component has considerably increased in size compared to previous examinations; this illustrates the possibility of malignant transformation as this enhancing component demonstrates the presence of some necrosis. Hyperintense FLAIR changes indicating tumor infiltration were also identified infiltrating the frontal, parietal, and temporal lobes (Fig. 4).

The extent of resection of insular gliomas depends on multiple variables, including morphological characteristics, anatomic location, and surrounding eloquent areas [4, 21, 34]. To understand further the complexity of this type of lesions, an anatomic classification of insular gliomas was introduced by Sanai et al. named the Berger-Sanai zone classification. In this classification, gliomas are divided into four zones depending on the location of the majority of the tumor: above the Sylvian fissure and anterior to the Foramen of Monro is classified as Zone 1; Zone 2 represents lesions located posterior to the Foramen of Monro and above the Sylvian fissure; Zones 3 and 4 are located below the Sylvian fissure, posterior and anterior to the Foramen of Monro, respectively [35]. In subsequent studies of the Berger-Sanai Classification, the greatest extent of resection is usually accomplished in Zones 1 and 4, and the smallest extent of resection is usually associated with Zone 2 [4, 21].

As noted in the preoperative MRI, the biggest component of this lesion was in zones 1 and 2, increasing the level of complexity of surgical intervention. Due to the extent of involvement of the frontal, parietal, and temporal lobe, it was decided to proceed with an awake craniotomy for cortical and subcortical mapping to identify and preserve eloquent areas on the right hemisphere. The morphological characteristics of the lesion, including the involvement of multiple lobes, oriented the decision to proceed with a transcortical approach for tumor resection [36].

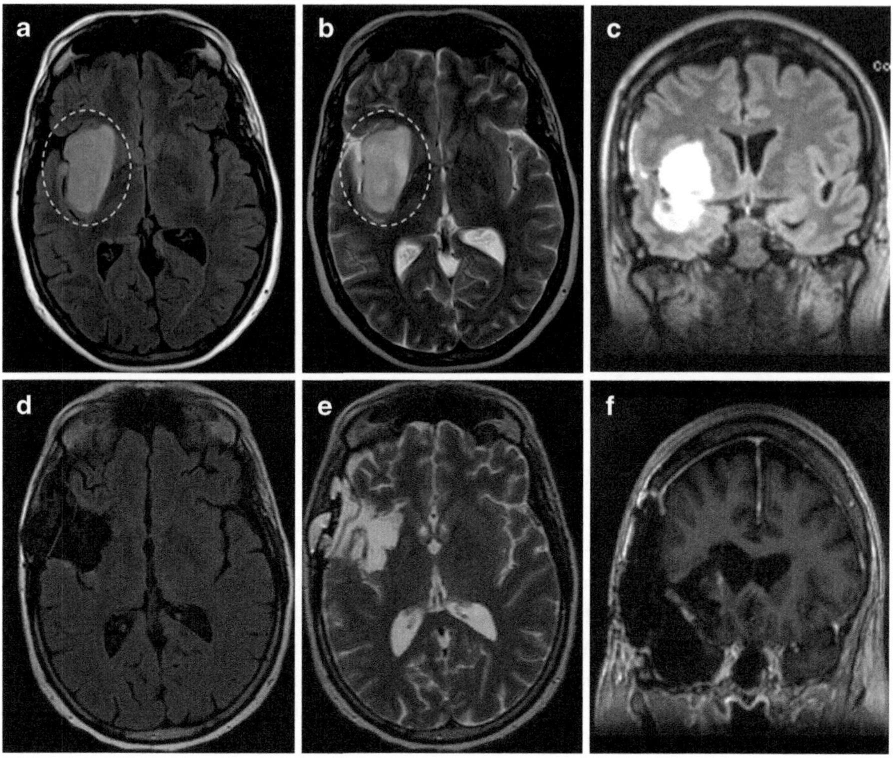

Fig. 4 (**a, b**) Axial views of a T2-FLAIR and T2-weighted MRI showing a right-sided hyperintense lesion (yellow-dotted circle) and adjacent superior temporal gyrus. (**c**) Coronal view of a T2-FLAIR MRI showing a right-insular hyperintense lesion extending into the right temporal lobe. (**d, e**) Axial view of a T2 FLAIR and T2-weighted MRI showing extent of resection of the right-insular glioma. (**f**) Coronal view of a T1 post-contrast MRI showing extent of resection of the right-temporal component of the tumor

9 Anatomy Approach: Transcortical

The patient underwent an awake, right-sided, pterional craniotomy for tumor resection. She was brought to the operative room and positioned supine, with the head slightly rotated to the left and fixed with a three-pin head clamp. The anesthesiologist conducted induction and kept the patient under conscious sedation after performing a bilateral scalp block. The previous incision was extended inferiorly and anteriorly, allowing greater access to the anterior frontal operculum and temporal floor. The cautery is used to dissect down, and a myocutaneous flap is mobilized anteriorly and posteriorly. Screws and plates from previous surgery and the bone flap were removed; another flap was formed anteriorly to allow for better exposure to the temporal tip and frontal operculum [36, 37].

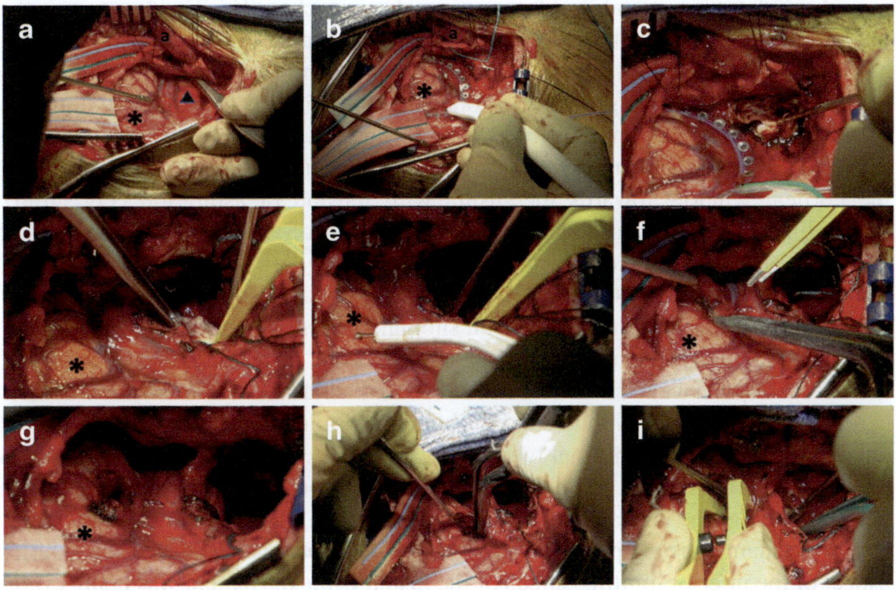

Fig. 5 Right-sided pterional craniotomy for insular glioma resection. (**a**) The frontal (asterisk) and temporal (arrowhead) lobes surrounding the Sylvian fissure are exposed after dural opening. (**b**) The 360o electrocorticography is again used alongside the Ojemann stimulator for brain mapping before proceeding with corticectomy. (**c, d**) Temporal lobectomy. Due to the extension of this tumor to the temporal lobe, the first portion of the surgery involved an anterior temporal lobectomy. (**e**) Transcortical approach for insular glioma resection. The right frontal lobe is mapped in order to identify a specific point for corticectomy. (**f**) The cortex is opened with the aid of the bipolar cautery. (**g**) Image showing the frontal and temporal windows to the insula. (**h**) The cortical window is further expanded towards the insula with the aid of aspiration and bipolar cautery. (**i**) Supra- and infra-sylvian resection cavities can be connected for better extent of tumor resection. a, indicates the anterior portion of the craniotomy; Asterisk, frontal lobe

Before dural opening, the patient is awakened and instructed to follow commands ahead of brain mapping procedures. The dura is opened and tacked up with sutures (Fig. 5a). Functional mapping of the area is conducted by direct stimulation from 2 to 3 to 4 mA in order to identify the lowest current threshold reproducing a functional response, at the same time that the patient performs multiple activities including number counting, executive function with alternation of letters and numbers, as well as identifying pictures and correlation (Fig. 5b) [33]. In this specific case, a positive motor speech deficit was elicited in the pre-motor region.

Due to the extension of the tumor to the temporal lobe, it was decided to start the approach with a right temporal corticectomy (Fig. 5c). First, a temporal dissection is performed, and a corticectomy is done through the middle temporal gyrus; this plane is further dissected down inferiorly to the temporal pole. The temporal lobectomy is accomplished, including the removal of the uncus (Fig. 5d). For a transcortical approach to the insula, three different cortical windows can be created:

trans-frontal, trans-parietal, and transtemporal approaches [36]. The decision on the best surgical approach is tailored based on tumor morphology, functional mapping, and surgical anatomy. In this case, a trans-frontal approach was performed due to the extent of the remaining lesions in zones 1 and 2 [35]. The frontal operculum was stimulated, and no responses were obtained (Fig. 5e). The cortex is opened with a combination of bipolar cautery and sharp dissection, further creating and extending the cortical window towards the insula with the aid of aspiration (Fig. 5f–h). Depending on the extent of the tumor, supra- and infra-Sylvian resection cavities or windows can be connected to aid tumor resection (Fig. 5i).

The insula is then cauterized with bipolar cautery and entered. The resection of the tumor is usually performed through subpial dissection, avoiding Sylvian vessel branches [36]. This lesion was noted to be fibrous and difficult to dissect, requiring the use of bipolar cautery to work around the insula and lenticulostriate vessels carefully; the use of the bipolar cautery should be limited to abnormal areas to avoid any potential damage to normal brain tissue. Tumor debulking was achieved with a combination of resection and electrical stimulation for subcortical mapping due to the proximity of the insula to deep white matter tracts, including motor and somatosensory thalamocortical pathways. The lesion was dissected down to the inferior aspect of the inferior limiting sulcus of the insula while the patient was being evaluated with multiple tasks. At this point, an intraoperative MRI was obtained, revealing residual lesion at the superior aspect of the tumor, and a decision was made to continue resecting the remaining component. In close relation to the posterior limb of the internal capsule, the most posterior portion of the tumor was considered high risk, and a decision was made not to dissect this area. The surgical cavity was irrigated, and hemostasis was achieved. The dura was closed, and the two bone flaps were approximated using a Y-shaped connector. The temporalis fascia was closed with interrupted sutures, and the skin was closed in layers.

After further pathological analysis, the lesion was consistent with an anaplastic astrocytoma (WHO grade III), and the patient received treatment with radiation therapy and concomitant temozolomide.

10 Conclusion

Surgical intervention for insular lesions is challenging due to their location deep to the Sylvian fissure, adjacent to highly eloquent areas. The ideal surgical approach is based on lesion location and preoperative imaging examination, including fMRI and DTI sequences. Awake craniotomy for cortical and subcortical mapping is preferred to identify and preserve the language, motor, and sensory functions.

References

1. Cunha-Cabral D, Silva SM, Alves H, Vaz RP, Pereira PA, Andrade JP. Neurosurgical anatomy of the insular cortex. Clin Neurol Neurosurg. 2019;186:105530.
2. Tanriover N, Rhoton AL Jr, Kawashima M, Ulm AJ, Yasuda A. Microsurgical anatomy of the insula and the sylvian fissure. J Neurosurg. 2004;100(5):891–922.
3. Bykanov AE, Pitskhelauri DI, Dobrovol'skiy GF, Shkarubo MA. Surgical anatomy of the insular cortex. Zh Vopr Neirokhir Im N N Burdenko. 2015;79(4):48–60.
4. Benet A, Hervey-Jumper SL, Sánchez JJ, Lawton MT, Berger MS. Surgical assessment of the insula. Part 1: surgical anatomy and morphometric analysis of the transsylvian and transcortical approaches to the insula. J Neurosurg. 2016;124(2):469–81.
5. Ribas EC, Yağmurlu K, de Oliveira E, Ribas GC, Rhoton A. Microsurgical anatomy of the central core of the brain. J Neurosurg. 2018;129(3):752–69.
6. Türe U, Yaşargil DC, Al-Mefty O, Yaşargil MG. Topographic anatomy of the insular region. J Neurosurg. 1999;90(4):720–33.
7. Warwick R, Williams PL. Gray's anatomy. Philadelphia: W. B. Saunders; 1973.
8. Uddin LQ, Nomi JS, Hébert-Seropian B, Ghaziri J, Boucher O. Structure and function of the human insula. J Clin Neurophysiol. 2017;34(4):300–6.
9. Stephani C, Fernandez-Baca Vaca G, Maciunas R, Koubeissi M, Lüders HO. Functional neuroanatomy of the insular lobe. Brain Struct Funct. 2011;216(2):137–49.
10. Kurth F, Zilles K, Fox PT, Laird AR, Eickhoff SB. A link between the systems: functional differentiation and integration within the human insula revealed by meta-analysis. Brain Struct Funct. 2010;214(5-6):519–34.
11. Guenot M, Isnard J, Sindou M. Surgical anatomy of the insula. Adv Tech Stand Neurosurg. 2004;29:265–88.
12. Duffau H, Capelle L, Lopes M, Faillot T, Sichez JP, Fohanno D. The insular lobe: physiopathological and surgical considerations. Neurosurgery. 2000;47(4):801–10. discussion 10-1.
13. Gibo H, Carver CC, Rhoton AL Jr, Lenkey C, Mitchell RJ. Microsurgical anatomy of the middle cerebral artery. J Neurosurg. 1981;54(2):151–69.
14. Marrone AC, Severino AG. Insular course of the branches of the middle cerebral artery. Folia Morphol (Praha). 1988;36(3):331–6.
15. Henderson F, Abdullah KG, Verma R, Brem S. Tractography and the connectome in neurosurgical treatment of gliomas: the premise, the progress, and the potential. Neurosurg Focus. 2020;48(2):E6.
16. O'Donnell LJ, Suter Y, Rigolo L, Kahali P, Zhang F, Norton I, et al. Automated white matter fiber tract identification in patients with brain tumors. Neuroimage Clin. 2017;13:138–53.
17. Çavdar S, Aydın AE, Algın O, Aydın S. The complex structure of the anterior white commissure of the human brain: fiber dissection and Tractography study. World Neurosurg. 2021;147:e111–e7.
18. Muftah Lahirish IA, Middlebrooks EH, Holanda VM, Batista-Quintero R, Maeda FL, Neto MR, et al. Comparison between transcortical and interhemispheric approaches to the atrium of lateral ventricle using combined white matter fiber dissections and magnetic resonance tractography. World Neurosurg. 2020;138:e478–e85.
19. Yagmurlu K, Vlasak AL, Rhoton AL Jr. Three-dimensional topographic fiber tract anatomy of the cerebrum. Neurosurgery. 2015;11(Suppl 2):274–305; discussion.
20. Eseonu CI, ReFaey K, Garcia O, Raghuraman G, Quinones-Hinojosa A. Volumetric analysis of extent of resection, survival, and surgical outcomes for insular Gliomas. World Neurosurg. 2017;103:265–74.

21. Hervey-Jumper SL, Li J, Osorio JA, Lau D, Molinaro AM, Benet A, et al. Surgical assessment of the insula. Part 2: validation of the Berger-Sanai zone classification system for predicting extent of glioma resection. J Neurosurg. 2016;124(2):482–8.
22. Hinojosa J, Gil-Robles S, Pascual B. Clinical considerations and surgical approaches for low-grade gliomas in deep hemispheric locations: insular lesions. Childs Nerv Syst. 2016;32(10):1875–93.
23. Duffau H, Capelle L. Preferential brain locations of low-grade gliomas. Cancer. 2004;100(12):2622–6.
24. Chevrier MC, Bard C, Guilbert F, Nguyen DK. Structural abnormalities in patients with insular/peri-insular epilepsy: spectrum, frequency, and pharmacoresistance. AJNR Am J Neuroradiol. 2013;34(11):2152–6.
25. Hervey-Jumper SL, Berger MS. Insular glioma surgery: an evolution of thought and practice. J Neurosurg. 2019;130(1):9–16.
26. Quiñones-Hinojosa A, Ojemann SG, Sanai N, Dillon WP, Berger MS. Preoperative correlation of intraoperative cortical mapping with magnetic resonance imaging landmarks to predict localization of the Broca area. J Neurosurg. 2003;99(2):311–8.
27. Walker JA, Quiñones-Hinojosa A, Berger MS. Intraoperative speech mapping in 17 bilingual patients undergoing resection of a mass lesion. Neurosurgery. 2004;54(1):113–7; discussion 8.
28. Safaee MM, Englot DJ, Han SJ, Lawton MT, Berger MS. The transsylvian approach for resection of insular gliomas: technical nuances of splitting the Sylvian fissure. J Neuro-Oncol. 2016;130(2):283–7.
29. Türe U, Yaşargil MG, Al-Mefty O, Yaşargil DC. Arteries of the insula. J Neurosurg. 2000;92(4):676–87.
30. Rey-Dios R, Cohen-Gadol AA. Technical nuances for surgery of insular gliomas: lessons learned. Neurosurg Focus. 2013;34(2):E6.
31. Simon M, Neuloh G, von Lehe M, Meyer B, Schramm J. Insular gliomas: the case for surgical management. J Neurosurg. 2009;110(4):685–95.
32. Gousias K, Markou M, Voulgaris S, Goussia A, Voulgari P, Bai M, et al. Descriptive epidemiology of cerebral gliomas in Northwest Greece and study of potential predisposing factors, 2005–2007. Neuroepidemiology. 2009;33(2):89–95.
33. Oberheim Bush NA, Hervey-Jumper SL, Berger MS. Management of glioblastoma, present and future. World Neurosurg. 2019;131:328–38.
34. Quinones-Hinojosa A, Raza SM, Ahmed I, Rincon-Torroella J, Chaichana K, Olivi A. Middle temporal gyrus versus inferior temporal gyrus transcortical approaches to high-grade astrocytomas in the mediobasal temporal lobe: a comparison of outcomes, functional restoration, and surgical considerations. Acta Neurochir Suppl. 2017;124:159–64.
35. Sanai N, Polley MY, Berger MS. Insular glioma resection: assessment of patient morbidity, survival, and tumor progression. J Neurosurg. 2010;112(1):1–9.
36. Roux A, Dezamis E, Trancart B, Pallud J. How I do it: trans-cortical approach for insular diffuse glioma. Acta Neurochir. 2020;162(12):3025–30.
37. Przybylowski CJ, Baranoski JF, So VM, Wilson J, Sanai N. Surgical morbidity of transsylvian versus transcortical approaches to insular gliomas. J Neurosurg. 2019;132(6):1731–8.

Surgical Anatomy of the Cerebellum

Jander Moreira Monteiro and Gustavo Rassier Isolan

1 Introduction

The cerebellum is part of the central nervous system located dorsally to the brainstem. It has an important role in movement coordination and equilibrium. Its fibers connect it to the brain, and spinal cord, passing through its peduncles to the brainstem. There are three peduncles: the superior cerebellar peduncle, which connects it to the midbrain; the middle cerebellar peduncle connects the cerebellum and the pons; and the inferior cerebellar peduncle connects the cerebellum to the medulla oblongata [1]. Besides the fourth ventricle, formed between the cerebellum and the pons, the peduncles form cisterns between them and the brainstem where the cranial nerves and important vessels pass, making surgical approaches to this region very challenging [1].

Superiorly, the cerebellum is separated from the inferior face of the occipital lobe by the cerebellar tentorium (*tentorial cerebelli*), which extends from the tentorial incisura (tentorial notch) to the transverse venous sinus and the temporal and occipital bones. It is laterally and posteriorly bounded by the petrous and mastoid portions of the temporal bone, a small part of the mastoid angle of the parietal bone, and the occipital bone. It has a close relation, inferiorly, to the foramen magnum, through which it can herniate in some conditions, as in intracranial hypertension. The cerebellum has three surfaces: tentorial, suboccipital, and petrosal, related to the tentorium, occipital, and petrosal bones, respectively [1].

J. M. Monteiro · G. R. Isolan (✉)
The Center for Advanced Neurology and Neurosurgery (CEANNE), Porto Alegre, RS, Brazil

Paraná Mackenzie Evangelic University, Curitiba, PR, Brazil

E. G. Figueiredo et al. (eds.), *Brain Anatomy and Neurosurgical Approaches*,
https://doi.org/10.1007/978-3-031-14820-0_8

2 Anatomical Divisions

The cerebellum can be divided into three parts: a single midline portion – the vermis, and two lateral masses – the hemispheres. The vermis is bounded anteriorly by the brainstem and laterally and posteriorly by the hemispheres, in an embracing disposition. As the hemispheres are larger than the vermis, they form a depression, the anterior and posterior cerebellar incisures.

Both the vermis and the hemispheres are divided by fissures in lobules. These lobules have no functional importance, but may be used as surgical landmarks. The fissures are horizontal and continuum in the whole cerebellar surface, although the lobules are named differently in the vermis and the hemispheres. The vermis lobules are lingula, central lobe, culmen, declive, folium, tuber, pyramid, uvula, and nodule from the most rostral to the most caudal. From the most rostral to the most caudal, the cerebellar hemisphere lobules are wings of the central lobule, anterior part of the quadrangular lobule, posterior part of the quadrangular lobule, superior semilunar lobule, inferior semilunar lobule, biventral lobule, tonsils, and the floculli. These lobules' disposition and the fissures that separate them will be better viewed in each corresponding surface [1].

3 Cerebellar Surfaces

The cerebellum surfaces are didactically divided according to the structure it faces. The tentorial surface faces the cerebellar tentorium, going from the tentorial incisure to the transverse venous sinus. It is retracted in a supra-cerebellar infratentorial approach (Fig. 1). The suboccipital surface extends between the transverse and sigmoid venous sinuses bilaterally and is exposed in a suboccipital craniotomy. The petrosal surface is located between the sigmoid venous sinus lateral and posterior and the brainstem, medial and anterior. Its exposure permits access to the cerebellar-pontine angle [1].

3.1 Tentorial Surface

The cerebellar tentorium shapes the tentorial surface. It goes from an anterior-medial portion, the highest point of the cerebellum called the apex, downward to the posterior-lateral border, at the same level as the transverse sinus. This sinus is the limit between the tentorial and suboccipital surfaces. It can be divided into postero-lateral, which is the largest part and has close relation to the transverse venous sinus, and posteromedial part, smaller comprising the cerebellar falx and posterior cerebellar incisure. The anterior border, which separates the tentorial and petrosal

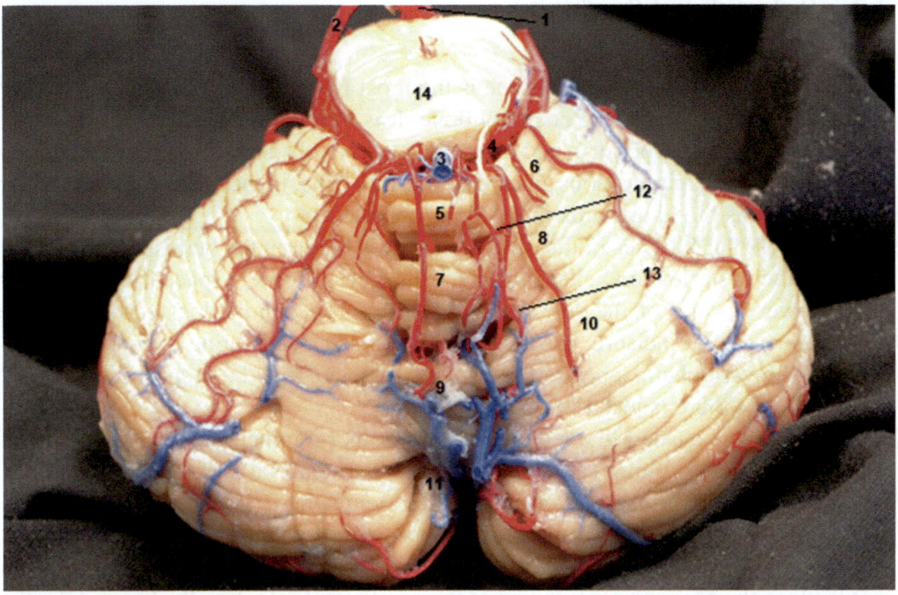

Fig. 1 Cerebellar tentorial surface. Notice that the cerebellar anatomical divisions are not well-defined (this distinction is even harder in the surgical field). (*1*) Basilar artery; (*2*) superior cerebellar artery; (*3*) precentral cerebellar vein; (*4*) anterior cerebellar incisure; (*5*) culmen; (*6*) quadrangular lobule; (*7*) declive; (*8*) simple lobule; (*9*) folium; (*10*) superior semilunar lobule; (*11*) posterior cerebellar incisure; (*12*) primary fissure; (*13*) post-clival fissure; (*14*) midbrain

surfaces, can also be divided into two parts: anterolateral and anteromedial. The anterolateral part of the anterior border is parallel to the petrous venous sinus. The anteromedial part faces the midbrain and forms the posterior wall of the fissure between the cerebellum and the midbrain, the cerebellar-mesencephalic fissure (Fig. 1) [1].

This surface has a smooth transition between the vermis and the hemispheres due to shallow vermohemispherics fissure, which separates these structures. As it goes downwards, especially in the suboccipital surface, this separation becomes more prominent, creating a posterior cerebellar incisure space, where lies the cerebellar falx. In the anterior portion, the "embrace" of the hemispheres to the vermis forms the anterior cerebellar incisure.

The cerebellar lobules that face the cerebellar tentorium are rostral to dorsal, the culmen and the quadrangular lobule, the declive and the simple lobule, and the folium superior r lobule, in the vermis and the hemisphere, respectively. The fissures that divide those lobules are the primary fissure, between culmen and declive, and the post-clival fissure, between declive and folium. The primary fissure also divides the tentorial surface into two parts: anterior and posterior. The tentorial fissures go from the midline to the anterolateral border of this surface and are a continuum with petrosal surface fissures [1].

3.2 Suboccipital Surface

The cerebellum's suboccipital surface is the most frequent approach to a cerebellar tumor, as it directly faces the occipital bone. It is limited by the transverse and sigmoid venous sinus, superiorly and laterally, respectively, and the foramen magnum inferiorly. The occipital venous sinus lies in the midline, in the base of the cerebellar falx, and drains blood upwards to the sinus confluence.

The cerebellar falx lies in the cerebellar recess between the hemispheres and the vermis – the posterior cerebellar incisure. This recess grows in space from the upper to the lower portion of the suboccipital surface. The vermis, located deep in the posterior cerebellar incisure, has a well-defined fissure separating it from the hemispheres, the vermohemispherics fissure. The opposite may be seen in the tentorial surface, where there is just a smooth transition.

The vermis is shaped like a diamond due to its prominent lobule, the pyramid. Above that lies the folium (in the transition to the tentorial surface) and the tuber. Downwards, the uvula and its corresponding hemispheric lobules, the tonsils, are arranged in a similar way to the oropharynx – from which they are named after. Right below this oropharynx-like structure lies the *vallecula cerebelli*, the recess that leads to the Magendie's foramen to the fourth ventricle. Finally, the nodule, the most caudal lobule, lies internally to the uvula and is not seen on this surface [1].

The vermis and hemispheric lobules of the suboccipital surface are the folium and the superior semilunar lobules, the tuber and the inferior semilunar lobules, the pyramid, and the biventral lobules, and finally, the uvula and the tonsils. This surface is divided into superior and inferior parts by the suboccipital fissure. This fissure is divided into vermian and hemispheric parts, although it is a continuum through the vermohemispheric fissure. The vermian part, called a pre-pyramidal fissure, separates the tuber and the pyramid, and the hemispheric part, called the pre-biventral fissure, separates the inferior semilunar biventral lobules. The other two fissures of this surface are the petrosal fissure, which comes from the petrosal surface and marks the limit between folium and superior semilunar lobules to the tuber and inferior semilunar lobules; and the tonsillobiventral fissure, which separates the biventral lobules and the tonsils [1].

The tonsils are pediculated structures projecting themselves to the caudal opening of the fourth ventricle. They are hemispheric ovoid-shaped lobules connected to the rest of the cerebellar hemisphere by a white matter bundle in their lateral part called the tonsillar peduncle. Besides the peduncle, the other surfaces are free. The inferior pole and posterior surface face the cisterna magna (where it herniates in intracranial hypertension or posterior fossa malformations). The superior part of the tonsils is divided from the medulla oblongata by the cerebellomedullary fissure. The tonsils face each other medially and are separated by the *vallecula cerebelli*. The ventral surface of the tonsils faces the tela choroidea, inferior medullary vellum, and nodule, which makes the lower half of the roof of the fourth ventricle. The superior pole of the tonsils faces the nodule, medially, and the biventral lobule, laterally. They are separated by the telovelotonsilar cleft [1].

3.3 Petrosal Surface

The petrosal surface faces the petrous bone, the brainstem, and the fourth ventricle. The hemispheric parts face the anterior bone wall, the petrous bone (Figs. 2 and 3). The vermian part has a midline depression, the anterior cerebellar incisure, which fits the brainstem, and the fourth ventricle. The fourth ventricle divides the vermis into rostral and caudal parts. Due to its presence, not all the hemispheres' lobules are connected by a continuous vermian part [1].

The lobules in the superior part are the lingula, the central lobule, and culmen, and the inferior lobules are the nodule and the uvula. The hemispheric surface is formed by the wings of the central lobule, the quadrangular lobule, the superior semilunar lobule, the inferior semilunar lobule, the biventral lobule the tonsil, and the floccule – all the hemispheric lobules can be seen in this surface. The superior and inferior semilunar lobules and the biventral lobules do not have a related vermian part on this surface. The most important fissure in this surface is the petrosal fissure, a continuum to the suboccipital surface. It divides the petrosal surface into superior and inferior, passing between the superior and inferior semilunar lobules [1].

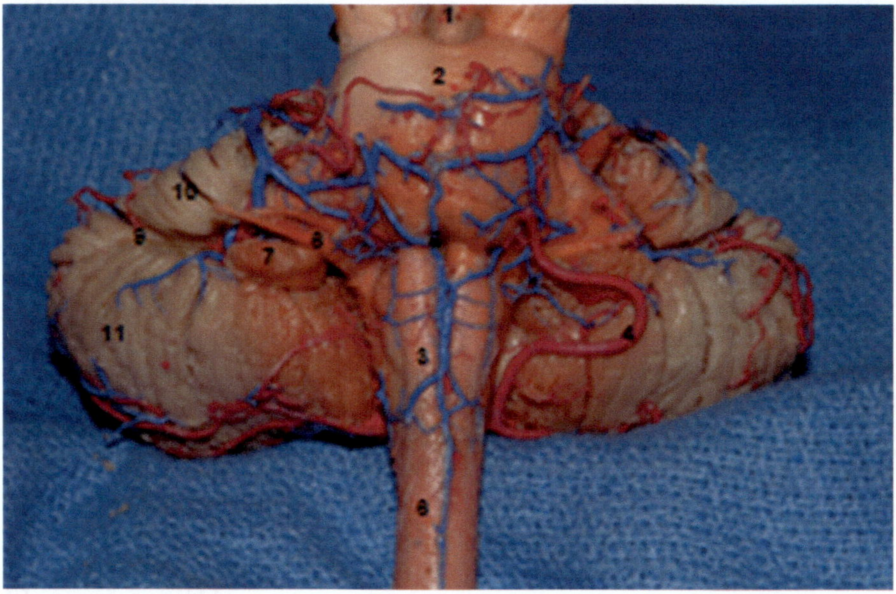

Fig. 2 Anterior view of the brainstem and the petrosal surface of the cerebellum. (*1*) Midbrain; (*2*) pons; (*3*) medulla oblongata; (*4*) posterior inferior cerebellar artery; (*5*) anterior pontomesencephalic–anterior medullary venous system; (*6*) spine cord; (*7*) flocculus; (*8*) VII–VIII cranial nerves; (*9*) petrosal fissure; (*10*) superior part of the petrosal surface; (*11*) inferior part of the petrosal surface

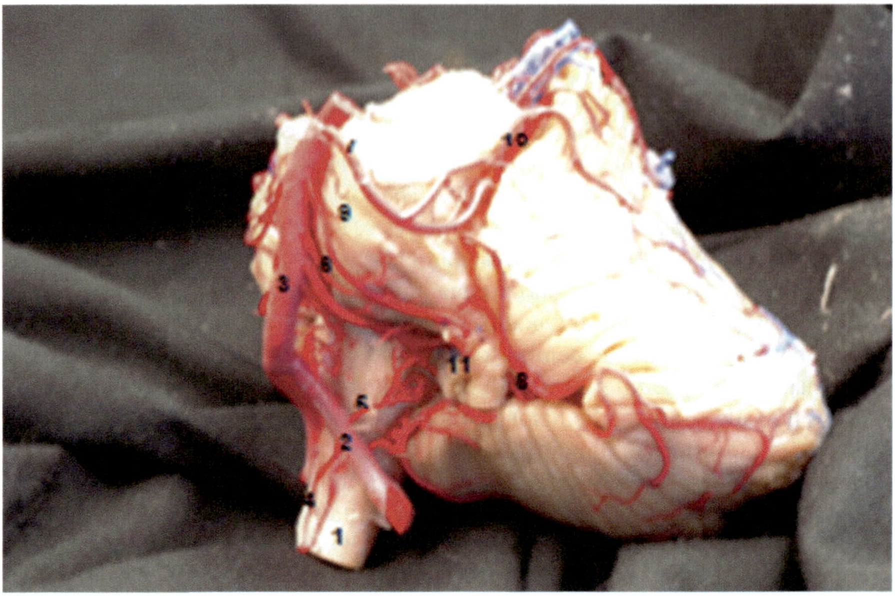

Fig. 3 Lateral view of the petrosal surface of the cerebellum. (*1*) Spinal cord; (*2*) vertebral artery; (*3*) basilar artery; (*4*) anterior spinal artery; (*5*) posterior inferior cerebellar artery; (*6*) anterior inferior cerebellar artery; (*7*) superior cerebellar artery; (*8*) petrosal fissure; (*9*) pons; (*10*) cerebellar-mesencephalic fissure; (*11*) flocculus

4 Cerebellar Nuclei and Fiber

There are four pair of cerebellar nuclei in the cerebellar white matter: dentate, emboliform, fastigial, and globose nuclei.

The fastigial nuclei are medial and have a close relation to the roof of the fourth ventricle. The dentate nuclei are bigger and more easily seen in anatomical specimens. Between the two lies the globose and emboliform nuclei.

The cerebellum has afferent and efferent fibers. The afferent fibers, arriving basically from the vestibular nuclei (pons and medulla oblongata), the pontine nuclei, and the spinal cord, make synapses in the cerebellar cortex, according to its function. The Purkinje cells in the cerebellar cortex make synapses to its corresponding cerebellar nuclei, from where the efferent fibers leave the cerebellum to its final synapses.

5 Surgical Corridors

In order to reach posterior cranial fossa structures and their many possible diseases, we must find surgical paths around the cerebellum. Even when a lesion (tumor) is located within the cerebellar parenchyma, knowing the way around the cerebellum should make its surgical access less likely to cause a neurological deficit, as one could plan the best-suited corticectomy, which could provide the best attack angle to the tumor and the best chance of complete resection. Therefore, anatomical knowledge of each cerebellar surface and its relations is fundamental in posterior cranial fossa surgery.

5.1 Telovelar Approach

The telovelar approach is very anatomical, as it does not enter the cerebellar parenchyma to reach the fourth ventricle. The patient must be positioned in a way that exposes the occipital region (semi-sitting, prone, or ¾ prone positions). Intraoperative neuromonitoring is important. A midline skin incision is made between the inium and the C2 spinous process, followed by a subperiosteal dissection of the soft-tissue plane. Exposing the asterium, bilaterally, the inium, superiorly, and the border of the foramen magnum inferiorly will provide an adequate view to the posterior fossa. Then, a midline suboccipital craniectomy is performed. Some authors suggest a C1–C2 laminectomy to achieve a better approach angle to the fourth ventricle, but that is not always necessary. Durotomy is usually Y-shaped [2, 3].

Using a surgical microscope, the arachnoid is opened sharply from the cisterna magna until the cerebellomedullary fissure is identified between cerebellar tonsils and the spinal cord. The posterior inferior cerebellar artery (PICA) must be freed from the arachnoid in the cerebellomedullary fissure bilaterally, and perforating arteries must be preserved. Dissection begins at the inferior edge of the tonsils, extending to the medial part and the adjacent uvula border so that the cerebellar tonsils can be mobilized laterally to expose the tela choroidea lower half of the roof of the fourth ventricle. The uvula can also be retracted superiorly. Next, the tela choroidea is sectioned from the Magendie's foramen up to the junction with the inferior medullary velum. This exposes the superolateral recess and provides access to the entire floor of the fourth ventricle. Big tumors can distort the region's anatomy and even protrude from the foramina of the fourth ventricle when large enough (Fig. 4) [2, 3].

After tumor resection (which exceeds this chapter's purpose), vigorous hemostasis is necessary due to the risk of postoperative bleeding and obstructive hydrocephaly. Brainstem manipulation itself can lead to swelling and obstructive hydrocephaly. It may be wise to use a postoperative external ventricular drain in the supratentorial compartment of the ventricular system to prevent postoperative morbidity and death. Watertight dural closure is mandatory to avoid CSF leakage and cranioplasty (preferably with autologous bone).

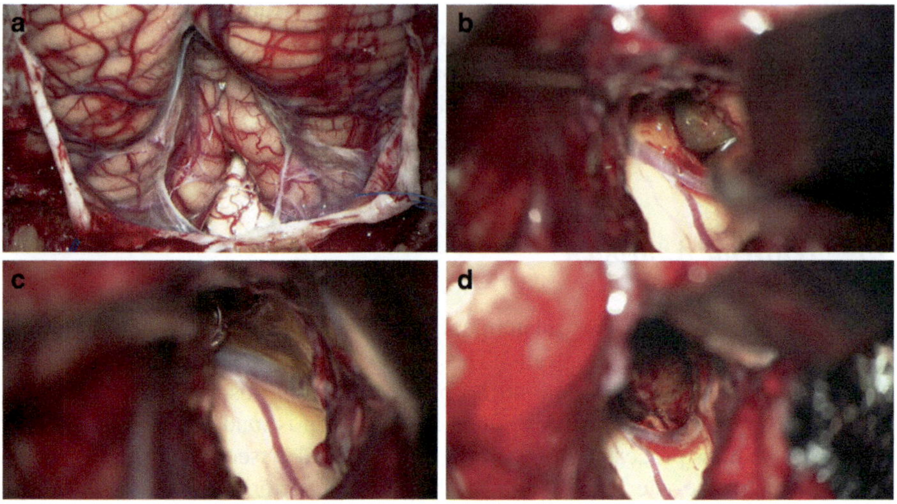

Fig. 4 Telovelar approach for a brainstem cavernous malformation. (**a**) Suboccipital craniotomy exposing the whole suboccipital surface of the cerebellum; (**b**) after dissection of the tonsils and the uvula, the view to the floor of the fourth ventricle; (**c**) Corticectomy and lesion dissection within the brainstem; (**d**) complete cavernous malformation resection

5.2 Retrosigmoid Approach

The retrosigmoid (or lateral suboccipital) approach is the most commonly used approach for cerebellar-pontine angle (CPA) tumors. It allows good exposure of the brainstem, cranial nerves (IV to XII), and when dealing with vestibular schwannomas, hearing, and facial motricity preservation.

The patient must be positioned with the retroauricular/mastoid region exposed. ¾ prone or semi-sitting positions are most used, but even a supine position is possible (if the patient has good neck mobility). Intraoperative neuromonitoring is crucial when treating tumors such as vestibular schwannomas, as mimic facial preservation is one of the main goals.

A retroauricular C-shaped (or inverted-J-shaped) skin incision is made, 2–3 cm behind the ear. Subperiosteal soft-tissue dissection is performed to expose the asterion. Classically, the asterion represents the intracranial junction between the transverse and sigmoid sinus, although it is not a precise landmark. Craniotomy starts at the asterion, extending medially and inferiorly, no more than 5 cm. The transverse and sigmoid sinus borders and their junction must be visible so that a durotomy can be made just parallel to them (C-shaped).

Using the surgical microscope, CSF must be drained (foramen magnum or CPA cisterns) so that the cerebellum can relax. A preoperative external lumbar drain is also a good strategy. With proper relaxation, the cerebellum is gently mobilized medially to expose the CPA nerves and vessels, as well as the CPA tumors. The durotomy should be more or less caudal, depending on the surgery's purpose. The superior petrosal vein (Dandy's vein) can hinder the approaching by blocking the view to the upper CPA region and limiting cerebellar retraction. Although its sectioning is usually harmless, authors suggest its preservation (Fig. 5).

After tumor resection, vigorous hemostasis is necessary. As all posterior fossa approaches, watertight dural closure is crucial to prevent CSF leakage. Cranioplasty with bone cement is made when an autologous bone is drilled in craniotomy, although the latter is always preferable.

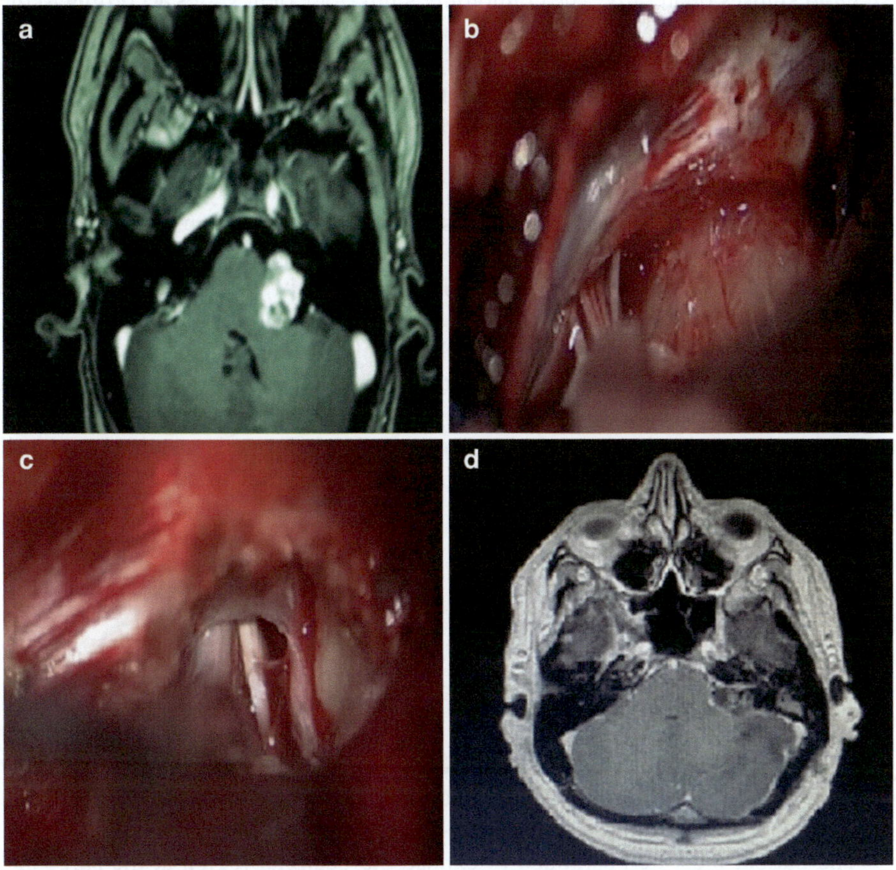

Fig. 5 Vestibular schwannoma in the left cerebellar-pontine angle (CPA). (**a**) Axial T1-weighted gadolinium-enhanced image showing vestibular schwannoma in the left CPA; (**b**) retrosigmoid approach. The cerebellum is mobilized medially. Large tumor distorts the region, but the lower cranial nerves are visible; (**c**) complete resection of the tumor, with opened internal acoustic meatus and identification of the facial nerve in the internal acoustic meatus; (**d**) postoperative axial T1-weighted gadolinium-enhanced image showing total tumor resection

5.3 Supracerebelar Infratentorial Approach

This approach provides an anatomical corridor to the pineal region, quadrigeminal plate, ambiens cistern, and surroundings.

The patient is preferably positioned in a semi-sitting position so that gravity can help "retract" the cerebellum. Besides, a good midline reference is helpful. Like the telovelar approach, an inium-to-C2 spinal process skin incision is made. The subperiosteal soft-tissue dissection is limited to the occipital bone, as there is no need to expose C1 and C2 posterior arch. The craniotomy must expose the transverse and sigmoid sinuses bilaterally. The craniotomy's inferior extent varies according to the need for CSF drainage in the foramen magnum cistern. An external lumbar drain may be welcome. C-shaped, curvilinear, or even straight durotomy can be performed, always near the inferior border of the transverse sinus. The dura-mater and transverse sinus can be retracted superiorly with simple sutures to enhance the surgeon's view [4, 5].

Using the microscope, the arachnoid attached to the cerebellum, and the tentorial edge are dissected to expose the tentorial surface of the cerebellum. Some veins also run from the cerebellar tentorial surface to the cerebellar tentorium and stand between the surgeon's view and the pineal region. It is always preferable to preserve those veins, but the small ones can be sectioned without any harm. For midline lesions in the pineal region, the precentral cerebellar vein can also obstruct the surgeon's view. However, in this case, a paramedian or lateral approach, rather than a purely median approach, is best suited to the vessel to be preserved (Figs. 6 and 7).

As in all posterior fossa surgeries, watertight dural closure is mandatory, as well as cranioplasty. In addition, postoperative care must be aware of the risk of hydrocephaly.

5.4 Transcortical Approach

When the tumor lies within the cerebellar parenchyma, we must perform a cortical approach. Due to cerebellar ontogenesis (biological development), the anatomical and functional divisions have no correspondence. That said, a corticectomy in the cerebellum will not lead to irreversible neurological deficits (cerebellar function is spread throughout the ontogenic divisions).

The patient must be positioned to expose the occipital region at the surgeon's preference. A midline inium-to-C2 spine process incision is made, although a smaller paramedian skin incision is an option for more lateral lesions. In this case, durotomy must be suited for the surgical objective, with no need to expose the venous sinuses or the CSF cisterns. Corticectomy must be as nearest to the lesion as possible.

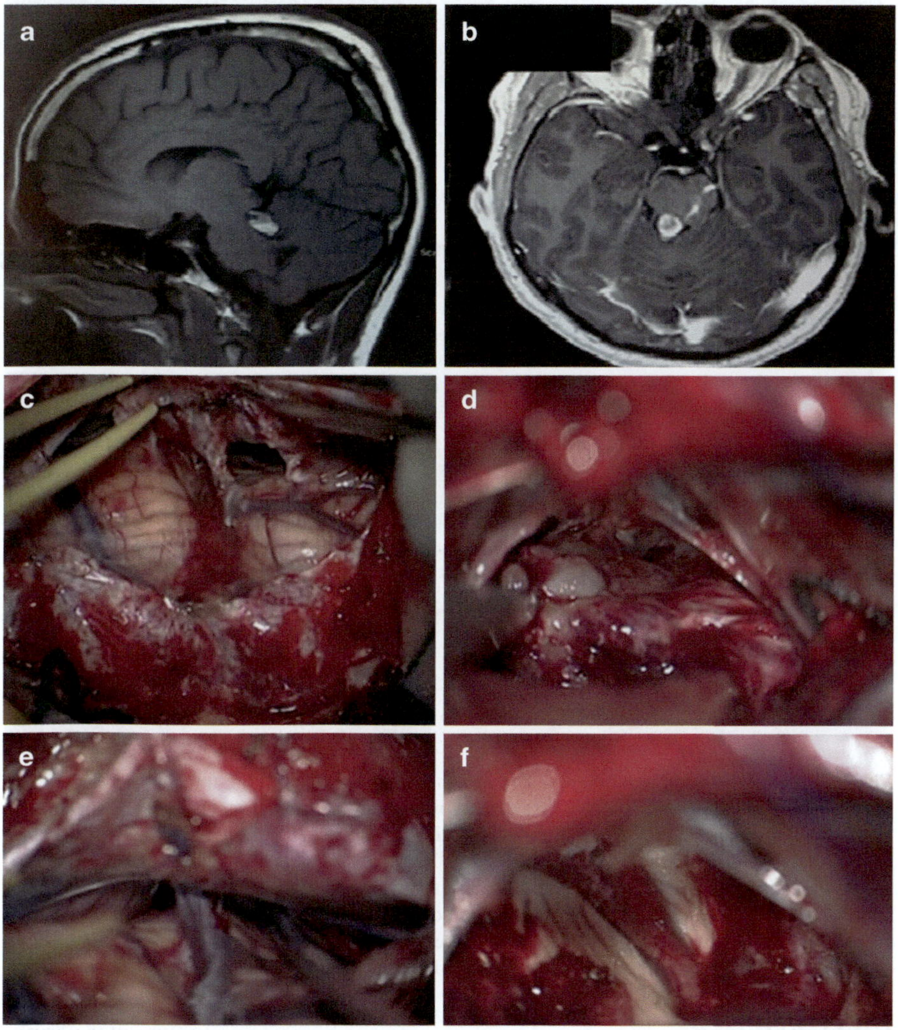

Fig. 6 Supracerebellar infratentorial approach for a pineal region tumor. (**a**) Sagittal T1-weighted gadolinium-enhanced image showing pineal region tumor with heterogeneous enhancement pattern; (**b**) axial T1-weighted gadolinium-enhanced image, same tumor; (**c**) "C-shaped" dural opening. CSF already drained with external lumbar drain; (**d**) Dissection of cerebellar tentorial surface to reach pineal tumor; (**e**) exposure of pineal tumor; (**f**) complete tumoral resection

As with all posterior fossa surgeries, vigorous hemostasis and watertight dural closure are mandatory. However, this approach carries considerable risk of postoperative swelling and obstructive hydrocephaly. Therefore, in some cases, external ventricular drain in the supratentorial ventricular system is welcome.

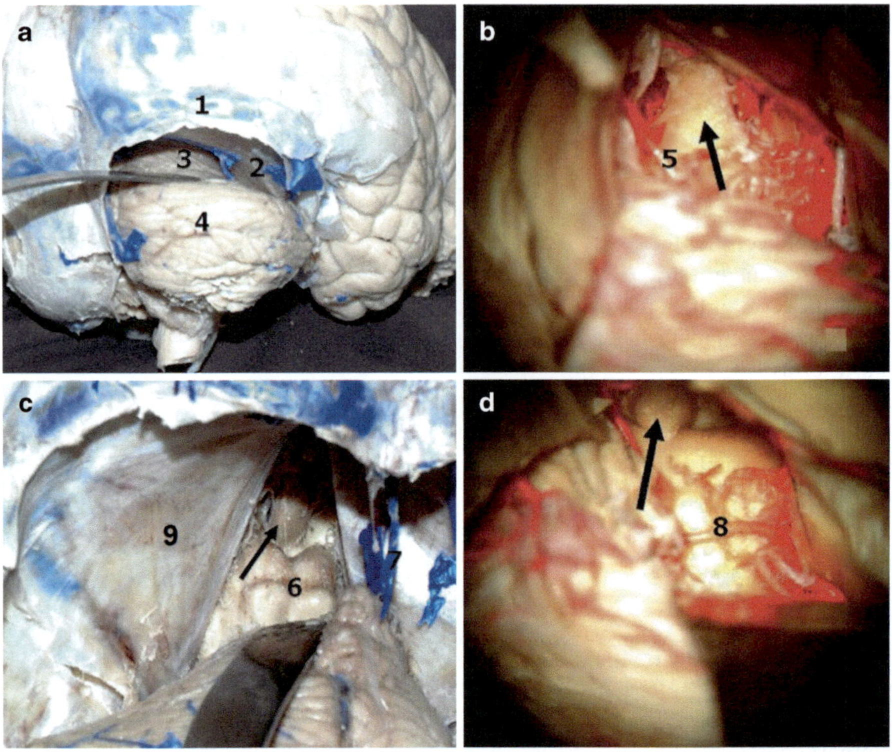

Fig. 7 Cadaveric dissection of the pineal region in a supracerebellar infratentorial approach (SIA). (**a**) macroscopic view of a posterior fossa durotomy with an inferior cerebellar retraction, exposing the cerebellar tentorial surface; (**b**) microscopic view of a midline SIA, showing the pineal gland; (**c**) macroscopic view of an SIA, exposing the quadrigeminal plate and the pineal gland; (**d**) microscopic view of an SIA with paramedian pineal gland exposure. (*1*) Transverse sinus; (*2*) cerebellar tentorium; (*3*) tentorial cerebellar surface; (*4*) suboccipital cerebellar surface; (*5*) medial posterior choroid artery; (*6*) quadrigeminal plate; (*7*) cerebellar bridging veins; (*8*) lateral posterior choroid artery; (*9*) cerebellar tentorium; (*pointer*) pineal gland

6 Conclusion

The cerebellar anatomical knowledge is very important to any surgeon who treats posterior fossa diseases. Knowing the way around the cerebellum is crucial for safer surgery, meaning a smaller chance of postoperative neurological deficits and hydrocephaly. Authors suggest extensive microsurgical laboratory training as the only acceptable way to prepare the surgeon to treat posterior fossa diseases, as the near relation between the brainstem, the posterior fossa vessels, and the cerebellum makes every mistake potentially catastrophic.

As the surgery itself, postoperative care is fundamental. Close attention to hydrocephaly and CSF leakage is mandatory, as both can turn the results of a very anatomical surgery into an unfortunate disabled patient. Posterior fossa surgery is a very illustrative example that neurosurgery is not restricted to the surgical act itself.

References

1. Rhoton AL Jr. Cranial anatomy and surgical approaches. Philadelphia: Lippincott Williams & Wilkins; 2003.
2. Tomasello F, Conti A, Angileri FF, Cardali S. Telo-velar approach to fourth-ventricle tumors: how I do it. Acta Neurochir. 2015;157(4):607–10. https://doi.org/10.1007/s00701-015-2358-z.
3. Liu JK, Dodson VN. Telovelar approach for microsurgical resection of fourth ventricular subependymoma arising from rhomboid fossa: operative video and technical nuances. Neurosurg Focus. 2019;1(2):V5.
4. La Pira B, Sorenson T, Quillis-Quesada V, Lanzino G. The paramedian supracerebellar infratentorial approach. Acta Neurochir. 2017;159(8):1529–32. https://doi.org/10.1007/s00701-017-3196-y.
5. Choque-Velasquez J, Resendiz-Nieves J, Jahromi BR, Colasanti R, Baluszek S, Muhammad S, Hernesniemi J. Midline and Paramedian Supracerebellar Infratentorial approach to the pineal region: a comparative clinical study in 112 patients. World Neurosurg. 2020;137:e194–207. https://doi.org/10.1016/j.wneu.2020.01.137.

Surgical Anatomy of the Brainstem

Yosef Dastagirzada, Akshay V. Save, and Daniel Cavalcanti

1 History of Brainstem Anatomy and Surgery

Our current understanding of the intricate structure that is the brainstem originates from centuries of dissections by brilliantly innovative neuroanatomists and subsequently by countless surgeon scientists. Claude Galen (c.129–216) has been described as the founder of experimental neurosurgery and is considered one of the first to write about the brainstem in his paper *On The Usefulness of the Parts* in 173 AD [1, 2]. Upon his dissections primarily in monkeys, he would begin to make crucial primitive observations that are now better understood anatomically. For example, he describes the intricate steps of a suboccipital craniectomy and how compression of the fourth ventricle led to severe impairment of consciousness [3, 4]. This led to the Greco-Roman scientists' interest in the physiologic role ("vital spirits") of the ventricles that would then be carried on by Leonardo Da Vinci (1452–1519) with his experiments of injecting melted wax into the ventricular system of an ox; one of the earliest creations of 3-dimensional anatomic modeling [5]. Andreas Vesalius (1514–1564) was one of the major contributors during the Golden Age of Italian Anatomy, with his accurate depictions of the brainstem and posterior fossa structures in one of the earliest works of human anatomy called *De Humani Corporis Fabrica* (1543) [6, 7].

Transitioning from macro-anatomical identification and descriptions, Thomas Willis (1621–1675) would deconstruct the brainstem, termed the *dorsalis medullae*

Y. Dastagirzada · A. V. Save
Department of Neurosurgery, NYU Langone Health, New York, NY, USA
e-mail: yosef.dastagirzada@nyulangone.org; akshay.save@nyulangone.org

D. Cavalcanti (✉)
Department of Neurosurgery, St. Vincent's Medical Center, Ayer Neuroscience Institute, Bridgeport, CT, USA

E. G. Figueiredo et al. (eds.), *Brain Anatomy and Neurosurgical Approaches*,
https://doi.org/10.1007/978-3-031-14820-0_9

educit initium ("beginning of the spinal cord") by Vesalius [7], into more detail (midbrain, pons, medulla) and shift his attention to their physiologic role. He would be described as the founder of neuropathology as he would perform dissections on patients that he would follow for years [8]. Aside from the circle of Willis that will perpetually carry his legacy, he is known for better delineating the cranial nerves, identifying critical structures such as the medullary pyramids, cerebellar peduncles, and acknowledging that the medulla oblongata had an intrinsic role in regulating the functions of various organs [8, 9]. A group of scientists called the *Naturophilosophen* would continue to lead the investigations in describing brainstem neurophysiology [10]. Gall and Spurzheim were responsible for a comprehensive dissection of cranial nerve tracts from their origin in the brainstem and the course of pyramidal fibers through the medulla [11, 12].

Aside from the earliest surgical interventions in the posterior fossa targeting fractures and war-related injuries, surgeon Sauveur François and his drainage of an abscess in 1768 is thought to be one of the earliest surgically treated lesions in this location [13]. Others include William Macewen (1848–1924), who described a successful posterior fossa surgery for a brain tumor in a 14-year-old child [14], Victor Horsley (1857–1916), describing decompressive craniectomy with his mastoid-to-mastoid skin incision [15, 16], and Thierry de Martel (1876–1940) who invented a surgical chair for sitting position posterior fossa surgery [10, 15]. Harvey Cushing (1869–1939) and Walter Dandy (1886–1946) would further revolutionize the surgical approach with protective subcapsular resection and more aggressive resections [1].

The tracing of this elegant structure's anatomic and surgical history allows for better comprehension of how a hidden, mysterious area has become one for which neurosurgeons continue to explore via safe entry zones (Table 1).

Table 1 Safe entry zones to the brainstem based on location

Brainstem location	Safe entry zone
Midbrain	A. Lateral Mesencephalic sulcus
	B. Anterior Mesencephalic zone
	C. Intercollicular region
Pons	A. Lateral Pontine zone
	B. Peritrigeminal zone
	C. Supratrigeminal zone
	D. Supracollicular/Infracollicular zones
	E. Median sulcus
Medulla	A. Anterolateral sulcus
	B. Posterior median sulcus
	C. Olivary zone
	D. Lateral medullary zone

2 The Midbrain

2.1 Anatomy & Neurophysiology

The midbrain connects the pons and the diencephalon and has many crucial neurophysiologic functions. It is the smallest segment of the brainstem, measuring 1.5 cm craniocaudally. In addition to relaying ascending and descending sensory and motor fibers, it also has connections to the cerebellum via mostly efferent axons through the superior cerebellar peduncles. The reticular activating system spans the midbrain and pons and regulates consciousness, wakefulness, and sleep. The midbrain also hosts the nuclei of the oculomotor and trochlear nerves and is involved in pupillary constriction and lens accommodation through the Edinger-Westphal nucleus.

The midbrain lies rostral to the pons and caudal to the diencephalon. Its caudal boundary is the superior pontine or mesencephalic-pons groove. It is commonly divided into three sections: the tectum, the tegmentum, and the basis pedunculi. The tectum, also known as the quadrigeminal plate, consists of two paired structures: the superior and inferior colliculi. The superior colliculus is involved in visual reflexes and rapid saccadic movements. Each superior colliculus projects to the ipsilateral lateral geniculate nucleus and the optic tracts through the superior brachium. Similarly, the inferior colliculus is involved in auditory processing, connecting to the medial geniculate nucleus through the inferior brachium. The cerebral aqueduct connects the third and fourth ventricles between the tectum and tegmentum. Periaqueductal grey matter plays a key role in pain suppression via enkephalin and serotonergic pathways resulting in the release of endogenous opioids. Ventral to the tectal plate and cerebral aqueduct lies the tegmentum. The tegmentum contains the reticular formation, ascending sensory fibers of the medial lemniscus, the red nucleus, the ventral tegmental area, and the substantia nigra. The reticular formation contains many fiber tracts involved in neuromodulation and, importantly, involves the ascending reticular activating system, which regulates consciousness and wakefulness. The reticular formation includes the dorsal raphe nuclei, which generate serotonin. The red nucleus is the origin of the rubrospinal tract, involved in the regulation and modulation of motor function through cerebellar circuits. The ventral tegmental area is the primary source of dopaminergic neurons and is an important part of reward circuitry. The substantia nigra also uses dopaminergic neurotransmitters to regulate motor control through the basal ganglia. Degeneration of this structure is notably found in Parkinson's disease. Finally, the basis pedunculi comprises the most ventral portion of the midbrain. Most importantly, it contains the descending motor corticobulbar and corticospinal tracts to continue the internal capsule. The corticobulbar tract lies medial to the corticospinal tract. The most medial portion of the cerebral peduncles hosts fibers of the corticopontine tract.

As mentioned previously, the major cranial nerve nuclei in the midbrain are the oculomotor and trochlear nerve nuclei. The oculomotor nerve innervates the superior rectus, inferior rectus, medial rectus, inferior oblique, and levator palpebrae muscles. The trochlear nerve is the only cranial nerve to exit the brainstem dorsally, where it then travels through the ambient cistern and into the orbital apex to innervate the superior oblique muscle, which depresses the eye when it is adducted. The Edinger-Westphal nucleus lies immediately dorsal to the oculomotor nucleus. It adjusts the lens during accommodation and mediates the pupillary light reflex, which constricts both pupils in response to light.

From a vasculature perspective, the blood supply to the midbrain is predominantly from the paired posterior cerebral arteries and the basilar artery. The basilar artery runs in the interpeduncular cistern before it transitions into the posterior cerebral arteries. Ventrally the blood supply is from the basilar artery, whereas dorsally and laterally, it is predominantly from branches of the posterior cerebral artery. The superior cerebellar artery also provides a small blood supply to the tectum.

2.2 *Safe Zones/Approaches*

2.2.1 Posterolateral Midbrain Approach: Lateral Mesencephalic Sulcus (LMS)

The lateral mesencephalic sulcus, localized by visualization of the lateral mesencephalic vein (critical anastomosis between the basal vein of Rosenthal and the petrosal system, i.e., supratentorial-infratentorial anastomosis) [17], can be used to access lesions in the posterolateral midbrain as it lies adjacent to the posterior ambient cistern [18]. It runs between the cerebral peduncle anteriorly and tegmentum posteriorly, bordered by the medial geniculate body above and pontomesencephalic sulcus below [19]. Surrounding vessels include the posterior P2 (P2P) segment of the posterior cerebral artery (PCA) superiorly and the cerebellomesencephalic segment of the superior cerebellar artery (SCA) inferiorly; the medial posterior choroidal artery crosses centrally. The trochlear nerve is also another localizing structure as it courses the LMS inferiorly.

The safe zone to this location is ideally between the substantia nigra anterolaterally, oculomotor fibers anteromedially, and the medial lemniscus posteriorly [20]. Ventral deviation could lead to damage to the substantia nigra or pyramidal tracts. In contrast, dorsal deviation could damage the sensory pathways of the medial lemniscus, oculomotor, and red nuclei, as well as the mesencephalic nucleus/tract of the trigeminal pathway. Studies have shown an average length of 9.6 mm and a working corridor length of 8.0 mm [21]. Surgical approaches to this location include the extreme lateral supracerebellar infratentorial (Fig. 1) and the subtemporal corridors.

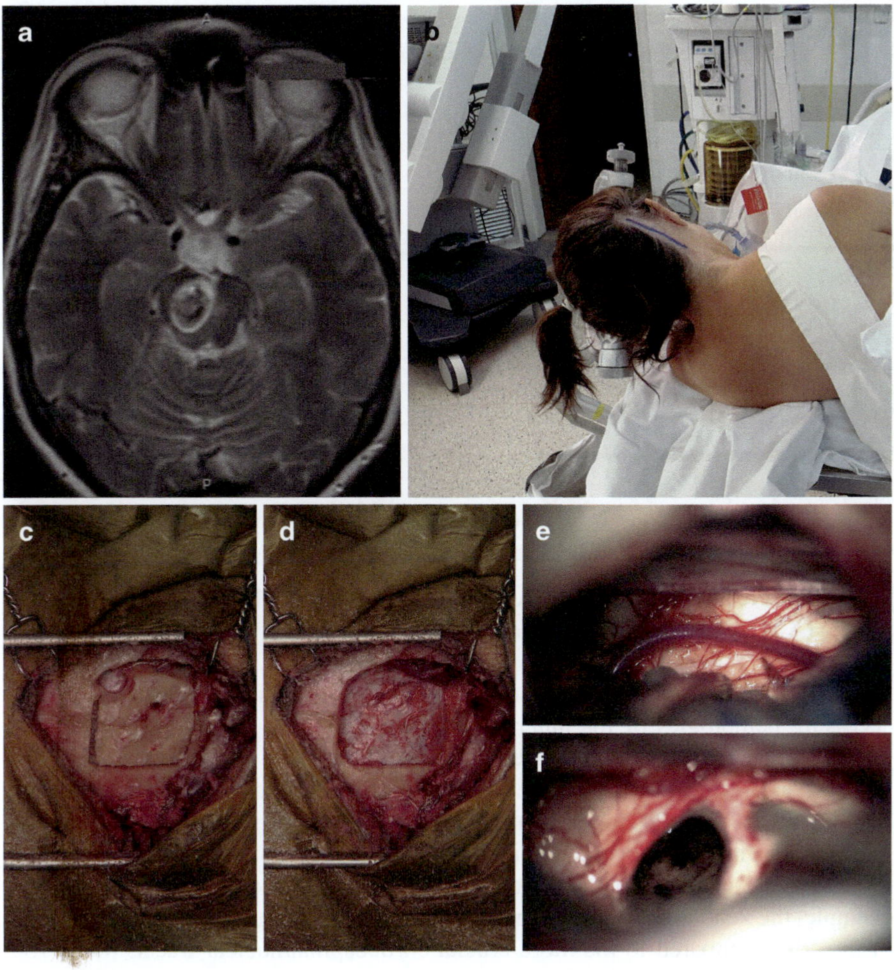

Fig. 1 The extreme lateral supracerebellar infratentorial approach is an ideal option for cavernous malformations that are located close to the posterolateral surface of the midbrain. (**a**) MRI of the brain, axial view, T2-weighted image, showing a heterogeneous rounded image centered in the right cerebral peduncle and tegmentum, approaching the surface of the brainstem at the right lateral mesencephalic sulcus. (**b**) Patient in the left lateral decubitus, with a linear retromastoid incision marked in order to proceed with a right retrosigmoid approach. The skin incision is placed approximately two fingerbreadths behind the pinna. (**c**, **d**) Right-sided retrosigmoid approach. A burr-hole is created immediately cranial to the asterion, on the parietomastoid suture. The craniotomy is tailored in a small rectangular shape with the posterior edge of the sigmoid sinus as well as the transverse-sigmoid junction carefully skeletonized. (**e**) Intraoperative exposure of the posterior aspect of the ambient cistern and the lateral segment of the quadrigeminal cistern. The right trochlear nerve along the superior cerebellar artery is seen crossing the cistern. The lateral mesencephalic sulcus was chosen as a safe entry zone. (**f**) Final view of the surgical cavity after gross total resection. (**g**) Two-year post-operative MRI revealing complete resection

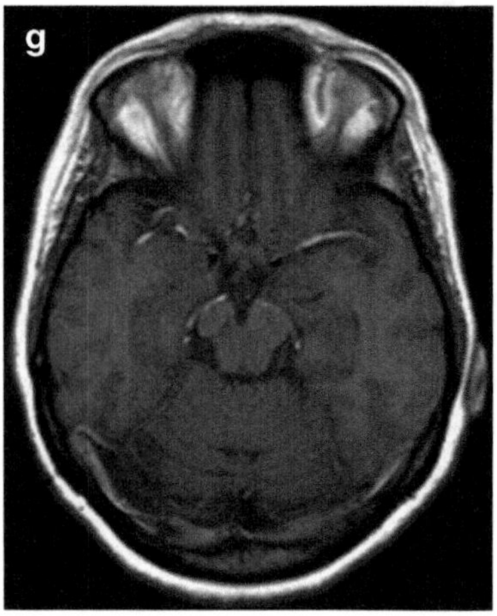

Fig. 1 (continued)

2.2.2 Anterolateral Midbrain Approaches: Anterior Mesencephalic Zone (AMZ)/Perioculomotor Zone

Understanding the anatomy of the AMZ in the crural cistern can be crucial to accessing pathologies of the anterolateral midbrain. It is located on the cerebral peduncle, specifically the crus cerebri (anterior portion of the peduncle). The corridor is a narrow space between the root exit zone of the oculomotor nerve medially (medial sulcus of the peduncle) and the corticospinal tracts laterally. The vasculature provides the craniocaudal boundaries, the PCA superiorly and the SCA inferiorly [22]. The approach takes advantage of the distribution of tracts in this area. For example, the zone is entered through frontopontine fibers in the ventral medial third of the crus, compared to the more crucial corticospinal/nuclear tracts, which routinely lie laterally [23, 24]. Prats-Galino et al. demonstrated that the red nucleus is located approximately 7.8 mm from the pial surface, which should be taken into consideration as oculomotor fibers curve around the medial boundary [25]. The group also noted that in about 40% of endoscopic dissections, perforators from the SCA obstructed the exposure. Of note, for pathologies in the anteromedial midbrain (medial to oculomotor), Spetzler et al. have proposed an interpeduncular fossa approach that is void of efferent motor fibers [26]. It is located between the mammillary bodies superiorly and the tip of the basilar artery inferiorly. These pathways can be accessed via classic pterional, orbitozygomatic, and subtemporal approaches.

2.2.3 Dorsal Midbrain Approaches: Intercollicular Region

Dorsal midbrain pathology is best accessed via the intercollicular region in the quadrigeminal cistern to access the area of the quadrigeminal plate/tectum. Landmarks in the area include the pineal gland superiorly and the paired superior and inferior colliculi. The safe entry zone is thought to be more tolerant of a neurotomy instead of the sparseness of fibers, specifically a vertical, midline entrance between the four colliculi [24, 27]. Other proposed entry points include supracollicular and infracollicular zones. The former's location, more specifically, has been described as the subpineal triangle (triangle of Obersteiner), immediately above the superior colliculi and below the posterior commissure [28]. The infracollicular zone is immediately inferior to the inferior colliculi and above the origin of the trochlear nerve [24]. The relationship to the cerebral aqueduct must be respected when considering these approaches as the mesencephalic nucleus/tract, oculomotor nucleus, and medial longitudinal fasciculus are at its depth [28]. Surgical approaches to this location include median or extreme lateral SCIT.

3 The Pons

3.1 Anatomy & Neurophysiology

The pons is the brainstem's middle section and lies between the midbrain and medulla. Dorsal to the pons is the fourth ventricle, followed by the cerebellum. The basilar artery travels directly anterior to the pons in the basilar groove. It is bounded by the mesencephalic-pons and pontomedullary sulci superiorly and inferiorly, respectively. The pons is often divided into dorsal and ventral section. The ventral section contains pontine nuclei and the descending corticospinal and corticopontine tracts. The dorsal section, also known as the tegmentum, comprises the reticular formation, the trapezoid body, and additional fiber pathways, which vary depending on the level of interest. The middle cerebellar peduncle is also found at the level of the pons. The middle cerebellar peduncle contains mostly afferent projection fibers from the corticopontocerebellar tract, which helps coordinate movement.

The cranial nerve nuclei of the pons include the trigeminal, abducens, facial, vestibular, and cochlear nerve nuclei. Due to the way cranial nerve nuclei migrate during development, the motor nuclei are all medial to the sensory nuclei, separated by the sulcus limitans, a preserved structure on the floor of the fourth ventricle.

As expected from its name, the trigeminal nerve has three components: the ophthalmic, maxillary, and mandibular nerves, which provide sensation to the face and oral cavity. The ophthalmic and maxillary nerves have only sensory components, whereas the mandibular nerve is a mixed sensory and motor nerve, innervating the lower face and mastication muscles. The trigeminal nerve has three distinct nuclei: spinal trigeminal, main trigeminal sensory, and mesencephalic trigeminal nuclei. Within the spinal trigeminal nuclei are three further subdivisions, the caudal nucleus, the interpolar nucleus, and the oral nucleus, which mediate facial pain, temperature, and itch. The main trigeminal sensory nucleus mediates touch sensation and oral mechanosensation. The mesencephalic trigeminal nucleus is involved with jaw proprioception using stretch receptors. A separate trigeminal motor nucleus gives rise to fibers innervating mastication.

Cranial nerves VI, VII, and VIII exit at the pontomedullary sulcus. The abducens nerve exits medially, where it travels through Dorello's canal underneath the petrosphenoidal ligament (Gruber's ligament) and innervates the lateral rectus muscle of the eye which is responsible for ocular abduction. The medial longitudinal fasciculus travels between the midbrain and pons, where it coordinates the movement of both eyes for smooth pursuit by integrating the activity of the oculomotor nerve with the contralateral abducens nerve. The facial nerve mediates all facial movement, a small amount of sensory territory in the posterior auricular region, and taste sensation in the anterior 2/3 of the tongue. It has a complicated trajectory, where it travels from the main nucleus to the ventricular floor, where it joins the abducens nucleus to form the facial colliculus. The intermediate nerve joins the motor component and mediates the sensory aspects of the facial nerve. The vestibulocochlear nerve mediates both hearing and balance. The cochlear nucleus is found in the rostral medulla but then travels superiorly to decussate in the trapezoid body in the dorsal pons. These nerve fibers travel to the superior olivary complex and then project to the inferior colliculus through the lateral lemniscus. There are four distinct vestibular nuclei: superior, inferior, lateral, and medial, which receive fibers from the semicircular canals, utricles, and saccules to mediate rotational and linear balance.

The anatomy of the rhomboid fossa is of special importance when considering neurosurgical approaches to pontine pathology. The median eminence divides the rhomboid fossa into left and right segments at the midline. As described previously, the sulcus limitans is another craniocaudal demarcation that divides the motor and sensory nuclei. The stria medullaris is a segment of transverse fibers that define the caudal boundary of the pons. The facial colliculus is a notable prominence on the floor of the fourth ventricle that contains facial nerve fibers and the abducens nucleus. The inferior component of the rhomboid fossa extends into the medulla and will be discussed subsequently.

The pons receives vascular supply from paramedian, short circumferential, and long circumferential branches of the basilar artery, and the superior cerebellar artery. Anteriorly and medially, the vasculature is predominantly from the basilar artery; laterally and posteriorly, there is a contribution from the superior cerebellar artery and the anterior inferior cerebellar artery. The posterior inferior cerebellar artery provides blood supply to the vestibular nuclei.

3.2 Safe Zones/Approaches

3.2.1 Ventrolateral Pontine Approaches: Lateral Pontine Zone/Peritrigeminal Zone/Supratrigeminal Zone

Baghai et al. [29, 30] originally reported a safe entry zone on the junction between the middle cerebellar peduncle and the pons and between the trigeminal root entry zone and the CN VII-VIII complex root entry zones. The main drawback of this safe zone is that vertical dissection is limited. The lateral pontine safe zone is better exposed using a retrosigmoid craniotomy (Fig. 2), then widely dissecting the cerebellopontine cistern and the petrosal fissure.

The peritrigeminal zone has been described to be medial to the trigeminal nerve and lateral to the pyramidal tract, between the emerging trigeminal and facial/vestibulocochlear complex (pontomedullary sulcus) in the prepontine cistern [31]. Recalde more elegantly evaluated it in 2008 via fiber dissections, showing approximately a depth of 11.2 mm (range 9.5–13.1 mm) to the trigeminal nuclei and a distance of 4.64 mm (range 3.8–5.6 mm) between the pyramidal tract and the trigeminal nerve [21]. When using this longitudinal corridor to access ventral pontine pathology, one must be considerate of the motor and sensory nuclei of the trigeminal nerve at the depth to avoid deficits [32]. Other considerations are the craniocaudal fibers of cranial nerves VI to VIII that are dorsal to these nuclei. One can consider the supratrigeminal zone if attempting to target pathologies slightly more lateral in the ventral pons. Hebb and Spetzler delineated the corridor immediately above the CNV root entry zone on the middle cerebellar peduncle, lateral to the descending motor tracts [33]. Superiorly, the fibers of CN III emerge.

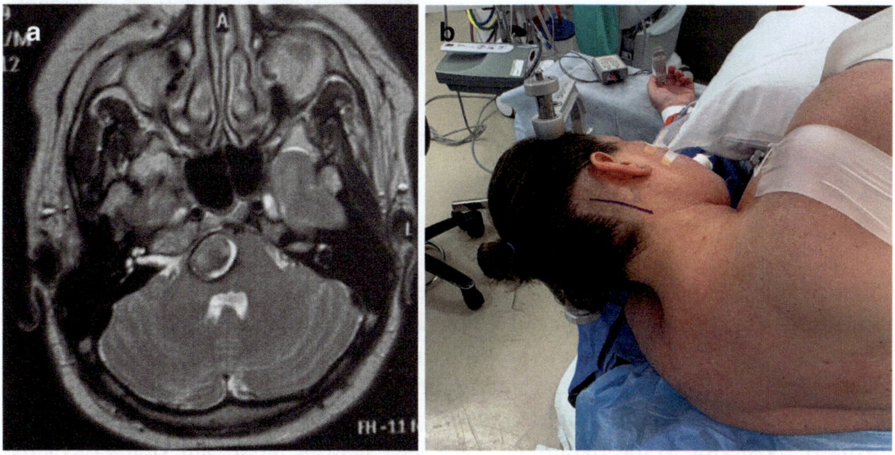

Fig. 2 The retrosigmoid craniotomy provides the neurosurgeon with access to important safe-entry zones including the lateral pontine zone. It is considered a workhorse approach in the management of several lesions involving the cerebellopontine angle. (**a**) MRI of the brain, axial view, T2-weighted image, showing a rounded heterogenous lesion centered in the right hemipons, with signal of various phases of hematoma, including the so-called hemosiderin rim. The bulk of the lesion is centered at the level of the internal acoustic meatus. (**b**) Patient is placed in the lateral decubitus. A right retromastoid incision is drawn approximately two fingerbreadths behind the pinna, starting from a point at the posterior projection of the superior pole of the pinna, extending down to a point approximately 1cm below the posterior projection of the mastoid tip. (**c**) A retrosigmoid craniotomy is completed. (**d**) Intraoperative exposure of the right cerebellopontine angle, with the lateral pontine zone located between the emergence of the cranial nerves VII and VIII. (**e**) Final aspect of the cerebellopontine angle partially showing the entry point. (**f**) Two-year follow-up MRI of the brain confirming complete resection

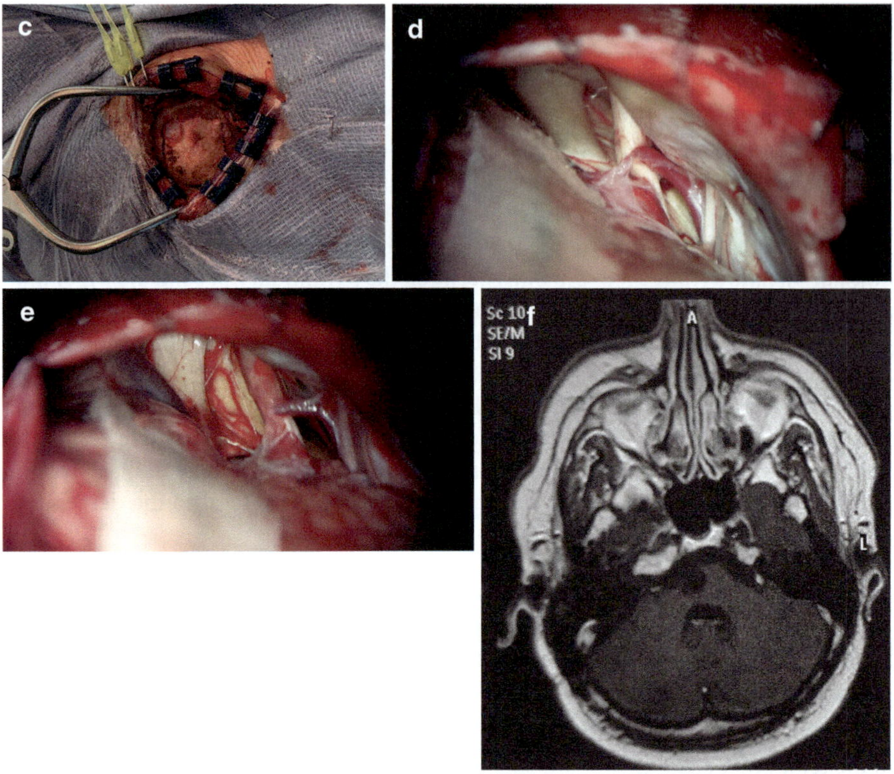

Fig. 2 (continued)

Alternatively, the supra-trigeminal safe entry zone constitutes another corridor utilized to manage anteriorly placed lesions in the upper pons located just above the trigeminal root entry zone on the middle cerebellar peduncle. Taking advantage of the posterolateral position of the middle cerebellar peduncle and the thick transverse pontine fibers, it is feasible to dissect along these fibers, medially or anteromedially, posterior to the pyramidal tract. The Kawase approach yields the ideal straight trajectory for this safe entry zone [34].

3.2.2 Dorsal Pontine Approaches: Supracollicular & Infracollicular Zones/Median Sulcus of the Fourth Ventricle (Interfacial)

Understanding the intricate anatomy of the rhomboid fossa, specifically the fourth ventricular floor, is crucial to conducting safe microsurgery in the dorsal pons. The facial colliculus, acoustic striae, sulcus limitans, are median sulcus are the most crucial landmarks to be aware of. The facial colliculus is lateral to the median sulcus, a midline structure that divides the rhomboid fossa, and is a landmark for the craniocaudal trajectory of the medial longitudinal fascicle (MLF) that runs in parallel. It is formed by the motor nerve fibers of CNVII, which loop over the ipsilateral abducens nucleus; the loop is responsible for the enlargement visualized intraoperatively [35, 36]. Literature reports a craniocaudal dimension of 3 mm and ventrodorsal elevation of 0.73 mm [37, 38]. Localization of this landmark can be done on MRI and by electrophysiologic methods used intraoperatively [39–42], as data have shown it to be indiscernible in about one third of the population [38].

The suprafacial and infrafacial triangles were first described by Kyoshima et al. [43] who studied the topographic anatomy of the colliculi and the potential neurological deficits when manipulating various parts of the fourth ventricular floor. The former is bordered by the MLF medially, the cerebellar peduncle laterally, facial nerve caudally, and the frenulum veli (crossing of the trochlear nerve) cranially, while the latter is defined by MLF medially, the stria medullaris inferiorly (hypoglossal and vagal trigones/nuclei below), and the facial nerve laterally. Dissection deeper than 4–5 mm has been shown to damage the medial lemniscus in both these approaches possibly [19]. The suprafacial approach's incision has been described as approximately 5 mm lateral to the median sulcus and about 1 cm in length, inferior to the cerebellar peduncle. The infrafacial approach incision is typically 5 mm lateral to the median sulcus, but above the stria medullaris and typically less than 1 cm in length [43]. Further studies have considered the significant variability of the anatomy of the striae medullaris and the possible damage to either the trigeminal motor nucleus or the nuclei of the lower cranial nerves when using the suprafacial or infrafacial triangles. Strauss et al. [42] then suggested a paramedian supracollicular zone measuring 13.8 (13.3–14.5) mm vertically between the facial colliculus and the decussation of CN IV, located approximately 0.6 mm from the midline. Additionally, the trigeminal motor nucleus limits the approach laterally, located 6.3 mm from the midline.

Finally, replacing the infrafacial triangle, a paramedian infracollicular zone was suggested between the projection of the facial nerve fibers on the facial colliculus cranially and the superior limits of the nucleus of the hypoglossal nerve and the dorsal nucleus of the vagal nerve caudally, extending approximately 9.2 (8.3–9.5) mm vertically, located approximately 0.3 mm from the midline.

A midline approach through the median sulcus of the fourth ventricle has been used, termed the interfacial approach. The oculomotor nucleus located in the midbrain superiorly and the abducens nucleus inferiorly define its borders, where there is the sparseness of crossing fibers [24]. Excessive lateral retraction could damage the MLF as it runs parallel to the median sulcus.

Figure 3 details the surgical anatomy of the median suboccipital approach, a workhorse in managing posterior fossa lesions, especially when involving the floor of the fourth ventricle.

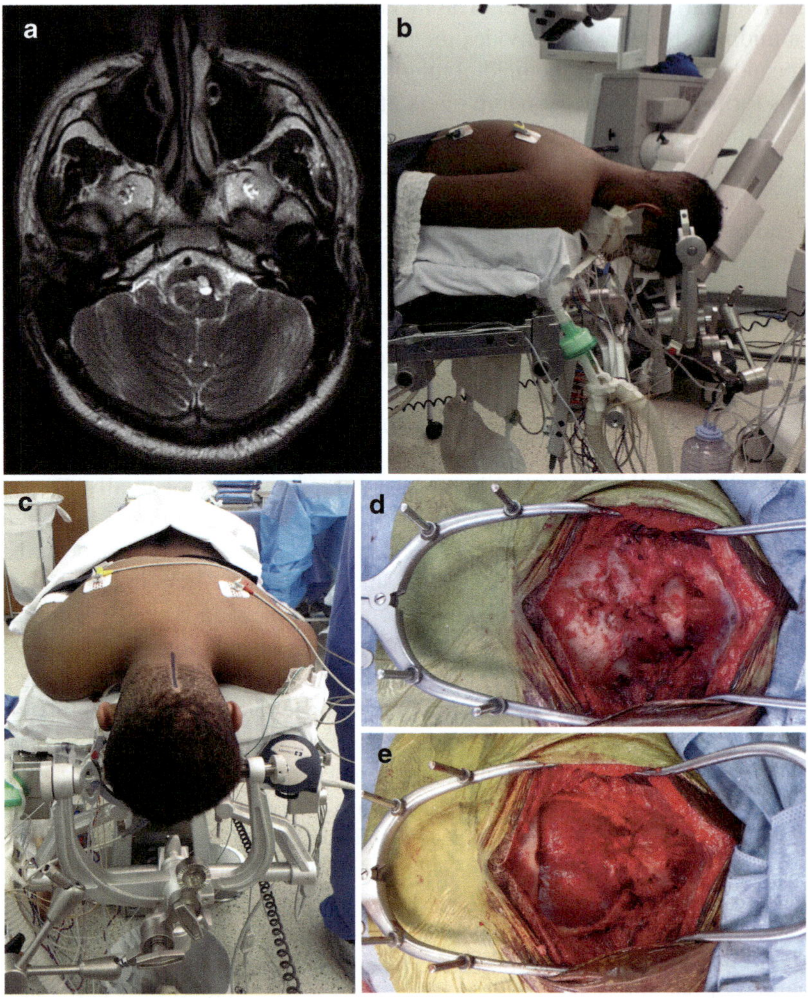

Fig. 3 The median suboccipital anatomy with telovelar dissection for dorsally located pontine cavernous malformation. (**a**) MRI of the brain, axial view, T2-weighted image, revealing an irregular rounded vascular lesion containing areas of bleeding at different stages and a hypointense peripheral rim, better seeing anterolaterally in the left. (**b**, **c**) Patient is carefully placed on prone position, chin flexed, with the head of the bed elevated to 30 degrees. Intraoperative monitoring is critical. With minimal hair trimming, a linear and median occipitocervical incision is marked. (**d**) The dissection is carried out to the deep muscular layers along the midline avascular fascial plane, with subperiosteal dissection of the occipital squama laterally. Being mindful of the anatomic relationship between the V3 segment and the posterior arch of C1, careful subperiosteal dissection of the posterior arch of C1 is also performed. (**e**) Completion of a median suboccipital craniotomy with one single midline burr hole. (**f**) Demonstration of the anatomy of the floor of the fourth ventricle; the main safe entry zones are demonstrated. (**g**) The dissection of the tela choroidea and the inferior medullary velum render the exposure of the rhomboid fossa and its landmarks, critical for selecting the right safe entry zone. (**h**) The median suboccipital anatomy. Three-year follow-up MRI of the brain demonstrating complete resection of the cavernous malformation

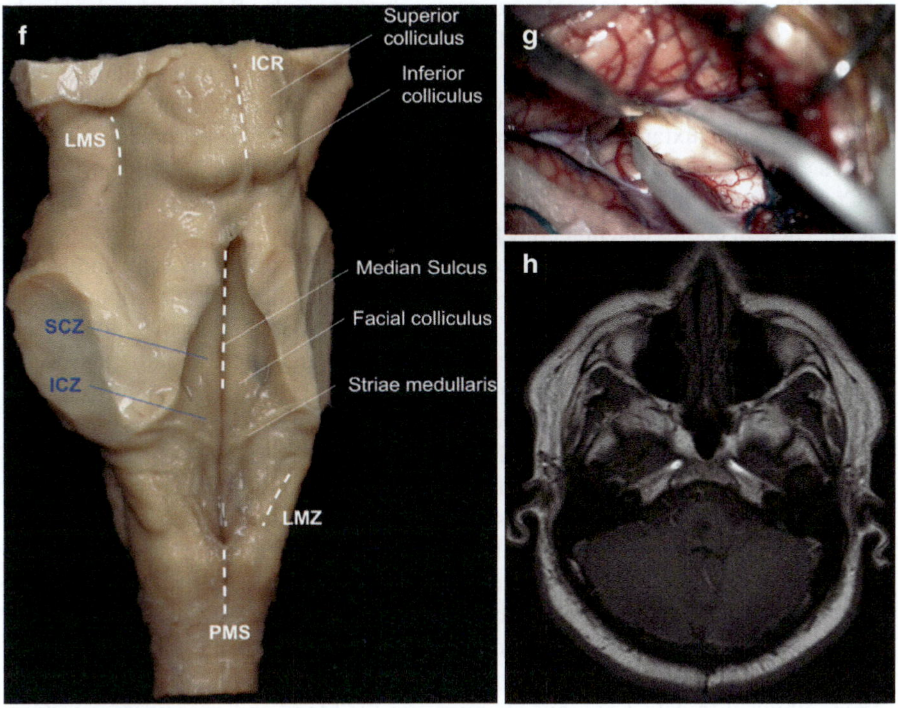

Fig. 3 (continued)

4 The Medulla

4.1 Anatomy & Neurophysiology

The medulla oblongata is the most caudal aspect of the brainstem, connecting the pons to the upper cervical spinal cord. The medulla regulates cardiovascular and pulmonary function, houses ascending and descending fiber tracts, includes the inferior olivary nucleus, and contains cranial nerve nuclei for the glossopharyngeal, vagus, and hypoglossal nerves. The spinal accessory nerve nucleus extends from the medulla to the C4-5 segments of the cervical spine. The dorsal column-medial lemniscus contains the fibers for light touch and proprioception in the medulla. Within this tract, the fibers for the lower extremities lie medial to those for the upper extremities. The pyramids lie ventrally and medially and comprise the descending corticospinal tract. Of note, both the dorsal column-medial lemniscus and corticospinal tract decussate in the medulla, with the corticospinal tract decussating caudal to the dorsal column-medial lemniscus. Lateral to the pyramids are the olives, made of a superior and inferior segment. The superior olivary tract is involved in the auditory processing pathway, while the inferior olivary tract is part of the olivocerebellar system. The inferior olivary nucleus contains the neuronal cell bodies for the climbing fibers that project to the cerebellum via the inferior cerebellar peduncle. The inferior cerebellar peduncle contains afferent and efferent fiber tracts and is heavily involved in integrating proprioceptive information with motor control of balance and posture. The juxtarestiform body in the inferior cerebellar peduncle also coordinates eye movements with the vestibular system.

Cranial nerves nine and ten have diverse functions, with corresponding distinct nuclei. The cell bodies that innervate taste sensation originate in the solitary nucleus. The neuronal cell bodies that innervate motor movement of the pharynx and larynx are housed in the nucleus ambiguous in the deeper medulla. The glossopharyngeal nerve innervates the pharynx and larynx via the stylopharyngeus muscle and receives taste sensation in the posterior 1/3 tongue and upper mouth. The vagus nerve is the longest autonomic nerve in the body. It provides the majority of the innervation of the pharynx and larynx, posterior/inferior pharyngeal taste sensation, parasympathetic innervation to multiple organs, and also provides sensory innervation of the external acoustic meatus. The dorsal nucleus of the vagus nerve is responsible for the parasympathetic response to the visceral organs.

The rhomboid fossa in the medulla extends inferiorly from the stria medullaris. Moving inferiorly from the stria medullaris are the hypoglossal trigone, vagal trigone, and the area postrema, respectively. The hypoglossal and vagal nerve nuclei approach the floor of the rhomboid fossa at the hypoglossal and vagal trigones. The area postrema is one of the circumventricular organs, where the lack of blood-brain barrier allows it to detect noxious chemicals in the blood and coordinate vomiting. The inferior-most apex of the rhomboid fossa is known as the obex.

The vascular supply of the medulla is largely supplied by paramedian and short circumferential branches of the vertebral arteries. The posterior inferior cerebellar artery provides blood supply to the dorsal and lateral regions of the medulla. A PICA artery occlusion results in a lateral medullary syndrome classically described as the loss of ipsilateral facial pain/temperature sensation, loss of contralateral limb and trunk pain/temperature sensation, ataxia vertigo, and dysphagia/dysphonia. The most caudal regions of the medulla at the cervicomedullary junction also receive vascular supply from the spinal arteries.

4.2 Safe Zone/Approaches

4.2.1 Anterior Medullary Approaches: Anterolateral Sulcus (ALS)/ Olivary Zone/Lateral Medullary Zone

Located along the preolivary sulcus between the caudal roots of the hypoglossal nerve superiorly and C1 rootlets inferiorly, the anterolateral sulcus approach can lead to anterolateral medullary pathologies [32]. It continues and eventually becomes the anterolateral sulcus of the spinal cord inferiorly. The ALS lies between the medullary pyramids (decussation point) and olive; therefore, surgery here can lead to stretch/damage to the corticospinal tracts. The posterolateral sulcus can be another entry point parallel to the postolivary sulcus. The zone lies anterior to the glossopharyngeal and vagus roots, between the olive and inferior cerebellar peduncle [21]. This approach has been deemed useful for exophytic lesions adjacent to the lateral medullary cistern [19]. As mentioned earlier, the olivary zone is situated between the zones (ALS/PLS). Safe dimensions for surgical dissection have been described at a depth of 4.7–6.9 mm and a length of 13.5 mm [21].

The lateral medullary zone ("inferior cerebellar peduncle approach") has been described for resecting dorsolateral medullary lesions [44]. It is described as a vertical incision in the inferior cerebellar peduncle, below the cochlear nuclei and posterior to the emergency of CNs IX and X. Surgically, one must search for the foramen of Luschka, as the zone is ventral to this landmark.

4.2.2 Posterior Medullary Approaches: Posterior Median Sulcus and Posterior Intermediate Sulcus

Akin to the interfacial pontine approach to the pons, the posterior median sulcus approach to the medulla takes advantage of this midline structure with sparse crossing fibers. The surgical corridor lies between the bilateral gracile fascicles/tubercles and the hypoglossal and dorsal vagal nuclei neighboring midline [28]. If you were to move laterally, one would discover the posterior intermediate sulcus, which runs between the gracile and cuneate tubercles and can serve as another access point to the posterior medulla [31].

5 Conclusion

Due to the densely packed tracts and nuclei in the brainstem, pathologies in this area tend to cause profound neurologic deficits. Primary intra-axial pathologies in the brainstem are most commonly glial neoplasms, cavernous malformations, and arteriovenous malformations. Glial neoplasms are broadly categorized as diffuse or focal and are further subdivided by location. Astrocytomas are the most common subtype of brainstem glioma, though primitive neuroendocrine tumors, lymphomas, ependymomas, gangliogliomas, and oligodendrogliomas can also occur in this region [45, 46]. Brainstem cavernomas are thought to make up between 4% and 35% of cavernous malformations [47, 48], with a bleed rate between 1% and 3% per year. Size of cavernoma has not been associated with an increase in bleeding risk; however, prior history of bleeds tended to increase the risk of rebleeding. Though widely considered inoperable in the past, modern surgical series have shown a 1.9% mortality rate from surgery and a 14% rate of permanent neurologic worsening. The majority of neurologic deficits from surgery are transient and must be balanced with the risk of rebleeding, causing devastating neurologic injury [49]. In experienced hands, brainstem arteriovenous malformations have surgical mortality of 7%, with a 14% rate of neurologic deterioration [50]. Lateral pontine and medullary were found to have the best functional outcomes, whereas anterior pontine and posterior midbrain AVMs had the highest rate of worsening neurologic function and death. Also to be considered in the differential diagnosis of brainstem lesions include metastatic lesions, intracranial abscesses, and other infectious etiologies.

In lieu of advancements in neuronavigation and instrumentation in the operating room (microscope, endoscope, neuromonitoring), the brainstem, previously known as "no man's land," has been well delineated anatomically by scientists and neurosurgeons in history, as described earlier in this chapter. The combination of surgical experience and comprehension of neurophysiology has allowed the advent of safe zones to this structure that harbors the control panel for nearly all human body functions. Our comprehensive review of the anatomy and safe entry zones is meant to allow for enhanced preparation when considering a surgical corridor to the *dorsalis medullae educit initium.*

References

1. Cucu AI, Turliuc S, Costea CF, et al. The brainstem and its neurosurgical history. Neurosurg Rev. 2021;44(6):3001–22. https://doi.org/10.1007/s10143-021-01496-3.
2. Swanson LW. Neuroanatomical terminology: a lexicon of classical origins and historical foundations. New York: Oxford University Press; 2003.
3. Viale G. The surgical approach to the posterior cranial fossa according to Galen. Oper Neurosurg. 2007;61(5):ONS399–403. https://doi.org/10.1227/01.neu.0000304000.05021.8b.
4. Freemon FR. Galens ideas on neurological function. J Hist Neurosci. 1994;3:263–71.
5. da Mota Gomes M. From the wax cast of brain ventricles (1508–9) by Leonardo da Vinci to air cast ventriculography (1918) by Walter E. Dandy. Rev Neurol. 2020;176(5):393–6. https://doi.org/10.1016/j.neurol.2019.11.005.
6. Catani M, Sandrone S. Brain renaissance: from Vesalius to modern neuroscience. Oxford: Oxford University Press; 2015.
7. Wijdicks EFM. Historical awareness of the brainstem: from a subsidiary structure to a vital center. Neurology. 2020;95(11):484–8. https://doi.org/10.1212/WNL.0000000000010504.
8. Molnár Z. ThomasWillis (1621–1675), the founder of clinicalneuroscience. Nat Rev Neurosci. 2004;5(4):329–35.
9. Arráez-Aybar LA, Navia-Álvarez P, Fuentes-Redondo T, Bueno-López JL. Thomas Willis, a pioneer in translational research in anatomy (on the 350th anniversary of Cerebri anatome). J Anat. 2015;226(3):289–300. https://doi.org/10.1111/joa.12273.
10. Martirosyan NL, Carotenuto A, Patel AA. History of brainstem surgery. In: Spetzler R, Kalani M, Lawton M, editors. Surgery of the brainstem. New York: Thieme; 2019.
11. Simpson D. Phrenology and the neurosciences: contributions of F. J. Gall and J. G. Spurzheim. ANZ J Surg. 2005;75(6):475–82. https://doi.org/10.1111/j.1445-2197.2005.03426.x.
12. Rawlings CE, Rossitch E. Franz Josef Gall and his contribution to neuroanatomy with emphasis on the brainstem. Surg Neurol. 1994;42:272–5.
13. Goodrich JT. History of posterior fossa tumor surgery. In: Özek MM, Cinalli G, Maixner W, Sainte-Rose C, editors. Posterior fossa tumors in children. Berlin: Springer International Publishing; 2015. p. 3–60.
14. Macewen W. Localization of cerebral lesions and antiseptic trephinning. Lancet. 1881;2:581–2.
15. Morcos JJ, Haines SJ. History of brain stem surgery. Neurosurg Clin N Am. 1993;4(3):357–65.
16. Michael P. Powell, Sir Victor Horsley at the birth of neurosurgery. Brain. 2016;139(2):631–4.

17. Cannizzaro D, Rammos SK, Peschillo S, et al. The lateral mesencephalic vein: surgical anatomy and its role in the drainage of tentorial dural arteriovenous fistulae. World Neurosurg. 2016;85:163–8.
18. Figueirido EG, Beer-Furlan A, Welling LC, et al. Microsurgical approaches to the ambient cistern region: an anatomic and qualitative study. World Neurosurg. 2016;87:584–90.
19. Yagmurlu K, Rhoton A, Tanriover N, et al. Three-dimensional microsurgical anatomy and the safe entry zones of the brainstem. Oper Neurosurg. 2014;10:602–20.
20. Cavalcanti D, Preul MC, Kalani Y, et al. Microsurgical anatomy of safe entry zones to the brainstem. J Neurosurg. 2016;124:1359–76.
21. Recalde RJ, Figueirido EG, de Oliveira E. Microsurgical anatomy of the safe entry zones on the anterolateral brainstem related to surgical approaches to cavernous malformation. Neurosurgery. 2008;62:9–17.
22. Yang Y, van Niftrik B, Ma X, et al. Analysis of safe entry zones into the brainstem. Neurosurg Rev. 2019;42:721–9.
23. Meyers R. The human frontocorticopontine tract: functional inconsequence of its surgical interruption. Neurology. 1951;5:341–56.
24. Bricolo A, Turazzi S. Surgery for gliomas and other mass lesions of the brainstem. Adv Tech Stand Neurosurg. 1995;22:261–341.
25. Weiss A, Perrini P, De Notaris M, et al. Endoscopic endonasal transclival approach to the ventral brainstem: anatomic study of the safe entry zones combining fiber dissection technique with 7 tesla magnetic resonance guided neuronavigation. Oper Neurosurg. 2019;16(2):239–49. https://doi.org/10.1093/ons/opy080.
26. Kalani MYS, Yagmurlu K, Spetzler RF. The interpeduncular fossa approach for resection of ventromedial midbrain lesions. J Neurosurg. 2018;128:834–9.
27. Kalani MYS, Yagmurlu K, Martirosyan NL, et al. Approach selection for intrinsic brainstem pathologies. J Neurosurg. 2016;125:1–12.
28. Meybodi AT, Hendricks B, Witten AJ, et al. Virtual exploration of safe entry zones in the brainstem: comprehensive definition and analysis of the operative approach. World Neurosurg. 2020;140:499–508.
29. Baghai P, Vries JK, Bechtel PC. Retromastoid approach for biopsy of brain stem tumors. Neurosurgery. 1982;10(5):574–9.
30. Cavalcanti DD, Catapano JS, Niemeyer FP. Using the lateral pontine safe entry zone for resection of deep-seated cavernous malformations in the lateral pons: 2-dimensional operative video. Oper Neurosurg. 2020;19(5):E518–9.
31. Giliberto G, Lanzino DJ, Diehn FE, et al. Brainstem cavernous malformations: anatomical, clinical, and surgical considerations. Neurosurg Focus. 2010;29(3):1–22.
32. Cantore G, Missori P, Santoro A. Cavernous angiomas of the brain stem: intra-axial anatomical pitfalls and surgical strategies. Surg Neurol. 1999;52:84–93.
33. Hebb MO, Spetzler RF. Lateral transpeduncular approach to intrinsic lesions of the rostral pons. Oper Neurosurg. 2010;66(1):26–9.
34. Zenonos GA, Fernandes-Cabral D, Nunez M, et al. The epitrigeminal approach to the brainstem. J Neurosurg. 2018;128(5):1512–21.
35. Takezawa K, Townsend G, Ghabriel M. The facial nerve: anatomy and associated disorders for oral health professionals. Odontology. 2018;106(2):103–16.
36. Yoo H, Mihaila DM. Neuroanatomy, facial colliculus. In: StatPearls. Treasure Island, FL: StatPearls Publishing; 2021. https://www.ncbi.nlm.nih.gov/books/NBK555907/.
37. Ogawa Y, Ito S, Makino T, et al. Flattened facial colliculus on magnetic resonance imaging in Machado-Joseph disease. Mov Disord. 2012;27:1041–6.
38. Bogucki J, Gielecki J, Czernicki Z. The anatomical aspects of a surgical approach through the floor of the fourth ventricle. Acta Neurochir. 1997;139:1014–9.
39. Uchida T, Kin T, Koike T, et al. Identifcation of the facial colliculus in two-dimensional and three dimensional images. Neurol Med Chir (Tokyo). 2021;61:376–84.

40. Suzuki K, Matsumoto M, Ohta M, et al. Experimental study for identification of the facial colliculus using electromyography and antidromic evoked potentials. Neurosurgery. 1997;41(5):1130–6. https://doi.org/10.1097/00006123-199711000-00021.
41. Katsuta T, Morioka T, Fuji K, et al. Physiologic localization of the facial colliculus during direct surgery on an intrinsic brainstem lesion. Neurosurgery. 1993;32:861–3.
42. Strauss C, Romstock J, Nimsky C, et al. Intraoperative identification of motor areas of the rhomboid fossa using direct stimulation. J Neurosurg. 1993;79:393–9.
43. Kyoshima K, Kobayashi S, Gibo H, et al. A study of safe entry zones via the floor of the fourth ventricle for brainstem lesions. J Neurosurg. 1993;78(6):987–93.
44. Deshmukh VR, Rangel-Castilla L, Spetzler RF. Lateral inferior cerebellar peduncle approach to dorsolateral medullary cavernous malformation. J Neurosurg. 2014;121:723–9.
45. Epstein F, McCleary EL. Intrinsic brainstem tumors of childhood: surgical indication. J Neurosurg. 1986;64:11–5.
46. Jallo GI, Biser-Rohrbaugh A, Freed D. Brainstem gliomas. Childs Nerv Syst. 2004;20(3):143–53. https://doi.org/10.1007/s00381-003-0870-6.
47. Gross BA. Bainstem cavernous malformations. Skull Base. 2009;64(5):805–18. https://doi.org/10.1227/01.NEU.0000343667.14177.72.
48. Arauz A, Patiño-Rodriguez HM, Chavarria-Medina M, Becerril M, Longo GM, Nathal E. Rebleeding and outcome in patients with symptomatic brain stem cavernomas. Cerebrovasc Dis. 2017;43(5–6):283–9. https://doi.org/10.1159/000463392.
49. Walcott BP, Choudhri O, Lawton MT. Brainstem cavernous malformations: natural history versus surgical management. J Clin Neurosci. 2016;32:164–5. https://doi.org/10.1016/j.jocn.2016.03.021.
50. Han SJ, Englot DJ, Kim H, et al. Brainstem arteriovenous malformations: anatomical subtypes, assessment of "occlusion in situ" technique, and microsurgical results. J Neurosurg. 2017;122(1):107–17. https://doi.org/10.3171/2014.8.JNS1483.

Part II
The Cisterns

Surgical Anatomy of the Sylvian Fissure

Pablo Augusto Rubino, Juan Santiago Bottan, and Román Arévalo

1 Introduction

Cerebral fissures are tight clefts between two lobes or hemispheres that are limited by the arachnoid membrane [1]. Cisterns are enlarged pockets within the arachnoid layer that fill the spaces where the brain folds itself [2]. Arachnoid cisterns are also formed around neurovascular structures such as cranial nerves and the cerebral vasculature [3]. Cerebrospinal fluid (CSF) flows through these structures. Fissures and cisterns are anatomical pathways for the surgeon to reach deeper regions and a protective sheath for delicate structures [4]. The Sylvian fissure is continued medially with a wider space, the Sylvian cistern. These terms can sometimes be used indistinctly by some authors. To avoid confusion, we refer to both as the **Sylvian** corridor. The Sylvian corridor is a wide and versatile pathway that allows access to the basal cisterns and their contents, avoiding transgression of the parenchyma or minimizing brain retraction [5].

Furthermore, the insula, central core, and mesial temporal region can also be exposed [6]. Knowledge of anatomy is mandatory for the microsurgeon, and a sharp, clean, arachnoidal dissection is the way to open these spaces and gently "creep into" the depths of the brain to finally access a lesion that needs to be taken care of. This chapter will attend to the anatomical aspects of the Sylvian corridor and the microsurgical nuances relevant to this structure.

P. A. Rubino (✉) · J. S. Bottan · R. Arévalo
Department of Neurosurgery, Hospital Aleman, Buenos Aires, Argentina

E. G. Figueiredo et al. (eds.), *Brain Anatomy and Neurosurgical Approaches*,
https://doi.org/10.1007/978-3-031-14820-0_10

2 History

The lateral cerebral fissure of the brain was originally described around the mid-1600s. However, it is uncertain if the first description is attributed to Sylvius or his disciple, Thomas Bartholin. Interestingly, Girolamo Fabrici d'Acquapendente illustrated the fissure 40 years before Sylvius's report, in *Tabulae Pictae* [7]. In the '60s, M.G. Yasargil led the transition from the lessons learned in H.G. Krayenbuhl's laboratory into the operating theater [8–12], introducing the operative microscope [13, 14], and thus permitting microsurgical exploration of structures such as the Sylvian fissure [15]. This allowed for revolutionary development in the surgical treatment of cerebral diseases such as aneurysms, tumors, and deep sitting lesions [9, 11, 12, 16].

3 Anatomy of the Sylvian Corridor

The Sylvian fissure is a deep sulcus that separates the frontoparietal and temporal opercula. It carries the middle cerebral artery M1, M2, and M3 branches within its depths. Thc M4 branches emerge on either side of the fissure to reach the lateral surface of the cerebral hemisphere. The disposition of the fissure is complex, starting at the orbitofrontal surface of the brain, lateral to the anterior clinoid process, and extends laterally in the convexity at the level of the supramarginal gyrus [17] (Fig. 1).

Concerning the cranial anatomical landmarks, the fissure runs at the level of the anterior squamous suture. The *anterior squamous point*, represented by the junction

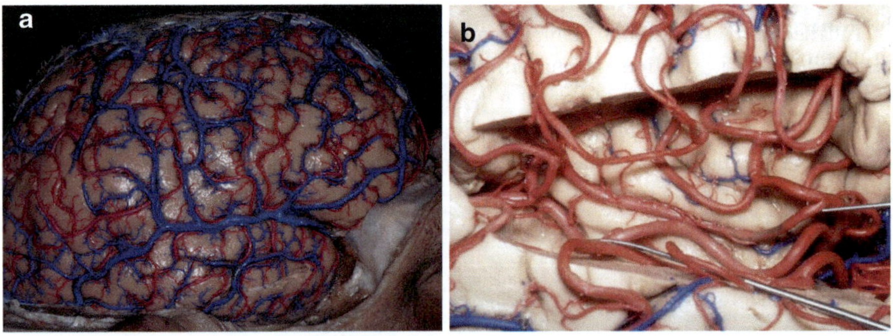

Fig. 1 The sylvian fissure. (**a**) In a lateral view of this right cerebral hemisphere. The disposition and extension of the sylvian fissure can be appreciated: commencing at the orbitofrontal surface of the brain, lateral to the anterior clinoid process and extending laterally in the convexity at the level of the supramarginal gyrus. The M4 branches of the middle cerebral artery (MCA), emerging from the fissure, and the superficial sylvian vein, with its tributaries, can be well appreciated. (**b**) The frontal and temporal opercula have been removed in this specimen and the insula is shown. The M2 and M3 branches of the MCA within the depths of the fissure and cistern

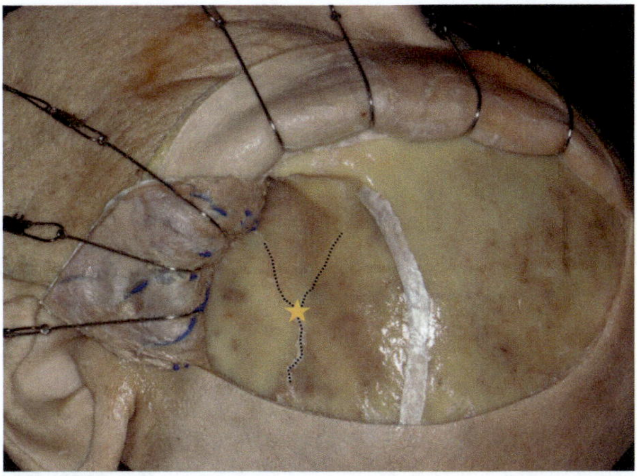

Fig. 2 The anterior squamous point is represented by the junction of the sphenoid wing, frontal and temporal bones, and is the topographical location of the anterior sylvian point, which divides the fissure in an anterior and a posterior rami

of the sphenoid wing, frontal and temporal bones, is the topographical location of the anterior Sylvian point, which in turn, is the most important surgical reference of the fissure [18] (Fig. 2). The anterior Sylvian point divides the fissure in anterior and posterior rami [17].

The stronger arachnoidal adhesions between the frontoparietal operculum and the temporal operculum give place to a broader space, deeper to the former, referred to as the **Sylvian cistern** (Fig. 3). This cistern is divided into two compartments: a proximal or **sphenoidal** compartment that harbors the M1 segment of the MCA and the distal or **operculoinsular** compartment containing the M2 and M3 branches [19]. The sphenoidal compartment separates the orbitofrontal cortex and anterior perforated substance from the temporal lobe. The operculoinsular compartment separates the frontal, parietal, and temporal operculum laterally and the insular surface medially. Both compartments are communicated at the level of the *limen insulae*. Thus, when the operculoinsular compartment is dissected in the laboratory or the OR, the insular surface represents "the floor" of this cavity.

The insula resembles a shallow three-sided pyramid with its base facing medially and the pyramid's tip being the insular apex. The short gyri are mostly located on the anterior facing side of the pyramid, while the long gyri represent the other two sides [20]. The shape of the insula is triangle-based anteriorly and with its tip pointing posteriorly. The anteroinferior angle of the insula is represented by the *limen insulae* [21]. The MCA's insular branches (M2) run in the Sylvian cistern and reach the edges of the insula (topographically, the limiting sulcus), and they loop, continuing as the M3 (opercular) branches, that run distally until reaching the Sylvian fissure more superficially, where they emerge as the M4 (cortical) branches [22] (Fig. 4).

The operculum covers the insula. The two upper thirds lay under the frontal and parietal lobes, represented by the inferior frontal gyrus (F3), while the lower third is covered by the superior temporal gyrus (T1) [23]. Therefore, a coronal cross-section of the Sylvian fissure and cistern will have an inverted "Y-" or "T-" shaped

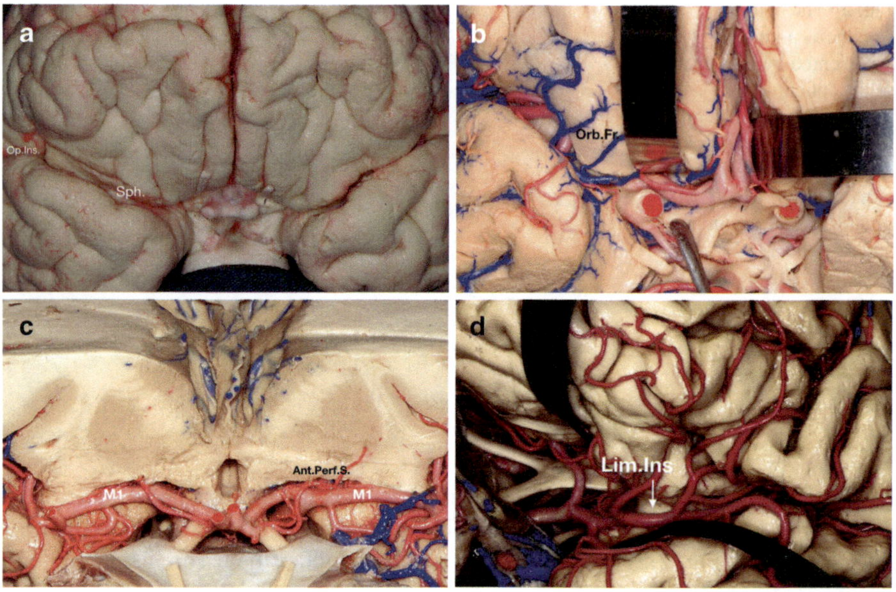

Fig. 3 Strong arachnoidal adhesions between the frontoparietal and the temporal opercula give place to the sylvian cistern. (**a**) This cistern is divided into two compartments: a proximal or sphenoidal compartment (Sph.), and a distal or operculoinsular compartment (Op.Ins.). (**b, c**) The sphenoidal compartment separates the orbitofrontal cortex (Orb.Fr.) and anterior perforated substance (Ant.Perf.S.) from the temporal lobe and contains the M1 segment of the MCA. (**d**) The operculo-insular compartment, which contains the M2 and M3 branches, separates the frontal, parietal, and temporal operculum laterally and the insular surface medially; the insular surface represents the floor of this cavity. Both compartments are communicated at the level of the limen insulae (Lim.Ins)

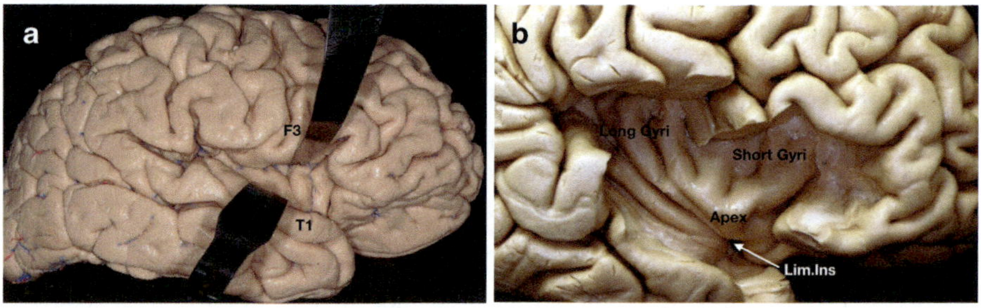

Fig. 4 Anatomy of the insula: In order to fully appreciate the anatomy of the insula, dissection and retraction of both, the frontoparietal and temporal operculum, is needed. (**a**) The insula is covered by the operculum. The two upper thirds lay under the frontal and parietal lobes, represented by the inferior frontal gyrus (F3), while the lower third is covered by the superior temporal gyrus (T1). (**b**) The insula resembles a shallow three-sided pyramid with its base facing medially and the tip of the pyramid being the insular apex, pointing laterally. The short gyri are mostly located in the anterior facing side of the pyramid while the long gyri represent the other two sides. The anteroinferior angle of the insula is represented by the limen insulae

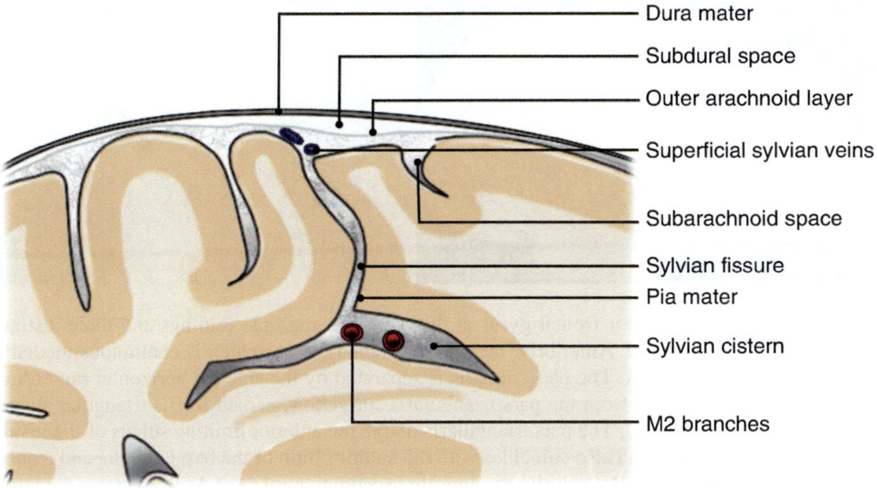

Fig. 5 Illustration of the disposition of the meninges and their relation to the sylvan fissure and cistern. Note the inverted "Y" or "T" shape of the cross section

disposition separating the three internal surfaces (Fig. 5). The Sylvian cistern entirely covers the superior surface of the temporal lobe. It is divided into two areas: anteriorly, the *planum polare*, posteriorly the *planum temporale,* and between both of them, the long oblique-running Heschl's gyrus [24].

The inferior frontal gyrus at this level has multiple foldings, and three distinct segments can be recognized. Anteriorly: the *pars orbitalis*, which is continuous medially with the orbitofrontal cortex. The pars orbitalis is separated by the horizontal rami of the Sylvian fissure from the *pars triangularis* (named for its triangular shape, with its tip facing inferiorly). The pars triangularis marks the anterior limiting sulcus of the insula in the fissure depth and coincides with the anterior limit of the basal ganglia and frontal horn of the lateral ventricle. Posteriorly: the anterior ascending rami separates the latter from the *pars opercularis*. This is a "U-" shaped segment that contains the most inferior aspect of the precentral sulcus. These structures are easily recognizable and are useful surgical landmarks [25] (Fig. 6). Broca's area usually resides in the superior half of the pars opercularis and the anterior half of the pars triangularis on the dominant hemisphere.

Across the fissure, and opposed to these structures, the superior temporal gyrus is mostly non-eloquent on its anterior part, but the posterior aspect of this convolution harbors Wernicke's area close to the supramarginal gyrus. As mentioned before, the superior temporal gyrus is continuous medially with the cisternal or superior surface of the temporal lobe. The inferior frontal gyrus correlates with the planum polare anteriorly. The precentral gyrus, which is frequently continuous with the posterior half of the pars opercularis, is a good landmark to identify the most external aspect of Heschel's gyrus. The postcentral gyrus and supramarginal gyrus are

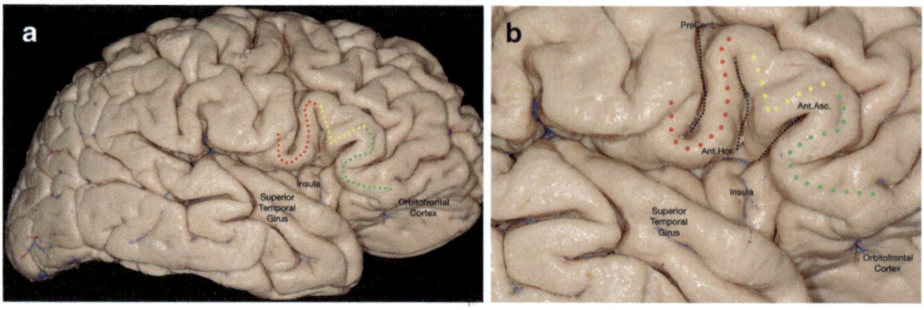

Fig. 6 (**a** and **b**) The inferior frontal gyrus at this level has multiple foldings and three distinct segments can be recognized. Anteriorly: the pars orbitalis (green), which is continuous medially with the orbitofrontal cortex. The pars orbitalis is separated by the anterior horizontal rami (Ant. Hor.) of the sylvian fissure from the pars triangularis (in yellow, named for its triangular shape, with its tip facing inferiorly). The pars triangularis marks the anterior limiting sulcus of the insula in the depth of the fissure and also coincides with the anterior limit of the basal ganglia and frontal horn of the lateral ventricle. Posteriorly: the anterior ascending rami (Ant.Asc) separates the latter from the pars opercularis. This is a "U" shaped segment which contains the most inferior aspect of the precentral sulcus (PreCent.)

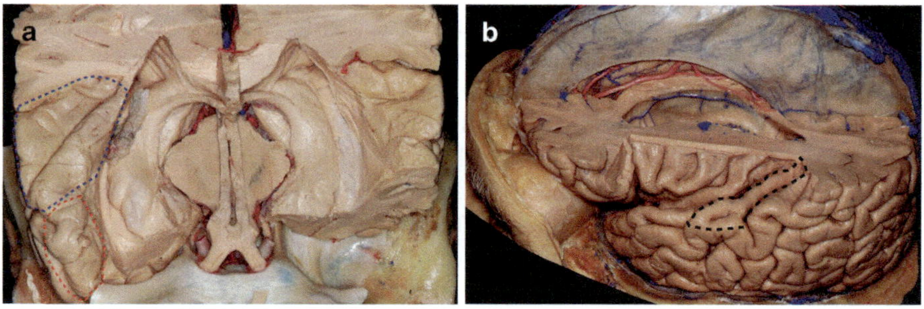

Fig. 7 The cisternal or superior surface of the temporal lobe: (**a**) the superior surface is continued laterally with the superior temporal gyrus. The inferior frontal gyrus correlates with the planum polare (red), anteriorly. The precentral gyrus, which is frequently continuous with the posterior half of the pars opercularis (ParsOp.), is a useful landmark to identify the most external aspect of Heschel's gyrus (**b**). The postcentral gyrus (PostC.) and supramarginal gyrus (SM) are coincidental with the planum temporale (**a**, blue)

coincidental with the planum temporale [26] (Fig. 7). Wernicke's area is usually seated at the level of the central lobule [27].

The superficial Sylvian veins are found in variable numbers, ranging from being absent in some cases to being a large venous complex with multiple veins of various sizes [28] (Fig. 8). However, the most common is to find 2–5 veins that typically drain anteriorly in the sphenoparietal sinus [29]. These veins are most frequently attached to the arachnoid over the fissure and run parallel to it, although they can be displaced inferiorly in some cases.

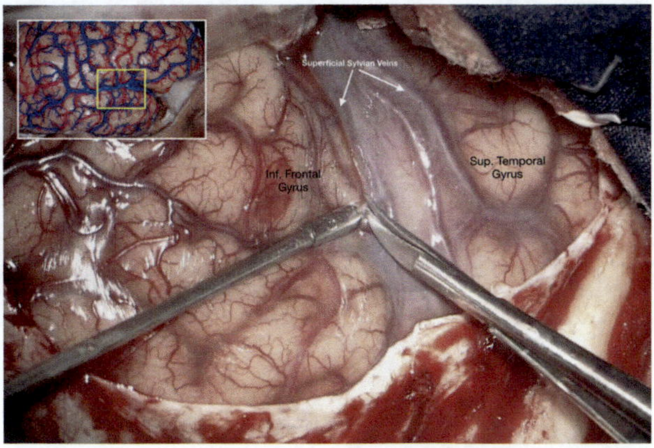

Fig. 8 The arachnoid membrane is incised by microscissors on the frontal side of the superficial sylvian veins at the level of the pars triangularis

4 Indications

The Sylvian corridor might be the most versatile approach in the microsurgeon's armamentarium (Fig. 9). When combined with the classic pterional craniotomy or extended variations of it [30, 31], this corridor can grant access to the sellar and parasellar regions [32], the anterior floor of the third ventricle via the *lamina terminalis* [33], the internal carotid artery and the carotid collar [34], the anterior clinoid process [35], the cavernous sinus [36], the interpeduncular cistern [37], the whole length of the middle cerebral artery [19], the A1 segment of the anterior cerebral artery and anterior communicating artery complex, the anterior perforated substance, the mesial temporal regions [26], the frontal and temporal horns of the lateral ventricle [38], the central core and basal ganglia [39], and the anterolateral surface of posterior fossa [40]. In addition, the sphenoidal compartment of the contralateral Sylvian cistern can be accessed as well from this corridor [41].

5 Surgical Technique

The "splitting" of the Sylvian fissure (Fig. 10) usually begins at the apex of the pars triangularis of the inferior frontal gyrus, the site where the brain parenchyma is most retracted and the arachnoid membrane can be incised safely [42, 43]. This landmark is known as the *anterior Sylvian point* [18]. The incision is carried out with a sharp instrument, either an arachnoid knife, micro scissors, or a number 11 blade, on the frontal side of the superficial temporal veins. These structures will be left attached to the temporal lobe to avoid damage by traction or the need to sacrifice them [19].

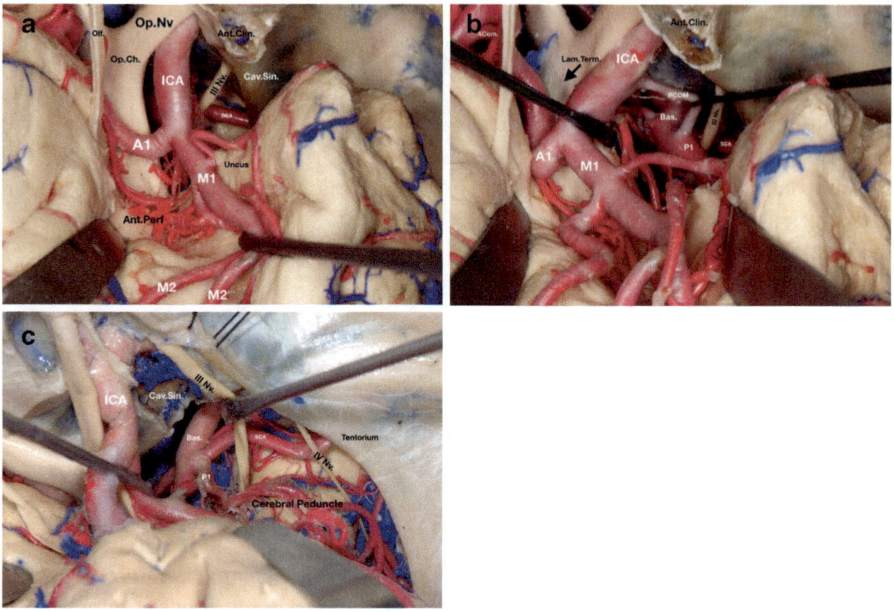

Fig. 9 The proximal sylvian fissure is dissected in this specimen. This corridor can grant access to the sellar and parasellar regions, the anterior floor of the third ventricle via the lamina terminalis (Lam.Term.), the internal carotid artery (ICA) and the anterior clinoid process (Ant.Clin), the cavernous sinus (Cav.Sin), the interpeduncular cistern, the whole length of the middle cerebral artery (M1–M2), the A1 segment of the anterior cerebral artery and anterior communicating artery complex (ACom), the anterior perforated substance (Ant.Perf), the mesial temporal regions, the frontal and temporal horns of the lateral ventricle, the central core and basal ganglia, and the anterolateral surface of posterior fossa. (**a, b**) Classic sylvian corridor. (**c**) The pretemporal corridor is an interesting way of complementing the sylvian corridor. It allows the surgeon to access the cavernous sinus and to perform an anterior petrosectomy in order to expose the anterolateral aspect of the brainstem

The arachnoid incision is wide enough to gently allow the bipolar forceps to spread the space between the temporal and frontal lobes. We recommend using small (0.5 mm) short non-stick bipolar forceps for the Sylvian fissure. The suction cannula can be used as a retractor as the arachnoid membranes are dissected. Using microscissors, the arachnoidal bands are cut, avoiding the vessels. Gentle counter traction with the suction cannula is useful to identify the avascular plane better to cut. It is advisable first to dissect in-depth, transitioning from the tight cleft of the fissure to the wider space of the Sylvian cistern. The dissection progresses gently until the M2 branches are identified. The "splitting" then continues proximally from "the inside to the outside," which usually makes the job of separating both lobes easier, especially in tighter fissures [19].

There are small superficial veins that cross as a bridge from one lip of the operculum to the other. These can be coagulated and divided. Arteries should be preserved and dissected carefully and only then displaced to the corresponding lobe. Unlike veins, M3–M4 branches don't cross over the fissure. They remain on their

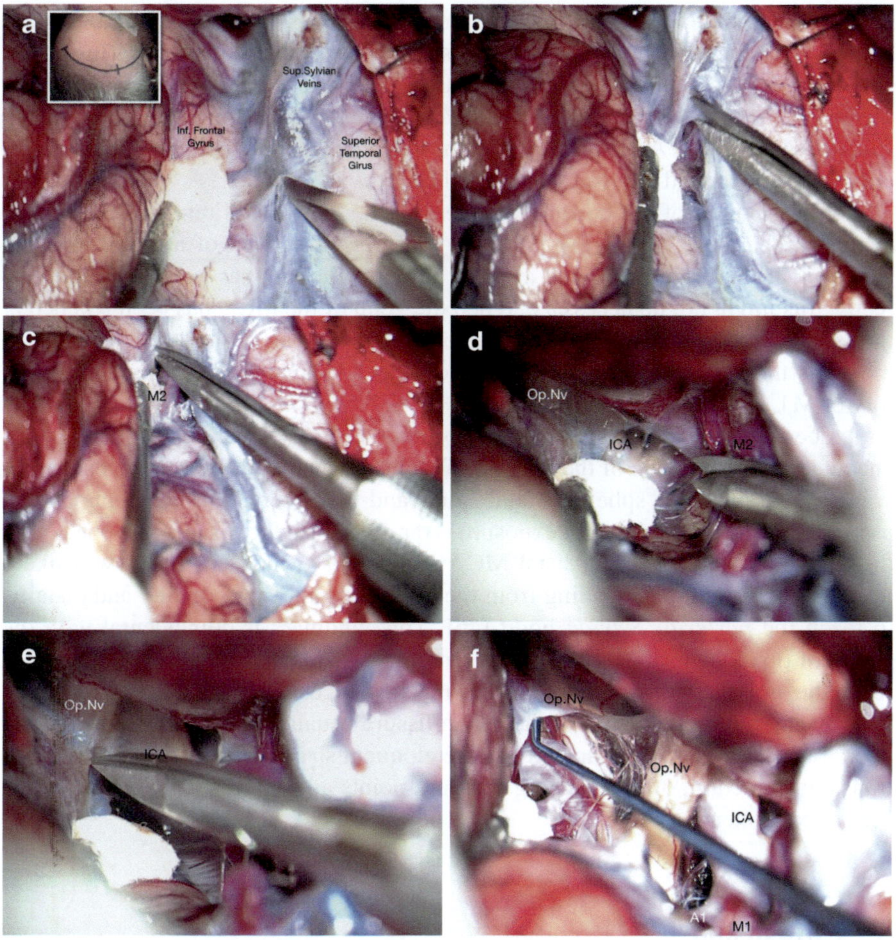

Fig. 10 Illustrative case. (**a**) The opening of the sylvian fissure usually begins at the anterior sylvian point where the brain parenchyma is most retracted and a safer incision can be made. (**b**) Using microscissors, the arachnoidal bands are cut avoiding any vessels. Gentle counter traction with the suction cannula is useful to better identify the avascular plane where to cut. (**c**) The dissection progresses gently until the M2 branches are identified. (**d**) The "splitting" then continues proximally following M1 to the carotid bifurcation. (**e, f**) In the final steps of the dissection, the complete opening of the sylvian corridor will allow the surgeon a full exposure of the neurovascular elements of the basal cisterns, which is, as already seen, a vast pathway to treat a grand variety of pathologies, both humoral and vascular. This will require the dissection of other arachnoidal membranes such as the suprasellar cistern (chiasmatic cistern), the carotid cistern and, if further depth is required, Liliequist membrane

corresponding lobe. Any bleeding from small veins or pial vessels should be meticulously coagulated, and the field must be kept clean at all times. Gentle compression with cottonoids will usually stop any minor venous oozing without the need for bipolar coagulation. In the case of ruptured aneurysms, keeping a clean surgical field, and identifying the structures under the microscope, can be difficult. Copious irrigation with saline can help clean some of the clots in the arachnoid. Certain authors have mentioned the benefits of using fibrinolytic agents such as urokinase in irrigation [44]. These can be dangerous in the presence of ruptured aneurysms; therefore, the surgeon should use these agents with caution and preferably away from the rupture site. Again, identifying the M2 branches and dissecting the perivascular spaces will lead the way to the basal cisterns [45].

As the opening of the corridor progresses, the frontal and temporal lobes will begin to separate. The surgeon can help with the placement of a brain retractor. However, if the brain is relaxed and the head positioning has been achieved correctly, this might not be necessary. When the genu of the MCA has been reached at the level of the limen insulae, the surgeon might need to switch to longer instruments as the dissection now will become deeper following the horizontal trajectory of M1.

A deep venous branch draining the orbital surface of the frontal lobe can be encountered in the depth of the sphenoidal compartment of the cistern. This vein or veins will drain in the sphenoparietal sinus, and its preservation is recommended, especially if they are large. Full exposure is rarely needed for vascular lesions within the basal cisterns. However, distal MCA aneurysms or insular tumors will likely need a wider opening, extending from the pars triangularis proximally and distally, to reach the *posterior Sylvian point* [46]. This landmark is coincidental with the posterior tip of the insula. For full exposure of the insula, the dissection will need to reach the anterior, superior, and inferior limiting sulcus [39].

Working within the depth of the silvan fissure usually requires retraction of the operculum. However, this can damage the eloquent structures [47]. Therefore, soft, transient retraction, often changing pressure points and sparing it whenever possible, is recommended.

6 Conclusion

The Sylvian corridor is a versatile avenue used by microdissection of the Sylvian fissure and cistern. It will grant access to a wide range of subdural, subarachnoid, and intraparenchymal lesions. Careful dissection and deep knowledge of anatomy are of capital importance for safer and more successful surgeries.

References

1. Shafique S, Rayi A. Anatomy, head and neck, subarachnoid space. In: StatPearls. Treasure Island, FL: StatPearls Publishing; 2021.
2. Adeeb N, Deep A, Griessenauer CJ, Mortazavi MM, Watanabe K, Loukas M, Tubbs RS, Cohen-Gadol AA. The intracranial arachnoid mater: a comprehensive review of its history, anatomy, imaging, and pathology. Childs Nerv Syst. 2013;29:17–33.

3. Lü J. Arachnoid membrane: the first and probably the last piece of the roadmap. Surg Radiol Anat. 2015;37:127–38.
4. Lü J, Zhu X-L. Cranial arachnoid membranes: some aspects of microsurgical anatomy. Clin Anat. 2007;20:502–11.
5. Ngando HM, Maslehaty H, Schreiber L, Blaeser K, Scholz M, Petridis AK. Anatomical configuration of the sylvian fissure and its influence on outcome after pterional approach for microsurgical aneurysm clipping. Surg Neurol Int. 2013;4:129.
6. Ribas EC, Yağmurlu K, de Oliveira E, Ribas GC, Rhoton A. Microsurgical anatomy of the central core of the brain. J Neurosurg. 2018;129:752–69.
7. Collice M, Collice R, Riva A. Who discovered the sylvian fissure? Neurosurgery. 2008;63:623–8. discussion 238.
8. Yasargil MG. Significance of microsurgery in brain surgery. Dtsch Med Wochenschr. 1969;94:1496–7.
9. Yaşargil MG. Intracranial microsurgery. Clin Neurosurg. 1970;17:250–6.
10. Yaşargil MG. Possibilities and limitations of vascular surgery of the central nervous system. Bull Schweiz Akad Med Wiss. 1969;24:487–93.
11. Yaşargil MG. Intracranial microsurgery. Proc R Soc Med. 1972;65:15–6.
12. Krayenbühl HA, Yaşargil MG, Flamm ES, Tew JMJ. Microsurgical treatment of intracranial saccular aneurysms. J Neurosurg. 1972;37:678–86.
13. Yaşargil MG, Krayenbühl H. The use of the binocular microscope in neurosurgery. Bibl Ophthalmol. 1970;81:62–5.
14. Krayenbühl H, Yasargil MG. The use of binocular microscopes in neurosurgery. Wien Z Nervenheilkd Grenzgeb. 1967;25:268–77.
15. Yasargil MG, Kasdaglis K, Jain KK, Weber HP. Anatomical observations of the subarachnoid cisterns of the brain during surgery. J Neurosurg. 1976;44:298–302.
16. Yasargil MG, Fox JL. The microsurgical approach to intracranial aneurysms. Surg Neurol. 1975;3:7–14.
17. Rhoton A. Rhoton's cranial anatomy and surgical approaches. Oxford: Oxford University Press; 2020.
18. Ribas GC, Ribas EC, Rodrigues CJ. The anterior sylvian point and the suprasylvian operculum. Neurosurg Focus. 2005;18:E2.
19. Yasargil MG. Microneurosurgery, volume I: microsurgical anatomy of the basal cisterns and vessels of the brain, diagnostic studies, general operative techniques . . . considerations of the intracranial aneurysms. New York: Thieme Medical Publishers.; 1984.
20. Bottan JS, Rubino PA, Lau JC, Macdougall KW, Parrent AG, Burneo JG, Steven DA. Robot-assisted insular depth electrode implantation through oblique trajectories: 3-dimensional anatomical nuances, technique, accuracy, and safety. Oper Neurosurg. 2020;18:278–83.
21. Guenot M, Isnard J, Sindou M. Surgical anatomy of the insula. Adv Tech Stand Neurosurg. 2004;29:265–88.
22. Türe U, Yaşargil MG, Al-Mefty O, Yaşargil DCH. Arteries of the insula. J Neurosurg. 2000;92(4):676–87. https://doi.org/10.3171/jns.2000.92.4.0676.
23. Tanriover N, Rhoton ALJ, Kawashima M, Ulm AJ, Yasuda A. Microsurgical anatomy of the insula and the sylvian fissure. J Neurosurg. 2004;100:891–922.
24. Wen HT, Rhoton AL, de Oliveira E, Cardoso AC, Tedeschi H, Baccanelli M, Marino R. Microsurgical anatomy of the temporal lobe: part 1: mesial temporal lobe anatomy and its vascular relationships as applied to amygdalohippocampectomy. Neurosurgery. 1999;45:549–91; discussion 591–2.
25. Muhammad S, Tanikawa R, Lawton M, Regli L, Niemelä M, Korja M. Microsurgical dissection of sylvian fissure-short technical videos of third generation cerebrovascular neurosurgeons. Acta Neurochir. 2019;161:1743–6.
26. Wen HT, Rhoton ALJ, de Oliveira E, Castro LHM, Figueiredo EG, Teixeira MJ. Microsurgical anatomy of the temporal lobe: part 2—sylvian fissure region and its clinical application. Neurosurgery. 2009;65:1–35. discussion 36.

27. Girvin JP. Temporal lobe surgery. In: Operative techniques in epilepsy surgery. Cham: Springer; 2015. p. 125–63.
28. Aydin IH, Tüzün Y, Takçi E, Kadioğlu HH, Kayaoğlu CR, Barlas E. The anatomical variations of sylvian veins and cisterns. Minim Invasive Neurosurg. 1997;40:68–73.
29. Oka K, Rhoton ALJ, Barry M, Rodriguez R. Microsurgical anatomy of the superficial veins of the cerebrum. Neurosurgery. 1985;17:711–48.
30. Chiarullo M, Voscoboinik DS, Taccone WV, Lafata JM, Rubino P, Lambre J. Abordaje pterional: alcances y técnica quirúrgica. Rev Arg Neuroc. 2014;28:156–61.
31. Taccone WV, Chiarullo MD, Voscoboinik DS, Caviglia MR, Bustamante JL, Medina L. Abordaje Fronto-Orbito-Cigomático en Dos Piezas: Indicaciones y Técnica Quirúrgica. Rev Arg Neuroc. 2013;27:119–23.
32. Krisht AF, Tindall GT. Pituitary disorders: comprehensive management. Baltimore: Lippincott Williams and Wilkins; 1999.
33. Apuzzo ML. Surgery of the third ventricle. Neurology. 1988;38(5):829. https://doi.org/10.1212/wnl.38.5.829-b.
34. Seoane E, Rhoton ALJ, de Oliveira E. Microsurgical anatomy of the dural collar (carotid collar) and rings around the clinoid segment of the internal carotid artery. Neurosurgery. 1998;42:866–9.
35. Krisht AF, Hsu SPC. Paraclinoid aneurysms. Part 1: superior (true ophthalmic) aneurysms. Contemp Neurosurg. 2008;30:1–6.
36. Krisht AF, Kadri PAS. Surgical clipping of complex basilar apex aneurysms: a strategy for successful outcome using the pretemporal transzygomatic transcavernous approach. Neurosurgery. 2005;56:261–73.
37. Seoane E, Tedeschi H, de Oliveira E, Wen HT, Rhoton ALJ. The pretemporal transcavernous approach to the interpeduncular and prepontine cisterns: microsurgical anatomy and technique application. Neurosurgery. 2000;46:891–9.
38. Ribas EC, Yagmurlu K, Wen HT, Rhoton AL. Microsurgical anatomy of the inferior limiting insular sulcus and the temporal stem. J Neurosurg. 2015;122(6):1263–73. https://doi.org/10.3171/2014.10.JNS141194.
39. Potts MB, Chang EF, Young WL, Lawton MT. Transsylvian-transinsular approaches to the insula and basal ganglia: operative techniques and results with vascular lesions. Neurosurgery. 2012;70:824–34; discussion 834.
40. Kawase T, Toya S, Shiobara R, Mine T. Transpetrosal approach for aneurysms of the lower basilar artery. J Neurosurg. 1985;63:857–61.
41. Rodríguez-Hernández A, Gabarrós A, Lawton MT. Contralateral clipping of middle cerebral artery aneurysms: rationale, indications, and surgical technique. Neurosurgery. 2012;71:114–6.
42. Cohen-Gadol AA. Atraumatic sylvian fissure split: nuances and pitfalls. Oper Neurosurg. 2020;18:217–24.
43. Tayebi Meybodi A, Borba Moreira L, Gandhi S, Preul MC, Lawton MT. Sylvian fissure splitting revisited: applied arachnoidal anatomy and proposition of a live practice model. J Clin Neurosci. 2019;61:235–42.
44. Tanikawa R. 2nd Rhoton Society Virtual Meeting and 8th International Zoomposium on Microneurosurgical Anatomy. Spec. Lect. Kamiyama's Instruments Surgeries. 2020. https://stanford.cloud-cme.com/course/courseoverview?P=3000&EID=34898.
45. Tanikawa R. Less invasive cistemal approach and removal of subarachnoid hematoma for the treatment of ruptured cerebral aneurysms. No Shinkei Geka. 2007;35:17–24.
46. Ribas GC, Yasuda A, Ribas EC, Nishikuni K, Rodrigues AJJ. Surgical anatomy of microneurosurgical sulcal key points. Neurosurgery. 2006;59:ONS177–210; discussion ONS210–1.
47. Safaee MM, Englot DJ, Han SJ, Lawton MT, Berger MS. The transsylvian approach for resection of insular gliomas: technical nuances of splitting the Sylvian fissure. J Neuro-Oncol. 2016;130:283–7.

Sylvian Fissure: Anatomical and a Clinical Correlations

Eduardo Carvalhal Ribas, Leila Maria da Róz, and Hung Tzu Wen

1 Sylvian Fissure and Cistern

The sylvian fissure (SyF) is the most prominent and complex fissure of the brain, promptly identifiable at the superolateral face of the brain, and harbors its underlying Sylvian cistern (SyC). Since 1976, Yasargil et al. [1] emphasized the importance of the SyF, describing in detail its microanatomy and related surgical approaches, to establish it as the main microneurosurgical corridor to the base of the brain. The transsylvian approach is possible through a pterional craniotomy and reaches anterior basal extrinsic lesions and frontobasal, mesial temporal, and insular intrinsic intracranial lesions. Other frontotemporal craniotomies, including pterional modifications, supraorbital and orbitozygomatic, can enhance basal approaches and expose the SyF [2, 3].

Although the cerebral surface is exposed "prima facie" to all scholars who open the cranium, attention, and description of its anatomy is relatively recent. The old Egyptians and Greek civilizations did not contribute deeply to its understanding, and Galenweres was later more concerned with the ventricular cavities [4, 5]. The first description of this lateral fissure is in the book *Casp. Bartolini Institutiones Anatomicae*, published in 1641 by Caspar Bartholin's son, Thomas [5]. He attributes its discovery to Franciscus de le Böe (1614–1672), who descended from a French family named de le Boë, but was born in Germany and spent much of his life in the Netherlands. He was also known as Dr. Sylvius and later personally described this fissure in his own text *Disputationem Medicarum*, published in 1663, stating:

E. C. Ribas · L. M. da Róz
Department of Neurosurgery, Hospital Israelita Albert Einstein, São Paulo, SP, Brazil

H. T. Wen (✉)
Department of Neurosurgery, Hospital das Clínicas, University of São Paulo Medical School, São Paulo, SP, Brazil

E. G. Figueiredo et al. (eds.), *Brain Anatomy and Neurosurgical Approaches*,
https://doi.org/10.1007/978-3-031-14820-0_11

"Particularly noticeable is the deep fissure or hiatus which begins at the roots of the eyes (*oculorum radices*) . . . it runs posteriorly above the temples as far as the roots of the brain stem (*medulla radices*). . . . It divides the cerebrum into an upper, larger part and a lower, smaller part" [6]. The painting of the lateral fissure is present in earlier works, as seen at the *Tabulae Pictae* made by the Italian anatomist Girolamo Fabrici D′ Acquapendente in 1600, but these plates lack text descriptions and were probably used as instructional aids during dissections [5].

From the embryological perspective, the primitive neural tube is first covered by a single layer of cells, which will form the intima pia, and later surrounded by a mesoderm derived tissue called *meninx primitive,* which cavitates to form the dura mater and arachnoid. By the end of the embryonic stages, the basal cisterns formation is correlated with regression of the arachnoid trabeculae [2]. The SyF results from the growth of the infoldings frontal, parietal, and temporal opercula, overlapping onto the insula and becoming narrow only in humans, in whom the frontoparietal operculum is particularly developed [4].

The SyC, found inside the SyF, is transitional between the basal cisterns and the subarachnoid space over the superolateral surface of the brain (Fig. 1). The frontotemporal arachnoid membrane covers the basal surface and the lateral surface of the cerebrum. At the SyF, this membrane bridges the antero-medial portion of the temporal pole (planum polare) to the posterior portion of the lateral, posterior, and medial orbital gyri. Also, it continues deeply and medially, where several cisterns confluence above the internal carotid artery bifurcation (carotid, chiasmatic, olfactory, lamina terminalis, Sylvian, crural, and interpeduncular). These multiple cisterns junction forms thickened bands of arachnoid from the area of the olfactory trigone to the lateral optic nerve and mesial temporal lobe, running across and completely enclosing the origins of the anterior and middle cerebral arteries. As the middle cerebral artery (MCA) continues laterally, it pierces the arachnoid and enters the SyC. The SyC begins narrow and extends laterally, under the anterior perforated substance, enlarging first at the limen insula to encompass the middle cerebral artery bifurcation and later spreading over all insular surface. It contains the MCA and its major branches, the origins of the lenticulostriate, temporopolar and anterior temporal arteries, the superficial and deep Sylvian veins (with insular branches) [2].

According to Gibo et al. [7], the SyF presents a superficial part, visible on the brain's surface, and the deep part, hidden below the surface.

The superficial part has a stem and three rami. The stem starts just lateral to the anterior perforated substance, between the lateral olfactory stria and the rhinal incisura or temporal incisura, and proceeds laterally and anteriorly toward the lateral surface of the brain [8]. This trajectory is located directly posterior to the sphenoid ridge, between the frontal and temporal lobes, and laterally reaches the anterior sylvian point (ASyP) [3, 9]. The average length of the sylvian stem is 39 mm (range 30–56 mm) [9], and rarely follows a straight line. Often the lateral orbital gyrus projects inferiorly, indenting and compressing the area of the temporal pole or proximal segment of the superior temporal gyrus, causing a C- or S-shaped course of the SyF stem [10].

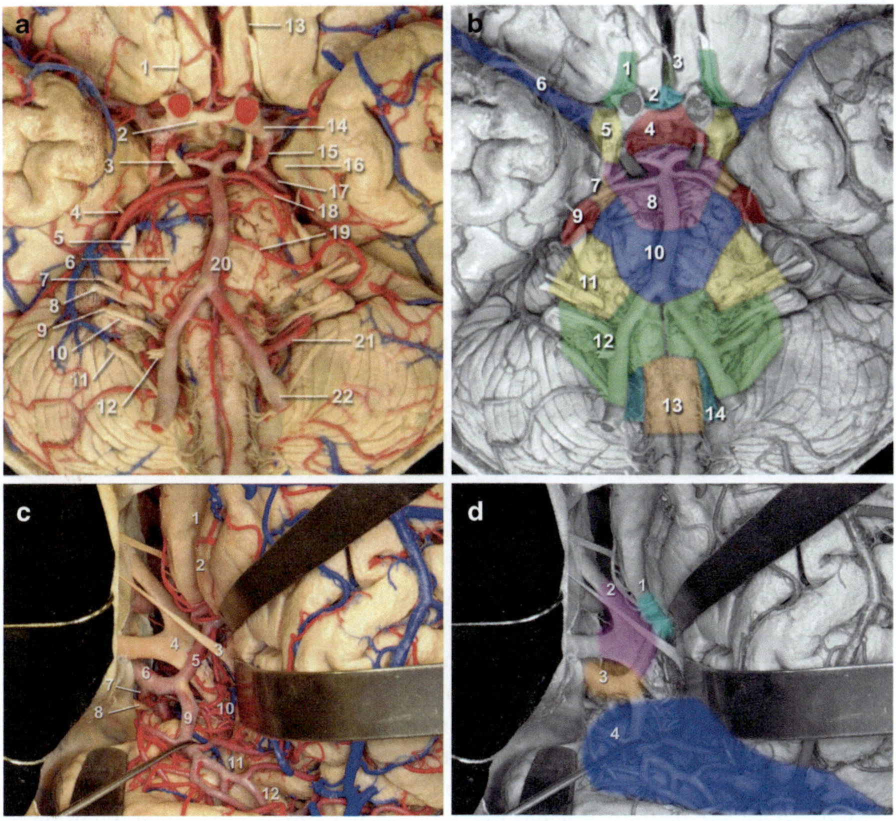

Fig. 1 Basal and sylvian cisterns. (**a**) Nervous and vascular structures related to the basal cisterns (anteroinferior view). (*1*) Callosal cistern, (*2*) lamina terminalis cistern, (*3*) carotid cistern, (*4*) sylvian cistern, (*5*) trigeminal nerve (V), (*6*) abducens nerve (VI), (*7*) facial nerve (VII), (*8*) vestibulocochlear nerve (VIII), (*9*) glossopharyngeal nerve (IX), (*10*) vagus nerve (X), (*11*) accessory nerve (XI), (*12*) hypoglossal nerve (XII), (*13*) olfactory sulcus, (*14*) internal carotid artery, (*15*) posterior communicating artery, (*16*) uncus, (*17*) posterior cerebral artery, (*18*) superior cerebellar artery, (*19*) anterior inferior cerebellar artery, (*20*) basilar artery, (*21*) posterior inferior cerebellar artery, (*22*) vertebral artery. (**b**) The basal and sylvian cisterns have been overlaid with colors. (*1*) lamina terminalis cistern, (*2*) callosal cistern, (*3*) olfactory cistern, (*4*) chiasmatic cistern, (*5*) carotid cistern, (*6*) sylvian cistern, (*7*) crural cistern, (*8*) interpeduncular cistern, (*9*) ambient cistern, (*10*) prepontine cistern, (*11*) superior cerebellar-pontine cistern, (*12*) inferior cerebellar-pontine cistern (lateral cerebello-medullary), (*13*) anterior spinal cistern, (*14*) posterior spinal cistern. (**c**) Nervous and vascular structures identified at a lateral view through a pterional approach. (*1*) interhemispheric fissure, (*2*) rectus gyrus, (*3*) olfactory tract (I), (*4*) chiasm (II), (*5*) anterior cerebral artery, (*6*) internal carotid artery, (*7*) posterior communicating artery, (*8*) oculomotor nerve (III), (*9*) middle cerebral artery - M1 segment, (*10*) lenticulostriate arteries, (*11*) limen insula, (*12*) middle cerebral artery - M2 segment. (**d**) Basal and sylvian cisterns, seen at a lateral view through a pterional approach, have been overlaid with colors. (*1*) Callosal cistern, (*2*) lamina terminalis cistern, (*3*) carotid cistern, (*4*) sylvian cistern

The ASyP was defined by Taylor and Haugton in their study of the topography of the convolutions and fissures of the brain published in 1900 [11], using the term Sylvian point as "the point where the main stem of the fissure of Sylvius reaches the outer aspect of the hemisphere," and later more deeply reviewed by Ribas et al. The ASyP is located inferior to the triangular part and anterior/inferior to the opercular part of the inferior frontal gyrus. Due to the usual retraction of the triangular part of the IFG concerning the SyF, the ASyP usually has a cisternal and enlarged appearance that facilitates its identification and can be used intraoperatively as an initial landmark to identify other important neural and sulcal structures along the fissure that are usually hidden by arachnoidal and vascular coverings.

From the ASyP, the stem divides into anterior horizontal, anterior ascending, and posterior rami. The two anterior rami divide the inferior frontal gyrus into the pars orbitalis, the pars triangularis, and the pars opercularis. The posterior ramus extends backward, separating the frontal and parietal superiorly from the temporal lobe inferiorly, and finishes at the posterior Sylvian point, where the posterior ascending ramus of the SyF originates and enters the supramarginal gyrus [3]. This segment measures 6 to 9 cm in length and courses in a slightly or moderately undulating line due to indentations of the frontal, parietal, and temporal gyri into the SyF [10]. The superior and inferior margins of the SyF constitute the frontoparietal and temporal operculum, which cover the superior and inferior aspects of the insula [3] (Fig. 2).

The deep compartment of the SyF on the basal surface encompasses a sphenoidal compartment that extends into the anterior insular cleft, a space between the posterior surface of the lateral and posterior orbital gyri and the anterior surface of the insula. The sphenoidal compartment is a narrow space posterior to the sphenoid ridge between the frontal and temporal lobes that communicates medially with the carotid cistern and laterally reaches the limen insula region, at the lateral margin of the anterior perforated substance [8]. It harbors the M1 segments of MCA, the extracerebral lateral lenticulostriate arteries, the deep Sylvian vein, and, occasionally, M2 segments, which may originate in the proximal or middle part of the M1 segment [10].

The sphenoidal compartment opens into the space that faces the anterior and lateral surfaces of the insula, which were named by Wen et al. as the anterior and lateral operculoinsular compartments, respectively [8]. Gibo et al. divided the anterior and posterior parts of the deep SyF as the sphenoidal and operculoinsular compartments, respectively [7]. Szikla et al. [12] also subdivided the SyF, naming superficial, intermediate, and deep planes, which are related to the superficial, opercular, and insular compartments of Gibo et al. [7], respectively [8].

More recently, Inoue et al. [13], described in detail the supratentorial arachnoidal membranes using microscope and endoscope views. More superficially, the outer arachnoid membranes surround the whole brain and, more deeply, the subarachnoid space is divided into cisterns by 12 inner arachnoid membranes: diencephalic, mesencephalic, medial carotid, intracarotid, intracrural, olfactory, medial, and lateral lamina terminalis, and proximal, medial, intermediate, and lateral Sylvian membranes. The proximal Sylvian membrane extends from the posterior orbital gyri to the anterior part of the uncus, and it separates the Sylvian and the carotid cisterns.

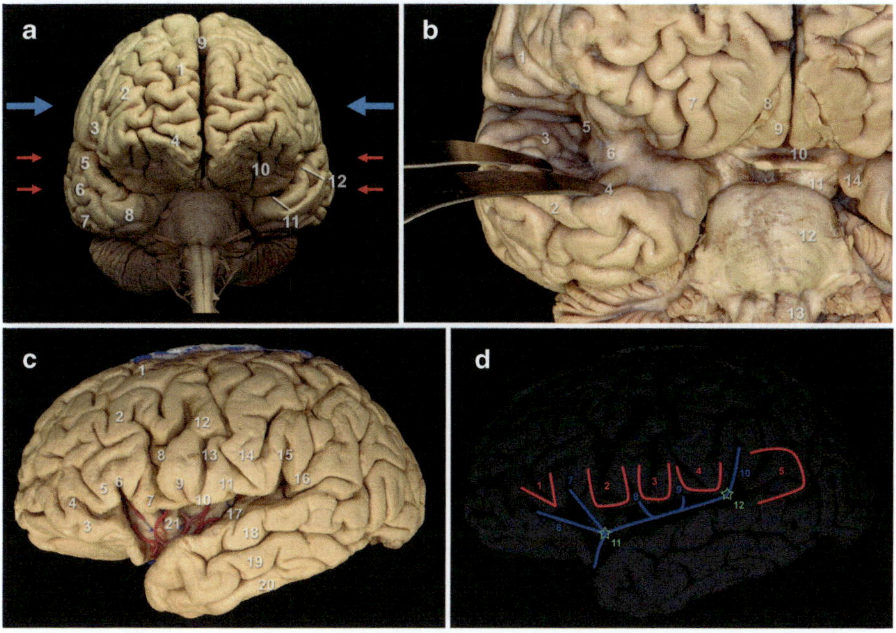

Fig. 2 Sylvian stem and opercula. (**a**) Anatomical brain dissection (frontal view): the largest transverse diameter of the brain in the suprasylvian region corresponds to the postcentral gyrus (blue large arrows). The largest transverse diameter of the brain in the infrasylvian region corresponds to the superior or middle temporal gyrus, in the coronal plane of the postcentral gyrus (small red arrows). (*1*) Superior frontal gyrus, (*2*) middle frontal gyrus, (*3*) inferior frontal gyrus, (*4*) frontal pole, (*5*) superior temporal gyrus, (*6*) middle temporal gyrus, (*7*) Inferior temporal gyrus, (*8*) temporal pole, (*9*) interhemispheric fissure, (*10*) orbital gyri, (*11*) sylvian fissure - stem, (*12*) sylvian fissure - lateral extension. (**b**) Magnified view of the sylvian stem (frontal view). (*1*) Suprasylvian operculum (inferior frontal gyrus), (*2*) infrasylvian operculum (superior temporal gyrus), (*3*) planum temporale, (*4*) planum polare, (*5*) insular surface, (*6*) limen insula, (*7*) orbital gyri, (*8*) olfactory tract, (*9*) rectus gyrus, (*10*) chiasm, (*11*) midbrain, (*12*) pons, (*13*) medulla, (*14*) uncus. (**c**) Anatomical brain dissection (lateral view). (*1*) Superior frontal gyrus, (*2*) middle frontal gyrus, (*3*) inferior frontal gyrus - pars orbitalis, (*4*) anterior horizontal ramus of sylvian fissure, (*5*) inferior frontal gyrus - pars triangularis, (*6*) anterior ascending ramus of sylvian fissure, (*7*) inferior frontal gyrus - pars opercularis, (*8*) precentral sulcus, (*9*) anterior subcentral ramus, (*10*) subcentral gyrus, (*11*) posterior subcentral ramus, (*12*) precentral gyrus, (*13*) central sulcus, (*14*) postcentral gyrus, (*15*) posterior ascending ramus of sylvian fissure, (*16*) supramarginal gyrus, (*17*) Heschl's gyrus, (*18*) superior temporal gyrus, (*19*) middle temporal gyrus, (*20*) inferior temporal gyrus, (*21*) insular surface. (**d**) The frontoparietal operculum extends from the anterior sylvian point (ASyP) to the posterior ascending branch of the SyF and can be understood as a series of convolutions roughly arranged, from anterior to posterior, in a V shape convolution, followed by three U-shaped convolutions and one C-shaped convolution. (*1*) Inferior frontal gyrus - pars triangularis, (*2*) inferior frontal gyrus - pars opercularis, (*3*) subcentral gyrus, (*4*) postcentral and supramarginal gyri connection, (*5*) supramarginal gyrus, (*6*) anterior horizontal ramus of sylvian fissure, (*7*) anterior ascending ramus of sylvian fissure, (*8*) anterior subcentral ramus, (*9*) posterior subcentral ramus, (*10*) posterior ascending ramus of sylvian fissure, (*11*) anterior sylvian point, (*12*) posterior sylvian point

The MCA originates at the carotid cistern and courses laterally, piercing this proximal Sylvian membrane and entering the SyC. The MCA surpasses the limen insula and spreads over the insular surface medially and the intermediate Sylvian membrane laterally, extending from the frontoparietal operculum to the upper surface of the temporal lobe along the medial edge of the superior and transverse temporal gyri. The lateral Sylvian membrane is found more superficially, connecting the frontoparietal and temporal opercula. The superficial Sylvian veins are sandwiched into the narrow space formed by the lateral Sylvian membrane (medially) and the outer arachnoid cerebral membrane (laterally). The medial Sylvian membrane extends inferomedially from the medial edge of the frontoparietal operculum and attaches to the insula.

Other and multiple interopercular sulci are found at the medial aspect of the opercula, and they are usually oblique and curved due to the indentations of the opponent gyri. Their usual depth has been related to each SyF part, tending to increase towards the distal SyF. The depth of the interopercular sulci is 10–20 mm at the proximal SyF part (between posterior medial orbital, posterior orbital, posterolateral orbital and suborbital gyri, and temporal pole, polar planum, and Schwalbe gyri), 25–40 mm at the SyF middle section (between subtriangular, subopercular, subprecentral gyri, and the opponent parts of the superior temporal gyrus) and increases to 35–50 mm at the distal SyF part (between the subcentral, the anterior, middle, and posterior transverse gyri of the inferior parietal lobe and the anterior and posterior transverse temporal gyri and temporal planum) [10].

The SyC is contained inside the SyF and assumes different shapes and sizes along its course. Proximally, inside the SyF deep compartment and named by anatomists "Vallecula" or preinsular sulcus [10], it measures 30 to 39 mm in length and 5 to 6 mm in width. The vallecula is covered laterally by the proximal medial part of the superior temporal gyrus (planum polare) and medially by the lateral orbital gyrus, which often protrudes and indents inside each other, conferring curving C or S shape to this SyF/SyC part. At the middle SyF part (insular section), the SyC increases to 6 to 7 cm in length, 5 to 6 cm in width, and 3 to 5 mm in depth, extending from the level of the limen insula to the posterior insular point. Yasargil et al. describes it as opening deeply, between the opercula laterally and the insular surface medially, to form 4 pouches: the anterior pouch, beneath the lateral orbital gyrus to the anterior limiting insular sulcus; the superior pouch, beneath the frontal operculum to the superior limiting insular sulcus; the posterior inferior, beneath the parietal operculum to the retroinsular fossa; and the inferior pouch, beneath the temporal operculum to the inferior limiting insular sulcus [10].

2 Fronto-Orbital Operculum

The fronto-orbital operculum is formed by the posterior orbital gyrus, the posterior portion of the lateral orbital gyrus, and the pars orbitalis of the inferior frontal gyrus. It covers the anterior surface of the insula, harbors inside the anterior operculum insular compartment of the SyF, and is related to the anterior limiting sulcus of the insula [8, 9]. The medial surface of the fronto-orbital operculum is formed by two suborbital gyri (superior and inferior), which are continuous with the accessory insular gyrus and the anterior surface of the anterior short gyrus [9].

The posterior orbital gyrus merges medially with the posterior portion of the medial orbital gyrus and laterally with the posterior portion of the lateral orbital gyrus, forming the posteromedial and posterolateral orbital lobules, respectively [9]. The posteromedial orbital lobule continues deeply with the transverse short insular gyrus [9].

3 Frontoparietal Operculum

The frontoparietal operculum covers the superior surface of the insula, harbors inside the lateral operculo-insular compartment of the SyF, and is related to the superior limiting sulcus of the insula [8, 9].

It extends from the anterior Sylvian point (ASyP) to the posterior ascending branch of the SyF. The suprasylvian structures can be understood as a series of convolutions roughly arranged, from anterior to posterior, in a V shape with its vertex constituted by the ASyP, followed by three U-shaped convolutions and one C-shaped convolution. The bottoms of the three U-shaped convolutions and their related sulcal extremities can be superior to the SyF or inside it (Fig. 2).

The inferior frontal gyrus (IFG) is located superiorly to the SyF, forming the most anterior segment of the frontoparietal operculum. The IFG merges anteriorly with the middle frontal gyrus and posteriorly is connected to the precentral gyrus. The anterior horizontal and anterior ascending rami of the SyF divides the IFG into three parts, from anterior to posterior: the pars orbitalis, the pars triangularis, and the pars opercularis. The anterior horizontal ramus can be understood as an extension of the superior limiting sulcus of the insula, and the anterior ascending ramus as an extension of the anterior limiting sulcus of the insula [9]. The insula sulci confluence of anterior and superior limiting sulci is identified as the “anterior insular point” [9].

The pars orbitalis is located anterior to the insula and is not seen as part of the frontoparietal operculum. The pars triangularis has a triangular shape, is often divided by a small descending branch of the inferior frontal sulcus, and is typically more retracted than the other two IFG parts, allowing the ASyP to have a cisternal appearance [3]. The pars opercularis is delimited anteriorly by the anterior ascending ramus of the SyF and posteriorly by the anterior subcentral ramus of the SyF. A U-shaped convolution is formed with the inferior most portion of the precentral sulcus inside. The pars opercularis anteriorly and its connection with the precentral gyrus posteriorly [3].

The precentral and postcentral gyri are connected superiorly and inferiorly, forming together an ellipse. The inferior connection, also called the subcentral gyrus, is delimited by the anterior and posterior subcentral rami of the SyF and presents as a U-shaped convolution with the inferior part of the central sulcus inside [3].

The third U-shaped convolution is composed of the arm connecting the postcentral and supramarginal gyri, harboring inside the inferior part of the postcentral sulcus and delimited anteriorly by the posterior subcentral ramus of the SyF and posteriorly by the posterior ascending ramus of the SyF [3].

The supramarginal gyrus can be understood as a posterior and superior continuation of the superior temporal gyrus. Their connection encircles the posterior end of the SyF, having the posterior ascending ramus of the SyF inside. This C-shaped convolution completes the frontoparietal operculum [3].

The medial aspect of each portion of the frontoparietal operculum is related to a particular portion of the insula. From anterior to posterior, the pars triangularis covers the anterior short insular gyrus; the pars opercularis covers the short insular sulcus, the middle short insular gyrus, and the posterior part of the anterior short insular gyrus; the precentral gyrus covers the middle short insular gyrus and the precentral insular sulcus; the subcentral gyrus covers the central insular sulcus; and the anterior transverse parietal gyrus covers the postcentral insular sulcus and the superior portion of the anterior and posterior long insular gyri, which are adjacent to the anterior transverse temporal gyrus. The middle transverse parietal gyrus covers the transverse temporal sulcus, and the posterior transverse parietal gyrus of the frontoparietal operculum and the temporal planum overlap and forms the medial aspect of the supramarginal gyrus [9].

4 Temporal Operculum

The temporal operculum is formed by the temporal pole, superior temporal gyrus, the inferior portion of the supramarginal gyrus covering the insula's inferior surface, and the anterior perforated substance [9]. The medial aspect of the operculum is the inferior wall of the SyF and comprises, from anterior to posterior, the planum polare, Heschl's gyri, and planum temporale.

The planum polare covers the inferior surface of the insula and is adjacent to the inferior limiting sulcus of the insula, along two-thirds of its length [9, 14]. Its convolutions are termed the "sulci and gyri of Schwalbe" [9]. The anterior part of the planum polare has its axis oriented lateromedially, particularly anterior to the pars triangularis. The posterior part, located between the anterior edge of the precentral gyrus and the Heschl's gyri, has its main axis oriented anteroposteriorly.

The Heschl's gyrus, designated as the most anterior transverse temporal gyrus or Heschl's gyri for the anterior and posterior transverse temporal gyri, is separated by the transverse temporal sulcus. Superficially, the Heschl's gyrus and the superior temporal gyrus usually join in the coronal plane at the level of the external acoustic meatus, where it can be promptly identified due to its characteristic hump [8].

Deeply, the Heschl's gyrus is directed posteriorly and covers the posterior third (the posterior portion) of the inferior limiting sulcus of the insula [9, 14].

The planum temporale constitutes the posterior portion of the inner surface of the temporal operculum and is usually composed of two transverse temporal gyri: the middle and the posterior [8]. The planum temporale is flat and, together with the Heschl's gyrus, form a triangular area with its apex pointing medially toward the retroinsular region, the posterior portion of the posterior limb and the retrolentiform parts of the internal capsule, and more deeply the atrium of the lateral ventricle (Fig. 3) [8].

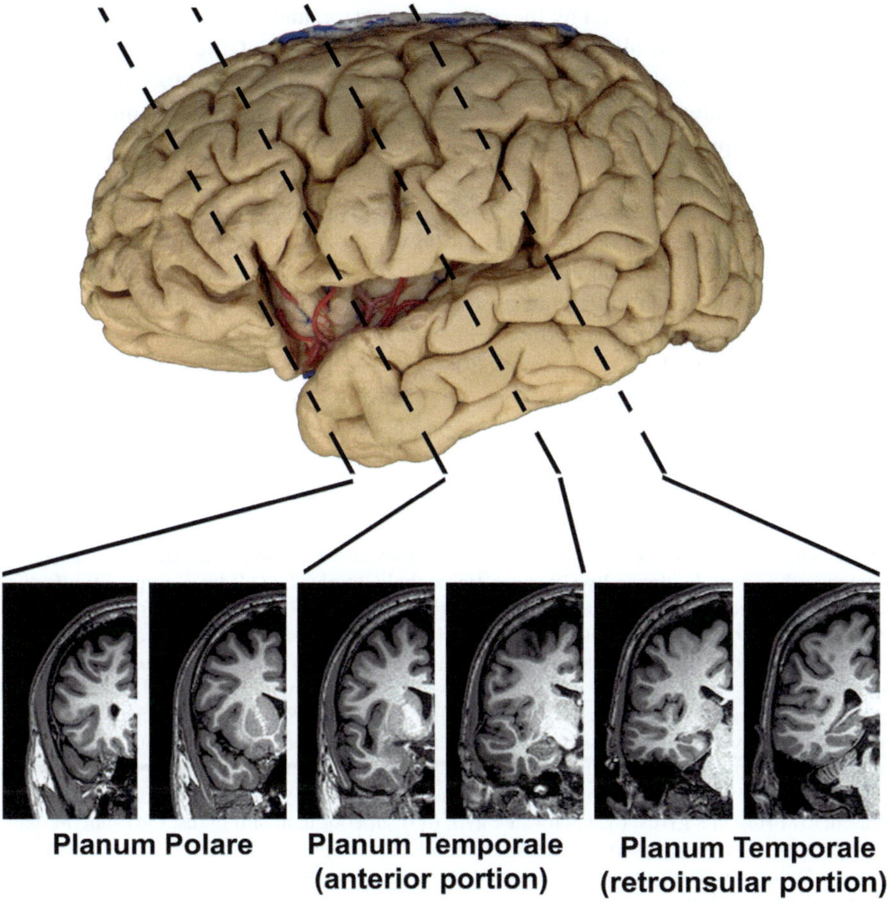

Fig. 3 Sylvian cistern morphology at coronal section (MRI). The sylvian cistern is oblique at the planum polare, horizontal with the insula at its depth at the anterior portion of the planum temporale and horizontal without the insula at its depth at the posterior portion of the planum temporale

5 Insula

The central core stands as a block on top of the brainstem, at the morphological center of the supratentorial compartment [15]. This solid block includes, from lateral to medial, the insular surface, the extreme capsule, the claustrum, the external capsule, the putamen, the globus pallidus, the internal capsule, the caudate nucleus, the stria terminalis, the septal region, and the thalamus.

The insular surface is the lateral aspect of the brain's central core and can be exposed surgically by dissecting the SyF and retracting its opercula (Fig. 4). This surface has a triangular shape, with its vertex located anteriorly and posteriorly at the limen insula. It is limited by the anterior, superior, and posterior insular limiting sulci (also referred to all together as the circular sulcus of the insula) [15]. The limen insula is a hook-like narrow structure formed by the olfactory cortex, connected to the anterior and inferior limiting sulci, defining at their meetings the "frontal limen point" and the "temporal limen point," respectively [15]. The superior limiting sulcus meets the anterior and inferior insular sulci, at regions referred to as the anterior insular point and posterior insular point, respectively [9].

The insula has lateral and anterior surfaces, and a central insular sulcus subdivides the lateral surface into anterior and posterior zones. The anterior zone is composed by the anterior, middle, and posterior short gyri that converge and fuse in the apex of the insula, which is the most convex or prominent portion of the insula. The posterior zone usually consists of one long gyrus that bifurcates posteriorly. The long gyrus is directed anteroinferior and reaches the most anteroinferior aspect of the insula pole but does not reach the insular apex [8].

Fig. 4 The insular surface and sylvian cistern. (**a**) Illustration of the insular surface, exposed by retracting the opercula, drawn by Giovanna Hespanhol Gütschow. (*1*) Rectus gyrus, (*2*) olfactory sulcus, (*3–6*) Medial: anterior: lateral and posterior orbital gyri, (*7*) orbital sulci, (*8*) posteromedial orbital lobule, (*9*) inferior frontal gyrus - pars orbitalis, (*10*) anterior horizontal ramus of sylvian fissure, (*11*) inferior frontal gyrus - pars triangularis, (*12*) anterior ascending ramus of sylvian fissure, (*13*) inferior frontal gyrus - pars opercularis, (*14*) inferior frontal sulcus, (*15*) middle frontal gyrus, (*16*) superior frontal sulcus, (*17*) superior frontal gyrus, (*18*) precentral sulcus, (*19*) precentral gyrus, (*20*) central sulcus, (*21*) postcentral gyrus, (*22*) anterior subcentral ramus, (*23*) subcentral gyrus, (*24*) posterior subcentral ramus, (*25*) postcentral sulcus, (*26*) superior parietal lobule, (*27*) intraparietal sulcus, (*28*) supramarginal gyrus, (*29*) angular gyrus, (*30*) intermediate sulcus of Jensen, (*31*) Heschl's gyrus, (*32*) Schwalbe gyri, (*33*) superior temporal gyrus, (*34*) superior temporal sulcus, (*35*) middle temporal gyrus, (*36*) inferior temporal sulcus, (*37*) inferior temporal gyrus, (*38–40*) superior: middle and interior occipital gyri, (*41*) transverse insular gyrus, (*42*) accessory insular gyrus, (*43*) short insular gyri, (*44*) insular apex, (*45*) central insular sulcus, (*46*) long insular gyri, (*47–49*) anterior: superior and inferior limiting insular sulci. (**b**) Illustration of the sylvian cistern compartments, exposed by retracting the opercula. (**c**) Illustration of the sylvian cistern pouches, exposed by retracting the opercula. (**d**) Anatomical insular dissection, after removing the opercula (lateral view). (*1*) Transverse insular gyrus, (*2*) accessory insular gyrus, (*3*) short insular gyri, (*4*) insular apex, (*5*) central insular sulcus, (*6*) long insular gyri, (*7–9*) anterior: superior and inferior limiting insular sulci, (*10*) limen insula, (*11*) Heschl's gyrus, (*12*) anterior insular point, (*13*) posterior insular point, (*14*) frontal limen point, (*15*) temporal limen point

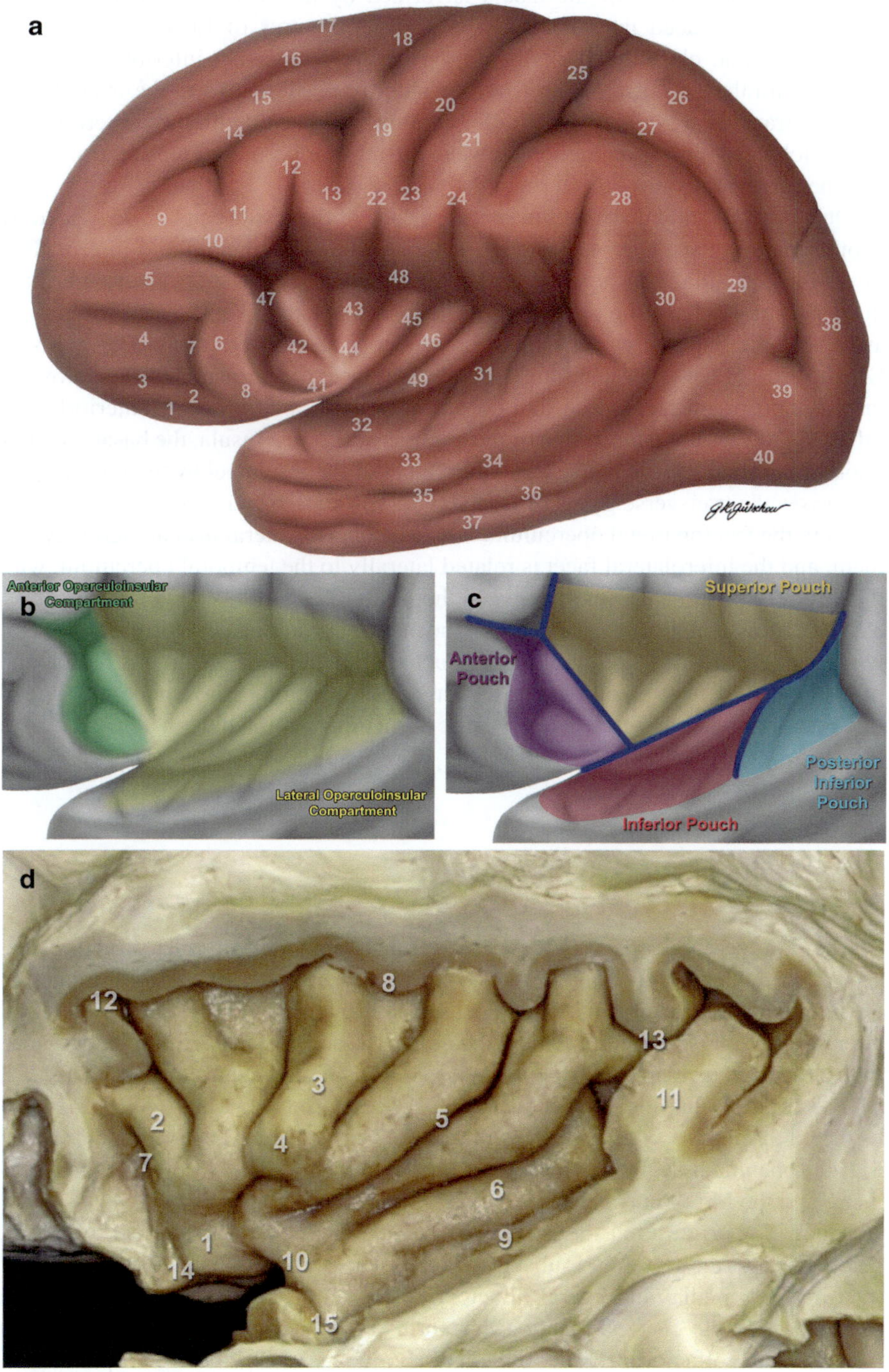
a
17
18
16
25
15
20
26
14
19
21
27
12
13
22
23
24
28
9
11
10
5
48
47
29
30
43
45
38
4
7
6
42
44
46
3
41
49
31
39
2
8
1
32
33
34
40
35
36
37
Anterior Operculoinsular Compartment
b
Lateral Operculoinsular Compartment
c
Superior Pouch
Anterior Pouch
Posterior Inferior Pouch
Inferior Pouch
d
12
8
13
3
11
2
5
4
7
6
9
1
10
14
15

The anterior surface of the insula is formed by the accessory gyrus, oriented vertically, and placed laterally, and the transverse gyrus of Eberstaller oriented transversely and placed medially. Both gyri converge together inferiorly and contribute, with the gyri originating from the anterior portion to form the insular pole [8]. The transverse gyrus of Eberstaller continues medially with the posteromedial orbital lobule, acting as a junction between the inferior portion of the anterior insula and the posterior fronto-orbital region.

Due to its pyramidal shape, the lateral surface of the insula can be subdivided by another perspective. The insular edge separates the superolateral and inferolateral facets of the insula. The superolateral facet is formed by the short insular gyri, the insular central sulcus, the upper part of the anterior long insular gyrus, the insular apex, and the edge between the anterior short gyrus and the transverse gyrus of Eberstaller. On the other hand, the inferolateral facet of the insula is formed by the inferior portion of the anterior and posterior long gyri that continue anteriorly and inferiorly toward the lateral edge of the anterior pole of the insula, the basal continuation of the short gyri of the superolateral facet of the insula below the insular edge comprises the transverse gyrus of Eberstaller. The superolateral facet is related laterally to the frontoparietal operculum, with the superior lateral insular cleft between them, and the inferolateral facet is related laterally to the temporal operculum, with the inferior lateral insular cleft between them [8].

The central core is directly medial to the insular surface and composed by several white matter pathways and grey matter nuclei. To better understand this region and possibly lead to more precise and safer surgeries, the central core can be subdivided into quadrants, each related to specific structures and white matter pathways [15, 16].

6 Arteries of the Sylvian Cistern

The middle cerebral artery (MCA) is the most important vascular structure that crosses through the Sylvian Fissure (SyF), mainly because it is the largest branch of the internal carotid artery (ICA), with a diameter roughly twice the anterior cerebral artery (ACA). It is also among the most common locations for cerebral aneurysms, with a reported incidence between 14.4% and 43% of all diagnosed aneurysms [17].

The surgical classification divides the MCA into four segments (M1–M4) from its origin through its final branches in the brain hemisphere. These segments are the M1 or sphenoidal segment, the M2 or insular segment, the M3 or opercular segment, and the M4 or cortical segment [7].

The MCA origin is at the medial end of the SyF, lateral to the optic chiasm, below the anterior perforated substance and posterior to the division of the olfactory tract into the medial and lateral olfactory striae. As the MCA courses laterally, it pierces a multiple cisterns junction that forms thickened bands of arachnoid from the area of the olfactory trigone to the lateral optic nerve and mesial temporal lobe, entering the SyF in its sphenoidal compartment. The M1 segment is found at the sphenoidal compartment of the SyF, coursing laterally below the anterior perforated substance and parallel, but approximately 1 cm posterior, to the sphenoid ridge. It extends from the carotid bifurcation to the limen of the insula or the genu of the MCA and can be subdivided into two portions. Its proximal half is related superiorly to the anterior perforated substance, posteriorly to the upper portion of the anteromedial surface of the uncus, inferiorly to the inferior portion of the anteromedial surface of the uncus, and anteriorly to the stem of the SyF, the frontotemporal arachnoid reflection, and the lesser wing of the sphenoid. The distal half of M1 is related superiorly to the inferior portion of the anterior surface of the insula, posteriorly to the inferior portion of the insular pole, inferiorly to the anterior portion of the planum polare, and anteriorly to the stem of the SyF, frontotemporal arachnoid reflection, and the lesser wing of the sphenoid. Along its path, M1 gives rise to perforating branches called lateral lenticulostriate arteries that supply subcortical structures of the brain like the basal ganglia and the internal capsule before entering the SyF. As MCA enters the SyF, it branches extensively, supplying the lateral surface of the brain's frontal, parietal, and temporal lobes (Figs. 5 and 6) [7, 18].

The M2 segment, also called insular, begins at the genu of MCA, where the trunks of MCA pass over the limen insula and ends at the circular sulcus of the insula. It sends off branches along its trajectory to supply all insular surfaces (they course in the anterior, superolateral, and inferolateral insular clefts). The M2 segment, during its trajectory on the lateral surface of the insula, sends off branches to the insula itself, the extreme capsule, and occasionally the claustrum and the external capsule. The lateral lenticulostriate arteries supply the structures of the central core of the hemisphere located medially to the claustrum. At the posterior portion of the insula, the M2 segment sends off branches to the corona radiata. As these arteries reach the limiting sulci of the insula (anterior, superior, and inferior), they continue over the medial aspect of the opercula as M3, opercular segment, coursing closely adherent to and over the surface of the frontoparietal and temporal opercula to reach the superficial part of the SyF. It then courses between the orbital operculum and the planum polare on the basal surface or between the frontal and parietal opercula above and the temporal operculum below on the lateral surface to exit the SyF. Once they exit the SyF, they become the M4, or cortical, segment, supplying lateral surface of frontal, parietal, and temporal lobes [7, 18].

MCA branches may also be classified as central and cortical branches; it has around 10 branches during its course through SyF. The *central branches* of the MCA arise from *M1 and M2 segments*. They are called the *lateral striate (lenticulostriate) arteries*. The lateral lenticulostriate arteries pierce the floor of the SyF and course as deep as to the external surface of the thalamus. They supply the striatum, much of the head and body of the caudate nucleus, and large portions of the lenticular nucleus and the external and internal capsules [7, 18].

The *cortical branches* of MCA supply most of the lateral surface of the brain, as the orbital, frontal, parietal, and temporal parts of the cerebral cortex. These branches gradually increase in size, with those originating from M1 being the shortest, while those originating from M4 are the longest.

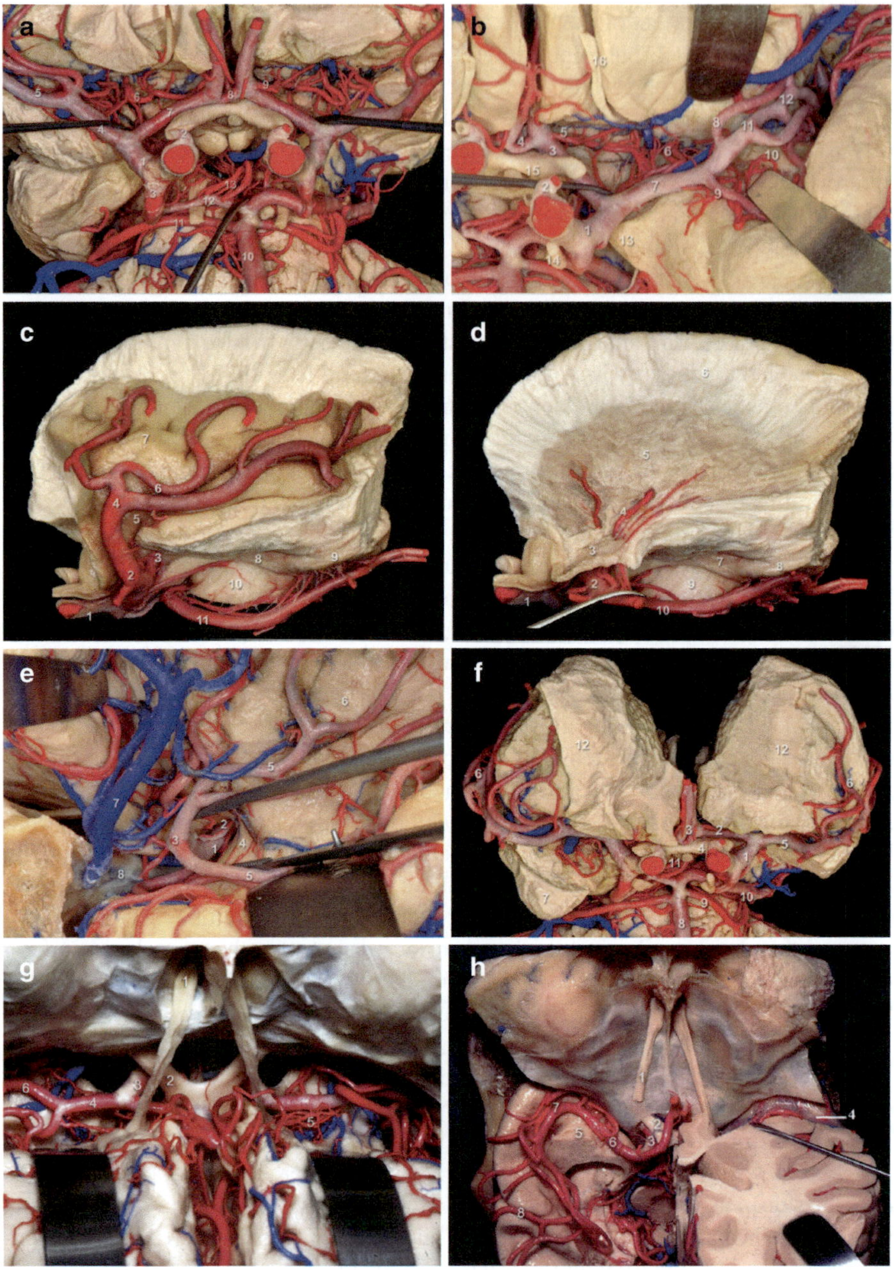
a
b
c
d
e
f
g
h

Fig. 5 Arterial vascularization of the sylvian cistern - proximal/sylvian stem. (**a**) Anatomical dissection of the circle of Willis (inferior view). (*1*) Internal carotid artery, (*2*) ophthalmic artery, (*3*) posterior communicating artery, (*4*) middle cerebral artery - M1 segment, (*5*) middle cerebral artery - M2 segment, (*6*) lenticulostriate arteries, (*7*) anterior cerebral artery, (*8*) anterior communicating artery, (*9*) recurrent artery of Heubner, (*10*) basilar artery, (*11*) superior cerebellar artery, (*12*) posterior cerebral artery, (*13*) thalamoperforating arteries. (**b**) The fronto-orbital surface, temporal pole and internal carotid artery are retracted to expose the sylvian stem. (*1*) Internal carotid artery, (*2*) ophthalmic artery, (*3*) anterior cerebral artery, (*4*) anterior communicating artery, (*5*) recurrent artery of Heubner, (*6*) lenticulostriate arteries, (*7*) middle cerebral artery - M1 segment, (*8*) frontal early branch, (*9*) temporal early branch, (*10*) limen insula, (*11*) middle cerebral artery bifurcation, (*12*) middle cerebral artery - M2 segment, (*13*) uncus, (*14*) oculomotor nerve (III), (*15*) chiasm, (*16*) olfactory tract (I). (**c**) central core of the brain (lateral view). (*1*) Internal carotid artery, (*2*) middle cerebral artery - M1 segment, (*3*) lenticulostriate arteries, (*4*) middle cerebral artery bifurcation, (*5*) limen insula, (*6*) middle cerebral artery - M2 segment, (*7*) insular apex, (*8*) optic tract, (*9*) lateral geniculate body, (*10*) midbrain, (*11*) posterior cerebral artery. (**d**) Central core of the brain dissection showing the lenticulostriate arteries passing through the anterior perforated substance and supplying the putamen (lateral view). (*1*) Internal carotid artery, (*2*) middle cerebral artery - M1 segment, (*3*) anterior perforated substance, (*4*) lenticulostriate arteries, (*5*) putamen, (*6*) internal capsule, (*7*) optic tract, (*8*) lateral geniculate body, (*9*) midbrain, (*10*) posterior cerebral artery. (**e**) Magnified view of the limen insula with the lateral lenticulostriate arteries. (*1*) Middle cerebral artery - M1 segment, (*2*) lenticulostriate arteries, (*3*) middle cerebral artery bifurcation, (*4*) limen insula, (*5*) middle cerebral artery - M2 segment, (*6*) insular surface, (*7*) superficial sylvian vein, (*8*) sphenoparietal sinus. (**f**) Central core of the brain dissection, showing the middle cerebral artery – M1 segment at its inferior aspect and the middle cerebral artery – M2 segment at its lateral aspect. (*1*) Internal carotid artery, (*2*) anterior cerebral artery, (*3*) anterior communicating artery, (*4*) chiasm, (*5*) middle cerebral artery - M1 segment, (*6*) middle cerebral artery - M2 segment, (*7*) hippocampus, (*8*) basilar artery, (*9*) superior cerebellar artery, (*10*) posterior cerebral artery, (*11*) thalamoperforating arteries, (*12*) central core. (**g**) Anatomical brain dissection (supero-anterior view). (*1*) Olfactory bulb (I), (*2*) optic nerve (II), (*3*) internal carotid artery, (*4*) fronto-temporal arachnoid reflection, (*5*) temporal stem, (*6*) middle cerebral artery - M2 segment. (**h**) Anatomical dissection showing the middle cerebral artery trajectory inside the sylvian cistern (supero-anterior view). (*1*) Olfactory bulb (I), (*2*) optic nerve (II), (*3*) internal carotid artery, (*4*) middle cerebral artery bifurcation, (*5*) lenticulostriate arteries, (*6*) middle cerebral artery - M1 segment, (*7*) middle cerebral artery - M2 segment, (*8*) middle cerebral artery - M3 segment

The cortical branches arising *before the bifurcation of the MCA* are often termed as the *early branches*. The cortical branches are named according to the region of the brain that they supply:

- Anterior temporal arteries arise from the M1 segment of MCA and vascularize the temporal pole of the brain, which is the most anterior aspect of the temporal lobe.
- The lateral frontobasal artery arises from the M2 segment of MCA and supplies the lateral part of the orbital surface of the frontal lobe and the inferior frontal gyrus.

The remainder of the cortical branches arises from the M4 segment of the MCA:

- The artery of the prefrontal sulcus supplies the anterior aspects of the inferior and middle frontal gyri.

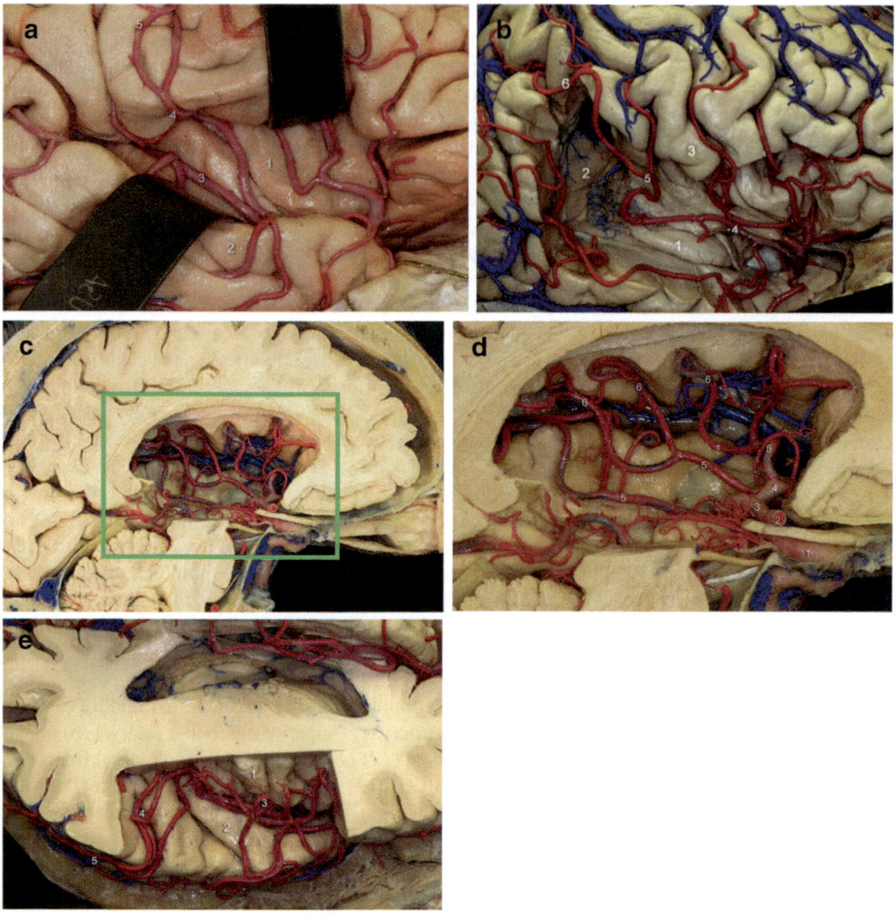

Fig. 6 Arterial vascularization of the sylvian cistern - distal/lateral. (**a**) The sylvian cistern is opened by retracting the opercula and the insular surface is exposes (lateral view). (*1*) Last short insular gyrus, (*2*) superior temporal gyrus, (*3*) middle cerebral artery – M2 segment, (*4*) middle cerebral artery – M3 segment, (*5*) middle cerebral artery – M4 segment. (**b**) Anatomical dissection of a right cerebral hemisphere, with removal of the superior temporal, middle temporal and supramarginal gyri, showing the temporal horn of the lateral ventricle opened and middle cerebral artery trajectory (lateral view). (*1*) Hippocampus, (*2*) *atrium* of lateral ventricle, (*3*) precentral gyrus, (*4*) middle cerebral artery – M2 segment, (*5*) middle cerebral artery – M3 segment, (*6*) middle cerebral artery – M4 segment. (**c**) Sagittal cut of left cerebral hemisphere after central core and insula removal to expose the middle cerebral artery trajectory (medial view) (a green rectangle delineates the next image to be explained). (**d**) Amplified view of the green rectangle area delineated at (**c**). (*1*) Internal carotid artery, (*2*) anterior cerebral artery, (*3*) middle cerebral artery – M1 segment, (*4*) middle cerebral artery bifurcation, (*5*) middle cerebral artery – M2 segment, (*6*) middle cerebral artery – M3 segment. (**e**) Anatomical axial cut of right cerebral hemisphere, after removal of frontoparietal operculum (superior view). (*1*) Insula, (*2*) Heschl's gyrus, (*3*) middle cerebral artery – M2 segment, (*4*) middle cerebral artery – M3 segment, (*5*) middle cerebral artery – M4 segment

- The artery of the precentral sulcus travels in the precentral sulcus to supply the posterior aspect of the inferior and middle frontal gyri, Broca's area, and the precentral gyrus, which contains the primary motor cortex for the head, upper limb, and trunk.
- The artery of the central sulcus travels within the central sulcus and contributes to the blood supply of the pre- and postcentral gyri.
- The artery of the postcentral sulcus travels in the postcentral sulcus to supply the anterior aspect of the parietal lobe and the postcentral gyrus, which contains the primary somatosensory cortex for the head, upper limbs, and trunk.
- A variety of parietal branches supply the lateral aspect of the parietal lobe, including the superior and inferior paracentral lobules.
- Angular artery supplies the angular and supramarginal gyri of the parietal lobe, the posterior part of the superior temporal gyrus, and the superior part of the lateral surface of the occipital lobe.
- Middle temporal branches supply the central aspect of the superior and middle temporal gyri and the primary auditory cortex, and Wernicke's area [7, 18].

Anatomical Variations – Anatomical variants of the MCA are uncommon. The MCA can be variable in its origin, but duplications, fenestrations, or accessory arteries are the most frequent. A duplicated MCA is only seen in less than 3% of individuals, but it may arise from the ICA, paralleling the main MCA and traveling towards the anterior temporal lobe to supply it. Up to 4% of people have an accessory MCA, which typically supplies the orbitofrontal aspect of the brain. It can arise from ICA (Type 1), A1 segment of ACA (Type 2), or A2 segment of ACA (Type 3) [7, 18].

7 Arteries: Angiographic View

One of the most difficult tasks when studying neuroanatomy is identifying the important information to neurosurgery since not all the important details to neuroanatomists are also important to neurosurgeons and vice versa. Examining angiographic images may help the young neurosurgeon learn anatomy, plan surgical strategies, and perform surgeries. Following are some details of SyF anatomy by angiographic study, but the first consideration is that angiographic images are 2D pictures representing a volumetric image. Nowadays, we have angiographic 3D images since modern angiography equipment allows 3D primary images, but it is always reliable to know how to interpret 2D images since they are the most easily accessible ones.

The neural and vascular structures along the trajectory of the M1 segment can be easily identified. Looking at the anteroposterior (AP) view of the carotid angiography, the neural structures located along the trajectory of M1 and M2 can be recognized. At the limen insulae, the M1 becomes M2 and turns around the pole of the insula. Thus, the insula is located medially to the M2, and the frontal, parietal, and temporal opercula are projected laterally to M2, between M2 and M4 segments, while the medial part of the temporal lobe is projected medially to the M2 (Figs. 7 and 8).

The several loops formed by the M2 when it reaches the superior circular sulcus of the insula can also be identified angiographically. These several loops approximately indicate the location of the superior circular sulcus of the insula. The first loop of the M2 is usually located more medially on angiography than the subsequent loops because the first loop usually is located in the anterior insular cleft.

The curvature and morphology of the M3 branches of the frontal and parietal opercula determine the location of that particular M3 branch. In the anterior frontal operculum, they loop shallowly, loop deeply in the central region, and then exit straight with a long course in the postcentral region. In the posterior parietal region, they exit straight with a short course [8, 12].

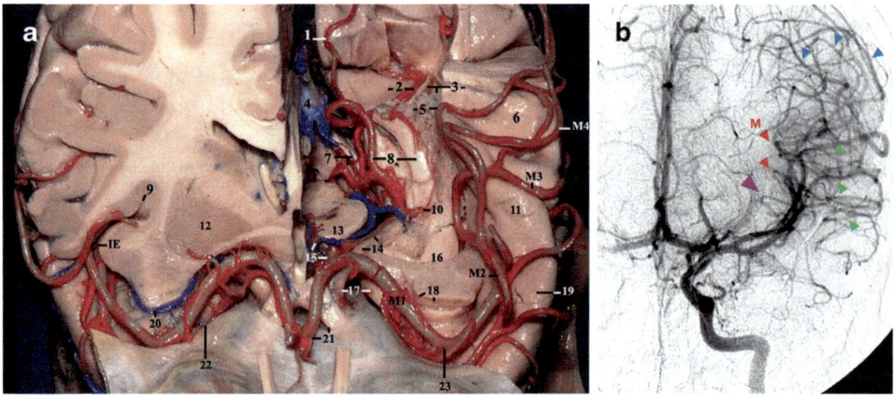

Fig. 7 Cerebral vascularization: Anatomy and angiography correlation – anterior view. (**a**) Anatomical brain dissection, with removal of brain portions to expose the main cerebral arteries trajectories (anterior view). (*1*) Parieto-occipital artery, (*2*) lingual gyrus and calcarine artery, (*3*) calcar avis: atrium: and posterior transverse temporal gyrus, (*4*) vein of Galen, (*5*) glomus of atrium and sylvian point, (*6*) middle transverse temporal gyrus, (*7*) tentorial edge and trochlear nerve, (*8*) P2P segment of the posterior cerebral artery: parahippocampal gyrus: and fornix, (*9*) superior limiting sulcus of the insula, (*10*) inferior choroidal point (entry point of the anterior choroidal artery in the temporal horn), (*11*) Heschl's gyrus, (*12*) lentiform nucleus, (*13*) crus cerebri, (*14*) apex and the posteromedial surface of the uncus and the anterior choroidal artery, (*15*) P1 segment of the posterior cerebral artery and the posterior communicating artery, (*16*) head of the hippocampus, (*17*) supraclinoid carotid artery and anteromedial surface of the uncus, (*18*) limen insulae and insular pole, (*19*) planum polare, (*20*) deep middle cerebral vein, (*21*) anterior cerebral artery and optic nerve, (*22*) lesser wing of the sphenoid, (*23*) genu of the MCA (M1 segment): M2: Insular segment of the MCA: M3: Opercular segment of the MCA: M4: Cortical segment of the MCA: IE: Insular edge. (**b**) Left Carotid angiography (AP view): The small red arrows indicate the M2 segment of the middle cerebral artery. The green arrows indicate the M3 segment of the middle cerebral artery over the *planum polare* and also over the *planum temporale*. The blue arrows indicate M4 segment of the middle cerebral artery over parietal area. The bigger pink arrow indicates the superior limiting sulcus of insula

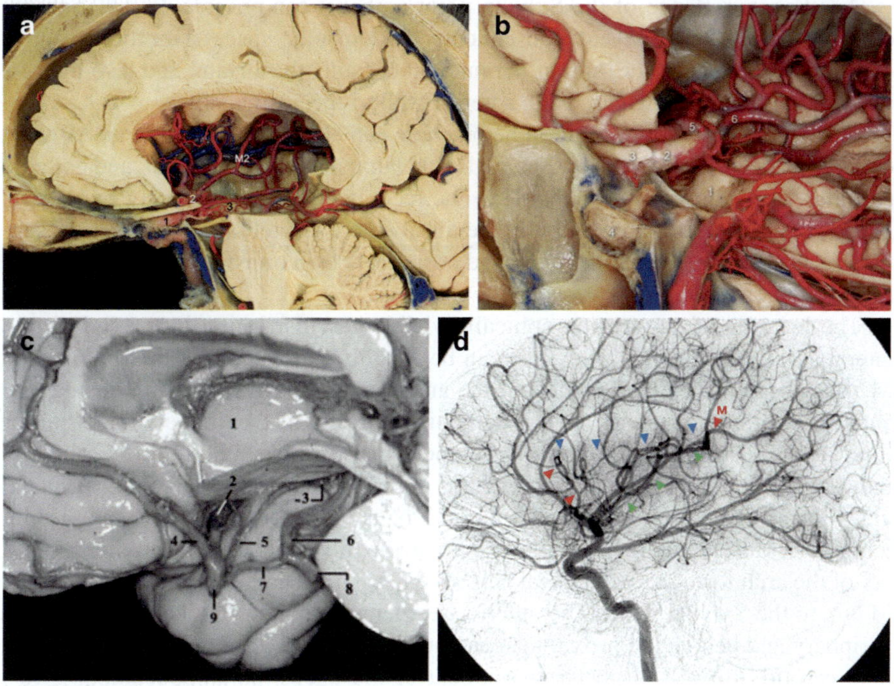

Fig. 8 Cerebral vascularization: Anatomy and angiography correlation – lateral view. (**a**) Sagittal cut of right cerebral hemisphere after central core and insula removal to display the trajectory of the middle cerebral artery (medial view). (*1*) Internal carotid artery (*2*) middle cerebral artery - M1 segment, (*3*) uncus and anterior choroidal artery passing over it. (**b**) Magnified view of a sagittal cut of right cerebral hemisphere after central core and insula removal (lateral view). (*1*) Uncus, (*2*) ICA, (*3*) optic nerve, (*4*) pituitary gland, (*5*) anterior cerebral artery, (*6*) middle cerebral artery. (**c**) Medial view of the right cerebral hemisphere with the crus cerebri removed to show the trajectory of the M1 segment in this projection. (*1*) Thalamus, (*2*) M1 segment of the MCA, (*3*) uncus and the inferior choroidal point, (*4*) anterior cerebral artery, (*5*) anterior choroidal artery, (*6*) P2A segment of posterior cerebral artery, (*7*) posterior communicating artery, (*8*) P1 segment of posterior cerebral artery, (*9*) supraclinoid carotid artery. (**d**) Carotid angiography (lateral view). The red arrows indicate anterior limiting sulcus. Blue arrows indicate superior limiting sulcus. The green arrows indicate inferior limiting sulcus. M, sylvian point

The last loop of the M3 segment that loops over the temporal operculum and exits the SyF constitutes an important angiographic landmark: the Sylvian point or the “M” point [9, 19–21]. The Sylvian point is usually located on the most medial aspect of the Heschl’s gyrus or even on the most medial aspect of the planum temporale. Besides being the last loop, the Sylvian point is usually the most medial loop on the temporal side since both Heschl’s gyrus and the planum temporale enclose the insula posteriorly and medially toward the atrium of the lateral ventricle, and this retroinsular region is located more medially than the anterior portion of the insula [8].

Another importance of the Sylvian point is that it can provide the location not only of Heschl's gyrus and consequently retroinsular space, but the posterior extremity of the posterior limb of the internal capsule, the atrium, and also the pulvinar of the thalamus, which is the anterior wall of the atrium.

The central core of the cerebral hemisphere can also be identified in an AP view of carotid angiography – the basal ganglia, thalamus, internal, external, and extreme capsules, and claustrum are located in the rectangle limited anteriorly by the A1, M1, and genu of MCA, medially by the distal anterior cerebral artery (A2–A5 segments), laterally by the M2 segment and posteriorly by a transverse line drawn from the Sylvian or M point.

The location of any angiographically visible vascular lesion in the SyF of the lateral surface of the cerebrum also can be roughly estimated: the first loop of the M2 usually indicates the location of the anterior limiting sulcus of the insula (the anterior limit of the insula), the Sylvian point indicates approximately the posterior limit of the insula, and the first M3 segment that displays a straight morphology indicates the location of the Heschl's gyrus or the planum temporale. Thus, in an AP view of carotid angiography, the temporal lobe is projected as follows: the planum temporale, Heschl's gyrus, and planum polare are all projected to the outer periphery of the arch formed by the distal half of M1, the genu and the whole extent of the M2 up to the Sylvian point. The anterior segment of uncus is projected at the outer periphery and behind the proximal segment of M1 and its apex behind the supraclinoid carotid artery. The posterior segment of the uncus, parahippocampal gyrus, temporal horn, and the hippocampus are all projected inside the square formed by the M1, the A1 anteriorly, the distal anterior cerebral artery medially, the M2 and the Sylvian point laterally, and an imaginary line traced horizontally from the Sylvian point toward the midline. The medial limit of the parahippocampal gyrus is projected approximately halfway between the distal ACA and M2 [8].

It is important to note that both frontal and occipital poles can be superimposed on MCA branches in AP view, but the largest transverse diameter of a normal brain (the suprasylvian lateral convexity) is located at the level of the postcentral gyrus. Thus, the most laterally located vessels over the lateral convexity are most likely running on the postcentral gyrus or supramarginal gyrus [8]. The same consideration may be done to the largest transverse diameter of the brain on the infrasylvian convexity, where we find the superior or middle temporal gyrus at the same coronal plane as the postcentral gyrus or supramarginal gyrus.

In the lateral view of the carotid angiography, the trajectory of the M1 segment may be difficult to trace since there is vessel superimposition. MCA loops are mainly composed of M2 and M3 segments over the insula and frontal, parietal, and temporal opercula, forming a "Sylvian triangle" [8].

The Sylvian triangle seams an upside-down right triangle, when the right angle is the junction between anterior and superior limiting sulci of the insula and the hypotenuse is the inferior limiting sulcus of the insula, it indicates not only the location of the insula but also approximately the location of the central core in angiography. In normal conditions, the operculum of the precentral gyrus covers the middle third and anterior portion of the posterior third of the insula. Thus, it is possible to identify approximately precentral gyrus location in angiography by analyzing the morphology of vascular loops [8].

Although most of the central core is inside the "Sylvian triangle," the head and body of the caudate nucleus are located above the lateral projection of the superior limiting sulcus of the insula. The lateral ventricle is also outside this triangle since the frontal horn is ahead and above the anterior limiting sulcus, the body is projected above the superior limiting sulcus, the atrium is projected behind the junction between the superior and inferior limiting sulcus of the insula, and the temporal horn is below the inferior limiting sulcus of the insula.

Also, in the lateral view of carotid angiography, the anterior portion of the planum polare of the temporal lobe and the genu of the MCA are projected anteriorly to the siphon of the ICA. The anterior segment of the uncus is projected at the level of ICA, the apex of the uncus is behind the carotid artery, and the temporal horn starts behind the carotid artery and is projected at the level of the middle temporal gyrus [8].

8 Veins of the Sylvian Cistern

The cerebral veins accompany arteries in the subarachnoid space, but unlike other parts of the body, cerebral veins do not have valves or contain muscle tissue. Thus, bidirectional flow is possible in them. They are divided into superficial, deep and veins of the brain stem and posterior fossa [22, 23].

The superficial veins may be divided into superior, middle, and inferior veins. The dominant vein of the middle group is the "superficial middle cerebral vein" (SMCV), which is also known as the Sylvian's vein due to its location in the SyF [22]. The superficial Sylvian vein usually arises at the posterior end of the Sylvian fissure and courses anteriorly and inferiorly along the lips of the fissure. The superficial Sylvian vein may arise as two trunks, but usually, they merge into one before emptying into the venous sinuses along the sphenoid ridge. It receives tributaries from the inferior part of the frontal lobe, superior temporal gyrus, and parietal opercula, and commonly anastomoses with the veins of Trolard and Labbé. The superficial Sylvian vein penetrates the arachnoid membrane that covers the anterior end of the SyF and joins the sphenoparietal sinus as it courses just below the medial part of the sphenoid ridge, or it may pass directly to the cavernous sinus [22, 24] (Fig. 9).

The superficial Sylvian vein may also leave SyF and courses around the temporal pole to reach dural sinuses in the middle fossa floor, empty into the superior petrosal sinus, or exit the intracranial cavity through the foramina in the sphenoid bone to reach the pterygoid plexus [24].

There are deep Sylvian veins that drain the insula and adjacent walls of the SyF [25]. The cisternal group of deep veins drains the area beginning anteriorly in front of the third ventricle and extends laterally into the SyF and backward to include the walls of the chiasmatic, interpeduncular, crural, ambient, and quadrigeminal cisterns. Cortical areas bordering the anterior incisural region, including the insula and

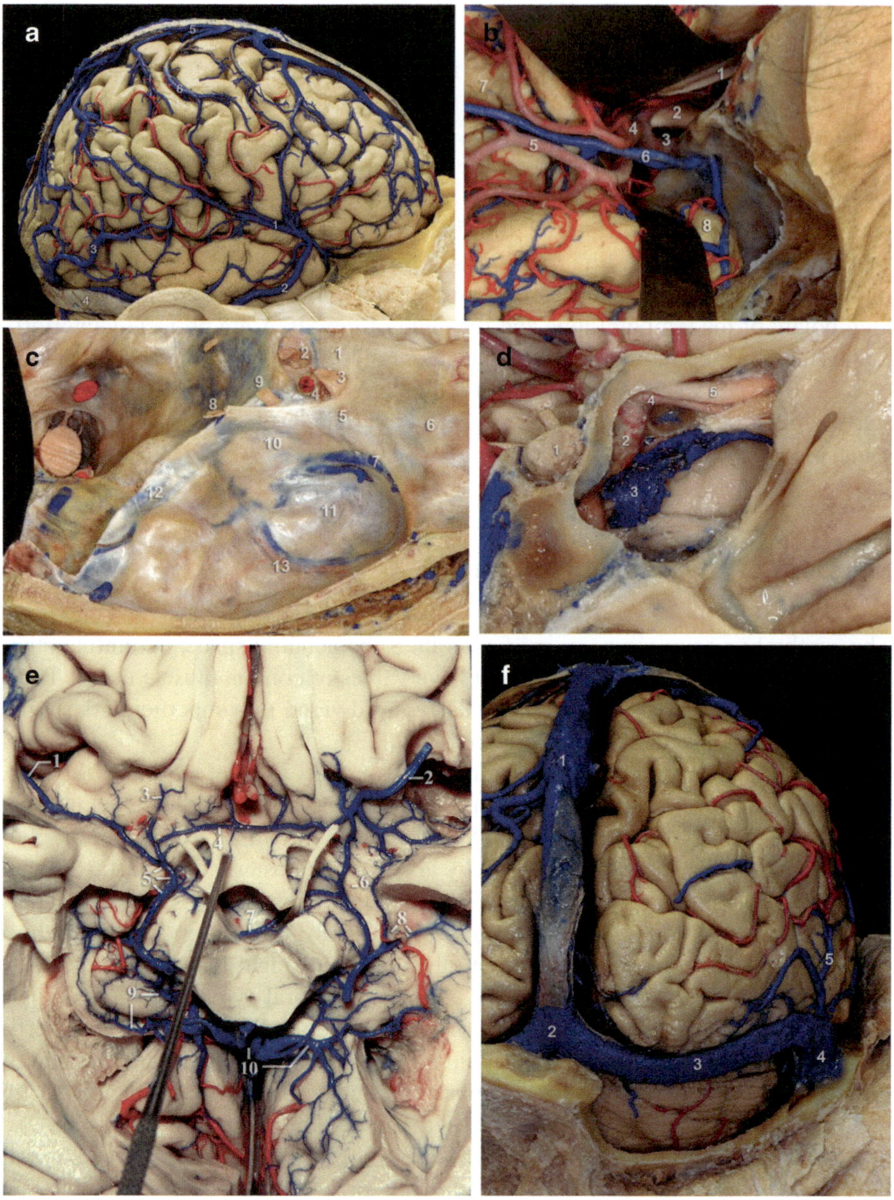
a
b
c
d
e
f

Fig. 9 Sylvian cistern venous drainage. (**a**) Anatomical brain dissection (lateral view). (*1*) Sylvian vein or superficial middle cerebral vein, (*2*) inferior cerebral vein, (*3*) inferior anastomotic vein of Labbé, (*4*) transverse sinus, (*5*) superior sagittal sinus, (*6*) superior anastomotic vein of Trolard. (**b**) Magnified view of the sylvian cistern, showing the superficial sylvian vein draining into the sphenoparietal sinus. (*1*) Olfactory tract (I), (*2*) chiasm, (*3*) internal carotid artery, (*4*) middle cerebral artery - M1 segment, (*5*) middle cerebral artery - M2 segment, (*6*) superficial sylvian vein, (*7*) insular surface, (*8*) temporal pole. (**c**) The middle fossa and the lateral wall of the cavernous sinus are exposed after the brain is removed. (*1*) Planum sphenoidale, (*2*) pituitary gland, (*3*) optic nerve (II), (*4*) internal carotid artery, (*5*) anterior clinoid, (*6*) anterior fossa, (*7*) sphenoparietal sinus, (*8*) trochlear nerve (IV), (*9*) oculomotor nerve (III), (*10*) cavernous sinus (lateral wall), (*11*) middle fossa, (*12*) superior petrosal sinus, (*13*) middle meningeal artery and vein. (**d**) The brain and sphenoidal sinus wall have been removed, exposing the internal carotid artery surrounded by cavernous sinus (medial view). (*1*) Pituitary gland, (*2*) internal carotid artery, (*3*) cavernous sinus, (*4*) ophthalmic artery, (*5*) optic nerve (II). (**e**) Deep drainage of sylvian cistern. The deep sylvian vein joins the anterior cerebral vein and form the basal vein of Rosenthal (inferior view). (*1*) Deep middle cerebral vein, (*2*) fronto-orbital vein, (*3*) olfactory vein, (*4*) anterior cerebral vein, (*5*) transition between the first (striate) and the second (peduncular) segments of basal vein, (*6*) optic tract, (*7*) peduncular vein, (*8*) inferior ventricular vein and inferior choroidal point, (*9*) posterior mesencephalic segment of basal vein and longitudinal hippocampal vein, (*10*) vein of Galen and internal occipital vein. (**f**) Final drainage of Labbé vein to transverse sinus and superior sagittal sinus to confluence of sinuses (posterior view). (*1*) Superior sagittal sinus, (*2*) confluence of sinuses, (*3*) transverse sinus, (*4*) sigmoid sinus, (*5*) inferior anastomotic vein of Labbé

the orbital surface of the frontal lobe, may drain to the basal vein. The insular veins are one of the major contributing groups to the first part of the basal vein, and they are named for their relationship to the insular sulci and gyri. The insula is situated deep to the frontoparietal and temporal opercula and has a pyramidal shape, as already explained in this chapter, which appears as an upside-down triangle in 2D angiography. Its apex or anterior pole is directed infero-medially toward the limen insula, which delineates the insula from the anterior perforated substance. The deep middle cerebral vein is formed by the union of the insular veins near the limen insula and passes medially across the anterior perforated substance, where it unites with the anterior cerebral vein to form the basal vein. The deep middle cerebral vein may receive the fronto-orbital, olfactory, anterior temporal, and a variable number of inferior striate veins as it courses under the anterior perforated substance. The deep middle cerebral vein, the anterior segment of the basal vein, or their tributaries may be connected by a bridging vein to the sphenoparietal or cavernous sinus. The anterior segment of the basal vein may be absent; in these cases, the deep middle cerebral vein unites with the olfactory, fronto-orbital, anterior cerebral, and inferior striate veins to form a trunk draining medially into the cavernous sinus or even laterally into the superficial veins along the SyF [25].

The veins of the insula drain predominantly through the deep middle cerebral vein into the basal vein. They include the anterior insular veins, which course on or near the anterior limiting sulcus; the precentral insular veins, which course anterior-inferiorly along the short gyri; the central insular vein, which courses antero-inferiorly along the central insular sulcus; and the posterior insular veins, which course on the long gyri near the posterior limiting sulcus. These veins commonly end by joining each other and draining directly into the deep middle cerebral vein. Others terminate by ascending or descending to drain into superficial cortical veins bordering the SyF [24, 25, 26, 27].

If the superficial Sylvian vein is too thin or absent, adjacent veins will support this drainage area. The veins arising on the upper lip of the SyF will join veins that empty into the superior sagittal sinus, and those on the lower lip will be directed to join the veins that empty into the sinuses below the temporal lobe [24].

Suppose the central segment of the superficial Sylvian vein is absent. In that case, the anterior segment will join the sinuses along the sphenoid ridge, and the posterior segment will join the anastomotic veins of Trolard and Labbé that drain into the superior sagittal and transverse sinuses [24].

Acknowledgments We thank the medical student and artist Giovanna Hespanhol Gütschow (Pontifícia Universidade Católica do Paraná - PUCPR - Brazil) for drawing Fig. 4a with great talent and effort.

References

1. Gazi Yaşargil M, Kasdaglis K, Jain KK, Weber H-P. Anatomical observations of the subarachnoid cisterns of the brain during surgery. J Neurosurg. 1976;44(3):298–302.
2. Yaşargil MG, Yaşargil MG. Microsurgical anatomy of the basal cisterns and vessels of the brain, diagnostic studies, general operative techniques, and pathological considerations of the intracranial aneurysms. Stuttgart: Thieme; 1984.
3. Ribas GC, Ribas EC, Rodrigues CJ. The anterior sylvian point and the suprasylvian operculum. Neurosurg Focus. 2005;18(6B):E2.
4. Ribas GC. The cerebral sulci and gyri. Neurosurg Focus. 2010;28(2):E2.
5. Collice M, Collice R, Riva A. Who discovered the sylvian fissure? Neurosurgery. 2008;63(4):623–8.
6. de le Boë Sylvius F. Disputationes medicarum pars prima, primarias corporis humani functiones naturales ex anatomicis, practicis et chymicis experimentis deductas complectens. Amsterdam: J van den Bergh; 1663.
7. Gibo H, Carver CC, Rhoton AL Jr, Lenkey C, Mitchell RJ. Microsurgical anatomy of the middle cerebral artery. J Neurosurg. 1981;54(2):151–69.
8. Wen HT, Rhoton AL Jr, De Oliveira E, Castro LHM, Figueiredo EG, Teixeira MJ. Microsurgical anatomy of the temporal lobe: part 2-sylvian fissure region and its clinical application. Neurosurgery. 2009;65(6 Suppl 1):1–36.
9. Türe U, Yaşargil DC, Al-Mefty O, Yaşargil MG. Topographic anatomy of the insular region. J Neurosurg. 1999;90(4):720–33.
10. Yaşargil M, Krisht A, Türe U, Al-Mefty O, Yaşargil D. Microsurgery of insular gliomas: part I—surgical anatomy of the sylvian cistern. Contemp Neurosurg. 2017;39(11):1–8.
11. Taylor E, Haugton W. Some recent researches on the topography of the convolutions and fissures of the brain. Trans R Acad Med Ireland. 1900;18:511–9.
12. Szikla G, Bouvier T, Hori T, Petrov V. The sylvian fissure. In: Angiography of the human brain cortex. Berlin: Springer; 1977. p. 101–25.
13. Inoue K, Seker A, Osawa S, Alencastro LF, Matsushima T, Rhoton AL Jr. Microsurgical and endoscopic anatomy of the supratentorial arachnoidal membranes and cisterns. Neurosurgery. 2009;65(4):644–65.

14. Ribas EC, Yagmurlu K, Wen HT, Rhoton AL Jr. Microsurgical anatomy of the inferior limiting insular sulcus and the temporal stem. J Neurosurg. 2015;122(6):1263–73.
15. Ribas EC, Yağmurlu K, de Oliveira E, Ribas GC, Rhoton AL Jr. Microsurgical anatomy of the central core of the brain. J Neurosurg. 2018;129(3):752–69.
16. Sanai N, Polley M-Y, Berger MS. Insular glioma resection: assessment of patient morbidity, survival, and tumor progression: clinical article. J Neurosurg. 2010;112(1):1–9.
17. Zaidat OO, Castonguay AC, Teleb MS, Asif K, Gheith A, Southwood C, et al. Middle cerebral artery aneurysm endovascular and surgical therapies: comprehensive literature review and local experience. Neurosurg Clin N Am. 2014;25(3):455–69.
18. Rhoton AL Jr. The supratentorial arteries. Neurosurgery. 2002;51(4 Suppl):S53–120.
19. Taveras JM. Neuroradiology. Baltimore: Williams & Wilkins; 1996.
20. Türe U, Yaşargil MG, Al-Mefty O, Yaşargil DC. Arteries of the insula. J Neurosurg. 2000;92(4):676–87.
21. Waddington M. Atlas of cerebral angiography with anatomic correlation. Boston: Little Brown and Company; 1974. p. 38–41.
22. Osborn A. Veins and venous sinuses. In: Harnsberger HR, Osborn AG, Macdonald AJ, Ross JS, Moore KR, Salzman KL, Wiggins RH, Davidson HC, Hamilton BE, Carrasco C, editors. Veins and venous sinuses. Utah: Amirsys; 2011. p. 1334–87.
23. Egemen E, Solaroglu I. Anatomy of cerebral veins and dural sinuses. In: Thomas AJ, Lo EH, Zhang JH, Biller J, Caplan LR, Leary MC, Yenari M, editors. Primer on cerebrovascular diseases. 2nd ed. London: Elsevier; 2017. p. 32–6.
24. Rhoton AL Jr. The cerebral veins. Neurosurgery. 2002;51(October):159–205.
25. Ono M, Rhoton AL Jr, Peace D, Rodriguez RJ. Microsurgical anatomy of the deep venous system of the brain. Neurosurgery. 1984;15(5):621–57.
26. Suzuki Y, Matsumoto K. Variations of the superficial middle cerebral vein: classification using three-dimensional CT angiography. Am J Neuroradiol. 2000;21(5):932–8.
27. Baker HL. The venous angiogram: a frequently overlooked phase of carotid angiography. Clin Neurosurg. 1964;10:130–50.

Surgical Anatomy of Interhemispheric Fissure

Tomokatsu Hori, Y. A. Alshebib, Hideki Shiramizu, and Seigo Matsuo

1 Introduction

Interhemispheric fissure (IF) is the surgical corridor to lesions of the anterior skull base, anterior and posterior third ventricle, and brain stem. Every neurosurgeon should know about the surgical anatomy of the fissure to approach these difficult lesions. At first, the long superior sagittal sinus (SSS) overrides the entire IF, which is very important to preserve at all costs. Although some old neurosurgeons advocate the possible sacrifice of the anterior third of SSS, the recent neurosurgical literature reported the remotion of the anterior third of SSS resulted in severe brain edema and eventually death of the patient who had a small falx meningioma situated at the anterior third of SSS [1].

2 Surgical Anatomy of IF

Concerning glioma invading IF, the usefulness of the VAC(CA), VPC(CP) line to delineate the pre-supplementary motor area (Pre SMA) (Fig. 1), SMA, and M1 of the contralateral lower extremity as reported by Talairach and Szikla, should be checked carefully [2].

T. Hori (✉) · Y. A. Alshebib
Department of Neurosurgery, Moriyama Neurological Center Hospital, Tokyo, Japan
e-mail: thori@moriyamaikai.or.jp

H. Shiramizu · S. Matsuo
Moriyama Memorial Hospital, Tokyo, Japan

E. G. Figueiredo et al. (eds.), *Brain Anatomy and Neurosurgical Approaches*,
https://doi.org/10.1007/978-3-031-14820-0_12

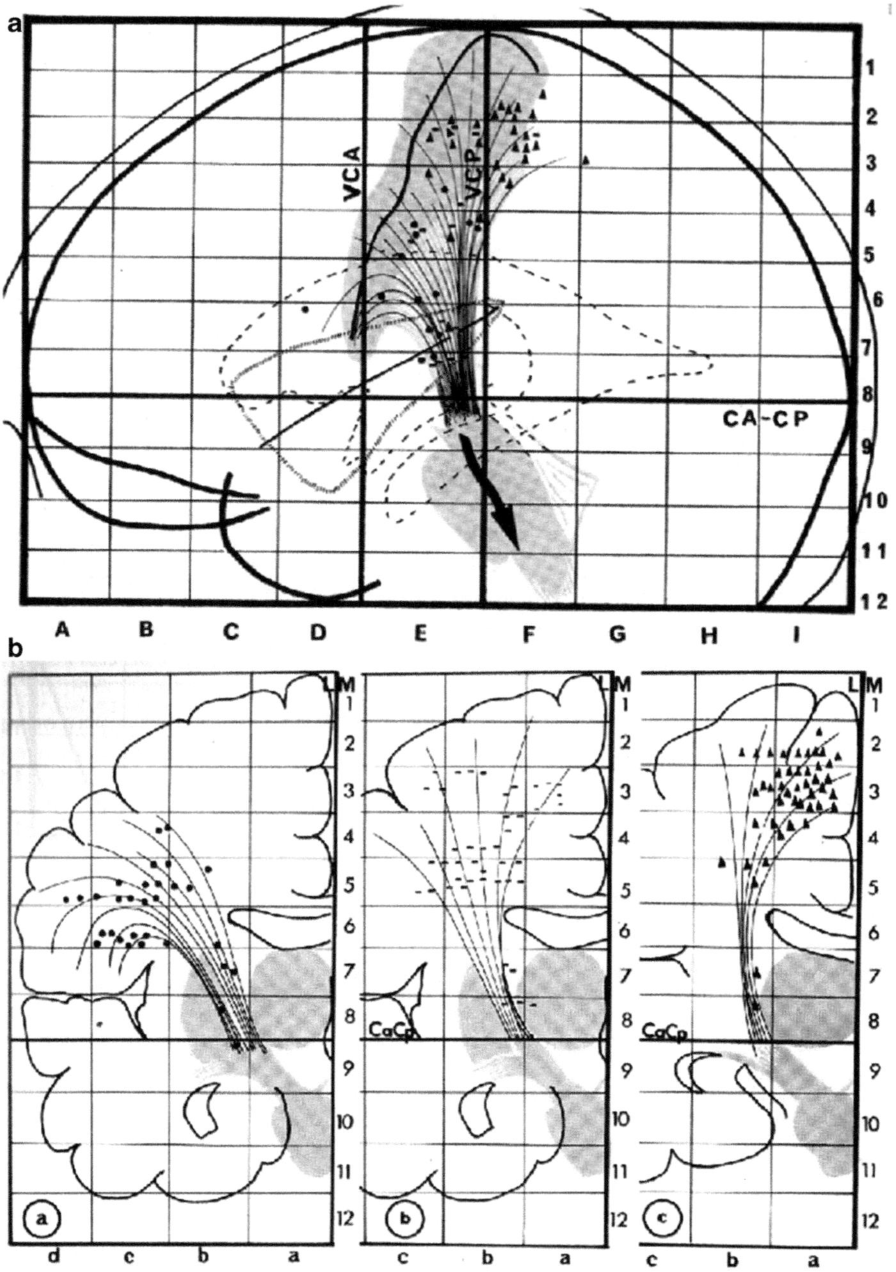

Fig. 1 (**a**, **b**) Triangular points indicate locations of positive motor responses of the contralateral lower limb, small transverse lines (**a**, **b**b) indicate areas of positive motor responses of the contralateral upper limb, and small open circles (**a**, **b**a) indicate locations of positive motor responses of contralateral face and tongue. Path of motor fibers in the white substance of the hemispheres between the cortex and the posterior limb of the internal capsule and topography in the standardized system of motor responses to low-frequency stimuli (1–10 V, 1–5 ms, 1 c/s) obtained during stereo EEG exploration of thirty cases of epilepsy are shown in Fig. 1**a**, **b**

Bereitschaftspotential (BP) was recorded directly from the right supplementary motor area proper (SMA-proper) and its rostral part by chronically implanted subdural electrodes in three patients with intractable focal motor seizure. Cortical electrical stimulation of the SMA-proper revealed the somatotopy as previously reported, and the supplementary negative motor area (SNMA) was identified just anterior to the SMA-proper in two of three cases.

In patient 1, eight kinds of simple movements, i.e., left and right middle finger extension (MF Ext), left arm abduction, left and right foot dorsiflexion (dorsiflex), left knee extension, tongue protrusion, and saccadic eye movement, were studied to record BP. In the SMA-proper, somatotopically distributed BP preceding

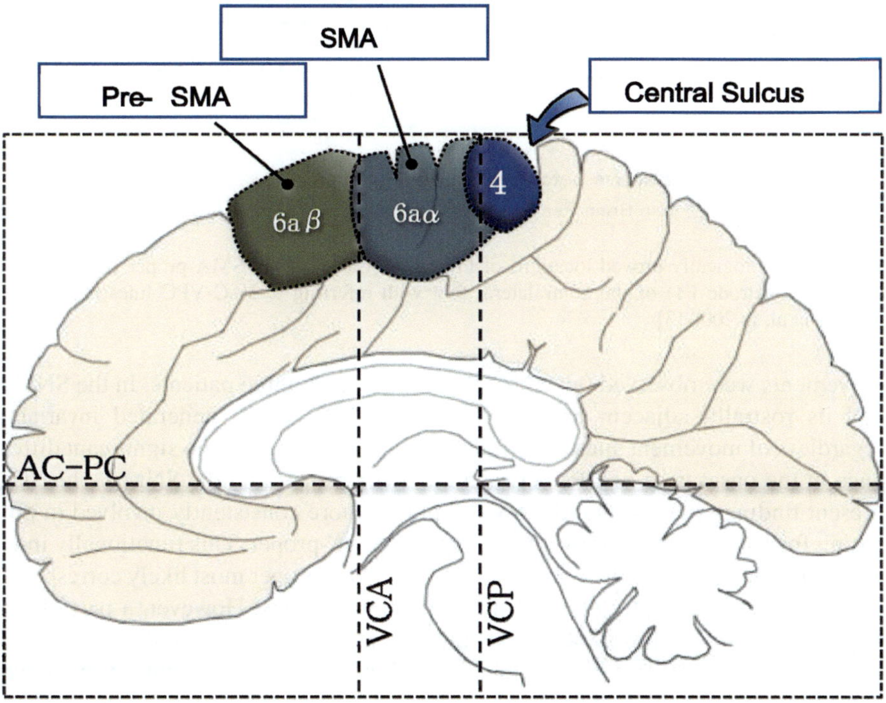

Fig. 2 AC-PC (CA-CP) line: Superior limits of the anterior commissure and inferior limit of posterior commissure are connected and named AC-PC line. The usual AC-PC line used in functional neurosurgery is the line connecting the posterior midpoint of the anterior commissure and the anterior midpoint of the posterior commissure. Just the anterior portion of the VCA line (vertical line to the AC-PC line at the posterior limit of AC) defined by 6aβ in this figure is the area of pre-supplementary motor area (Pre SMA). The area between the VCA and VCP defined by 6aα is the supplementary motor area (SMA). Just posterior part of VCP is area 4 (M1 of foot area), and the blue arrow indicates the medial beginning of the central sulcus. This picture demonstrates the location of Pre SMA(6aβ), SMA(6aα), and Area 4 in terms of VAC-VPC lines, and the blue arrow indicates the central sulcus of the interhemispheric fissure (Fig. 2)

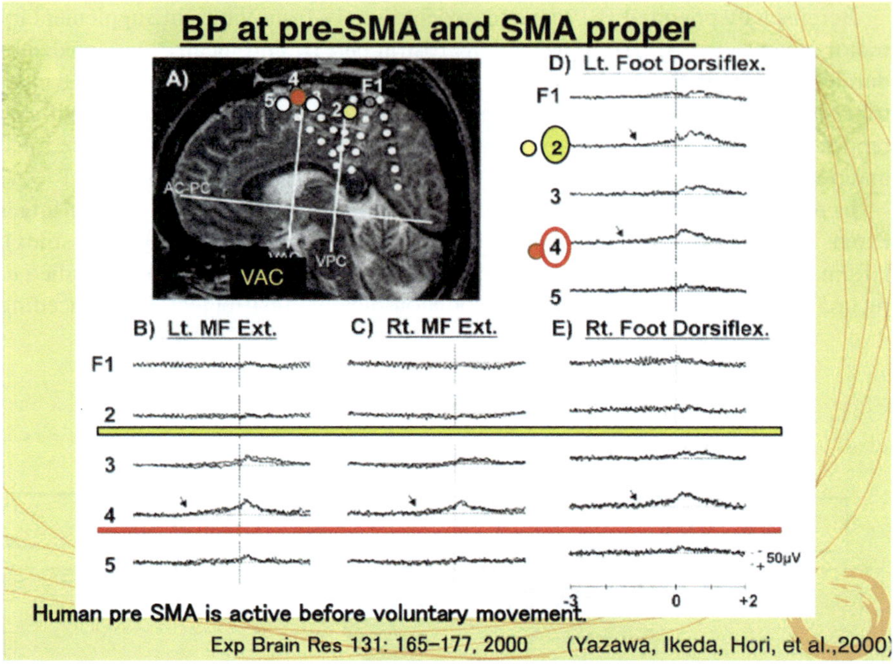

Fig. 3 Physiologically proved locations of Pre-SMA (Electrode 4), SMA proper (Electrode 2), and M1 (Electrode F1) of the contralateral foot with referring to VAC-VPC lines reported by Yazawa et al. in 2000 [3]

movements were observed (electrode 2 in Fig. 3) in all three patients. In the SNMA and its rostrally adjacent areas, "SNMA-plus" BPs were generated invariably regardless of movement sites (electrode 4 in Fig. 3). There was no significant difference in the onset time of BPs between the SMA-proper and the SNMA-plus. The present findings suggest that the SNMA-plus is more consistently involved in preparing for various simple movements than the SMA-proper. This functionally independent region (SNMA-plus) just rostral to the SMA-proper most likely corresponds to a Pre SMA, originally defined in non-human primates. However, a part of this area elicited the inhibition of various movements by cortical stimulation. Since it generated BPs regardless of movement sites, it may play a higher role in the movement preparatory process than the SMA-proper. F1 is the M1 of the foot area. These findings are very important for glioma surgery invading the medial, middle third of IF [3].

Surgical management and strategies for the SMA gliomas with epilepsy were previously described [4–6]. The following are our preoperative evaluations. The steps include functional magnetic resonance imaging (fMRI), interictal dipole tracing (DT), subdural electrodes mapping, measurements of movement-related cortical potential (MRCP), and the use of the intraoperative open MRI under awake craniotomy. Six patients with SMA glioma who presented with epilepsy underwent surgery after the mapping procedures and are now seizure-free. Combinations of preoperative (fMRI, subdural electrodes mapping) and intraoperative mapping allow exact localization and identification of the critical functional areas. Early postoperative deficits in motor and speech function were profound, although patients recovered rapidly. It is concluded that the step of mapping procedures plays an important role in managing SMA glioma with epilepsy surgery.

We performed the following steps for SMA gliomas with epilepsy [4].

2.1 Step 1: Non-invasive Step

The aim was to evaluate the relationship between tumor locations and functional zones using the Talairach standard brain atlas. The hand motor area, central sulcus, speech area, and other functional areas (auditory and visual) are identified as our standard fMRI mapping methods. It has been generally accepted that activation sites on fMRI correlate well with the motor cortex's intraoperative electrical stimulation. To map and perform non-invasive epileptic foci exploration, EEG dipole was traced. In patients with temporal lobe epilepsy, we have shown its usefulness in determining the epileptic zones. These non-invasive steps are important for the limitations of the extent of electrode placement. However, non-invasive mapping is not enough to identify epileptogenic foci and the exact map of eloquent zones. Thus, chronic subdural electrode implantations and intraoperative direct cortical stimulations have been carried out.

2.2 Step 2: Invasive Step

The subdural electrode mapping allows identification of the sensorimotor area using SEP and the SMA using the MRCP. In this step, the standard studies of the chronic subdural mappings are performed, where analysis of the critical functional areas, epileptogenic zone, and irritative zones becomes possible. This process includes video-EEG recording and cortical electric stimulations.

2.3 Step 3: Resection Surgery

In some cases, awake craniotomy and open MRI were used. Functional mapping and 3-D update navigation can aid in identifying the location of sensorimotor and speech function on the critical areas and irritative zones. The intraoperative identification of functionally eloquent cortex is considered by many to be the best available method for a safe resection with minimizing morbidity. Intraoperative ECS remains an easy, reliable, and secure method for functional localization of essential regions to maximize a safe resection of brain tumors.

The steps of multimodality mapping were feasible to identify the eloquent cortex. In particular, the final steps using intraoperative MRI and 3-D navigation systems were important. We have not predicted which patients are more likely to experience acute or chronic postoperative defects following medial frontal lobere sections, including the SMA. Thus, further studies of the SMA functions and recovery of the surgical removal of the SMA are needed.

The next section will discuss our various approaches to remove deep-seated lesions from the anterior, middle, and finally posterior IF.

2.3.1 Anterior Interhemispheric (AIH) Approach for the Tumors in and around the Third Ventricle

Surgical management of the tumors in and around the anterior third ventricle, including craniopharyngioma (CRP), hypothalamic glioma (HTG), meningioma, suprasellar large pituitary adenoma, hypothalamic hamartoma, atypical teratoma, and metastatic tumor are still controversial [7].

Generally, tumors extending deeply into the third ventricle can be removed by the anterior transcallosal approach. Using this approach, the anterior part of the tumor invading the optic chiasm and anterior cerebral artery complex is difficult to remove sufficiently. In our experiences, interhemispheric approach with or without opening of lamina terminalis is the preferred way to remove such tumors under direct control. The posterosuperiorly extended tumor is usually not firmly attached to the surrounding structures and thus readily removed from the narrow lamina terminalis (LT) space without compromising the posterosuperior third ventricular structures and veins. Concerning the interhemispheric approach, some neurosurgeons advocate the transbasal approach because the lesion of the posterior part of the third ventricle can be controlled by direct vision [8]. However, opening the frontal sinus will increase infection, and osteotomy of the frontal base bone will add cosmetic problems.

Since 1990, the author (TH) has adopted a simple anterior interhemispheric approach (AIH) with or without opening the LT for 100 consecutive cases having the lesion in and around the anterior third ventricle without major postoperative deficits except for optic-hypothalamic-pituitary axis impairment. Our approach is characterized as follows: (1) No frontal sinus opening was needed, (2) No permanent olfactory nerve impairment at least in one side, (3) Narrow space between the bridging veins, usually less than 20 mm, was sufficient to remove the lesion using the dissection technique of bridging veins, (4) Anterior communicating artery can be preserved during resection of the lesion, (5) No mammillary body or fornix impairment were noticed post-operatively, (6) The midline important vascular system within prepeduncular cistern has always been in direct vision and under control without injury. The small feeding vessels going to the lesion can be identified, coagulated, and cut. In all 100 cases, no permanent morbidity and mortality related to this approach have been observed except for intrinsic complications associated with the hypothalamic-pituitary hormonal axis and visual system.

Illustrative Case

The authors will demonstrate operative pictures of AIH with LT opening to remove the posterior two-thirds of hypothalamic hamartoma of a 10-year-old boy presented with intractable gelastic seizure and precocious puberty. Preoperative MRI demonstrated Valdueza Type IIb hamartoma attached to the right hypothalamic wall (Fig. 4).

Many approaches for resection of lesions in and around the anterior third ventricle have been described. However, surgical management of such lesions remains technically challenging. Among these approaches, AIH approach with or without the trans-lamina terminalis (trans-LT) opening seems to be the least invasive procedure for managing tumors in and around the third ventricle. If the tumor is not invading the dorsal part of the anterior third ventricle, the lesion can be safely resected using the AIH approach with or without the use of trans-LT approach. Although we prefer using this technique, large tumors invading the dorsal part of the

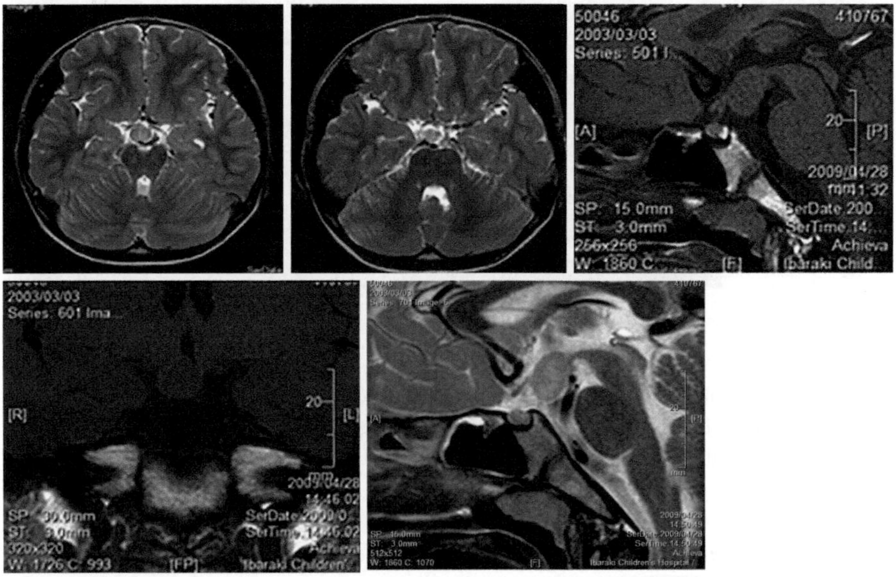

Fig. 4 Upper left: AIH approach and LT (red arrow) was exposed. Upper right: Red arrow indicates hypothalamic hamartoma attached to the right side of the hypothalamus. Lower left: Green arrow indicates hypothalamic hamartoma, and a depth electrode is inserted into the hamartoma to record EEG from the tumor. Lower right: Red arrow indicates the last piece of hamartoma, and basilar artery was exposed. And finally, posterior 2/3 of the hamartoma was removed, and 1/3 of the anterior part covered by the chiasm was left in place. The operative plan was to remove posterior 2/3 of hamartoma completely from the right hypothalamic tissue to control the intractable gelastic seizure. Immediately after the operation, the gelastic seizures were completely stopped, and higher cognitive function returned to a normal level, and precocious puberty was also controlled. Now more than 12 years was passed without any seizure, and the patient is free from antiepileptic medication. *HH* Hypothalamic hamartoma

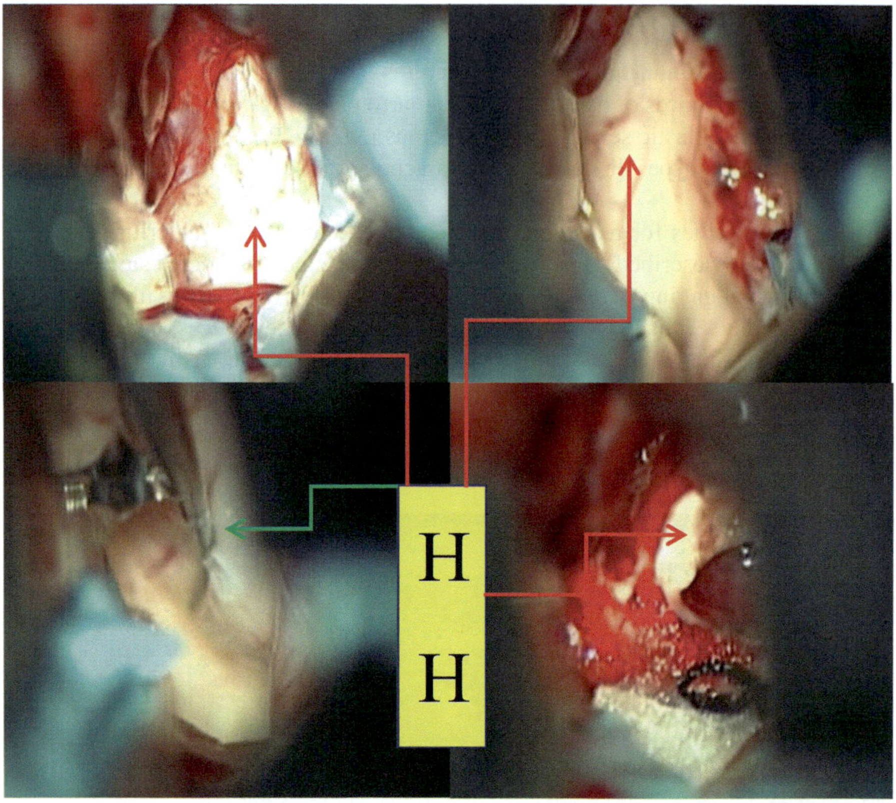

Fig. 4 (continued)

anterior third ventricle are difficult to manage by AIH alone because this part is hidden within the anterior portion of the corpus callosum. This fact creates a significant obstacle to operate on such lesions [9].

2.3.2 AIH Approach with Anterior Callosal (AC) Section

To the best of our knowledge, no cases of ACS have been reported for large tumors in and around the third ventricle. Therefore, we have added a simple anterior callosal section (ACS) to AIH with or without LT section to manage these tumors, with excellent results. In our experience, no postoperative mortality or major postoperative neurological deficits have been observed. AIH with ACS appears to be a very simple and useful method for managing tumors invading the dorsal part of the anterior third ventricle without substantial morbidity. We report four tumors using this technique in 2013 [9]. A large meningioma operation using this technique will be presented.

Illustrative Case

The patient had no deficits and showed improved higher cognitive functioning after surgery. The pathological diagnosis revealed atypical meningioma (WHO grade I). The patient was discharged to his home after a short rehabilitation period and has since returned to work. Six months later, MRI showed no sign of regrowth (Fig. 5d–f).

Ten years have passed since his last operation, and his postoperative MRIs with Gd enhancement showed the small fragment of residual tumors without further growth. The patient is leading normal social life with antiepileptic medication.

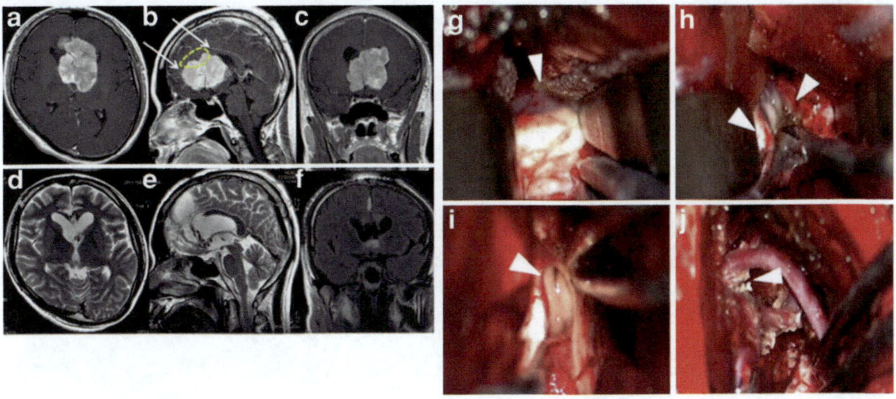

Fig. 5 A 33-year-old man found that his ability to concentrate was becoming gradually impaired. He developed epilepsy and was transferred to another emergency hospital. There, a giant tumor was seen around the third ventricle. The patient was referred to our hospital for treatment. MRI obtained at consultation demonstrated a giant enhanced homogenous tumor (60 × 65 × 65 mm) that involved bilateral A2A3 in the midpoint (**a**–**c**). The patient had no deficits and showed improved higher cognitive functioning after surgery. The pathological diagnosis revealed atypical meningioma (WHO grade I). The patient was discharged to his home after a short rehabilitation period and has since returned to work. Six months later, MRI showed no sign of regrowth (**d**–**f**). Skin incision, craniotomy, and meticulous dissection via AIH approach were performed. Although the surface of the tumor was soft and removable, the actual substance of the tumor core has very hard and strongly adhered to ACA midway between A2 and A3 portion bilaterally. We repeated coagulation and cutting of the feeders and meticulously dissected the tumor piecemeal. There were many small feeders all around the tumor. Eventually, we decided on a second-look operation in which the posterior part of the tumor was removed. A second surgery was performed 3 weeks later. First, we divided the anterior corpus callosum between the bilateral pericallosal arteries, visualized the tumor's posterosuperior part, and dissected the tumor in a piecemeal fashion (**g**, white arrowhead indicates the tumor and anterior corpus callosum.). The compressed fornix and anterior commissure were inspected (**h**, **i**). Finally, near-complete removal of the tumor was achieved using electrocoagulation for the residual part of the A2A3 adhesion (**j**)

2.3.3 Anterior Interhemispheric Transcallosal Transventricular Approach

Illustrative Cases

Large Neurocytoma (Fig. 6)

Large Ventricular Meningioma (Fig. 7)

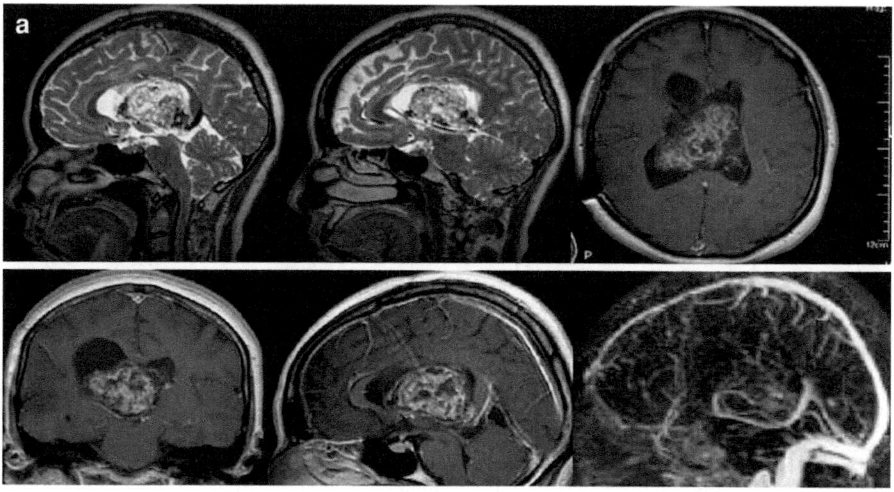

Fig. 6 A 52-year-old woman has been operated on by another neurosurgeon using right transcortical approach resulting in partial removal and VP shunt placement. (**a**) MRI showing brain tumor. Further postoperative growth of the tumor forced her to consult us, and MRI demonstrated a large neurocytoma invading lateral ventricles bilaterally. The tumor was completely removed by AIH with a transcallosal transventricular approach. Panel **b** left upper column shows the operative image of the tumor in the lateral ventricle. Finally, total removal with preservation of the right thalamostriate vein has been done (Panel **b** left below). The right corner of the MRI shows complete removal of the tumor without further recurrence, even 10 years after the final operation. For such a large size of neurocytoma, the authors prefer an anterior interhemispheric transcallosal transventricular approach rather than a transcortical one because the important intraventricular veins such as bilateral thalamostriate veins can be better preserved than the transcortical approach. The patient is free from epilepsy without antiepileptic medication and leading a normal social life

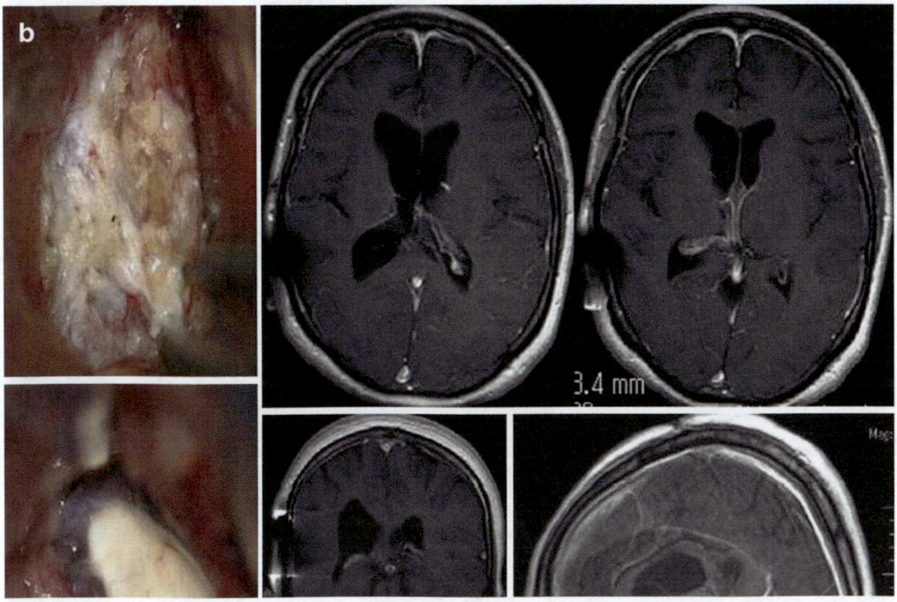

Fig. 6 (continued)

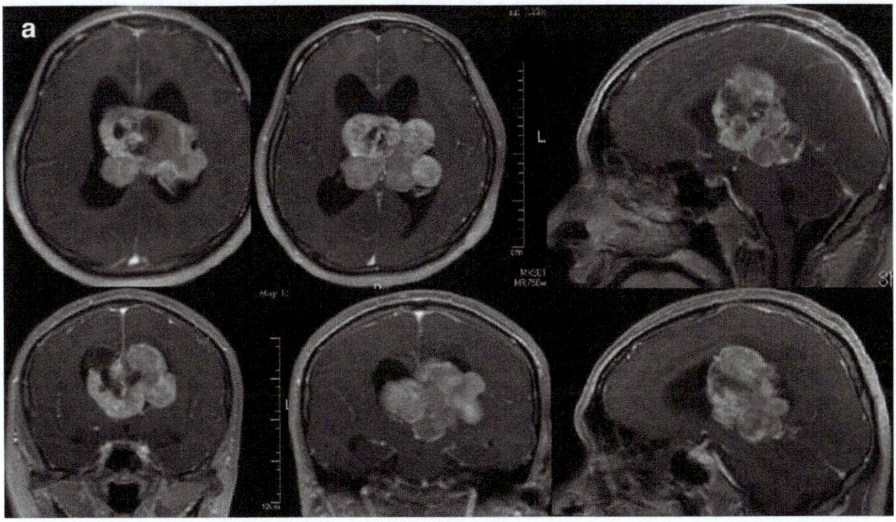

Fig. 7 A 36-year-old man with large bilateral intraventricular meningioma with intracranial hypertension. Panel **a** depicts the preoperative MR images, and Panel **b** the postoperative MRI images after multiple surgeries. The small residual tumor has been irradiated by cyberknife, but further small growth of the tumor obliged us to operate again, and finally, total removal has been accomplished. The patient is complaining of right spastic hemiparesis, which is improved by Botox injection

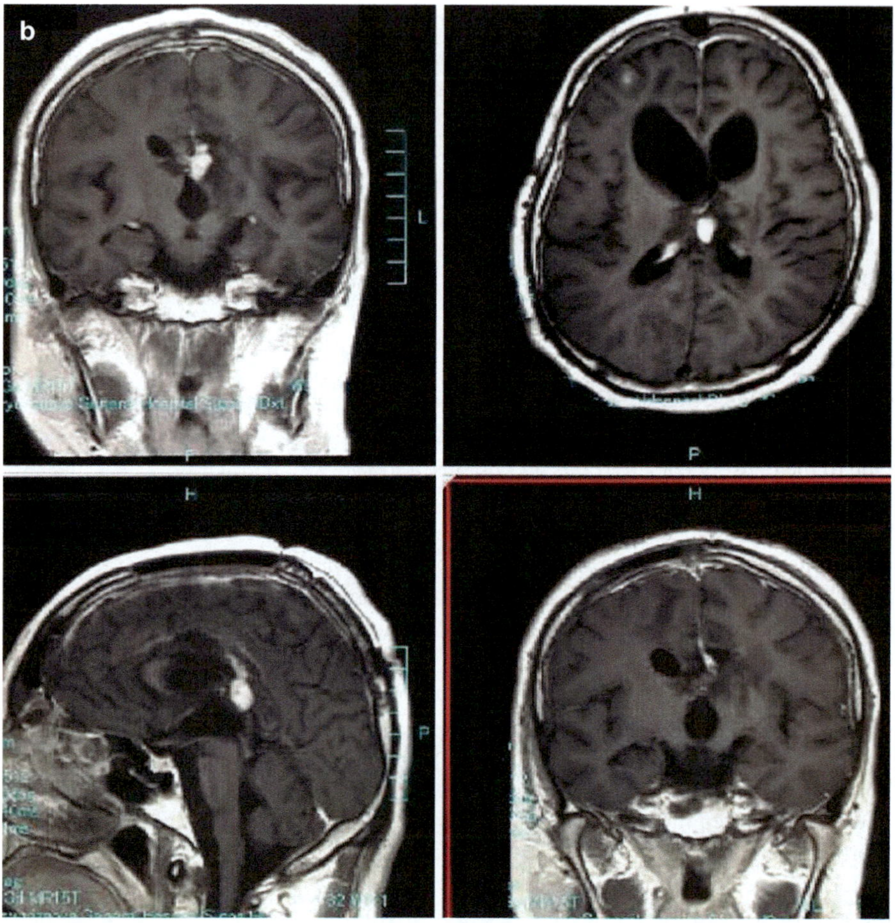

Fig. 7 (continued)

2.3.4 Interhemispheric Transcallosal Trans-hippocampal Commissure Trans-internal Cerebral Vein Approach

Midbrain Cavernous Malformation (CM) (Fig. 8)

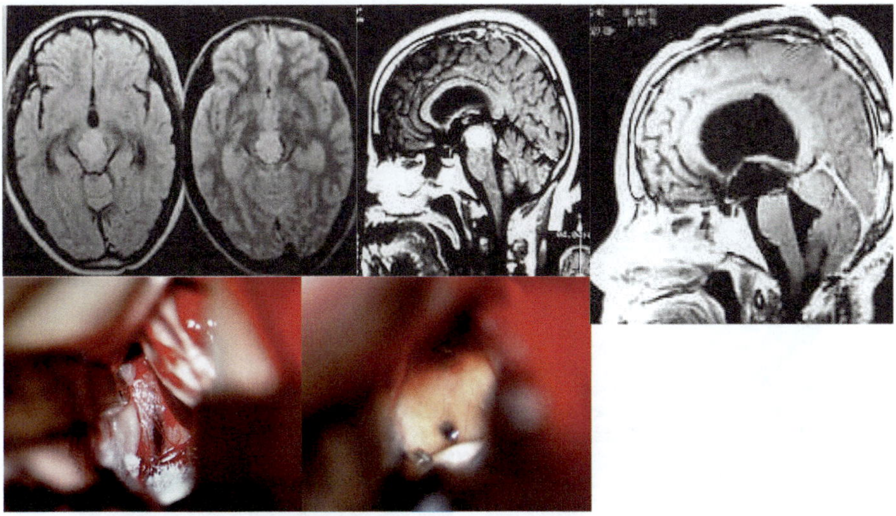

Fig. 8 A 46-year-old woman presented with bilateral oculomotor palsy and quadriparesis. Upper left, center MRI (T1 & T2WI axial plane), and right Gd-enhanced T1WI sagittal images showing the extent of the lesion before the first operation. The lower left picture during operation shows the tela choroidea of the third ventricle and bilateral internal cerebral veins in the roof of the third ventricle. After incising the midline tela choroidea of the third ventricle roof, the floor of the ventricle with small dark discoloration in front of the posterior commissure was detected (lower middle). The depth of the surgical field and hemorrhage from CM obliged the surgeon to stop surgery by partial removal. Finally, using infratentorial supracerebellar approach, the midbrain CM was resected (upper right sagittal MRI), and the postoperative condition of this patient is Glasgow Coma Scale 14

Occipital Transtentorial (OTT) Approach for Midbrain CM

A case of midbrain CM causing bilateral oculomotor palsy and right side dominant quadriparesis is presented. MRI showed that the maximum axial size of the CM was 22 mm, crossing the midline, with developing venous anomaly, and the CMs were multiple (culmen, right cerebellar peduncle, and left peri-third ventricle wall) (Fig. 9).

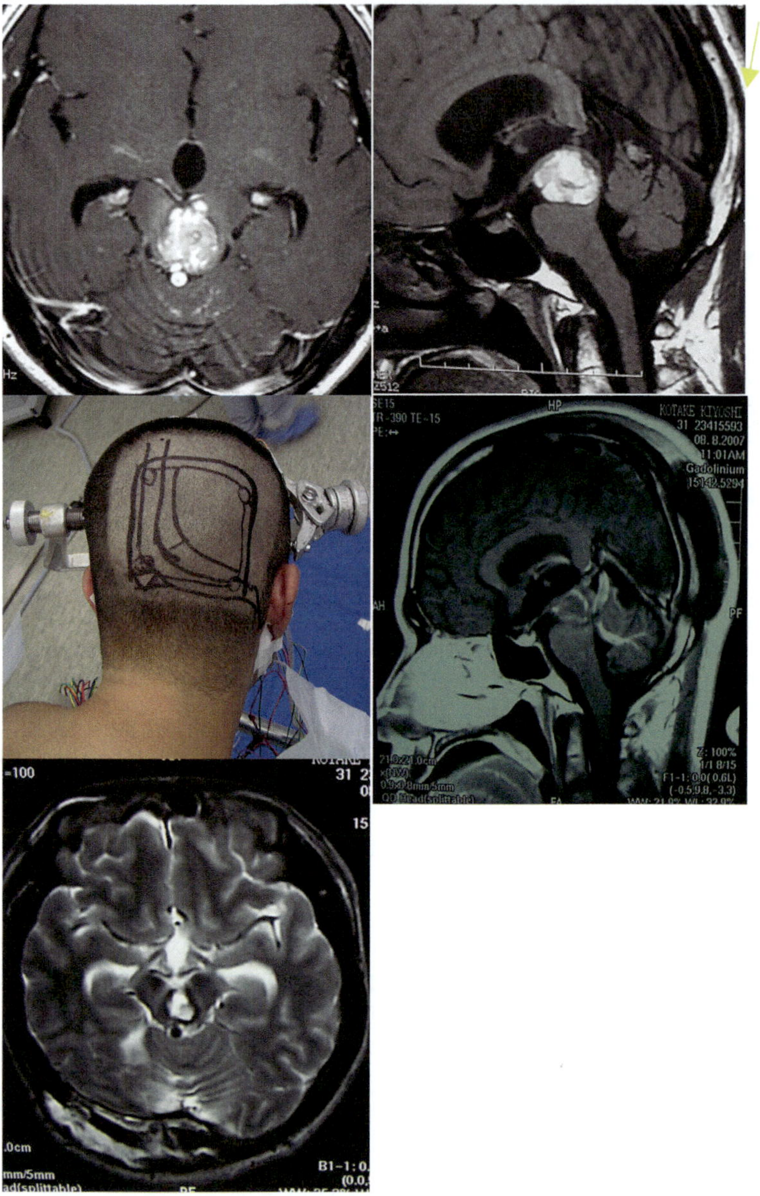

Fig. 9 Upper left: axial Gd-enhanced T1WI and the upper right: sagittal MRI T1WI shows midbrain and culmen (yellow arrow) CMs, both of which lesions were removed using the occipital transtentorial approach. Middle left: head position, skin incision, craniotomy, and dural incision of this patient. Middle right: After a small incision of the exposed dorsal surface of the right quadrigeminal plate, the CM with hematoma was removed with meticulous coagulation and cut of surrounding vessels. Postoperative T2 (lower left) axial and T1 sagittal with Gd enhancement (middle right) showing complete removal of both CM (midbrain and culmen) with a preserved venous anomaly. The patient could not return to work because of persistent diplopia caused by bilateral oculomotor palsy

3 Discussion

3.1 SMA Lesion

We performed the following steps in SMA surgery in patients with gliomas and epilepsy [4].

Step 1: Non-invasive step
Step 2: Invasive step
Step 3: Resection surgery

The intraoperative identification of functionally eloquent cortex is considered by many to be the best available method for a safe resection to minimize morbidity. Intraoperative ECS remains an easy, reliable, and safe method for functional localization to maximize a safe resection of brain tumors.

The steps of multimodality mapping were feasible to identify the eloquent cortex. In particular, the final steps using intraoperative MRI and 3-D navigation systems were important.

It is difficult to predict which patients are likely to experience acute or chronic postoperative defects following the medial frontal lobe, namely IF resections, which include the SMA. Further studies of the SMA functions and recovery of the surgical removal of the SMA are needed.

3.2 Various IF Approaches to Remove Deep-Seated Lesions

Lesions growing primarily within the third ventricle or extending into the ventricle include HTGs, pituitary adenomas, CRPs, meningiomas, cavernous angiomas, and, rarely, arteriovenous malformations. The surgical management of these lesions represents a significant challenge. Whereas intrasellar or intracisternal tumors located in the subdiaphragmatic region can be successfully treated via a transnasal transsphenoidal approach, neoplasms that protrudes from the sellar-suprasellar area into the third or lateral ventricle or septum pellucidum present particularly difficulty, with a significant risk of damaging the optic pathway and the hypothalamus. The transcallosal approach was frequently used in such cases. Subfrontal access through the anterior cranial fossa can also be done via a transfrontal sinus approach (if the frontal sinuses are sufficiently large), suprasinus transfrontal approach, frontobasal approach, or pterional approach [10–25].

The present chapter described a modified version of the traditional frontal AIH approach combined with a trans-LT approach, which prevents the need for opening the frontal sinus, dividing the falx, and sectioning the bridging veins. In suprasellar lesions that displace the third ventricle inferoposteriorly, the AIH approach provides a wide operative field. Consequently, good operative results without significant damage to brain tissue can be obtained. The AIH trans-LT approach is suitable for lesions located in the anterior part of the third ventricle, especially for those that grow anteriorly from the line joining the anterior ridge of the foramen of Monro and the cerebral aqueduct. For lesions of the pineal region, complete excision is not possible by opening only the LT.

It should be noted that this access is facilitated when the optic nerves are short. This interhemispheric lamina terminalis approach provides a good view of the structures of the infundibulo-hypophyseal axis. It usually does not require strong retraction of the frontal lobes, thus preserving the olfactory tracts. The first step of interhemispheric dissection is directed toward the frontal base, and the olfactory tract on the side of the approach is dissected from the frontal base to avoid injury. In contrast, the contralateral olfactory tract is left in place without dissection. The LT offers relatively safe access to the inferior part of the third ventricle, where the tumor is attached to the tuber cinereum and is readily exposed by a standard subfrontal or pterional approach. The LT is a soft, thin, white matter structure located in the inferior part (inferior two-thirds) of the anterior ventricular wall, between the optic tracts, proceeding from the anterior commissure to the posterior limit of the chiasm. It is crossed by the anterior cerebral arteries and the anterior communicating artery. Opening the LT permits observation of the tumor and access to the third ventricle. It is important to distinguish the LT from the thinned-out medial border of the optic tract and the posterior limit of the chiasm. The supraoptic nuclei and the columns of the fornix lie in the anterior wall of the hypothalamus immediately dorsal to the optic chiasm and lateral to the LT. The organum vasculosum of the LT, which is implicated in body fluid homeostasis and reproduction, lies beneath the anterior commissure in the midline of the LT. Both the pterional-transsylvian and subfrontal approaches provide exposure of the LT, through which the anteroinferior portion of the third ventricle can be accessed. Still, the AIH approach is more straightforward than the pterional-transsylvian approach.

Suppose total removal of the tumor is the aim of the surgery. In that case, postoperative hormonal disturbances decline significantly compared with the preoperative results, as demonstrated in CRP cases. If subtotal removal is the aim of the surgery, significant postoperative decline is not observed, as shown in the HTG group. As reported by Ohashi et al. [26], the operative strategy of CRP has been changed from total removal to subtotal removal with invaded hypothalamic tissue should be intact. Postoperative cyber-knife treatment for residual CRP is effective for consecutive 32 CRP without recurrence.

Meningiomas and pituitary adenomas are not primary third ventricle tumors. However, in our experience, the adhesion of posteriorly extended large adenomas is relatively easy to dissect without injury to the anterior hypothalamus using the AIH approach. Meningiomas invading the medial side of the optic canal can also be safely dissected and resected with or without optic canal drilling. The disadvantage of the AIH approach is the narrow corridor relative to other surgical approaches to the region in and around the anterior third ventricle, which may be a daunting task for less experienced neurosurgeons, where brain retraction seems to be greater than that in other approaches.

For large or giant lesions in and around the anterior third ventricle, in particular, these approaches alone may not permit sufficient removal. The superior part of the tumor disappears with the subfrontal and pterional approaches with trans-LT alone. With the transcortical–transventricular approach and traditional AIH approaches alone, the anterior part of the tumor invading the optic chiasm that adheres to the anterior cerebral artery complex disappears [10–25].

Consequently, we adopted the anterior callosal (AC) section in the modified AIH for managing large four tumors removal with excellent results. The AC section combined with the modified AIH provides sufficient visualization of the lesion invading the dorsal part of the anterior third ventricle. Using the combined AIH with AC, the dorsal aspect of the tumor was visualized and sufficiently removed as depicted in the respective illustrative case. The patient demonstrated improvement in short-term memory and cognitive functioning after surgery because the fornix and anterior commissure compression had been resolved. Ultimately, the AC section did not cause any major problem concerning postoperative cognitive function. The AC section itself appears to be a safe and useful procedure (Fig. 5g, h, i). To the best of our knowledge, there are few papers in the literature reporting AC (or genu) section so far, compared to many reports using the traditional transcallosal approach.

It remains controversial whether the region around the anterior wall of the foramen of Monro should be considered an untouchable area. However, the anterior commissure must be preserved since it is a significant commissural connection between the temporomesial regions and part of the frontal area. In our experience, the fornices and anterior commissure can be preserved (Fig. 5f, g). Unfortunately, the patient in case 1 already had severe short-term memory impairment before surgery. Furthermore, it was difficult to inspect the anterior commissure by microscopy due to tumor invasion.

Improving his short-term memory disturbance was also difficult. However, 2 months after surgery, he scored 2 on the modified Rankin Scale and was discharged. In all cases shown in this chapter, AIH combined with AC or AIF was successfully used to remove tumors without any major complications, especially of memory. Rosenfeld et al. report AC with AIF [23] for managing hypothalamic hamartomas. To the best of our knowledge, there are no cases of AIH with AC or AIF for managing large lesions in the third ventricle. Taken together, AIH with AC appears to have overcome the disadvantages of AIH for managing large lesions in and around the anterior third ventricle.

4 Conclusion

IF is a relatively safe surgical corridor to lesions in and around the anterior third ventricle, lateral ventricles, posterior third ventricle, pineal lesions, and brain stem cavernous malformations with no significant immediate or late complications.

Neurosurgeons should know the detail of physiological and anatomical landmarks, and physiological mapping using subdural and depth electrode recording should be done intraoperative MRI or awake craniotomy to improve operative results and lessen the postoperative adverse events. IF should also be used to approach various lesions in and around the third ventricle.

References

1. Kinjyo T, Mukawa J, Miyagi K, et al. Fatal venous infarction due to obliteration of the anterior third of the superior sagittal sinus in a falx meningioma. In: Hakuba A, editor. Surgery of the intracranial venous system. Berlin: Springer; 1996. p. 260–3.
2. Talairach J, Szikla G, Tournoux P, Prossalentis A, Bords-Ferrer M, Covello L, Iacob M, Mempel E. Atlas d'anatomie stereotaxique du telencephale: etudes anatomo-radilologiques. Paris: Masson & Cie; 1967. p. 239–1.
3. Yazawa S, Ikeda A, Kunieda T, Ohara S, Mima T, Nagamine T, Taki W, Kimura J, Hori T, Shibasaki H. Human presupplementary motor area is active before voluntary movements: subdural recording of bereitschaftspotential from medial frontal cortex. Exp Brain Res. 2000;131:165–77.
4. Yamane F, Muragaki Y, Maruyama T, Okada Y, Iseki H, Ikeda A, Homma I, Hori T. Preoperative mapping for patients with supplementary motor area epilepsy: multimodality brain mapping. Psychiatry Clin Neurosci. 2004;58:S16–21.
5. Muragaki Y, Maruyama T, Kawamata T, Yamane F, Nakamura R, Kubo O, Takakura K, Hori T. Usefulness of intraoperative magnetic resonance imaging for glioma surgery. Acta Neurochir. 2006;98(Suppl):67–75.
6. Hori T. Brain neuronavigation for deep-seated targets. In: Sindou M, editor. Practical handbook of neurosurgery: from leading neurosurgeons, Intracranial tumors, intraoperative explorations, pediatrics, vol. 2. Wien: Springer; 2009. p. 427–48.
7. Hori T, Kawamata T, Amano K, Aihara Y, Ono M, Miki N. Anterior interhemispheric approach for 100 tumors in and around the anterior third ventricle. Neurosurgery. 2010;66(3 Suppl Operative):65–74. https://doi.org/10.1227/01.neu.0000365550.84124.b.
8. Fujitsu K, Sekino T, Sakata K, Kawasaki T. Basal interfalcine approach through a frontal sinusotomy with vein and nerve preservation: technical note. J Neurosurg. 1994;80(3):575–9.
9. Shiramizu H, Hori T, Matsuo S, Niimura K, Yoshimoto H, Ishida A, Asakuno K, Yuzawa M. Anterior callosal section is useful for the removal of large tumors invading the dorsal part of the anterior third ventricle: operative technique and results. Neurosurg Rev. 2013;36:467–75. https://doi.org/10.1007/s10143-013-0455-0.
10. Alshail E, Rutka JT, Becker LE, Hoffman HJ. Optic chiasmatic hypo thalamic glioma. Brain Pathol. 1997;7(2):799–806.
11. Couldwell WT, Weiss MH, Rabb C, Liu JK, Apfelbaum RI, Fukushima T. Variations on the standard transsphenoidal approach to the sellar region, with emphasis on the extended approaches and parasellar approaches: surgical experience in 105 cases. Neurosurgery. 2004;55(3):539–50.
12. Dehdashti AR, de Tribolet N. Frontobasal interhemispheric trans-lamina terminalis approach for suprasellar lesions. Neurosurgery. 2005;56(2 Suppl):418–24.
13. Kupers RC, Fortin A, Astrup J, Gjedde A, Ptito M. Recovery of anterograde amnesia in a case of craniopharyngioma. Arch Neurol. 2004;61(12):1948–52.
14. Maira G, Anile C, Colosimo C, Cabezas D. Craniopharyngiomas of the third ventricle: trans-lamina terminalis approach. Neurosurgery. 2000;47(4):857–65.
15. Samii M. Technical aspects of excision of giant basal tumors with third ventricular involvement. In: Apuzzo ML, editor. Surgery of the third ventricle. Baltimore: Williams & Wilkins; 1987. p. 684–97.

16. Suzuki J, Katakura R, Mori T. Interhemispheric approach through the lamina terminalis to tumors of the anterior part of the third ventricle. Surg Neurol. 1984;22(2):157–63.
17. Karavitaki N, Cudlip S, Adams CB, Wass JA. Craniopharyngiomas. Endocr Rev. 2006;27(4):371–97.
18. Pedreira CC, Stargatt R, Maroulis H, et al. Health related quality of life and psychological outcome in patients treated for craniopharyngioma in childhood. J Pediatr Endocrinol Metab. 2006;19(1):15–24.
19. Kawamata T, Kubo O, Hori T. Histological findings at the boundary of craniopharyngiomas. Brain Tumor Pathol. 2005;22(2):75–8.
20. Pasquier B, Péoc'h M, Morrison AL, et al. Chordoid glioma of the third ventricle: a report of two new cases, with further evidence supporting an ependymal differentiation, and review of the literature. Am J Surg Pathol. 2002;26(10):1330–42.
21. Apuzzo ML, Chikovani OK, Gott PS, Teng EL, Zee CS, Giannotta SL, Weiss MH. Transcallosal, interfornicial approaches for lesions affecting the third ventricle: surgical considerations and consequences. Neurosurgery. 1982;10(5):547–54.
22. Mazza M, Di Rienzo A, Costagliola C, Roncone R, Casacchia M, Ricci A, Galzio RJ. The interhemispheric transcallosal-transversal approach to the lesions of the anterior and middle third ventricle: surgical validity and neuropsychological evaluation of the outcome. Brain Cogn. 2004;55(3):525–34. https://doi.org/10.1016/j.bandc.2004.03.005.
23. Rosenfeld JV, Freeman JL, Harvey AS. Operative technique: the anterior transcallosal transseptal interforniceal approach to the third ventricle and resection of hypothalamic hamartomas. J Clin Neurosci. 2004;11(7):738–44. https://doi.org/10.1016/j.jocn.2004.03.008.
24. Shapiro S, Rodgers R, Shah M, Fulkerson D, Campbell RL. Interhemispheric transcallosal subchoroidal fornix-sparing craniotomy for total resection of colloid cysts of the third ventricle. J Neurosurg. 2009;110(1):112–5. https://doi.org/10.3171/2008.4.17495.
25. Winkler PA, Weis S, Wenger E, Herzog C, Dahl A, Reulen HJ. Transcallosal approach to the third ventricle: normative morphometric data based on magnetic resonance imaging scans, with special reference to the fornix and forniceal insertion. Neurosurgery. 1999;45(2):309–17.
26. Ohhashi G, Miyazaki S, Ikeda H, Hori T. Postoperative long-term outcomes of patient with craniopharyngioma based on cyberknife treatment. Cureus. 2020;12(3):e7207. https://doi.org/10.7759/cureus.7207.

Surgical Anatomy of the Quadrigeminal Cistern and Pineal Gland

Leonardo Christiaan Welling, Nicollas Nunes Rabelo, Raphael Bertani, Bruno Henrique Dallo Gallo, and Eberval Gadelha Figueiredo

1 Introduction

The pineal gland is a small, pine cone-shaped structure named "konareion" by Galen's Greek physician. It measures approximately 7 × 6 × 3 mm and is predominantly composed of pinealocytes (similar to retinal photoreceptor cells) divided by highly vascular septa [1, 2]. It is the only organ whose diverse functions are not fully understood.

Pineal tumors represent 1% of all central nervous system neoplastic lesions (3% of tumors in adults and 8% in children) [3]. Due to early involvement of the cerebral aqueduct and secondary hydrocephalus (Fig. 1), intracranial hypertension syndrome occurs in 87% of cases. On average, 11 months before diagnosis, the symptoms are compatible with intracranial hypertension. Eye movement disorders, represented by compression of the mesencephalic structures, and Parinaud's syndrome, can be seen in 76% of cases. Cerebellar signs are less frequent and are diagnosed in 56% of patients [4, 5].

Since the beginning of the twentieth century, pineal region surgery has progressively evolved, reflecting an understanding of tumor behavior. Also, the evolution of

L. C. Welling (✉)
Department of Neurological Surgery, State University of Ponta Grossa, Ponta Grossa, PR, Brazil

N. N. Rabelo
Department of Neurological Surgery, Atenas Medical School, Passos, MG, Brazil

R. Bertani · E. G. Figueiredo
Department of Neurological Surgery, University of Sao Paulo, Sao Paulo, SP, Brazil

B. H. D. Gallo
School of Medicine, Pontificia Universidade Católica do Paraná, Curitiba, PR, Brazil

E. G. Figueiredo et al. (eds.), *Brain Anatomy and Neurosurgical Approaches*, https://doi.org/10.1007/978-3-031-14820-0_13

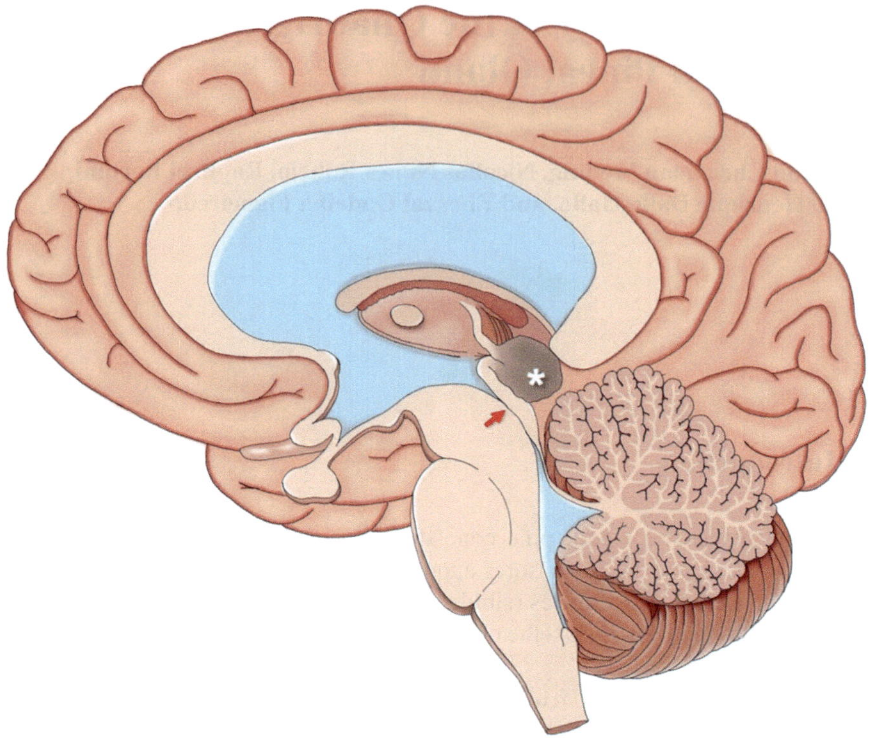

Fig. 1 Illustration representing a pineal tumor (*) causing aqueductal stenosis (red arrow) and, consequently, hydrocephalus

microsurgery and complementary therapies, whether radiotherapy or chemotherapy. Furthermore, it became evident that the management of pineal lesions must be multidisciplinary [6].

In 1913, the transcallosal interhemispheric approach was one of the first descriptions for pineal tumor resection. Dandy modified this approach in the 1920s; however, from 1910 to 1930, the results were unsatisfactory, mainly due to the lack of microscopy and adequate lighting in an environment with many anatomical challenges [7]. Van Wagenen described the transcortical transventricular approach in 1931. Horrax, on the other hand, described the resection of the occipital lobe to allow the exposure of large tumors [8]. Even with advances, perioperative mortality was close to 70% in the 1950s [6].

In the 1960s, Poppen et al. described the occipital transtentorial approach. However, radiotherapy was the first-line treatment since the risks were lower than surgery. It is observed that only half of the pineal tumors are susceptible to radiotherapy, and patients with low-grade lesions often have received whole-brain radiation [9].

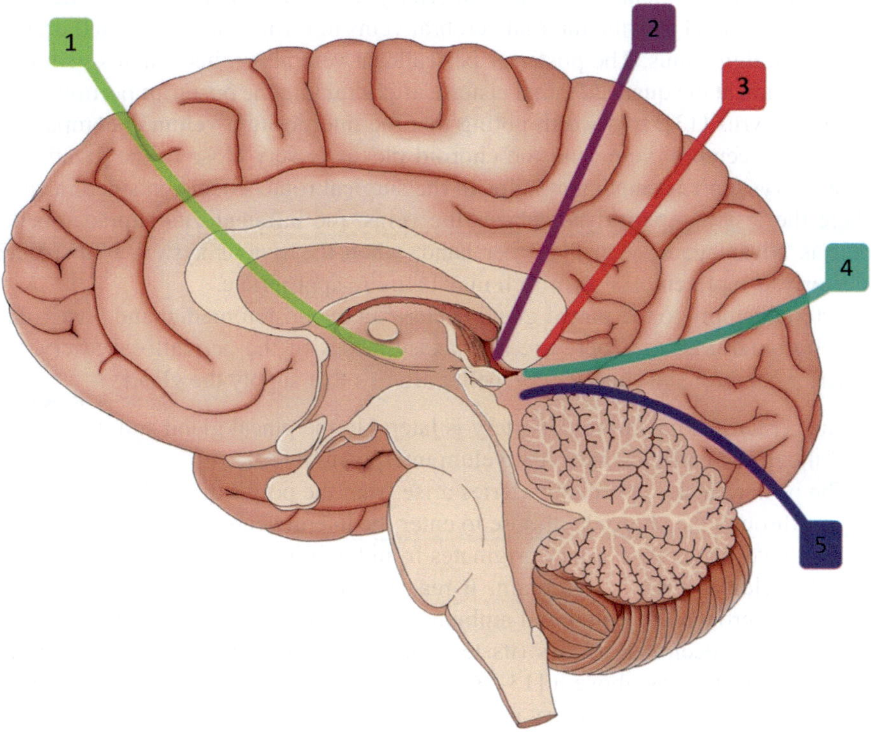

Fig. 2 Illustration showing possible surgical approaches to the pineal region: (*1*) Endoscopic approach for select cases; (*2, 3*) Less used interhemispheric approaches: transcallosal approach by Dandy [2] and retrocallosal approach [3]; (*4*) Occipital transtentorial approach (Poppen); and (*5*) Supracerebellar infratentorial

In this context, the need for a tissue diagnosis became evident but without high morbidity and mortality. The infratentorial supracerebellar approach was then improved, and from the 1970s onwards, the approach to pineal gland pathologies became safer [10, 11]. The first reports of stereotactic biopsies and radiosurgery became evident [12]. Surgical approaches are illustrated in Fig. 2.

2 Anatomy

The pineal gland is an extra-axial, encapsulated, deep structure in the geometric center of the brain. It is located between the thalamic pulvinars, in the posterior incisural space, and between the posterior midbrain and the tentorial apex; its ventral limit is the posterior commissure, the dorsal limit is the habenular commissure, and the superior limit is the corpus callosum splenium. It is, therefore, positioned inferiorly to the anterior portion of splenium, posteriorly to the third ventricle, superiorly to the collicular plate, and anteriorly to the vein of Galen [13]. The basal veins of Rosenthal join the internal cerebral veins to form Galen's vein and drain it into the straight sinus. The posterior pericallosal vein also joins Galen's vein near the splenium, at the quadrigeminal cistern, after draining the posterior portion of the cingulate gyrus [13]. It is worth noting that the interpositum velum encompassing the internal cerebral veins and the choroid plexus is very close to the gland. The drainage system's arrangement helps in the surgical route's decision since in cases where the lesion superiorly displaces the veins, the infratentorial supracerebellar route is advantageous. On the other hand, when the venous system is displaced inferiorly, a supratentorial approach may be more appropriate.

Arterial supply to the pineal gland occurs via the posteromedial and posterolateral choroidal arteries. There are also anastomoses of the pericallosal artery, the posterior cerebral and superior cerebellar arteries that supply the gland.

- The posteromedial choroidal artery is lateral to the pineal gland and, in the roof of the third ventricle, enters the velum interpositum [13].
- The posterolateral choroidal arteries arise from the posterior cerebral artery and pass through the choroidal fissure to enter the lateral ventricle [13].
- The posterior cerebral artery originates from the bifurcation of the basilar artery and circles around the midbrain. It reaches the posterior incisural space after passing through the crural and ambiens cisterns. It bifurcates into the calcarine and parieto-occipital arteries (its terminal branches) near the free edge of the tentorium, crossing above it [13–16].
- The superior cerebellar artery arises from the basilar trunk and circumvents the brainstem inferiorly to the oculomotor nerve. It can arise as single or duplicated trunks. It courses inferiorly to the trochlear nerve and superior to the trigeminal nerve, entering the cerebellomesencephalic fissure [14]. It supplies the structures below the upper limit of the inferior colliculus [15].

Most pineal lesions are infratentorial in origin and invade the posterior part of the third ventricle. As progression occurs, the lesion may reach the thalamus and the dorsal surface of the quadrigeminal lamina. Malignant lesions of glial origin may invade the thalamus and brainstem, compromising tumor resectability [17, 18].

3 Surgical Indications

A great variety of neoplastic and non-neoplastic lesions affect the pineal region. Approximately 50% of these lesions are sensitive to radiotherapy, and in this scenario, the surgical indication depends on the biopsy pathological findings, hydrocephalus, and tumor markers. Magnetic resonance imaging findings are not reliable in predicting the histological lesion subtype. However, it is instrumental for surgical planning to understand the relationship between the lesion and the anatomical structures and the feasibility of total resection [9, 19].

Due to the morbidity associated with the surgical topography and the large proportion of lesions susceptible to adjuvant therapies, non-invasive methods to obtain a definitive diagnosis should always be performed. The detection of human chorionic gonadotropin, alpha-fetus protein, or both, identified in the serum or the cerebrospinal fluid is highly suggestive of germinative origin tumors, for which resection or biopsy is not indicated, and the combination of radio and chemotherapy is enough. Cerebrospinal fluid (CSF) should preferably be obtained at an external ventricular shunt placement or during a third ventriculostomy. If there is no need for hydrocephalus treatment, a lumbar puncture can be performed, provided that small volumes are slowly removed due to the risk of herniation syndromes [19].

4 Hydrocephalus Management

Many patients with pineal tumors have their initial manifestation related to hydrocephalus (headache, reduced level of consciousness, and visual acuity), and sometimes there is a need to place an external ventricular shunt until adequate surgical planning is carried out.

Even concerning pineal masses that compress the aqueduct or posterior portion of the third ventricle, the symptoms of hydrocephalus can be relatively acute or more chronic. The last one may demonstrate signs of ventriculomegaly that are partly due to slow-growing lesions. Regardless of patient symptoms, the presence of moderate hydrocephalus on preoperative images justifies the proposition of temporary or permanent cerebrospinal fluid diversion.

Patients may not depend on the CSF diversion procedures in the case of small well-encapsulated tumors with mild hydrocephalus. In cases of moderate hydrocephalus, the ideal approach is an endoscopic third ventriculostomy. The third ventricle floor opening is done simultaneously as the biopsy of the pineal lesion is performed. In situations where there is a significant distortion of the third ventricle floor due to tumor dimensions and the relationship with the basilar artery is not favorable, performing a ventriculoperitoneal shunt is an alternative. However, CSF flow diversion is the last option since higher infection rates, overdrainage syndrome, subdural hematoma, and peritoneal dissemination of neoplastic cells are reported [20].

5 Patient Positioning

Patients can be placed in more varied positions, and considering the approaches to be exposed, the nuances of sitting, lateral (including ¾ prone), and prone positions, each with their respective advantages and disadvantages, are described.

5.1 Sitting (or Semi-sitting) Position

For the infratentorial supracerebellar approach, the sitting position is an optimal configuration. Gravity minimizes the accumulation of blood in the surgical cavity and facilitates tumor dissection from the venous system (when it is superiorly displaced). The risks are air embolism, pneumocephalus, subdural hematoma due to cortical collapse, and parasagittal bridge veins rupture in the supratentorial compartment.

Air embolism occurs when a vein is opened, and its internal pressure is negative due to gravitational effects on central venous pressure. The hemodynamics effects are secondary to the amount of air entering the venous system. The actual incidence of this complication is difficult to measure because diagnostic methods are very different. The most common devices for this are the transesophageal Doppler and the transesophageal echocardiogram. Also, a drop in expired carbon dioxide levels in the capnography can be helpful for diagnosis. Although capnography is less sensitive and indirect, it reflects the cardiovascular repercussions.

The best way to minimize air embolism occurrence is to be meticulous with hemostasis during initial surgical steps, from the skin to bone. It is crucial to maintain a clean surgical field, saline irrigation during coagulation, use bone wax properly, and keep central venous pressure (CVP) elevated at the expensc of volume expansion. Some authors suggest using high end-expiratory pressure (PEEP) to increase CVP, but their results are inconclusive, and it increases the chance of paradoxical embolism if there is an undiagnosed patent foramen ovale preoperatively.

If air embolism occurs, an increment in oxygen supply raises the O_2-partial pressure and decreases the arterial nitrogen partial pressure. This action increases the nitrogen diffusion gradient from the air bubbles into the circulation, thus accelerating their reabsorption. On the other hand, nitrous oxide, a common agent in inhalational anesthesia, can diffuse from the blood to the air bubbles, increasing in size. In any suspicion of air embolism, discontinuing inhalational anesthesia with nitrous oxide is the first measure.

In the most dramatic situations, when massive embolism occurs, air aspiration through the central venous catheter may be attempted just after removing the patient from the sitting position with operating table manipulation (i.e., quickly lowering the headboard and aspirating the central venous catheter).

Even within meticulous anesthetic care concerning the position, arterial hypotension is frequent because there is an accumulation of blood in the lower limbs, which is responsible for the mean arterial pressure decrease. Macroglossia occurs when the head is subjected to accentuated flexion and makes the venous and lymphatic tongue drainage difficult. Also, with accentuated head flexion, there is a risk of quadriplegia, as the combination of hypotension with flexion and compression over the anterior spinal artery can lead to catastrophic results.

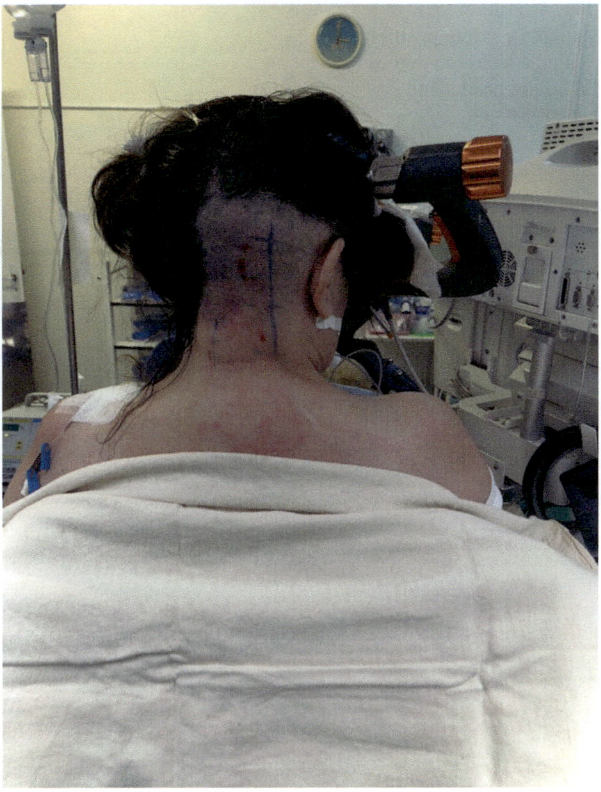

Fig. 3 Posterior view of a patient in semisitting position. Hair was removed to show a straight incision preoperative marking

The placement in a sitting position relies on an assistant stabilizing the patient's head in the Mayfield fixator (Figs. 3 and 4). The operating table inclination must be such that the tentorium becomes parallel to the floor. The legs should be elevated to facilitate venous return. The distance between the chin and the sternum must be at least 4 cm, so venous return from the cephalic segment is not compromised. The patient's pressure points must be padded to avoid abrasive skin injuries or even postoperative pressure neuropathy. The surgical microscope must be adjusted for a horizontal view of the operative field [21].

In the authors' opinion, the sitting position offers optimal operative conditions for pineal tumors. Despite the numerous complications reported, hypotension is most frequently observed, although transient and without clinical repercussions. As much as air embolism scares untrained teams, the interaction between neurosurgeons and anesthesiologists can avoid potentially unfavorable outcomes or minimize their risks.

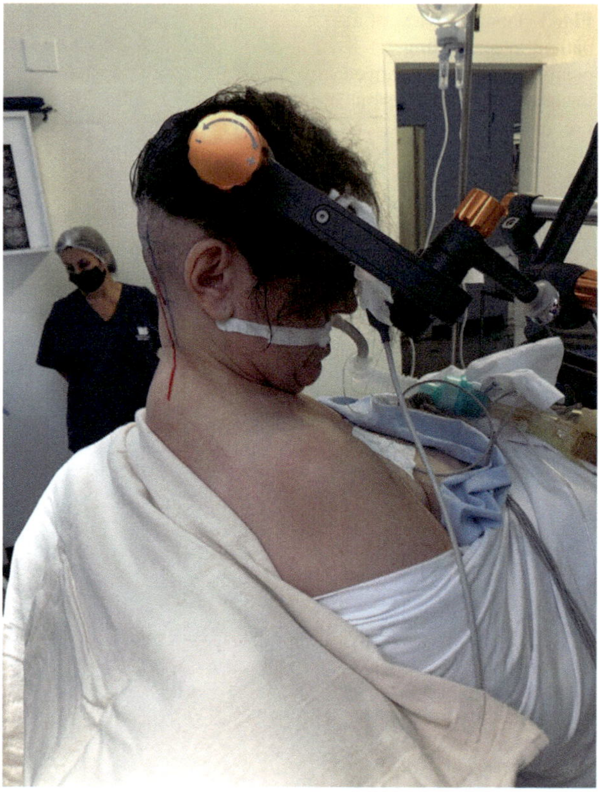

Fig. 4 Lateral view of a patient in sitting position, with the head secured by a Mayfield headclamp

5.2 *Lateral and Park Bench Position (¾ Prone)*

The lateral decubitus is usually performed with the patient lying on the right side, allowing the non-dominant hemisphere to relax downward and away from the falx cerebri to create a surgical corridor towards the pineal. For most approaches, the head should be elevated 30° above horizontal in the sagittal plane. On the other hand, the patient's nose should be about 30° towards the floor if the occipital transtentorial approach is chosen. The Park bench (or ¾ prone) position is a variant of the lateral position. The legs should be flexed and pressure points minimized with pillows. A pad should be placed under the right arm to minimize the chances of brachial plexus injury, and a support pad should be under the left hemithorax (with the patient supported on it). The use of adhesive tapes that help support the patient on the table is beneficial, as it allows lateral movements of the table during surgery to assist in the microscopic viewing angles.

Unlike the sitting position, the lateral positions and their variants cause less surgeon fatigue [22, 23]. In the author's opinion, it is feasible to get an adequate brain

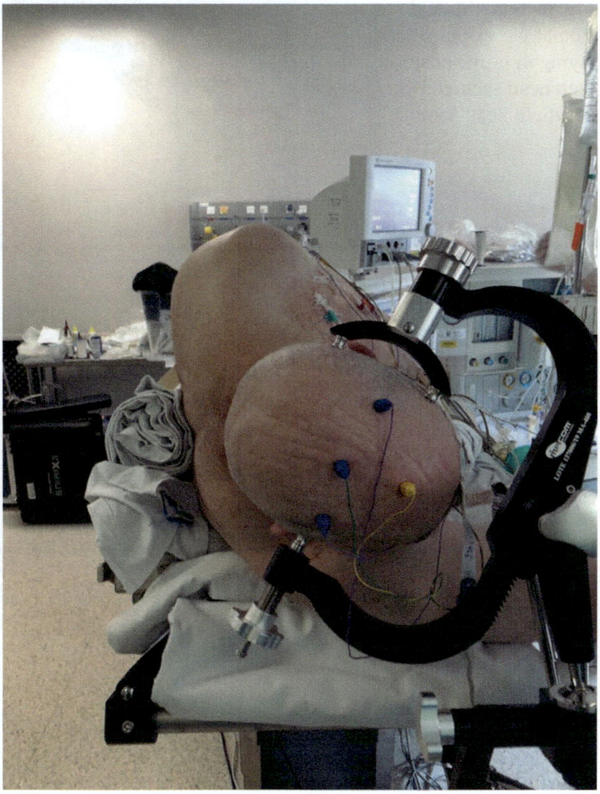

Fig. 5 Frontal view of a patient in Park-Bench position, with head secured by a Mayfield headclamp

relaxation in the lateral position and 30° head rotation, which can be considered a "combination" of the lateral and park-bench position (Fig. 5). When necessary and if not contraindicated, an external lumbar drain is advised.

5.3 Prone Position

The patient lays in a prone position, directly facing the floor (Figs. 6 and 7). This position is simple, safe, and brings with it several advantages. The first is that it is comfortable for the surgeon, but it can be challenging to perform the procedure sitting in a chair because of the surgical field depth. It is also useful when two surgeons work simultaneously in a face-to-face configuration. A slight extension of the head by 15° produces the variation known as the Concorde position (Fig. 6). This is not a good position for the supracerebellar and infratentorial pathways; on the other hand, supratentorial approaches are viable [21].

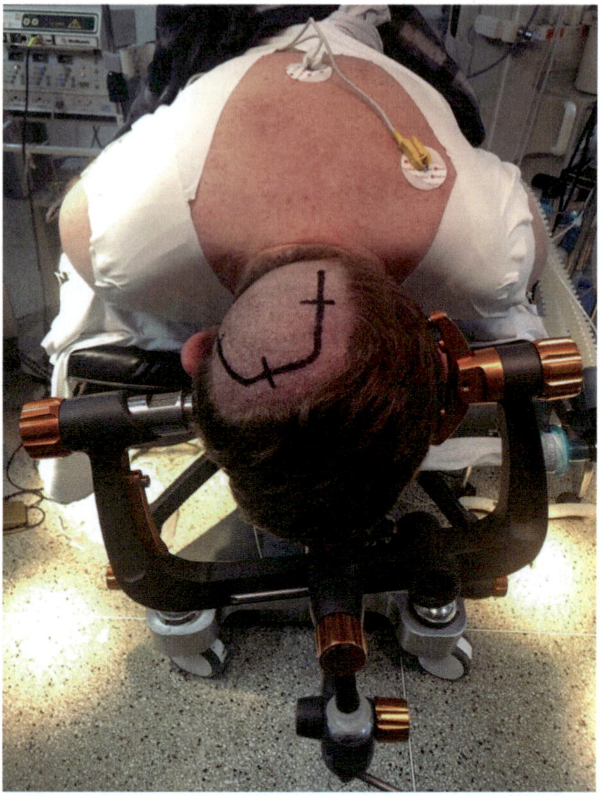

Fig. 6 Superior and frontal view of a patient laying in prone position, with head secured by a Mayfield headclamp

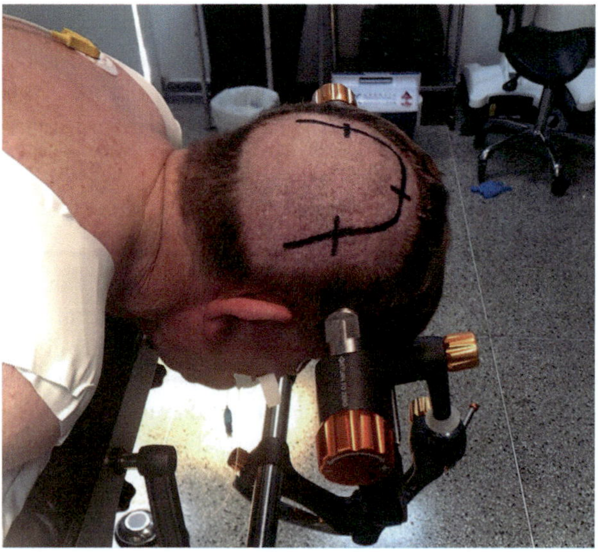

Fig. 7 Lateral view of a patient in prone position with a 15-degree extension of the head, the Concorde position

6 Surgical Approaches

6.1 Median and Paramedian Infratentorial Supracerebellar

This approach is preferably performed with the patient in a sitting or semi-sitting position. If an external ventricular drain (EVD) is required, the Frazier point can be easily accessed with the patient in this position. The skin incision is made from the external occipital protuberance to the C4 spinous process. The sharp dissection follows the nuchal ligament to minimize bleeding from muscles and works as a midline guide. The initial trepanation can be done over the superior sagittal sinus (SSS) above the torcula or lateral to the SSS. The other burr holes can be made over the transverse sinuses, one on each side. The last hole can be positioned in the midline, about 1 cm above the foramen magnum.

The holes arrangement does not necessarily have to be as exposed and varies between the surgeons. After the craniotomy, both transverse sinuses should be exposed. The dural opening follows a Y- or U-shaped fashion, and its reflection allows the direct visualization of the entire tentorium [11, 21]. The dural flap, attached cranially, should be lightly tractioned to minimize venous sinus thrombosis incidence. Concerning the inferior craniotomy extension and dural opening, it is observed that leaves a thin layer of 1 cm of bone and 1.5 cm of the dura works as a support against the cerebellum descent and some bridging veins traction from the tentorium.

The next step is to create the infratentorial corridor for the microsurgical approach. The bridging veins in the pineal trajectory are carefully coagulated. The coagulation of the more lateral vessels should be avoided, which do not interfere with the surgical corridor. There are no concerns if adequate venous collateral circulation exists; however, any vein that does not interfere with the surgical view should be spared. Sometimes, protecting the exposed cerebellum dorsal surface with cotton pads is necessary [24–26].

As the surgeon goes deeper into the surgical field, the anterior cerebellar vermis bridging veins become more visible as the retraction increases. The arachnoid membrane covering the quadrigeminal lamina can be sectioned with minimal cauterization since it is an avascular plane. Once the precentral cerebellar vein is identified, it also can be cauterized and sectioned. The remaining veins must be meticulously preserved, as the impairment of deep venous drainage can lead to catastrophic results.

After identifying the tumor mass, the microscope light must have its direction adjusted to the center of the lesion because if it remains parallel to the tentorium, the direction assumed is to the vein of Galen (Fig. 8).

Once the posterior part of the tumor is identified, its central portion is coagulated and opened with microscissors. The samples are obtained for intraoperative pathology that may be useful in whether to proceed with surgical resection. The tumor debulking is completed with simple aspiration or with ultrasonic aspiration. As the

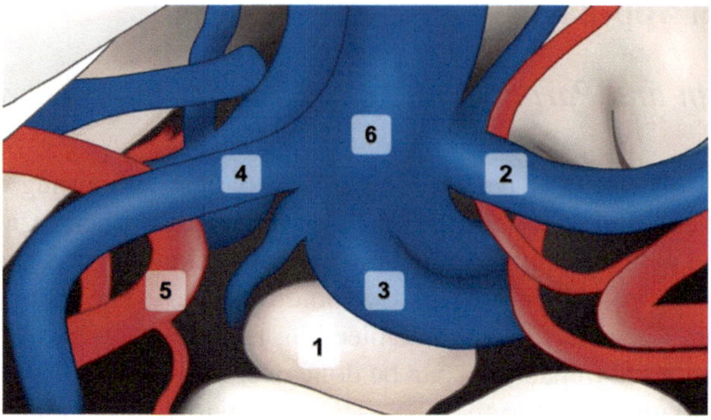

Fig. 8 Illustration showing the microsurgical anatomy of a posterior view of the posterior fossa, as would be seen in a cadaveric specimen, showing the following structures: (*1*) Pineal gland; (*2*) Internal occipital vein; (*3, 4*): Basal veins of Rosenthal; (*5*) P3 segment of the posterior cerebral artery; and (*6*) Vein of Galen

tumoral volume is reduced, its capsule can be easily detached from the thalamus and midbrain. The dissection continues until the third ventricle is found.

After this step, the lower part of the lesion is carefully separated from the brainstem. This is often the most challenging part; however, superior tumor retraction allows the dissection under direct visualization. If there is a brainstem invasion, removal progression must be carefully considered. Despite studies demonstrating that survival is directly related to resection extension, such results are not uniformly accepted in the literature, especially in glial histopathology, where the outcome results are controversial.

The upper part of the lesion is the last to be removed. Dissection of the velum interpositum and the deep venous system is performed because reduced tumor size is helpful for its dissection from fragile venous walls.

After tumor removal, a comprehensive view of the third ventricle is recommended. Micromirrors (or even endoscopy) are used to inspect the lower portion of the tumor field. Meticulous hemostasis is done with bipolar rather than hemostatic agents, as these can float in the ventricle and cause obstruction of its outflow. The dura coverings are hermetically closed, and the bone flap is replaced and fixed with mini plates [24, 25]. At the end of the procedure, and if extubating the patient is chosen while still in the operating room, it should be performed under pressure control (if necessary, sodium nitroprusside or nitroglycerin can be used) and under cough avoidance if possible.

A variant of the median approach currently used by the authors is the paramedian approach. Yasargil et al. described that this route could be used to reach the posterior incisural space and clip superior cerebellar artery aneurysms [27]. Based on cadaveric and surgical studies, it was observed that the posterior incisural space could also be well approached by an oblique trajectory [28–30].

Ueyama et al. demonstrated that the midline approach requires sectioning all vermian veins and at least one hemispherical vein. In contrast, the paramedian approach spares the vermian vessels. The sectioning of at least one hemispherical vein in the midline access is based on 59% of these being located in the first third of the dorsal cerebellum hemisphere, so the venous infarct chance is theoretically higher in the midline corridor [31].

A venous MRI and tumor projection can be helpful in the preoperative analysis to realize which is the ideal side of craniotomy when opting for the paramedian approach. Kulwin et al., in a 10 patients series, demonstrated excellent results, and this proposal is to avoid the cerebellar culmen obstruction in the surgeon's vision and minimizes the cerebellar retraction. In addition, the working distance to the lesion is shorter than the midline route and is ergonomically better for the surgeon [32].

6.2 *Transcallosal Interhemispheric*

The transcallosal interhemispheric approach provides a surgical corridor between the falx cerebri and the parieto-occipital junction. Dandy's first studies, which evaluated the arrangement of cortical bridging veins between the hemisphere and the sagittal sinus, allowed the improvement of this surgical corridor as an option to remove pathologies located in the pineal and adjacent areas.

This approach can be performed with the patient in the prone, park-bench, or even semi-sitting position. A linear incision is made [6, 21], and the craniotomy dimension depends on the tumor size and its arrangement inside the third ventricle. Preferably, a wide craniotomy allows flexibility in the microscopical vision angles and reduces the possibility of bridging veins lesion that could harm the surgical view. The right side is preferred for the approach, and the craniotomy should expose the entire sagittal sinus and even one or 2 cm of the contralateral hemisphere. The dura mater is opened to the superior sagittal sinus in a U-shaped fashion. As with the supracerebellar infratentorial approach, excessive dural traction in its opening should be avoided to reduce venous sinus thrombosis.

Depending on the chosen position, especially in those where the gravity does not allow the hemisphere to move away from the falx cerebri, the Layla or Greenberg retractors expand the surgical corridor. The maneuver, as mentioned above, enables the lesion to be placed within the center of the surgeon's field. The adhesions between the falx cerebri and the cingulate gyrus are gently separated.

Next, the corpus callosum is identified, which has a whitish appearance. The pericallosal arteries are a reference and can be retracted together to one side or farther apart. Approximately 2 cm in the corpus callosum opening are necessary. The splenium part of the corpus callosum is avoided due, at least theoretical, to disconnection syndromes.

If necessary, the tentorium or falx cerebri can be sectioned to enlarge the surgical lesion visualization. Once the corpus callosum is crossed, the deep venous system may already be evident on the dorsal surface of the tumor. The reports of the sacrifice of deep system veins and how many of these can be sacrificed are very controversial, and as there is no defined pattern, the more veins that are preserved, the lower the complications chances.

The other steps of the microsurgical approach follow the patterns previously described.

6.3 Occipital Transtentorial

The occipital transtentorial approach is one of the variations of the supratentorial pathways to the pineal gland. Its first descriptions were made by Poppen et al., using the parasagittal occipital route, with exposure of the superior sagittal sinus and elevation of the occipital lobe to have an interhemispheric and suboccipital corridor [33]. Jamieson et al. modified the approach by placing the patient in lateral decubitus, making a craniotomy exposing the transverse and sagittal sinuses, retracting the occipital lobe laterally, and excising the tentorium through the interhemispheric corridor. This formed the basis of the contemporary interhemispheric occipital approach [10].

According to Moshel et al., the transtentorial approach can be performed interhemispherically or laterally. Preferably, the patient is placed in a prone or park bench position. As the trajectory to the lesion is oblique, it may confuse surgeons unfamiliar with this route. The tentorium section offers excellent exposure to the quadrigeminal plate, which is favorable in tumors that extend inferiorly. In the interhemispheric modality, the craniotomy is located over the occipital lobe on the right and exposes the superior sagittal sinus. In the lateral transtentorial occipital modality, there is no need to expose the superior sagittal sinus, but on the other hand, the exposition of the ipsilateral transverse sinus [34].

When opting for the interhemispheric transtentorial occipital approach, the ¾ prone position permits gravity to the occipital lobe to decrease, mainly because there are no bridging veins in this region that could act as anchors against the gravity effects. In the lateral occipital transtentorial variation, a slightly flexed prone position with the head is useful to move the occipital lobe away from the tentorium. A mannitol infusion and external ventricular drainage can be helpful to minimize the hemianopsia risks if the occipital lobe retraction is necessary for the classical interhemispheric approach and, less commonly, in the more lateral pathways.

Eventually, intraoperative Doppler may help identify the straight sinus location. Under microscopy, after the straight sinus is identified, the tentorium is sectioned adjacent to it. If more space is needed, the falx cerebri also be sectioned.

As the tumor and the quadrigeminal cistern are identified, the tumor removal follows the microneurosurgery principles previously described [8, 33].

6.4 Transcortical Transventricular

The transcortical transventricular approach uses a direct route through the lateral ventricle (after a corticectomy has been performed). The brain entry point must be chosen by stereotaxis or neuronavigation. Its use is very restricted, and one of the few advantages is in tumors that project to the lateral ventricle [35].

6.5 Endoscopic Approaches

The use of neuroendoscopy in pineal tumors was first described in the late 1990s and was used both by the transventricular and extracerebral routes and sometimes as a complement to classical microscopic surgery [36].

Initially, the purpose of endoscopy was to treat obstructive hydrocephalus, but in certain situations, with more anterior holes anterior to the coronal suture, an endoscopic debulking of the tumor lesion is feasible.

After general anesthesia, the patient is fixed in the Mayfield clamp without rotation (i.e., 0°). After asepsis and antisepsis surgical protocol, a burr hole is placed 3 cm anterior to the coronary suture. This anteriorization aims to reach the anterior face of the tumor lesion by the third ventricle route. If neuronavigation is available, there will be more precision in choosing the ideal trajectory. After the corticectomy, the endoscope cannula is implanted perpendicularly to the cortex towards the frontal horn of the right lateral ventricle. It is entered with an endoscope with a 30° lens and crosses the foramen of Monro, respecting widely known anatomical limits. Coagulation must be performed meticulously, and samples of the lesion collected for histopathological analysis when identifying the tumor lesion.

The endoscope can also be used in the infratentorial supracerebellar approach, even with small craniotomies. After entering this surgical corridor, the assistant surgeon maintains the light and the vision of the structures in the posterior incisural space. The senior surgeon, following microsurgical teachings, progresses in its dissection toward the pineal gland. It is observed that the instruments do not pass through the endoscope portal, and the operative technique requires that it be performed bimanually.

As it is considered a minimally invasive procedure, the endoscopic approach has advantages. The first is using narrow surgical corridors which respect the bridging veins. Besides, the endoscope bypasses these veins that work as obstacles to the tumor dissection. In addition, small craniotomy, minimal brain retraction, and excellent direct tumor vision reveal promising alternatives to conventional microsurgical techniques.

On the other hand, the method has some limitations, whether related to the endoscopic procedure or the tumor itself. In the first case, the monohand endoscopy, the instruments are manipulated inside the endoscope portal; and there are a limited tumor dissection and hemostasis. When performed bimanually, the numerous moments that the instruments pass through the surgical corridor before reaching the primary target increases the injury risk to the cerebellum and the surrounded veins. The limitations imposed by the tumor are excessive bleeding, fibrous tumoral consistency, and lesions larger than 2.5 cm.

To minimize the complications, careful evaluation of the preoperative exams considering the distance from the cortex to the frontal horn, the position of the basilar artery tip concerning the third ventricle floor, anterior tumor protrusion into the third ventricle, a superior extension of the tumor below the splenium of the corpus callosum.

Based on intraoperative conditions, the best surgical strategy should constantly be reviewed. Whether an isolated biopsy or lesion debulking, never traction the tumor, perform rigorous hemostasis, and be prepared for open surgery conversion if necessary.

7 Postoperative Care and Complications

Corticosteroids should be maintained at high doses for the first few days and progressively reduced as the patient's clinical condition improves. During the early postoperative period, anticonvulsants are strongly recommended, especially in supratentorial approaches [35]. After the early pos operative, the anticonvulsants should be tapered off.

Confusion and drowsiness are common in sitting position patients, mainly due to the postoperative pneumocranium that is observed frequently. In addition, subdural hematoma and ventricular collapse with CSF overdrainage may occur during surgery. Air embolism is rare in the postoperative period and is usually identified during surgery with a decrease in expired carbon dioxide detected in the capnography. The detection of air bubbles by Doppler also confirms the air embolism [37].

Other complications are observed in supratentorial approaches, such as hemianopia due to retraction of the occipital lobe [38], contralateral sensory or motor deficits due to parietal lobe traction, or venous infarction after inadvertent retraction or sectioning of bridging veins.

The first 72 h are the most important, and any neurological changes should be promptly investigated with radiological imaging (CT scan or MRI) to exclude hemorrhage, hydrocephalus, or tension pneumocephalus. Among these, hemorrhage can be catastrophic, especially in those patients with malignant and partially resected tumors.

Venous infarctions are also very deleterious, occur later, and are often unpredictable. They are related to the inability of the previous healthy venous system that has not been manipulated or coagulated to support the venous drainage from the others who be sectioned. If venous infarction occurs in the brainstem, the neurological prognosis may worsen depending on its extension [21, 35, 37].

In patients with external ventricular drainage or ventriculoperitoneal shunts, the dysfunction can occur due to postoperative debris obstructing the valve or catheters. If the EVD is placed simultaneously with the tumor removal, the shunt dependency is evaluated in the first 3 days, so EVD is avoided for more extended periods.

Patients should be advised that extraocular movement disorders are frequent during this period, with vertical gaze and ocular convergence being the most affected. They are usually transient, but eventually, they can last up to months. It is observed, mainly in the vertical view, that a residual limitation can remain indefinitely, however, with little clinical significance. In addition, the greater the preoperative impairment, the lower the recovery chances [6, 37].

Gait ataxia is frequent and resolves within the first few days. The occurrence of akinetic mutism and cognitive dysfunction is also observed, but with a much lower incidence. Other more severe complications are directly related to the involvement of the brainstem, either by the long tracts or the cranial nerves nuclei injury [35].

On the other hand, as most patients with lesions in the pineal gland are relatively young, cardiological and respiratory complications are less frequent.

It is worth noting that early ambulation and physical therapy are very important at this stage. In addition, a brain MRI should be performed within the first 48 h to evaluate the correct degree of tumor resection [25, 35].

8 Conclusions

The wide range of possible diagnoses makes it imperative to establish a tissue analysis in patients with pineal tumors. Decisions about adjuvant therapies, prognosis, and follow-up depend on the histopathological diagnosis. Stereotactic methods are valuable in obtaining a tissue sample with less aggressiveness. Better anatomical knowledge, combined with technological advances, has allowed more aggressive resections, and the long-term prognosis is excellent for benign tumors; in cases of malignant tumors, the response to treatment is also much better than in the past.

References

1. Sumida M, Barkovich AJ, Newton TH. Development of the pineal gland: measurement with MR. AJNR Am J Neuroradiol. 1996;17(2):233–6.
2. Macchi MM, Bruce JN. Human pineal physiology and functional significance of melatonin. Front Neuroendocrinol. 2004;25(3–4):177–95.
3. Blakeley JO, Grossman SA. Management of pineal region tumors. Curr Treat Options in Oncol. 2006;7(6):505–16.
4. Konovalov AN, Pitskhelauri DI. Principles of treatment of the pineal region tumors. Surg Neurol. 2003;59(4):250–68.
5. Gaillard F, Jones J. Masses of the pineal region: clinical presentation and radiographic features. Postgrad Med J. 2010;86(1020):597–607.
6. Kennedy BC, Bruce JN. Surgical approaches to the pineal region. Neurosurg Clin N Am. 2011;22(3):367–80.
7. Dandy WE. Operative experience in cases of pineal tumor. Arch Surg. 1936;33(1):19.
8. Horrax G. Extirpation of a huge pinealoma from a patient with pubertas praecox: a new operative approach. Arch Neurol Psychiatr. 1937;37(2):385–97.
9. Regis J, Bouillot P, Rouby-Volot F, Figarella-Branger D, Dufour H, Peragut JC. Pineal region tumors and the role of stereotactic biopsy: review of the mortality, morbidity, and diagnostic rates in 370 cases. Neurosurgery. 1996;39(5):907–12; discussion 912–914.
10. Jamieson KG. Excision of pineal tumors. J Neurosurg. 1971;35(5):550–3.
11. Stein BM. The infratentorial supracerebellar approach to pineal lesions. J Neurosurg. 1971;35(2):197–202.

12. Pecker J, Scarabin JM, Vallee B, Brucher JM. Treatment in tumours of the pineal region: value of sterotaxic biopsy. Surg Neurol. 1979;12(4):341–8.
13. Matsuo S, Baydin S, Güngör A, et al. Midline and off-midline infratentorial supracerebellar approaches to the pineal gland. J Neurosurg. 2017;126:1984–94. https://doi.org/10.3171/2016.7.JNS16277.
14. Rhoton AL Jr. The cerebellar arteries. Neurosurgery. 2000;47(3 Suppl):S29–68.
15. Rhoton AL Jr. Tentorial incisura. Neurosurgery. 2000;47(3 Suppl):S131–53.
16. Zeal AA, Rhoton AL Jr. Microsurgical anatomy of the posterior cerebral artery. J Neurosurg. 1978;48:534–59.
17. Erlich SS, Apuzzo MLJ. The pineal gland: anatomy, physiology, and clinical significance. J Neurosurg. 1985;63(3):321–41.
18. Yamamoto I, Kageyama N. Microsurgical anatomy of the pineal region. J Neurosurg. 1980;53(2):205–21.
19. Choi JU, Kim DS, Chung SS, Kim TS. Treatment of germ cell tumors in the pineal region. Childs Nerv Syst. 1998;14(1–2):41–8.
20. Goodman RR. Magnetic resonance imaging-directed stereotactic endoscopic third ventriculostomy. Neurosurgery. 1993;32(6):1043–7; discussion 1047.
21. Bruce JN, Ogden AT. Surgical strategies for treating patients with pineal region tumors. J Neuro-Oncol. 2004;69(1–3):221–36.
22. Ausman JI, Malik GM, Dujovny M, Mann R. Three-quarter prone approach to the pineal-tentorial region. Surg Neurol. 1988;29(4):298–306.
23. Brotchi J, Levivier M, Raftopoulos C, Dewitte O, Pirotte B, Noterman J. Three-quarter prone approach to the pineal-tentorial region. Report of seven cases. Acta Neurochir Suppl (Wien). 1991;53:144–7.
24. Tate M, Sughrue ME, Rutkowski MJ, Kane AJ, Aranda D, McClinton L, et al. The long-term postsurgical prognosis of patients with pineoblastoma: long-term survival in pineoblastoma. Cancer. 2012;118(1):173–9.
25. Farnia B, Allen PK, Brown PD, Khatua S, Levine NB, Li J, et al. Clinical outcomes and patterns of failure in Pineoblastoma: a 30-year, single-institution retrospective review. World Neurosurg. 2014;82(6):1232–41.
26. Yu T, Sun X, Wang J, Ren X, Lin N, Lin S. Twenty-seven cases of pineal parenchymal tumours of intermediate differentiation: mitotic count, Ki-67 labelling index and extent of resection predict prognosis. J Neurol Neurosurg Psychiatry. 2016;87(4):386–95.
27. Yasargil M. Microneurosurgery, vol. 1. New York: Thieme; 1987.
28. Ammirati M, Bernardo A, Musumeci A, Bricolo A. Comparison of different infratentorial-supracerebellar approaches to the posterior and middle incisural space: a cadaveric study. J Neurosurg. 2002;97(4):922–8.
29. Vishteh AG, David CA, Marciano FF, Coscarella E, Spetzler RF. Extreme lateral supracerebellar infratentorial approach to the posterolateral mesencephalon: technique and clinical experience. Neurosurgery. 2000;46(2):384–8; discussion 388–9.
30. Van den Bergh R. Lateral-paramedian infratentorial approach in lateral decubitus for pineal tumours. Clin Neurol Neurosurg. 1990;92(4):311–6.
31. Ueyama T, Al-Mefty O, Tamaki N. Bridging veins on the tentorial surface of the cerebellum: a microsurgical anatomic study and operative considerations. Neurosurgery. 1998;43(5):1137–45.
32. Kulwin C, Matsushima K, Malekpour M, Cohen-Gadol AA. Lateral supracerebellar infratentorial approach for microsurgical resection of large midline pineal region tumors: techniques to expand the operative corridor. J Neurosurg. 2016;124(1):269–76.
33. Poppen JL. The right occipital approach to a Pinealoma. J Neurosurg. 1966;25(6):706–10.

34. Moshel YA, Parker EC, Kelly PJ. Occipital transtentorial approach to the precentral cerebellar fissure and posterior incisural space. Neurosurgery. 2009;65(3):554–64; discussion 564.
35. Sonabend AM, Bowden S, Bruce JN. Microsurgical resection of pineal region tumors. J Neuro-Oncol. 2016;130(2):351–66.
36. Robinson S, Cohen AR. The role of neuroendoscopy in the treatment of pineal region tumors. Surg Neurol. 1997;48(4):360–5; discussion 365–7.
37. Bruce JN, Stein BM. Surgical management of pineal region tumors. Acta Neurochir. 1995;134(3–4):130–5.
38. Nazzaro JM, Shults WT, Neuwelt EA. Neuro-ophthalmological function of patients with pineal region tumors approached transtentorially in the semisitting position. J Neurosurg. 1992;76(5):746–51.

Surgical Anatomy of Cerebellopontine Cistern

Ricardo Ramina, Gustavo Simiano Jung, Erasmo Barros da Silva Jr, and Rogerio S. Clemente

1 Introduction

Subarachnoid cisterns are compartments of the subarachnoid space filled with cerebrospinal fluid (CSF), separating the arachnoid membrane and the pia mater. These spaces contain fibers and many trabeculae [1]. The subarachnoid cisterns are interconnected, and their patency is important for CSF circulation. Cranial nerves, arteries, and veins pass through them. Knowledge of the anatomy of their neural and vascular structures is important in planning and performing intracranial surgery. Opening basal cisterns to release CSF reduces the intracranial pressure allowing better identification of the anatomic structures. The subarachnoid cisterns are divided into two groups: supratentorial and infratentorial. The posterior fossa cisterns are unpaired and paired. Unpaired are the following cisterns: interpeduncular, prepontine, premedullary, quadrigeminal, and the *cisterna magna*. Paired cisterns are the cerebellopontine and cerebellomedullary (Figs. 1 and 2) [2].

This chapter will discuss the surgical anatomy, radiologic findings of the cerebellopontine cistern, and the surgical approaches to different pathologies encountered in this region.

R. Ramina (✉) · G. S. Jung · E. B. da Silva Jr · R. S. Clemente
Neurosurgical Department, Neurological Institute of Curitiba, Curitiba, PR, Brazil
e-mail: ramina@inc-neuro.com.br

E. G. Figueiredo et al. (eds.), *Brain Anatomy and Neurosurgical Approaches*,
https://doi.org/10.1007/978-3-031-14820-0_14

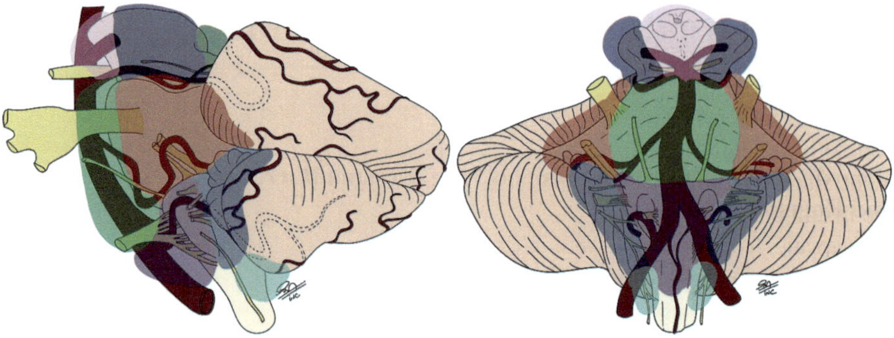

Fig. 1 Drawings showing the cerebellopontine cisterns (brawn), the prepontine cistern (green), and the cerebellomedullary cisterns (gray)

Fig. 2 Surgical picture showing the cerebellopontine cistern filled with CSF and the cranial nerves VII and VII at the internal auditory meatus. The cranial nerves IX, X, and XI at the jugular foramen

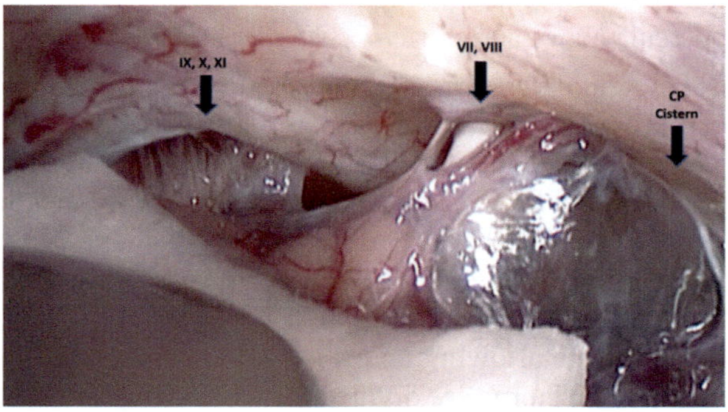

2 History

Key and Retzius first studied the anatomy of the subarachnoid cistern in 1875 [3]. In 1959, the neuroradiologist Liliequist described the cistern based on pneumographs and anatomical studies [4]. Other studies were made by Lewtas NA [5], Amundsen P [6], Yasargil MG [1], and Matsuno H [2].

The first successful complete cerebellopontine angle tumor removal is credited to Charles Ballance in 1894. The tumor was removed with the finger, and the patient recovered from surgery and was alive at least 18 years after surgery [7]. Victor Horsley, Fedor Krause, Harvey Cushing, Walter Dandy, and other pioneers contributed to the CPA surgery [8]. In 1939, Olivecrona presented a rate of facial nerve preservation of 65% in the surgical treatment of vestibular schwannomas; this result was extraordinary for that time [9]. William House using the surgical microscope and microsurgical techniques through the translabyrinthine approach, reported excellent facial nerve preservation rates and low operative mortality rate [10–13]. In 1965, Rand and Kurze, using the suboccipital transmeatal approach and microdissection, removed vestibular schwannomas totally with a high rate of preservation of facial, vestibular, and cochlear nerves [14]. More recently, Madjid Samii, in a large series of patients submitted to vestibular schwannoma removal, showed rates of anatomical facial preservation of approximately 93% (100% in smaller tumors) and improved rates of hearing preservation with minimal mortality rates (below 1%) [15, 16].

3 Surgical Anatomy

The cerebellopontine cistern is on the posterior portion of the petrous bone between the anterolateral surface of the pons and the cerebellum (Figs. 2 and 3). The pontomesencephalic membrane splits this cistern superiorly from the ambient cistern. The pontomesencephalic membrane is attached to the brainstem and intersects the oculomotor nerve anteriorly. Inferiorly, the cerebellopontine cistern is separated from the cerebellomedullary cistern by the lateral pontomedullary membrane, which lies between the cranial nerves VII, VIII, and IX. The anterior pontine membrane separates the cerebellopontine cistern medially from the prepontine cistern through the anterior pontine membrane. Laterally, the cerebellopontine cistern extends to the edge of the cerebellar surface [2, 17].

The cranial nerves running through the cerebellopontine cistern are the trigeminal nerve, the abducens nerve, the facial nerve, and vestibulocochlear nerve (Fig. 4). The trigeminal nerve originates about halfway between the lower and upper lateral portion of the pons and runs in the superolateral part of the cerebellopontine cistern. This nerve has two roots: the larger one is the sensory root located laterally and a narrow motor root located superior and medially to the sensory root. The *abducens* nerve originates in the pontomedullary junction and passes anterior to the pontine membrane. The facial and vestibulocochlear nerves originate in the pontomedullary junction running in the inferior part of the cerebellopontine cistern (Figs. 5 and 6).

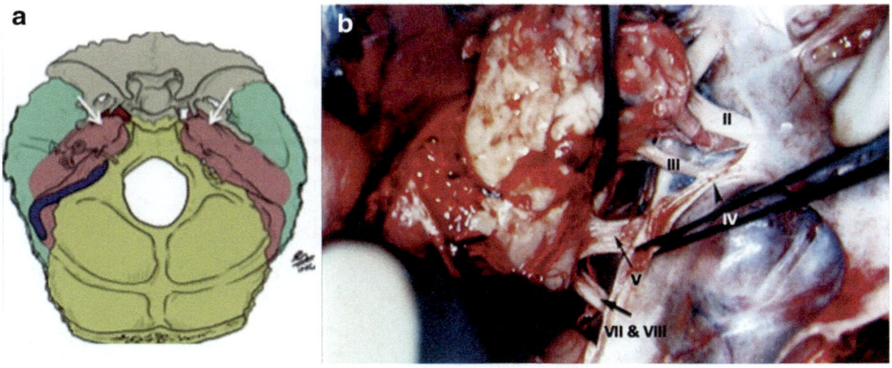

Fig. 3 (**a**) Drawing showing the relationship of the posterior portion of the petrous bone (brawn) and the structures of the internal auditory canal. (**b**) Anatomical preparation with the cranial nerves V, VII and VIII in the cerebellopontine angle

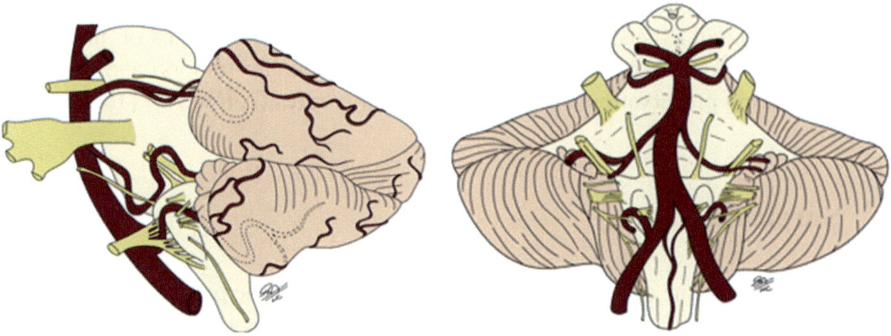

Fig. 4 Drawing showing the cranial nerves at the cerebellopontine angle

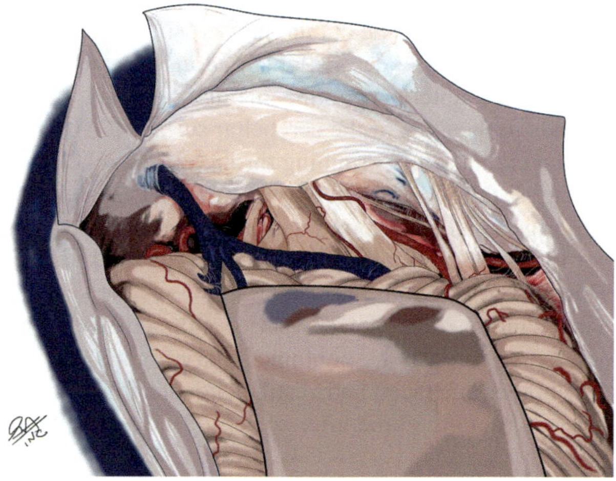

Fig. 5 Drawings showing the cranial nerves V, VII, and VII in the cerebellopontine angle and the basilar artery with its branches (AICA, SCA, and PCA)

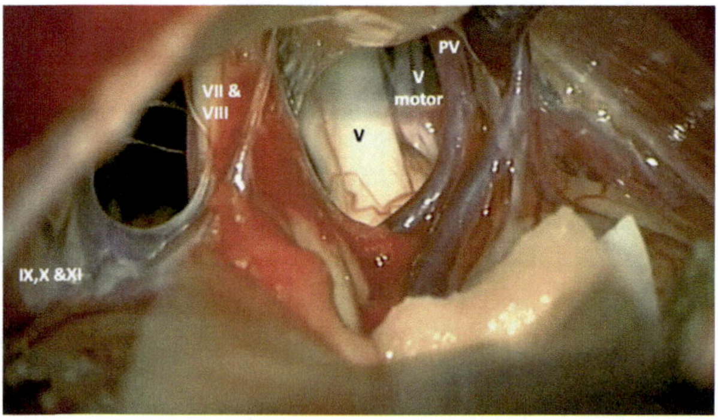

Fig. 6 Surgical view of the CPA after opening the arachnoid of the cerebellopontine cistern. PV, petrosal vein; V cranial nerve with its motor portion, VII and VIII cranial nerves and caudal cranial nerves (IX, X, and XI) covered by arachnoid

The facial nerve is usually hidden by the choroid plexus and the flocculus (Fig. 7). This portion of the facial nerve (subpial) is the point of vascular conflict of hemifacial spasm. The facial nerve arises 2–3 mm anterior to the root entry zone of the vestibulocochlear nerve. The lateral recess of the fourth ventricle (foramen of Luschka) is inferior and posterior to the root entry zones of the facial and vestibulocochlear nerve. In the cerebellopontine cistern, the facial nerve is located anteromedially, and the vestibulocochlear nerve is posterolateral. The outer arachnoid membrane goes into the internal auditory canal around the VII and VIII cranial nerves. Behind these nerves is the *flocculus*, which covers the aperture of the lateral recess of the fourth ventricle. The facial nerve is anterosuperior in the internal auditory canal, the cochlear nerve anteroinferior, the superior vestibular nerve is posterosuperior, and the inferior vestibular nerve is posteroinferior (Fig. 8).

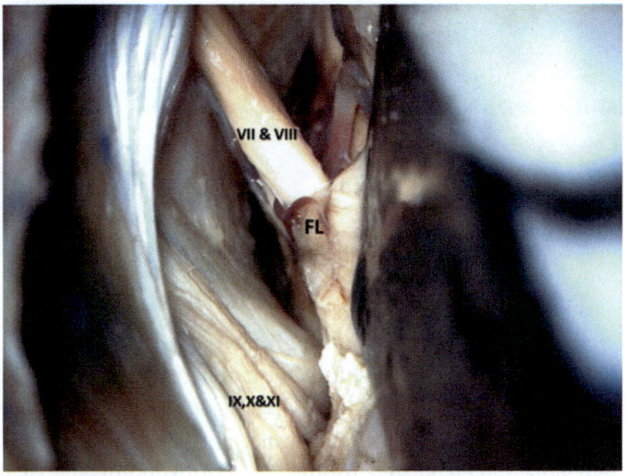

Fig. 7 Anatomical preparation showing the cranial nerves VII and VIII hidden by the flocculus

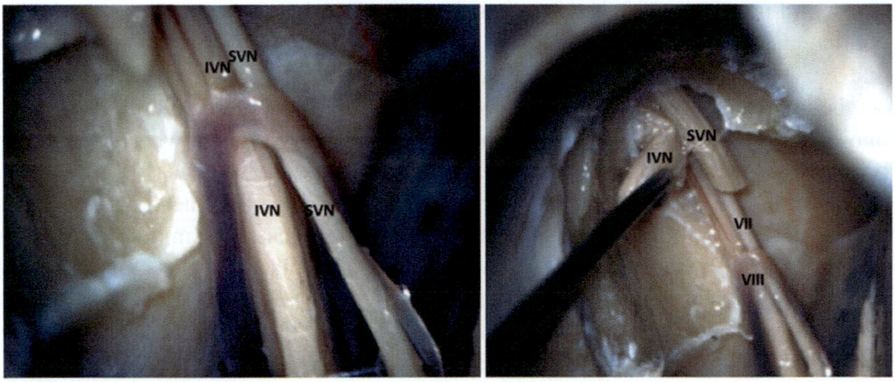

Fig. 8 Anatomical preparation. Left: Inferior (IVN) and superior (SVN) vestibular nerves in the CPA and in the internal auditory canal. Right: The IVN and the SVN are cut exposing the VII and VIII cranial nerves

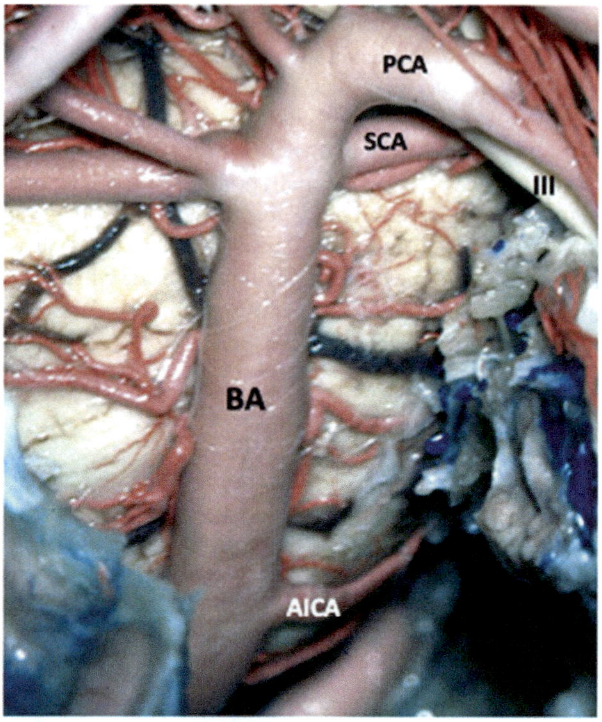

Fig. 9 Anatomical preparation showing the main arteries coursing in the CPA. PCA, posterior cerebral artery; SCA, superior cerebellar artery; AICA, anterior inferior cerebellar artery; BA, basilar artery and its perforator branches. III- oculomotor nerve

The main arteries coursing through the cerebellopontine cistern are the superior cerebellar artery (SCA) and the anterior inferior cerebellar artery (AICA) (Fig. 9). The SCA arises from the basilar SCA runs through the junction of the oculomotor nerve and the pontine membrane, above the trigeminal nerve, and below the trochlear nerve. The SCA supplies the tentorial surface of the cerebellum, lateral to the vermis. The AICA originates from the basilar artery at the pontine level in most cases, and in some cases, it arises from the vertebral artery. It enters the cerebellopontine cistern inferiorly through the anterior pontine membrane, close to the cranial nerves VII and VIII (usually inferior to these nerves) to reach the middle cerebellar peduncle supplying the petrosal surface of the cerebellum. The AICA may form a vascular loop close to the internal auditory meatus where two major branches are identified: the subarcuate and the labyrinthine arteries (Fig. 10).

The major veins crossing this cistern are the transverse pontine veins, the veins of the cerebellopontine fissure, pontomedullary sulcus, and middle cerebellar peduncle [2]. The transverse pontine veins join the tributaries of the superior petrosal *sinus* at the cerebellopontine cistern (Figs. 4 and 6). The superior petrosal vein (Dandy's vein) is formed by veins draining the cerebellum and brainstem into the superior petrosal sinus posterior to the trigeminal nerve.

The cerebellopontine angle (CPA) has the brainstem's medial boundary, the *cerebellum* as the posterior boundary, and its roof. Laterally is limited by the temporal bone, and its floor is formed by the cranial nerves IX, X, and XI. This space is filled with CSF. The cranial nerves 4th to the 11th are within the CPA [18] (Fig. 6).

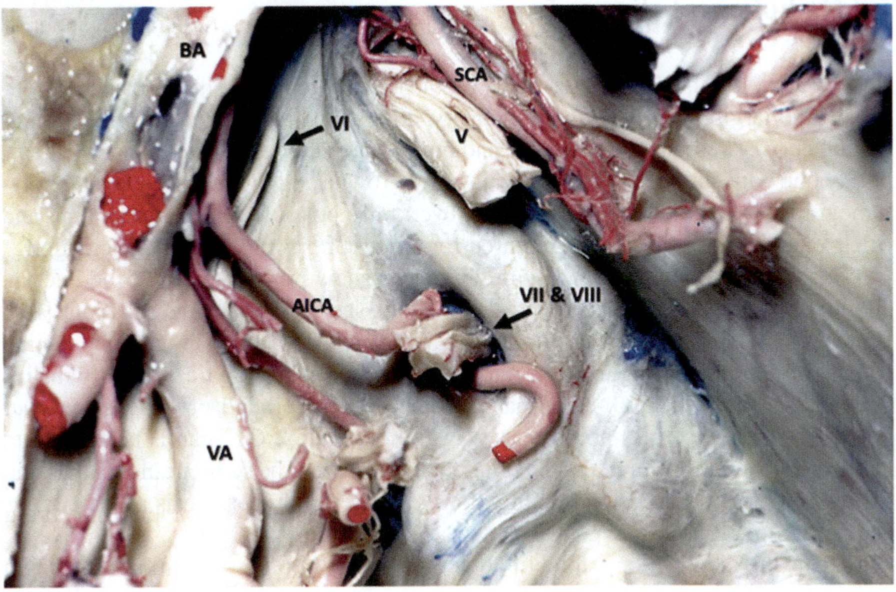

Fig. 10 Anatomical specimen demonstrating the relationship of the main CPA arteries (SCA, AICA) and the cranial nerves

4 Pathologies and Clinical Symptoms

Cerebellopontine angle tumors account for 5% to 10% of all intracranial neoplasms. Most CPA tumors are benign and vestibular schwannomas (acoustic neuromas) account for over 80%. Meningiomas are the second more common (3–10%), followed by epidermoid cysts (2–4%) and other rarer tumors: chordomas, chondrosarcomas, arachnoid or neuroenteric cysts, metastases, schwannomas of the trigeminal, facial, and caudal cranial nerves. Brainstem gliomas, ependymomas, medulloblastomas, tumors of the endolymphatic sac, and papillomas of the choroid plexus may involve the CPA secondarily. Aneurysms, brainstem cavernomas, and arteriovenous malformations are the vascular lesion encountered in this region. The presenting symptoms may be vague and nonspecific. Benign lesions usually present progressive and long-standing symptoms so that they may remain undetected for long periods. Patients with malignant lesions typically offer a short history of pain and multiple cranial nerves. Cranial compression nerves by arteries and veins (trigeminal neuralgia, hemifacial spasm, and glossopharyngeal neuralgia) are surgical pathologies found in the CPA [19–22]. Clinical symptoms are related to the etiology, involvement of cranial nerves, mass effect, and compression of the brainstem. High-frequency hearing loss and tinnitus are the most frequent symptoms in patients with vestibular schwannomas.

5 Radiological Diagnosis

Radiological diagnosis requires analysis of precise location, site of origin, extensions and margins, density, signal intensity, contrast enhancement, identification of cranial nerves, vessels, jugular bulb, and bone abnormalities. Evaluation of the temporal bone with standard radiography includes Towne and Stenver's views and polytomography. Accurate diagnosis is, however, only possible with CT scanning and MRI. Usually, both studies are required for more definitive diagnosis elucidation.

CT scans with bone window with thin slices show bone changes of the internal auditory canal (IAC), petrous bone, mastoid emissary veins, and calcifications [23]. Bone erosion with smooth margins suggests the presence of a slow-growing benign lesion. Infections and aggressive tumors may present osteolytic lesions with ill-defined margins and moth-eaten patterns. The IAC is eroded or widened in over 70% of vestibular schwannomas cases. A high-resolution CT scan is performed to delineate labyrinthine air cells, the relationship between the semicircular canals and the internal auditory meatus. Meningiomas may present hyperostosis of the petrous apex or temporal bone.

MRI is the diagnostic method of choice for all CPA tumors. Contrast enhancing lesions usually are schwannomas, meningiomas, chordomas, paragangliomas, chondrosarcomas, hemangioblastomas, metastases, and medulloblastomas. Non-enhancing tumors are the epidermoid cysts, arachnoid and neuroenteric cysts, low-grade gliomas, and other less frequent lesions [24]. Vestibular schwannomas are isointense on T1-weighted sequences and hyperintense on T2-weighted sequences. They enhance intensely and may present cystic portions. Vestibular schwannomas usually extend into the IAC. Meningiomas on T1-weighted are frequently isointense or hypointense to the brain parenchyma and hyperintense on T2-weighted MRI studies. Meningiomas have broad contact with the petrous bone, tentorium, or clivus. Invasion of the IAC is rare, and meningiomas arising primarily within the IAC are very rare. The "dural tail signal" (enhancement along the dura) is not pathognomonic of meningiomas but is frequently observed. Calcifications and hyperostosis may be present in meningiomas and are very rare in schwannomas. Chordomas and chondrosarcomas are heterogeneous, and bone erosion is observed [25, 26]. Paragangliomas enhance intensely and extend into the ear, jugular bulb, jugular vein, and neck [27, 28]. A precise radiological diagnosis is essential for surgical planning.

Digital angiography is performed when an aneurysm is suspected and for preoperative embolization (e.g., paragangliomas and other highly vascularized tumors).

6 Surgical Approaches

Careful preoperative evaluation is critical in minimizing intraoperative and postoperative complications. Preoperative imaging studies may define the exact size and location of the lesion.

Different surgical approaches may be used to treat tumors and other lesions in the CPA, and each has its advantages and disadvantages [29]. Selection of surgical technique is based on several factors: patient's age, general clinical and neurological status, type of lesion, tumor size and extensions (posterior fossa, middle fossa, IAC, and cavernous sinus), site of origin, and surgeon's experience. The surgical approach should provide sufficient exposure to the CPA and its related structures. It should permit adequate pathology exposure with no or very low morbidity, protect the neural and vascular structures, and permit adequate reconstruction of the dura and skull base. The CPA may be surgically approached through the posterior cranial fossa, through the petrous bone, through the middle fossa, or with a combination of approaches. Continuous neurophysiological monitoring is performed throughout the surgery, from the positioning of the patient to the skin closure. Somatosensory evoked potentials, electromyography of the facial nerve, monitoring of the brainstem auditory evoked potentials, oculomotor, trochlear, *abducens*, and caudal cranial nerves are used according to tumor extension and clinical presentation. During surgery, continuous exchange between the neurosurgeon and the neurophysiologist is of fundamental importance.

The main operative approaches are the retrosigmoid, translabyrinthine, presigmoid, retrolabyrinthine petrosectomy, total petrosectomy, transcochlear, and middle fossa.

The retrosigmoid approach introduced by Fedor Krause (1903) is one of the most frequently used approaches (Fig. 11). It is simple, safe, fast, and associated with a very low procedure-related morbidity rate [30–32]. The patient may be placed in a semi-sitting position, lateral or park-bench position, or supine position

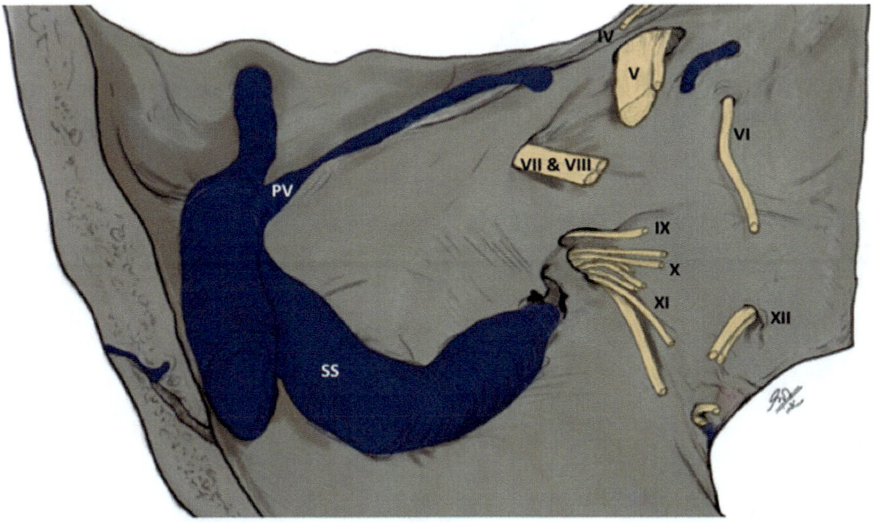

Fig. 11 Drawing showing the Sigmoid sinus (SS), the petrosal vein (PV) and the cranial nerves IV, V, VI, VII, and VIII in the CPA, as well as the cranial nerves IX, X, and XI in the jugular foramen and the XII cranial nerve

[33]. Every surgical position has pros and cons (Fig. 12). The semi-sitting position has the risk of venous air embolism, paradoxical air embolism, tension pneumocephalus, and circulatory instability. However, experienced anesthesiologists can minimize these risks using transesophageal echocardiography or the combined monitoring of end-tidal carbon dioxide and precordial Doppler [34–36]. Patients placed in a semi-sitting position should have their legs raised to or above the level of the heart, this will increase the venous pressure, and the risk of air embolism is reduced. In the last 20 years, we have been using the supine position in all surgeries through the retrosigmoid approach (Fig. 13) [33]. The retrosigmoid approach provides access to all levels of the CPA and neighboring regions. Meckel's cave and posterior portion of the middle cranial fossa and posterior part of the cavernous sinus may be accessed by opening the tentorium and removing the suprameatal tubercle [37]. All sizes of vestibular schwannomas may be removed through the retrosigmoid/transmeatal approach. It offers very good control of the neurovascular structures of the posterior fossa, preservation of hearing, and, if necessary,

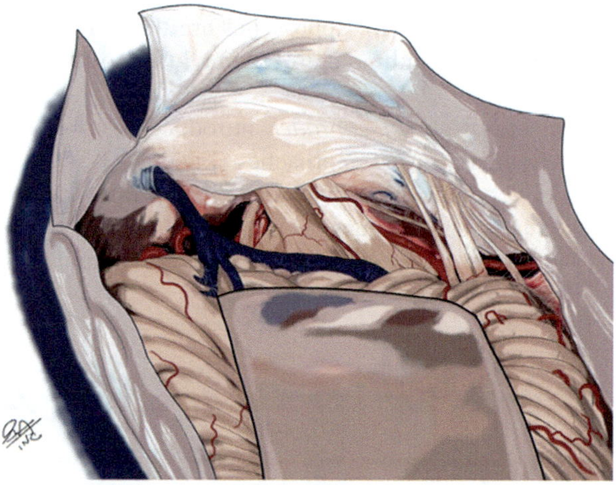

Fig. 12 Drawing of retrosigmoid approach (right side) exposing the anatomical structures in the CPA region

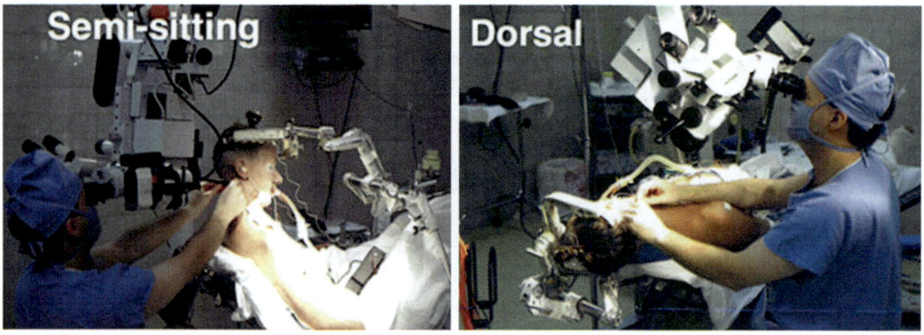

Fig. 13 Surgical pictures of patient positioning for retrosigmoid approach. Semi-sitting and dorsal (supine)

reconstruction of the facial nerve. Meningiomas, epidermoid cysts, vascular decompression of cranial nerves, and other pathologies may be adequately addressed by this approach. Endoscopy is very useful for removing small intracanalicular or intralabyrinthine tumors (Fig. 14), identifying cranial nerves and vessels, closing opened mastoid cells during drilling of the IAC, and checking vascular structures compressing cranial nerves (Figs. 15, 16, 17 and 18).

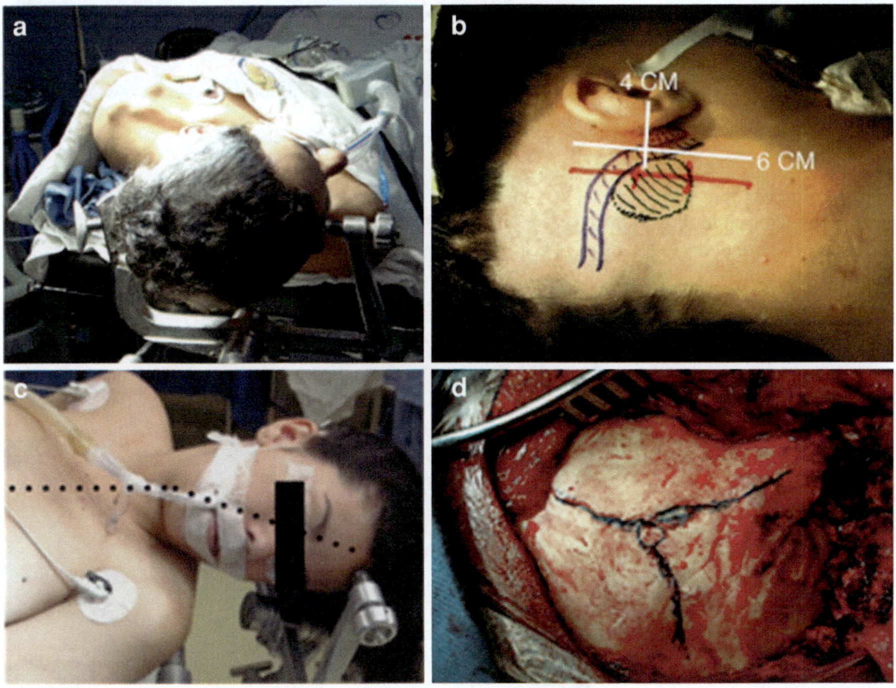

Fig. 14 (**a, b**) Surgical position for the retrosigmoid approach. (**c**) Surgical incision. (**d**) Asterium as marker for the burr hole

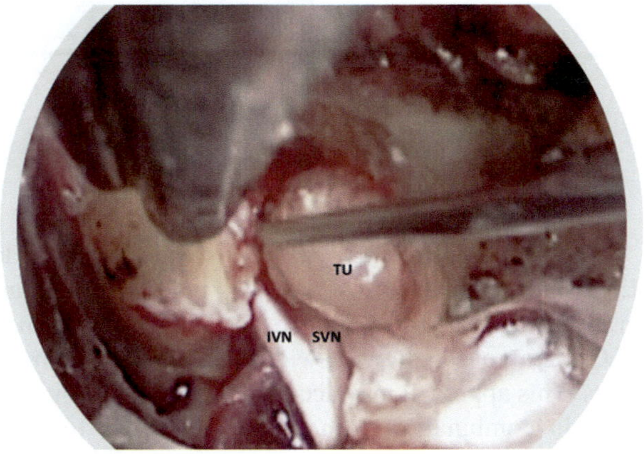

Fig. 15 Endoscopic removal of a small vestibular schwannoma (TU), inferior (IVN), and superior(SVN) vestibular nerves

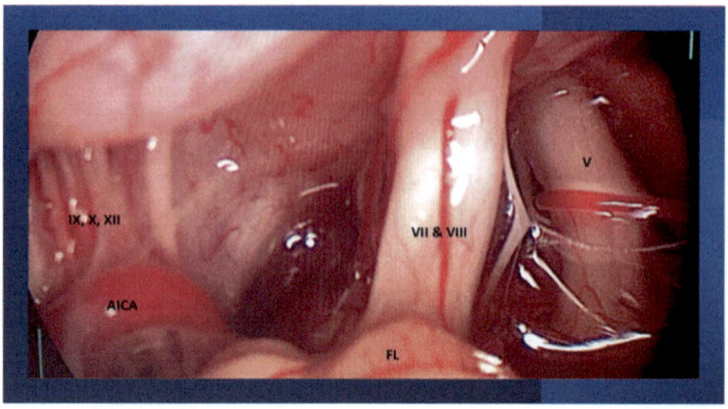

Fig. 16 Endoscopic view of the cranial nerves (V, VII & VIII and IX, X, XI), AICA and flocculus (FL)

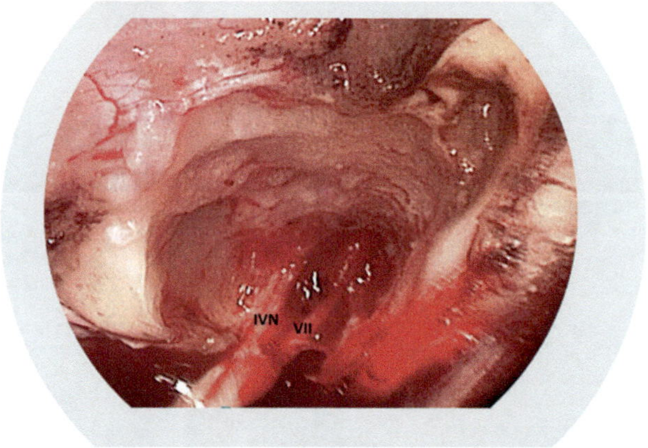

Fig. 17 Endoscopic view within the internal auditory canal after removal of a small vestibular schwannoma from the superior vestibular nerve. Inferior vestibular nerve (IVN) and facial nerve (VII)

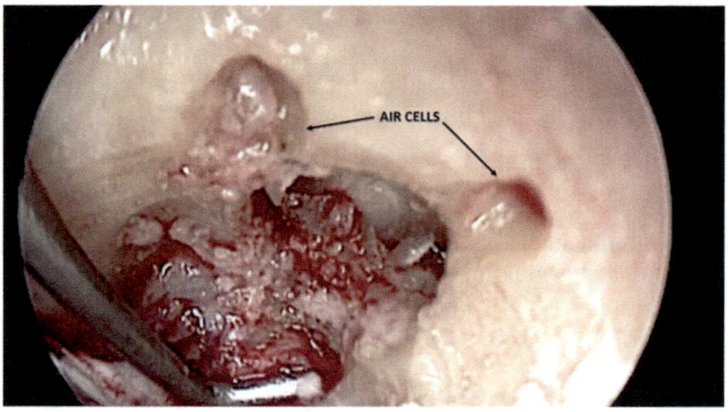

Fig. 18 Endoscopic identification of opened air cells in the internal auditory canal

The translabyrinthine approach introduced by Rudolf Panse in 1904 is used by ENT surgeons and neurosurgeons to remove vestibular schwannomas when preoperative hearing is lost [38–40]. The semicircular canals and the lateral wall of the IAC are removed, allowing wide exposure of the intrameatal and extrameatal tumor parts. This approach may access different pathologies of the petrous bone. It is also used in combination with other approaches to removing cranial base tumors.

Water-tight dura reconstruction is difficult, and an additional abdominal incision is usually needed to remove fat graft to occlude the surgical field and avoid CSF-fistula.

The middle fossa approach (extra of intradural) is utilized in cases of small vestibular schwannomas, trigeminal schwannomas, cysts, and other petrous apex tumors with extensions to the CPA up to the VII & VIII complex [41, 42]. An extended middle fossa approach with resection of the petrous apex behind the horizontal segment of the petrous internal carotid artery medial to the IAC and incision of the *tentorium* (Kawase approach) may be used for larger lesions [43]. This approach exposes the petrous apex intradural and extradural. It lead dissection of the second and third divisions of the trigeminal nerve, the Gasserian ganglion, the petrosal portion of the internal carotid artery, and the meatal and petrosal portions of the facial nerve. Retraction of the temporal lobe is necessary to expose the petrous apex.

Petrous approaches [44–47] with partial or complete resection of the petrous bone has the advantage of shortening the distance to reach the tumor and minimizing brain retraction.

The presigmoid approach allows supra and infratentorial exposure without resection of the labyrinth and preservation of hearing (Fig. 19). It requires, however, more bone drilling increasing the surgical time and may have higher approach-related morbidity.

The hearing may also be preserved by the retrolabyrinthine approach (removal of the bone between the sigmoid sinus and the semicircular canals).

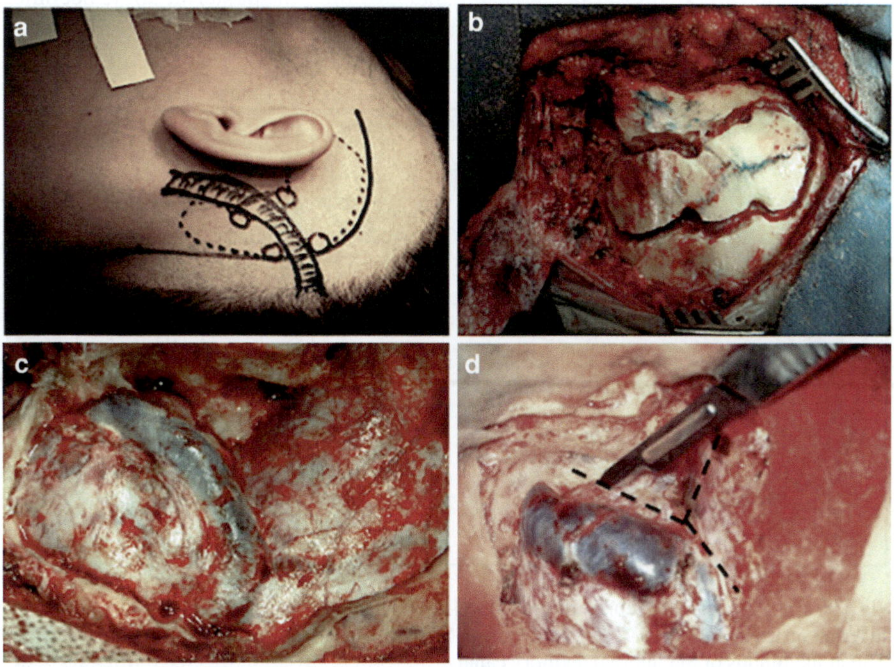

Fig. 19 Presigmoid approach. (**a**) Skin incision and planning the craniotomy. (**b**) Craniotomy. (**c**) Surgical exposure of the sigmoid and transverse sinuses, the middle and posterior fossa dura. (**d**) Dura opening in front of the sigmoid and tranverse sinuses, extending to the middle fossa

Total petrosectomy and the transcochlear approach are used if the hearing is already lost. This surgical access permits a wide exposure of the structures anterior to the IAC and petroclival region [48].

According to our experience, the approaches to CPA should be simple, permit adequate control of the important neurovascular structures of this region, reconstruction of the posterior fossa cranial base, and be associated with no or very-low approach-related morbidity. The retrosigmoid is the more used approach to the CPA. It allows surgical exposure from the Gasserian ganglion and posterior portion of the middle fossa (with removing the suprameatal tubercle and opening of the tentorium) to the cerebellomedullary cistern.

7 Case Studies

7.1 Case 1: Small Vestibular Schwannoma

A 42-year-old patient female was admitted to our hospital with a 3-years history of headaches, tinnitus, and progressive hearing loss on the right ear. MRI showed a small (T2) vestibular schwannoma on the right CPA (Fig. 19). The tumor was radically removed through a right retrosigmoid approach to preserve the facial nerve and hearing (Figs. 20 and 21).

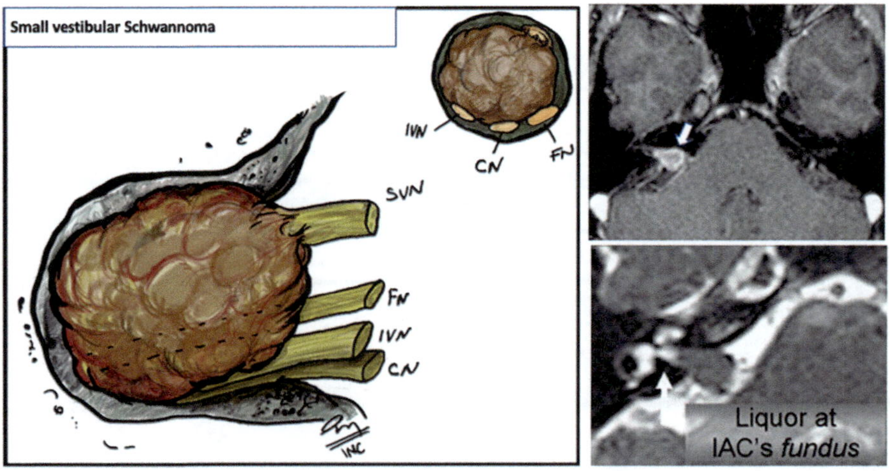

Fig. 20 Case 1. Drawing representing a small vestibular schwannoma that originates from the superior vestibular nerve (SVN). MRI of a right side vestibular schwannoma with the presence of CSF at the fundus of the internal auditory canal

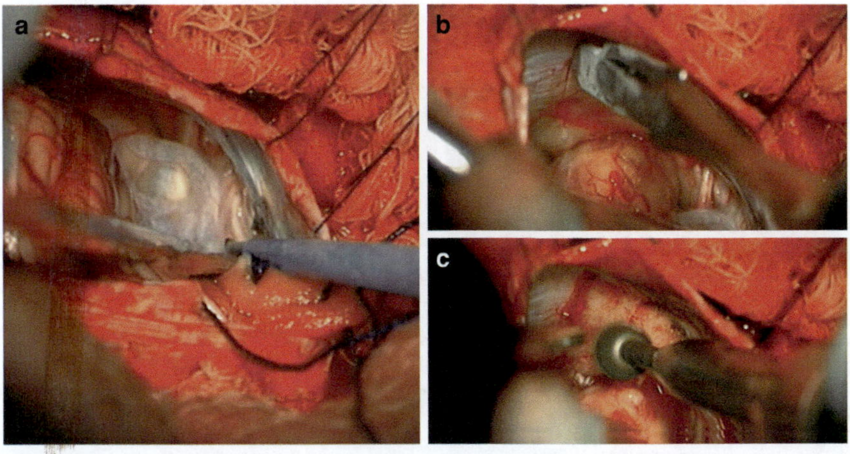

Fig. 21 Case 1. (**a**) Surgical exposition of the right cerebellopontine cistern through the retrosigmoid approach. (**b**) Opening of the dura mater over the internal auditory canal. (**c**) Drilling of the internal auditory canal

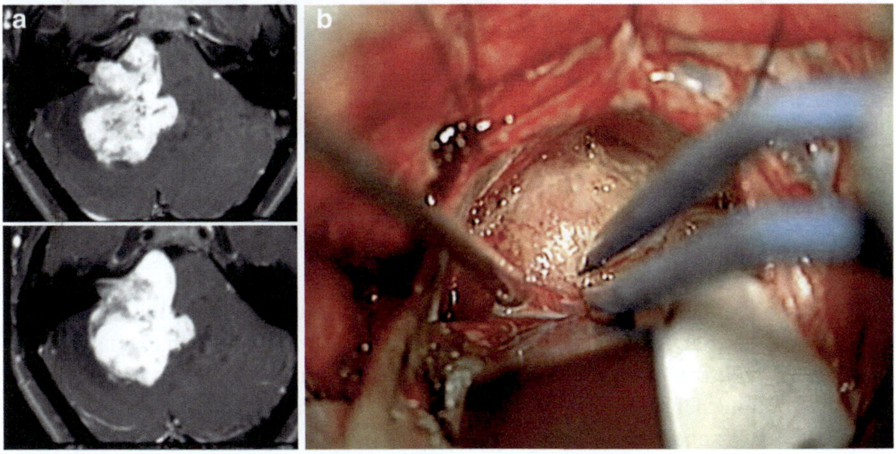

Fig. 22 Case 2. (**a**) MRI (T1-wGd) demonstrating a giant vestibular schwannoma in the right CPA. (**b**) Surgical approach to the tumor through the retrosigmoid approach

7.2 *Case 2: Large Vestibular Schwannoma*

This 45-year-old female reported a history of progressive hearing loss for 5 years. She became deaf on the right ear 2 years before admission to our department and developed gait difficulty and facial numbness in the last 12 months. An MRI disclosed a very large right-sided vestibular schwannoma (T4B) with compression of the IV ventricle (Fig. 22). Radical tumor removal was accomplished through the retrosigmoid-transmeatal approach with preservation of the facial nerve (Fig. 23).

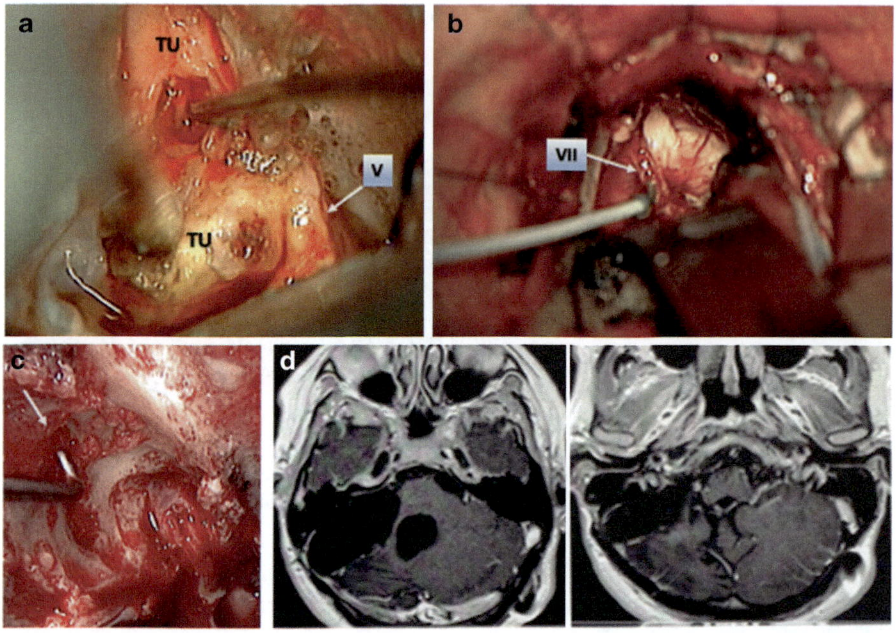

Fig. 23 Case 2. (**a**) Dissection of the trigeminal nerve (V) from tumor capsule (TU). (**b**) Electric stimulation of facial nerve after complete removal of the VS. (**c**) Endoscopic identification of opening air cells (hook). (**d**) Postoperative MRI demonstrating radical removal of the tumor

7.3 *Case 3: Cerebellopontine Angle Meningioma*

A 55-year-old male was admitted due to hearing loss and tinnitus of the left ear. The hearing was normal. MRI revealed a left side CPA meningioma with extensions into the internal auditory canal (Fig. 24). Total removal of the tumor was possible after opening the internal auditory canal and the jugular foramen. The cranial nerves V, VII, VIII, IX, X, and XI were preserved (Fig. 25).

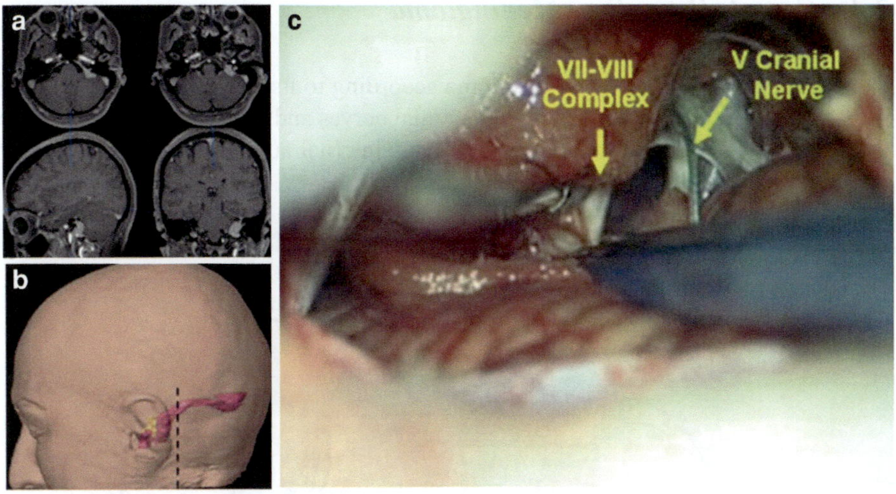

Fig. 24 Case 3. (**a**) MRI (T1-wGd) showing a left-side meningioma in the CPA. (**b**) Planned skin incision for the retrosigmoid approach. (**c**) Surgical view with exposition of cranial nerves V, VII, and VIII

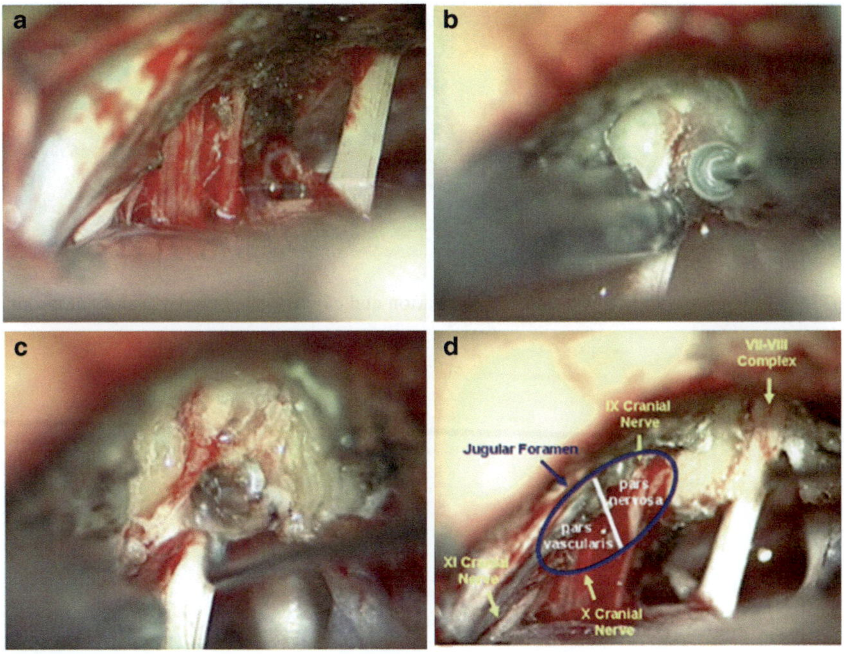

Fig. 25 Case 3. (**a**) Surgical view after radical removal of the intradural portion of the tumor. (**b**) Drilling the internal auditory canal. (**c**) Tumor removal from the intracanalicular portion. (**d**) Radical removal of the meningioma from the CPA, internal auditory canal and jugular foramen

7.4 Case 4: Petroclival Meningioma

Our definition of petroclival meningioma according to its implantation is shown in Fig. 26. These tumors are medial to the cranial nerves and may extend to the middle fossa, cavernous sinus, Meckel's cave, clivus, internal auditory canal, and jugular foramen involving the lower cranial nerves.

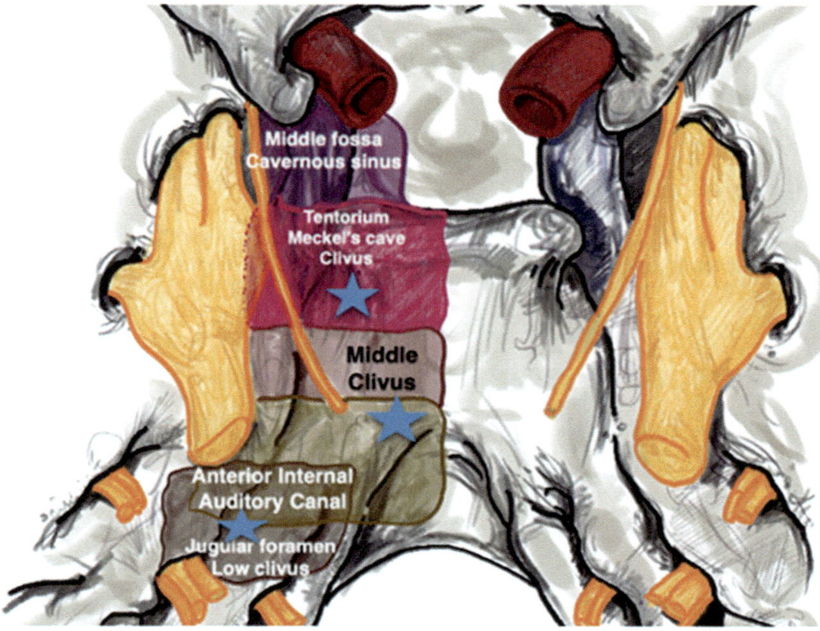

Fig. 26 Drawing showing the areas of implantation and extensions of petroclival meningiomas

The case of a 33-year-old female who complained for 3 years of headache, tinnitus, and decreasing hearing on the left ear is presented. In the last 1½ years, she developed gait difficulty and ataxia. MRI showed a very large left-sided petroclival meningioma (Fig. 27). This tumor was radically removed using a presigmoid approach. Cranial nerves IV, V, VII, and VIII were preserved (Fig. 28).

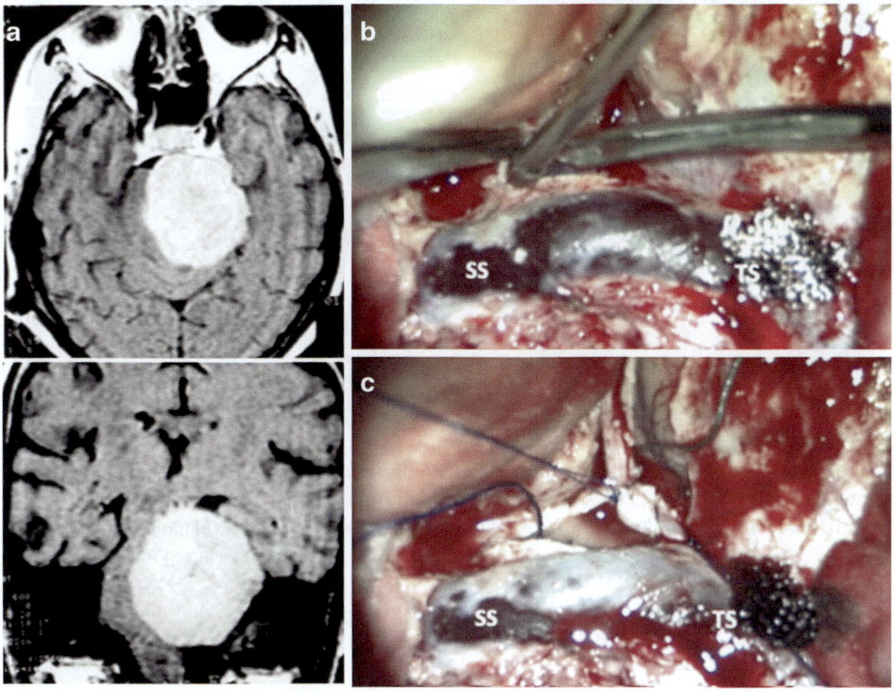

Fig. 27 Case 4. (**a**) MRI (T1-wGd) of a large left side petroclival meningioma. (**b**) Presigmoid approach with exposure of the sigmoid (SS) and transverse sinuses (TS). (**c**) Opening of the dura mater in front of the sigmoid sinus (SS) and in the middle fossa

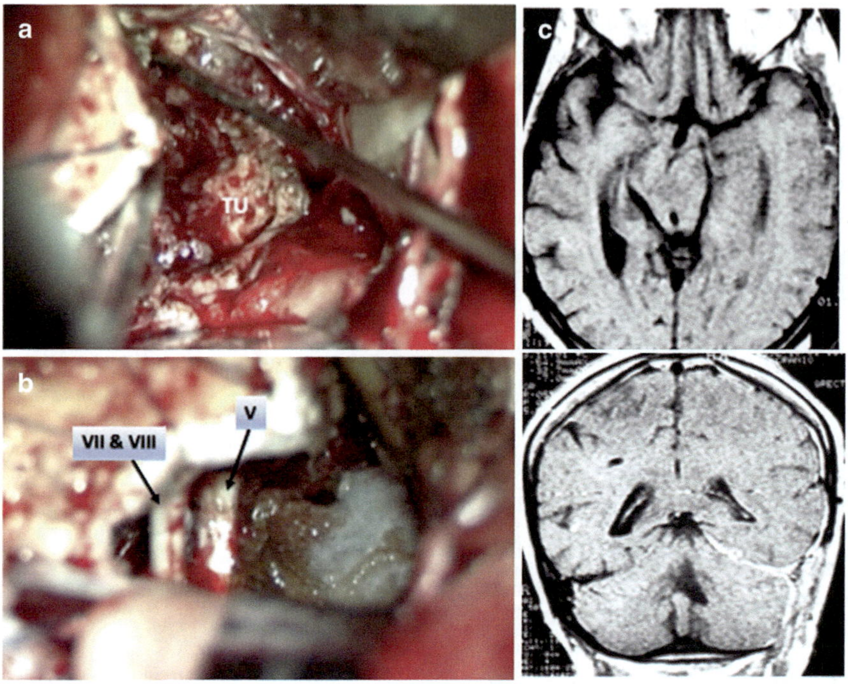

Fig. 28 Case 4. (**a**) Presigmoid approach to the CPA and middle fossa showing the meningioma in the cerebellopontine angle (TU). (**b**) After radical removal of the tumor the cranial nerves V, VII, and VIII are identified. (**c**) Postoperative MRI (T1-wGd) demonstrating radical removal of the tumor

7.5 *Case 5: Epidermoid Cyst*

This 43-year-old female patient complained of decreased hearing and facial pain on the right side. An MRI disclosed a large epidermoid cyst occupying the whole cerebellopontine cistern (Fig. 29). Radical removal with the cyst capsule was possible through the retrosigmoid approach, and all involved cranial nerves were preserved (Fig. 30).

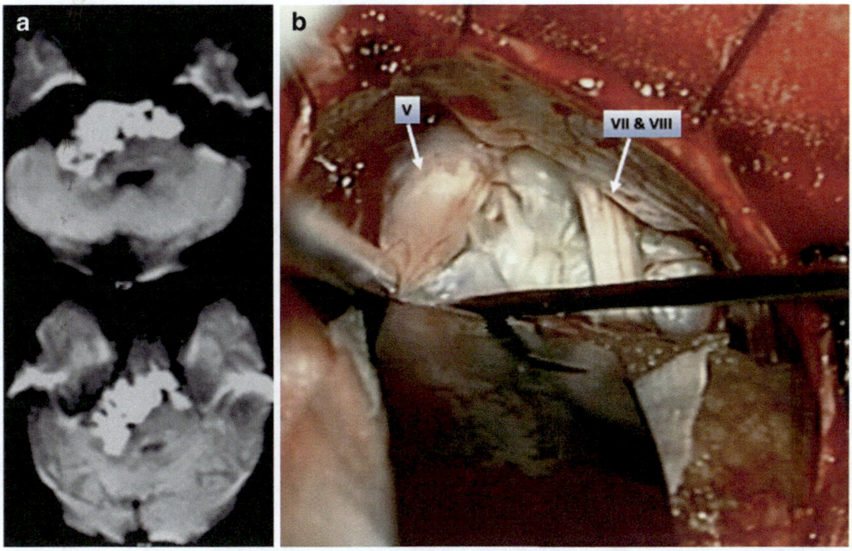

Fig. 29 Case 5. (**a**) MRI (diffusion scan) showing a large epidermoid cyst occupying the CPA and prepontine region. (**b**) Retrosigmoid approach showing epidermoid cyst medial to cranial nerves V, VII, and VIII

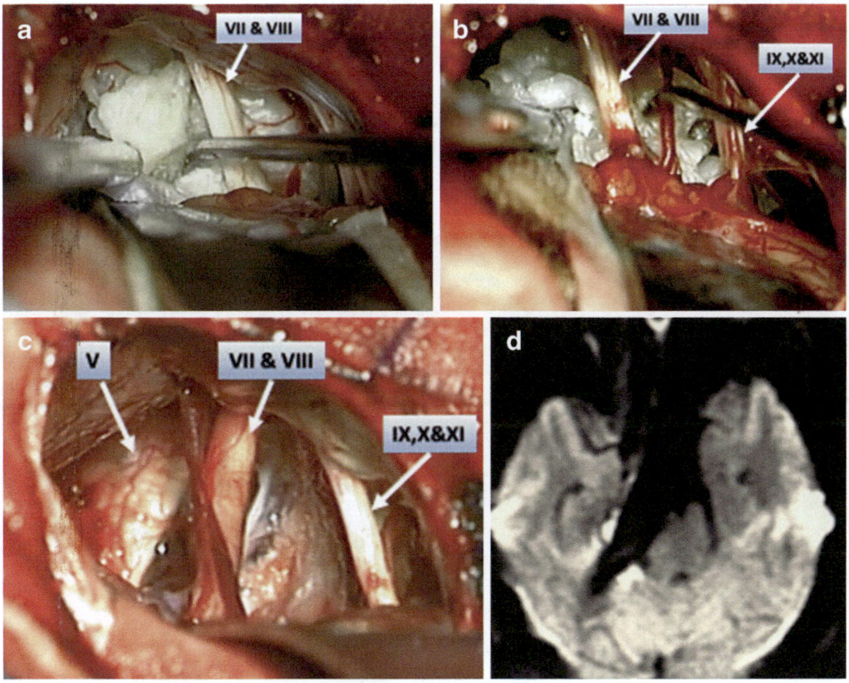

Fig. 30 Case 5. (**a**) Dissection of epidermoid cyst capsule from cranial nerves VII and VIII. (**b**) Epidermoid cyst medial to cranial nerves VII, VII, IX, X, and XI. (**c**) After radical removal the CPA and prepontine region are free from tumor. (**d**) MRI (diffusion scan) demonstrating radical removal of the lesion

7.6 *Case 6: Trigeminal Neuralgia*

A 63-year-old female presented a history of refractory left side hemifacial pain (V2 division of the trigeminal nerve) that started 4 years before. MRI showed a vascular compression of the trigeminal nerve at its entry zone at the brainstem (Fig. 31). Vascular decompression of the trigeminal nerve with interposition of a Teflon implant was carried out through the retrosigmoid approach (Fig. 32).

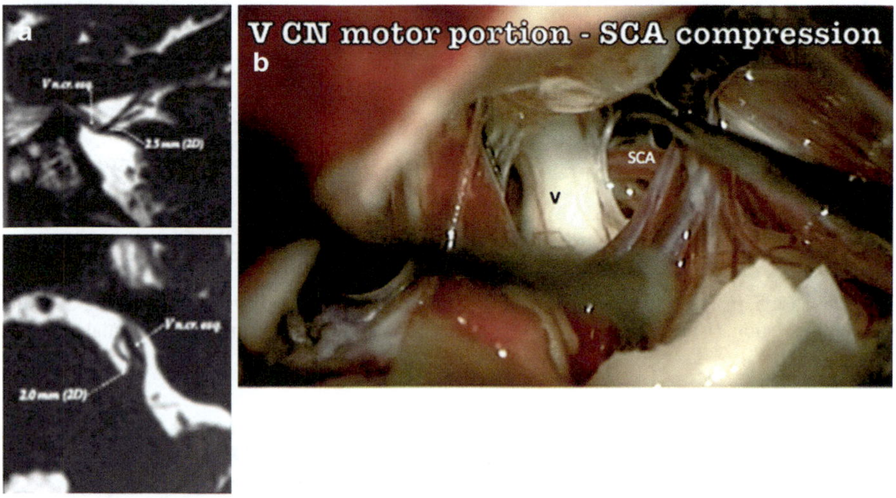

Fig. 31 Case 6. (**a**) MRI (T2-w) of a patient with left side trigeminal neuralgia. The trigeminal nerve (V) is compressed by the superior cerebellar artery. (**b**) Surgical view showing the trigeminal nerve (V) and the superior cerebellar artery (SCA)

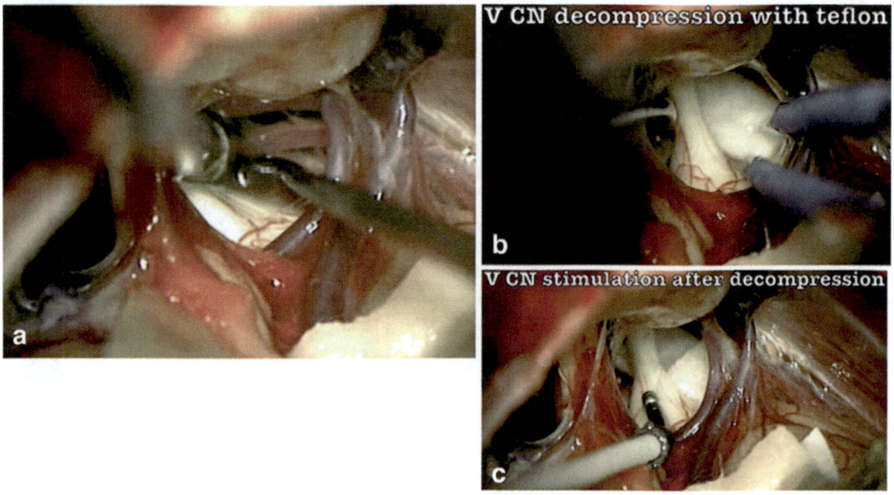

Fig. 32 Case 6. (**a**) The SCA is dissected from the trigeminal nerve at its entry-zone. (**b**) A small piece of Teflon in interposed between the nerve and the artery. (**c**) After vascular decompression, the trigeminal nerve is electric stimulated

8 Complications and How to Avoid

CSF fistula is a common complication after surgical removal of tumors and other pathologies involving the temporal bone and mastoid. It may occur with a frequency of 2% to 25% [49–51]. It may occur through the mastoid cells or externally through the skin (Table 1). Usually, it appears immediately or a few days after surgery, but it may develop in a delayed fashion. CSF fistulas may present a "low-flow" or "high-flow." A thin-cut CT scan may demonstrate the area of leakage. High-flow CSF fistulas may be associated with pneumocephalus. In these cases, an immediate surgical exploration may be necessary. Endoscopy is very useful for identifying and sealing opened air cells within the internal auditory canal after drilling the internal auditory meatus through the retrosigmoid approach. If postoperative CSF-fistula develops, these opened air cells may be occluded through the mastoid (Table 1).

Dysfunction of cranial nerves is another possible complication after surgery of pathologies in the cerebellopontine cistern. Intraoperative monitoring is of fundamental importance to avoid damage to cranial nerves and brainstem. Postoperative facial nerve palsy after removing vestibular schwannomas or other lesions should be carefully treated with artificial tears and other measures for eye protection. Special care is needed in patients with facial analgesia and corneal reflex absent (trigeminal nerve deficit) to avoid corneal ulceration. In these cases, an early tarsorrhaphy should be performed. Lower cranial nerves neuropathy needs early identification and management. Patients with tumors requiring extensive surgical dissection around the caudal cranial nerves should be extubated in the ICU only after they wake up from anesthesia and a positive swallowing testing. Aspiration pneumonia is a very dangerous complication and may be the cause of postoperative mortality. Modified *barium* or direct laryngoscopy may be used to diagnose deficits of these nerves. Oral intake is allowed only after the normal function of swallowing is observed. Postoperative infection may occur due to intraoperative contamination, previous infection in the mastoid or ear, or postoperative CSF-fistula. Early recognition and adequate management are necessary to avoid meningitis.

Table 1 CSF-fistula types and management

CSF-fistula types	Prevention and management
Through mastoid cells	Prevention: First, water-tight dura closure. Fill all opened pneumatized cells with muscle graft or fat graft or bone wax and fibrin glue
	Management: Lumbar drainage, acetazolamide for 3 days. If it fails, reoperation
Subcutaneous	Prevention: Water-tight dural closure, if not possible, fascia or muscle flap rotation
	Management: Compressive dressing and lumbar drainage. In case of failure, reoperation

9 Pearls and Tips

1. The cerebellopontine region and the CPA have a complex anatomy with the presence of important neurovascular structures. A thorough understanding of this complex anatomy is essential to obtain a successful surgery with minimal complications.
2. Many pathologies may arise in this region requiring different treatment strategies.
3. Clinical examination and careful review of neuroradiological findings will determine the location of pathology, the differential diagnosis and help to delineate the best surgical approach.
4. Selection of the appropriate surgical strategy should be individualized for each case.
5. Surgical approach should be simple, avoid brain retraction and damage to healthy tissues, and preserve neurovascular structures.
6. Choice of surgical approaches depends on many factors according to etiology and extension of the lesion, involvement of cranial nerves, and surgeon's experience.
7. Intraoperative monitoring of cranial nerves and brainstem is very important during surgery in this area.
8. Adequate skull base reconstruction with water-tight dura closure will avoid postoperative CSF leak and achieve good cosmetic result.
9. Anticipating, avoiding, recognizing, and managing early surgical complications will improve outcomes and patients' quality of life.

References

1. Yasargil MG, Kasdaglis K, Jain KK, Weber HP. Anatomical observations of the subarachnoid cisterns of the brain during surgery. J Neurosurg. 1976;44(3):298–302.
2. Matsuno H, Rhoton AL Jr, Peace D. Microsurgical anatomy of the posterior fossa cisterns. Neurosurgery. 1988;23(1):58–80.
3. Key A, Retzius G. Studien in der Anatomie des Nervensystems und des Bindegewebes. Stockholm: Samson & Wallin; 1875.
4. Liliequist B. The subarachnoid cisterns. An anatomic and roentgenologic study. Acta Radiol Suppl. 1959;185:1–108.
5. Lewtas NA, Jefferson AA. The carotid cistern. A source of diagnostic difficulties with suprasellar extensions of pituitary adenomata. Acta Radiol Diagn (Stockh). 1966;5:675–90.
6. Amundsen P, Newton TH. Subarachnoid cisterns. In: Newton TH, Potts DG, editors. Radiology of the skull and brain, Ventricles and cisterns, vol. 4. Great Neck, NY: MediBooks; 1978. p. 3588–711.
7. Ballance CA. Some points in the surgery of the brain and its membranes. 2nd ed. London: Macmillan; 1908.

8. Koerbel A, Gharabaghi A, Safavi-Abbasi S, Tatagiba M, Samii M. Evolution of vestibular schwannoma surgery: the long journey to current success. Neurosurg Focus. 2005;18(4):e10.
9. Olivecrona H. Acoustic tumors. In: Winther K, Krabbe K, editors. III Congrès Neurologique International, Copenhague, 21/8–25/8 1939. Comptes Rendus des Séances. Copenhague: Einar Munksgaard; 1939. p. 761–71.
10. Crabtree JA, House WF. Transtemporal bone microsurgical removal of acoustic neuromas X-ray diagnosis of acoustic neuromas. Arch Otolaryngol. 1964;80:695–7.
11. Hitselberger WE, Raney AA. Transtemporal bone microsurgical removal of acoustic neuromas. Neurosurgical thoughts comparing suboccipital and tranlabyrinthine approaches. Arch Otolaryngol. 1964;80:754–6.
12. House WF, Hitselberger WE, Knouf EG. Transtemporal bone microsurgical removal of acoustic neuromas. Postoperative complications. Arch Otolaryngol. 1964;80:742–5.
13. House WF. Transtemporal bone microsurgical removal of acoustic neuromas. Evolution of transtemporal bone removal of acoustic tumors. Arch Otolaryngol. 1964;80:731–42.
14. Rand RW, Kurze T. Preservation of vestibular, cochlear, and facial nerves during microsurgical removal of acoustic tumors. Report of two cases. J Neurosurg. 1968;28(2):158–61.
15. Samii M, Gerganov VM, Samii A. Functional outcome after complete surgical removal of giant vestibular schwannomas. J Neurosurg. 2010;112(4):860–7.
16. Samii M, Gerganov V, Samii A. Improved preservation of hearing and facial nerve function in vestibular schwannoma surgery via the retrosigmoid approach in a series of 200 patients. J Neurosurg. 2006;105(4):527–35.
17. Rhoton AL Jr. The cerebellopontine angle and posterior fossa cranial nerves by the retrosigmoid approach. Neurosurgery. 2000;47(3 Suppl):S93–129.
18. Rhoton AL Jr. The posterior cranial fossa: microsurgical anatomy and surgical approaches. Neurosurgery. 2000;47(3 Suppl):S5–6.
19. Brackmann DE, Bartels LJ. Rare tumors of the cerebellopontine angle. Otolaryngol Head Neck Surg (1979). 1980;88(5):555–9.
20. Hitselberger WE, House WF. Tumors of the cerebellopontine angle. Tumors of the cerebellopontine angle. Arch Otolaryngol. 1964;80(6):720–31.
21. Mattei TA, Goulart CR, Lima JS, Ramina R. Differential diagnosis and surgical management of cerebellopontine angle cystic lesions. J Bras Neurocirurg. 2011;22(3):66–71.
22. Ramina R, Mattei TA, Sória MG, da Silva EB Jr, Leal AG, Neto MC, et al. Surgical management of trigeminal schwannomas. Neurosurg Focus. 2008;25(6):E6.
23. Roser F, Ebner FH, Ernemann U, Tatagiba M, Ramina K. Improved CT imaging for mastoid emissary vein visualization prior to posterior fossa approaches. J Neurol Surg A Cent Eur Neurosurg. 2016;77(6):511–4.
24. Moura da Silva LF Jr, Buffon VA, Coelho Neto M, Ramina R. Non-schwannomatosis lesions of the internal acoustic meatus-a diagnostic challenge and management: a series report of nine cases. Neurosurg Rev. 2015;38(4):641–8.
25. Bonneville F, Savatovsky J, Chiras J. Imaging of cerebellopontine angle lesions: an update. Part 1: enhancing extra-axial lesions. Eur Radiol. 2007;17(10):2472–82.
26. Bonneville F, Savatovsky J, Chiras J. Imaging of cerebellopontine angle lesions: an update. Part 2: intra-axial lesions, skull base lesions that may invade the CPA region, and non-enhancing extra-axial lesions. Eur Radiol. 2007;17(11):2908–20.
27. Ramina R, Maniglia JJ, Fernandes YB, Paschoal JR, Pfeilsticker LN, Coelho NM. Tumors of the jugular foramen: diagnosis and management. Neurosurgery. 2005;57(1 Suppl):59–68.
28. Ramina R, Tatagiba MS. Radiological diagnosis. In: Ramina R, Tatagiba MS, editors. Tumors of the jugular foramen. Cham: Springer International Publishing; 2017. p. 51–62.

29. Samii M, Gerganov V. Approaches to the cerebellopontine angle. In: Samii M, Gerganov V, editors. Surgery of Cerebellopontine lesions. Berlin: Springer-Verlag; 2013. p. 115–45.
30. Lang J Jr, Samii A. Retrosigmoidal approach to the posterior cranial fossa. An anatomical study. Acta Neurochir. 1991;111(3–4):147–53.
31. Ojemann RG. Retrosigmoid approach to acoustic neuroma (vestibular schwannoma). Neurosurgery. 2001;48(3):553–8.
32. Ramina R, Constanzo F, da Silva EB Jr, Coelho Neto M. Retrosigmoid transmeatal approach with 360-degree drilling of the internal auditory canal for the resection of intracanalicular meningioma. J Neurol Surg B Skull Base. 2019;80(Suppl 3):S311.
33. Cardoso AC, Fernandes YB, Ramina R, Borges G. Acoustic neuroma (vestibular schwannoma): surgical results on 240 patients operated on dorsal decubitus position. Arq Neuropsiquiatr. 2007;65(3A):605–9.
34. Engelhardt M, Folkers W, Brenke C, Scholz M, Harders A, Fidorra H, et al. Neurosurgical operations with the patient in sitting position: analysis of risk factors using transcranial doppler sonography. Br J Anaesth. 2006;96(4):467–72.
35. Gildenberg PL, O'Brien RP, Britt WJ, Frost EA. The efficacy of doppler monitoring for the detection of venous air embolism. J Neurosurg. 1981;54(1):75–8.
36. Porter JM, Pidgeon C, Cunningham AJ. The sitting position in neurosurgery: a critical appraisal. Br J Anaesth. 1999;82(1):117–28.
37. Samii M, Tatagiba M, Carvalho GA. Retrosigmoid intradural suprameatal approach to Meckel's cave and the middle fossa: surgical technique and outcome. J Neurosurg. 2000;92(2):235–41.
38. Brackmann DE, Green JD Jr. Translabyrinthine approach for acoustic tumor removal. Neurosurg Clin N Am. 2008;19(2):251–64.
39. Briggs RJ, Luxford WM, Atkins JS Jr, Hitselberger WE. Translabyrinthine removal of large acoustic neuromas. Neurosurgery. 1994;34(5):785–90.
40. House WF. Translabyrinthine approach. In: House WF, Luetje CM, editors. Acoustic tumors, Management, vol. 2. Baltimore: University Park Press; 1979. p. 43–87.
41. House WF, Hitselberger WE, Horn KL. The middle fossa transpetrous approach to the anterior-superior cerebellopontine angle. Am J Otol. 1986;7(1):1–4.
42. House WF, Shelton C. Middle fossa approach for acoustic tumor removal. 1992. Neurosurg Clin N Am. 2008;19(2):279–88.
43. Kawase T, Shiobara R, Toya S. Middle fossa transpetrosal-transtentorial approaches for petroclival meningiomas. Selective pyramid resection and radicality. Acta Neurochir. 1994;129(3–4):113–20.
44. Brackmann DE. Translabyrinthine/transcochlear approaches. In: Sekhar LN, Janecka IP, editors. Surgery of cranial base tumors. New York: Raven Press; 1993. p. 351–65.
45. Ramina R, Neto C, Nogueira GF, da Silva EB Jr. Tumors of the petrous apex. In: Hanna EY, DeMonte F, editors. Comprehensive management of skull base tumors. 2nd ed. New York: Springer; 2021. p. 393–419.
46. Rhoton AL Jr. The temporal bone and transtemporal approaches. Neurosurgery. 2000;47(3 Suppl):S211–65.
47. Sekhar LN, Schessel DA, Bucur SD, Raso JL, Wright DC. Partial labyrinthectomy petrous apicectomy approach to neoplastic and vascular lesions of the petroclival area. Neurosurgery. 1999;44(3):537–50.
48. House WF, Hitselberger WE. The transcochlear approach to the skull base. Arch Otolaryngol. 1976;102(6):334–42.
49. Fayad JN, Schwartz MS, Slattery WH, Brackmann DE. Prevention and treatment of cerebrospinal fluid leak after translabyrinthine acoustic tumor removal. Otol Neurotol. 2007;28(3):387–90.

50. Sampath P, Rini D, Long DM. Microanatomical variations in the cerebellopontine angle associated with vestibular schwannomas (acoustic neuromas): a retrospective study of 1006 consecutive cases. J Neurosurg. 2000;92(1):70–8.
51. Stieglitz LH, Giordano M, Gerganov VM, Samii A, Samii M, Lüdemann WO. How obliteration of petrosal air cells by vestibular schwannoma influences the risk of postoperative CSF fistula. Clin Neurol Neurosurg. 2011;113(9):746–51.

Surgical Anatomy for Auditory Brainstem Implantation

Steffen K. Rosahl

1 Introduction

Profound deafness is a condition that can isolate people in social environments. Some profoundly deaf patients are born without a functioning auditory nerve or severely ossified cochleae, and there are others, like those suffering from neurofibromatosis type 2, lost their auditory nerve to bilateral growth of vestibular schwannomas.

These conditions are known as neural hearing loss or retrocochlear hearing loss. Currently, the only option to provide sound access for these patients is electrodes placed higher up in the auditory pathway that can relay sound-driven electrical impulses. The neuro-electronic interfaces that connect to these electrodes have been named "auditory brainstem implants" (ABI) if they stimulate the cochlear nucleus or "auditory midbrain implants" (AMI) if they connect to the midbrain nuclei of the hearing pathway. Midbrain implantation is exceedingly rare and involves penetrating electrodes, but the cochlear nucleus is a more popular target for hearing restoration in cases of non-functional cochlear nerves.

Surgeons who are assigned to implant electrodes over the cochlear nucleus in the lateral recess of the fourth ventricle will appreciate the guidance offered by the following chapter, but so will be microneurosurgeon who tackle tumors originating from or invading the lateral recess.

S. K. Rosahl (✉)
Department of Neurosurgery, Helios Klinikum, Erfurt, Germany

Albert-Ludwigs-University, Freiburg, Germany
e-mail: steffen.rosahl@helios-gesundheit.de

E. G. Figueiredo et al. (eds.), *Brain Anatomy and Neurosurgical Approaches*,
https://doi.org/10.1007/978-3-031-14820-0_15

2 The Cochlear Nerve

Patients to be implanted with an ABI lack a functional cochlear nerve. In some of them, however, the stump of this nerve can still serve as a guiding structure for the placement of neuroelectronic interfaces to the cochlear nucleus.

The cochlear nerve fibers originate from the nerve cell bodies of the spiral ganglion in the modiolus of the cochlea. Up to 95% of these fibers originate from the inner hair cells, the sensory receptor cells for the mechanical impulses primarily generated when sounds hit the eardrum and transmitted via the middle ear structures to the cochlea [1].

The cochlear nerve trunk exits the modiolus to run into the internal auditory canal. The peripheral segment of the cochlear nerve joins the vestibular nerve to form the vestibulocochlear nerve (eighth cranial nerve) at the meatal part of the internal auditory canal. The two portions of the nerve have a distinct and quite constant spatial relationship: in the internal auditory canal, the cochlear portion lies anterior to become located inferior to the vestibular nerve in the middle of the nerve and remains there until its entrance into the brainstem [2].

In general, the cochlear nerve and inferior vestibular nerve course beneath the superior vestibular nerve in the IAC and cerebellopontine cistern [3].

Glial tissue surrounds the nerve in the inner ear canal. Close to its exit, the meatus, the glia encasement changes into by Schwann[1] cells ("Obersteiner-Redlich-Zone").

The length of the cochlear nerve from the glial-Schwann junction to the brainstem is 10–13 mm in human males and 7–10 mm in females [4].

It contains 32–41,000 fibers with a diameter of 2–3 µm, nearly all myelinated for fast conduction [1]. The fibers reach the ventral cochlear nucleus (VCN) at its ventromedial surface. Each axon splits into an ascending and a descending fiber at the root entry zone [5]. Ascending fibers run dorsolaterally into the VCN, descending fibers run caudal and dorsal straight through this portion and diverge into the dorsal cochlear nucleus (DCN). In contrast to other mammals, the human DCN does not have a laminar structure since it is entirely penetrated by these descending fibers, forming a rather homogeneous plexus with the neurons [5].

3 Cochlear Nuclei

3.1 Histological Anatomy

Earlier, Lorente de No distinguished a compact ventral and a smaller, longer portion of the cochlear nucleus [6]. On the cellular level, this subclassification has been confirmed [7–11].

[1] Theodor Ambrose Hubert Schwann, German Anatomist and Physiologist, 1810–1882

The cochlear nerve runs through the VCN, resulting in a secondary descriptive division into a "nucleus cochlearis ventralis superior" (NCVS) and "inferior" (NCVI) [9, 12–14].

Animal and human research use a different nomenclature because of differences in the spatial orientation of the nuclei in the brainstem: "anterior" in the animal refers to "superior" in the human. The same is true for "inferior" and "posterior."

Moreover, there is no "pars anteroventralis" and no "pars posteroventralis" in humans [11].

The volume of the cochlear nucleus almost triples over the first five decades of life and then decreases again by about 1/3 until toward the end of life [15].

From a surgical perspective, the inferior cerebellar peduncle, a large neural fiber connection between the pons and the cerebellum, borders on the most ventral portion of the cochlear nucleus (Fig. 1) [5, 11, 17].

The floccular peduncle of the cerebellar flocculus forms part of the lateral border of the VCN. The VCN is partially crossed by myelinated fibers of the pontobulbar body, a neuronal relay site with connections to visual and auditory regions of the cerebellum [11, 18, 19].

The 0.5 mm-thin cell layer can also overlap the dorsal aspect of the DCN [11].

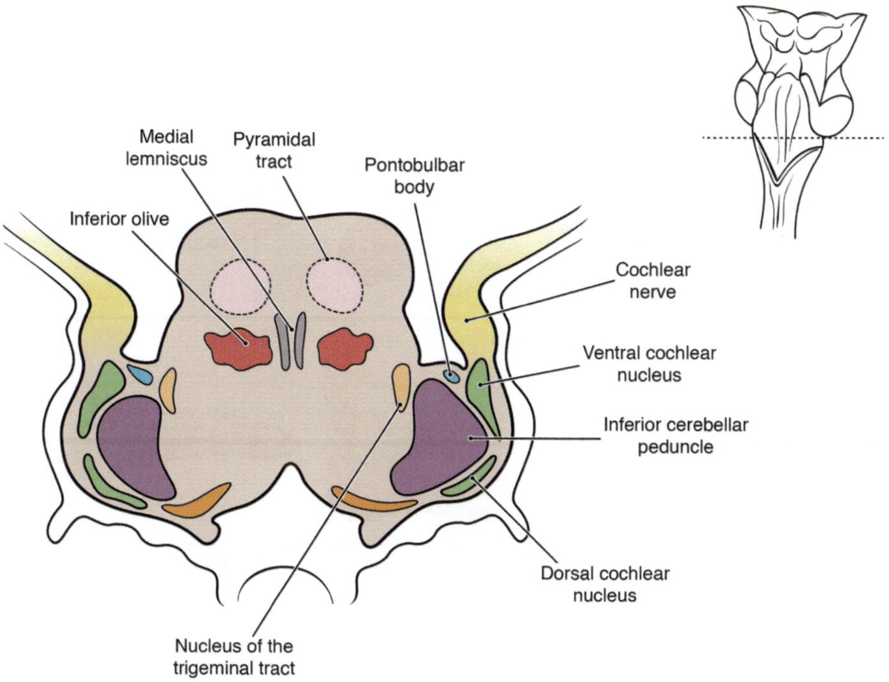

Fig. 1 Schematic axial section through the brainstem at the level of the entry zone of the cochlear nerve (permission granted from [16]). Upper-right insert: Location of the axial section shown in A at the pontomesencephalic junction (broken line)

For several reasons, the real dimensions of the human cochlear nucleus are difficult to assess. In histological sections, the borders of the VCN and DCN are not delineated. Histological fixation ultimately leads to shrinkage of specimen in the range of 10–17%, more pronounced in longitudinal than transversal direction, so that measurements need to be multiplied by a correction factor [15, 20–22].

The nucleus is hardly detectable on magnetic resonance imaging, but its length and width on a horizontal plane have been estimated at 8 × 3 mm [14].

A human cadaver study of 20 brainstem specimens (33 nuclei) found a striking inward rotation of the ventral cochlear nucleus to the longitudinal axis of the brainstem so that the rostral half of the ventral cochlear nucleus has several millimeters to the surface of the brainstem (Fig. 2).

In the same study, the overall extensions of the cochlear nucleus complex (CNC) were 8.01 × 1.53 × 3.76 mm (length × width × height) with respect to the intrinsic axis of the complex, with standard deviations between 0.2 and 1.21 mm. The VCN becomes wider in more rostrally located regions of the brainstem, and its longitudinal axis (length) was longer in more caudally located sections.

Individual variability of the dimensions of the cochlear nucleus was high, with the smallest and largest values for maximum dimensions and maximum surface depths varying by a factor of 3.

If the DCN was small in width and length, the VCN was too. There were no significant side differences for length and height of the cochlear nucleus, except for the

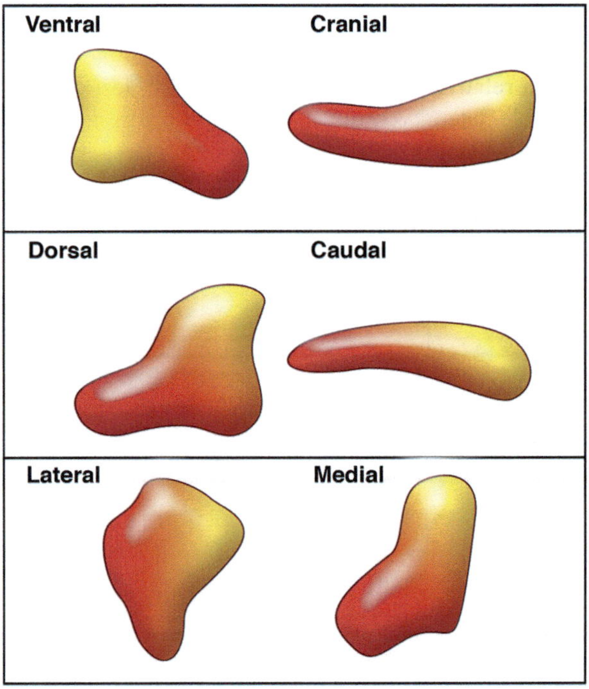

Fig. 2 Distance of the ventral cochlear nucleus (VCN) to the surface of the brainstem (permission granted from [16]). Left: Minimal surface depth (mean ± standard error) of the VCN. The broken line represents the first histological section of the facial nerve. Right: Three-dimensional rendering of the cochlear nucleus complex on the right side (view from above) concerning an ABI surface electrode. Note the distance of the rostral part of the VCN from a surface electrode placed in the lateral recess of the fourth ventricle (double arrow)

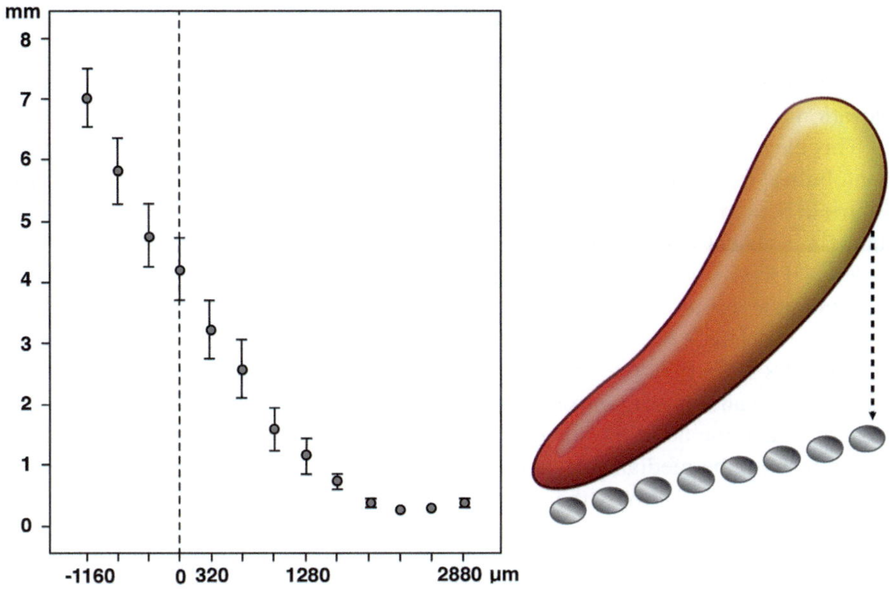

Fig. 3 Aspects of a three-dimensional, true-to-scale model based on the measurements in about 1000 histological sections (permission granted from [16])

mean length of the DCN in the sagittal plane. The DCN was longer than the right (2.8 vs. 4.14 mm) in this plane, but this difference was not statistically significant.

In-molded three-dimensional renderings of the CNC, it has a distorted X-like or boot-like shape viewed from the side and looks like a wedge from above (Fig. 3) [22–24].

The model was molded from plasticine, photographed from several angles, and uploaded onto a graphic computer to smooth of the surfaces and re-coloring. Views are spatially rotated according to the axes of the brainstem.

With advancing age, the CNC shrinks without reducing the number of cells [25, 26].

3.2 *Surgical Anatomy*

Only a small portion of the ventral human cochlear nucleus lies superficially [11]. The cochlear nucleus itself is not visible on the surface of the human brainstem.

However, some intrinsic anatomical structures correlate to outer landmarks such as the "torus" and the "auditory tubercle" [27–30].

The cochlear nuclei form part of the wall of the lateral recess of the fourth ventricle, and except for a tiny caudal portion, the dorsal surface of the DCN is almost entirely accessible from the surface of this region of the brainstem [13, 14, 31].

Jacob et al. measured this accessible area inside the lateral recess to be 7.5×2.5 mm [32].

The DCN lies back to back with the cerebellar peduncle and contributes a slight elevation on the surface of the brainstem - the “auditory tubercle” [33].

This elevation, however, is probably only partially formed by the DCN, with the pontobulbar body, the vestibular nucleus, and the ependymal lining of the fourth ventricle being responsible for another portion [16, 22, 24]. Moreover, the auditory tubercle is absent in each case [29].

Even if the regional anatomy is undisturbed by the presence of a tumor, the auditory tubercle is not a reliable surgical landmark. Still, the “visible area of the cochlear nucleus” on the surface of the brainstem has been reported to be $11.7 \pm 2.7 \times 3.1 \pm 0.7$ mm [28], and the length range of the tubercle was reported to be 3.5–9 mm and its height (rostrocaudal extension) 1.2–3 mm [29]. The mean distance between the auditory tubercle and another visible structure on the surface of the brainstem, the median sulcus, has been measured to be 6.5–6.9 mm [29].

In histological sections, the mean distance between the DCN and the median sulcus is about 10 mm, and the auditory tubercle extends medial to the DCN [22].

The term “Nuclear torus” refers to a more pronounced eminence that curves around the brainstem’s posterior lateral border on the medulla oblongata’s surface. It is thought to be created by the DCN, the lower part of the VCN, and the root entry zone of the vestibulocochlear nerve. The mean length of the torus has been measured to 12.8 mm using a caliper [22].

Altogether, extrinsic “landmarks” like the “auditory tubercle” and the “torus” appear only vaguely correlated to the extensions of the CNC in histological sections. However, both extrinsic and intrinsic measurements share a common feature in their individual variability.

Abe et al. noted that the VCN may reach the entrance of the lateral recess of the fourth ventricle (foramen of Luschka) [33]. Since it is difficult to define a clear border between DCN and VCN and most of the latter is located deep inside the brainstem, it is unlikely to cause much of an outer landmark on the surface.

The distance of the ventral cochlear nucleus to the facial nerve is more important because this nerve usually is well visible during surgery in this area, even in tumor cases. The DCN is located below the facial nerve exit zone in most cases, but the more deeply located VCN extends right across the apparent origin of the facial nerve. The cochlear nerve enters the middle portion of the VCN [25]. Penetrating electrodes that target the primary auditory neurons of the ventral cochlear nucleus may thus penetrate facial nerve fibers inside the brainstem [16].

The most important surgical landmark for implanting surface electrodes into the lateral recess is the foramen of Luschkae. Komune et al. [34] have described the anatomy around this entrance into the lateral recess (Fig. 4). In their report, the “rhomboid lip” marks the ventral border of the foramen. This surface structure is probably consistent with the lateral part of the tenia of the choroid plexus [23–25]

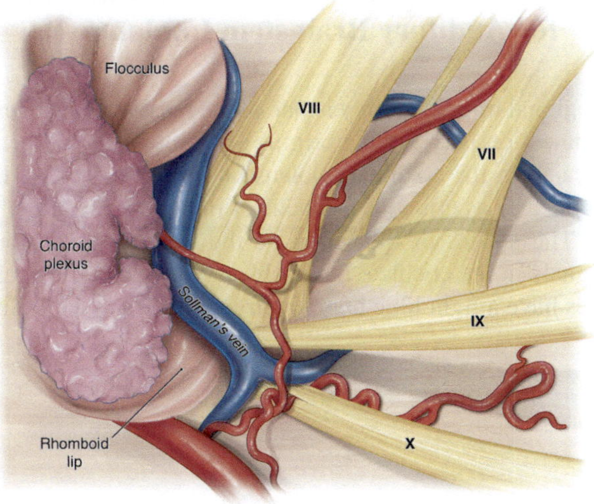

Fig. 4 Surgical anatomy of the brainstem region around the entrance to the lateral recess (foramen of Luschka) (permission granted from [16])

Terr and Edgerton at the House Ear Institute in Los Angeles showed that the terminal part of the cochlear nerve lies close to the line of attachment of the tenia of the choroid plexus (inferior "medullary velum" of the fourth ventricle) in human cadaveric specimens [13, 35].

The glossopharyngeal nerve guides the approach to the foramen of Luschka. It also serves as an estimate of the trajectory for implanting a surface ABI electrode array.

At the entrance of the lateral recess, a small vein runs under the choroid plexus at the surface between the cochlear nucleus and the pontobulbar body in most cases. It has first been described by the German neurosurgeon Wolf-Peter Sollmann who emphasized that this vein is straight and not tortuous like most other vessels in that region and that it is enormously helpful to pinpoint the exact placement of ABI surface electrode carriers (Fig. 4, personal communication) [23].

One of the pioneers of human auditory brainstem implantation, the neuro otologist Derald Brackmann, has coined the term, "Peters vein" in honor of Sollmann's contribution. In formalin-fixed specimens, the foramen of Luschka has a size of about 3.5 × 2 mm [28, 29].

The choroid plexus has been retracted medially to expose Sollmann's vein, a very small, straight-running vessel that is help the right location for electrode carriers of auditory brainstem implants.

CN VIII vestibulocochlear nerve

CN VII facial nerve

CN IX glossopharyngeal nerve

CN X vagus nerve

4 The Cochlear Nucleus as an Anatomic Interface for Auditory Brainstem Implants

Regional microanatomy is one of the reasons that despite a presumed point-to-point connection between the origins of auditory nerve fibers at the cochlea and neurons located in the cochlear nucleus of the brainstem, auditory brainstem implants are not nearly as effective as cochlear implants (Fig. 5).

The shades of blue indicate the frequency distribution (tonotopy) of auditory signals in the cochlea, the cochlear nerve, and the cochlear nucleus. High frequencies have a light blue hue, low frequencies are dark blue [16].

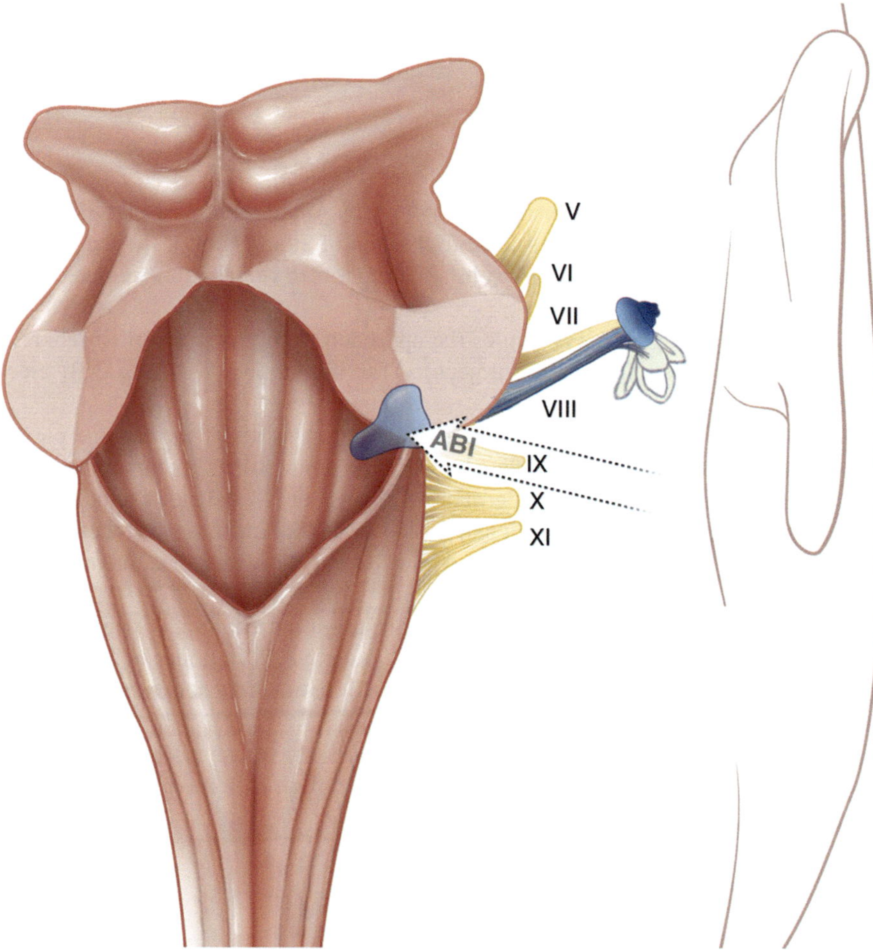

Fig. 5 Functional anatomy of the brainstem, vestibulocochlear structures in the petrous bone, and the cranial nerves in the cerebellopontine angle from a retrosigmoid perspective (permission requested from [16])

There are five main difficulties for interfacing the cochlear nucleus with an ABI electrode carrier:

1. There are very few landmarks helping to guide a surgeon to the cochlear nucleus.
2. The human cochlear nucleus, especially the VCN, has a complex shape that cannot be correlated to external anatomical landmarks.
3. The size and spatial orientation of the cochlear nucleus complex vary individually.
4. The primary terminals of the cochlear nerve in the VCN are located at a depth of several millimeters. Intrinsic fibers of the facial nerve cross the VCN, so penetrating electrodes risk damaging these fibers.
5. Large may compromise the cochlear nuclei and displace them.

The latter assumption is sustained by the fact that the best results with ABIs are often obtained in deaf patients with non-tumor pathologies, albeit with a wide variation in speech recognition from 10% to 100% [36, 37].

ABI surface electrodes reach the cochlear nucleus at its caudal end, where it is about 0.8 mm thin and only about 0.5 mm deep to the surface. They overlap about one- to two-thirds of the nucleus area that projects itself to the lateral brainstem border and preferably stimulate the whole DCN and about 1/3 of the length of the VCN. This means that electrical stimulation mostly does not directly reach the brainstem's primary terminals of cochlear nerve fibers. The maximal portion of the CNC reached by electrical stimulation at the surface includes an area of about 3.4 mm + (1/3 of 4.6 mm) × 3.8 mm = 4.9 mm × 3.8 mm. Electrodes outside this small area cannot be expected to stimulate the cochlear nucleus at reasonable charge densities.

While the glossopharyngeal nerve, the lateral recess with its choroid plexus, Sollman's vein, and trial stimulation at target sites are sufficient to guide surface electrode implantation, clear landmarks, and trajectories for penetrating implants are lacking. Therefore, most of the ventral portion of the CN is still out of reach.

To reach this target with penetrating electrodes, the entry point will have to be cranial to the lateral recess at the level of the pontocerebellar junction and at the lateral vertex of the brainstem, a site that roughly correlates with the middle portion of the VCN. The trajectory would have to be obliquely upwards (from caudal to rostral), avoiding the exit zone and the intra-axial course of the facial nerve.

Side effects caused by the electrical stimulation cannot be entirely avoided because of the close spatial relationship of the electrodes to the rubrospinal tract, the tectospinal tract, the pontobulbar body, the trigeminal tract, the second motor neuron of the long sensory tracts, the middle and inferior cerebellar peduncle, the vestibular nucleus, and the glossopharyngeal nerve [38–41].

Individual approaches probably depend on further improvement of magnetic resonance imaging of the CNC.

References

1. Spoendlin H, Schrott A. Analysis of the human auditory nerve. Hear Res. 1989;43(1):25–38.
2. Kunel'skaya NL, Yatskovsky AN, Mishchenko VV. Microanatomy of the cranial segment of the vestibulocochlear nerve. Possible correlations with the symptoms of neurovascular compression syndrome. Vestn Otorinolaringol. 2016;81(1):25–8.
3. Ryu H, Tanaka T, Yamamoto S, Uemura K, Takehara Y, Isoda H. Magnetic resonance cisternography used to determine precise topography of the facial nerve and three components of the eighth cranial nerve in the internal auditory canal and cerebellopontine cistern. J Neurosurg. 1999;90(4):624–34.
4. Lang J. Skull base and related structures. 2nd ed. Stuttgart: Schattauer Verlag; 2001.
5. Moore JK. Cochlear nuclei: relationship to the auditory nerve. In: Altschuler RA, Hoffmann DW, Bobbin RP, editors. Neurobiology of hearing: the cochlea. New York: Raven Press; 1986. p. 283–301.
6. Lorente de No R. Anatomy of the eighth nerve. III. General plan of the primary cochlear nuclei. Laryngoscope. 1933;43:327–50.
7. Bacsik RD, Strominger NL. The cytoarchitecture of the human anteroventral cochlear nucleus. J Comp Neurol. 1973;147(2):281–9.
8. Brawer JR, Morest DK, Kane EC. The neuronal architecture of the cochlear nucleus of the cat. J Comp Neurol. 1974;155(3):251–300.
9. Luxon LM. The anatomy and pathology of the central auditory pathways. Br J Audiol. 1981;15(1):31–40.
10. Mobley JP, Huang J, Moore JK, McCreery DB. Three-dimensional modeling of human brain stem structures for an auditory brain stem implant. Ann Otol Rhinol Laryngol Suppl. 1995;166:30–1.
11. Moore JK, Osen KK. The cochlear nuclei in man. Am J Anat. 1979;154(3):393–418.
12. Dublin WB. The cochlear nuclei revisited. Otolaryngol Head Neck Surg. 1982;90:744–60.
13. Terr LI, Edgerton BJ. Surface topography of the cochlear nuclei in humans: two- and three-dimensional analysis. Hear Res. 1985;17(1):51–9.
14. Terr LI, Sinha UK, House WF. Anatomical relationships of the cochlear nuclei and the pontobulbar body: possible significance for neuroprosthesis placement. Laryngoscope. 1987;97(9):1009–11.
15. Konigsmark BW, Murphy EA. Volume of the ventral cochlear nucleus in man: its relationship to neuronal population and age. J Neuropathol Exp Neurol. 1972;31(2):304–16.
16. Rosahl SK. Neuroanatomy and physiology relevant to auditory brainstem implants. In: Wilkinson EP, Schwartz MS, editors. Auditory brainstem implants. New York: Thieme Medical Publishers; 2021:3–13.
17. Moore JK. The human auditory brain stem: a comparative view. Hear Res. 1987;29(1):1–32.
18. Terr LI, House WF. Neurons of the inferior medullary velum in the cerebellopontine angle. Ann Otol Rhinol Laryngol. 1988;97(1):52–4.
19. Terr LI, Sinha UK. Three-dimensional computer-aided reconstruction of the pontobulbar body. Am J Otol. 1987;8(5):432–5.
20. Moore JK, Niparko JK, Perazzo LM, Miller MR, Linthicum FH. Effect of adult-onset deafness on the human central auditory system. Ann Otol Rhinol Laryngol. 1997;106(5):385–90.
21. Quester R, Schroder R. The shrinkage of the human brain stem during formalin fixation and embedding in paraffin. J Neurosci Methods. 1997;75(1):81–9.
22. Rosahl SK, Rosahl S. No easy target: anatomic constraints of electrodes interfacing the human cochlear nucleus. Neurosurgery. 2013;72(1 Suppl Operative):58–64; discussion 5.
23. Rosahl SK, Rosahl S. Letter: anatomy and auditory brainstem implants. Neurosurgery. 2016;78(4):E601–2.
24. Rosahl SK, Rosahl S, Walter GF, Hussein S, Matthies C, Samii M. Cochlear region of the brainstem. J Neurosurg. 2000;93(4):724–9.

25. Rosahl S. Ausdehnung, Lage und Form des menschlichen Nucleus cochlearis. Freiburg: Albert-Ludwigs-Universität; 2008.
26. Shepherd RK, Hardie NA. Deafness-induced changes in the auditory pathway: implications for cochlear implants. Audiol Neurootol. 2001;6(6):305–18.
27. Klose AK, Sollmann WP. Landmarks for implantation at the cochlear nucleus—a microanatomical study. Zentralbl Neurochir. 1998;(Suppl.1):61.
28. Klose AK, Sollmann WP. Anatomical variations of landmarks for implantation at the cochlear nucleus. J Laryngol Otol Suppl. 2000;27:8–10.
29. Quester R, Schroder R. Topographic anatomy of the cochlear nuclear region at the floor of the fourth ventricle in humans. J Neurosurg. 1999;91(3):466–76.
30. Schwalbe G. Lehrbuch der Neurologie. Jena: Gustav Fischer; 1881.
31. McElveen JT Jr, Hitselberger WE, House WF. Surgical accessibility of the cochlear nuclear complex in man: surgical landmarks. Otolaryngol Head Neck Surg. 1987;96(2):135–40.
32. Jacob U, Mrosack B, Gerhardt HJ, Staudt J. The surgical approach to the cochlear nucleus area. Anat Anz. 1991;173(2):93–100.
33. Abe H, Rhoton AL Jr. Microsurgical anatomy of the cochlear nuclei. Neurosurgery. 2006;58(4):728–39.
34. Komune N, Yagmurlu K, Matsuo S, Miki K, Abe H, Rhoton AL Jr. Auditory brainstem implantation: anatomy and approaches. Neurosurgery. 2015;11(Suppl. 2):306–20; discussion 20–1.
35. Terr LI, Edgerton BJ. Three-dimensional reconstruction of the cochlear nuclear complex in humans. Arch Otolaryngol. 1985;111(8):495–501.
36. Choi JY, Song MH, Jeon JH, Lee WS, Chang JW. Early surgical results of auditory brainstem implantation in nontumor patients. Laryngoscope. 2011;121(12):2610–8.
37. Colletti V, Shannon R, Carner M, Veronese S, Colletti L. Outcomes in nontumor adults fitted with the auditory brainstem implant: 10 years' experience. Otol Neurotol. 2009;30(5):614–8.
38. Edgerton BJ, House WF, Hitselberger W. Hearing by cochlear nucleus stimulation in humans. Ann Otol Rhinol Laryngol Suppl. 1982;91(2 Pt 3):117–24.
39. Matthies C, Thomas S, Moshrefi M, Lesinski-Schiedat A, Frohne C, Battmer R-D, et al. Auditory brainstem implants: current neurosurgical experiences and perspective. J Laryngol Otol. 2000;114(Suppl. 27):32–7.
40. Shannon RV, Fayad J, Moore J, Lo WW, Otto S, Nelson RA, et al. Auditory brainstem implant: II. Postsurgical issues and performance. Otolaryngol Head Neck Surg. 1993;108(6):634–42.
41. Sollmann WP, Laszig R, Marangos N. Surgical experiences in 58 cases using the nucleus 22 multichannel auditory brainstem implant. J Laryngol Otol Suppl. 2000;27:23–6.

Part III
The Central Core

Surgical Anatomy of the White Fiber Tracts

Richard Gonzalo Párraga and Armando Morales

1 General Considerations

The white matter within the central nervous system is located in different areas. At the cerebral and cerebellum levels, its distribution shows similarity: we can see a peripheral zone and central area, limited by cortical gray matter and deep basal nuclei. At the level of the brain stem, the white matter is located predominantly in its anterior portions (base of the cerebral peduncle, the base of the pons, and the bulb pyramid), compared to a smaller proportion of fibers, in the texture that we have seen in the tectum and tegmentum, respectively. The white matter is in the peripheral portion at the spinal cord level, forming tracts and fascicles inside the spinal cords [1].

2 White Substance in the Cerebral Hemispheres

The white matter bulk of the cerebral hemispheres is composed of myelinated bundles of fibers, denominated fiber tracts or fasciculi, and they are categorized based on their course and connections, as classically proposed by Theodor Meynert (1833–1891) in 1872 [2, 3], who divided these fibers into three groups: (1) association fibers: this links functional areas in the **same hemisphere;** (2) commissural fibers: which join homologous areas, not necessarily symmetrical **between the hemispheres;** (3) projection fibers enter or leave the telencephalic, **relating the cortex to the brainstem and the spinal cord**.

R. G. Párraga (✉) · A. Morales
Department of Cerebrovascular and Skull Base, Institute of Neurosurgery Bolivia (INEB), Cochabamba, Bolivia
e-mail: richardparraga@ineb.com.bo

E. G. Figueiredo et al. (eds.), *Brain Anatomy and Neurosurgical Approaches*,
https://doi.org/10.1007/978-3-031-14820-0_16

3 White Matter Sublevels

The white matter substance can be subdivided into two distinct zones, the peripheral zone (gyral) and the central zone (capsular) (Fig. 1) [1]. The large peripheral zone consists of white matter between the cortex and the periventricular matrix of the lateral ventricles. The smaller central zone consists of the projection fibers surrounding the lateral border of the lentiform nucleus (external capsule) and the V-shaped projection fiber bundles, located lateral to the lentiform nucleus and lateral to the caudate nucleus and thalamus (internal capsule) [1].

3.1 Association Fibers

These nerve fibers essentially connect several cortical regions within the same hemisphere and can be divided into short and long [3–5]:

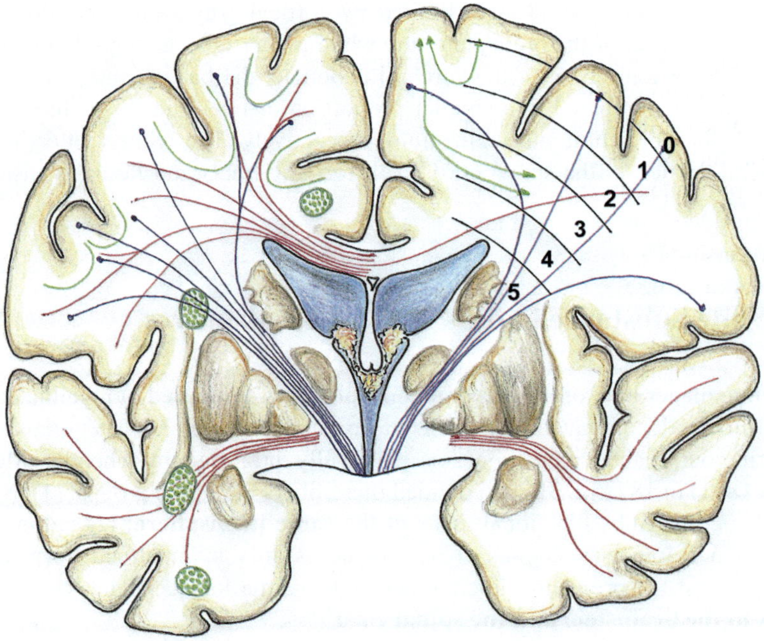

Fig. 1 Schematic image of a coronal view of the brain, showing its connection fiber (Adapted by Yasargil MG. Microneurosurgery Vol IVA: CNS Tumors: Surgical Anatomy, Neuropathology, Neuroradiology, Neurophysiology, Clinical Considerations, Operability, and Treatment Options. Stuttgart: Georg Thieme-Verlag; 1994). Blue Projection fibers; Green Association fibers; Red Commissural fibers. (*0*) Córtex; (*1*) subcortical; (*2*) subgyral; (*3*) gyral; (*4*) lobar; (*5*) capsular, pyramidal

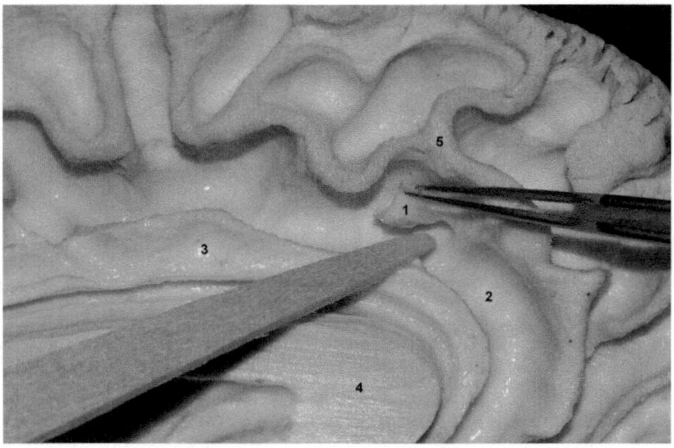

Fig. 2 Magnified view of the medial surface of the left hemisphere, where the removal of a short association fiber located in the depth of the sulcus of the cingulated joins the cingulated with the superior frontal gyrus. (*1*) Arcuate fibers; (*2*) depth of the sulcus; (*3*) cingulum; (*4*) corpus callosum; (*5*) vertex of the gyrus

Short association fibers: Also called arcuate *fibers of Arnold* are fibers in the shape of a "U" of Meynert [6]; the typical fibers are born on the vertex or the side of the gyrus and end at the vertex or the end of the next gyrus, after having surrounded the intermediate sulcus, joining to the adjacent gyri (Fig. 2) [6–9].

Long association fibers connect more widely separated gyri within the same hemisphere and are grouped into bundles.

Next, we describe the different fascicles according to their location and distribution, both the superolateral, medial, and inferior or basal surfaces of the brain:

3.1.1 In the Superolateral Surface of the Brain

Superior Longitudinal Fasciculus

Reil y Autenrieth (1809, 1812) were the first to describe the superior longitudinal fasciculus as a group of fibers located in the white matter of the temporal, parietal lobes and around the Sylvian fissure. Burdach (1819–1826) and subsequently Dejerine (1895) described this fiber system in detail as a tract that is located around the Sylvian fissure, above and lateral-wise to the lentiform nucleus and the insula and connects the posterior temporal lobe with the frontal lobe (Fig. 3). They named this section "fasciculus arcuatus" (arcuate fasciculus), and it is considered as a part of the superior longitudinal fasciculus (SLF), using the terms "superior longitudinal fasciculus" or "fasciculus arcuatus" in their descriptions [10, 11] are interchangeable.

In contrast to the classic descriptions, recent studies of diffusion tensor images (DTI) tractography demonstrated that the arcuate fasciculus is a subdivision of the superior longitudinal fasciculus. In addition, neuroimaging experiments with primates and humans have revealed that the superior longitudinal fasciculus is a complex system of the brain association fibers composed of three segments: (1) Fronto-parietal or horizontal segment, (2) Temporo-parietal or vertical segment, and (3) Temporo-frontal segment or arcuate fasciculus [12, 13].

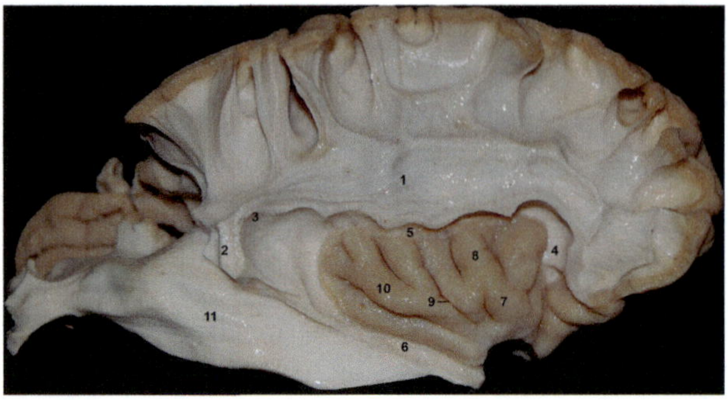

Fig. 3 Lateral view of the right cerebral hemisphere where the cerebral cortex and the short association fibers were removed. Exposing the superior longitudinal fasciculus represents the largest system of the intrahemispheric association in the superior lateral surface of each cerebral hemisphere. (*1*) Fronto-parietal or horizontal segment of the superior longitudinal fasciculus; (*2*) temporo-parietal or vertical segment of the superior longitudinal fasciculus; (*3*) temporo-frontal segment or arcuate fasciculus; (*4*) anterior limiting sulcus of the insula; (*5*) superior limiting sulcus of the insula; (*6*) inferior limiting sulcus of the insula; (*7*) insular apex; (*8*) short gyri of the insula; (*9*) central sulcus of the insula; (*10*) long gyrus of the insula; and (*11*) optic radiation

The frontoparietal or horizontal segment originates in the inferior parietal lobe, at the angular and supramarginal gyrus level, inside the white matter of the front and parietal operculum lateral to the arcuate fasciculus. Finally, it ends in the posterior and inferior frontal lobe, at the end level of the precentral gyrus and the posterior portion of the inferior frontal gyrus (Broca's territory) [13]. Isotope studies of non-human primate brains, as well as DTI analyses of human brains, revealed that this frontoparietal segment is not one single fiber tract but can be divided into three dorsal to ventral components in the white matter of the parietal and frontal lobes [12, 14]: the SLF I, II, and III. The SLF I originates from the dorsal superior parietal lobe and the medial parietal lobe (precuneus), runs through the white matter of the superior parietal and frontal regions, and terminates at the premotor and prefrontal cortex (dorsal parts of areas 6, 8, and 9 and the supplementary motor area).

The SLF II originates from the posterior portion of the inferior parietal lobe (angular gyrus), runs through the white matter's central core above the insula's superior limiting sulcus, and terminates in the dorsal premotor and prefrontal regions. The SLF III originates at the anterior portion of the inferior parietal lobe (supramarginal gyrus), runs through the opercular white matter of the parietal and frontal lobes, and terminates at the ventral premotor and prefrontal cortex (Broca's territory). This third portion of the SLF corresponds to the horizontal segment previously described [12, 15].

The temporoparietal or vertical segment of the SLF originates in the posterior portion of the superior and middle temporal gyrus (Wernicke's s territory). The fibers turn vertically, running parallel and lateral to the arcuate fasciculus, and terminate at the inferior parietal lobe [2, 12, 13].

The temporo-frontal segment corresponds to the classical arcuate fasciculus and is a long white matter tract that directly connects the posterior temporal lobe with the posterior frontal lobe. The posterior projection of the arcuate fasciculus is not limited to a well-defined anatomical territory with precise landmarks, and it mostly encompasses the medial and posterior portions of the superior, middle, and inferior temporal gyri. Then, the fibers converge in a single tract that arches around the caudal end of the Sylvian fissure, running within the white matter of the parietal and frontal operculum. This fascicle runs parallel and medial to the two superficial tracts previously described and lateral to the corticospinal tract [2, 12, 13].

In summary, such recent data suggest two parallel pathways connecting temporal and frontal regions: (1) a direct pathway corresponding to the classical arcuate fasciculus and (2) an indirect pathway running parallel and lateral to the direct pathway and consisting of two segments, an anterior or horizontal segment linking Broca's territory with the inferior parietal lobe; a posterior or vertical segment linking the inferior parietal lobe with Wernicke's territory. The SLF has been central to the neurobiological interpretation of higher brain functions, especially language disorders. The present subdivision of the SLF into direct and indirect segments highlights the importance of the inferior parietal cortex as a separate primary language area with dense connections to the classical language areas [12].

The superior longitudinal fasciculus is then composed of an indirect pathway that links the Broca area with Wernicke's area through the inferior parietal lobule along with its horizontal and vertical segments, and by a direct and deeper pathway which corresponds to the arcuate fasciculus, with both being related more particularly with language functions.

Topographically, it has an inverted C shape; on the surface, it is constituted by short fibers and deep by long fibers, constituting the largest intra-hemisphere association system on the superolateral surface of each cerebral hemisphere [3–5, 16]. To identify this fasciculus, it is necessary to remove the gray matter and the short-arched fibers from the middle, inferior frontal gyri, the supramarginal gyrus, angular gyrus, and middle temporal gyrus. Removing the front orbital, front parietal, and temporal operculum is necessary to expose them more easily (Fig. 3). The vertex that forms the change of direction of the fibers between horizontal and vertical is in the depth of the supramarginal gyrus (Fig. 4). It is important to mention that the vertical segment of the superior longitudinal fasciculus covers the sagittal stratum (Figs. 5 and 6).

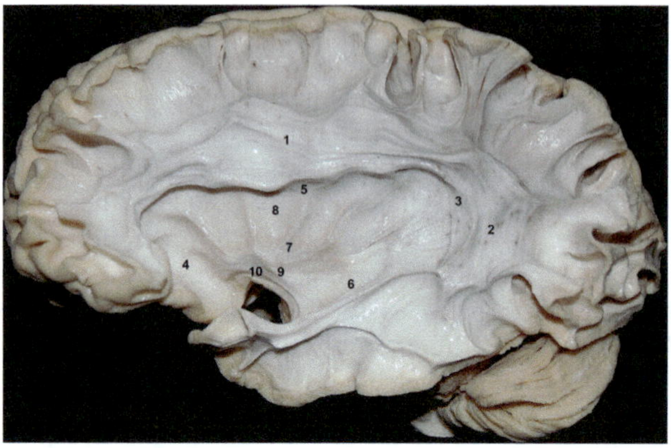

Fig. 4 Lateral view of the left cerebral hemisphere where the cerebral cortex and the short association fibers were removed, view of the superior longitudinal fasciculus that topographically has the inverted C shape around the lateral sulcus. (*1*) Fronto-parietal or horizontal segment of the superior longitudinal fasciculus; (*2*) temporo-parietal or vertical segment of the superior longitudinal fasciculus; (*3*) temporo-frontal segment or arcuate fasciculus; (*4*) anterior limiting sulcus of the insula; (*5*) superior limiting sulcus of the insula; (*6*) inferior limiting sulcus of the insula; (*7*) claustrum; (*8*) external capsule; (*9*) occipitofrontal fasciculus; and (*10*) uncinate fasciculus

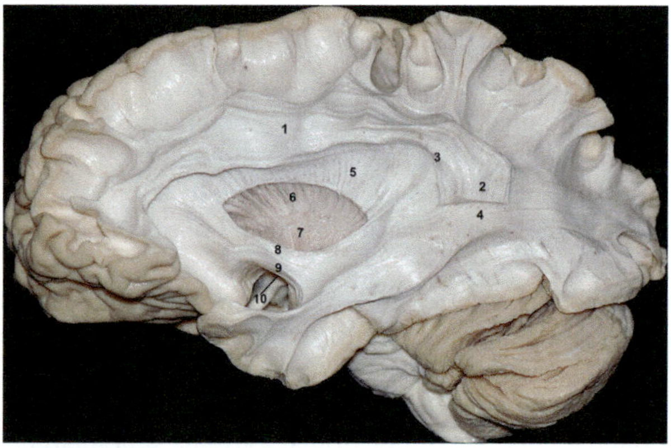

Fig. 5 Lateral view of the left cerebral hemisphere. The vertical portion of the superior longitudinal fasciculus was removed to expose the Sagittal stratum. (*1*) Fronto-parietal or horizontal segment of the superior longitudinal fasciculus; (*2*) temporo-parietal or vertical segment of the superior longitudinal fasciculus; (*3*) temporo-frontal segment or arcuate fasciculus; (*4*) sagittal stratum; (*5*) corona radiata; (*6*) internal capsule; (*7*) globus pallidus; (*8*) occipitofrontal fasciculus; (*9*) uncinate fasciculus; and (*10*) limen of the insula

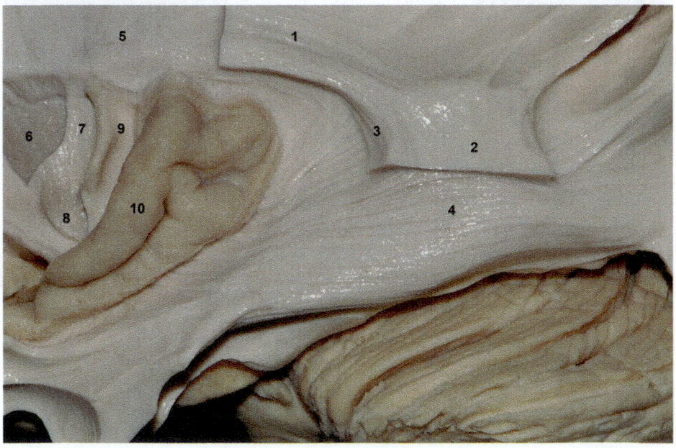

Fig. 6 Magnified view of the left cerebral hemisphere. The vertical portion of the superior longitudinal fasciculus was removed to expose the Sagittal stratum. (*1*) Fronto-parietal or horizontal segment of the superior longitudinal fasciculus; (*2*) temporo-parietal or vertical segment of the superior longitudinal fasciculus; (*3*) temporo-frontal segment or arcuate fasciculus; (*4*) sagittal stratum; (*5*) corona radiata; (*6*) putamen; (*7*) external capsule; (*8*) claustrum; (*9*) extreme capsule; and (*10*) long gyrus of the insula

Uncinate Fasciculus

The uncinate fasciculus is a hook-shaped associative bundle that connects the anteromedial temporal lobe (superior, middle, and inferior temporal gyri, cortical nuclei of the amygdala) with the orbitofrontal region (medial and posterior orbital cortex, gyrus rectus, and subcallosal area) [17]. Its fibers constitute a well-defined tract along the temporal stem. It occupies its anterior one-third immediately posterior to the limen insulae and anterior to the inferior fronto-occipital fasciculus, underneath the most anterior aspect of the inferior limiting sulcus of the insula. Both the uncinate and the inferior fronto-occipital fasciculi intermingle with the most ventral fibers of the extreme and external capsules [2, 13, 17].

The uncinate fasciculus is a ventral association fascicle divided into three segments: temporal, insular, and frontal (Figs. 7 and 8) [17]. The temporal segment originated in the cortical nucleus of the amygdala (areas 28, 34, and 36) and the anterior portion of the three gyri of the superolateral of the temporal surface (areas 20 and 38). These fibers come together to form a solid tract and make a portion for the anterior temporal stem, in the deeper part of the middle temporal gyrus and front of the temporal horn. The insular segment is located on the limen insula, under the

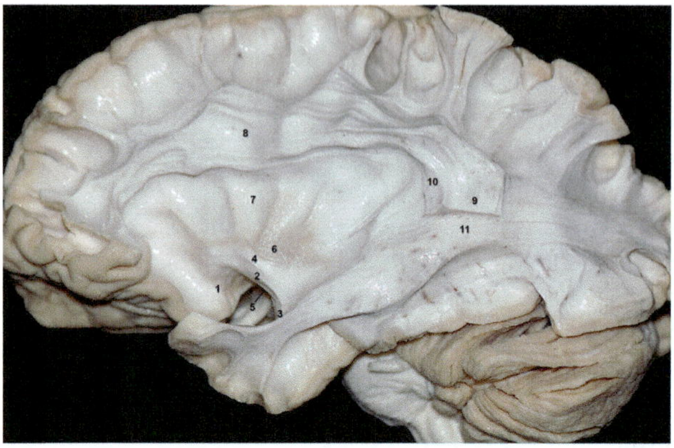

Fig. 7 Lateral view of the left cerebral hemisphere where the three segments of the uncinate fasciculus are exposed, and its relationship with the fronto-occipital fasciculus. (*1*) Uncinate fasciculus, frontal portion; (*2*) uncinate fasciculus, insular portion; (*3*) uncinate fasciculus, temporal portion; (*4*) fronto-occipital fasciculus; (*5*) limen of the insula; (*6*) claustrum; (*7*) external capsule; (*8*) fronto-parietal or horizontal segment of the superior longitudinal fasciculus; (*9*) temporo-parietal or vertical segment of the superior longitudinal fasciculus; (*10*) temporo-frontal segment or arcuate fasciculus; and (*11*) sagittal stratum

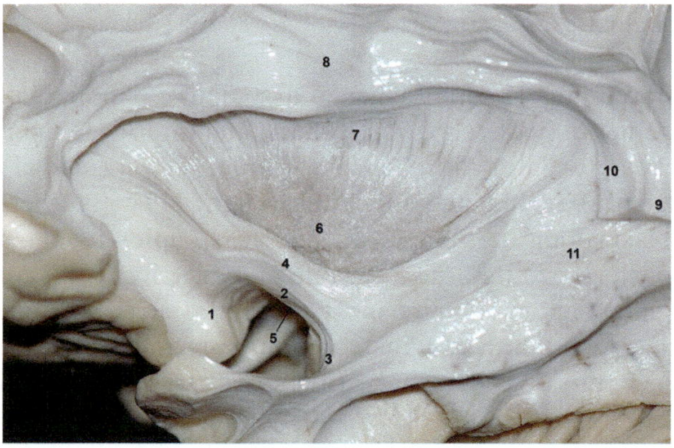

Fig. 8 Magnified view of the anatomic specimen where the external capsule was removed, preserving the putamen. The uncinate fasciculus and fronto-occipital fasciculus are observed without a precise limit between them. (*1*) Uncinate fasciculus, frontal portion; (*2*) uncinate fasciculus, insular portion; (*3*) uncinate fasciculus, temporal portion; (*4*) fronto-occipital fasciculus; (*5*) limen of the insula; (*6*) putamen; (*7*) corona radiata; (*8*) fronto-parietal or horizontal segment of the superior longitudinal fasciculus; (*9*) temporo-parietal or vertical segment of the superior longitudinal fasciculus; (*10*) temporo-frontal segment or arcuate fasciculus; and (*11*) sagittal stratum

putamen and the claustrum (Figs. 9 and 10). This segment has 2 to 5 mm of width and 3 to 7 mm of height, [12] here the fibers of the uncinate fasciculus go through the extreme capsule and the external, anterior inferior to the fronto-occipital fasciculus without a precise limit between them. The frontal segment has the shape of

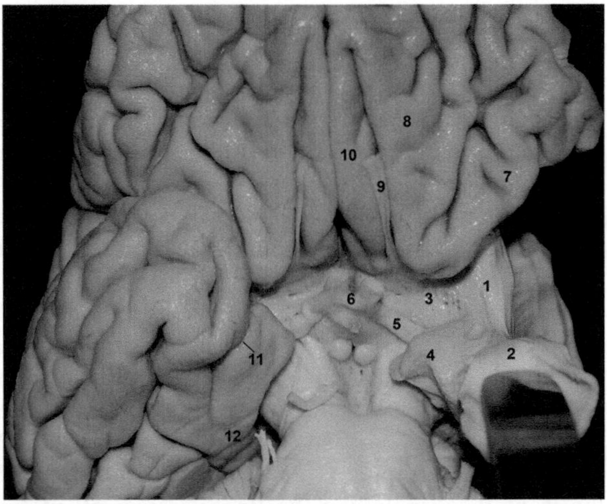

Fig. 9 The basal surface of the brain. The insular portion of the uncinate fasciculus is exposed. (*1*) Uncinate fasciculus, insular portion; (*2*) uncinate fasciculus, temporal portion; (*3*) anterior perforated substance; (*4*) uncus; (*5*) optic tract; (*6*) optic chiasm; (*7*) posterior orbital gyri; (*8*) medial orbital gyri; (*9*) olfactory tract; (*10*) gyrus rectus; (*11*) rhinal sulcus; and (*12*) parahippocampal gyrus

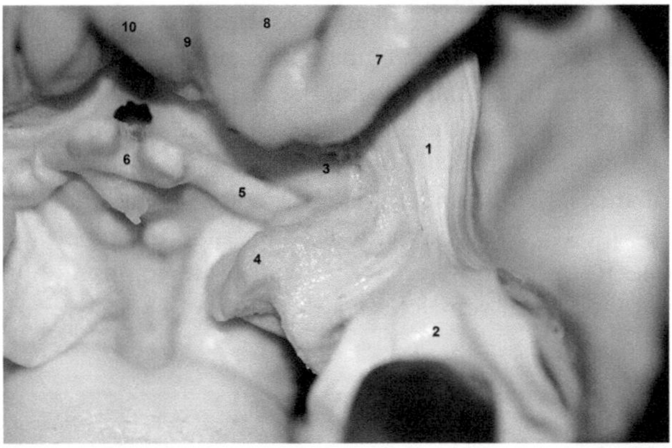

Fig. 10 The basal surface of the brain. Magnified view of the same anatomical specimen. (*1*) Uncinate fasciculus, insular portion; (*2*) uncinate fasciculus, temporal portion; (*3*) anterior perforated substance; (*4*) uncus; (*5*) optic tract; (*6*) optic chiasm; (*7*) posterior orbital gyri; (*8*) medial orbital gyri; (*9*) olfactory tract; and (*10*) gyrus rectus

a fan in the white matter in the front orbital portion and has a horizontal orientation in the gyrus rectus (area 11), in the medial and lateral orbital gyrus (area 12), and the sub callosum area (area 25) [12, 15].

The uncinate fasciculus fibers connect the cortical nucleus of the amygdala and the uncus with the subcallosum region and the pole of the superior, middle, and inferior temporal gyri with the gyrus rectus and the medial, lateral orbital gyri, as well as the orbital parts of the inferior frontal gyrus [6, 7, 12, 15, 18–23].

Fronto-Occipital Fasciculus

The fronto-occipital fasciculus are fibers that communicate the frontal lobe (fronto-orbital cortex and the prefrontal region), with the posterolateral region of the temporal lobe and the occipital lobe [2, 9]. In 1909, Curran first described this fascicle by postmortem fiber dissection [24]. Since then, many other authors have used white-matter dissection to elucidate the anatomical course of this fascicle. Recently, DTI studies and studies combining fiber dissection with DTI described the main course of the fronto-occipital fasciculus at the level of the limen insulae, the roof of the temporal horn, and within the anterior and middle temporal lobe [10, 25]. Morphologically, this set of fibers are equivalent to a fan that is doubly connected by an istmo situated deeply to the limen insulae, next to the uncinate fasciculus fibers open anteriorly in the frontal lobe, and they open later along with the lateral aspects of the temporal and occipital lobes (Figs. 11, 12, and 13) [11].

The fronto-occipital fasciculus is extended from the superolateral surface of the front lobe, including the Broca's area; all of these fibers converge in one unique band that passes throughout the inferior lateral lentiform nucleus, between extreme and the external capsule, in the basal portion of the insula and of the claustrum, the immediate superior uncinate fasciculus [15]. When this reaches the temporal lobe, it opens like a fan, having the largest fibers in the posterior direction, passing to make part of the sagittal stratum. It is also associated with the inferior longitudinal fasciculus that interconnects the occipital cortex (lingula, lateral occipital surface, and cuneus) with the superior, middle, and inferior temporal gyri cortex [3, 15, 24]. The parallel course and the superficial plane of the optical radiations form part of the external wall of the temporal horn, atrium, and occipital horn of the lateral ventricles.

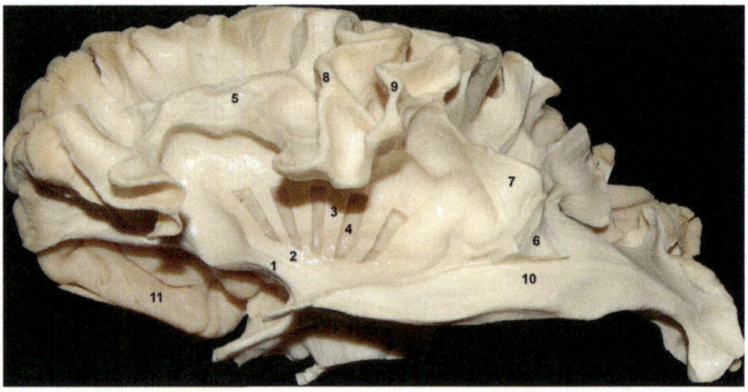

Fig. 11 In lateral view of the left hemisphere, the fronto-occipital fasciculus is exposed, that it communicates the frontal lobe (front-orbital cortex and the prefrontal region) with the posterolateral region of the temporal and the occipital lobes. (*1*) Uncinate fasciculus; (*2*) fronto-occipital fasciculus; (*3*) external capsule; (*4*) putamen; (*5*) fronto-parietal or horizontal segment of the superior longitudinal fasciculus; (*6*) temporo-parietal or vertical segment of the superior longitudinal fasciculus; (*7*) temporo-frontal segment or arcuate fasciculus; (*8*) precentral gyrus; (*9*) postcentral gyrus; (*10*) sagittal stratum; and (*11*) gyrus rectus

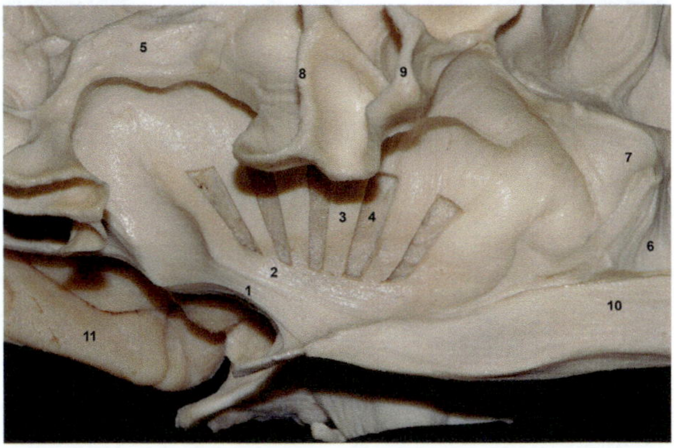

Fig. 12 Magnified view of the same anatomical specimen, the fronto-occipital fasciculus is exposed, that it communicates the frontal lobe (front-orbital cortex and the prefrontal region), with the posterolateral region of the temporal and occipital lobes. (*1*) Uncinate fasciculus; (*2*) fronto-occipital fasciculus; (*3*) external capsule; (*4*) putamen; (*5*) fronto-parietal or horizontal segment of the superior longitudinal fasciculus; (*6*) temporo-parietal or vertical segment of the superior longitudinal fasciculus; (*7*) temporo-frontal segment or arcuate fasciculus; (*8*) precentral gyrus; (*9*) postcentral gyrus; (*10*) sagittal stratum; and (*11*) gyrus rectus

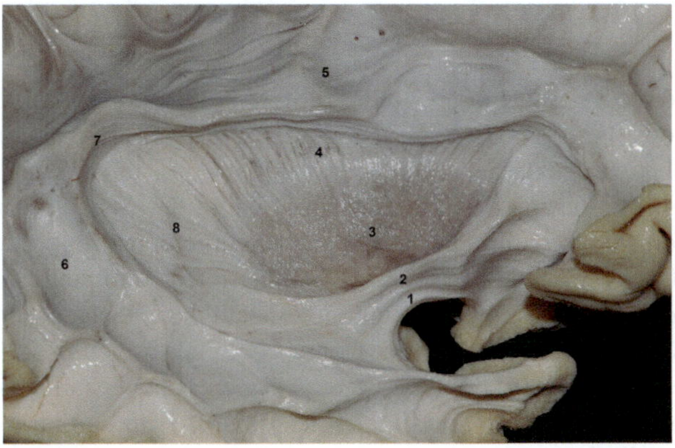

Fig. 13 Lateral view of the right cerebral hemisphere, the fronto-occipital fasciculus is exposed. (*1*) Uncinate fasciculus; (*2*) fronto-occipital fasciculus; (*3*) putamen; (*4*) corona radiata; (*5*) fronto-parietal or horizontal segment of the superior longitudinal fasciculus; (*6*) temporo-parietal or vertical segment of the superior longitudinal fasciculus; (*7*) temporo-frontal segment or arcuate fasciculus; and (*8*) sagittal stratum

Recent tractography studies DTI have described the frontal connections of the fronto-occipital fasciculus with the prefrontal cortex and the fronto-orbital cortex. The anatomical course of the front occipital fascicle in the temporal stem has been studied, besides its relationship with uncinate fasciculus; this last one goes through the third anterior part of the temporal stem. Meanwhile, the fronto-occipital fasciculus goes through the third posterior parts of the temporal stem, even though they rotate in

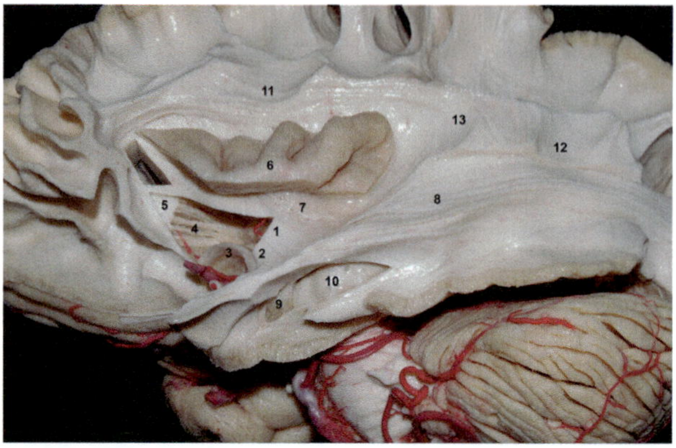

Fig. 14 Lateral view of the left cerebral hemisphere. The course identifies the fronto-occipital fasciculus in the temporal stem and its relationship with the uncinate fasciculus; this last one goes through the previous third of the temporal stem. Meanwhile, the fronto-occipital fasciculus goes through the two-thirds posterior parts of the temporal stem. Besides, they both turn in the posterior direction and run above the roof of the temporal horn. (*1*) Fronto-occipital fasciculus; (*2*) uncinate fasciculus; (*3*) limen of the insula; (*4*) internal capsule, anterior limb; (*5*) corona radiata; (*6*) insula; (*7*) claustrum; (*8*) sagittal stratum; (*9*) amygdala; (*10*) head of the hippocampus; (*11*) fronto-parietal or horizontal segment of the superior longitudinal fasciculus; (*12*) temporo-parietal or vertical segment of the superior longitudinal fasciculus; and (*13*) temporo-frontal segment or arcuate fasciculus

the posterior direction and run above the roof of the temporal horn, the superior to optical radiations (Fig. 14). The hearing radiations, the claustrum opercula fibers, and insulo-opercular of the external capsules and extreme pass through the temporal stem above of the fronto-occipital fasciculus, meanwhile the [5] radiations pass underneath. The uncinate fasciculus crosses the anterior temporal stem portion in the same place as the fronto-occipital fasciculus. The anterior commissural and the inferior thalamic pedicle cross the temporal stem underneath the uncinate fasciculus [13].

In the middle temporal region, the fronto-occipital fasciculus is located on the roof of the temporal horn, above and laterals of the optical radiations, and at the middle part of the inferior longitudinal fasciculus. It is important to mention that there is no exact limit between the fronto-occipital and uncinate fasciculus [7, 9, 12, 21, 26]. According to the findings of studies based on subcortical brain mapping by electrical stimulation during awake neurosurgery, while the superior longitudinal fasciculus is more particularly related to phonological aspects of language, the inferior fronto-occipital fasciculus is more related to its semantic aspect [10].

Inferior Longitudinal Fasciculus

The inferior longitudinal fasciculus was described for the first time by Burdach in 1822; it is constituted by a group of fibers of the anteroposterior direction, which connects sagittally in the occipital lobe on the point of the temporal lobe, on the long part of the basal portion of each brain's hemisphere. It is disposed laterally to the optical radiations, parallel to the temporal horn and in the deep part of the fusiform gyrus or occipitotemporal gyri (Figs. 15 and 16) [6–10].

Recent studies of the DTI demonstrated that the inferior longitudinal fasciculus is composed of a direct and indirect pathway [1]. The indirect pathway, the occipitotemporal projection system, consists of U-shaped fibers connecting adjacent gyri at the inferior temporal and occipital convexity. The direct pathway is composed of long association fibers located medial to the short fibers.

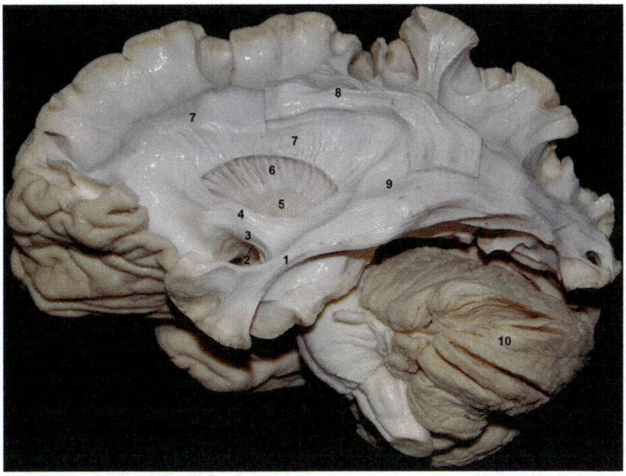

Fig. 15 Inferolateral view of the anatomical specimen. It is identified that the inferior longitudinal fasciculus, which is constituted by a group of fibers that goes to the anteroposterior direction, connects sagittally the occipital lobe to the tip of the temporal lobe, along the basal portion of each cerebral hemisphere. (*1*) Inferior longitudinal fasciculus; (*2*) limen of the insula; (*3*) uncinate fasciculus; (*4*) fronto-occipital fasciculus; (*5*) globus pallidus; (*6*) internal capsule; (*7*) corona radiata; (*8*) superior longitudinal fasciculus; (*9*) sagittal stratum; and (*10*) cerebellar hemisphere

Fig. 16 Magnified view of the same anatomical specimen. (*1*) Inferior longitudinal fasciculus; (*2*) limen of the insula; (*3*) uncinate fasciculus; (*4*) fronto-occipital fasciculus; (*5*) globus pallidus; (*6*) internal capsule; (*7*) corona radiata; (*8*) superior longitudinal fasciculus; (*9*) sagittal stratum; and (*10*) head of the hippocampus

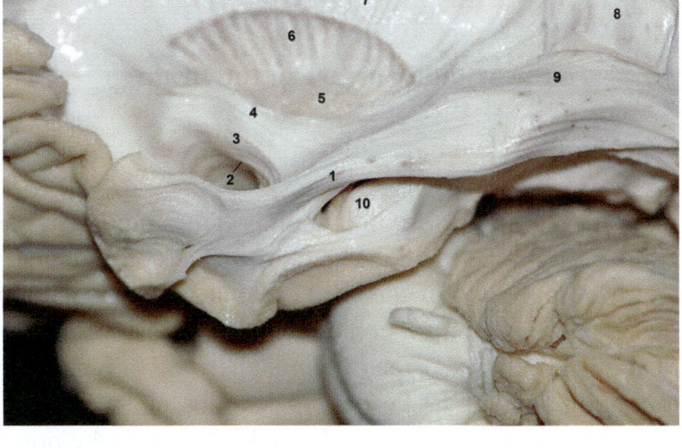

Despite numerous DTI studies analyzing the course of the direct segment of the inferior longitudinal fasciculus, controversy exists about the anterior cortical termination of this fascicle. There have been described connections with the anterior portion of the superior, middle, and inferior temporal gyri on the lateral surface of the temporal lobe, the fusiform gyrus, parahippocampal gyrus, the amygdala, and the hippocampus [1, 10, 14, 27]. At the anterior portion of the temporal horn of the lateral ventricle, these fibers gather in a single bundle running laterally and inferiorly to the lateral wall of the temporal horn. At this level, the inferior longitudinal fasciculus is located lateral and below the optic radiations, whereas the fronto-occipital fasciculus runs medial and above the optic radiations. Thus, the roof of the ventricle is a good anatomical landmark to distinguish the inferior longitudinal fasciculus (below) from the fronto-occipital fasciculus (above) [10, 28]. The atrium of the lateral ventricle is lateral to the sagittal stratum and medial to the arcuate fasciculus and the temporoparietal segment of the superior longitudinal fasciculus. Posteriorly, terminates on the convexity surface of the occipital pole, posterior lingual gyrus, posterior fusiform gyrus, and the cuneus [1, 10].

Subcallosal Fasciculus

Onufrowicz first described the subcallosal fasciculus in human brains in 1887. This bundle originates at the fronto-mesial precentral structures (the supplementary motor area and the cingulum). Then, it runs in a vertical direction passing through the white matter surrounding the lateral angle of the frontal horn of the ventricle. It terminates at the head of the caudate nucleus [4, 10, 29, 30]. On occasions, it has been confused with the fronto-occipital superior fasciculus because Muratoff described it in 1893. Since both have common connections after subsequent publications, it was evidenced that these fibers were independent. In the anterior frontal cuts, the system can be separated into three components: a superior horizontal part is placed under the corpus callosum; a lateral part that descends and encloses the angle of the corona radiata and the corpus callosum; and an inferior part adjacent to the basal ganglia [10].

3.1.2 In the Middle Surface of the Brain

Cingulum (Rodete Fasciculus)

The cingulum is also nominated as only cíngulo; it has a semi-angular shape and courses along the medial aspect of the cerebral hemisphere, following the curve of and forming much of the white matter within the cingulate gyrus and the parahippocampal gyrus, and it constitutes the largest intra-hemispheric association system of the mid-portion of each hemisphere of the brain (Fig. 17). The cingulum is a longitudinal compact fasciculus running above and parallel to the corpus callosum (Fig. 18), connecting the prefrontal lobes with the posterior cortices and the hippocampus. The cingulum receives fibers from the anterior thalamic nucleus, superior

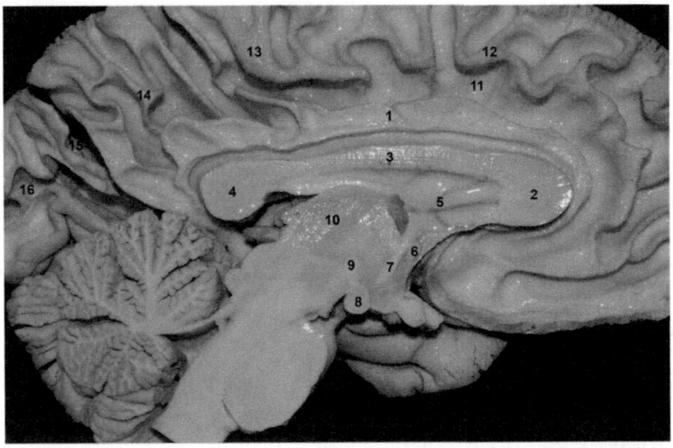

Fig. 17 Medial view of the left hemisphere, all the cerebral cortex was removed. The cingulum is exposed; it extends inside the cingulate gyrus's white matter and the parahippocampal gyrus. (*1*) Cingulum; (*2*) genu of the corpus callosum; (*3*) body of the corpus callosum; (*4*) splenium of the corpus callosum; (*5*) septum pellucidum; (*6*) anterior commissure; (*7*) column of the fornix; (*8*) mamillary body; (*9*) mamillothalamic tract; (*10*) thalamus; (*11*) cingulate sulcus; (*12*) superior frontal gyrus; (*13*) paracentral lobule; (*14*) precuneus; (*15*) parieto-occipital sulcus; (*16*) calcarine sulcus; (*17*) corpus callosum

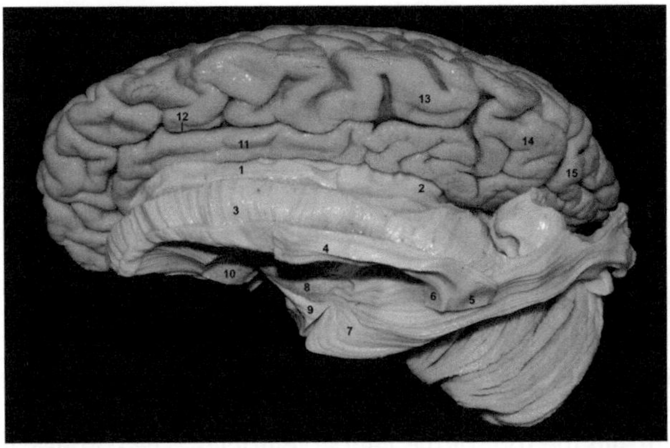

Fig. 18 Superior view of the brain. The left hemisphere was partially removed to expose the cingulum and its relationship with the corpus callosum. (*1*) Cingulum; (*2*) splenium of the corpus callosum; (*3*) corona radiata; (*4*) fronto-parietal or horizontal segment of the superior longitudinal fasciculus; (*5*) temporo-parietal or vertical segment of the superior longitudinal fasciculus; (*6*) temporo-frontal segment or arcuate fasciculus; (*7*) optic radiation; (*8*) globus pallidus; (*9*) anterior commissure; (*10*) nucleus accumbens; (*11*) cingulate gyrus; (*12*) cingulate sulcus; (*13*) paracentral lobule; (*14*) precuneus; and (*15*) cuneus

frontal gyrus, paracentral lobule, and precuneus. Fibers coming from the precuneus greatly contribute to the enlargement of the cingulum [6–9].

Rostrally, the cingulum curves anteriorly in front of the genu of the corpus callosum and ends in the subcallosal gyrus, also known as the paraolfactory area of Broca and paraterminal gyrus. Caudally, the cingulum crosses the back of the fibers of the forceps major, covers the inferior lip of the anterior portion of the calcarine sulcus, and continues toward the anterior parahippocampal region, ending in the presubiculum and entorhinal cortex [10], at the moment that it surrounds the splenium of the corpus callosum, the cingulum receives a fasciculus of reinforcement from the precuneus and lingula. The union of these three contingents of fibers, istmo of the cingulum, precuneus, and the língula, forms the parahippocampal gyrus.

3.2 *Commissural Fibers*

They are interhemispheric association fibers; they are transversely disposed of, crossing the middle line and finishing in the homologous regions of the cortex of both brain hemispheres. The commissural fibers of the brain form the corpus callosum, the anterior commissure, the commissure of the fornix, the posterior commissure, and the habenular commissure, being the last two already constitute diencephalic structures [6, 11].

3.2.1 The Corpus Callosum

The corpus callosum represents the largest commissural of the encephalon, composed of nearly 200 million fibers, that form a sheet of a white substance, in a quadrilateral shape, transversely from one hemisphere to another, it is a wide commissure or better yet, a vast system of association that gathers between both halves of the brain in non-symmetrical points of the cortex [6]. With the exceptions of the anterior portions of the temporal lobes that are together by the anterior commissure, the primary visual areas, and the sensitive somatic area parts that do not connect inter hemispherical. Its fibers continue through sectors 0, 1, 2, 3, and 4 of the gyral section, cross the middle line, and go through sectors 4, 3, 2, 1, and 0 of the gyral section of the opposite side (Fig. 1) [31].

The corpus callosum is located between the hemispheres on the floor of the longitudinal fissure and the roof of the lateral ventricles. The corpus callosum, which forms the largest part of the ventricular walls, contributes to the wall of each of the five parts of the lateral ventricle. Its anterior half is situated in the midline deep to the upper part of the inferior frontal gyrus. The corpus callosum posterior part, also known as the splenium, is situated deep to the supramarginal gyrus and the lower third of the pre- and postcentral gyrus. The corpus callosum has five parts: two anterior parts, the genu and rostrum; a central part, the body; and two posterior parts, the splenium and tapetum (Fig. 19) [7].

The curved anterior part, the genu, wraps around and forms the anterior wall and adjacent part of the roof of the frontal horn. The genu blends below into the rostrum, a thin tapered portion that forms the floor of the frontal horn and is continuous downward, in front of the anterior commissure, with the lamina terminalis. The genu gives rise to a large fiber tract, the forceps minor, which forms the anterior wall

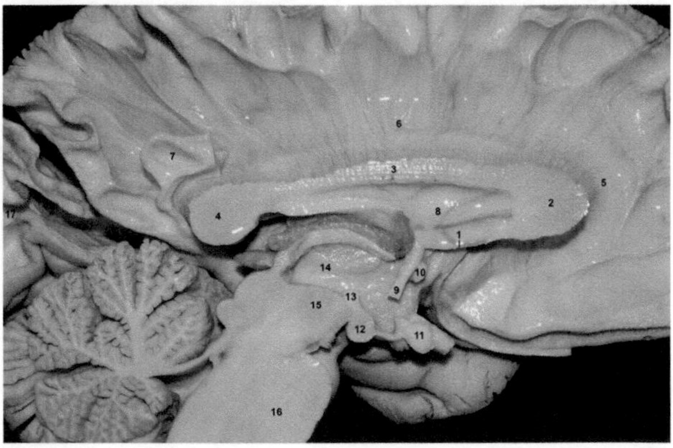

Fig. 19 In medial view of the left hemisphere, the five segments of the corpus callosum are exposed: two anterior segments (genu and rostrum), a central segment (body), and two posterior segments (splenium and tapetum). (*1*) Rostrum of the corpus callosum; (*2*) genu of the corpus callosum; (*3*) body of the corpus callosum; (*4*) splenium of the corpus callosum; (*5*) forceps minor; (*6*) corpus callosum; (*7*) cingulum; (*8*) septum pellucidum; (*9*) column of the fornix; (*10*) anterior commissure; (*11*) optic chiasm; (*12*) mamillary body; (*13*) mamillothalamic tract; (*14*) thalamus; (*15*) red nucleus; (*16*) pons; and (*17*) calcarine sulcus

of the frontal horn and interconnects the frontal lobes. The forceps minor sweeps obliquely forward and laterally, as does the anterior wall of the frontal horn. The genu blends posteriorly into the midportion, the body, located above the body of the lateral ventricle.

The central segment (body) is located above the lateral ventricle body, and it possesses two surfaces (Fig. 20): superior, convex from front to back, flat or slightly concave in the transversal sense, it occasionally presents, on the middle line, a longitudinal sulcus, a vestige of the raphe of the corpus callosum. On each side of this sulcus, it is possible to observe two small longitudinal cords, denominated medial longitudinal stria white tracts or Lancisi nerves [6], of white color and variable dimensions, usually 1 mm wide. Laterally of these tracts, the lateral longitudinal stria is located gray tracts or taeniae tectae [6], the cingulate gyrus covers them, and they have a grayish color, throughout the medial border, it is linked to the medial tract by a thin veil of gray matter, where it is given the name of indicium griseum. Its inferior surface is convex in the transversal sense. It is fasciculate in a transverse direction; in the middle line, it inserts the septum pellucidum; from behind, it goes into contact with the fornix and hippocampal (psalterium or the lira of David) [6]. Laterally, it forms the lateral ventricle roof (Fig. 21) [6, 7].

The posterior segment (splenium or the rodete of Reil) [6], the thick, rounded posterior end, is situated dorsal to the pineal body and the upper part of the medial wall of the atrium, forming a superior lip of the brain slit of Bichat. It is 6 to 7 cm from the occipital pole. The splenium gives rise to a large tract, the forceps major, which forms a prominence called the bulb in the upper part of the medial wall of the atrium and occipital horn as it sweeps posteriorly to interconnect the occipital lobes.

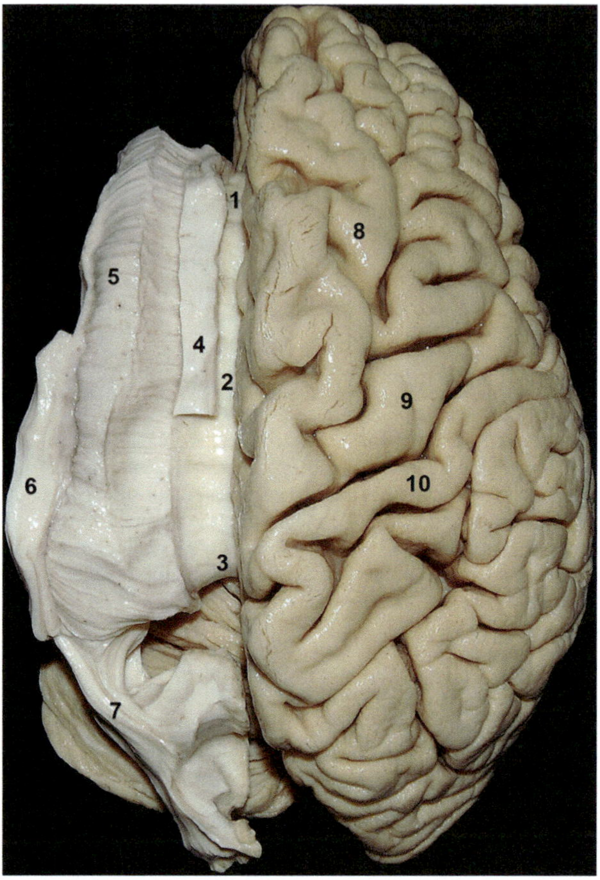

Fig. 20 Superior view of the brain, exposing the superior surface of the corpus callosum above the lateral ventricle body. Its superior surface is convex from front to the back, flat or slightly concave in the transversal sense. On occasions, it presents a longitudinal sulcus in the medial line, the vestige of the raphe of the corpus callosum. On each side of this groove, two small longitudinal cords, denominated Medial longitudinal stria (white tract or nerves of Lancisi), are visualized. Laterally, these tracts are found in the lateral longitudinal stria (gray tracts or taeniae tectae), covered by the cingulate gyrus, and have a grayish color. (*1*) Genu of the corpus callosum; (*2*) body of the corpus callosum; (*3*) splenium of the corpus callosum; (*4*) cingulum; (*5*) corona radiata; (*6*) superior longitudinal fasciculus; (*7*) optic radiation; (*8*) superior frontal gyrus; (*9*) precentral gyrus; and (*10*) postcentral gyrus

Another fiber tract, the tapetum, which arises in the posterior part of the body and splenium, sweeps laterally and inferiorly to form the roof and lateral wall of the atrium and the temporal and occipital horns. When bending posteromedially, the splenium fibers form a prominence in the upper part of the medial atrium wall and the occipital horn called the bulb of the corpus callosum [7, 11]. The other tract is from fibers denominated tapetum. It originates from the posterior portion of the body and the splenium. It projects laterally and inferiorly, forming the roof and the

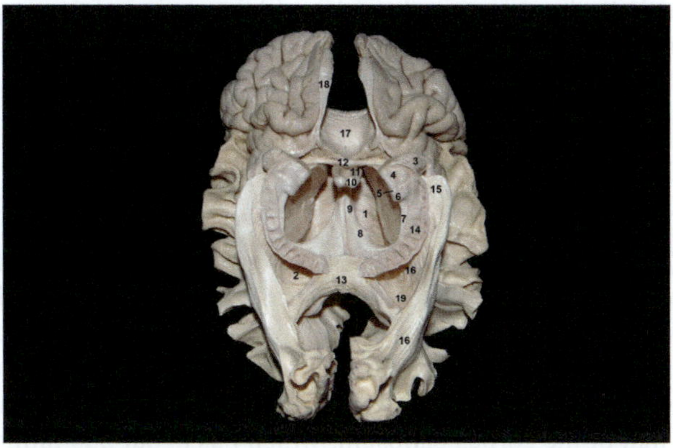

Fig. 21 Basal view of the brain, the brainstem, and the thalamus was removed in addition to having been removed the parahippocampal gyrus. The ependyma was removed, exposing the inferior surface of the corpus callosum convex in the transversal sense. It is fasciculate in the transversal direction; in the middle line, it inserts the septum pellucidum; from behind, it goes into intimate contact with the fornix and hippocampal commissure. Laterally, it forms the roof of the lateral ventricle. (*1*) Corpus callosum; (*2*) atrium of the lateral ventricle; (*3*) amygdala; (*4*) uncinate gyrus; (*5*) limbus of giacomini; (*6*) intralimbic gyrus; (*7*) fimbria; (*8*) crus of the fornix; (*9*) corpus of the fornix; (*10*) mamillary body; (*11*) column of the fornix; (*12*) anterior commissure; (*13*) splenium of the corpus callosum; (*14*) dentate gyrus; (*15*) meyer's loop; (*16*) optic radiation; (*17*) rostrum of the corpus callosum; (*18*) olfactory tract; and (*19*) tapetum

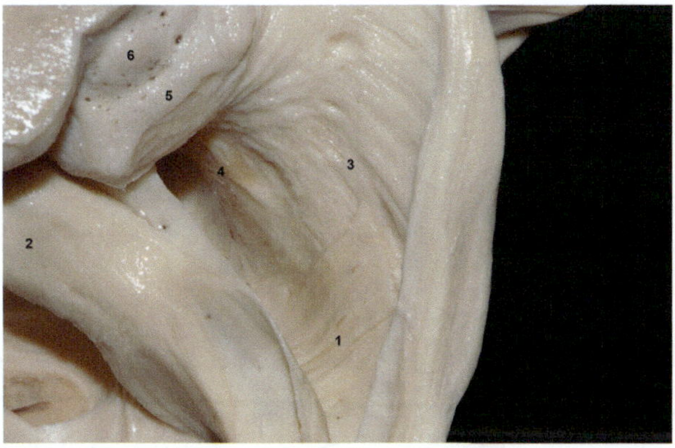

Fig. 22 Basal view of the brain. Magnified view of the atrium of the lateral ventricle, exposing the relationship of the tapetum with optical radiation. (*1*) Tapetum; (*2*) splenium of the corpus callosum; (*3*) optic radiation; (*4*) the tail of the caudate nucleus; (*5*) pulvinar of the thalamus; and (*6*) medial geniculate body

lateral wall of the atrium, temporal, and the occipital horn. The tapetum separates the fibers of the optic radiations from the temporal horn and the atrium. The cingulate gyrus surrounds and is separated from the corpus callosum by the callosal sulcus (Fig. 22) [6–8, 31, 32].

Morphologically, the corpus callosum must be comprehended by a group of transversal fibers, that while they cross the medial line, opens in a fan way to reach the different points of the brain convexity. The body fibers can also be comprehended through the assumed conformation of a butterfly moving its wings, where the body of the butterfly corresponds to the middle portion of the corpus callosum, and the movement of each wing is the distribution of the fibers in each of the brain hemisphere [11].

3.2.2 The Anterior Commissure

The anterior commissure is constituted by a group of transversal fibers, which mainly connect the temporal poles. The anterior commissure is made by an anterior and medial part and by its lateral parts or posterior extensions that go through basally in each one of the brain's hemispheres (Fig. 23). It has an average diameter of 4 mm [33], and it harbors about 3.5 million fibers, but with an overall axon density 2.4 times that of the corpus callosum, which has about 200 million fibers. The optic nerve, which has a similar diameter, harbors 1.2 million fibers [34].

Its anterior part crosses the inferior medial line of the rostrum of the corpus callosum, in the front of the fornix columns, and it has the superior portion of the terminal sheet stuck more to the medial segment, that of the same time, it forms a prominent indentation inside the third ventricle, after under the interventricular foramen. A small beam of fibers is curved to the front of the anterior perforated substance and the olfactory trigon in this segment. This fascicle originated medially in the fiber portion of the internal capsule's anterior arm that goes in front of the anterior commissure. It connects the olfactory system, which creates the real

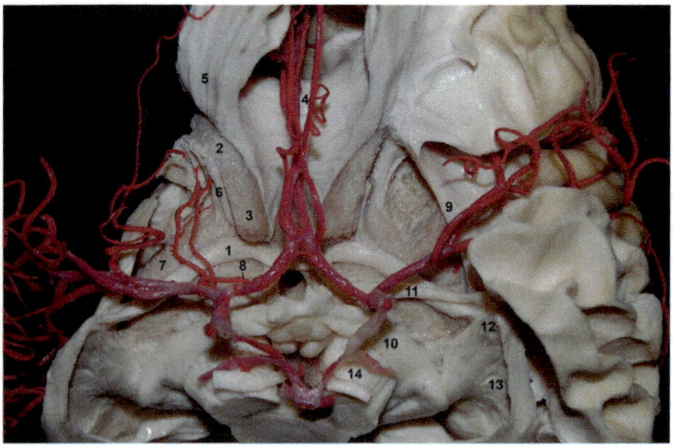

Fig. 23 Anteroinferior view of the brain injected with resin. The anterior commissure is identified, formed by an anterior and medial part, and another lateral part or posterior extension that goes through basally on each brain's hemispheres. It has the shape of a bicycle handlebar. (*1*) Anterior commissure; (*2*) head of the caudate nucleus; (*3*) nucleus accumbens; (*4*) genu of the corpus callosum; (*5*) forceps minor; (*6*) internal capsule, anterior limb; (*7*) globus pallidus; (*8*) heubner's artery; (*9*) fronto-occipital fasciculus; (*10*) optic tract; (*11*) diagonal band; (*12*) meyer's loop; (*13*) optic radiation; and (*14*) crus cerebri

olfactory commissure [6, 11]. After this, it continues its lateral journey and a little posterior to cross the inferior anterior surface of the globus pallidus, where it stays (Channel of Gratiolet) [6] (Fig. 24). Laterally, it is disposed superiorly to the amygdala, perpendicular to the anterior bundle of the optical radiation, and posteromedial to the uncinate fasciculus. This structure resembles a bicycle handlebar.

The lateral extension of the anterior commissure, at the limen insulae level, starts to gain diameter and to turn flat until it enters the white matter of the temporal lobe where it opens in a radiated shape, its anterior fibers that correspond to the major contingent of fibers of the anterior commissure. These fibers direct posteriorly to join and mix with the fronto-occipital fasciculus and the optical radiation forming the stratum sagittal. Its posterior fibers join the uncinate fasciculus, and it has an anterior direction to the temporal pole (Fig. 24) [6–9, 11, 21, 35, 36].

It is important to remember that the anterior commissure constitutes, on each side, the posterior limit of the region called the ventral striatum. That shelters at the time the acumens nucleus [11].

In humans, the anterior commissure seems to be related to the interhemispheric transfer of olfactory, gustative, visual, and auditory information [33]. The anterior commissure–posterior commissure line (AC–PC line) is a very important landmark in stereotactic atlases for the proper localization of neuroanatomical targets utilized in stereotactic neurosurgical procedures [11, 33].

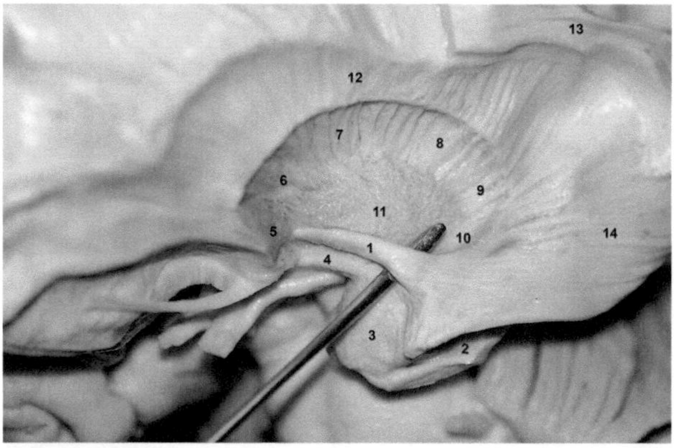

Fig. 24 Lateral view of the brain. Sequential dissection of the superior lateral surface of the left hemisphere. The lateral extension of the anterior commissure, at the level of the limen insulae it starts to increase in diameter, and it turns, until it enters the white matter of the temporal lobe where it opens in a radiated shape, its anterior fibers that correspond to the major contingent of fibers of the anterior commissure, it directs posteriorly, to join and mix with the fronto-occipital fasciculus and the optical radiation forming the sagittal stratum. Its posterior fibers join the uncinate fasciculus, and it has an anterior direction to the temporal pole. (*1*) Anterior commissure; (*2*) head of the hippocampus; (*3*) amygdala; (*4*) ventral amygdalofugal fibers; (*5*) nucleus accumbens; (*6*) internal capsule, anterior limb; (*7*) internal capsule, genu; (*8*) internal capsule, posterior limb; (*9*) internal capsule, retrolenticular part; (*10*) internal capsule, sublenticular part; (*11*) globus pallidus; (*12*) corona radiata; (*13*) superior longitudinal fasciculus; and (*14*) stratum sagittal

3.2.3 Fornix (Hippocampal Commissure)

The fornix is also known as the brain trigon of the four-pillared vault of Winslow [6]. It is a white substance sheet located in the middle line under the corpus callosum. The fornix is the main efferent system of the hippocampus, starting as the posterior extension of the fimbria. It is divided into three portions (Fig. 25): crura, body, and columns. The crura of the fornix runs in a rostromedial direction just below the splenium of the corpus callosum, forming the anterior wall of the lateral ventricle. At the level of the body, both fornices run together in the midline along the superomedial border of the thalami, in the medial wall of the body of the lateral ventricle. Fornices fuse below the body of the corpus callosum in the commissure of the fornix [6, 7, 10].

The fornix has a format of a C structure that wraps up the thalamus; it extends from the hippocampus and the mamillary (Fig. 25). It originates from the floor of the temporal horn of the ventricle surface, next to the ventricle surface of the hippocampus, starting from the fibers that are joined throughout the medial edge of the hippocampus. These nerve fibers are the alveo, a thin sheet of white matter that covers the ventricle surface of the hippocampus; and then converges to form the fimbria. The fimbria is separated from the dentate gyrus of the fimbriodentate sulcus and accompanies the side lateral geniculate body, separated from this and the optical radiations and auditory by the

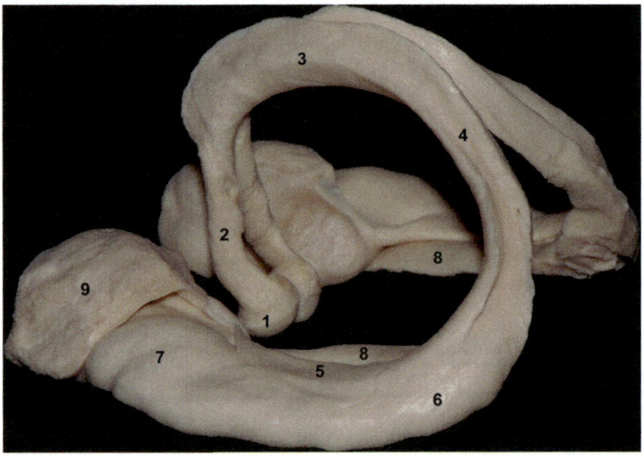

Fig. 25 The lateral view of the fornix, a C format structure that wraps up the thalamus, extends from the hippocampus to the mamillary body and has three parts: crura, body, and columns. (*1*) Mamillary body; (*2*) column of the fornix; (*3*) body of the fornix; (*4*) crus of the fornix; (*5*) fimbria; (*6*) body of the hippocampus; (*7*) head of the hippocampus; (*8*) subiculum; (*9*) amygdala; and (*10*) dentate gyrus

choroidal fissure. The fimbriae of both sides increase in thickness. When it reaches the extreme posterior part of the hippocampus, they arched forward to form the crura of the fornix, circling the posterior surface of the pulvinar of the thalamus, in the medial part of the atrium, right away an orientation arc is formed in the superior medial area, which is projected for the inferior surface of the splenium of the corpus callosum. Under the splenium, a fine fiber sheet denominated hippocampus commissure (also fibers of the lira of David or psalterium) [6] connect the medial edges to the crura of the fornix and functionally connects both hippocampus formations. At the junction of the atrium with the body of the lateral ventricle, both cruras meet, forming the body of the fornix that goes above the thalamus and under the septum pellucidum in the inferior portion of the medial wall of the lateral ventricle body. In the anterior margin of the thalamus, the body of the fornix is divided into two columns that form an arc throughout the superior anterior margin of the foramen of Monroe. Right away, they melt to the lateral walls of the third ventricle as they pass back of the anterior commissure, projecting down until reaching the outer side and posterior of the mamillary bodies [6–9]. From the internal nucleus of the mamillary body, it gives off a common trunk of fibers, constituted by the fascicle of Vicq-d Ázyr and the fascicle of Gudden [6]. From these two fascicles, the first one reaches the thalamus and the second one the mesencephalic tegmentum after passing above the red nucleus. Not all the fibers of the fornix are directed to the mamillary body. When the fornix column deviates down to address the mamillary body, a special fascicle is revealed from its previous face, denominated Precommissural fornix or the olfactory fasciculus of the Ammon's horn [6]. This fasciculus follows the posterior inferior border of the septum pellucidum, and it passes in front of the anterior commissure. It reaches the base of the brain between the rostrum of the corpus callosum and the anterior perforated substance [6].

The body and crus are located deep to the lower part of the pre- and postcentral gyri, and the fimbria is located deep to the lower part of the superior temporal gyrus. All of its parts are located deep to the posterior part of the insula. In the body of the lateral ventricle, the body of the fornix is in the lower part of the medial wall; in the atrium, the crus of the fornix is in the medial part of the anterior wall; and in the temporal horn, the fimbria of the fornix is in the medial part of the floor [7].

The inner edge of the fornix forms the outer border of the choroidal fissure, the cleft between the thalamus and the fornix, along which the choroid plexus in the lateral ventricle attaches. The choroidal fissure is a C-shaped arc extending from Monroe's foramen through the body, atrium, and temporal horn of the lateral ventricle (Fig. 26) [7, 37–39].

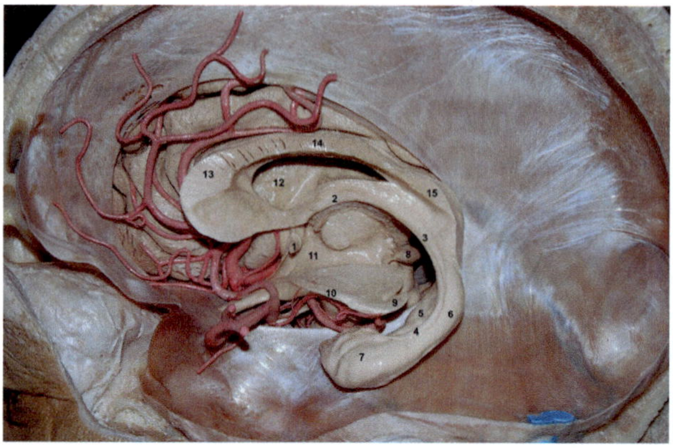

Fig. 26 Lateral view of the head. The relationship of the fornix with the choroidal fissure and ventricular cavities is exposed. (*1*) Anterior commissure; (*2*) body of the fornix; (*3*) crus of the fornix; (*4*) fimbria; (*5*) subiculum; (*6*) body of the hippocampus; (*7*) head of the hippocampus; (*8*) pineal body; (*9*) lateral geniculate body; (*10*) optic tract; (*11*) third ventricle; (*12*) frontal horn of the lateral ventricle; (*13*) genu of the corpus callosum; (*14*) body of the corpus callosum; and (*15*) splenium of the corpus callosum

3.2.4 Posterior Commissure

The posterior commissure lies below the pineal recess of the third ventricle, crossing the midline along the caudal lamina of the pineal stalk, and corresponds to the upper aspect of the aqueduct opening within the third ventricle.

It is a complex bundle that contains both decussating and commissural fibers that connect diencephalic and mesencephalic nuclei (interstitial and dorsal nuclei of the posterior commissure located within the periventricular gray matter, nucleus of Darkschewitsch of the periaqueductal gray matter, interstitial nucleus of Cajal located at the rostral end of the oculomotor nucleus and closely linked with the medial longitudinal fasciculus, and posterior thalamic, pretectal, tectal, and habenular nuclei), mostly still anatomically and functionally incompletely understood [11, 34].

3.2.5 Commissure Habenular

The habenular commissure lies between both habenulas, which are small protuberances of both thalami located at the distal ends of both striae medullaris that course across the superior part of the medial surface of both thalami, within the posterior aspect of the lateral wall of the third ventricle.

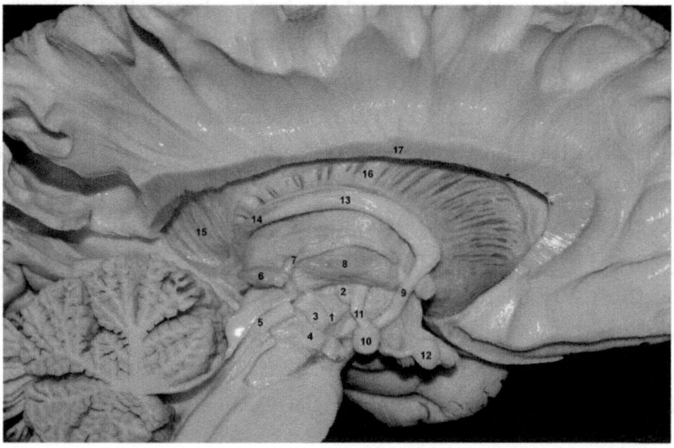

Fig. 27 Medial view of the thalamus and the midbrain. The Meynert's retroreflexus fasciculus is evidenced, which originates in the habenular nuclei. It descends throughout the inner face of the thalamus, medially passes to the red nucleus and ends in a small cellular accumulation situated in the posterior perforated substance, the interpeduncular ganglion. (*1*) Meynert's retroreflexus fasciculus; (*2*) subthalamic nucleus; (*3*) red nucleus; (*4*) oculomotor nerve; (*5*) cerebral aqueduct; (*6*) pineal body; (*7*) taenia thalami; (*8*) thalamus; (*9*) column of the fornix; (*10*) mamillary body; (*11*) mamillothalamic tract; (*12*) optic chiasm; (*13*) body of the fornix; (*14*) crus of the fornix; (*15*) tapetum; (*16*) thalamic radiation; and (*17*) corpus callosum

Since the pineal gland is superiorly attached to the habenular commissure and inferiorly attached to the posterior commissure, the pineal recess of the third ventricle is located between their two commissures [11].

As with the posterior commissure, the habenular commissure contains both decussating fibers (tectohabenular) and commissural fibers (predominantly connecting the habenular nuclei) [34]. The habenular nuclei receive many afferents from olfactory inputs from the septal nuclei, the amygdala nuclei, and the hippocampus, through the medullary stria of the thalamus [6–8], from the habenular nuclei come to the fiber that constitutes the Meynert's retroreflexus fasciculus (Fig. 27) [6]. This fasciculus descends throughout the inner face of the thalamus, it passes medially to the red nucleus, and it ends in a small cellular accumulation situated in the posterior perforated substance of the interpeduncular nucleus of the mesencephalon and to the rostral salivatory nucleus on the floor of the fourth ventricle to activate reflex salivation [6, 11].

3.2.6 Septum Pellucidum

The septum pellucidum stretches across the interval between the anterior parts of the corpus callosum and the body of the fornix. It is composed of paired laminae and separates the frontal horns and bodies of the lateral ventricles in the midline. In the frontal horn, the septum pellucidum is attached to the rostrum of the corpus callosum below, the genu anteriorly, and the body above. In the body of the lateral ventricle, the septum is attached to the body of the corpus callosum above and the body of the fornix below. The septum pellucidum disappears posteriorly where the body of the fornix meets the splenium. There may be a cavity, the cavum septum pellucidum, in the midline between the laminae of the septum pellucidum [6, 7].

The three angles are: anterior, posterior, and inferior. The anterior angle corresponds to the genu of the corpus callosum and its blunt. The posterior angle becomes pointy between the body of the corpus callosum and the body of the fornix, and it goes backward with the name of the septum tail, and it ends later in the point where the body of the fornix joins the splenium of the corpus callosum. The inferior angle corresponds to the superior edge of the anterior commissure [6–8].

3.3 *Projection Fibers*

The projection fibers comprise the corticofugal and corticopetal fibers that connect the cerebral cortex with the corpus striatum, the thalamus, cerebellum, brainstem, and spinal cord. Underneath the cerebral cortex, these fibers converge to form the corona radiata, which intersects the commissural fibers and is continuous with the internal capsule.

Topographically, it represents a vast cone whose base corresponds to the cortex and the vertice at the lower end of the spinal cord. These fibers, since their origin they crisscross with the rest of the fibers of association and the commissure to form the centrum semiovale of Vieussens [6]. Above the level of the thalamus, these projection fibers are arranged in a radiating pattern called the corona radiata or great (sun of Reil) [6] (Figs. 28, 29, and 30). The corona radiata is continuous caudally with the more compact internal capsule, which fibers constitute the cerebral peduncle. The internal capsule is a thick mass of white matter bounded laterally by the lentiform nucleus and medially by the caudate nucleus and the thalamus.

According to this definition, projection fibers are [11]:

1. Cortico-striatal fibers, system quota Cortico-striatal-pallido-thalamus-cortical.
2. Thalamic peduncle is constituted by the corticothalamic fibers systems and for their reciprocal systems of the thalamocortical fibers systems.
3. Corticopontine fibers that originated in different cortical areas and are directed to the pontine nuclei, from where they are projected to the cerebellum.
4. Corticobulbar fibers and corticospinal fibers, which is a set are denominated pyramidal tract.
5. Corticoreticular fibers originate in the motor and somatosensitive cortex, and they project to nuclei of the brainstem's reticular formation.
6. The set of longitudinal fibers that constitute the fornix and connect mainly each hippocampus to the ipsilateral mammillary body [11].

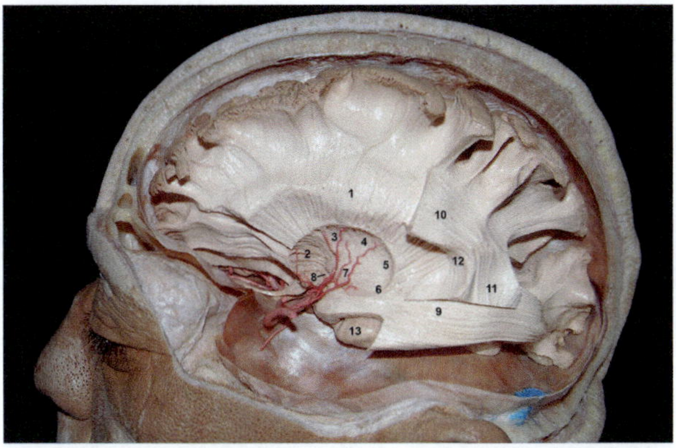

Fig. 28 Lateral view of the left hemisphere. The cortex of the brain was removed and partially the superior longitudinal fasciculus. The projection fibers are exposed. They are a set of afferent and efferent fibers of the brain's cortex that connect it to the base of the nucleus, the thalamus, and the spinal cord. Topographically, it represents a vast cone whose base corresponds to the cortex and the vertice at the lower end of the spinal cord. These fibers, since their origin they crisscross with the rest of the fibers of association and the commissure to form the centrum semiovale. Above the upper edge of the putamen, the projection fibers are arranged in a radial form, thus receiving the denomination of corona radiata. (*1*) Corona radiata; (*2*) internal capsule, anterior limb; (*3*) internal capsule, genu; (*4*) internal capsule, posterior limb; (*5*) internal capsule, retrolenticular part; (*6*) internal capsule, sublenticular part; (*7*) globus pallidus; (*8*) anterior commissure; (*9*) sagittal stratum; (*10*) fronto-parietal or horizontal segment of the superior longitudinal fasciculus; (*11*) temporo-parietal or vertical segment of the superior longitudinal fasciculus; (*12*) temporo-frontal segment or arcuate fasciculus; and (*13*) head of the hippocampus

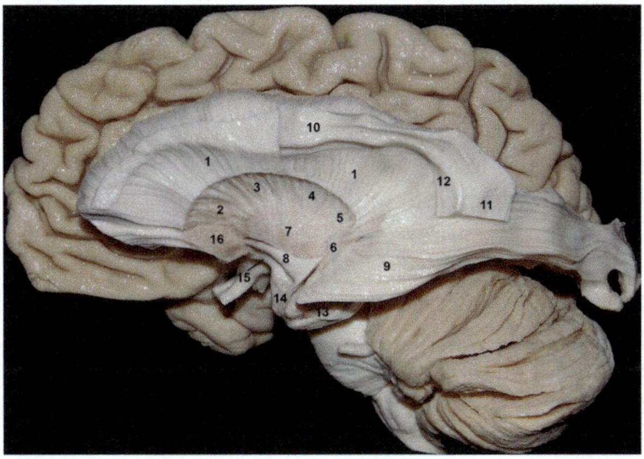

Fig. 29 Lateral view of the left hemisphere, the anatomical specimen is dissected with the Kingler's Technique. The horizontally and the vertical portion of the superior longitudinal fascicle was removed partially to identify the projection fibers that make up the internal capsule, in addition to having removed the putamen. Topographically the fibers and the internal capsule are divided into five parts. (*1*) Corona radiata; (*2*) internal capsule, anterior limb; (*3*) internal capsule, genu; (*4*) internal capsule, posterior limb; (*5*) internal capsule, retrolenticular part; (*6*) internal capsule, sublenticular part; (*7*) globus pallidus; (*8*) anterior commissure; (*9*) optic radiation; (*10*) fronto-parietal or horizontal segment of the superior longitudinal fasciculus; (*11*) temporo-parietal or vertical segment of the superior longitudinal fasciculus; (*12*) temporo-frontal segment or arcuate fasciculus; (*13*) head of the hippocampus; (*14*) amygdala; (*15*) diagonal band; and (*16*) nucleus accumbens

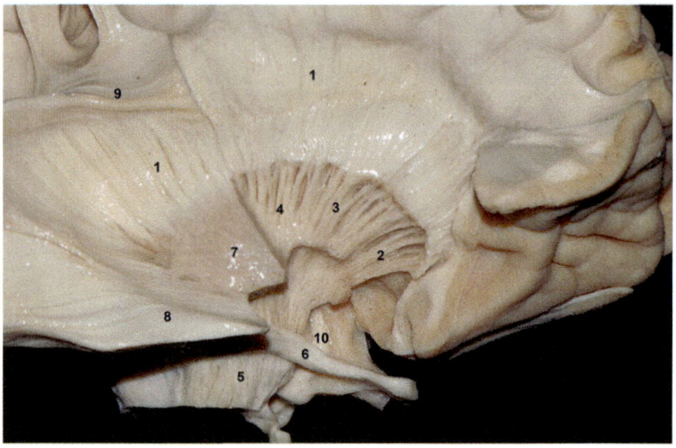

Fig. 30 Lateral view of the right hemisphere, the anatomical specimen is dissected with the Kingler's Technique. The horizontal portion of the superior longitudinal fascicle was partially removed to identify the projection fibers that make up the corona radiata to remove the putamen. (*1*) Corona radiata; (*2*) internal capsule, anterior limb; (*3*) internal capsule, genu; (*4*) internal capsule, posterior limb; (*5*) crus cerebri; (*6*) optic tract; (*7*) putamen; (*8*) optic radiation; (*9*) superior longitudinal fasciculus; and (*10*) column of the fornix

3.3.1 Internal Capsule

The internal capsule of Reil [6] consists of a thicker white substance, limited laterally by the lentiform nucleus and medially by the caudate nucleus and the thalamus. The projection fibers are arranged to progressively converge from the cerebral cortex constituting in conjunction with the various association fibers, the semioval center or Corona radiata; when the upper limit of the base of the nuclei is reached (lentiform nucleus and caudate nucleus), they become part of the internal capsule, in the inferior part to the lentiform nucleus, these form the base of the cerebral peduncle [7, 11]. From the morphological point of view, these fibers constitute the internal capsule of the lentiform nucleus, in opposition to its external capsule, arranged laterally to the nucleus.

The internal capsule shelters only project fibers and, given the conceptualization that defines it, place them between the upper and lower limits of the lentiform nucleus. These corticostriatal fibers connect the cerebral cortex with the putamen, and the caudate nucleus does not enter its constitution by reaching those nuclei along its upper edge.

Because it is constituted by fibers arranged between the cerebral cortex superiorly and the cerebral peduncle and the thalamus at the inferior part, the internal capsule has the shape of a fan that opens from the bottom to the top [11]. Foville compared it to a stalk whose fruits were represented by the nuclei of the base; layering both of its visible portions in an axial cut of the hemisphere made it comparable, according to Gratiolet, to an open-cut, turbinate's that includes the lentiform nucleus [6].

In resemblance to what happened along with the phylogeny's embryological development of the thalamus and the neocortical portion of the brain hemisphere, what morphologically implies the division of the original stratum so that the nucleus is maintained medially and the putamen laterally. As a result of this separation, the cells remain among their fibers due to the internal capsule interposition. Once the division of the stratum occurs to a lesser extent in its previous portion, the anterior part of the internal capsule, it usually has fewer portions; it is thinner and fenestrated by cellular striate bridges that communicate the caudate nucleus with the putamen. Given the final fan-shaped conformation of the internal capsule, the most anterior and basal portion of the stratum does not become divided, corresponding to the striate of the nucleus accumbens (Figs. 28, 29, and 30) [11, 40].

Given the conformation of the head of the caudate nucleus and the thalamus, and the previous provision, of the first related to the second, the internal capsule that is arranged laterally, those structures present the form of a "V" in the axial cuts, with its vertex disposed medially and between the two structures and aiming for Monroe's interventricular foramen.

Due to the topography of internal capsule fibers, it is divided into five parts [7, 11, 31]: (1) The Anterior limb is located between the head of the caudate nucleus, medially and the lentiform nucleus, in the lateral part. (2) The genu is in the vertex region between the head of the caudate nucleus, putamen, and thalamus, therefore lateral and adjacent to the interventricular foramen. (3) The posterior limb is located between the thalamus medially and the lentiform nucleus laterally. (4) The retro lenticular part, which is subsequently disposed to the lentiform nucleus. (5) The sublenticular part, which is disposed inferiorly to the lentiform nucleus (Fig. 29).

1. The anterior limb, also called lenticulocaudate segment, is mainly formed by horizontal fibers that pass between the caudate nucleus and the lentiform nucleus and converge towards the anterior end of the thalamus. These fibers are dissociated by the internuclear gray substance, where the bridges stretch between the head of the caudate nucleus and the putamen. It is approximately 2 cm shorter than the posterior segment, and it is interrupted down and forward by the gray substance bridge that joins the putamen to the caudate nucleus (nucleus accumbens) [6]. It is directed obliquely from the back to the front and from inside to out. It contains frontopontine fibers, which originated in the frontal cortex and synapse with the pontine nuclei, whose axons are directed to the opposite cerebellum hemisphere by the middle of the cerebral peduncle. It is also conformed by the thalamus to the anterior radiation, which connects the anterior and medial nucleus of the thalamus with the cortex of the frontal lobe [6, 7, 31].
2. The genu is only visible in the axial cuts; its appearance and constitution vary according to the resection level. It contains corticonuclear fibers, which originate mainly from the motor cortex (area 4 of Brodmann), and they target the motor of the nuclei of the cranial nerves. By occupying the knee of the internal capsule, it is also called the geniculate fasciculus. The most anterior fibers of the superior thalamic radiation also extend into the genu (Fig. 29) [6, 7, 9, 31].
3. The posterior limb, also called the optical lenticular segment [6], measures from 3 to 4 cm. Its fibers are generally oblique downwards and inwards [6]. It converges towards the base of the cerebral peduncle: the most anterior fibers are oblique downward, inward, and in the back; the middle ones go directly down

and forward and contain frontopontine fibers, as well as corticospinal fibers for the motor nuclei of the brainstem and the superior and inferior extremities. The fibers directed for the superior limbs are located closer to the knee, in contrast to the fibers directed to the inferior limbs. The precentral gyrus is located superficially to the posterior limb of the internal capsule [7]; It is also composed of corticorubral fibers, pallidothalamic fibers, and superior thalamic radiation (for the premotor, motor, and somatic sensory cortex) (Fig. 29) [6, 7, 9, 31].

Considering its situation in the thalamic region and the subthalamic region: (a) In the thalamic region, they pass between the body of the caudate nucleus and the superior edge of the putamen. The fibers descend between the outer face of the thalamus and the inner face of the lentiform nucleus. (b) In the subthalamic region is limited by the lenticular loop fibers, and in the back, by the lateral reticulated body and a thin gray layer that belongs to the reticule of the thalamus, in the middle is in high and low relation: (1) The zona incerta separates it from the thalamic fasciculus of Forel and the lower face of the thalamus; (2) With the lenticular fascicule of Forel; (3) With the subthalamic nucleus of Luys [6].

4. The retrolenticular part consists of fascicules that seemed to be crossed and direct horizontally from out to in, between the caudate nucleus tail and the putamen's posterior edge to reach the thalamic pulvinar by its external face [6]. It is made up of parietopontine and occipitopontine fibers, by the fibers of the occipital cortex that are directed to the superior colic and the pretectal region, and for the posterior thalamic radiation that mainly includes optical radiation, and interconnections between the cortex of the parietal and occipital lobes by the posterior portion of the thalamus (especially the pulvinar) (Fig. 29) [6, 7, 9, 31].
5. The sublenticular part, this segment continues the preceding one below the lenticular nucleus. Remember that this area forms a triangular leaf that is part of the roof of the temporal horn of the lateral ventricle, whose anterior vertex is related to the amygdala and whose base corresponds to the lower part of the retro lenticular segment (Fig. 29) [6]. It contains the temporopontine fibers and acoustic radiation fibers that from the medial geniculate body are directed to Heschl's transverse gyrus (areas 41 y 42 of Brodmann) and the adjacent areas of the superior temporal gyrus. It still contains part of the fibers of the optical radiation, which extends from the lateral geniculate body and the pulvinar of the thalamus to the lips of the calcarine sulcus [6, 7, 9, 31].

References

1. Catani M, Jones DK, Donato R, Ffytche DH. Occipito-temporal connections in the human brain. Brain. 2003;129:2093–107.
2. Catani M, Thiebaut de Schotten M. A diffusion tensor imaging tractography atlas for virtual in vivo dissections. Cortex. 2008;44(8):1105–32.
3. Crosby EC, Humphery T, Lauer EW. Correlative anatomy of the nervous system. New York: MacMillan; 1962.

4. Duffau H. The anatomo-functional connectivity of language revisited. New insights provided by electrostimulation and tractography. Neuropsychologia. 2008;46:927–34.
5. Duvernoy HM. The human brain: surface, three-dimensional sectional anatomy and MRI, and blood supply. Vienna: Springer-Verlag; 1999.
6. Testut L, Latarjet A. Tratado de Anatomía Humana. 9th ed. Barcelona: Editorial Salvat; 1997.
7. Rhoton AL Jr. Cranial anatomy and surgical approaches. Schaumburg: Lippincott Williams & Wilkins; 2003.
8. Snell R. Neuroanatomía Clínica. 4th ed. Buenos Aires: Editorial Panamericana; 1999.
9. Türe U, Yasargil MG, Friedman AH, Al-Mefty O. Fiber dissection technique: lateral aspect of the brain. Neurosurgery. 2000;47:417–27.
10. Duffau H. Brain mapping: from neural basis of cognition to surgical applications. Wien: Springer; 2011.
11. Ribas GC. Applied cranial-cerebral anatomy brain architecture and anatomically oriented microneurosurgery. Cambridge University Press; 2018.
12. Ebeling U, von Cramon D. Topography of the uncinate fascicle and adjacent temporal fiber tracts. Acta Neurochir. 1992;115:143–8.
13. Fernández-Miranda F, Rhoton AL, Álvarez-Linera J. Three-dimensional microsurgical and tractographic anatomy of the white matter of the human brain. Neurosurgery. 2008;62(SHC Suppl 3):SHC989–SHC1027.
14. Fernandez-Miranda JC, Rhoton AL Jr, Kakizawa Y, Choi C, Alvarez-Linera J. The claustrum and its projection system in the human brain: a microsurgical and tractographic anatomical study. J Neurosurg. 2008;108:764–74.
15. Kier EL, Staib LH, Davis LM, Bronen RA. MR imaging of the temporal stem: anatomic dissection tractography of the uncinate fasciculus, inferior occipitofrontal fasciculus, and Meyer's loop of the optic radiation. AJNR Am J Neuroradiol. 2004;25:677–91.
16. Rubino P, Rhoton AL Jr, Tong X, Oliveira E. Three-dimensional relationships of the optic radiation. Neurosurgery. 2005;57(Suppl 4):ONS219–27.
17. Klingler J. Development of the macroscopic preparation of the brain through the process of freezing [in German]. Schweiz Arch Neurol Psychiatr. 1935;36:247–56.
18. Klingler J, Gloor P. The connections of the amygdala and the anterior temporal cortex in the human brain. J Comp Neurol. 1960;115:333–69.
19. Ludwig E, Klingler J. Atlas cerebri humani. S. Karger: Basel; 1956.
20. Nieuwenhuys R, Voogd J, Huijzen C. The human central nervous system. A synopsis and atlas. 3rd ed. Berlin: Springer; 1988.
21. Peltier J, Travers N, Destrieux C, Velut S. Optic radiations: a microsurgical anatomical study. J Neurosurg. 2006;105:294–300.
22. Primbam KH, Lennox MA, Dunsmore RH. Some connections of the orbito-fronto-temporal, limbic and hippocampal areas of Macaca mulatta. J Neurophysiol. 1950;13:127–35.
23. Quensel F. Ueber den Stabkranz des menschlichen Stirnhirnes. Folia Neuro-Biol. 1910;4:319–34.
24. Curran EJ. A new association fiber tract in the cerebrum. With remarks on the fiber tract dissection method of studying the brain. J Comp Neurol. 1909;19:645–56.
25. Wang F, Sun T, Li XG, Liu NJ. Diffusion tensor tractography of the temporal stem on the inferior limiting sulcus. J Neurosurg. 2008;108:775–81.
26. Párraga RG, Ribas GC, Welling LC, Alves RV, de Oliveira E. Microsurgical anatomy of the optic radiation and related fibers in 3-dimensional images. Neurosurgery. 2012;71(1 Suppl Operative):160–71.
27. Schmahmann JD, Pandya DN. Fiber pathways of the brain. New York: Oxford University Press; 2006.
28. Mandonnet E, Nouet A, Gatignol P, Capelle L, Duffau H. Does the left inferior longitudinal fasciculus play a role in language? A brain stimulation study. Brain. 2007;130:623–9.
29. Makris N, Kennedy DN, McInerney S, Sorensen AG, Wang R, Caviness VS Jr, Pandya DN. Segmentation of subcomponents within the superior longitudinal fascicle in humans: a quantitative, in vivo, DT-MRI study. Cereb Cortex. 2005;15:854–69.

30. Makris N, Papadimitriou GM, Kaiser JR, Sorg S, Kennedy DN, Pandya DN. Delineation of the middle longitudinal fascicle in humans: a quantitative, in vivo, DT-MRI study. Cereb Cortex. 2009;19:777–85.
31. Yasargil MG. Microneurosurgery, volume IVA: CNS tumors: surgical anatomy, neuropathology, neuroradiology, neurophysiology, clinical considerations, operability, treatment options. Stuttgart: Georg Thieme; 1996.
32. Yasargil MG. Microneurosurgery, volume IVB: CNS tumors. Stuttgart: Georg Thieme; 1996.
33. Peltier J, Verclytte S, Delmaire C, et al. Microsurgical anatomy of the anterior commissure: correlations with diffusion tensor imaging fiber tracking and clinical relevance. Neurosurgery. 2011;69:ONS241–7.
34. Williams PL, Warwick R, editors. Gray's anatomy. 36th ed. Philadelphia: WB Saunders; 1980.
35. Latarjet A, Ruiz Liard A. Anatomía humana. 3rd ed. Madrid: Editorial Panamericana; 1997.
36. Peuskens D, Van Loon J, Van Calenbergh F. Anatomy of the anterior temporal lobe and the frontotemporal region demonstrated by fiber dissection. Neurosurgery. 2004;55:1174–84.
37. Nagata S, Rhoton AL Jr, Barry M. Microsurgical anatomy of the choroidal fissure. Surg Neurol. 1988;30:3–59.
38. Timurkaynak E, Rhoton AL Jr, Barry M. Microsurgical anatomy and operative approaches to the lateral ventricles. Neurosurgery. 1986;19:685–723.
39. Wen HT, Rhoton AL Jr, de Oliveira EP. Transchoroidal approach to the third ventricle: an anatomic study of the choroidal fissure and its clinical application. Neurosurgery. 1998;42:1205–19.
40. Türe U, Yasargil D, Al-Mefty O, Yasargil G. Topographic anatomy of the insular region. J Neurosurg. 1999;90:720–33.

Surgical Anatomy of the Basal Ganglia and Thalamus

Vanessa Milanese Holanda Zimpel, Erik Middlebrooks, and Natally Santiago

1 Introduction

Deep brain stimulation (DBS) of the subthalamic nucleus (STN) or globus pallidus internus (GPi) is a proven therapeutic modality for Parkinson's disease (PD) and dystonia [1]. Many randomized clinical studies have shown an effective suppression of PD motor symptoms when stimulating either target: STN or GPi [2, 3]. However, incomplete understanding of the basal ganglia neuroanatomy and the therapeutic mechanisms of DBS impede personalization of this therapy and may contribute to suboptimal outcomes.

In 1990, Delong et al. elucidated the basal ganglia pathways in monkeys. They proposed that the basal ganglia exercise excitatory and inhibitory control over the cerebral cortex. The direct pathway, via putaminal inhibitory projections to GPi, decreases pallidothalamic inhibition and facilitates cortical motor activity. In contrast, with putaminal projections via GPe then STN, the indirect pathway sends excitatory projections to GPi, increasing pallidothalamic inhibition and inhibiting motor-cortical activity [4, 5].

V. M. H. Zimpel (✉)
Microsurgical Anatomy Dr. Evandro de Oliveira Laboratory, BP—A Beneficência Portuguesa de São Paulo, São Paulo, SP, Brazil

Department of Neurosurgery, Mayo Clinic, Jacksonville, FL, USA

E. Middlebrooks
Department of Radiology, Mayo Clinic, Jacksonville, FL, USA

N. Santiago
Hospital Alipio Correa Neto, São Paulo, SP, Brazil

BP—A Beneficência Portuguesa de São Paulo, São Paulo, SP, Brazil

E. G. Figueiredo et al. (eds.), *Brain Anatomy and Neurosurgical Approaches*,
https://doi.org/10.1007/978-3-031-14820-0_17

The STN is a topographically organized homogeneous nucleus that has been functionally divided into inconsistent subdivisions over the last century [6]. The tripartite subdivision of the STN into the motor, limbic, and associative divisions has been questioned, and an organization of the STN without strict anatomical boundaries has been proposed [7]. Precise knowledge of the STN connections may provide the basis for optimal partitioning of the STN based on function.

Because its volume is substantially larger than that of the STN, the functional circuitry of the GPi is less densely packed, and surrounding functional circuits are further away and less likely to be affected by spreading current from therapeutic DBS. One predictable advantage of GPi–DBS over STN–DBS is that fewer unintended stimulation-induced side-effects are observed with GPi stimulation.

The GPi is more variable in its position relative to the mid-commissural point than the STN, which may explain the more inconsistent results of early DBS practitioners—who used primarily indirect targeting—when targeting GPi as opposed to STN. If modern, direct targeting methods are employed to localize the GPi, however, the increased volume of the GPi compared to that of the STN makes it is easier to stimulate only the pathologic somatosensory circuitry in the posterolateral portion of the GPi without spreading current into surrounding structures that produce unintended effects (e.g., the internal capsule, located posterior and medial to the GPi; the limbic and associative circuitry of the anteromedial GPi). The larger volume of the GPi requires a larger volume of neural tissue activation to produce a similar therapeutic benefit to that seen with STN stimulation, resulting in the higher energy demands and shorter battery life associated with GPi stimulation.

Overall, both STN and GPi have proven to be very effective DBS targets for PD [2, 8, 9]. The safety profile of GPi–DBS is superior to that of STN, but more energy is required in the GPi to produce optimal therapeutic results. Both targets also have other advantages and disadvantages that should be considered when choosing an optimal DBS target for a given patient [10]. The GPi is considered the primary output structure of the basal ganglia and is known to have an increase in its neuronal activity before the onset of PD motor symptoms [11]. GPi–DBS is effectively used to abolish dyskinesias in patients with treatment-resistant hyperkinetic movements.

Both targets may produce tremor suppression, the most common and often debilitating symptom associated with many neurological disorders, such as PD, dystonia, and essential tremor (ET). The collective evidence regarding ET proposes that the neuromodulation of the cerebellum–thalamocortical pathway is fundamental to control the tremor network dysfunction [12].

Neurosurgical ablative procedures have provided successful symptomatic relief from medically refractory tremors; nevertheless, bilateral thalamotomies were related to substantial side effects, such as dysarthria, dysphagia, cognitive impairment, and gait problems [13]. DBS emerged as a promising alternative in 1991 for ET. It offered tremor suppression while avoiding the usual complications of thalamotomies. One of the most widely targeted brain regions for tremor treatment is the ventral intermediate nucleus (VIM) and the thalamus's ventral oral nucleus (VO) [12]. The effect of VO stimulation has also been clinically shown to improve tremor control in patients whose VIM–DBS initially failed. In these cases, a VO rescue lead was placed anterior to the previously placed VIM lead [14].

Finally, the ventral anterior limb of the internal capsule and the adjacent ventral striatum junction (VC/VS) has been used as a target for DBS for the treatment of highly resistant obsessive–compulsive disorder (OCD) and treatment-resistant major depressive disorder, showing encouraging results [15].

2 Microsurgical Anatomy

The STN is an almond-shaped nucleus situated at an oblique angle to the anteroposterior axis of the subthalamus [16], anterolaterally to the red nucleus (Fig. 1). The zona incerta is located superiorly to the STN, immediately superior to the lenticular fasciculus. DBS of the caudal zona incerta region has been reported to be efficacious in the suppression of essential and other forms of tremor, particularly those with prominent proximal limb involvement, as well as in PD and possibly dystonia.

The corticospinal fibers course through the cerebral peduncle ventrally and laterally to the STN before ascending and merging into the posterior limb of the internal capsule. The medial lemniscus passes posterolaterally to the STN, ascending from the gracile and cuneate tubercles to the thalamus and dividing the brainstem into ventral and dorsal parts (Fig. 2).

The optic tract runs inferolateral to the STN along its path toward the lateral geniculate body (Fig. 2). The oculomotor nerve is located medial and inferior to the STN. Some clinical studies have reported that STN–DBS can result in altered oculomotor function [17]. Stimulation-induced deconjugate gaze suggests a lead position that is excessively medial and deep.

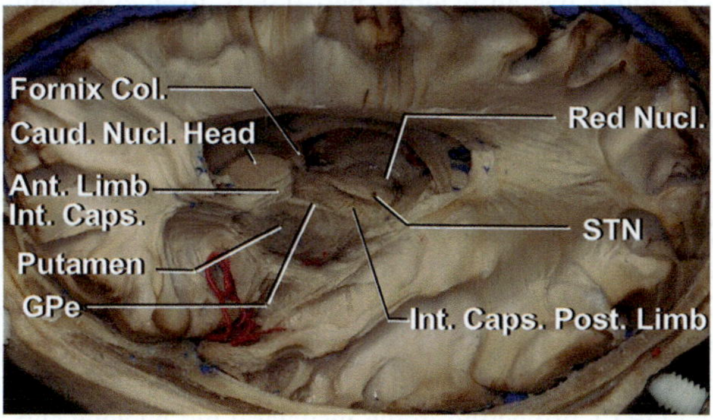

Fig. 1 Detailed anatomy demonstrating the STN relationships to the GPi, red nucleus, and posterior limb of the internal capsule. *Caud.*, caudate; *Conn.*, connections; *GPe*, Globus Pallidus Externus; *GPi*, Globus Pallidus Internus; *Int. Caps. Ant. Limb*, anterior limb of the internal capsule; *Int. Caps. Post. Limb*, posterior limb of the internal capsule; Nucl., nucleus; *STN*, subthalamic nucleus

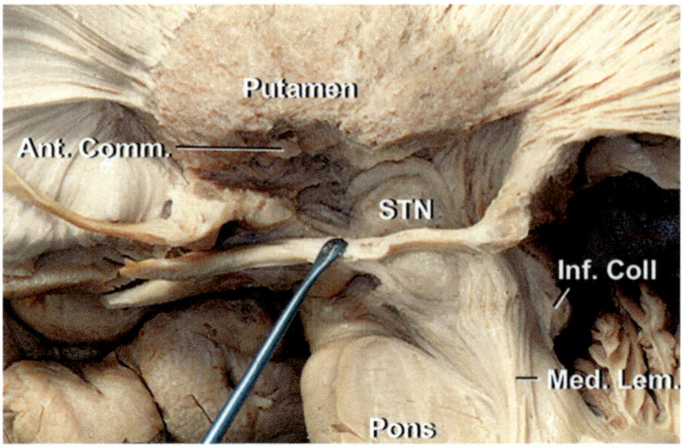

Fig. 2 Lateral view of the STN and putamen showing their relationship with the optic tract and anterior commissure. *Ant. Comm.*, anterior commissure; *Inf Coll.*, inferior colliculus; *Lat. Gen.*, lateral geniculate; *Med. Lem.*, medial lemniscus; *STN*, subthalamic nucleus

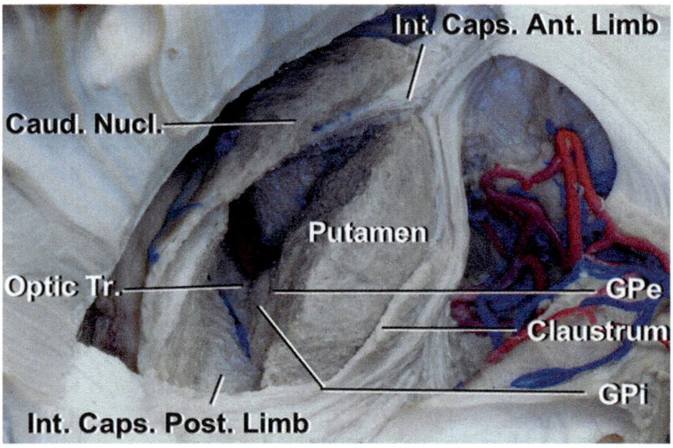

Fig. 3 Superior white matter dissection exposing the lateral ventricles, caudate nucleus, putamen, and GPi. *Caud.*, caudate; *GPe*, Globus Pallidus Externus; *GPi*, Globus Pallidus Internus; *Int. Caps. Ant. Limb*, anterior limb of the internal capsule; Nucl., nucleus; *STN*, subthalamic nucleus

The globus pallidus has an external part (GPe), located laterally, that extends anteriorly and medially to the genu of the internal capsule, and an internal part (GPi), located medially, that abuts the posterior limb of the internal capsule along its medial border (Fig. 3]. The optic tract passes along the inferomedial border of the posterior GPi on its path from the chiasm to the lateral geniculate body of the midbrain (Fig. 4). These are important anatomic associations during GPi–DBS. The globus pallidus has primarily inhibitory control over movement [18].

The VIM and the VO of the thalamus have connections to the dentatorubrothalamic tract (DRTT). The DRTT originates in the contralateral dentate nucleus and decussates in the tegmentum toward the ipsilateral red nucleus before reaching the thalamus in the neighborhood of the VIM (Fig. 5) [19]. More recently, nondecussating fibers of the DRTT projecting to the ipsilateral red nucleus and thalamus have been described, which may act as a substrate for the involvement of each cerebellar hemisphere in bilateral limb movements. The DRTT has been suggested as the neural substrate for stimulation-induced tremor reduction [20].

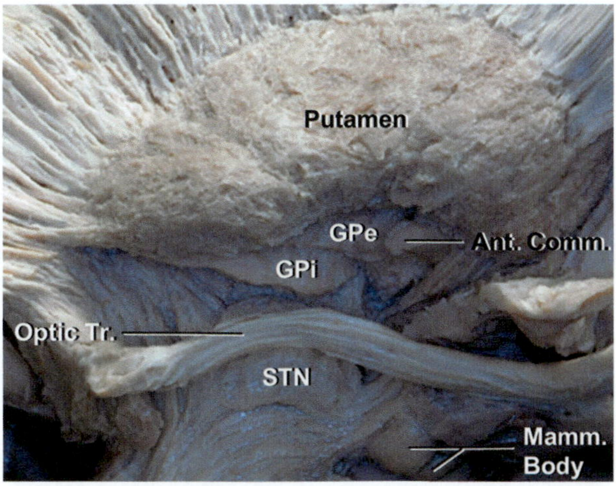

Fig. 4 Lateral white matter dissection of the STN and GPI and their relationship with the optic tract. *Ant. Comm.*, anterior commissure; *GPe*, Globus Pallidus Externus; *GPi*, Globus Pallidus Internus; *Mamm*, mammillary; *Opt. Tr.*, optic tract; *STN*, subthalamic nucleus

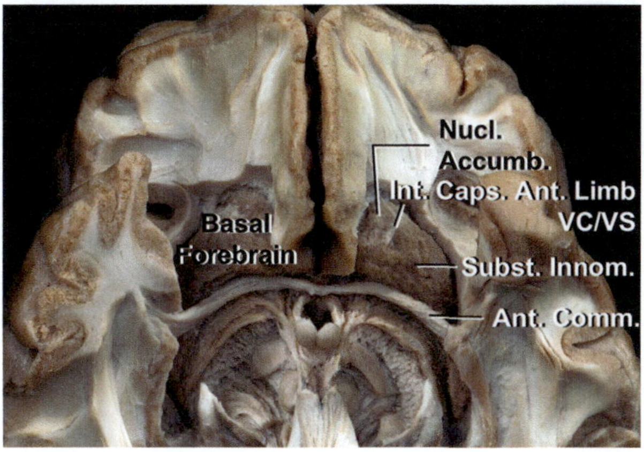

Fig. 5 Inferior view of the nucleus accumbens, VC/VS, and substantia innominata relationships. *Accumb*, accumbens; *Ant. Comm.*, anterior commissure; *Int. Caps. Ant. Limb*, anterior limb of the internal capsule; Nucl., nucleus; *Subst. Innom.*, substantia innominata; *STN*, subthalamic nucleus

The anterior limb of the internal capsule in humans corresponds to a bundle that carries ascending and descending fibers from the orbitofrontal cortex (OFC) and the anterior cingulate cortex. The fibers are organized in such a way that the ventral portion of the VC carries fibers from the ventromedial region of the prefrontal cortex and OFC [21].

The VS includes the nucleus accumbens (NAc), medioventral portion of the caudate and putamen, olfactory tubercle, and anterior perforated substance (Fig. 6). The term VS highlights the ventral extension of this portion to the basal surface of the brain. The VS and striatal network are part of the reward system and support goal-motivated behavior. Unlike the dorsal portion of the striatum, which is well-demarcated and more easily identified, the VS lacks well-defined anatomical limits

Fig. 6 Inferior view of the brain after removal of fibers of the brainstem and cerebellum showing the dentate nucleus. The DRTT arises in the contralateral dentate nucleus and decussates toward the ipsilateral red nucleus, before reaching the thalamus in the neighborhood of the VIM nucleus. *Accumb*, accumbens; *Med.*, medial; Nucl., nucleus; *VIM*, ventral intermedius; *VC*, ventral caudal; *Tr.*, tract

for its delimitation [22]. The VC/VS is a complex but also promising surgical target for DBS [23]. Reviewing tracts around the VC/VS for DBS treatments since the therapeutic efficacy of DBS is thought to be mediated by modulation of activity within both the gray and the white matter surrounding the target. It is of main importance to understand it anatomically and functionally for better achievements in surgical procedures.

3 Discussion

Targeting methods for functional neurosurgical interventions have improved substantially with the exploitation of recent technological advances. Indirect targeting—whereby the position of a neuroanatomical target relative to the mid-commissural point in a stereotactic brain atlas is used to predict its location in the brain of a given patient—while useful as a starting point for DBS targeting, should now be considered obsolete as a stand-alone targeting technique [24]. Various strategies to improve upon indirect atlas targeting have been developed by stereotactic surgeons over the years to account for the error attributable to neuroanatomical variability from patient to patient.

Modern imaging modalities (T2 MRI [25], FGATIR MRI [26], and DTI tractography [27, 28]) may provide more direct visualization of anatomic structures of interest than ever before, facilitating direct targeting, which provides more precise identification of intended brain targets in a given individual. Direct targeting is also enhanced using advanced targeting software that provides three-dimensional image reconstruction and incorporates neuroanatomical atlas overlays that can be deformed to precisely adapt to a given patient's visible anatomy.

A high-quality deformable digital brain atlas enables the modern stereotactic surgeon to more accurately infer the position of important neuroanatomical structures or boundaries that may be imperfectly visualized even with the best available imaging [26]. As presented in this chapter, a solid understanding of the neuroanatomical relationships of the STN, GPi, ViM, VO, and VC/VS and their surrounding structures is a prerequisite to optimal DBS targeting.

The continued shift in the understanding of functional brain disorders from traditional localizationist theories to widely distributed "circuitopathies"has been driven by advances in noninvasive brain mapping, namely, diffusion-based tractography–MRI (DT–MRI) and resting-state functional connectivity (rs-fMRI). There is converging evidence that many traditional targets in DBS surgery likely exhibit their effect either directly or indirectly by modulation of adjacent white matter tracts. Such evidence has led to a consolidation of several targets, such as VIM and caudal ZI, likely related to the dentato-rubro-thalamic tract and anterior STN and vALIC mediated through limbic and associative STN hyper direct pathways. The following paragraphs discuss the most recent evidence regarding connectivity-based targeting and programming for common DBS targets.

Connectomics has been well-studied in targeting the VIM and PSA for tremor suppression. Indeed, multiple studies have evaluated the role of connectivity-based segmentation of the ventral thalamus as a target for functional neurosurgery [12, 29–33]. The summation of studies to date has led to the consolidation of the idea that tremor suppression is largely based on modulation of the dentato-rubro-thalamic tract (DRTT), which traverses traditional targets of cZI and VIM, as well as the posterior border of VOp [34]. Ataxia is an expected adverse effect of ET DBS, in every case to some degree. Although the mechanism for this phenomenon has not been fully elucidated, it is thought that the tremor circuit is the same circuit largely responsible for cerebellar modulation of coordinated movement [13].

STN–DBS is a complex topic covering multiple indications; however, most connectomes' research has been related to PD and OCD. Connectivity studies have supported the concept of a tripartite functional gradient within the STN made up of a motor division posterolateral, a middle associative division, and limbic division anteromedially [35, 36]. Such functional division has been shown to correlate with specific white matter connections that correlate with improvements in rigidity, tremor, and bradykinesia [37]. Specifically, improvement in bradykinesia was associated with SMA connectivity, while rigidity improvement correlated with both SMA and prefrontal regions. Meanwhile, tremor improvement correlated most with primary motor cortex connectivity [37].

Functional and structural connectivity has also been shown to be an independent predictor of improvement in PD after STN–DBS and predict postoperative motor scores within 15% [38]. Others have also employed machine learning to perform a connectivity-based "monopolar review" to predict effect electrodes based solely on connectivity data with 84.9% accuracy [39]. Likewise, this approach was able to predict settings that would produce stimulation-induced dyskinesia in another study

[40]. The summation of these studies provides an intriguing framework for how connectivity data may be utilized in the future to predict optimal settings for symptom control while reducing the incidence of side effects.

STN–DBS may disrupt the pathologic synchronization of β oscillations in the motor network (thereby facilitating movement) by altering the timing of motor cortex firing through orthodromic stimulation of somatosensory afferents to the STN that are most concentrated in the dorsolateral STN [41]. An improved understanding of the various effects of STN–DBS can be obtained by studying STN microanatomy and the connections between the STN, GPi, GPe, and putamen, as demonstrated in the dissections presented here.

Studies have also shown OCD to be responsive to STN–DBS and other targets, such as NAc, ALIC, and VS. The mechanism of action between these targets has historically been debated. For instance, a tract originally referred to as the superolateral branch of the medial forebrain bundle (slMFB) was described as the ideal DBS target; however, no additional evidence of this tract's existence is available outside of few DTI studies [42]. Importantly, the proper MFB lies in a more ventral location and does not traverse ALIC at all. Conversely, the anterior thalamic radiations have been shown to correlate with greater improvement in OCD with no effect with stimulation of the MFB proper [43]. Interestingly, the tract previously described as slMFB is closely correlated in space with the associative hyper direct STN fibers [44]. Given the known treatment effect of anteromedial STN stimulation in OCD would be suggested that ALIC–DBS may in part be related to stimulation of these hyper direct fibers rather than any portion of the MFB, and the slMFB may simply be false fiber connections in DTI. Indeed, the consolidation of many OCD targets, including STN, slMFB, ITP, ALIC, MD/VA, and VC/VS, reveals proximity to these STN-frontal lobe hyper direct tracts supporting a theory of common network modulation in these targets.

The GPi has been less studied by functional or structural connectivity. For evaluation of GPi–DBS in PD, connectivity to the primary motor cortex, and to a lesser degree, the supplemental motor area and premotor cortex were predictive of UPDRS-III motor score improvement. GPi connectivity has also been evaluated in DBS for dystonia [45]. Similar to PD, improvement in primary generalized dystonia was associated with greater connectivity to the sensorimotor network [45, 46].

References

1. Obeso JA, Olanow CW, Rodriguez-Oroz MC, et al. Deep-brain stimulation of the subthalamic nucleus or the pars interna of the globus pallidus in Parkinson's disease. N Engl J Med. 2001;345:956–63.
2. Okun MS, Fernandez HH, Wu SS, et al. Cognition and mood in Parkinson's disease in subthalamic nucleus versus globus pallidus interna deep brain stimulation: the COMPARE trial. Ann Neurol. 2009;65:586–95.
3. Odekerken VJ, van Laar T, Staal MJ, et al. Subthalamic nucleus versus globus pallidus bilateral deep brain stimulation for advanced Parkinson's disease (NSTAPS study): a randomised controlled trial. Lancet Neurol. 2013;12:37–44.

4. Wichmann T, Delong MR. Deep-brain stimulation for basal ganglia disorders. Basal Ganglia. 2011;1:65–77.
5. DeLong MR. Primate models of movement disorders of basal ganglia origin. Trends Neurosci. 1990;13:281–5.
6. Keuken MC, Uylings HB, Geyer S, et al. Are there three subdivisions in the primate subthalamic nucleus? Front Neuroanat. 2012;6:14.
7. Alkemade A, Forstmann BU. Do we need to revise the tripartite subdivision hypothesis of the human subthalamic nucleus (STN)? NeuroImage. 2014;95:326–9.
8. Combs HL, Folley BS, Berry DT, et al. Cognition and depression following deep brain stimulation of the subthalamic nucleus and globus pallidus pars internus in Parkinson's disease: a meta-analysis. Neuropsychol Rev. 2015;25:439–54.
9. Follett KA, Weaver FM, Stern M, et al. Pallidal versus subthalamic deep-brain stimulation for Parkinson's disease. N Engl J Med. 2010;362:2077–91.
10. Williams NR, Foote KD, Okun MS. STN vs. GPi deep brain stimulation: translating the rematch into clinical practice. Mov Disord Clin Pract. 2014;1:24–35.
11. Wichmann T, Dostrovsky JO. Pathological basal ganglia activity in movement disorders. Neuroscience. 2011;198:232–44.
12. Middlebrooks EH, Tuna IS, Almeida L, et al. Structural connectivity-based segmentation of the thalamus and prediction of tremor improvement following thalamic deep brain stimulation of the ventral intermediate nucleus. Neuroimage Clin. 2018;20:1266–73.
13. Wong JK, Hess CW, Almeida L, et al. Deep brain stimulation in essential tremor: targets, technology, and a comprehensive review of clinical outcomes. Expert Rev Neurother. 2020;20:319–31.
14. Oyama G, Foote KD, Hwynn N, et al. Rescue leads: a salvage technique for selected patients with a suboptimal response to standard DBS therapy. Parkinsonism Relat Disord. 2011;17:451–5.
15. Malone DA Jr, Dougherty DD, Rezai AR, et al. Deep brain stimulation of the ventral capsule/ventral striatum for treatment-resistant depression. Biol Psychiatry. 2009;65:267–75.
16. Yagmurlu K, Rhoton AL Jr, Tanriover N, et al. Three-dimensional microsurgical anatomy and the safe entry zones of the brainstem. Neurosurgery. 2014;10(Suppl 4):602–19; discussion 619–620.
17. Fridley J, Adams G, Sun P, et al. Effect of subthalamic nucleus or globus pallidus interna stimulation on oculomotor function in patients with Parkinson's disease. Stereotact Funct Neurosurg. 2013;91:113–21.
18. Yagmurlu K, Vlasak AL, Rhoton AL Jr. Three-dimensional topographic fiber tract anatomy of the cerebrum. Neurosurgery. 2015;11(Suppl 2):274–305; discussion 305.
19. Gallay MN, Jeanmonod D, Liu J, et al. Human pallidothalamic and cerebellothalamic tracts: anatomical basis for functional stereotactic neurosurgery. Brain Struct Funct. 2008;212:443–63.
20. Yang AI, Buch VP, Heman-Ackah SM, et al. Thalamic deep brain stimulation for essential tremor: relation of the dentatorubrothalamic tract with stimulation parameters. World Neurosurg. 2020;137:e89–97.
21. Jbabdi S, Lehman JF, Haber SN, et al. Human and monkey ventral prefrontal fibers use the same organizational principles to reach their targets: tracing versus tractography. J Neurosci. 2013;33:3190–201.
22. Baydin S, Yagmurlu K, Tanriover N, et al. Microsurgical and fiber tract anatomy of the nucleus accumbens. Oper Neurosurg. 2016;12:269–88.
23. Fudge JL, Haber SN. Defining the caudal ventral striatum in primates: cellular and histochemical features. J Neurosci. 2002;22:10078–82.
24. Holanda VM, Okun MS, Middlebrooks EH, et al. Postmortem dissections of common targets for lesion and deep brain stimulation surgeries. Neurosurgery. 2020;86(6):860–72.
25. Rabie A, Verhagen Metman L, Slavin KV. Using "functional" target coordinates of the subthalamic nucleus to assess the indirect and direct methods of the preoperative planning: do the anatomical and functional targets coincide? Brain Sci. 2016;6:65.

26. Sudhyadhom A, Haq IU, Foote KD, et al. A high resolution and high contrast MRI for differentiation of subcortical structures for DBS targeting: the fast gray matter acquisition T1 inversion recovery (FGATIR). NeuroImage. 2009;47(Suppl 2):T44–52.
27. Alho A, Hamani C, Alho EJL, et al. Magnetic resonance diffusion tensor imaging for the pedunculopontine nucleus: proof of concept and histological correlation. Brain Struct Funct. 2017;222:2547–58.
28. Sajonz BE, Amtage F, Reinacher PC, et al. Deep brain stimulation for tremor tractographic versus traditional (DISTINCT): study protocol of a randomized controlled feasibility trial. JMIR Res Protoc. 2016;5:e244.
29. Kim W, Sharim J, Tenn S, et al. Diffusion tractography imaging-guided frameless linear accelerator stereotactic radiosurgical thalamotomy for tremor: case report. J Neurosurg. 2018;128:215–21.
30. Pouratian N, Zheng Z, Bari AA, et al. Multi-institutional evaluation of deep brain stimulation targeting using probabilistic connectivity-based thalamic segmentation. J Neurosurg. 2011;115:995–1004.
31. Tsolaki E, Downes A, Speier W, et al. The potential value of probabilistic tractography-based for MR-guided focused ultrasound thalamotomy for essential tremor. Neuroimage Clin. 2018;17:1019–27.
32. Akram H, Dayal V, Mahlknecht P, et al. Connectivity derived thalamic segmentation in deep brain stimulation for tremor. Neuroimage Clin. 2018;18:130–42.
33. Al-Fatly B, Ewert S, Kubler D, et al. Connectivity profile of thalamic deep brain stimulation to effectively treat essential tremor. Brain. 2019;142:3086–98.
34. Calabrese E, Hickey P, Hulette C, et al. Postmortem diffusion MRI of the human brainstem and thalamus for deep brain stimulator electrode localization. Hum Brain Mapp. 2015;36:3167–78.
35. Parent A, Hazrati L-N. Functional anatomy of the basal ganglia. II. The place of subthalamic nucleus and external pallidium in basal ganglia circuitry. Brain Res Rev. 1995;20:128–54.
36. Accolla EA, Ruiz MH, Horn A, et al. Brain networks modulated by subthalamic nucleus deep brain stimulation. Brain. 2016;139:2503–15.
37. Akram H, Sotiropoulos SN, Jbabdi S, et al. Subthalamic deep brain stimulation sweet spots and hyperdirect cortical connectivity in Parkinson's disease. NeuroImage. 2017;158:332–45.
38. Horn A, Reich M, Vorwerk J, et al. Connectivity predicts deep brain stimulation outcome in Parkinson disease. Ann Neurol. 2017;82:67–78.
39. Lin H, Na P, Zhang D, et al. Brain connectivity markers for the identification of effective contacts in subthalamic nucleus deep brain stimulation. Hum Brain Mapp. 2020;41(8):2028–36.
40. Tsuboi T, Charbel M, Peterside DT, et al. Pallidal connectivity profiling of stimulation-induced dyskinesia in Parkinson's disease. Mov Disord. 2021;36:380–8.
41. Karas PJ, Mikell CB, Christian E, et al. Deep brain stimulation: a mechanistic and clinical update. Neurosurg Focus. 2013;35:E1.
42. Coenen VA, Panksepp J, Hurwitz TA, et al. Human medial forebrain bundle (MFB) and anterior thalamic radiation (ATR): imaging of two major subcortical pathways and the dynamic balance of opposite affects in understanding depression. J Neuropsychiatry Clin Neurosci. 2012;24:223–36.
43. Baldermann JC, Melzer C, Zapf A, et al. Connectivity profile predictive of effective deep brain stimulation in obsessive-compulsive disorder. Biol Psychiatry. 2019;85:735–43.
44. Middlebrooks EH, Domingo RA, Vivas-Buitrago T, et al. Neuroimaging advances in deep brain stimulation: review of indications, anatomy, and brain connectomics. AJNR Am J Neuroradiol. 2020;41(9):1558–68.
45. Okromelidze L, Tsuboi T, Eisinger RS, et al. Functional and structural connectivity patterns associated with clinical outcomes in deep brain stimulation of the globus pallidus internus for generalized dystonia. AJNR Am J Neuroradiol. 2020;41:508–14.
46. Rozanski VE, Vollmar C, Cunha JP, et al. Connectivity patterns of pallidal DBS electrodes in focal dystonia: a diffusion tensor tractography study. NeuroImage. 2014;84:435–42.

Part IV
The Ventricles

Surgical Anatomy of the Lateral Ventricles

Richard Gonzalo Párraga

1 Introduction

Embryologically, the lateral ventricles are constituted by remaining internal spaces of the forebrain vesicle, which corresponds to the most superior dilation of the primitive neural tube. Throughout embryogenesis, the duplication of the forebrain vesicle and the subsequent C-folding of the entire telencephalon cause each lateral ventricle and its related deeper structures to be arranged on the diencephalon, around each thalamus, which it constitutes the morphological center of each cerebral hemisphere [1–4]. From a morphological point of view, it should be noted that each thalamus corresponds to an ovoid diencephalic mass that is arranged in continuity to each half of the midbrain, forming the top of each half of the brain stem and on which each cerebral hemisphere is attached [5].

Currently, the identification of the deeper brain structures in neuroimaging examinations is made from the initial observation of the ventricular compartments, and with the advent of modern microneurosurgical and neuroendoscopic techniques, the ventricular cavities have become important access routes neurosurgical to deep brain lesions [6–12].

For this chapter, ten cerebral hemispheres were studied, which were fixed in 10% formalin for 40 days [13], four cerebral hemispheres after being fixed in formalin were frozen at −10 to −15° for 14 days and were dissected with the Klingler technique [14–16].

R. G. Párraga (✉)
Cerebrovascular & Skull Base Surgery, Institute of Neurosurgery Bolivia (INEB), Cochabamba, Cercado, Bolivia
e-mail: richardparraga@ineb.com.bo

E. G. Figueiredo et al. (eds.), *Brain Anatomy and Neurosurgical Approaches*,
https://doi.org/10.1007/978-3-031-14820-0_18

2 Lateral Ventricle

The thalamus is located in the center of the lateral ventricle. Each lateral ventricle wraps around the superior, inferior, and posterior surfaces of the thalamus. Each thalamus is characterized as an ovoid mass, as each lateral ventricle is surrounded by a C-shaped, made up of compartments communicated with each other, but which receive different names according to their arrangement in relation to the thalamus itself: (1) the frontal horn, corresponds to the ventricular portion located anterior to the thalamus; (2) The ventricular body, arranged on the thalamus; (3) the atrium or ventricular trigone, located posterior to the thalamus; (4) the temporal horn, arranged below the thalamus; and (5) the occipital horn corresponds to a posterior extension of the ventricular atrium (Figs. 1, 2, 3, and 4) [7].

Each of these five parts has two walls, medial and lateral, as well as a ceiling and a floor. The frontal, temporal horn, and the atrium have an anterior wall. In addition to the thalamus, the structures that delimit each lateral ventricle are the hippocampus, fornix, caudate nucleus, amygdala, corpus callosum, and septum pellucidum (Fig. 5).

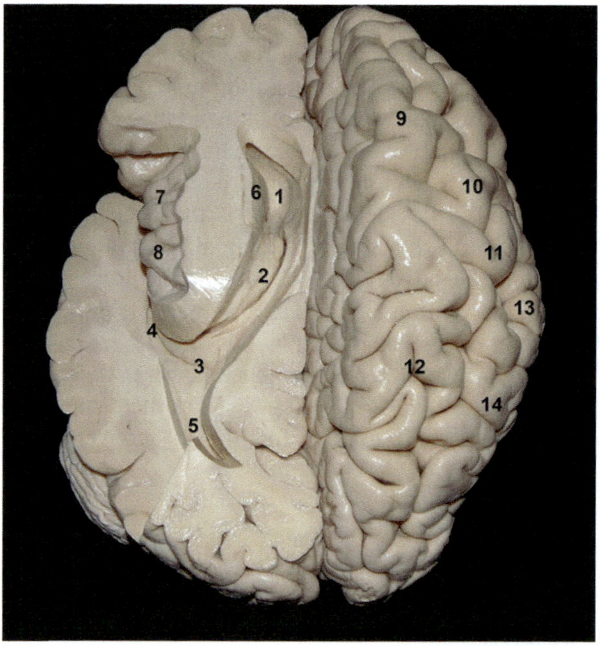

Fig. 1 Superior view of the brain, left cerebral hemisphere has been removed, to expose ventricle cavities. (1) Frontal horn of the lateral ventricle; (2) Body of the lateral ventricle; (3) Atrium; (4) Temporal horn; (5) Occipital horn; (6) Head of the caudate nucleus; (7) Short gyri of the insula; (8) Long gyrus of the insula; (9) Superior frontal gyrus; (10) Precentral gyrus; (11) Postcentral gyrus; (12) Superior parietal lobule; (13) Supramarginal gyrus; (14) Angular gyrus

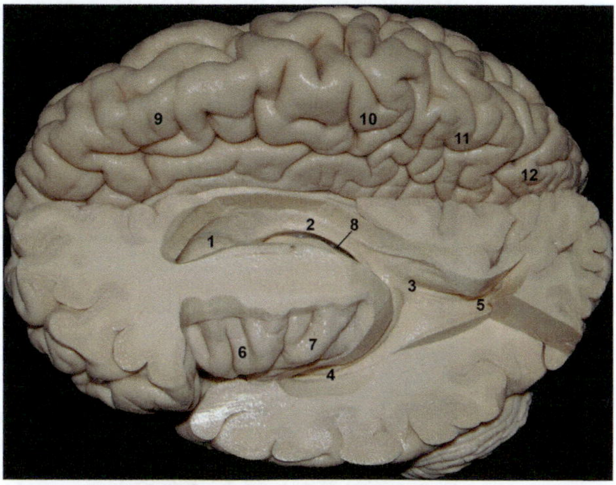

Fig. 2 Superolateral view of the brain, same anatomical specimen (1) Frontal horn of the lateral ventricle; (2) Body of the lateral ventricle; (3) Atrium; (4) Temporal horn; (5) Occipital horn; (6) Short gyri of the insula; (7) Long gyrus of the insula; (8) Choroidal fissure; (9) Superior frontal gyrus; (10) Paracentral lobule; (11) Precuneus; (12) Cuneus

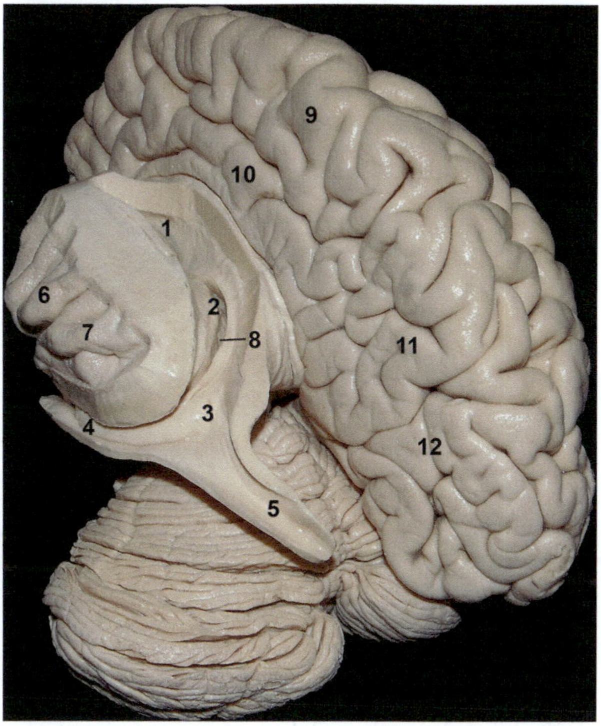

Fig. 3 Posterolateral view. Anatomical specimen where the left cerebral hemisphere was removed, preserving the central block and the lateral ventricle. (1) Frontal horn of the lateral ventricle; (2) Body of the lateral ventricle; (3) Atrium; (4) Temporal horn; (5) Occipital horn; (6) Short gyri of the insula; (7) Long gyrus of the insula; (8) Choroidal fissure; (9) Superior frontal gyrus; (10) Cingulate gyrus; (11) Precuneus; (12) Cuneus

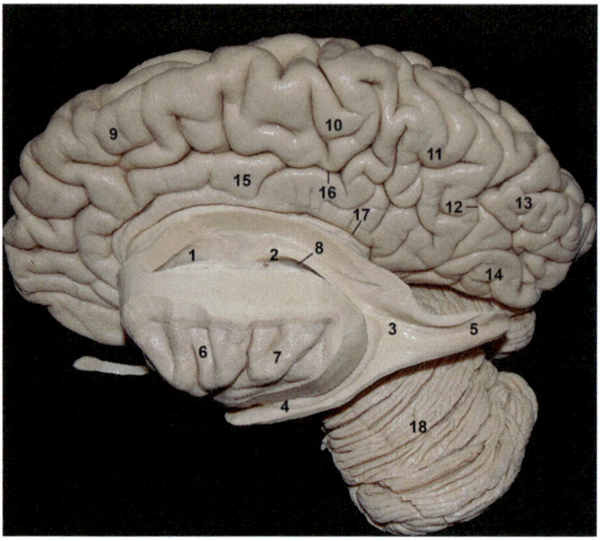

Fig. 4 Lateral view of the same anatomical specimen as the previous image. (1) Frontal horn of the lateral ventricle; (2) Body of the lateral ventricle; (3) Atrium; (4) Temporal horn; (5) Occipital horn; (6) Short gyri of the insula; (7) Long gyrus of the insula; (8) Choroidal fissure; (9) Superior frontal gyrus; (10) Paracentral lobule; (11) Precuneus; (12) Parieto-occipital sulcus; (13) Cuneus; (14) Medial occipitotemporal (lingual) gyrus; (15) Cingulate gyrus; (16) Cingulate sulcus; (17) Sulcus of the corpus callosum; (18) Cerebellum

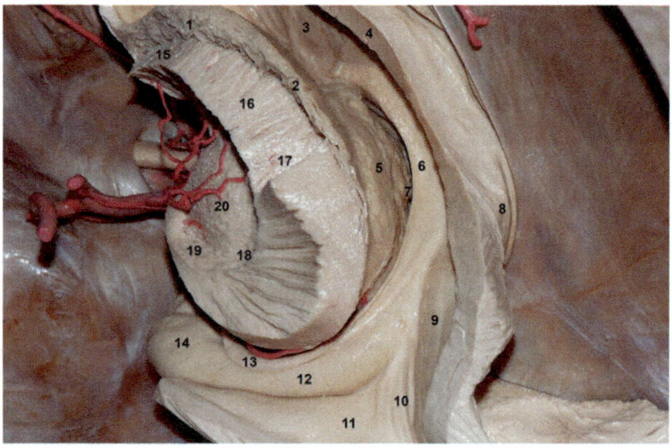

Fig. 5 Posterosuperior view, anatomical specimen dissected with the Klingler technique, preserved nuclei of the base. The relationship of these nuclei with the ventricular cavity is exposed. (1) Head of caudate nucleus; (2) Body of caudate nucleus; (3) Septum pellucidum; (4) Trunk of the corpus callosum; (5) Thalamus; (6) Body of the fornix; (7) Choroidal fissure; (8) Splenium of corpus callosum; (9) Bulb of the corpus callosum; (10) Calcar avis; (11) Collateral trigone; (12) Body of the hippocampus; (13) Fimbria; (14) Head of the hippocampus; (15) Internal capsule, anterior limb; (16) Internal capsule, genu; (17) Internal capsule, posterior limb; (18) Internal capsule, retrolenticular part; (19) Internal capsule, sublenticular part; (20) Globus pallidus

2.1 Front Horn

It is located anterior to the interventricular foramen, it has a medial wall, formed by the septum pellucidum; an anterior wall and a roof formed by the knee of the corpus callosum; a lateral wall, formed by the head of the caudate nucleus; and a narrow floor, formed by the rostrum of the corpus callosum (Figs. 6 and 7).

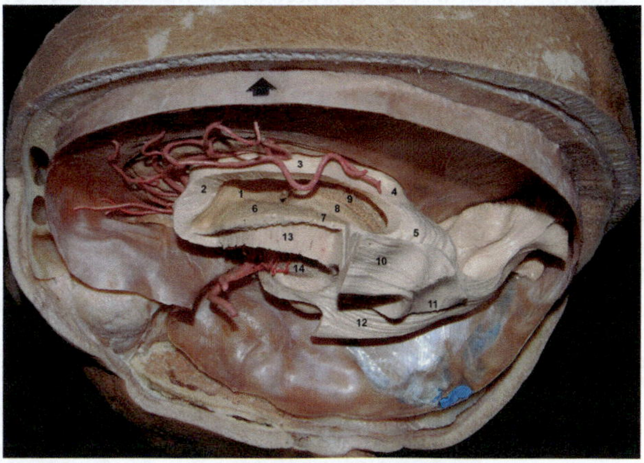

Fig. 6 Superior view of a head. The left cerebral hemisphere was removed, partially preserving the walls of the ventricular cavities. (1) Rostrum of corpus callosum; (2) Genu of corpus callosum; (3) Body of corpus callosum; (4) Splenium of corpus callosum; (5) Tapetum; (6) Head of caudate nucleus; (7) Body of caudate nucleus; (8) Thalamus; (9) Choroid plexus; (10) Fronto-parietal or horizontal segment of the superior longitudinal fasciculus; (11) Temporo-parietal or vertical segment of the superior longitudinal fasciculus; (12) Sagittal stratum; (13) Internal capsule; (14) Globus pallidus

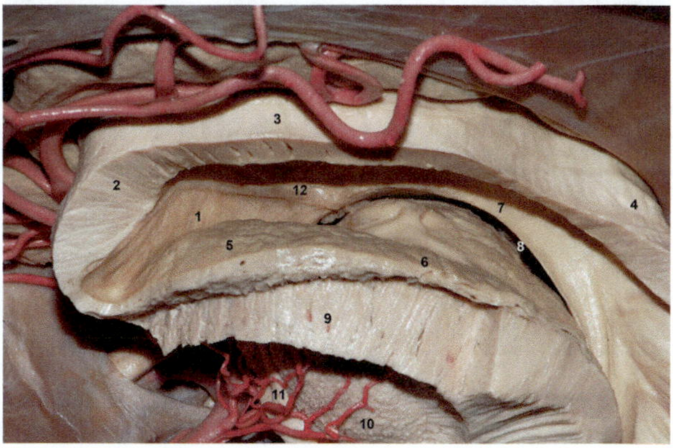

Fig. 7 Magnified view of the same anatomical specimen as the previous figure. (1) Rostrum of corpus callosum; (2) Genu of corpus callosum; (3) Body of corpus callosum; (4) Splenium of corpus callosum; (5) Head of caudate nucleus; (6) Body of caudate nucleus; (7) Fornix; (8) Choroidal fissure; (9) Internal capsule; (10) Globus pallidus; (11) Anterior commissure; (12) Septum pellucidum

2.2 Body

It extends from the posterior border of the interventricular foramen to the point where the septum pellucidum disappears and the corpus callosum and fornix meet. The roof is formed by the corpus callosum (Figs. 8 and 9); the medial wall is formed by the septum pellucidum, above, and by the body of the fornix, below; the side wall is formed by the body of the caudate nucleus; and the floor by the thalamus. The caudate nucleus and the thalamus are separated by the thalamostriate groove, through which the thalamostriate vein and the terminal stria run (Figs. 10, 11, and 12) [6].

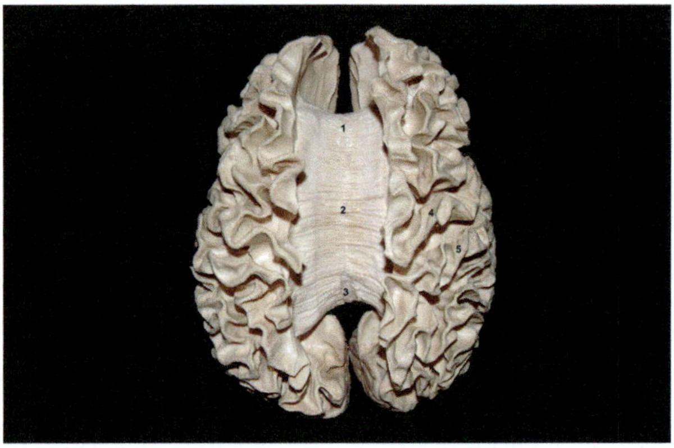

Fig. 8 Superior view of a brain dissected with the Klingler technique. The corpus callosum is exposed. (1) Genu of the corpus callosum; (2) Trunk of the corpus callosum; (3) Splenium of the corpus callosum; (4) Precentral gyrus; (5) Postcentral gyrus

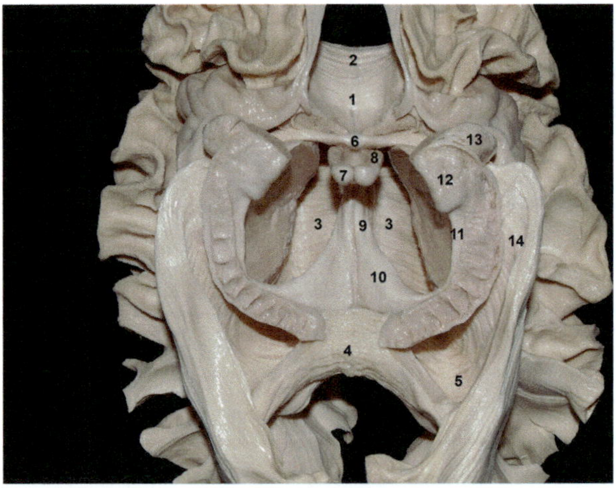

Fig. 9 Inferior view of a brain, where the brain stem and partially the thalamus were removed, to show demonstrating that the corpus callosum forms the roof of the lateral ventricles. (1) Rostrum of corpus callosum; (2) Genu of corpus callosum; (3) Body of corpus callosum; (4) Splenium of corpus callosum; (5) Tapetum; (6) Anterior commissure; (7) Mamillary body; (8) Column of the fornix; (9) Body of the fornix; (10) Crus of the fornix; (11) Fimbria; (12) Head of the hippocampus; (13) Amygdala; (14) Optic radiation

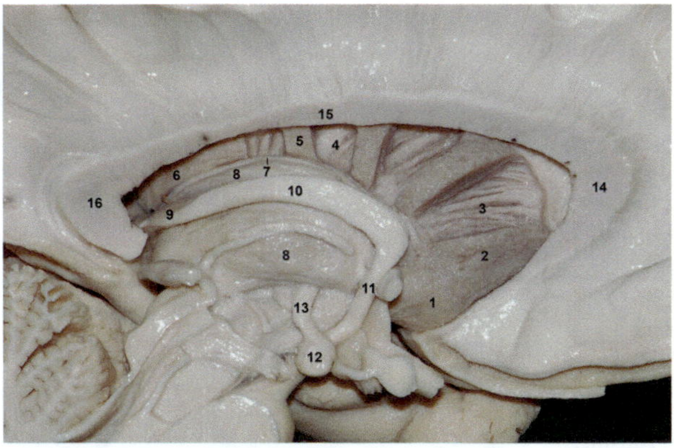

Fig. 10 Medial view of the left cerebral hemisphere, dissected with the Klingler technique. The structures that make up the lateral wall of the lateral ventricle are exposed. (1) Nucleus accumbens; (2) Head of caudate nucleus; (3) Anterior thalamic radiation; (4) Superior thalamic radiation; (5) Body of caudate nucleus; (6) Tail of caudate nucleus; (7) Stria terminalis; (8) Thalamus; (9) Crus of the fornix; (10) Body of the fornix; (11) Column of the fornix; (12) Mamillary body; (13) Mammillothalamic tract; (14) Genu of the corpus callosum; (15) Body of the corpus callosum; (16) Splenium of the corpus callosum

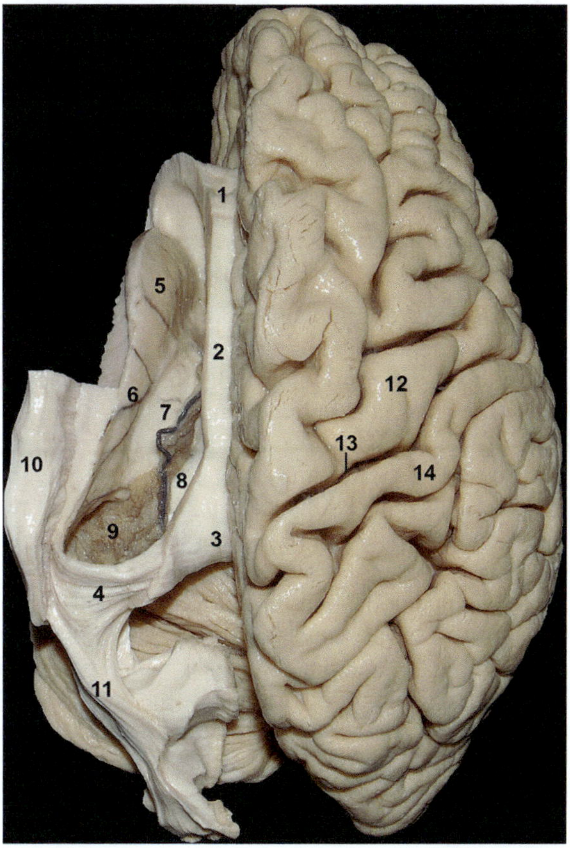

Fig. 11 Superior view. The left cerebral hemisphere was partially removed. The structures that form the floor of the frontal horn and the body of the lateral ventricle are exposed. (1) Genu of the corpus callosum; (2) Body of the corpus callosum; (3) Splenium of the corpus callosum; (4) Tapetum; (5) Head of caudate nucleus; (6) Body of caudate nucleus; (7) Thalamus; (8) Fornix; (9) Glomus of the choroid plexus; (10) Superior longitudinal fasciculus; (11) Optic radiation; (12) Precentral gyrus; (13) Central sulcus; (14) Postcentral gyrus

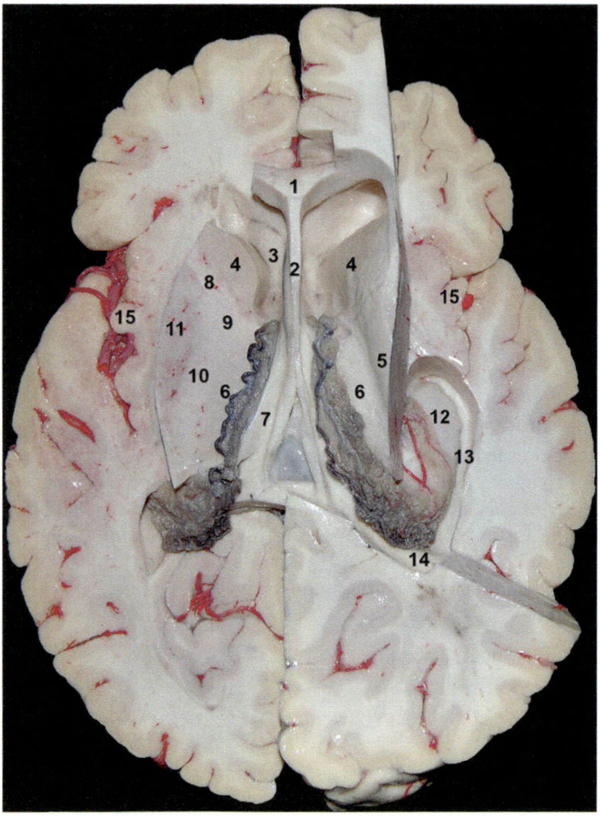

Fig. 12 Superior view of a brain. An axial section of both cerebral hemispheres was made at the level of the apex of the insula. (1) Genu of the corpus callosum; (2) Septum pellucidum; (3) Rostrum of the corpus callosum; (4) Head of caudate nucleus; (5) Body of caudate nucleus; (6) Thalamus; (7) Body of the fornix; (8) Internal capsule, anterior limb; (9) Internal capsule, genu; (10) Internal capsule, posterior limb; (11) Lentiform nucleus; (12) Head of the hippocampus; (13) Temporal horn of the lateral ventricle; (14) Atrium of lateral ventricle; (15) Insula

2.3 *Atrium and Occipital Horn*

The atrium and occipital horn form a triangular cavity with a posterior vertex, in the occipital lobe, and an anterior base in relation to the pulvinar of the thalamus. The atrium opens anteriorly above the thalamus, into the body of the lateral ventricle; below thalamus in temporal horn; and later it communicates with the occipital horn (Figs. 13 and 14) [14].

The roof of the atrium is formed by the body, splenium, and the tapetum of the corpus callosum. The medial wall is formed by two horizontal prominences: An upper one called the bulb of the corpus callosum, made up of a group of fibers called the greater forceps. A lower one called calcar avis, which is the intraventricular representation of the bottom of the calcarine fissure. The lateral wall has an anterior part, formed by the caudate nucleus, and a posterior part, formed by the tapetum (Fig. 9) [6]. The anterior wall has a medial portion, formed by the cross of the fornix; and a lateral portion, formed by the pulvinar of the thalamus. The floor is made up of the collateral trigone, which is the intraventricular representation of the bottom of the collateral sulcus.

The occipital horn extends posteriorly to the atrium to the occipital lobe although its size is variable and may even be absent.

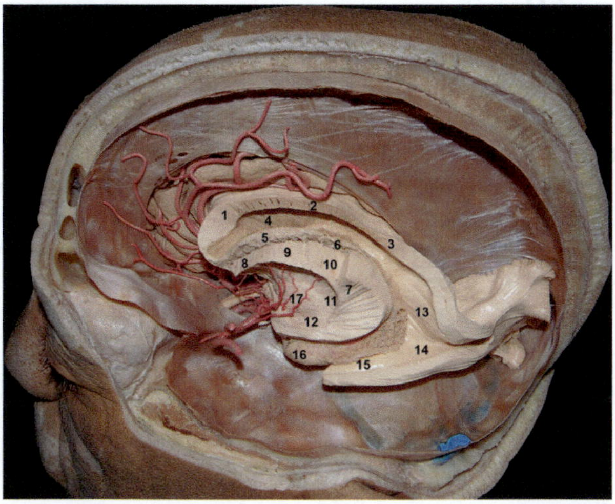

Fig. 13 Lateral view of a head. The arrangement of the C-shaped ventricular cavities around the thalamus is demonstrated. (1) Genu of the corpus callosum; (2) Body of the corpus callosum; (3) Splenium of the corpus callosum; (4) Septum pellucidum; (5) Head of caudate nucleus; (6) Body of caudate nucleus; (7) Corona radiata; (8) Internal capsule, anterior limb; (9) Internal capsule, genu; (10) Internal capsule, posterior limb; (11) Internal capsule, retrolenticular part; (12) Internal capsule, sublenticular part; (13) Atrium of lateral ventricle; (14) Collateral trigone; (15) Temporal horn of the lateral ventricle; (16) Amygdala

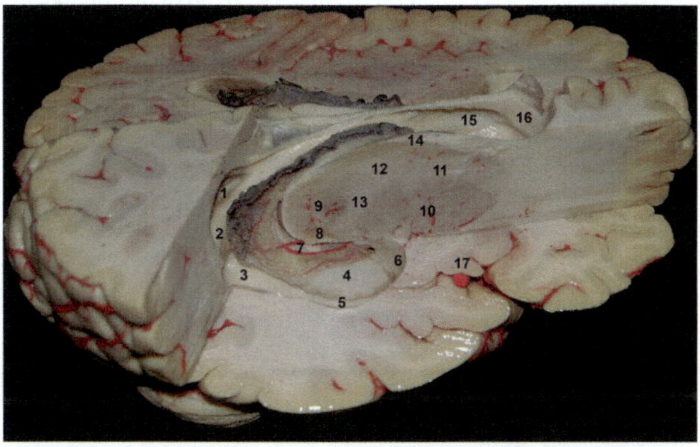

Fig. 14 Lateral view. The roof and the lateral wall of the right temporal horn as well as have been removed. The hippocampus proper occupies the medial part of the floor of the temporal horn and can be divided into three parts: head, body, and tail. The head of the hippocampus is the anterior and largest part of the hippocampus; it is directed anteriorly and then medially. (1) Bulb of the corpus callosum; (2) Calcar avis; (3) Collateral trigone; (4) Head of the hippocampus; (5) Temporal horn of the lateral ventricle; (6) Amygdala; (7) Fimbria; (8) Lateral geniculate body; (9) Thalamus; (10) Lentiform nucleus; (11) Internal capsule, anterior limb; (12) Internal capsule, genu; (13) Internal capsule, posterior limb; (14) Head of caudate nucleus; (15) Septum pellucidum; (16) Genu of the corpus callosum. (17) Insula

2.4 Temporal Horn

It extends from the atrium, below the pulvinar, to the medial portion of the temporal lobe, ending in its anterior wall located behind the amygdala. Its floor is formed medially by the hippocampus and laterally by the collateral eminence, a prominence that is the intraventricular representation of the bottom of the collateral sulcus that separates the parahippocampal and occipitotemporal gyri, on the lower surface of the temporal lobe (Figs. 15, 16, and 17). The roof in its medial portion is formed

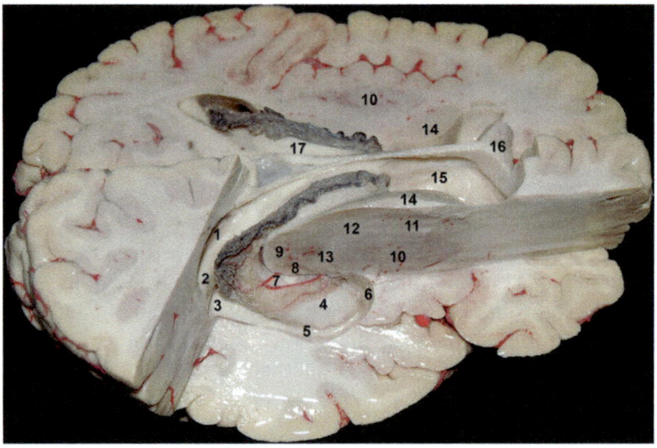

Fig. 15 Superior view of the same anatomical specimen as the previous image. (1) Bulb of the corpus callosum; (2) Calcar avis; (3) Collateral trigone; (4) Head of the hippocampus; (5) Temporal horn of the lateral ventricle; (6) Amygdala; (7) Fimbria; (8) Lateral geniculate body; (9) Thalamus; (10) Lentiform nucleus; (11) Internal capsule, anterior limb; (12) Internal capsule, genu; (13) Internal capsule, posterior limb; (14) Head of caudate nucleus; (15) Septum pellucidum; (16) Genu of the corpus callosum; (17) Body of the fornix

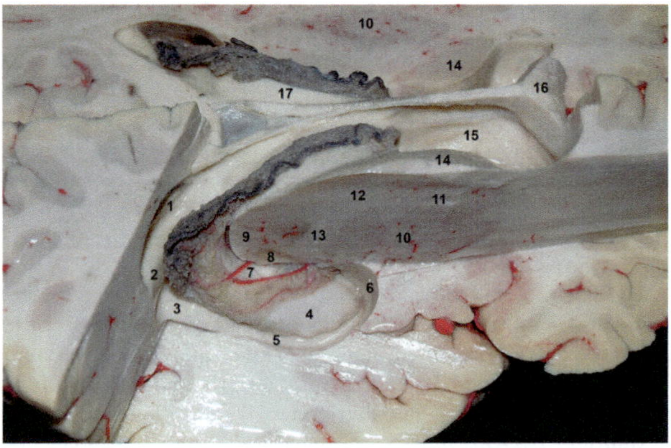

Fig. 16 Magnified view of the same anatomical specimen. (1) Bulb of the corpus callosum; (2) Calcar avis; (3) Collateral trigone; (4) Head of the hippocampus; (5) Temporal horn of the lateral ventricle; (6) Amygdala; (7) Fimbria; (8) Lateral geniculate body; (9) Thalamus; (10) Lentiform nucleus; (11) Internal capsule, anterior limb; (12) Internal capsule, genu; (13) Internal capsule, posterior limb; (14) Head of caudate nucleus; (15) Septum pellucidum; (16) Genu of the corpus callosum; (17) Body of the fornix

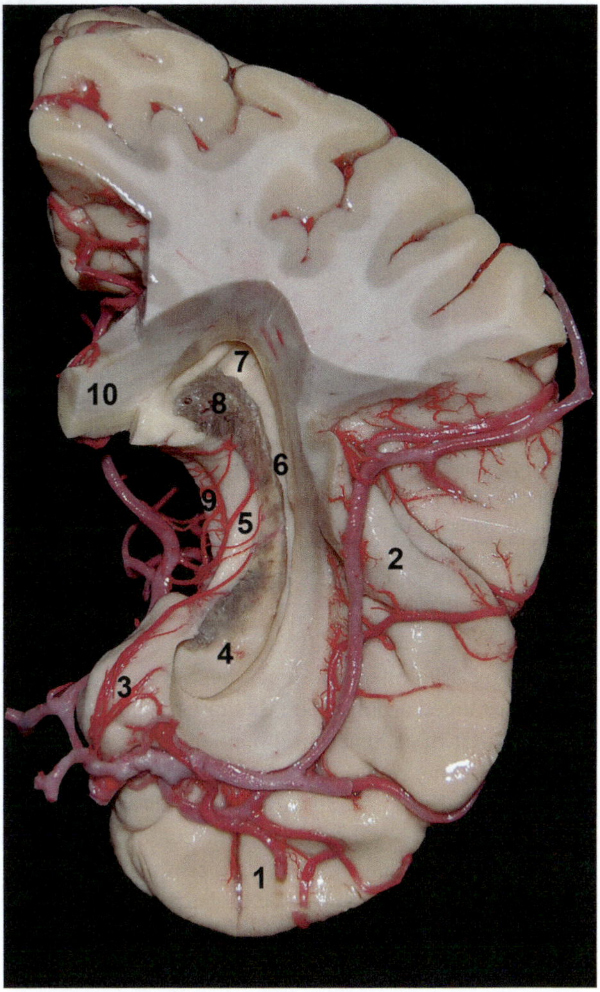

Fig. 17 Superior view of the left temporal lobe. It has been disconnected medially from the thalamus through the choroidal fissure, anteriorly and medially from the globus pallidus, anteriorly and laterally from the insula, and superiorly from the temporal stem. The superior surface is also called the opercular surface, and it is the temporal operculum of the sylvian fissure. It presents three morphologically distinct parts: the planum polare, Heschl's gyrus, and the planum temporal (usually formed by the middle and posterior transverse temporal gyri). (1) Planum polare; (2) Heschl's gyrus; (3) Uncus; (4) Head of the hippocampus; (5) Fimbria; (6) Temporal horn of the lateral ventricle; (7) Collateral trigone; (8) Glomus of the choroid plexus; (9) Dentate gyrus; (10) Splenium of the corpus callosum

by the lower surface of the thalamus and the tail of the caudate nucleus. The lateral portion of the roof is formed by the tapetum of the corpus callosum, which radiates inferiorly to later form the lateral wall of the temporal horn. The tapetum separates the temporal horn from optical radiation.

3 Discussion

Embryologically, the forebrain vesicle, around the fifth week, divides into two secondary vesicles: (1) the telencephalon, with its primitive cerebral hemispheres, and (2) The diencephalon, which develops the optic vesicles. The cavity in each cerebral hemisphere is known as the lateral ventricle [1, 2, 17, 18]. Throughout embryogenesis, the duplication of the forebrain vesicle and the subsequent C-folding of the

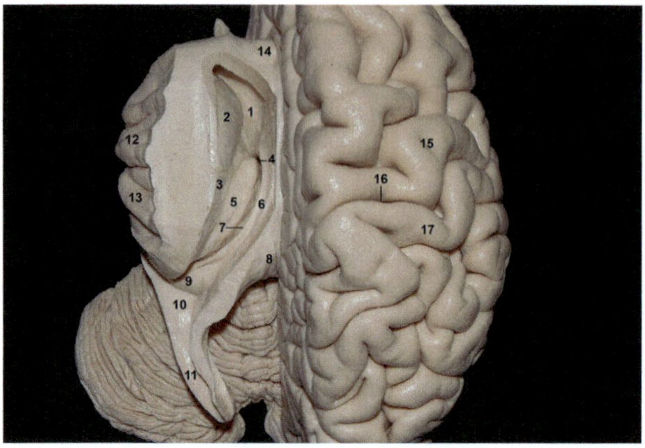

Fig. 18 Anatomical specimen where the left cerebral hemisphere was removed, preserving the central block and the lateral ventricle. (1) Rostrum of the corpus callosum; (2) Head of the caudate nucleus; (3) Body of the caudate nucleus; (4) Foramen of Monro; (5) Thalamus; (6) Fornix; (7) Choroidal fissure; (8) Splenium of the corpus callosum; (9) Body of the hippocampus; (10) Collateral trigone; (11) Occipital horn of the lateral ventricle; (12) Short gyri of the insula; (13) Long gyrus of the insula; (14) Genu of the corpus callosum; (15) Precentral gyrus; (16) Central sulcus; (17) Postcentral gyrus

entire telencephalon, causes each lateral ventricle, and the deep structures related to it, to be arranged around each thalamus, which by its Once, it becomes the morphological center of each cerebral hemisphere (Fig. 18).

The following radial folds of the cerebral surface around each thalamus give rise to the cerebral grooves and gyrations, determining with which the grooves of the superolateral and basal surface of the cerebral hemispheres point to the nearest ventricular cavity, which constitutes a useful feature for intra-operative topographic orientation from brain surface exposure [8, 19].

4 Surgical Applications

The choice of the surgical approach for lesions located in the lateral ventricles depends on the side of the origin of the lesion, its growth pattern, its location, and the existence or not of ventricular obstruction. Lesions located within the anterior portion of the lateral ventricle are reached by the transcallosal anterior and transcortical anterior approaches. The anterior transcallosal approach is useful for lesions located in the frontal horn and the body of the lateral ventricle. The transcortical approach is facilitated when the ventricle is dilated [7].

The temporal cortex that surrounds the temporal horn and the ventricular atrium has an important functional organization, including language, vision, memory, and music processing [20]. Therefore, temporal lobe surgery implies a precise knowledge of the intrinsic architecture of the brain (Fig. 19).

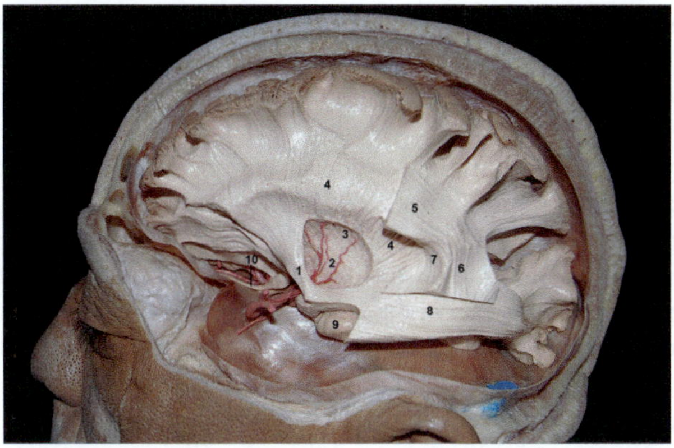

Fig. 19 Lateral view of the left cerebral hemisphere. Specimen dissected with the Klingler technique where the path of the association and projection fibers that make up the white matter of the brain is identified. (1) Occipitofrontal fasciculus; (2) Globus pallidus; (3) Internal capsule; (4) Corona radiata; (5) Fronto-parietal or horizontal segment of the superior longitudinal fasciculus; (6) Temporo-parietal or vertical segment of the superior longitudinal fasciculus; (7) Temporo-frontal segment or arcuate fasciculus; (8) Sagittal stratum; (9) Head of the hippocampus; (10) Olfactory tract

4.1 Approaches to the Temporal Horn and Mesial Structures

The approaches to these structures were grouped into three groups: the first, lateral approaches through the superior, middle, and inferior temporal gyri. The second, with a subtemporal approach through the fusiform or parahippocampal gyrus. The third, the Transsylvian approach.

Different approaches have been described for resection of the temporal lobe, from the amygdalohippocampectomy first described by Niemeyer in 1958 through an incision in the median temporal gyrus, which was only accepted as a treatment technique for patients with refractory epilepsy until 1980 mesial temporal lobe [21]. Median temporal gyrus approaches cause contralateral quadrantanopia. At the level of the median temporal gyrus, the lateral wall of the temporal horn and atrium is 22–26 mm from the cortical surface and at the basal level 10–14 mm from the floor of the middle fossa. Seven percent of the patients where this corridor is used develop visual deficits [20, 22, 23]. In addition, the median temporal gyrus approach carries at risk for Wernicke's aphasia in the dominant hemisphere since the sensory language area is in the superior and middle temporal gyrus approximately 5–6 cm behind the temporal pole. Olivier in 1992 [24] describes the approach via the superior temporal sulcus for surgical treatment of epilepsy.

In 1993, Hori et al. [25] described the approach through the fusiform gyrus through a subtemporal route. Park et al. [26] in 1996 described the subtemporal transparahippocampal approach for amygdalohippocampectomy in temporal lobe epilepsy surgery. Make a cortical incision at the uncus 1–1.5 cm posterior to the point where the third cranial nerve crosses the border of the tentorium. The temporal horn and anterior hippocampus are exposed through an incision of the parahippocampal gyrus 2–3 cm anterior hippocampus is removed subpially. They report homonymous quadrantanopia in a single patient out of seven and memory disorders in another.

The transsylvian transventricular approach proposed by Yasargil et al. [27–29], for selective amygdalohippocampectomy, with the opening of the sylvian fissure until identifying the lower limiting sulcus of the insula and making an opening of the same behind the limen of the insula until it enters the temporal horn. The advantage of this approach is to preserve the language area without damaging the cortical surface. The disadvantage that it requires a more complex surgical technique and that the incision of the temporal stem injures the uncinate fascicle, the anterior commissure, and part of the Meyer's loop.

In 1998, Vajkoczy et al. [30] proposed the transsylvian, transcisternal approach for mesial block resection in cases of hippocampal sclerosis. The medial surface of the parahippocampal gyrus and the rhinal sulcus are exposed. With in bloc resection of the anterior hippocampus and tonsil. Preserves the lateral basal surface of the temporal bone. 9% of patients present with third nerve palsy and three percent superior quadrantanopia.

Coopens et al. [31] describe the anteromedial approach to the temporal horn to avoid damage from optic radiation. Using the transsylvian approach, to identify the uncus and the anterior portion of the parahippocampal gyrus, a corticectomy is performed in the 3-cm piriformis cortex to enter the amygdala and resect it, medial to the uncinate fascicle, limen of the insula, and optic radiation.

Choi et al. [32] identified a safe triangular area below the floor of the sylvian fissure to access the temporal horn. With an incision at the level of the limen of the insula 5 mm posterior adjacent to the inferior insular sulcus, it is extended 10–20 mm posteriorly causing less damage to the Meyer's loop and the optical radiation. This triangular area is located between the medial border of the Meyer's loop and lateral to the optic tract, the base at the level of the limen of the insula, and the apex at the level of the anterior border of the lateral geniculate body. In the anterior portion of the triangle is the amygdala. The approach in this area passes through the tonsil to the temporal horn.

Yeni et al. [33] studied 30 patients who underwent selective amygdalohippocampectomy for secondary generalized complex partial seizures that did not respond to medical treatment, all of whom had normal visual fields examination. In the postoperative period, ten patients [36.6%] presented superior quadrantanopia, due to injury to the anterior beam of optical radiation [Meyer's loop]. In contrast, Yasargil [34] reports only 1 patient out of 73, with superior quadrantanopia after selective amygdalohippocampectomy.

4.2 Approaches to the Ventricular Atrium

Fornari et al. [35] describe the approach to the ventricular atrium through the superior parietal gyrus, making an incision 1 cm posterior to the postcentral sulcus, and extending posteriorly from 4 to 5 cm, it is widely used but causes neurological deficit, including apraxia, acalculia, visual deficit with homonymous hemianopia due to lesion of the three beams of optical radiation.

Nagata and Sasaki [36] propose the trans-sulcal, lateral insular approach for the ventricular atrium, through a horizontal incision of 15 mm in the posterior surface of the insula at the level of the Heschl's transverse gyrus indicating that it is not provoked optical radiation injury.

Yasargil [37, 38] describes the posterior interhemispheric, parieto-occipital approach to the ventricular atrium, generally to its medial wall. Make an incision through the pre-wedge, anterior to the parieto-occipital groove, avoiding damaging the optic radiation and the visual cortex. Rhoton [7] describes the approach to the atrium through an incision in the posterior part of the cingulate, lateral to the splenium of the corpus callosum, without compromising the optical radiation.

Muftah Lahirish et al. [39] taking aspects such as neuroanatomical dissection, tractography with fasciculus at risk and prospective clinical symptoms, they concluded that the interhemispheric parieto-occipital approach is the best option with less morbidity for the approach to the ventricular atrium.

Mahaney et al. [40] report that the medial interhemispheric parieto-occipital approach to the ventricular atrium avoids damage from optic radiation and the calcarine cortex. It proposes as entry point for the twist of the cingulate 5 mm superior and 5 mm posterior of the falcotentorial junction.

Ribas et al. [41] propose the superior parietal approach for the atrium through the anterior portion of the intraparietal sulcus at the junction with the postcentral sulcus. Similarly, when it comes to the non-dominant hemisphere, he recommends approaching the atrium through the posterior portion of the superior temporal sulcus.

5 Conclusion

Currently, the identification of the deeper brain structures in neuroimaging examinations is made from the initial observation of the ventricular compartments.

The study of neuroanatomy with anatomical specimens allows obtaining a deep understanding of the three-dimensional relationship of the external and internal morphology of the brain. It will allow us to choose the safest access routes in injuries that compromise the ventricular cavities.

References

1. Testut L, Latarjet A. Tratado de anatomía humana, vol. 2. 9th ed. Barcelona: Salvat; 1997.
2. Snell R. Neuroanatomía clínica. 4th ed. Argentina Bs. As: Panamericana; 1999.
3. Machado A. Neuroanatomia funcional. 2th ed. São Paulo: Atheneu; 1993.
4. Sociedade Brasileira de Anatomia. Anatomical Terminology. São Paulo: Manole; 2001.
5. Lockard I. Desk reference for neuroanatomy: a guide to essential terms. New York: Springer; 1977.
6. Ribas GC, Ribas EC, Rodrigues AJ Jr. The brain, the tridimensional vision, and the techniques to obtain stereoscopic images. Rev Med (Sao Paulo). 2006;85:78–90.
7. Rhoton AL Jr. Cranial anatomy and surgical approaches. Schaumburg: Lippincott Williams & Wilkins; 2003.
8. Ribas GC. Microanatomia cirúrgica dos pontos-chaves dos sulcos e giros cerebrais (tese livre-docência). São Paulo: Faculdade de Medicina, Universidade de São Paulo; 2005.
9. Timurkaynak E, Rhoton AL Jr, Barry M. Microsurgical anatomy and operative approaches to the lateral ventricles. Neurosurgery. 1986;19:685–23.
10. Wen HT, Rhoton AL Jr, Oliveira E. Transchoroidal approach to the third ventricle: an anatomic study of the choroidal fissure and its clinical application. Neurosurgery. 1998;42:1205–19.
11. Wen HT, Rhoton AL Jr, Oliveira E, Cardoso ACC, Tedeschi H, Baccanelli M, et al. Microsurgical anatomy of the temporal lobe: part I: mesial temporal lobe anatomy and its vascular relationships and applied to amygdalohippocampectomy. Neurosurgery. 1999;45:549–92.
12. Yasargil MG. Microneurosurgery: Vol IVA CNS Turnors: surgical Anatorny, neuropathology, neuroradiology, neurophysiology, clinical considerations, operability, treatment options. Stuttgart: Georg Thieme; 1996.
13. Sanan A, Abdel Aziz KM, Janjua R, Van Loveren H, Keller J. Colored silicone injection for use in neurosurgical dissection. Neurosurgery. 1999;45:1267–74.
14. Klingler J. Development of the macroscopic preparation of the brain through the process of freezing [in German]. Schweiz Arch Neurol Psychiatr. 1935;36:247–56.
15. Ludwig E, Klingler J. Atlas Cerebri Humani. Basel: S. Karger; 1956.
16. Türe U, Yasargil MG, Friedman AH, Al-Mefty O. Fiber dissection technique: lateral aspect of the brain. Neurosurgery. 2000;47:417–27.
17. Ribas GC. Telencéfalo. In: Meneses MS, editor. Neuroanatomia aplicada. Rio de Janeiro: Guanabara Koogan; 1999. p. 237–72.
18. Williams PL, Warwick R, editors. Gray's anatomy. 36th ed. Philadelphia: Saunders; 1980.
19. Harkey HL, Al-Mefty O, Haines DE, Smith RR. The surgical anatomy of the cerebral sulci. Neurosurgery. 1989;24:651–4.
20. Peltier J, Travers N, Destrieux C, Velut S. Optic radiations: a microsurgical anatomical study. J Neurosurg. 2006;105:294–300.
21. Niemeyer P. The transventricular amygdala-hippocampectomy in temporal lobe epilepsy. In: Baldwin M, Bailey P, editors. The temporal lobe epilepsy. Springfield, IL: Charles C Thomas; 1958. p. 461–82.
22. D'Angelo VA, Galarza M, Catapano D, Monte V, Bisceglia M, Carosi I. Lateral ventricle tumors: surgical strategies according to tumor origin and development—a series of 72 cases. Neurosurgery. 2005;56(Suppl 1):36–45.
23. Ebeling U, Reulen HJ. Neurosurgical topography of the optic radiation in the temporal lobe. Acta Neurochir. 1988;92:29–36.
24. Olivier A. Relevance of removal of limbic structures in surgery for temporal lobe epilepsy. Can J Neurol Sci. 1991;18(Suppl 4):628–35.
25. Hori T, Tabuchi S, Kurosaki M. Subtemporal amygdalohippocampectomy for treating medically intractable temporal lobe epilepsy. Neurosurgery. 1993;33:50–7.
26. Park TS, Bourgeois BF, Silbergeld DL, Dodson WE. Subtemporal transparahippocampal amygdalohippocampectomy for surgical treatment of mesial temporal lobe epilepsy. Neurosurg Focus. 1996;85:1172–6.

27. Wieser HG, Yasargil MG. Selective amygdalohippocampectomy as a surgical treatment of mesiobasal limbic epilepsy. Surg Neurol. 1982;17:445.
28. Yasargil MG. Impact of temporal lobe surgery. J Neurosurg. 2004;101:725–38.
29. Yasargil MG, Teddy PJ, Roth P, Selective amygdalo-hippocampectomy. Operative anatomy and surgical technique. Adv Tech Stand Neurosurg. 1985;12:93–123.
30. Vajkoczy P, Krakow K, Stodieck S, Pohlmann-Eden B, Schmiedek P. Modified approach for the selective treatment of temporal lobe epilepsy: transsylvian-transcisternal mesial en bloc resection. J Neurosurg. 1998;88:855–62.
31. Coppens J, Mahaney K, Abdulrauf S. An anteromedial approach to the temporal horn to avoid injury to the optic radiation fibers and uncinate fasciculus: anatomical and technical note. Neurosurg Focus. 2005;18(6b):E3.
32. Choi C, Rubino P, Fernandez-Miranda J, Rhoton A Jr. Meyer's loop and the optic radiations in the transsylvian approach to the mediobasal temporal lobe. Neurosurgery. 2006;59(Ons Suppl 4):ONS-228–ONS-36.
33. Yeni S, Tanriover N, Uyanik O, Onur M. Visual field defects in selective amygdalohippocampectomy for hippocampal sclerosis: the fate of meyer's loop during the transsylvian approach to the temporal horn. Neurosurgery. 2008;63:507–15.
34. Yaşargil MG, Krayenbühl N, Roth P, Hsu S, Yaşargil DC. The selective amygdalohippocampectomy for intractable temporal limbic seizures. J Neurosurg 2009.
35. Fornari M, Savoiardo M, Morello G, Solero CL. Meningiomas of the lateral ventricles. Neuroradiological and surgical considerations in 18 cases. J Neurosurg. 1981;54:64–74.
36. Nagata S, Sasaki T. Lateral transsulcal approach to asymptomatic trigonal meningiomas with correlative microsurgical anatomy: technical case report. Neurosurgery. 2005;56(Suppl 2):E438.
37. Yasargil MG. Microneurosurgery: microneurosurgery of CNS Tumors, vol. IVB. Stuttgart: Georg Thieme; 1996.
38. Yasargil MG, Türe U, Yasargil DC. Surgical anatomy of supratentorial midline lesions. Neurosurg Focus. 2005;18(6B):E1.
39. Muftah Lahirish IA, Middlebrooks EH, Holanda VM, Batista-Quintero R, Maeda FL, Neto MR, et al. Comparison between transcortical and interhemispheric approaches to the atrium of lateral ventricle using combined white matter fiber dissections and magnetic resonance tractography. World Neurosurg. 2020;138:e478–85.
40. Mahaney K, Abdulrauf S. Anatomic relationship of the optic radiations to the atrium of the lateral ventricle: description of a novel entry point to the trigone. Neurosurgery. 2008;63(ONS Suppl 2):ONS195–ONS203.
41. Ribas GC, Yasuda A, Ribas EC, Nishikuni K, Rodrigues AJ Jr. Surgical anatomy of microneurosurgical sulcal key points. Neurosurgery. 2006;59(Suppl 2):ONS177–ONS211.

Surgical Anatomy of the Third Ventricle

Joyce Koueik, Sima Sayyahmelli, and Mustafa K. Başkaya

1 Introduction

The surgical management of third ventricle (TV) tumors remains a challenge to neurosurgeons, since these are deep-seated tumors in a difficult to reach location, and are surrounded by vital neurovascular structures. The appropriate surgical approach depends on the tumor's location, origin, size, extension, and its laterality (Fig. 1). There are three broad approaches to the TV: anterior, lateral, and posterior. All three approaches require navigating through normal neural tissue. Thus, when choosing the right approach, one might choose a longer trajectory corridor to minimize brain retraction or risk injury to vital structures. In this chapter, we review the most important open surgical approaches with an emphasis on their potential key surgical pitfalls and nuances.

J. Koueik · S. Sayyahmelli · M. K. Başkaya (✉)
Department of Neurological Surgery, University of Wisconsin Hospital and Clinics, Madison, WI, USA
e-mail: koueik@neurosurgery.wisc.edu; sayyahmelli@neurosurgery.wisc.edu; m.baskaya@neurosurg.wisc.edu; baskaya@neurosurgery.wisc.edu

E. G. Figueiredo et al. (eds.), *Brain Anatomy and Neurosurgical Approaches*, https://doi.org/10.1007/978-3-031-14820-0_19

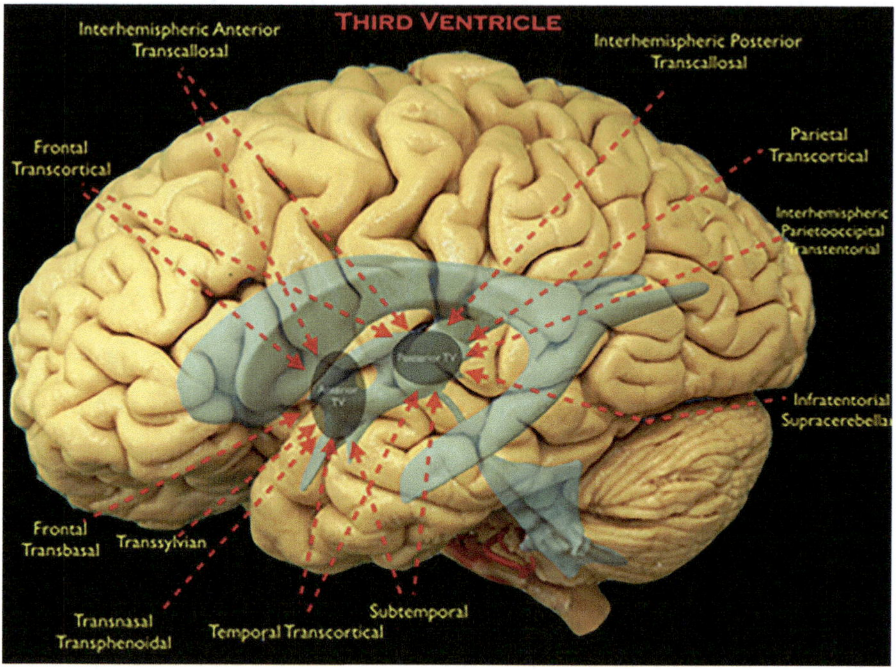

Fig. 1 Open surgical approaches to the third ventricle. (*1*) Anterior approaches include trancallosal, subtemporal, subfrontal, subchiasmatic, and trans-optico-carotid. (*2*) Lateral approaches include the subtemporal and pterional. (*3*) Posterior approaches include transcortical, transcallosal, occipital transtentorial, and supracerebellar infra-tentorial [2]. (This figure first appeared in Baskaya, 2016; Journal of Neuro-oncology)

2 Anterior Approaches

The anterior wall of the TV is formed in the rostral-to-caudal direction by the interventricular foramen (IVF), the anterior commissure, the lamina terminalis (LT), the optic recess, and the optic chiasm (Fig. 2). The anterior cerebral arteries pass anterior to that. The anterior roof is formed from the anterior to posterior direction by the IVF, the columns of fornix, the tela-choroida (which encloses the internal cerebral vein (ICV) and the medial posterior choroidal artery), the hippocampal commissure, the habenular commissure, and the supra-pineal recess [1]. Thus, access to the anterior TV requires traversing many critical structures.

After accessing the lateral ventricles using the interhemispheric anterior transcallosal approach (IATcA) or the frontal transcortical approach (FTA), one can reach the TV via: 1) the transforaminal, 2) the inter-forniceal, 3) the trans-choroidal, and 4) the sub-choroidal approaches (Fig. 3) [2–5].

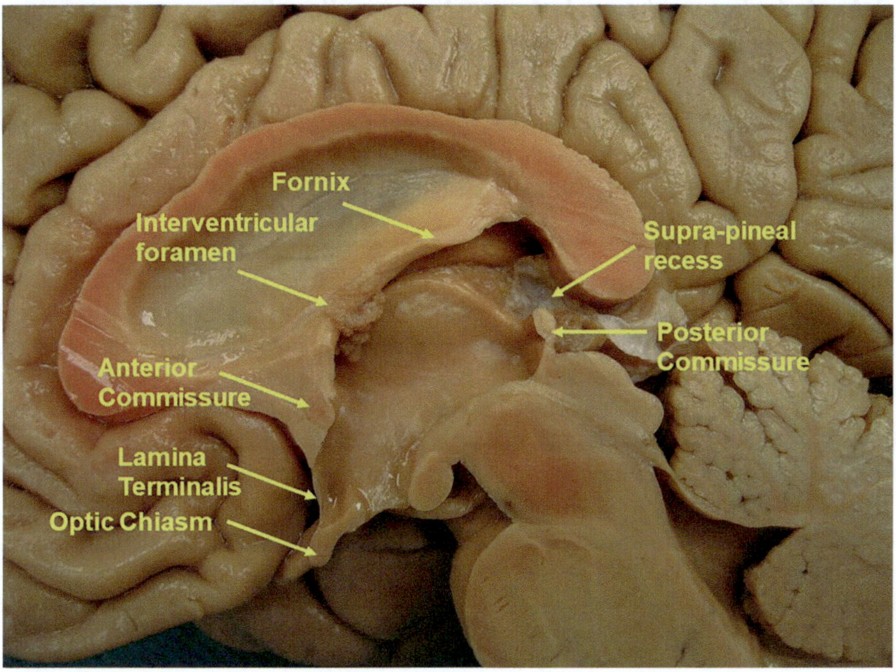

Fig. 2 Cadaveric view of the third ventricle anterior and posterior borders. Cadaveric view of the anterior and posterior third ventricle borders

Fig. 3 Approaches to reach the third ventricle. Cadaveric view shows that after accessing the lateral ventricles using the interhemispheric anterior trans-callosal approach (IATcA) or the frontal transcortical approach (FTA), one can reach the TV via these four approaches: (*1*) transforaminal (TF), (*2*) sub-choroidal (SC), (*3*) trans-choroidal (TC), and (*4*) inter-forniceal (IF)

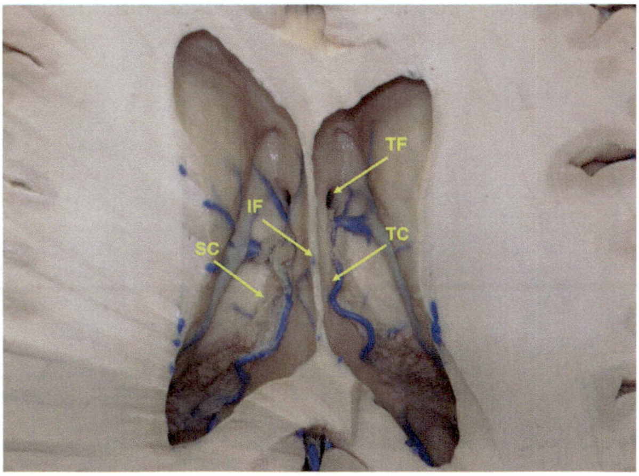

2.1 Interhemispheric Anterior Trans-callosal Approach (IATcA)

The distance to the TV via the IATcA is shorter that the FTA and is associated with less risk of post-operative seizures, contralateral hemiparesis, and porencephaly [6]. This approach can provide superior visualization of the entire TV, depending on which corridor is used.

The anterior interhemispheric approach requires meticulous dissection of the interhemispheric fissure to separate the falx cerebri from the medial surface of the superior frontal gyrus. Identification and preservation of the callosomarginal and peri-callosal arteries during the dissection are essential. At the depth of the fissure, the corpus callosum is identified by its pearly white appearance. A 1–2 cm long callosotomy is made in order to enter the lateral ventricle. It is important to use anatomic landmarks such as the thalamostriate vein and choroid plexus (CP) to enter the correct ventricle. Complications of this approach include injury to the superior frontal and cingulate gyri, along with injury to the peri-callosal arteries and bridging veins. Additional complications include injury to the genu of the internal capsule causing hemiplegia. Disconnection syndrome may also result from the callosotomy, while injury to the deep draining veins can cause basal nuclei and thalamic infarctions [7] (Fig. 4).

2.1.1 Transforaminal Approach

The transforaminal approach provides access to the TV through a natural opening and provides good visualization of small tumors in the anterior TV. Larger tumors

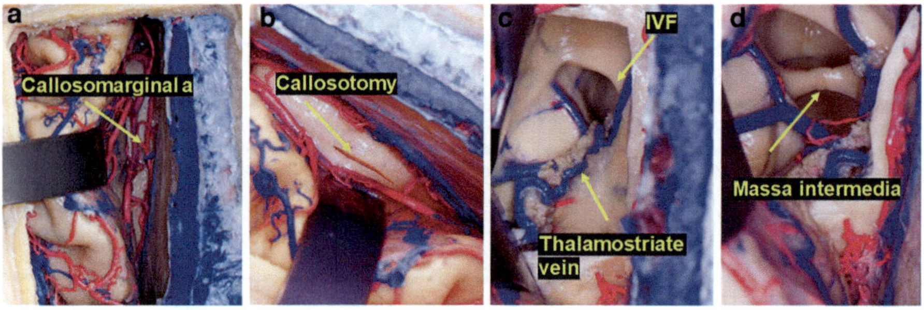

Fig. 4 Performing the anterior interhemispheric trans-callosal approach in a cadaver. (**a**) While dissecting the interhemispheric fissure to separate the falx cerebri from the medial surface of the superior frontal gyrus, the callosomarginal arteries are first identified and preserved followed by the peri-callosal arteries. (**b**) 1–2 cm callosotomy is made. (**c**) Right interventricular foramen is identified. Note the thalamostriate vein to the right of the choroid plexus. (**d**) Massa intermedia, also called the inter-thalamic adhesion, is observed in this case

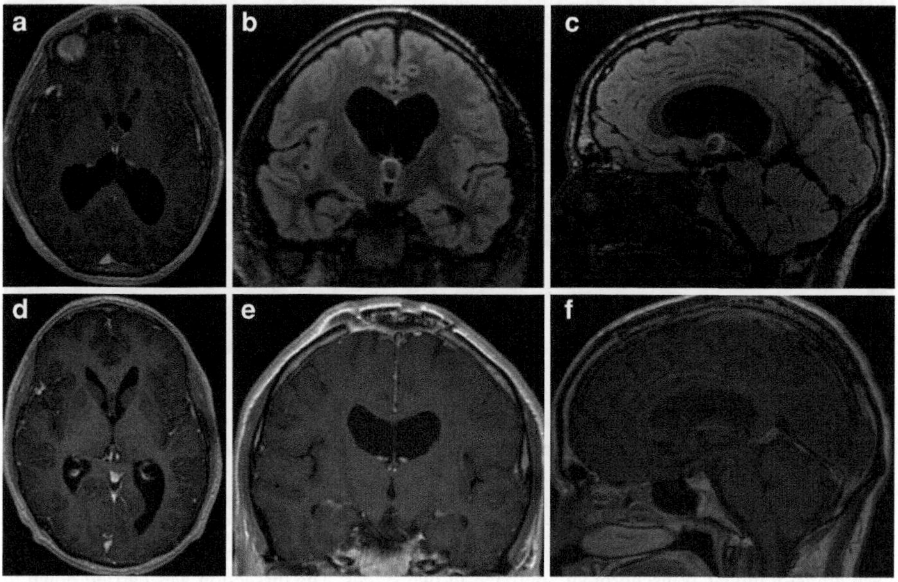

Fig. 5 Anterior interhemispheric trans-callosal transforaminal approach case history. A 33-year-old man presented with a 2-year history of intermittent right-sided hearing loss. He underwent magnetic resonance imaging (MRI) which revealed an incidentally found 1.0 × 1.3 cm colloid cyst within the foramen of Monroe causing hydrocephalus (**a–c**). He underwent microsurgical gross total resection of the colloid cyst via bifrontal craniotomy and interhemispheric transcallosal approach. The surgery and postoperative course were uneventful. At 18-month follow-up, MRI showed no residual or recurrent tumor (**d–f**) and that the patient was continuing to do well

might require extending this opening posteriorly by cutting the ipsilateral fornix or dividing the thalamostriate vein. The former has a significant risk of memory problems, while the latter has the risk of developing hemiplegia, mutism, drowsiness, hemorrhagic infarct in the basal nuclei, and even death [8, 9] (Fig. 5—case illustration).

2.1.2 Inter-forniceal Approach

The inter-forniceal approach provides access to the anterior and middle portions of the TC by dividing the midline forniceal raphe and opening the roof of the TV between the forniceal bodies up to 1.5 cm in length. The septum pellucidum can be a helpful anatomic landmark if there is no midline shift. A septum cavum can also minimize manipulations of the forniceal columns. This approach carries the risk of bilateral forniceal damage with resultant extensive memory problems, thus rendering this approach out of favor. In addition, there is a risk of injury to the ICVs internal cerebral veins and posterior choroidal arteries [2, 10].

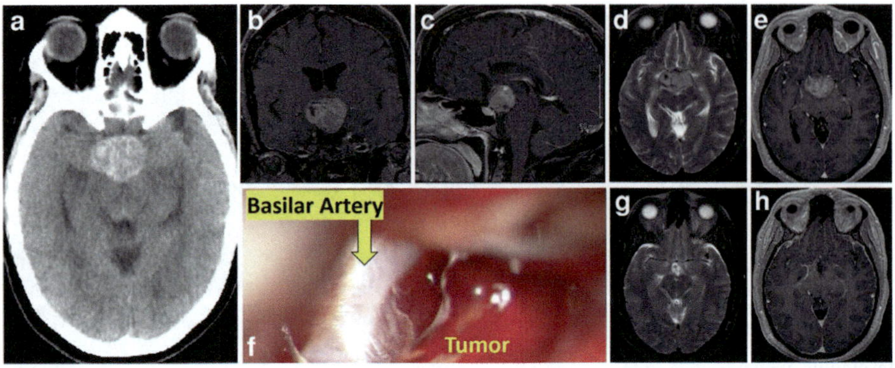

Fig. 6 Anterior interhemispheric trans-callosal trans-choroidal approach case history. A 42-year-old woman presented with sudden onset of severe headache and depressed level of consciousness. She had short-term memory problems and excessive urination for the last 2 weeks. Computed tomography (CT) scan of the head showed a hemorrhagic mass in the TV and suprasellar cisterns (**a**). MRI revealed a hemorrhagic enhancing 2.7 × 3.4 × 2.9 cm suprasellar and third ventricular mass (**b–e**). Ventriculostomy was inserted for obstructive hydrocephalus. She underwent microsurgical gross total resection via central midline craniotomy, and anterior interhemispheric transcallosal–transforaminal–transchoroidal approach in two stages. The dissection was carried down to the floor of the TV, where the basilar artery was identified, and a microsurgical third ventriculostomy was done by tumor removal (**f**). The histopathology was compatible with chordoid glioma of the TV, WHO grade II. The surgery and postoperative course were uneventful. She continued to have persistent diabetes insipidus which was treated with Desmopressin (DDAVP). She has completed the radiation therapy. Postoperative MRI showed no residual or recurrent tumor (**g, h**) [27]. (This figure first appeared in Baskaya, 2018; J Neurol Surg B Skull Base)

2.1.3 Trans-choroidal Approach

The choroid fissure lies between the floor of the lateral ventricle and the thalamus, and is just superior to the roof of the TV. Using this approach, one can access the middle to posterior portions of the TV. Once the choroidal fissure is open, the CP is freed from the fornix to expose the velum interpositum which contains the ICV. The ICV should be preserved as injury to this can result in venous infarct of the deep nuclei [3, 7, 11] (Fig. 6—case illustration).

2.1.4 Sub-choroidal Approach

The sub-choroidal approach is similar to the trans-choroidal approach, except that with the sub-choroidal approach, the CP is retracted medially, thus away from thalamus. This necessitates coagulating and dividing the thalamostriate vein, thereby opening a corridor between the thalamus and the CP. Even though the fornix is well-protected in this approach, this approach carries the risk of injury to the thalamus, the stria medullaris thalami, to the anterior and superior thalamic veins, and the thalamostriate vein and choroidal arteries, thereby rendering the sub-choroidal approach out of favor [11, 12].

2.2 *Sub-frontal Approach*

The sub-frontal approach is useful for small anterior TV tumors that do not extend superiorly or posteriorly. It has several modifications: 1) trans-lamina terminalis, 2) optic-carotid, 3) subchiasmatic, and 4) transnasal and transsphenoidal approaches [13, 14].

2.2.1 Trans-lamina Terminalis

The trans-lamina terminalis modification is useful when the chiasm is prefixed, and the LT is stretched by the tumor [15, 16].

2.2.2 Opticocarotid Approach

The opticocarotid modification is most useful when the chiasm is fixed and the tumor is extending antero-superiorly [13, 15, 17].

2.2.3 Subchiasmatic Approach

The subchiasmatic approach modification is best when the optic chiasm is fixed or post-fixed and the tumor is located anteriorly with inferior extension as this approach provides limited viewing of the IVF and the roof of the TV. Craniopharyngiomas extending into the TV are the most common tumors removed via this approach [13, 15, 17].

3 Lateral Approaches

The lateral walls of the TV are formed by the thalamus and hypothalamus [18].

3.1 *Sub-temporal Approach*

The sub-temporal approach is indicated when the tumor is located lateral to the sella turcica or extending into the middle cranial fossa. Usually, these tumors are medial to the perforating branches of the posterior communicating artery; thus, it is often inevitable that these vessels are sacrificed [16, 18].

3.2 Trans-sylvian Approach

After a wide dissection of the sylvian fissure and opening of the LT, the trans-sylvian approach provides access to the anterior TV. This approach is commonly used for TV craniopharyngiomas [1, 19, 20].

4 Posterior Approaches

The posterior wall is formed in the rostral-to-caudal direction by the splenium of the corpus callosum, the supra-pineal recess, the pineal gland, the pineal recess, the posterior commissure, and the cerebral aqueduct. Three approaches are commonly used to access tumors in the posterior part of the TV: 1) supra-cerebellar infra-tentorial (ScItA), 2) interhemispheric posterior trans-callosal approach (IPTcA), and 3) occipital trans-tentorial approach (OTtA) [21–23].

4.1 Supra-cerebellar Infra-tentorial (ScItA)

The Sclta provides access to areas from the transverse fissure of the cerebellum, the quadrigeminal plate, the medial upper cerebellar peduncle, the pineal region, and the posterior TV. However, this surgical corridor to the TV is long and narrow and is thus not suitable for tumors extending superiorly and laterally, since the exposure is limited by the angle of the tentorium. During this approach, careful dissection is required to avoid injury of the vein of Galen, the basal vein of Rosenthal, the ICVs, and the precentral cerebella vein, as well as the petrosal and dorsal cerebellar bridging veins [22, 24] (Fig. 7—case illustration).

4.2 Interhemispheric Posterior Trans-callosal Approach

The interhemispheric posterior trans-callosal approach is similar to the ScItA; however, a callosotomy through the posterior CC is performed. This is indicated when the tumor is located in the posterior TV and is extending superiorly toward the splenium of the CC. Here, the veins are mobilized posteriorly. Transecting the posterior CC can involve injury to the habenular and posterior commissure and thus result in memory problems and disconnection syndrome [23, 25, 26].

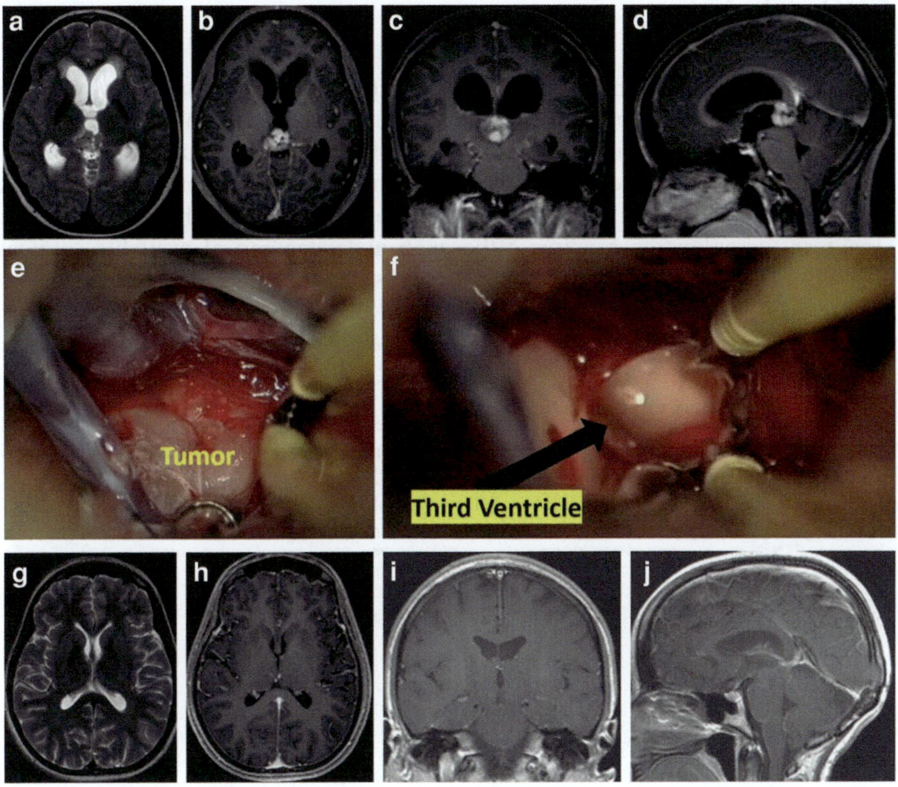

Fig. 7 Supracerebellar infratentorial approach case history. A 27-year-old woman with history of hydrocephalus who underwent endoscopic third ventriculostomy and biopsy of the papillary tumor of the pineal region at an outside hospital, 2 years prior to presenting to out institution. Due to the enlargement of the tumor on follow-up studies, she was referred for surgical resection. MRI revealed an enhancing, partially cystic mass in the pineal region measuring 2.4 × 2.3 × 2.1 cm, extending into the TV (**a–d**). She underwent microsurgical gross total resection (**e, f**) via midline supracerebellar infratentorial approach under intraoperative monitoring with somatosensory evoked potentials (SSEPs) and motor evoked potentials (MEPs). The histopathology was compatible with the papillary tumor of the pineal region. The surgery and postoperative course were uneventful and neurologic examination including the extraocular eye movements remained intact. She completed the radiation therapy. At the 5-year follow-up appointment, MRI showed no residual or recurrent tumor (**g–j**) and she was still doing well [28]. (This figure first appeared in Baskaya, 2018; Oper Neurosurg (Hagerstown). 2018;15(6):E87)

4.3 *Occipital Trans-tentorial Approach*

The occipital trans-tentorial approach is indicated for tumors in the pineal region and posterior TV with supra-tentorial extension. This approach does not require a callosotomy; however, there is high risk of injury to the tectum and the pulvinar as well as the occipital lobes if retraction is performed [21].

5 Conclusions

Surgical excision is an important predictor of the outcome for tumors within the ventricular system. The origin, type, location and size of the tumor, the age of the patient and their co-morbidities, any limitations in positioning, and the tumor patho-anatomy should all be carefully considered when choosing the appropriate approach for resecting TV tumors. Achieving a gross total resection of the tumor without significant complication requires a thorough understanding of available surgical approaches and their relative advantages and disadvantages.

Acknowledgements Figure 1 was previously published in Microsurgical resection of tumors of the lateral and third ventricles: Operative corridors for difficult-to-reach lesions. J Neurooncol. 2016;130(2):331-40.

Figure 6 was previously published in Gross Total Resection of Chordoid Glioma of the Third Ventricle via Anterior Interhemispheric Transcallosal Transforaminal Approach at Two Stages. J Neurol Surg B Skull Base. 2018; 79(Suppl 3):S281-S282.

Figure 7 was previously published in Microsurgical resection of posterior third ventricular/ pineal region papillary tumor via supracerebellar infratentorial approach: 3-dimensional operative video. Oper Neurosurg 2018; 15(6):E87.

Disclosure The authors report no financial conflicts of interest.

References

1. Rhoton AL. The lateral and third ventricles. In: Rhoton's cranial anatomy and surgical approaches. Oxford: Oxford University Press; 2003. p. 207–71.
2. Cikla U, Swanson KI, Tumturk A, Keser N, Uluc K, Cohen-Gadol A, et al. Microsurgical resection of tumors of the lateral and third ventricles: operative corridors for difficult-to-reach lesions. J Neuro-Oncol. 2016;130(2):331–40.
3. Rhoton AL Jr, Yamamoto I, Peace DA. Microsurgery of the third ventricle: part 2. Operative approaches. Neurosurgery. 1981;8(3):357–73.
4. Timurkaynak E, Izci Y, Acar F. Transcavum septum pellucidum interforniceal approach for the colloid cyst of the third ventricle operative nuance. Surg Neurol. 2006;66(5):544–7; discussion 7.
5. Winkler PA, Ilmberger J, Krishnan KG, Reulen HJ. Transcallosal interforniceal-transforaminal approach for removing lesions occupying the third ventricular space: clinical and neuropsychological results. Neurosurgery. 2000;46(4):879–88; discussion 88–90.
6. Danaila L, Radoi M. Surgery of tumors of the third ventricle region. Chirurgia (Bucur). 2013;108(4):456–62.
7. Yasargil MG, Abdulrauf SI. Surgery of intraventricular tumors. Neurosurgery. 2008;62(6 Suppl 3):1029–40; discussion 40–1.
8. Lavyne MH, Patterson RH Jr. Subchoroidal trans-velum interpositum approach to mid-third ventricular tumors. Neurosurgery. 1983;12(1):86–94.
9. Ture U, Yasargil MG, Al-Mefty O. The transcallosal-transforaminal approach to the third ventricle with regard to the venous variations in this region. J Neurosurg. 1997;87(5):706–15.

10. Fornari M, Savoiardo M, Morello G, Solero CL. Meningiomas of the lateral ventricles. Neuroradiological and surgical considerations in 18 cases. J Neurosurg. 1981;54(1):64–74.
11. Strugar J, Piepmeier JM. Approaches to lateral and third ventricle tumors. In: Schmidek HH, Sweet WH, editors. Operative neurosurgical techniques: indications, methods, and results. Philadelphia: WB Saunders; 2000. p. 837–51.
12. Cossu M, Lubinu F, Orunesu G, Pau A, Sehrbundt Viale E, Sini MG, et al. Subchoroidal approach to the third ventricle. Microsurgical anatomy. Surg Neurol. 1984;21(4):325–31.
13. Herrmann HD, Winkler D, Westphal M. Treatment of tumours of the pineal region and posterior part of the third ventricle. Acta Neurochir. 1992;116(2–4):137–46.
14. Shucart WA, Stein BM. Transcallosal approach to the anterior ventricular system. Neurosurgery. 1978;3(3):339–43.
15. Johnson RR, Baehring J, Piepmeier J. Surgery for third ventricular tumors. Neurosurg Q. 2003;13(3):207–25.
16. Oi SSA, Samii M. Operative techniques for tumors in the third ventricle. Op Tech Neurosurg. 2003;6(4):205–14.
17. Kulwin C, Chan D, Ting J, Hattab EM, Cohen-Gadol AA. Endoscopic endonasal transplanum transtuberculum resection of a large solid choroid plexus papilloma of the third ventricle. J Clin Neurosci. 2014;21(7):1263–6.
18. Piepmeier JM, Westerveld M, Spencer DD, Sass KJ. Surgical management of intraventricular tumors of the lateral ventricles. In: Schmidek HH, Sweet WH, editors. Operative neurosurgical techniques: indications, methods, and results. Philadelphia: WB Saunders; 1995. p. 725–38.
19. Yasargil MG. Microneurosurgery. Microneurosurgery of CNS tumors, vol. IVB. Stuttgart: Georg Thieme Verlag; 1996. p. 38–42, 56–7, 63–5, 313–23.
20. Yasargil MG, Ture U, Yasargil DC. Impact of temporal lobe surgery. J Neurosurg. 2004;101(5):725–38.
21. Fukui M, Natori Y, Matsushima T, Nishio S, Ikezaki K. Operative approaches to the pineal region tumors. Childs Nerv Syst. 1998;14(1–2):49–52.
22. Laborde G, Gilsbach JM, Harders A, Seeger W. Experience with the infratentorial supracerebellar approach in lesions of the quadrigeminal region, posterior third ventricle, culmen cerebelli, and cerebellar peduncle. Acta Neurochir. 1992;114(3–4):135–8.
23. Schijman E. Microsurgical anatomy of the transcallosal approach to the ventricular system, pineal region and basal ganglia. Childs Nerv Syst. 1989;5(4):212–9.
24. Lozier AP, Bruce JN. Surgical approaches to posterior third ventricular tumors. Neurosurg Clin N Am. 2003;14(4):527–45.
25. Benes V. Advantages and disadvantages of the transcallosal approach to the III ventricle. Childs Nerv Syst. 1990;6(8):437–9.
26. Jia W, Ma Z, Liu IY, Zhang Y, Jia G, Wan W. Transcallosal interforniceal approach to pineal region tumors in 150 children. J Neurosurg Pediatr. 2011;7(1):98–103.
27. Dogan I, Ucer M, Baskaya MK. Gross total resection of chordoid glioma of the third ventricle via anterior interhemispheric transcallosal transforaminal approach at two stages. J Neurol Surg B Skull Base. 2018;79(Suppl 3):S281–S2.
28. Sayyahmelli S, Dogan I, Baskaya MK. Microsurgical resection of posterior third ventricular/pineal region papillary tumor via supracerebellar infratentorial approach: 3-dimensional operative video. Oper Neurosurg. 2018;15(6):E87.

Surgical Anatomy of the Fourth Ventricle

Dan Zimelewicz Oberman, Matías Baldoncini, Alvaro Campero, and Pablo Ajler

1 Introduction

The fourth ventricle is a midline cavity between the brainstem and the cerebellum [1]. It is composed of several important internal structural elements of nuclei and fasciculi related to prominences or depression on the surface of the floor of the fourth ventricle, which must be identified by the neurosurgeon [2–5]. Knowledge of the anatomy of the IV ventricle is required to plan the most appropriate approaches, choose the safest entry zones, and minimize possible surgical complications.

2 Anatomy

The fourth ventricle is a broad, tent-shaped midline cavity between the cerebellum and the brainstem. It is connected rostrally through the aqueduct with the third ventricle, caudally through the foramen of Magendie with the cisterna magna, and laterally through the foramina of Luschka with the cerebellopontine angles (CPA) [1, 6].

The roof expands laterally and posteriorly from its narrow rostral end just below the aqueduct to the fastigium and lateral recess level. The floor of this tent-shaped

D. Z. Oberman (✉)
Neurosurgery Department, Hospital de Força Aérea do Galeão, Rio de Janeiro, RJ, Brazil

M. Baldoncini
Department of Neurological Surgery, Hospital San Fernando, Buenos Aires, Argentina

A. Campero
Neurosurgery Department, Hospital Italiano de Buenos Aires, Buenos Aires, Argentina

P. Ajler
Neurosurgery Department, Padilla Hospital, Tucumán, Argentina

E. G. Figueiredo et al. (eds.), *Brain Anatomy and Neurosurgical Approaches*, https://doi.org/10.1007/978-3-031-14820-0_20

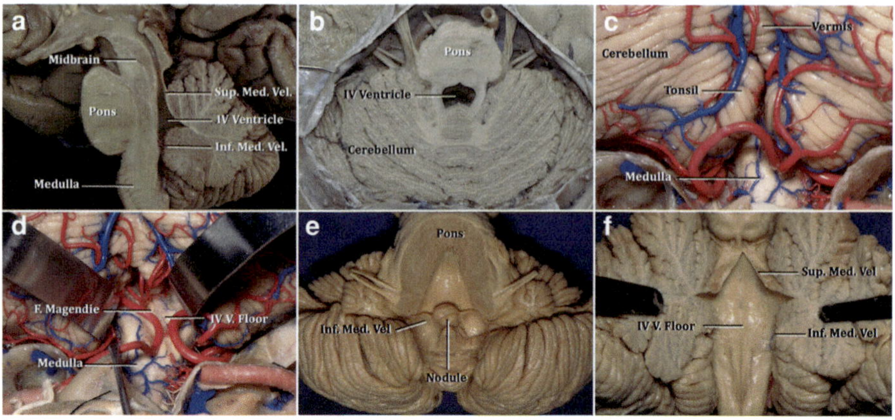

Fig. 1 (**a**) Lateral view of the brainstem and cerebellum. The superior and inferior medullary velum resembles a tent with its apex, the fastigium, pointing posteriorly. (**b**) Superior axial view of cross-section at the level of the pons. (**c**, **d**) Posterior view of the midline suboccipital craniotomy. The cerebellomedullary fissure is adherent to the tonsil. The medullotonsillar space is located between the medulla and the tonsil. Opening the cerebellomedullary cistern and gently retracting cerebellar tonsils laterally exposes the tela choroidea, the foramen of Magendie, and both PICAs. (**e**) Anterior view of the fourth ventricle. A portion of the pons and the whole medulla have been removed to provide this view into the roof of the fourth ventricle. Inferior medullary velum is suspended from the inferior roof of the fourth ventricle, which sweeps laterally from the surface of the nodule and blends into the flocculus at the level of the lateral recess. (**f**) Posterior view of the floor of fourth ventricle. The cerebellar peduncles have been sectioned and the vermis of the cerebellum removed in order to expose the floor of the fourth ventricle. The floor of the IV ventricle is formed by the dorsal surfaces of the pons and the rostral medulla

cavity faces anteriorly, and the pons and the medulla form it. The apex of the roof, the fastigium, divides it into superior and inferior parts. The superior part is distinctly different from the inferior part in that the inferior part is formed largely by thin membranous layers, and thicker neural structures form the superior part (Fig. 1).

3 Fourth Ventricle Roof

3.1 Upper Ventricular Roof

The ventricular surface of the superior half of the roof is formed by two parts, medially by the superior medullary velum and laterally by the ventricular surface of the superior cerebellar peduncles. The rostral portion of the ventricular surface of each lateral wall is formed by the medial surface of the superior cerebellar peduncle. The superior portion of the roof is formed by the lingula of the vermis, which adheres to the outer surface of the superior medullary velum, and is bordered on each side by the superior cerebellar peduncles. The superior peduncle fibers arise in the dentate nucleus located lateral to the fastigium.

3.2 *Inferior Ventricular Roof*

The nodule, tela choroidea, and inferior medullary velum formed the inferior ventricular roof. They are located in the upper portion of the cerebellomedullary fissure, the complex cleft that extends superiorly between the cerebellum and the medulla and is intimately related to the inferior half of the roof of the fourth ventricle.

The ventricular surface of the lower half of the roof is divided into a cranial portion, formed by the nodule and inferior medullary velum, and a caudal portion, formed by the tela choroidea. The inferior medullary velum is a membranous layer connecting the nodule and the flocculus. It is separated from the superior pole of both tonsils by narrow rostral extensions of the cerebellomedullary fissure. The inferior medullary velum is continuous at the level of the fastigium with the superior medullary velum. Caudally, it is attached to the tela choroidea.

The tela choroidea sweeps inferiorly from the telovelar junction around the superior pole of each tonsil to its attachment to the inferolateral edges of the floor along narrow white ridges, the taeniae, which meet at the obex. Cranially, the taeniae turn in a lateral direction over the inferior cerebellar peduncles and pass horizontally along the inferior borders of the lateral recesses. The tela choroidea does not completely enclose the inferior half of the fourth ventricle. However, it has three openings into the subarachnoid space: the paired foramina of Luschka, located at the outer margins of the lateral recesses, and the foramen of Magendie, located in the midline at the caudal tip of the fourth ventricle.

4 Fourth Ventricle Floor

The floor of the fourth ventricle has a rhomboid shape and is limited superolaterally by the superior cerebellar peduncles and inferolaterally by the inferior cerebellar peduncles and gracile cuneate tubercles (Fig. 2) [1, 7]. It presents a strip between the lower margin of the cerebellar peduncles and the site of attachment of the tela choroidea; this strip, called the junctional part, is characterized by the striae medullary that extend into the lateral recesses. The floor of the fourth ventricle is created by the pons on its superior two-thirds and by the medulla on its inferior one-third. Its cranial apex is directed to the cerebral aqueduct, Its caudal tip, directed toward the obex. The medullary part of the floor and its lateral angles open through the lateral recesses and foramina of Luschka into the CPA.

The floor of the fourth ventricle is divided into three parts [1]: (1) a superior or pontine part; (2) an intermediate or junctional part; and (3) an inferior or medullary part:

- Pontine part: it has a triangular shape. Its apex is located at the cerebral aqueduct; its base is represented by an imaginary line connecting the lower margin of the cerebellar peduncles, and its lateral limits are formed by the ventricular surface of the cerebral peduncles.

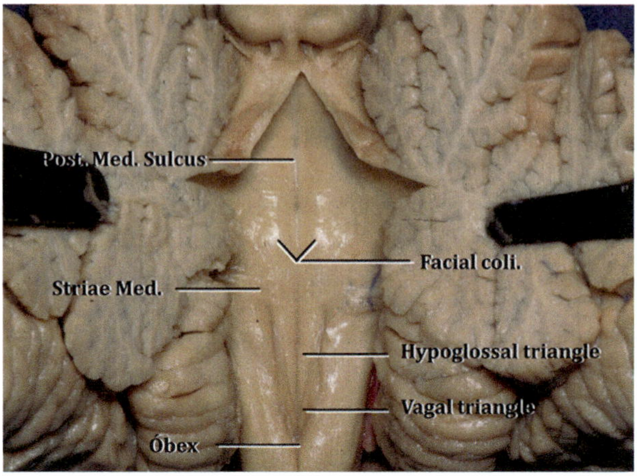

Fig. 2 Superior view of the floor of the fourth ventricle. The cerebellum, superior and inferior medullary velums have been removed to reveal the floor of the fourth ventricle. Coli: Coliculli; Med: median, medullary

- Junctional part: it is the strip between the lower margin of the cerebellar peduncles and the site of attachment of the tela choroidea to the taeniae just below the lateral recess. The intermediate part extends into the lateral recesses.
- Medullary part: the inferior part has a triangular shape and is limited laterally by the taeniae that mark the floor's inferolateral margin. Its caudal tip, the obex, is anterior to the foramen of Magendie.

Longitudinally, the floor of the fourth ventricle is divided from the rostral to the caudal tip into symmetrical halves by the median sulcus. The sulcus limitans, another longitudinal sulcus, divide each half of the floor into a raised median strip called the median eminence and a lateral strip called the area vestibular. The motor nuclei of the cranial nerves are located medial to the sulcus limitans, and the sensory nuclei are located lateral to it.

In the pontine part, each median eminence contains two rounded prominences, facial colliculi, located on each side of the median sulcus. The facial colliculi are limited laterally by the superior fovea, a dimple formed by the sulcus limitans. In medullary part, it presents the configuration of a feather, or pen nib, called the calamus scriptorius, with three triangular areas overlying the hypoglossal and vagus nuclei (hypoglossal and vagal trigones), and the area postrema; just lateral to the hypoglossal trigone, the sulcus limitans sulcus, presents another dimple called the inferior fovea. At the junctional part, the sulcus limitans is discontinuous. The locus ceruleus is located at the upper end of the sulcus limitans. The vestibular area is widest in the intermediate part of the floor, where it forms a rounded elevation that extends into the lateral recess, is crossed by the striae medullary, that crosses the midportion of the floor, and overlies the vestibular nuclei. The auditory tubercle, a prominence in the lateral portion of the vestibular area, overlies the dorsal cochlear nucleus and the cochlear part of the vestibulocochlear nerve.

5 Cerebellar–Brainstem Fissures

The cerebellum wraps around the posterior surface of the brainstem to create three cerebellar–brainstem fissures: cerebellomesencephalic, cerebellopontine, and cerebello-medullary related to the roof and lateral recesses of the fourth ventricle.

The suboccipital surface of the cerebellum, the surface that borders the cerebellomedullary fissure and faces the occipital bone, has a deep vertical depression, the posterior cerebellar incisura, into which the vermis is folded between the hemispheres, the cerebellomedullary fissure. This fissure is one of the most complex fissures in the brain, and is through this fissure; the fourth ventricle is accessed.

The cerebellomedullary fissure extends superiorly between the cerebellum and medulla. The ventral wall is formed by the posterior surface of the medulla, the inferior medullary velum, and the tela choroidea. The dorsal wall is formed by the uvula and nodule of the vermis medially and the tonsils and biventral lobules laterally. The cerebellomedullary fissure is separated from the fourth ventricle by the tela choroidea and inferior medullary vellum and communicates with the ventricle through the foramen of Magendie. The cerebellomedullary fissure also communicates below the lateral recess and around the foramen of Luschka with the cerebellopontine fissure. CNs IX–XII are located in the cerebellomedullary cistern just ventrolateral-to-the lateral margin of the cerebellum–medullary fissure. The cisternal space of cerebemedullary fissure has been divided into three spaces: the supratonsillar space rostrally, uvulo-tonsillar space medially, and medullotonsillar space ventrally, and faces the cisterna magna dorsally and caudally.

6 Lateral Recesses

The lateral recesses are narrow curved pouches extending laterally below the cerebellar peduncles, which open through the foramina of Luschka into the CPA. It has a ventral, rostral, caudal, and posterior wall. The ventral wall of each lateral recess is formed by the junctional part of the floor and the rhomboid lip, a sheetlike layer of neural tissue that extends laterally from the floor and unites with the tela choroidea to form a pouch at the outer extremity of the lateral recess. The rostral wall of each lateral recess is formed by the caudal margin of the cerebellar peduncles. The inferior cerebellar peduncle courses upward in the floor, forming the ventral wall of the lateral recess, and turns posteriorly at the lower portion of the pons to form the ventricular surface of the rostral wall of the recess. The caudal wall is formed by the tela choroidea, which stretches from the taenia and attaches to the edge of the peduncle of the flocculus. The biventral lobule is dorsal to the lateral recess. The flocculus extends laterally from the superior edge of the outer extremity of the lateral recess. The rootlets of the glossopharyngeal and vagus nerves arise ventrally, and the facial nerve rostral, to the choroid plexus, which extends through the lateral recess and the foramen Luschka into the CPA.

7 Choroidal Plexus

The choroid plexus in the fourth ventricle is located in the roof and lateral recesses. It is composed of paired inverted L-shaped fringes that arise on the ventricular surface of the tela choroidea. Each of the paired fringes of the plexus has a longitudinal limb, the medial segment, which stretches from the nodule level to the foramen of Magendie. A transverse limb, the lateral segment, originates from the rostral end of the medial segment and extends parallel to the telovelar junction through the lateral recesses and the foramen of Luschka into the CPA. The entire structure presents the form of a letter T, the vertical limb of which is double.

Each medial segment is subdivided into a rostral or nodular part and a caudal or tonsillar part. The nodular parts are widest at their junction with the lateral segments. The tonsillar parts are anterior to the tonsils and extend inferiorly through the foramen of Magendie. The rostral and caudal ends of the medial segments are often fused. Each lateral segment is subdivided into a medial or peduncular part and a lateral or floccular part. The peduncular part forms a narrow fringe that is continuous with the rostral part of the medial segment and is attached to the tela choroidea covering the lateral recess inferior to the cerebellar peduncles. The floccular part is continuous with the peduncular part at the lateral margin of the cerebellar peduncles and protrudes through the foramen of Luschka into the CPA below the flocculus.

8 Vascular Relationships

Each wall of the fourth ventricle has surgically important arterial relationships. There are three cerebellar arteries, which comprise the superior cerebellar artery, which is intimately related to the superior half of the roof; the anterior–inferior cerebellar artery (AICA) is intimately related to the lateral recess and the foramen of Luschka, and the basilar and vertebral arteries give rise to many perforating branches that reach the floor of the fourth ventricle; and the posterior–inferior cerebellar artery (PICA), which is intimately related to the inferior half of the roof. They traverse the brainstem, cerebellar peduncles, and cerebellar–brainstem fissures, respectively, before reaching the cerebellar surface. The principal surgical approach to the fourth ventricle is through the suboccipital surface, exposing the PICA, where it dips into the cerebellomedullary fissure and terminates by supplying the suboccipital surface of the cerebellum.

The PICA is the most inconstant cerebellar artery regarding its course and distribution area. It usually originates from the posterior or lateral margin of the vertebral artery, near the olive, courses around the medulla oblongata from the anterior to the posterior aspects of the brainstem to reach near the foramen of Magendie. It runs near the roots of low cranial nerves, first passing the hypoglossal nerve above, below, or through its roots, and then, the glossopharyngeal, vagus, and accessory nerves usually located below or at the same level as this artery. Then, it passes through the cerebellomedullary fissure between the tonsil and the posterior part of the roof of the fourth ventricle, reaches the cerebellar hemispheric and vermian surfaces, and finally supplies the suboccipital cerebellar surface.

The artery has been divided into five segments. The PICA segment coursing in the cleft between the tonsil on one side and the tela and velum on the opposite side is called the "telovelotonsillar segment." This segment of PICA, which forms a superior loop, which forms a convex rostral curve in its course around the rostral pole of the tonsil, is also referred to as either the "cranial" or "supratonsillar loop." The apex of the cranial loop faces the inferior medullary velum. From this PICA segment, the choroidal branches to the tela and choroid plexus arise. The segment, which passes across the posterior medulla, often forms a caudally convex loop that coincides with the caudal pole of the tonsil, but it may also course superior or inferior to the caudal pole the tonsil without forming a loop. Most PICAs bifurcate into a medial and a lateral trunk in their passage around the tonsil. The medial trunk ascends to supply the vermis and the adjacent hemisphere, and the lateral trunk passes laterally over the tonsil to supply most of the hemispheric and tonsillar surfaces. The AICA main trunks are infrequently exposed during an operation directed through the cerebellomedullary fissure, a course near the foramen of Luschka, where they extend small choroidal branches to the tela and choroid plexus in the lateral recess.

The vein of the cerebellomedullary fissure originates on the lateral edge of the nodule and uvula, courses laterally near or along the junction between the inferior medullary velum and the tela choroidea, and dorsal or ventral to the flocculus, to reach the CPA. The vein of the inferior cerebellar peduncle courses on the peduncle parallel and several millimeters lateral to the inferolateral edge of the fourth ventricle, from the foramen of Magendie to the lateral recess. The cerebellomedullary fissure and inferior cerebellar peduncle veins drain into the CPA through the communication between the cerebellomedullary and cerebellopontine fissures empty into the vein of the middle cerebellar peduncle near the lateral end of the pontomedullary sulcus.

9 Approaches to the Fourth Ventricle

Lesions of the fourth ventricle have posed a special challenge to neurosurgeons because of the severe deficits that may follow injury to the ventricular walls and floor [8]. Historically, tumors of the fourth ventricle were operated on either by splitting the vermis, removing a part of the cerebellum, or dissecting the cerebellomedullary fissure [6, 9]. In the past, a common approach to the fourth ventricle and lateral recess has consisted of splitting the vermis on the suboccipital surface or removing a portion of one cerebellar hemisphere [8, 10, 11]. However, vermian lesions may cause equilibratory disturbances with truncal ataxia, staggering gait, oscillation of the head and trunk, nystagmus, and cerebellar mutism [12–14]. Because of this, the development of new routes was necessary.

In examining the clefts and walls of the cerebellomedullary fissure, the microsurgical anatomy of the posterior fossa revealed detailed information for neurosurgeons. Several authors reported [15–17] that opening both the tela choroidea and inferior medullary velum and extending the telar opening laterally toward the foramen of Luschka exposes the ventricle cavity from aqueduct to obex and laterally to the peduncular surface bordering the recess. The telovelar approach was initially described by Matsushima et al. [1] and has been extensively studied by other authors. They have found that precise dissection and opening of the cerebellomedullary, which is the entrance of the ventricle, offer a sufficient operative view not only of the medial region of the ventricle but also of the lateral part of the fissure and the deep area toward the cerebral aqueduct, without splitting the vermis. The inferior medullary velum, another paper-thin layer, can also be opened if the tela opening does not provide adequate exposure. Opening the inferior medullary velum accesses the latter areas and the superior half of the roof. Extending the telar opening laterally toward the foramen of Luschka opens the lateral recess and exposes the peduncular surfaces bordering the recess [1, 6]. If better exposure of the superolateral recess is needed, then removing the ipsilateral tonsil will provide a direct view of this area. However, care should be taken while performing this approach, since the dentate nucleus is just above the superior pole of the tonsil [6, 9, 16]. One of the main limitations of the telovelar approach is to access deep rostral tumor attachment compared with the transvermian approach, which was confirmed in cadaveric dissections [18–20].

In addition, cadaveric and clinical studies related to lesions involving the fourth ventricle classified exposure to the fourth ventricle into three types: extensive opening, lateral wall opening, and lateral recess opening [17]. Tanriover et al. [14] reported the microanatomy and exposures gained through the trans-cerebellomedullary and trans-vermian approaches. Both approaches provide access to the entire width of the floor of the fourth ventricle. The major difference between the two approaches is the exposure of the lateral recess and the foramen of the Luschka. Even in the midline suboccipital approach, the trans-cerebellum medullary approach exposes the lateral and superolateral recesses and the foramen of the Luschka. It allows observation of the lateral end of the fourth ventricular lesion. The trans-vermian approach, which offers an incision through at least the lower third of the vermis, affords a modest increase of the operator's working angle compared to the trans-cerebellum medullary approach when accessing the rostral half of the fourth ventricle. In addition, the trans-cerebellomedullary approach can access the lateral part of the fourth ventricle [13, 14]. In addition, the recent use of endoscopy for cerebellomedullary fissure opening surgery has been investigated, and more development of more advanced equipment will improve this method of surgery in the future [21].

10 Case Examples

10.1 Case 1 Subependymoma

A 33-year-old man with no relevant clinical record was admitted with a 5 day history of oppressive occipital headache and posterior cervical pain. Neurologic examination revealed photophobia and neck stiffness without additional neurological focus. An unenhanced CT scan demonstrates a well-circumscribed isointense mass located in the inferior portion of the fourth ventricle with a spontaneously hyperdense acute extratumoral hemorrhage in the cisterna magna (Fig. 1). Contrast-enhanced magnetic resonance imaging (MRI) demonstrated a well-delimited non-enhanced tumor, hypointense T1-weighted and hyperintense on T2-weighted images, involving the floor of the fourth ventricle extending caudally into the cervical spinal canal via the foramen magnum (Fig. 3). The patient was operated on immediately through a midline suboccipital craniectomy and a telovelotonsillar approach. The cisterna magna was opened, and hemorrhagic CSF under high pressure was aspirated sequentially. A large blood clot was removed and a firm, well-delimited, yellow–white partially cyst-lobulated, macroscopically hypovascular lesion was recognized and completely excised, which was attached to the right lateral recess and the floor of the fourth ventricle previous devascularization from

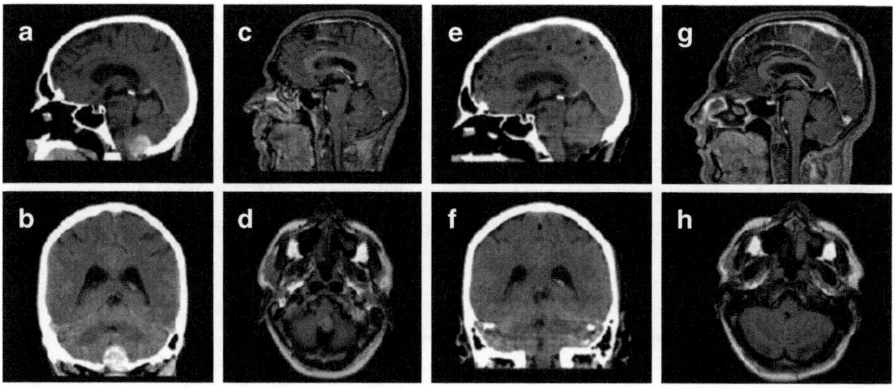

Fig. 3 (**a**, **b**) Shows sagittal and coronal CT views demonstrating acute extra-tumoral hemorrhage between the cisterna magna and the atlas. (**c**, **d**) Represents sagittal and axial views of a non-enhancing fourth ventricle tumor on T1-weighted contrast MRI sequences. (**e**, **f**) shows sagittal and coronal postoperative CT views demonstrating surgical approach and hemorrhage resolution. (**g**, **h**) Represents T1-weighted contrast MRI sagittal and axial views demonstrating complete tumoral resection. *CT* computed tomography, *MRI* magnetic resonance imaging

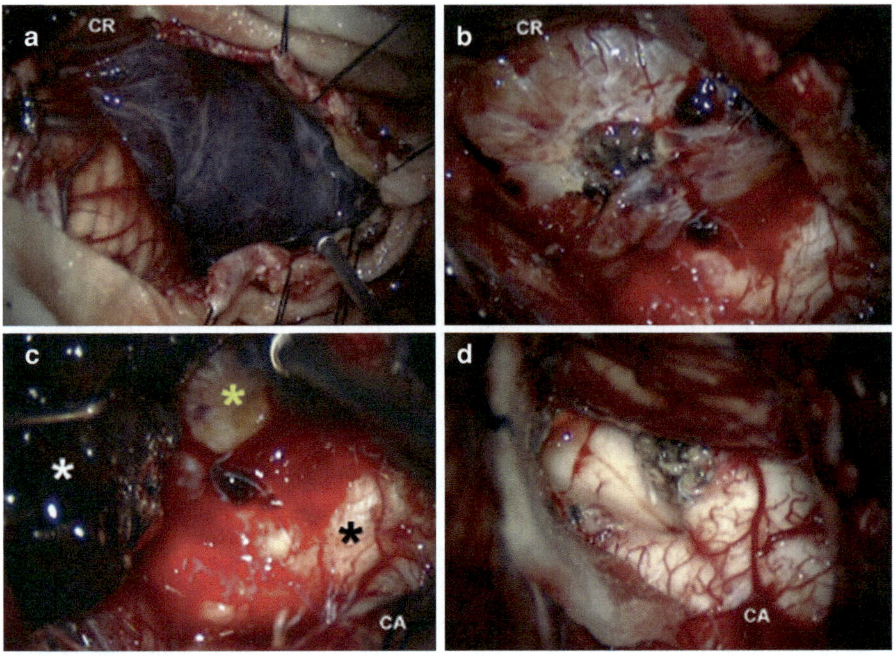

Fig. 4 (**a**) intraoperative photo demonstrate subarachnoid hemorrhage (CR, cranial). (**b**) white asterisk shows large blood clot, yellow revealed fourth ventricle tumor and the black asterisk mark the brain stem (CA, caudal). (**c**) Represents tumor site of hemorrhage. (**d**) Shows complete tumor excision and the floor of fourth ventricle

small branches of the right PICA (Fig. 4). Intra- and postoperative immunohistopathological examination revealed a subependymoma. The patient has a normal postoperative course and was discharged on the fifth postoperative day.

10.2 Case 2 Hemangioblastoma

A 38-year-old woman presented with a history of nausea, vomiting, atypical headache, and loss of consciousness. MRI revealed hydrocephalus, and a well-circumscribed contrast enhancement disclosed a nonhomogeneous lobulated mass filling the cavity of the fourth ventricle, without sharp demarcation toward the surrounding tissue of the cerebellar convexity. The patient was treated by external ventricular drainage and microsurgical telovelar approach and resection of the lesion, which was compatible with anaplastic hemangioblastoma (Fig. 5). The patient had a normal postoperative course and was discharged on the seventh postoperative day. A postoperative MRI study confirmed complete resection of the tumor with normal CSF circulation.

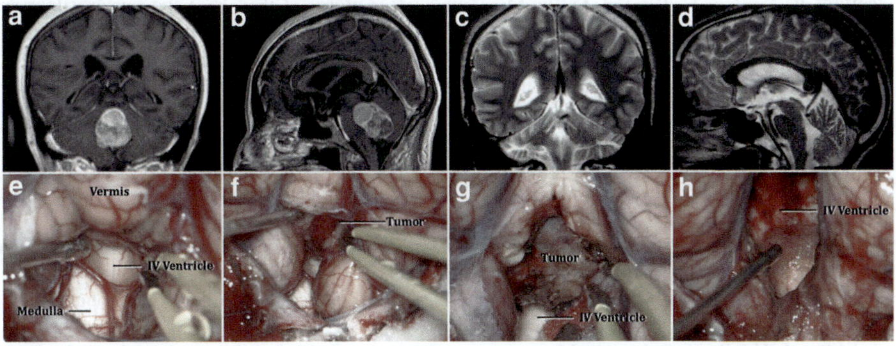

Fig. 5 (**a**, **b**) Shows coronal and sagittal views demonstrating enhancing fourth ventricle tumor on T1-weighted contrast MRI sequences. (**c**, **d**) Shows coronal and sagittal postoperative MRI demonstrating complete tumor resection. (**e**) Intraoperative photo of the fourth ventricle. (**f**, **g**) The tonsils were dissected and retracted laterally, exposing the tumor. (**h**) Shows complete tumor excision of the lesion and the floor of fourth ventricle

11 Conclusions

The fourth ventricle is an important anatomical space. Opening cerebellomedullary fissure provides wide exposure of the brainstem surfaces, segments of the cerebellar arteries and veins, and vital neural structures hidden by the cerebellum. The operative outcome depends on a clear knowledge of the microsurgical anatomic relationships between the neurovascular structures in the operative corridors to the fourth ventricle.

References

1. Matsushima T, Rhoton AL, Lenkey C. Microsurgery of the fourth ventricle: part 1. Neurosurgery. 1982;11(5):631–67. https://doi.org/10.1227/00006123-198211000-00008.
2. Párraga RG, Possatti LL, Alves RV, Ribas GC, Türe U, de Oliveira E. Microsurgical anatomy and internal architecture of the brainstem in 3D images: surgical considerations. J Neurosurg. 2016;124:1377–95.
3. Abla AA, Lekovic GP, Turner JD, de Oliveira JG, Porter R, Spetzler RF. Advances in the treatment and outcome of brainstem cavernous malformation surgery: a single-center case series of 300 surgically treated patients. Neurosurgery. 2011;68:403–14; discussion 414–5.
4. Abla AA, Turner JD, Mitha AP, Lekovic G, Spetzler RF. Surgical approaches to brainstem cavernous malformations. Neurosurg Focus. 2010;29:E8.
5. Bertalanffy H, Benes L, Miyazawa T, Alberti O, Siegel AM, Sure U. Cerebral cavernomas in the adult. Review of the literature and analysis of 72 surgically treated patients. Neurosurg Rev. 2002;25:1–53; discussion 54–5.
6. Matsushima T, Fukui M, Inoue T, Natori Y, Baba T, Fujii K. Microsurgical and magnetic resonance imaging anatomy of the cerebellomedullary fissure and its application during fourth ventricle surgery. Neurosurgery. 1992;30(3):325–30. https://doi.org/10.1227/00006123-199203000-00003.

7. Fujii K, Lenkey C, Rhoton AL Jr. Microsurgical anatomy of the choroidal arteries. Fourth ventricle and cerebellopontine angles. J Neurosurg. 1980;52:504–24.
8. Kellogg JX, Piatt JH Jr. Resection of fourth ventricle tumors without splitting the vermis: the cerebellomedullary fissure approach. Pediatr Neurosurg. 1997;27:28–33. https://doi.org/10.1159/000121221.
9. Matsushima T. Microsurgical anatomy of the cerebellomedullary fissure and variations of the transcerebellomedullary fissure approach. In: Microsurgical anatomy and surgery of the posterior cranial fossa. Tokyo: Springer; 2015. p. 73–99. https://doi.org/10.1007/978-4-431-54183-7_7.
10. Erşahin Y, Mutluer S, Cağli S, Duman Y. Cerebellar mutism: report of seven cases and review of the literature. Neurosurgery. 1996;38:60–5;discussion 66.
11. Herb E, Thyen U. Mutism after cerebellar medulloblastoma surgery. Neuropediatrics. 1992;23:144–6.
12. Miller J, Hdeib A, Cohen A. Management of tumors of the fourth ventricle. In: Schmidek and Sweet operative neurosurgical techniques: indications, methods, and results, vol. 1. 6th ed. Philadelphia: Elsevier; 2012. p. 367–97. https://doi.org/10.1016/b978-1-4160-6839-6.10031-0.
13. Deshmukh VR, Figueiredo EG, Deshmukh P, Crawford NR, Preul MC, Spetzler RF. Quantification and comparison of telovelar and transvermian approaches to the fourth ventricle. Neurosurgery. 2006;58:ONS202–6; discussion ONS206–7.
14. Tanriover N, Ulm AJ, Rhoton AL Jr, Yasuda A. Comparison of the transvermian and telovelar approaches to the fourth ventricle. J Neurosurg. 2004;101:484–98.
15. Matsushima K, Yagmurlu K, Kohno M, Rhoton AL Jr. Anatomy and approaches along the cerebellar-brainstem fissures. J Neurosurg. 2016;124:248–63.
16. Matsushima T, Inoue T, Inamura T, Natori Y, Ikezaki K, Fukui M. Transcerebellomedullary fissure approach with special reference to methods of dissecting the fissure. J Neurosurg. 2001;94:257–64.
17. Matsushima T. Microsurgical anatomy and surgery of the posterior cranial fossa: surgical approaches and procedures based on anatomical study. Tokyo: Springer; 2015.
18. Eissa E. The role of telovelar approach in fourth ventricular surgery: a new perspective. Turk Neurosurg. 2018;28(4):523–9. https://doi.org/10.5137/1019-5149.jtn.21209-17.1.
19. Han S, Wang Z, Wang Y, Wu A. Transcerebellomedullary fissure approach to lesions of the fourth ventricle: less is more? Acta Neurochir. 2013;155:1011–6.
20. Tomasello F, Conti A, Cardali S, La Torre D, Angileri FF. Telovelar approach to fourth ventricle tumors: highlights and limitations. World Neurosurg. 2015;83:1141–7.
21. Yang L, Zhang H, Wang X, Yan Z, Chen L, Ji X, et al. Midline suboccipital endoscopic transcerebellomedullary fissure keyhole approach. J Craniofac Surg. 2017;28:1603–6.

Part V
The Skull Base

Surgical Anatomy of the Anterior Fossa

Bradley Kolb and Andre Beer-Furlan

1 Anterior Cranial Fossa Anatomy

The anterior cranial fossa forms the bony floor beneath the olfactory bulbs and tracts, as well as the basal frontal lobe. It also functions as the roof of the nasal cavity and the orbits. Its posterior border is the sphenoid ridge of the lesser wing of the sphenoid bone laterally and the limbus of the sphenoid bone in the midline.

The ethmoid bone forms the anterior aspect of the anterior cranial fossa in the midline. This spongy, highly pneumatized bone has four parts, two lateral masses known as labyrinths, the vertical perpendicular plate, and the horizontal cribriform plate. The cribriform plate presents an olfactory fossa on its superior surface, over which the olfactory bulbs lie. The crista galli protrudes upward in the midline, forming the attachment point for the falx cerebri.

The posterior aspect of the crista galli articulates with the frontal crest of the frontal bone, and this articulation forms the so-called foramen cecum, which in some cases is occupied by a bridging vein that connects a nasal emissary vein to the superior sagittal sinus. The anterior-most aspect of the cribriform plate houses the two nasal slits, foramina through which the anterior ethmoidal nerves pass. The vertical plate of the ethmoid bone forms the superior anterior part of the nasal

B. Kolb
Department of Neurological Surgery, Rush University Medical Center, Chicago, IL, USA

A. Beer-Furlan (✉)
Department of Neurosurgical Oncology, Moffitt Cancer Center, Tampa, FL, USA

E. G. Figueiredo et al. (eds.), *Brain Anatomy and Neurosurgical Approaches*,
https://doi.org/10.1007/978-3-031-14820-0_21

septum, articulating posteriorly with the vomer via the sphenoid crest and articulating anteriorly with the nasal bone crests.

The lateral masses of the ethmoid bone, also known as the labyrinths, form the medial wall of the orbit via the lamina papyracea and part of the lateral wall of the nasal cavity via the superior and middle turbinates [1]. The ethmoidal sinuses sit beneath the cribriform plate and the orbits, and anterior to the sphenoid sinuses, and offer a corridor to the floor of the anterior cranial fossa. The anterior and posterior ethmoidal foramen transmits the anterior and posterior ethmoidal arteries, veins, and nerves. The anterior and posterior ethmoidal arteries are branches of the ophthalmic artery. The anterior ethmoidal nerve is a continuation of the nasociliary nerve, which arises from the ophthalmic division of the trigeminal nerve after it has given off its other branches, which includes the posterior ethmoidal nerve. The falcine artery, a branch of the anterior ethmoidal artery, supplies the dura of the anterior cranial fossa and often represents an important blood supply to meningiomas in this region.

The olfactory epithelium is found in the superior aspect of the nasal cavity, on the superior turbinates, the nasal septum including the perpendicular plate of the ethmoid, and cribriform plate, also known as the horizontal plate. The olfactory epithelium is critical in extended endoscopic endonasal approaches (EEA) in patients with preserved olfaction, as some portion of it must be preserved to avoid anosmia.

The posterior border of the anterior cranial fossa in the midline is formed by the chiasmatic sulcus, the shallow groove stretching between each optic foramen that forms the surface for the optic chiasm. This structure is also variously known as the optic groove and the prechiasmatic sulcus. The anterior ridge of the chiasmatic sulcus is known as the superior limbus sphenoidale. The posterior ridge of the chiasmatic sulcus is known as the posterior limbus sphenoidale. Immediately anterior to the chiasmatic sulcus is the planum sphenoidale. This structure forms the roof of the sphenoid sinuses and is alternatively known as the jugum sphenoidale [2]. The anterior-most aspect of the planum sphenoidale articulates with the ethmoid bone via a structure known as the ethmoidal spine. The lateral extension of the planum sphenoidale consists of the anterior clinoid processes. The sphenoid ridge forms the posterior aspect of the anterior cranial fossa, and its most lateral extension terminates at the front-lateral surface of the skull as the pterion (Fig. 1).

The more lateral aspect of the anterior cranial fossa floor is formed by the portion of the frontal bone known as the orbital roof, with the orbit and its soft tissue contents immediately inferior. The orbital roof presents bony protrusions known as jugea, limiting sub frontal access to anterior skull base pathology and should be drilled down to reduce retraction on the frontal lobe (Fig. 2).

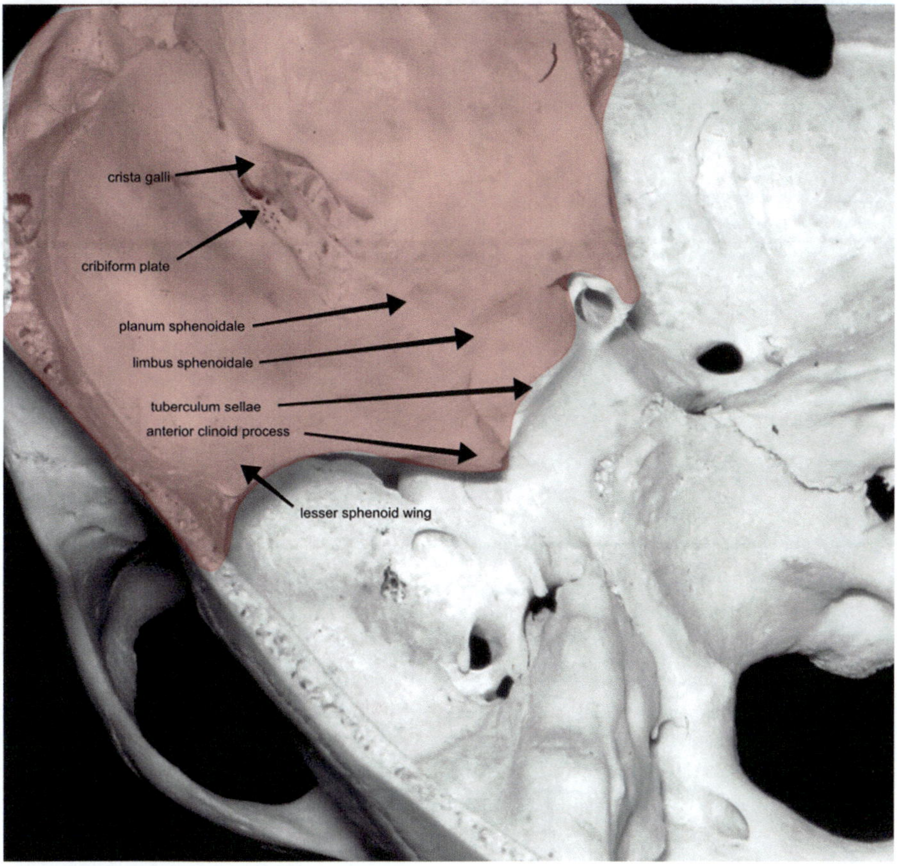

Fig. 1 Skull base bony anatomy highlighting the limits of the anterior cranial fossa and its associated bone landmarks. (*Courtesy from Alberto Capel Cardoso, MD, PhD*)

Major intracranial arteries associated with the anterior cranial fossa floor include the medial orbitofrontal artery, typically a branch of the second division of the anterior cerebral artery, and the lateral orbitofrontal artery, typically a branch of the superior division of the middle cerebral artery. Both arteries provide the blood supply to the orbital frontal cortex, and the medial orbitofrontal artery supplies blood to the olfactory bulb and tracts [3].

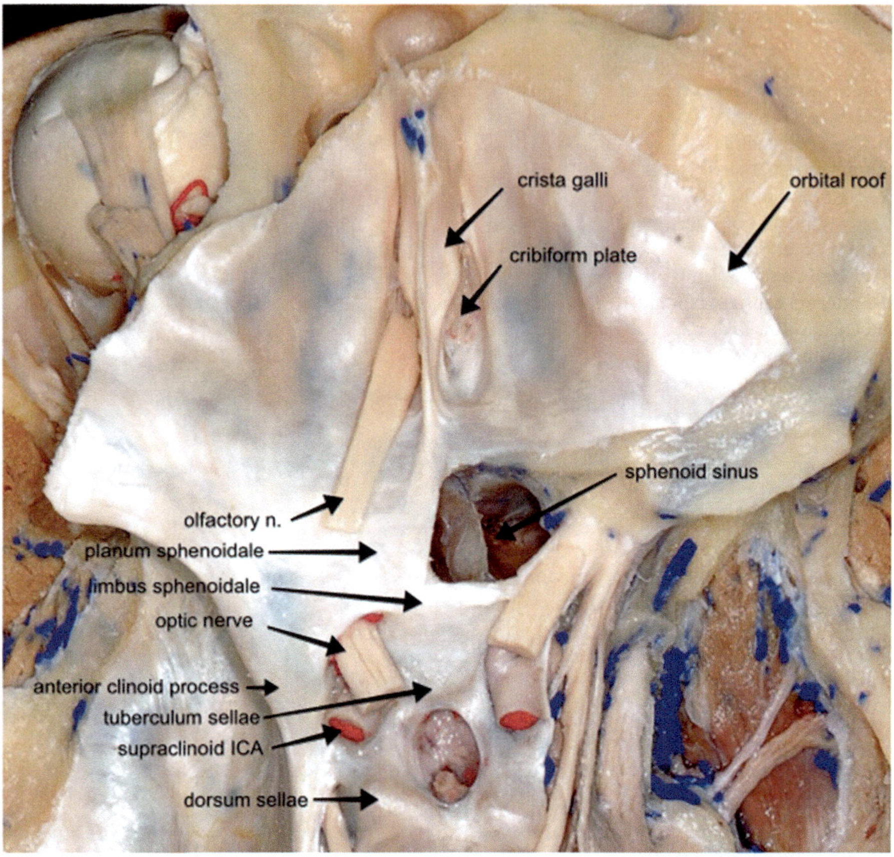

Fig. 2 Anterior skull base anatomy demonstrating the relationship of bone, dura mater and cranial nerves. The image also highlights location of the orbit and sphenoid sinus in relation to the inner surface of the anterior cranial fossa. (*Courtesy from Alberto Capel Cardoso, MD, PhD*)

2 Surgical Approaches and Techniques

2.1 Bifrontal Craniotomy

The bifrontal craniotomy involves a coronal scalp incision, the so-called Souttar incision. It starts just above the zygoma and about 1 cm in front of the tragus and is carried through to the galea without incising the pericranium, the temporalis, or its overlying fascia. The scalp flap is raised, and the plane between the galea and the pericranium is developed. The pericranium is raised as a separate pedicled flap after careful interfascial dissection over the temporalis muscle, which preserves the facial nerve and the blood supply from the supratrochlear supraorbital, and zygomatico-temporal branches of the superficial temporal artery. It is important to dissect and preserve the supraorbital artery pedicle of the pericranial flap and expose the bilateral superior orbital rims. This maneuver allows the inferior cut of the bifrontal craniotomy to be flush with the anterior skull base, and usually, additional orbitotomy is unnecessary. If needed, additional exposure may be obtained through an orbital bar osteotomy with or without a nasal bone osteotomy.

Detachment of the medial canthal ligament with medial orbital osteotomies extends the approach further, and the addition of complete lateral orbital wall removal to the inferior orbital fissure maximizes the exposure. Regardless of whether additional osteotomies are used, closure necessarily involves cranialization of the frontal sinus, and the pedicled pericranial flap harvested during opening can be affixed to the inferior frontal dura. For lesions whose removal results in a bony defect in the anterior cranial fossa floor, the defect can usually be closed with a vascularized tissue flap (mucoperiosteal or pericranial). Nevertheless, some authors advocate using a titanium mesh implant or a split-thickness bone graft [4].

The bifrontal craniotomy approach to the anterior skull base, when used to treat large extradural anterior skull base tumors, such as a giant olfactory groove or planum sphenoidale meningiomas, is typically understood to involve the sacrifice of olfaction. Some authors have argued for a unilateral frontal–orbital approach with attempted preservation of at least the contralateral olfactory nerve when treating these large anterior skull base meningiomas [5]. In the case of extensive bifrontal–transbasal approaches, Spetzler and colleagues have also described a technique in which circumferential osteotomies are made around the cribriform plate, allowing for the maintenance of its attachment to the nasal mucosa and the frontal lobe dura [6].

Additional drawbacks to the bifrontal craniotomy as an approach for anterior skull base pathology are the necessity of frontal lobe retraction when gaining access to the basal cisterns to allow brain relaxation. In order to retract the frontal lobes, the anterior one-third of the superior sagittal sinus must also be ligated, which has been associated with venous stroke in rare patients. In addition, violation of the frontal sinuses is unavoidable in bifrontal craniotomy approaches to the anterior skull base, and this necessitates reconstruction to prevent postoperative cerebrospinal fluid (CSF) leak. This can be time-consuming, and unsuccessful reconstruction will result in CSF leak and possible meningitis. Finally, but importantly, from a cosmetic standpoint, bifrontal craniotomies are more challenging to reconstruct and could result in suboptimal cosmetic results impacting patient quality of life.

2.2 *Frontolateral Craniotomy*

The frontolateral craniotomy, also known as the pterional craniotomy, and its derivative, the lateral supraorbital craniotomy, allow for a more laterally based attack to anterior skull base pathology when compared to the corridor afforded by the traditional bifrontal craniotomy (Fig. 3).

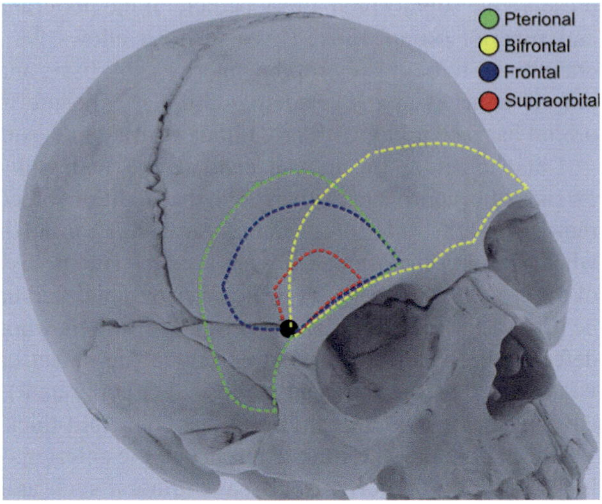

Fig. 3 Most common craniotomies to approach the anterior cranial fossa

A curvilinear incision is made behind the hairline, exposing the lateral orbital rim and the frontozygomatic suture, which roughly approximates the floor of the anterior cranial fossa. The area of the McCarty keyhole, defined as 5 mm behind and 7 mm above the frontozygomatic suture, is exposed by subperiosteal dissection of the temporalis muscle, which preserves the frontalis branches of the facial nerve. A single burrhole, either at the keyhole or, in the case of the "lateral supraorbital" approach, at the superior temporal line is typically sufficient, and a bone flap is raised, extending medially usually to about the mid pupillary line, and posteriorly about 2–4 cm, depending on the size of the lesion in question [7, 8].

The advantages of this approach compared to a bifrontal craniotomy include early access to the basal cisterns without brain retraction, which allows for CSF drainage and frontal lobe relaxation. For large tumors, a unilateral frontal/front lateral approach limits any frontal lobe retraction to one side, compared to the bifrontal craniotomy, which risks bilateral frontal lobe retraction injury. The frontal/front lateral operative corridor also allows early identification of the optic apparatus and the bilateral carotid arteries. Finally, compared to the bifrontal craniotomy, the frontal sinus may be avoided, minimizing the risk for postoperative CSF leak.

One chief downside of the frontolateral approach to anterior skull base pathology is that for many tumors, especially large tumors, the angle of attack places the contralateral ACA and its branches, together with the contralateral olfactory nerve, at a depth of dissection, meaning that these structures are often not identified until later in the surgery. The anterior falx may also prevent visualization of the contralateral side and cutting the falx may be necessary to access the contralateral extent of the tumor. The extent to which the traditional fronto-lateral craniotomy exposes the temporal lobe and more posterior and superior aspects of the frontal lobe is also traditionally thought to be unnecessary when approaching the anterior skull base. Nevertheless, a unilateral frontal/front lateral craniotomy may be tailored depending on tumor size and extension into surrounding compartments.

2.3 Supraorbital Craniotomy

The supraorbital craniotomy represents a minimally invasive "keyhole" approach to lesions of the anterior cranial fossa (Fig. 4). Like the fronto-lateral (pterional craniotomy), a burrhole is placed at the MacCarty keyhole. However, rather than extending superior and posterior to expose the frontal lobe, superior temporal lobe, and a large amount of the Sylvian fissure, the bone flap in the supraorbital craniotomy extends superiorly and medially, creating a smaller, more frontally oriented angle of approach to the anterior cranial fossa.

The supraorbital craniotomy, as its name suggests, is centered over the brow and takes advantage of the so-called "keyhole" effect, allowing for access to tumors larger than the craniotomy itself by maximizing working distance and working angles [9]. Because the supraorbital craniotomy is small relative to the pathology it seeks to treat, proper head positioning is essential to maximize exposure. The head should be elevated above cardiac level and then retroflexed about 20–30° to allow gravity retraction of the frontal lobes. Because the medial extent of the craniotomy is limited by the supraorbital foramen and the frontal sinus, the head should be rotated laterally between 40° and 60°, with the more anterior the tumor, the more lateral the rotation of the head [9].

Although the supraorbital craniotomy can be made by employing a traditional bicoronal (Souttar) or fronto lateral incision, it has become customary to approach this craniotomy through the so-called "eyebrow" incision, centered either just inside

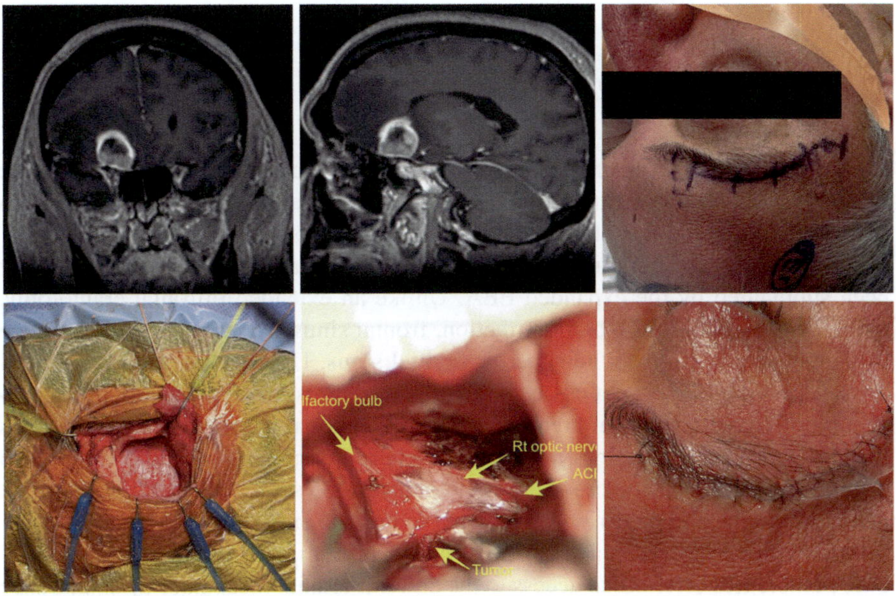

Fig. 4 Keyhole supraorbital craniotomy to approach a right anterior cranial fossa dural based metastasis. Note the sufficient exposure despite the small craniotomy and brain swelling

the eyebrow itself or just superior to it and stretching from just lateral to the supraorbital foramen laterally to the end of the eyebrow [9–12]. It is important to stay lateral to the supraorbital foramen to avoid injury to the supraorbital nerve, which provides sensory innervation to the forehead and brow. Similarly, extending the incision just lateral to the edge of the eyebrow, medial to the frontal process of the zygoma, ensures protection of the superficial temporal branch of the facial nerve, which courses in a superior direction about 1 cm behind the frontal process. The incision is carried down to just above the pericranium, and a single myocutaneous flap is raised and retracted using fishhooks.

A pericranial flap can be raised if there is a concern for violation of the frontal sinus or if the need for an anterior skull base repair is envisioned. The temporalis muscle is mobilized laterally, and a burrhole is placed in the region of the MacCarty keyhole, making sure to orient the drill away from the orbit and towards the anterior cranial fossa floor unless orbital access is desired to treat the pathology in question. The lateral extension of the craniotomy is typically to the superior temporal line. The medial extension of the craniotomy can travel medially as far as dictated by pre-op imaging, and intra-op navigation suggests the need for access to the tumor but should avoid violation of the frontal sinus. The superior craniotomy extension is dictated by the location and size of the lesion. However, it is important to ensure that the base of the craniotomy is flush with the anterior cranial fossa floor, as this will allow both for minimal brain retraction when traveling through the sub-frontal corridor and a maximum working angle to approach the superior aspect of the lesion. Once the bone flap is raised, any bony ridges or irregularities on the anterior cranial fossa floor can be drilled down to provide additional subfrontal access.

2.4 Endoscopic Endonasal Approaches

Historically, access to anterior skull base pathology occurred either through a large bifrontal craniotomy or through one of many craniofacial approaches, each with considerable disturbance of facial soft tissue and bony structures. Trans-oral, trans-labial, trans-septal, lateral rhinotomy, and trans-maxillary approaches have largely been supplanted by the expanded EEA. Unlike an extended bifrontal craniotomy, the EEA does not require brain retraction, frontal sinus reconstruction, or sacrifice of the anterior portion of the superior sagittal sinus. Unlike extended craniofacial approaches, the EEA is not cosmetically disfiguring, is not associated with significant blood loss, and does not require complex rotational flap-based reconstructions for closure. With endoscopes and other specialty instruments, this basic approach can be tailored to provide broad access to many of the skull base pathologies of the anterior, middle, and posterior fossa. In particular, the extended endoscopic endonasal transcribiform, transplanum, and transtuberculum sellae approaches to skull base pathologies offer broad access to the anterior cranial fossa floor by an entrance through the ethmoidal and sphenoidal sinuses.

Access to the anterior cranial fossa through the EEA involves resection of the roof of the nasal cavity, which consists of the ethmoid air cells and the cribriform plate. The most anterior ethmoid air cells, the so-called *agger nasi* cells, are contained within the lacrimal bone, anterior–superior to the middle turbinate, and anterior, lateral, and inferior to the frontal recess of the frontal sinus. Posterior to the frontal recess and immediately lateral to the medial wall of the orbit, the so-called ethmoid bulla represents the largest and most consistent anterior ethmoid air cells. The anterior ethmoidal artery can be found along the posterior wall of the frontal recess, where it arises from the medial wall of the orbit after branching off the ophthalmic artery. The medial wall of the orbit, formed by the lamina papyracea of the ethmoid bone, represents the lateral most border of the EEA corridor to the anterior cranial fossa. The posterior ethmoidal artery also arises from the medial wall of the orbit after branching from the ophthalmic artery, but as its name suggests, is found more posteriorly in the nasal cavity, specifically about 8 mm from the point, where the optic nerve exits the optic canal, at the angle formed by the ethmoidal roof and the medial wall of the orbit as it enters the posterior ethmoidal canal [13].

For the unilateral transcribiform EEA, which can be used to remove small tumors of the anterior cranial fossa and repair encephaloceles and other CSF leaks, the head is placed in pins, elevated about 30°, and turned slightly to the right. The uncinate process is removed, and an incision is carried down inferiorly and laterally to the maxillary sinus ostium. Then, the middle turbinate is resected at its base, which is the perpendicular plate of the ethmoid bone. Removal of the uncinate process and the middle turbinate exposes the anterior and posterior ethmoidal air cells, then drilled out [14].

The bilateral transcribiform EEA, which approaches large anterior cranial fossa meningiomas and sinonasal malignancies with intracranial extension, involves similar positioning to the unilateral approach, but starts with a submucosal resection the cartilaginous septum and removal of its attachment to the perpendicular plate of the ethmoid. The uncinate process is removed, followed by a total ethmoidectomy, in which the anterior and posterior ethmoid air cells are removed A superior septectomy is performed anterior to the cribriform plate, and a frontal sinusotomy exposes the posterior frontal sinus wall. Authors favoring the Lothrop procedure in the bilateral transcribiform EEA note that it identified the anterior extent of bony resection, provides a natural shelf for stabilizing graft material during skull base reconstruction, and minimizes postoperative iatrogenic mucocele formation thanks to the wide opening of the frontal sinuses that it provides [15]. Bilateral sphenoidotomy is also performed to identify the planum sphenoidale and optic nerves, which serve as landmarks to delimit the posterior boundary of the anterior cranial fossa. The key endoscopic landmark for the optic nerves is the medial optico-carotid recess, which sits at the junction of the parasellar carotid canal and the optic canal. The carotid artery and optic nerve identification are especially important while performing posterior ethmoidectomy in patients with an Onodi cell, a large posterior ethmoid air cell superior and lateral to the sphenoid sinus, like the carotid artery and optic nerve, often run in the lateral wall of these air cells [15].

Following exposure of the anterior cranial fossa floor through total ethmoidectomy and sphenoidotomy, the cribriform plate is drilled off from the posterior wall of the frontal sinus and posteriorly to include the planum sphenoidale, taken care to identify and ligate the anterior and posterior ethmoidal arteries in the process.

Transplanum or transtuberculum EEA may be performed in a more focused fashion when there is no need to expose the most anterior aspect of the anterior cranial fossa and olfactory groove. In those scenarios, only removal of posterior ethmoid cells and opening of the sphenoid sinus is needed, allowing for preservation of the olfaction by preserving the cribriform plate.

3 Pearls and Tips (Pathology Specific)

3.1 Olfactory Groove (OG) Meningiomas

- It is important to differentiate an anterior falx meningioma from an OG meningioma. The anterior falx meningioma is a convexity tumor with its attachment at the most anterior aspect of the frontal bone/posterior table of the frontal sinus. Hence, a bifrontal craniotomy is usually required to resect it
- In general, approach selection in OG meningiomas should be guided by olfaction status, tumor extension, and vessel encasement
- The transcranial approach should be favored in patients with preserved olfaction, since the loss of smell is certain in a transcribiform EEA
- The ipsilateral groove between the medial orbit and the cribriform plate is usually a blind spot during frontal or supraorbital unilateral craniotomies for OG meningiomas. The assistance of angled endoscope/instruments may be needed to reach the caudal aspect of an ipsilateral OG (Fig. 5)
- EEA should be favored in OG meningiomas extending into the sinonasal cavity. Anatomical limitations of the endoscopic approach are anterior tumor extension

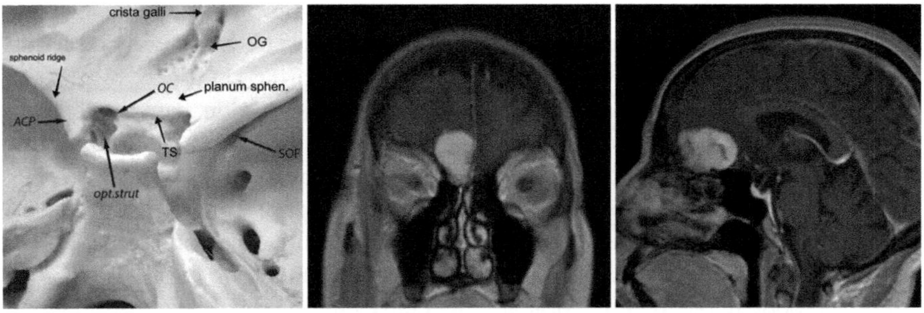

Fig. 5 Anterior skull base anatomical landmarks with emphasis at the cribiform plate. MRI of an OG meningioma highlighting the location of blind spot (the ipsilateral groove between the medial orbit and the cribriform plate) when approaching this tumor through a frontal or supraorbital unilateral craniotomy

at the posterior table of the frontal sinus and laterally over the orbital roof past the midline point of the orbit
- Vessel encasement is a relative contraindication for EEA resection of OG meningiomas. Similarly, when the anterior cerebral arteries are encased, a transcranial approach usually represents a safer alternative
- In OG meningiomas, the ethmoidal arteries are usually hypertrophic due to the presence of the tumor. It is important to identify, coagulate, and divide the arteries as early as possible to reduce overall blood loss. Ethmoidal arteries that are not properly controlled can result in stump retraction and cause intraorbital hematomas. This complication is traditionally associated with the EEA, but it is important also to realize that it can be seen in transcranial approaches

3.2 Planum Sphenoidale and Tuberculum Sellae Meningiomas

- The distinction between meningiomas of the planum sphenoidale and the tuberculum sella can typically be made based on radiographic features, such as the site of most significant hyperostosis and the point, where the tumor blood supply converges at the base of the mass. For larger tumors, the distinction may be more challenging
- In general, planum sphenoidale meningiomas may be safely resected through multiple approaches. We tend to favor transplanum EEA in patients without olfaction and in patients, where the planum sphenoidale meningioma fails to extend anteriorly to the cribriform plate. A keyhole supraorbital craniotomy is favored in tumors with dural attachment extending laterally over the roof of the orbit or to the anterior clinoid process
- We favor resection of most tuberculum sellae meningiomas through a transturberculum/transplanum EEA, since this approach allows for complete tumor resection while minimizing optic apparatus manipulation. A transcranial approach is favored in situations, where the dural attachment of the tumor extensions over the anterior clinoid process or laterally beyond the supraclinoid ICA
- Regardless of approach selection, preoperative imaging assessment regarding optic canal invasion is crucial. Most tuberculum sellae meningiomas will invade the optic canal medial to the optic nerve. The EEA offers a critical advantage here, as it allows for early optic canal decompression and removal of the intracanalicular tumor without significant optic nerve manipulation
- When approaching a tuberculum sellae meningioma transcranially, we tend to favor a keyhole supraorbital craniotomy. It is important to assess the angle of the skull base formed by the planum sphenoidale and the sella and the caudal extension of the tumor in the sagittal views. If the angle is obtuse or flat, the tumor can typically be resected through a supraorbital craniotomy without issue. If the angle is acute, the caudal reach of the supraorbital craniotomy may be limited, necessitating either angled endoscopies and instruments or a frontolateral craniotomy to reach the most caudal aspect of the tumor

3.3 Ethmoidal Dural Arteriovenous Fistula

- Surgical ligation of ethmoidal DAVF has been shown to have superior rates of complete obliteration and 30-day good outcomes [16]
- Although endovascular treatment of ethmoidal DAVF is possible, there are risks, including unintended embolization of the central retinal artery causing blindness, the difficulty with selective catheterization of small anterior circulation arterial branches, and finally, unintended reflux of embolic material into the cerebral circulation [17]
- Microsurgical treatment of ethmoidal DAVF requires exposure of the anterior cranial fossa floor in the midline to identify the fistulous point, which typically drains into the sagittal sinus. This may be achieved through a bifrontal or unilateral frontal/supraorbital craniotomy that allows sufficient access to the anterior cranial fossa floor base
- An understanding of the angiographic anatomy is crucial for identifying the fistulous point and surgical planning, particularly in small craniotomies. A keyhole supraorbital approach is suitable for ipsilateral basally located fistulas, with the resulting surgical corridor almost perpendicular to the falx
- If the ipsilateral fistula is higher into the interhemispheric fissure, a unilateral frontal craniotomy that extends to the midline is more suitable
- Bifrontal craniotomies provide wide exposure to the surgical field, and there is less need for tailored planning. They are required in rare cases of bilateral fistulas

3.4 Anterior Cranial Fossa Meningocele

- These lesions are best approached using a unilateral EEA
- Fluorescein dye injection through a preoperative lumbar puncture helps visualize the exact location of the CSF leaks using endoscopy
- If the meningocele is seen emanating from the cribriform plate medial to the middle turbinate, the middle turbinate can be left intact, and the ethmoid air cells do not need to be entered. On the other hand, if the CSF leak is identified lateral to the middle turbinate, it must be removed, and the ethmoidal air cells must be drilled to identify further and treat the meningocele [14]
- The dura and any associated herniated brain (encephalocele) are resected, and a graft is used to close the defect

3.5 Sinonasal Malignancies with Intracranial Extension

- Sinonasal malignancies are locally destructive lesions that can invade the anterior cranial fossa by destroying the ethmoid and sphenoid bones. These lesions can often involve the orbits and the ventral skull base as well

- Sinonasal anatomy is grossly distorted or destroyed by these malignancies, especially in lesions large enough to invade the anterior cranial fossa floor. This usually compromises nasal cavity flap options for skull base reconstruction
- The main advantages of EEA for these lesions include avoidance of scarring and facial incisions, avoidance of maxillofacial skeletal dissection and translocation, with lower overall approach-related morbidity. Nevertheless, it is important to understand the limitations of EEA and not to compromise the oncological goals of a resection
- Regardless of approach, the dural margins and olfactory bulbs should always be sent to ensure an oncologic resection
- Open or transfacial approaches are usually required when there is tumor involvement of the skin or subcutaneous tissue, the anterior table of the frontal sinus, deep orbital invasion, lateral supra-orbital extension, or extensive dural and brain parenchymal invasion
- In some scenarios, a combined cranio-endoscopic approach, in which an EEA from below is performed with a bifrontal craniotomy from above, can obviate the need for a more radical transfacial approach

4 Conclusions

The location of the anterior cranial fossa, its anatomical structure, and associated pathologies lend themselves to traditional open craniotomies, minimally invasive transcranial approaches through a supra-orbital "keyhole" craniotomy, and EEA across the cribriform plate and planum sphenoidale. As with all other areas of skull base surgery, appropriate approach selection involves careful consideration of patient presentation, preoperative imaging, and the overall goals of the surgery. In this respect, the various minimally invasive approaches outlined above are only useful insofar as they maximize the effectiveness of the surgery. As always, approach selection in anterior skull base surgery should ultimately be made with the patient's best interests in mind.

References

1. Rhoton AL Jr. The anterior and middle cranial base. Neurosurgery. 2002;51:S273–302.
2. Chmielewski PP. New terminologia anatomica: cranium and extracranial bones of head. Folia Morphol (Warsz). 2021;80(3):477–86.
3. Mavridis IN, Kalamatianos T, Koutsarnakis C, Stranjalis G. The microsurgical anatomy of the orbitofrontal arteries. World Neurosurg. 2016;89:309–19.
4. Feiz-Erfan I, et al. Proposed classification for the transbasal approach and its modifications. Skull Base. 2008;18:29–47.
5. Ung TH, et al. Preservation of olfaction in anterior midline skull base meningiomas: a comprehensive approach. Acta Neurochir. 2019;161:729–35.

6. Spetzler RF, Herman JM, Beals S, Joganic E, Milligan J. Preservation of olfaction in anterior craniofacial approaches. J Neurosurg. 1993;79:48–52.
7. Romani R, Laakso A, Kangasniemi M, Niemelä M, Hernesniemi J. Lateral supraorbital approach applied to tuberculum sellae meningiomas: experience with 52 consecutive patients. Neurosurgery. 2012;70:1504–18. ; discussion 1518–9.
8. Downes AE, Freeman JL, Ryan Ormond D, Lillehei KO, Samy Youssef A. Unilateral tailored fronto-orbital approach for giant olfactory groove meningiomas: technical nuances. World Neurosurg. 2015;84:1166–73.
9. Zador Z, Gnanalingham K. Eyebrow craniotomy for anterior skull base lesions: how I do it. Acta Neurochir. 2013;155:99–106.
10. Delashaw JB Jr, Jane JA, Kassell NF, Luce C. Supraorbital craniotomy by fracture of the anterior orbital roof. Technical note. J Neurosurg. 1993;79:615–8.
11. Reisch R, Marcus HJ, Kockro RA, Ulrich NH. The supraorbital keyhole approach: how I do it. Acta Neurochir. 2015;157:979–83.
12. Dlouhy BJ, Chae MP, Teo C. The supraorbital eyebrow approach in children: clinical outcomes, cosmetic results, and complications. J Neurosurg Pediatr. 2015;15:12–9.
13. de Notaris M, et al. Endoscopic endonasal approach to the ethmoidal planum: anatomic study. Neurosurg Rev. 2008;31:309–17.
14. Greenfield JP, et al. Endoscopic endonasal transethmoidal transcribriform transfovea ethmoidalis approach to the anterior cranial fossa and skull base. Neurosurgery. 2010;66:883–92. ; discussion 892.
15. Liu JK, Christiano LD, Patel SK, Tubbs RS, Eloy JA. Surgical nuances for removal of olfactory groove meningiomas using the endoscopic endonasal transcribriform approach. Neurosurg Focus. 2011;30:E3.
16. Giannopoulos S, et al. Treatment of ethmoidal dural arteriovenous fistulas: a meta-analysis comparing endovascular versus surgical treatment. World Neurosurg. 2019;128:593–599.e1.
17. Lawton MT, Chun J, Wilson CB, Halbach VV. Ethmoidal dural arteriovenous fistulae: an assessment of surgical and endovascular management. Neurosurgery. 1999;45:805–10. ; discussion 810–1.

Surgical Anatomy of the Orbit

Claudio Henrique F. Vidal, Caetano J. Coimbra, Cristina Baracuhy de Melo, Grant Gilliland, Breno J. C. de Lima, Hugo N. A. Coelho, Camila B. M. Muniz, and Ricardo M. C. Aragão

1 Overview

The orbits are anatomically complex cavities filled by: ocular globes, adipose tissue, muscle tissue, glands, vascular structures, and nerves. The orbits communicate with the anterior and middle cranial fossa, temporal fossa, pterygopalatine fossa, and paranasal sinuses [1].

Several pathologies can affect the orbit, including infections, inflammations, traumas, and tumors. Kronlein, in 1888 apud Lee et al. [2], described for the first time a lateral orbitotomy for resection of posterior orbital tumors. Since then, numerous transorbital, transcranial, and endoscopic approaches have been developed recently. These techniques can be used alone or in combination. In addition, intraoperative devices for locating lesions, such as neuronavigation, were incorporated, increasing the accuracy of these approaches and reducing tissue damage.

Supplementary Information The online version contains supplementary material available at https://doi.org/10.1007/978-3-031-14820-0_21.

C. H. F. Vidal · H. N. A. Coelho
Neurosurgery Department, Hospital Getúlio Vargas, Recife, PE, Brazil

C. J. Coimbra (✉)
Skull Base Surgery Center, Baylor University Medical Center, Dallas, TX, USA

Minimally Invasive Brain Surgery Center, Medical City Hospital, Dallas, TX, USA

C. B. de Melo
Department of Ophthalmology, Hospital das Clínicas (UFPE), Recife, PE, Brazil

B. J. C. de Lima · C. B. M. Muniz · R. M. C. Aragão
Department of Otorhinolaryngology, Hospital das Clínicas (UFPE), Recife, PE, Brazil

G. Gilliland
Skull Base Surgery Center, Baylor University Medical Center, Dallas, TX, USA

E. G. Figueiredo et al. (eds.), *Brain Anatomy and Neurosurgical Approaches*,
https://doi.org/10.1007/978-3-031-14820-0_22

Transfacial approaches classically used to address medial orbital lesions have been largely replaced by endonasal approaches. However, each of these techniques must be used after individual analysis of the case and under the medical team's experience.

Considering the possibility of intracranial, nasosinusal, and orofacial extensions, the modern surgical approach to the orbit brings together several specialties, including neurosurgery, ophthalmology, otolaryngology, plastic surgery, and oral–maxillofacial surgery.

2 Pertinent Anatomical Considerations

The orbit is a pyramidal bone structure that harbors vision apparatus structures. The orbital apex is located posteriorly, and the base of the pyramid is located anteriorly [3]. The orbital apex is the area between the orbit and the intracranial cavity. It includes the foramina communicating the intracranial and orbital compartments. The foramina in the orbital apex include the optic canal (OC), the superior orbital fissure (SOF), and the inferior orbital fissure (IOF). The structures of the orbits that are of special interest to the neurosurgeon are:

1. The OC measures approximately 5 mm in diameter and 1 cm in length and is directed from medial to lateral. It is located in the medial portion of the orbital apex and within the confines of the sphenoid bone [3]. The OC contains the optic nerve (ON), the ophthalmic artery, and the postganglionic sympathetic nerves of the internal carotid plexus. The OC is bordered medially by the body of the sphenoid bone, superiorly by the upper root of the lesser sphenoid wing, inferolaterally by the lower root of the lesser sphenoid wing or "optic strut," and laterally by the anterior clinoid process (ACP) [4].
2. The SOF represents a bone gap between the greater and lesser sphenoid wings. It is located superolaterally to the OC and extends inferiorly to a point immediately medial to the OC [3]. The SOF connects the middle cranial fossa to the orbit and is separated from the OC and foramen rotundum by the optic and maxillary struts, respectively [5].
3. The IOF is the inferolateral continuation of the SOF. This fissure connects the orbit floor with the pterygopalatine and infratemporal fossa [3].
4. The annulus of Zinn is a funnel-shaped fibrotendinous tissue (Fig. 1) located at the orbital apex. It gives rise to all four extraocular rectus muscles and the superior oblique muscle [5, 6]. The annulus of Zinn and the extraocular muscles form a cone that delimits the intraconal and extraconal compartments in the orbital apex [6]. The annulus is attached to the ON dorsally, and it is divided by a dural band into two other compartments [6]. The medial compartment contains the ON and the ophthalmic artery, while the lateral compartment or oculomotor foramen contains the superior and inferior divisions of the oculomotor nerve and the

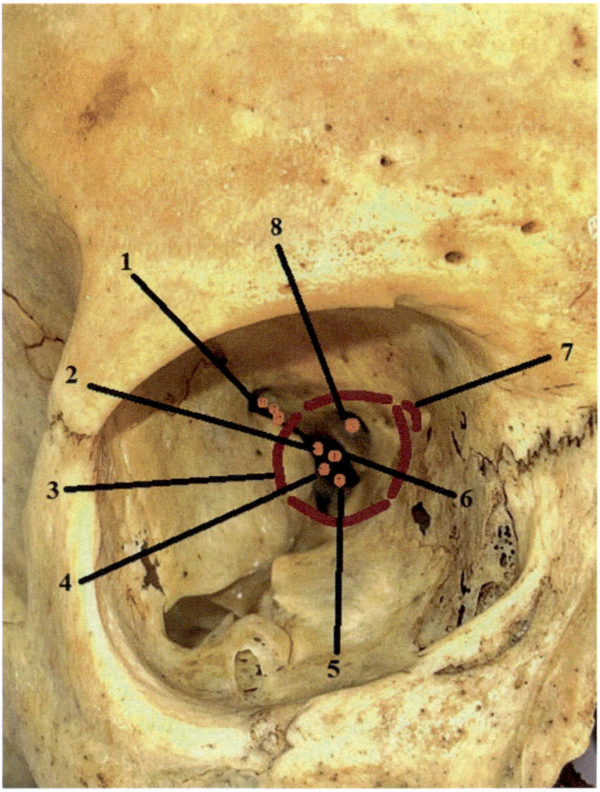

Fig. 1 Orbital anatomy on the right side: (*1*) extraconal segment of the superior orbital fissure with the lacrimal, frontal, and trochlear nerves; (*2*) superior division of the oculomotor nerve, within the intraconal segment of the superior orbital fissure; (*3*) annulus of Zinn with representation of the insertion of the superior, inferior, medial, and lateral rectus muscles in the region of the orbital apex; (*4*) abducent nerve; (*5*) inferior division of the oculomotor nerve; (*6*) nasociliary nerve; (*7*) insertion of the superior oblique muscle at the orbital apex; and (*8*) optical nerve within the optical canal

nerves abducens and nasociliary. The superior and inferior ophthalmic vein and the lacrimal, frontal, and trochlear nerves pass through the SOF but in an extraconal position [5, 6].

The ON is divided into four segments: intraocular, intraorbital, intracanalicular, and intracranial. The intraocular segment is beyond the scope of the chapter. The intraorbital segment is located within the intraconal space and extends from the globe to the orbital apex. The intracanalicular segment is located above the ophthalmic artery as it travels in the OC along the medial portion of the ACP, between the roof of the OC and the optic strut. The intraorbital and intracanalicular segments are covered by pia mater, arachnoid, and dura mater [5]. The intracranial segment is in the ON cistern in the subarachnoid space and has a close relation to the internal carotid artery (ICA).

The ACP is a triangular bony prominence formed by the medial and posterior end of the lesser sphenoid wing [5]. Its superior root forms the roof of the OC and continues with the tuberculum sellae (TS) medially, while its inferior root forms the optic strut and represents the floor of the OC. The optic strut separates the OC from

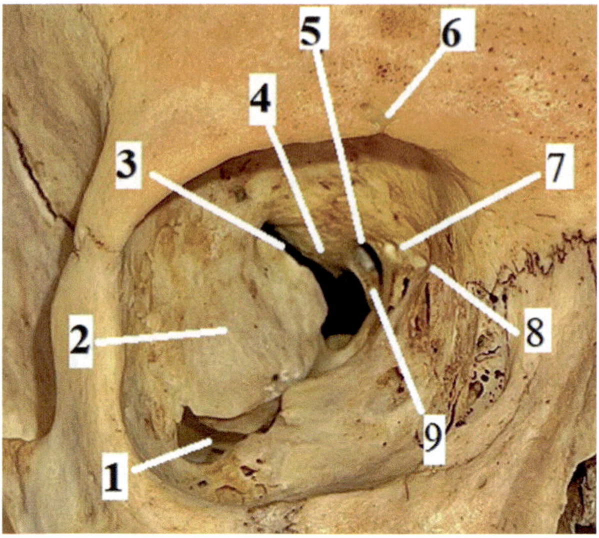

Fig. 2 Orbital anatomy on the right side: (*1*) inferior orbital fissure; (*2*) greater wing of the sphenoid bone; (*3*) superior orbital fissure; (*4*) lesser wing of the sphenoid bone; (*5*) optic foramen; (*6*) supraorbital notch; (*7*) posterior ethmoidal foramen; (*8*) anterior ethmoidal foramen; and (*9*) optic strut

the SOF (Fig. 2). Important dural folds are related to the ACP. The distal dural ring is related to the upper surface of the ACP and forms a dural continuum with the falciform ligament in the proximal OC and the sellar diaphragm. The proximal dural ring is related to the inferior surface of the ACP and represents the upper limit of the cavernous segment of the ICA. The clinoidal segment of the ICA is located between these two dural rings and posteriorly to the optic strut. The ON is medial to ACP, and the oculomotor nerve travels in the inferolateral aspect of the ACP before it passes through the annulus of Zinn.

The orbital roof is the bone boundary between the anterior cranial fossa and the orbit. The supraorbital foramen is located in the medial third of the superior orbital ridge (Fig. 2). The supraorbital nerve, supraorbital artery, and vein pass through this foramen and provide blood supply and sensation to the forehead. The supratrochlear nerve exits medial to the supraorbital nerve [3].

3 Epidemiology of Orbital Lesions

A wide spectrum of tumoral and pseudotumoral lesions can affect the orbital space, being primary or secondary in relation to the origin. Shields et al., reviewing 1264 cases of orbital tumors, found that the prevalence of orbital lesions is relatively constant throughout life (0–18 years: 23%; 19–59 years: 44%; and 60–92 years: 33%) [7]. Despite this, the incidence of malignant tumors is higher in older individuals. In this review, 64% of the masses were benign pathologies (0–18: 80%; 19–59: 73%; and 60–92: 42%) and 36% were malignant tumors (0–18: 20%; 19–59: 27%; and 60–92: 58%).

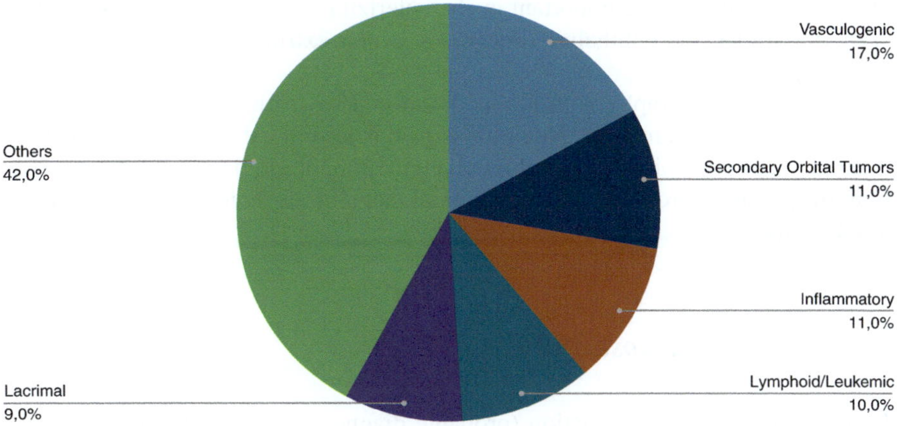

Fig. 3 Distribution of types of orbital lesions in an ophthalmology reference center [7]

Figure 3 [7] shows the distribution of orbital lesion types. The most common are vasculogenic lesions, 17% of the total, with the following distribution: 36% cavernous hemangioma, 25% lymphangioma, and 17% capillary hemangioma. Secondary orbital tumors (primary tumors from nearby regions that extend to the orbit) accounted for 11% of the cases, of which: 29% were uveal melanoma, and 13% were conjunctival melanoma. Inflammatory lesions, excluding those of lacrimal origin, also accounted for 11% of the cases, and the most prevalent were: 74% pseudotumor/non-granulomatous idiopathic inflammation and 10% infection. Lymphoid or leukemic origin, excluding those of lacrimal gland, accounted for 10% of the cases; the most prevalent were: 73% non-Hodgkin lymphoma and 9% atypical lymphoid hyperplasia. Lacrimal apparatus lesions were 9% of the cases, of which 34% were malignant lesions, mainly non-Hodgkin's lymphoma, and 66% were benign, mainly dacryoadenitis/pseudotumor. The remaining 42% are subdivided into 13 other types of lesions, such as meningiomas, ON lesions, peripheral nerve lesions, bony lesions, cartilaginous lesions, metastatic tumors, among others

4 Preoperative Evaluation

4.1 Radiological Evaluation

Adequate surgical planning involves extensive radiological evaluation. Contrast-enhanced magnetic resonance imaging (MRI) of the skull and orbits is necessary for the proper characterization of the lesion. The addition of angio-MRI techniques not only aids in neuronavigation-guided approaches but also enables the differential diagnosis of vascular lesions and the study of the patency of the cranial base vessels in cases of tumors in this topography with a secondary invasion of the orbit. The other MRI sequences are important in characterizing the involvement of structures, such as the ON, the ophthalmic artery, the ocular extrinsic musculature, and the orbital apex region.

Computed tomography with a bone window is essential when planning endoscopic endonasal approaches. Attention is paid to defining the relationship with the maxillary and sphenoid sinuses, ethmoidal labyrinth, in addition to the characterization of possible problems, such as nasal septum deviations that narrow the nasal cavity.

4.2 Surgery Purpose and Possibilities

Based on the imaging evaluation (probable diagnosis, lesion extension, and relationship with neurovascular structures), the objective of the surgery will be defined. It may vary from complete resection, with or without safety margins, to partial resection for decompression or even biopsy with a diagnostic purpose [8]. The type of approach to be used will also depend on this surgical proposal and the surgical team's experience and may include traditional direct transorbital approaches, primary transorbital endoscopy, endonasal endoscopic approaches, multiportal endoscopic approaches (endonasal + transorbital), or transcranial approaches. The main point for choosing adequate surgical access is the topography of the lesion and its relationship with the surrounding neurovascular structures.

Pathologies involving the superior and anterior region of the orbit can be accessed through an anterior orbitotomy. Localized lesions of the lateral aspect of the orbit can be approached through lateral orbitotomy or transcranial access (frontotemporal craniotomy). On the other hand, pathologies involving the medial and inferior region of the orbit can be addressed through medial orbitotomy or endonasal endoscopy. Each access provides a limited visual field for the orbital approach, but together they can provide 360° access for lesion resection [9].

5 Transorbital Approaches

The purpose of the chapter is not to exhaust the theme, since there are countless variations and possibilities, but only to stimulate readers' interest based on some illustrative cases.

The upper and lower lids maintain their consistency due to a dense layer of connective tissue that ends at the periosteum of the orbital rim, the orbital septum. The upper eyelid mixes with the levator tendon of the upper eyelid and in the lower eyelid, with the capsulopalpebral fascia, 5 mm inferior to the tarsal plate. It also has continuity with the medial and lateral canthal ligaments [3].

5.1 Superior Orbitotomy

5.1.1 Upper Eyelid Crease

It represents an easy access route for affections of the roof of the orbit, superior extraconal compartment, superior medial wall of the orbit, and lacrimal gland. Its advantages are shorter surgical time, less tissue manipulation, good esthetic results, and less risk of complications than the transcranial approach. This approach can be associated with detachment of the trochlea of the superior oblique muscle and removal of the superior orbital rim, representing, in the latter situation, an orbitocranial extension that allows exposure of the dura mater of the frontal lobe [10]. The trochlea, medially, and the lacrimal gland, laterally, are important anatomical elements. Indications for this approach are lacrimal gland tumors; resection or biopsy of extraconal tumors (hemangiomas, dermoid cyst, lymphomas, and neurofibromas); drainage of abscess and superior mucocele; removal of intraorbital foreign body; bone tumors; and fractures of the roof of the orbit.

Surgical Technique [11, 12]

- Horizontal incision of the skin in the region of the upper eyelid crease (8–10 mm above the eyelid margin), from the medial corner to the superior lacrimal punctum (Fig. 4)
- The skin, subcutaneous, and pre-septal orbicularis muscle are opened. Apposition of retractor hooks on the upper eyelid and cauterization of vessels with bipolar
- Dissection of the transseptal plane with blunt scissors, type Westcott or Stevens, toward the orbital roof with supraperiosteal exposure of the superior edge of the frontal bone. Avoid manipulation or sectioning of the elevator muscle below the septum and pre-aponeurotic fat
- Opening the periosteum from the frontal bone at the orbital rim and detach this periosteum
- Insertion of a malleable Jaeger spatula into the extraconal space toward the globe to retract the orbital contents and expose the subperiosteal space
- Identification of the trochlea, which can eventually be detached if the removal of the lesion requires a larger field of exposure
- Periorbital opening with Westcott scissors, exposure of the lateral region of the orbit, visualization of the main lacrimal gland, and extraconal fat dissection to remove the lesion
- Systematic cauterization of vessels and not injuring the zygomatic–temporal and lacrimal nerves responsible for parasympathetic tear production
- After removing the lesion, the periorbita does not need to be sutured; only the periosteum is reapproximated and the skin. The suture can be done with continuous or separate stitches. If excess skin is on the upper eyelid, a strip of skin can be excised simultaneously

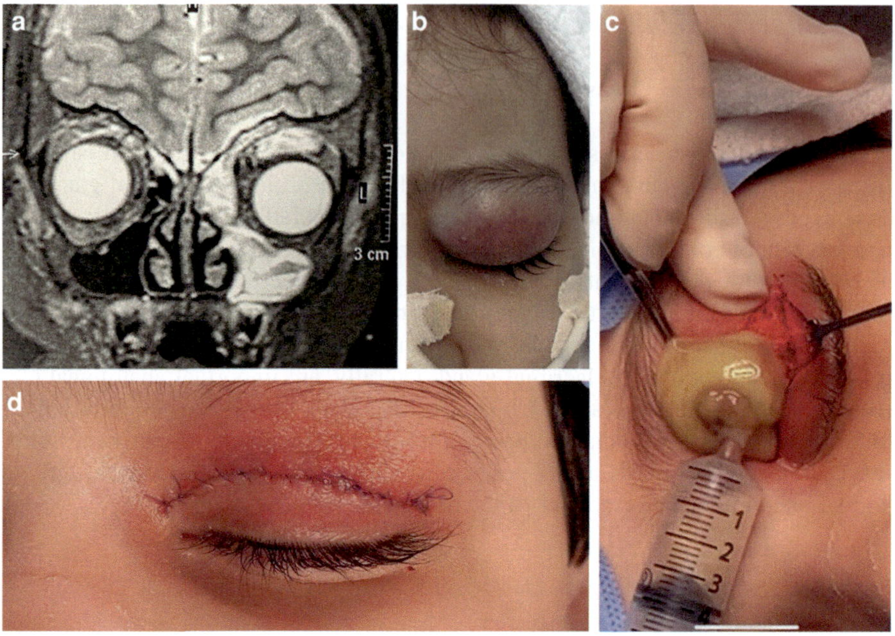

Fig. 4 (**a**) Coronal T2-weighted MRI showing maxillary and ethmoidal sinusopathy on the left, and abscess in the upper region of the left orbit; (**b**) left orbital cellulitis; (**c**) superior orbitotomy with incision via the eyelid sulcus and drainage of the abscess; and (**d**) continuous suture in the eyelid crease. There was clearly a reduction in local volume

5.2 *Medial Orbitotomy*

The medial approach must consider the following anatomical structures: common canaliculus, lacrimal sac, and the ethmoid foramen that mark the transition between the ethmoid and the frontal bone.

5.2.1 Transcaruncular Approach

Indications: resection or biopsy of extra and intraconal tumors; drainage of abscesses and medial mucocele; fractures of the medial wall of the orbit; orbital decompression for the treatment of Graves' ophthalmopathy; and intraorbital foreign body removal.

Surgical Technique [11–13]

- The upper and lower eyelids, close to the medial angle, are retracted with traction sutures or can be retracted with the aid of retractors (Fig. 5). Using delicate scissors, a curvilinear incision is made in the precaruncular region between the car-

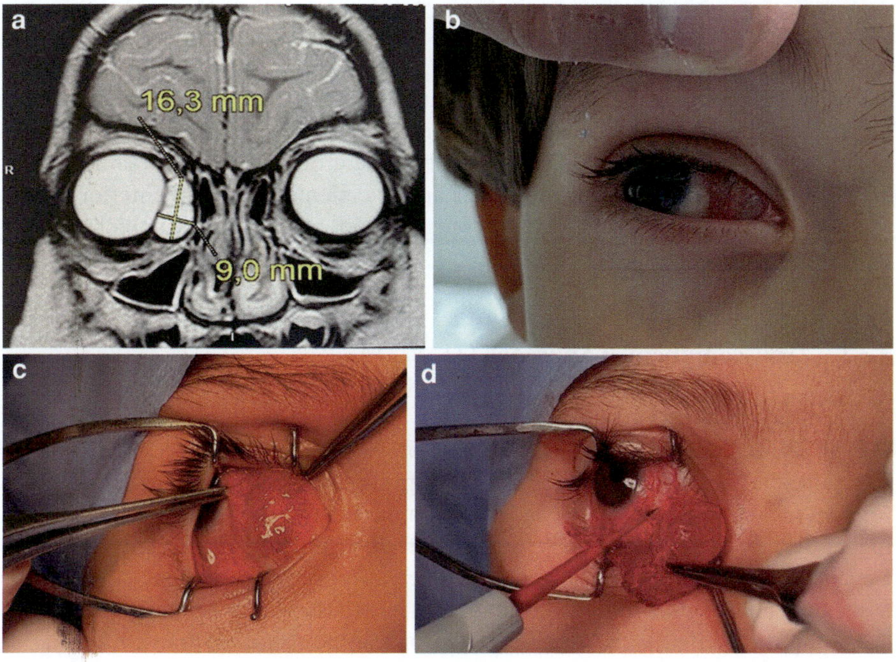

Fig. 5 (**a**) Axial T2-weighted MRI demonstrating hyperintense medial extraconal lesion; (**b**) medial conjunctival lesion with lateral deviation of the right eye; (**c**) apposition of the eyelid retractors and exposure of the lesion and surgical field; and (**d**) tumor dissection and resection via the precaruncular transconjunctival approach

uncle and the semilunar fold. The caruncle is laterally displaced. The incision is extended vertically, about 12–15 mm, avoiding compromising the canaliculus and lacrimal sac.

- Loose areolar tissue is dissected following the surface of Horner's muscle to the posterior lacrimal crest, where the muscle inserts. A retractor is inserted behind the posterior lacrimal crest to aid exposure.
- An incision is made in the periorbita along the superior lacrimal crest with Westcott scissors. Then, a subperiosteal dissection is performed to expose the medial wall of the orbit.
- The anterior and posterior ethmoid arteries indicate the superior extension of the surgical field, while the inferior oblique muscle gives the inferior limit. The ethmoid arteries can be divided after coagulation, and the superior oblique muscle can be de-inserted from the trochlea to improve superior and posteromedial exposure.
- In medial orbital decompression, the ethmoid osteotomy is performed with a delicate osteotome. Underlying ethmoid cells, as well as the mucosa, must also be removed.
- The periorbita is usually not closed, whereas absorbable sutures are placed in the conjunctiva and caruncle.

5.3 *Inferior Orbitotomy*

There are three relevant anatomical structures on the orbital floor: the infraorbital nerve canal, the inferior oblique muscle, and the perforating branchs of the infraorbital artery. Indications for this route are: treatment of orbital floor fractures; incisional biopsy of ON tumors; incisional and excisional biopsy of inferior orbital tumors; drainage of inferior orbital abscesses; and orbital floor decompression in Graves' ophthalmopathy.

5.3.1 **Transconjunctival Approach** [11, 12, 14]

Transconjunctival incisions to the medial wall and orbital floor are widely used, mainly by oculoplastic surgeons. They allow wide exposure without leaving an apparent scar, in addition to a lower risk of postoperative complications compared to accesses that use skin incisions. For example, the transpalpebral approach has a higher risk of ectropion, with retraction and exposure of the sclera, whereas the Lynch approach leaves an apparent facial scar. Transconjunctival incisions can be associated with canthotomy or cantholysis in accessing the orbital floor, and in accessing the medial wall, a vertical incision can be made via transcaruncular or precaruncular region.

The endonasal endoscopic approach can be performed via a transethmoidal approach to the medial wall, or transantral approach to the orbital floor, with results similar to the direct approaches through orbitotomies. Endoscopic techniques have advantages in globe injuries, such as perforations and ruptures, as they avoid using retractors that compress the globe.

Surgical Technique

- The transconjunctival approach of the inferior fornix to access the inferior orbital rim can be performed via the preseptal or retroseptal approach. The latter will be described, because it is the most used. Tension sutures should be performed on the inferior eyelid to aid its eversion.
- The inferior fornix is identified below the tarsal conjunctiva to initiate the incision. This extends across the conjunctiva in a lateromedial direction to expose the fat compartment behind the orbital septum.
- With the aid of retractors, the inferior eyelid is pulled. The periorbital incision is performed posteriorly to the infraorbital rim until the periosteum is exposed, retracted with a detacher (Fig. 6).
- When broad exposure along the lateral orbital edge is required, transconjunctival access can be combined with a lateral canthotomy and cantholysis, exposing the entire lateral orbital rim to a level about 10 mm above the frontozygomatic suture.

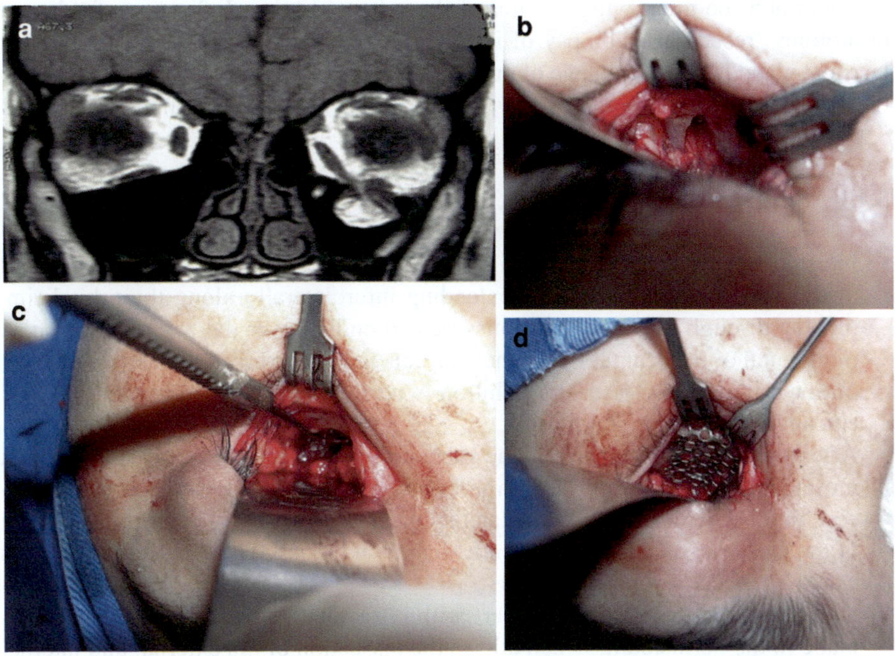

Fig. 6 (**a**) Coronal T1-weighted MRI demonstrating soft tissue herniation and muscle entrapment in blowout fracture of left orbital floor; (**b**) transconjunctival exposure with identification of bone floor defect; (**c**) removal of inferior rectus muscle trapped in the fracture; and (**d**) attachment of titanium mesh recomposing the bone defect in the floor

- Closure of the periosteum or periorbita is usually unnecessary unless lateral canthotomy is performed, requiring simultaneous canthopexy. The conjunctival incision is closed with continuous suture with 6-0 absorbable sutures. Conjunctival closure reestablishes the normal anatomical position of the lower eyelid. Care must not incorporate the orbital septum at closure, leading to postoperative entropion.

5.4 Lateral Orbitotomy

The classic incisions of Kronlein and Stallard–Wright have been less applied nowadays. Currently, the approach to the lateral wall has been more frequent through the superior eyelid sulcus, through the Berke incision (behind the lateral or transcantal corner), or through the swing eyelid.

The main anatomical elements of the lateral wall are the frontozygomatic suture, the zygomatic–facial, and temporal neurovascular bundles; the lateral canthal ligament and the great wing portion of the sphenoid bone. Indications for lateral orbitotomy: resection or biopsy of intra and extraconal tumors lateral to the ON; lacrimal gland lesions; drainage of orbital abscesses; deep lateral decompression for the treatment of Graves' orbitopathy; and treatment of orbitozygomatic fractures.

5.4.1 Surgical Technique [11, 12, 15]

- Incision marking is performed in a sinusoidal shape, starting in the lateral third of the upper eyelid sulcus and extending inferolaterally along the orbital rim, passing through the lateral commissure and ending over the zygomatic arch. A transcantal or retrocantal Berke incision can also be performed. Then, local infiltration with a vasoconstrictor solution is performed along the marking, lateral epicanthus, and temporal muscle.
- After making the skin incision, the subcutaneous tissue and the orbicularis oculi muscle are pulled apart. The skin incision should not go beyond the posterolateral limit above 2.4 cm from the lateral epicanthus to preserve the frontal branch of the facial nerve and the zygomatic–temporal branch. The course of the nerves can be mapped using a stimulator.
- Underlying tissues are pulled apart with retractors. Then, a vertical incision is made in the periosteum, which allows the dissection of the external and internal walls of the orbit, with the subperiosteal release of the temporal fossa and orbital cavity (superior, lateral, and inferior regions). The loosening of the temporal muscles increases the access exposure area. During dissection of the inner surface of the orbit, the lateral canthal ligament is often disinserted, and in intraconal lesions, the lateral rectus muscle can also be disinserted to facilitate removal of the lesion.
- Lateral orbitotomy is performed with the aid of a reciprocating or piezoelectric saw, with two cuts being made, one superior at the level of the frontozygomatic suture and the other inferior, immediately above the anterior insertion of the zygomatic arch. The bone fragment is fractured and removed (Fig. 7). The cuts are made in a V-shape to aid reconstruction. During bone cuts, the orbital content must be protected and irrigated to prevent heat transfer to the globe. Part of the greater wing of the sphenoid can be removed using the high-speed drill to increase exposure and facilitate lateral orbital decompression.
- The lateral periorbita is exposed and can be opened above or below, depending on the pathology to be addressed, allowing access to the extra and intraconal compartments.
- In case of a dural lesion, it must be corrected primarily or with the aid of grafts and fibrin glue. The periorbita must be closed using sutures, the interposition of autologous fat, or synthetic dural graft to prevent extrusion of its contents and optimize fistula repair. Bone reconstruction can be done by fixing the bone flap with miniplates and screws, which will provide good esthetic results.

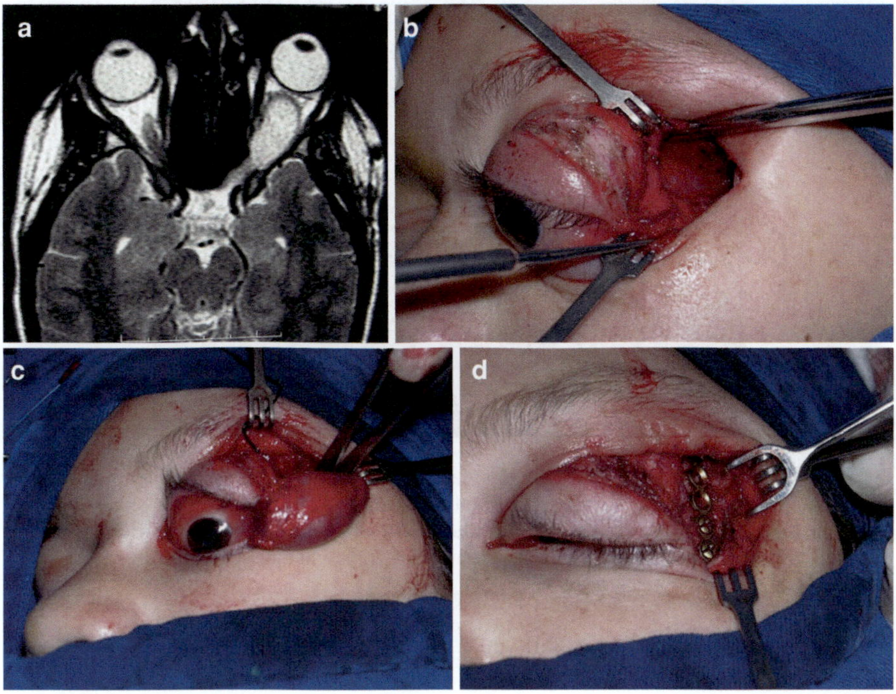

Fig. 7 (**a**) Axial T2-weighted MRI shows hyperintense fusiform intraconal lesion with mild anterior displacement of the left globe; (**b**) lateral orbitotomy via extended upper eyelid crease; (**c**) after the osteomy had been made, the lesion was dissected and completely resection was achieved. Histopathological examination was compatible with a pilocytic astrocytoma of the optic nerve; and (**d**) osteosynthesis with a miniplate on the lateral rim

6 Endonasal Endoscopic Medial Approach

Endonasal endoscopic access represents a safe and effective way to approach intra- and extraconal orbital lesions, medial to the ON, without the need for an external incision and with preservation of the orbicularis musculature, canthal ligament, and lacrimal apparatus [16]. The nasal septum can be partially removed, and the sinus cavities enlarged, allowing an adequate operative field without compromising the nasosinusal physiology [17]. A wide opening of the ethmoid, maxillary and sphenoid sinuses must be performed for an adequate characterization of the limits of the orbit and the skull base (Fig. 8), avoiding unwanted complications [18].

Navigation systems evolve rapidly and play an important role in surgical outcomes and patient safety. Image-guided navigation aids in tumor location and avoids nearby neurovascular structures, but it has limitations. The systems do not provide precise spatial information during the differentiation of soft tissues, especially in the orbital surgical space, where structures are closely superimposed [19].

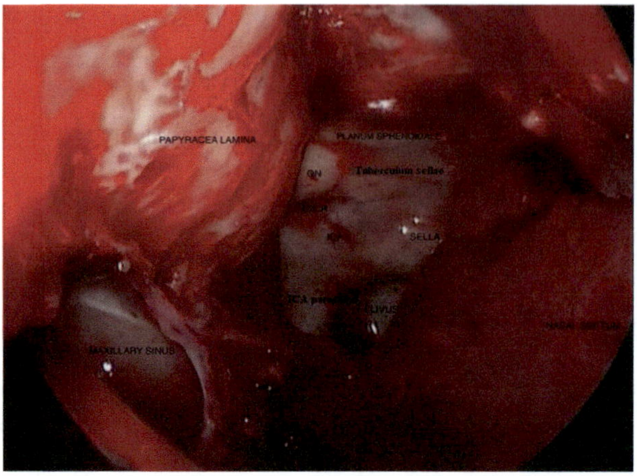

Fig. 8 Endonasal endoscopic view after complete ethmoidectomy on the right side, wide opening of the right maxillary sinus, and opening on the right of the anterior wall of the sphenoid sinus. Important landmarks of the skull base are identifiable, and the internal carotid artery can be observed in its paraclival and parasellar segments. LOCR, lateral optic-carotid recess; ICA, internal carotid artery; ON, optical nerve

Intraorbital fat can obliterate the view of the surgical field, making access to the desired structures difficult. Removal or cauterization of intraorbital fat, whether extraconal or intraconal, should be avoided. This fat should be just displaced from the field of vision. Excessive removal of extraconal fat can lead to enophthalmia [9]. Exceptionally, the intraconal fat can be cauterized, but its resection is inadvisable due to the risk of damage to neurovascular structures. Intraconal fat, extrinsic ocular musculature, and neurovascular structures, as they do not allow resections, represent anatomical obstacles to the endonasal endoscopic approach [20].

6.1 Surgical Technique

The endoscopic endonasal view of the orbit demands wide opening of the maxillary sinus, as anatomical repair of the orbital floor, complete anteroposterior ethmoidectomy for exposure and removal of the lamina papyracea and exposure of the medial aspect of the orbit, and wide sphenoidotomy to assist in the localization and approach to the orbital apex. Once the lamina papyracea and its periosteum (periorbita) are removed, they are opened, and the extraconal space is entered.

6.1.1 Step 1: Removal of the Anterior Wall of the Sphenoid Sinus

Complete removal of the anterior wall of the sphenoid sinus makes it possible to identify the transition between the medial wall of the orbit, represented by the lamina papyracea, and the lateral wall of the sphenoid.

6.1.2 Step 2: Removal of the Lamina Papyracea

After complete exposure of the medial wall of the orbit, the lamina papyracea is removed from anterior to posterior, with the posterior limit being just 1 cm anterior to the OC (Fig. 9). The anterior limit of the dissection corresponds to the junction of the lacrimal sac with the lamina papyracea, represented by the frontal process of the maxilla. Using angled curettes, it is possible to create an anterior opening, and with a "Freer detacher," carefully perform the dissection and removal of the lamina papyracea without incising the underlying periorbita. If there is a need to enlarge the operative field, a posterior septal window can be made [21].

6.1.3 Step 3: Opening of the Periorbita

The opening of the periorbita is done through an anterior vertical incision associated with a second perpendicular incision (see Video 1), with consequent exposure of the extraconal intraorbital fat, which is gently dissected until there is adequate exposure of the medial rectus muscle. This muscle represents the main anatomical barrier of the endonasal approach to remove intraconal lesions. From this moment forward, it is recommended that the surgery be performed by four hands, with the assistant surgeon holding the endoscope and retracting the medial rectus muscle, allowing to main surgeon freedom, in both hands, for tumor dissection.

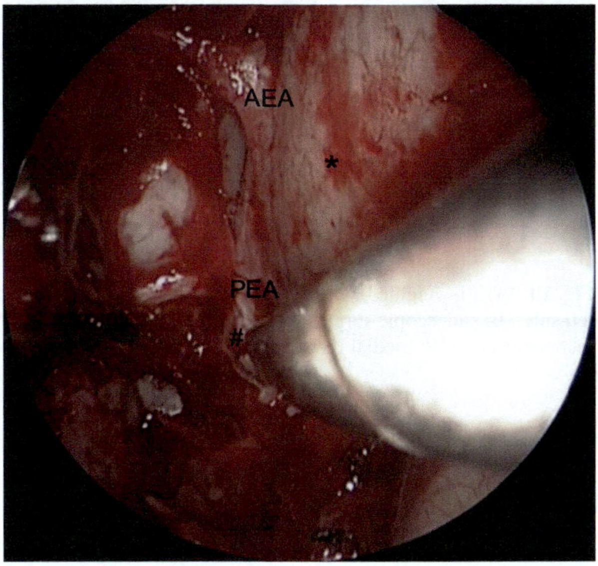

Fig. 9 Endoscopic endonasal view of the left periorbita after removal of the lamina papyracea. *, Periorbita; #, Orbital Apex; AEA, anterior ethmoidal artery; PEA, posterior ethmoidal artery

6.1.4 Step 4: Intraconal Dissection

The retraction of the medial rectus muscle can be superior or inferior, creating an inferior or superior orbital window, respectively. When possible, dissection through the inferior orbital window (between the medial and inferior rectus muscles) is preferable, displacing the medial rectus superiorly (Fig. 10—see Video 1), due to the scarcity of neurovascular structures. When the medial rectum is pulled downward, there is a risk of rupture of the ethmoid neurovascular bundles, which can cause bleeding and difficulties in the progression of the surgery.

As previously mentioned, intraorbital fat, and especially intraconal fat, should not be removed or cauterized to avoid postoperative enophthalmos and inadvertent damage to the ophthalmic artery [20]. Endonasal endoscopic access to the orbit is limited to lesions medial to the ON, and this nerve can be identified up to the region of the orbital apex. Using lenses with an angle of 30° promotes excellent visualization of the orbital apex, allowing the removal or biopsy of tumors from this topography with low complication rates (Fig. 11—see Video 2). Postoperative diplopia and enophthalmos occur in less than 15% of patients [22].

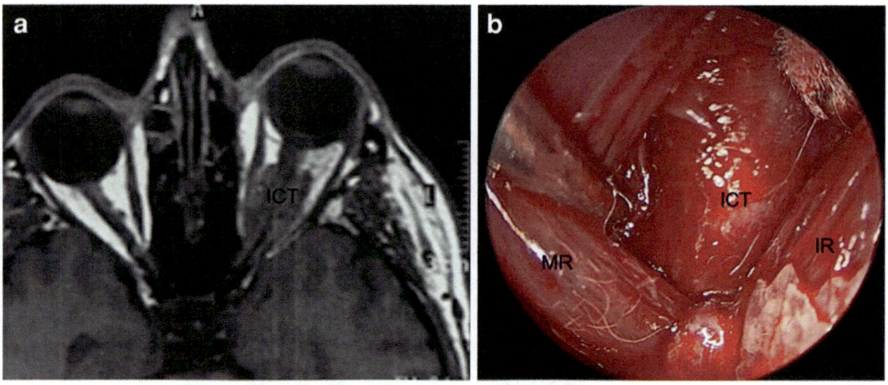

Fig. 10 (**a**) T1-weighted MRI demonstrating an intraconal lesion involving the optic nerve on the left side; (**b**) endoscopic surgical view of the inferior window after superior retraction of the medial rectus muscle. MR, medial rectus; IR, inferior rectus; ICT, intraconal tumor

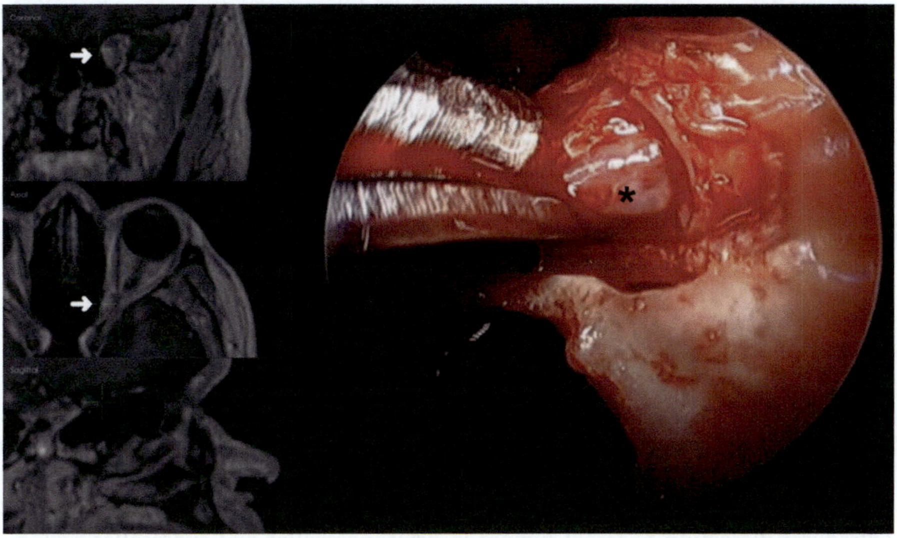

Fig. 11 On the left, a contrast-enhancing lesion (white arrow) can be seen in different sections of a T1-weighted MRI, located medial-to-medial rectus muscle (extraconal) at the orbital apex. On the right, a surgical photograph after opening the lamina papyracea and the periorbita, with the lesion being dissected. *, orbital apex cavernous hemangioma

7 Transcranial Approaches

Neurosurgeons classically use cranio-orbital approaches to approach lesions extending from the intracranial space to the orbit, primary orbital lesions located in the posterior third of the orbit [4], and intracranial tumors of the suprasellar and parasellar regions.

Transcranial approaches to the orbit can be divided into superior and posterolateral approaches. The superior approach is widely used in neurosurgical practice and was initially systematized by Joseph Maroon [4, 6]. It includes a frontotemporal craniotomy with the orbital rim and roof of the orbit. The frontotemporal–orbital bone flap is removed in one piece. This approach uses a bicoronal scalp incision to allow access to the medial aspect of the orbit and often requires penetration of the frontal sinus. In the posterolateral or pterional approach, removing the sphenoid bone allows a better working angle for lesions around the SOF and OC and lesions with intradural extension. With transcranial approaches, the trochlear nerve is quite vulnerable to injury [6], as it is an extraconal structure that crosses the orbit in a lateral-to-medial direction and immediately below the periorbita.

The authors use a customized focus fronto-orbito-zygomatic approach (FOCA) to the orbit in their practice. This approach, as discussed below, combines the advantages of superior and posterolateral transcranial approaches. Bone removal and dural elevation are performed according to the location of each particular type of tumor, primarily involving the orbital apex, the intracranial compartment alone, or the orbital apex and intracranial compartment.

7.1 Focus Fronto-Orbito-Zygomatic Approach (FOCA)

7.1.1 Concept and Rationale

The FOCA was conceptualized by the senior author (CC) to minimize the traditional fronto-orbito-zygomatic approah and yet to allow direct extradural access to skull base tumors. The FOCA uses an extended eyebrow incision or the traditional fronto-temporal scalp incision, a small focus fronto-orbitozygomatic bone flap, and extradural dissection to target tumors in orbit, parasellar and suprasellar regions. Extradural dissection to the origin of the tumor avoids the need for early Sylvian fissure opening, brain dissection, and the use of mechanical retraction. More importantly, this minimalist approach still permits extradural access for selected skull base tumors and orbit without direct manipulation or undue retraction of the brain parenchyma.

7.1.2 Surgical Technique

Step 1: Patient Position and Skin Incision

The patient is placed in a supine position, with the dorsum elevated and the neck extended. The head is secured to the skeletal head pin frame and turned 15°–30° to the opposite side. All surgical dissections are carried out under magnification with the surgical microscope.

The FOCA uses an eyebrow incision extending laterally in a curve fashion toward the zygoma. The incision starts in the lateral third of the eyebrow after the supraorbital notch to avoid injury to the supraorbital nerve and curves laterally following the eyebrow and frontal process of the zygomatic bone (Fig. 12). At the lateral canthus, the incision is directed posteriorly for 1.50 cm. The incision on the zygomatic arch and eyebrow avoids injury to the frontalis branches of the facial nerve. The skin incision in the eyebrow is carried out obliquely to minimize injury to the hair follicle and hair loss. The frontal branch of the facial nerve goes horizontally to the orbit at the height of 2.8 cm from the lateral canthus [23], reducing the risk of its integrity.

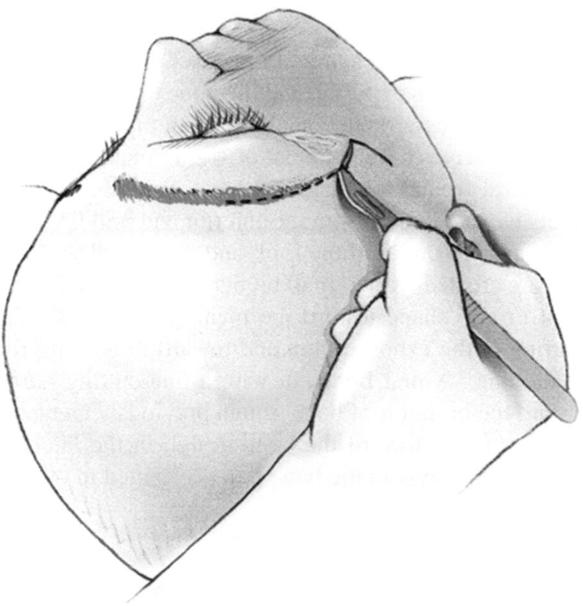

Fig. 12 Skin incision on the right side: The incision starts in the lateral third of the eyebrow (after the supraorbital notch), curving inferiorly through the lateral edge of the orbit, and then it is directed posteriorly for approximately 1.5 cm from the lateral canthus

Step 2: Soft Tissue Dissection and Muscle Mobilization

After deepening the plane, the osseous surface of the orbital rim is exposed, followed by a subperiosteal dissection of the frontal and temporal surfaces. The dissection follows the tissue plane, and the loose areolar tissue plane can increase the wound retraction capacity. Since only the zygomatic process of the frontal bone is exposed during the approach, the frontal branch of the facial nerve is protected by the temporoparietal fascia [24, 25].

The skin incision continues through the orbicularis muscle and the periosteum of the orbital rim, the zygomatic process of the frontal bone, and the frontal process of the zygomatic bone. The anterior position of the periosteum incision avoids injury to the frontal branches of the facial nerve. Subperiosteal dissection is carried out anteriorly and posteriorly for exposure to the orbital rim (Fig. 13). The frontal branch of the facial nerve remains protected by the temporoparietal fascia [24, 25]. After mobilization and retraction of the temporal muscle and the temporoparietal fascia, adequate exposure of the keyhole area (pterion) is obtained.

A periorbital dissection is then carried out to separate the eye globe from its roof and lateral wall.

Step 3: Bone Dissection

The bone dissection starts at the MacCarty's keyhole with the high-speed electric drill. The dura is exposed, and the bone dissection continues laterally to include the pterion, the lesser sphenoidal wing, and the anterior temporal squama. This bone removal creates a bone defect of 4–5 cm and allows visualization of the dura over the lateral frontal lobe and anterior temporal lobe (Fig. 14a). The dura is progressively elevated away from the anterior and middle fossa skull base, and further drilling is carried out until the roof, and the lateral wall of the orbit is visualized. Bone cuts are created with a small bit burr over the roof of the orbit and lateral wall of the orbit in a V shape toward the meningo-orbital band. A small frontal craniotomy starting at the exposed dura and toward the orbital rim is carried with the electric craniotome. A first bone cut with a fine cutting burr is created in the orbital rim toward the orbital roof bone trough previously created. A second bone cut is created in the zygoma toward the bone trough in the lateral wall of the orbit. The small fronto-orbito-zygomatic bone flap is elevated in one piece (Fig. 14b).

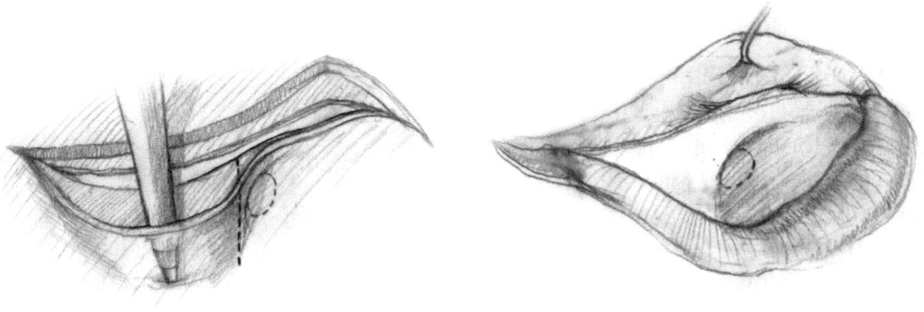

Fig. 13 Subperiosteal dissection of the frontal and temporal surfaces on the right side with pterion exposure after soft tissue retraction

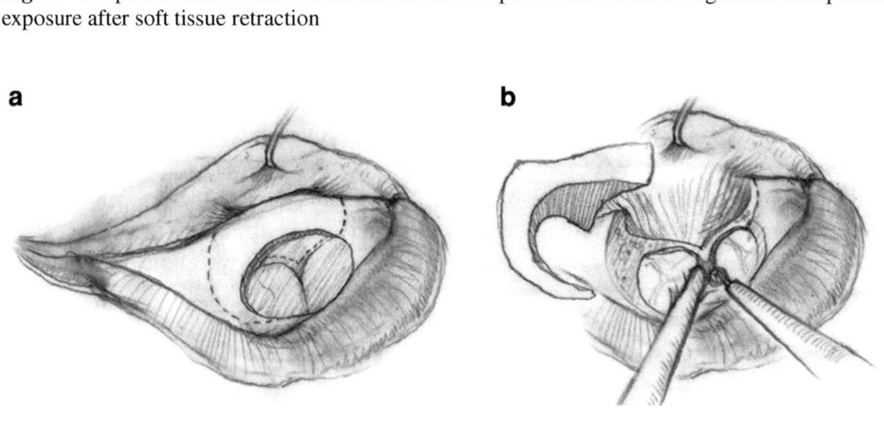

Fig. 14 (**a**) Area of exposure of the dura mater of the frontal and temporal lobes after craniectomy. The dotted lines show the locations of the bone cuts; (**b**) piece of bone which includes part of the orbital rim, the zygomatic process of the frontal bone and the orbital roof. Additional bone must be removed using a high-speed drill at the lesser wing of the sphenoid bone

Step 4: Epidural Dissection

After bone flap elevation, epidural dissection continues in the middle fossa skull base, and the dura of the temporal lobe is progressively elevated from the SOF starting at the meningo-orbital band. The dural fold that enters through the SOF, toward the orbit, comprises two distinct layers. The section of the lateral portion of this fold is carried out using a cutting instrument, and after identifying the cleavage plane between the layers, the so-called "dural peeling" is made by retracting the external leaflet [26]. The dissection continues posteriorly, and the dura of the temporal lobe is elevated from the dura propria of the anterior cavernous sinus. The dissection is continued laterally, and the dura is peeled away from the maxillary branch of the trigeminal nerve (V2) at the foramen rotundum. The dura of the frontal lobe is further elevated from the anterior skull base fossa with exposure to the roof of the OC and ACP. This epidural dissection is tailored to each specific location of the lesion to allow access to the origin of the tumor without undue brain manipulation.

Step 5: Removal of the Orbital Roof, the Roof of the Optical Canal, and Anterior Clinoid Process

The roof of the OC is removed with the high-speed electric drill and provides identification of the orbital apex with bone decompression of the ON. The orbital roof can be removed from the orbital apex to the orbital rim. Removing the orbital roof is necessary for surgery of tumors extending from the intracranial space to the orbit or primary orbital lesions. The extension of bone removal of the orbital roof is tailored to the needs and location of the orbital lesion.

Removal of the ACP when associated with resection of the roof of the OC, the opening of the falciform ligament, and the opening of the "distal ring" greatly increases the capacity to mobilize the ON and the ICA in the vicinity of the skull base. Anterior clinoidectomy is regularly applied in transcranial surgeries for parasellar tumors (meningiomas, adenomas, and craniopharyngiomas) regardless of orbital invasion.

The ACP is closely related to the oculomotor nerve laterally, the clinoidal segment of the ICA, the dural rings, and the ON medially. Opening the roof of the OC prior to the anterior clinoidectomy provides safe identification of the medial border of the ACP and protects the ON during the clinoid resection. The superior cortical surface of the ACP is first drilled, and then, its center is cored out. The ACP is then disconnected from the body of the sphenoidal bone after drilling of the optic strut. Next, the inferior cortex is carefully thinned out, and a thin shell is then elevated from the roof of the cavernous sinus inferiorly and the oculomotor nerve inferolaterally. A dense oculo-carotid membrane protects the ICA and the oculomotor nerve during anterior clinoidectomy, and it corresponds to the superior wall of the cavernous sinus. Extradural anterior clinoidectomy with the opening of the OC allows early identification of the ON and ICA (Figs. 15 and 16). Opening of the falciform ligament and dural sheath provides decompression of the ON.

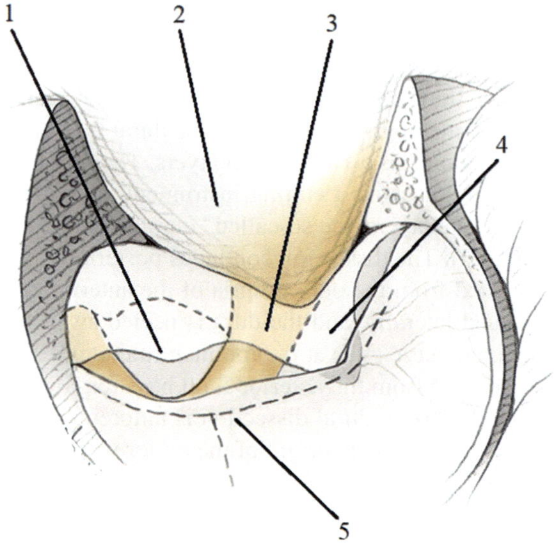

Fig. 15 Dotted lines demarcate the following structures: (*1*) sphenoidal limen at the tuberculum sellae; (*2*) orbit; (*3*) roof of the optical canal; (*4*) anterior clinoid process; and (*5*) incision lines of the dura mater

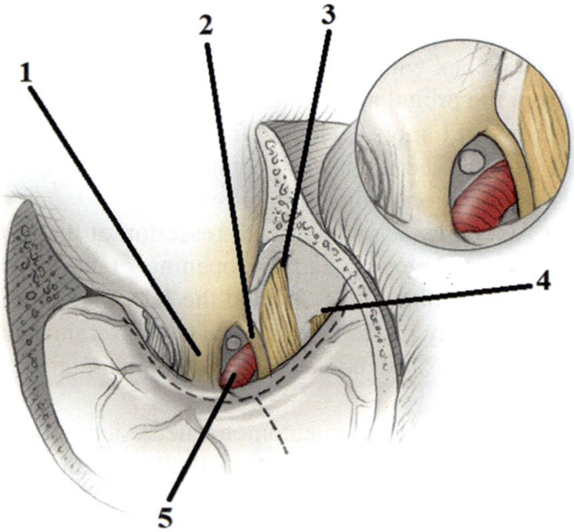

Fig. 16 Exposed structures after extradural removal of the anterior clinoid process and optic roof canal: (*1*) intracanalicular segment of the optical nerve; (*2*) the oculomotor nerve running through the upper wall of the cavernous sinus and closely related to the anterior clinoid process, laterally; (*3*) maxillary branch of the trigeminal nerve at the foramen rotundum; (*4*) mandibular branch of the trigeminal nerve at the foramen ovale; and (*5*) clinoid segment of the internal carotid artery

In the case of tumors in the suprasellar region, such as TS meningioma, the extradural dissection is extended medially to the OC toward the hyperostosis in the planum sphenoidale and the TS to allow extradural access to the origin of the tumor (Fig. 15).

Step 6: Dural Opening

Using the concept of "extradural target navigation" [27], the dura mater is opened as close as possible to the site of the lesion at the skull base, reducing parenchymal exposure and damage. For tumors in the suprasellar and parasellar region, the dura opening and entry into the subarachnoid space starts medial to the falciform ligament and continues over the OC dura and laterally along the dura margins of the resected ACP. For tumors in the planum sphenoidal or the TS, the dura opening continues toward the opposite side to allow access to the dural origin of the tumor. The dura opening is extended laterally for tumors in the ACP and the middle cranial fossa. Soon after entry into the subarachnoid space, the dura of the nerve sheath of the ON is opened, and the tumor in its medial or lateral aspect is removed for early decompression of the compromised ON prior to any tumor manipulation (Fig. 17).

The intradural dissection will be processed in a customized way for each type of lesion, but the principle of working through these small dural openings requires progressive tumor debulking and smooth retraction of the edges of the lesion to the surgeon's field of vision.

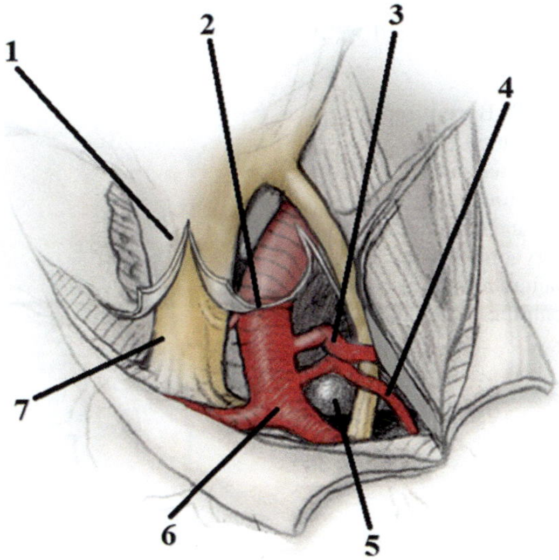

Fig. 17 Structures visualized after dural opening and resection of the anterior clinoid process and the roof of the optic canal: (*1*) opened falciform ligament; (*2*) distal dural ring; (*3*) posterior communicating artery; (*4*) anterior choroidal artery; (*5*) posterior clinoid process; (*6*) internal carotid artery bifurcation; and (*7*) intracranial segment of the optic nerve in the subarachnoid space

Step 7: Dural, Bone, and Soft Tissue Reconstruction

The dural defect in the skull base created by the resection of the infiltrated dura is repaired with a non-suturable synthetic dural substitute, placed as an inlay. The use of fibrin glue is recommended. The previously removed piece of bone is then replaced and secured with miniplates and screws to the surrounding frontal bone and frontal process of the zygomatic bone. A titanium mesh is used to repair the cortical defect created by the bone removed (Fig. 18). The remainder of the closure should include all tissue planes, including aponeurotic galea, subcutaneous, and skin, reducing dead spaces and the formation of collections. No drain is used.

7.2 *Clinical Cases*

7.2.1 Case # 1

A 56-year-old female presented with visual loss, diplopia, pain, and axial proptosis on the right side. The MRI showed an intraconal mass with homogeneous contrast enhancement at the orbital apex and slight anterior displacement on the globe (Fig. 19a, b).

With the use of FOCA, the roof of the OC and the ACP were removed extradurally. The falciform ligament was opened, and a tumor was seen below the ON in its intracanalicular segment (Fig. 19c).

The periorbita was opened with a sickle knife (Fig. 19d), and the superior and inferior edges were tagged with a 4-0 silk suture to increase exposure. The lateral rectus was looped with a vessel loop and pulled inferiorly to expose the intraconal

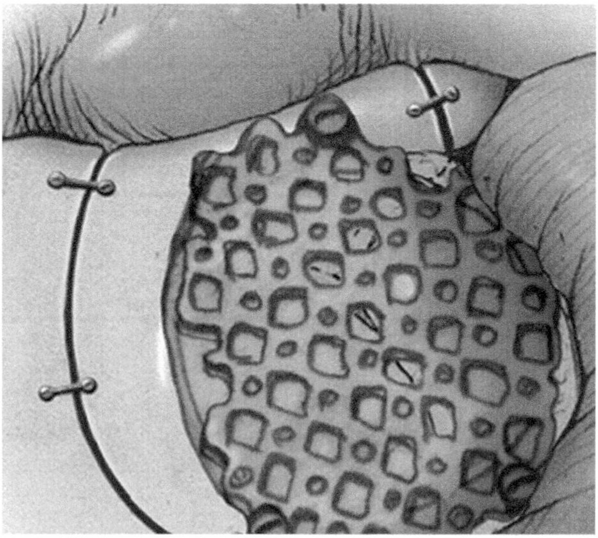

Fig. 18 Bone is replaced and secured with miniplates and screws to the surrounding frontal bone and frontal process of the zygomatic bone. A titanium mesh is used to repair the cortical defect

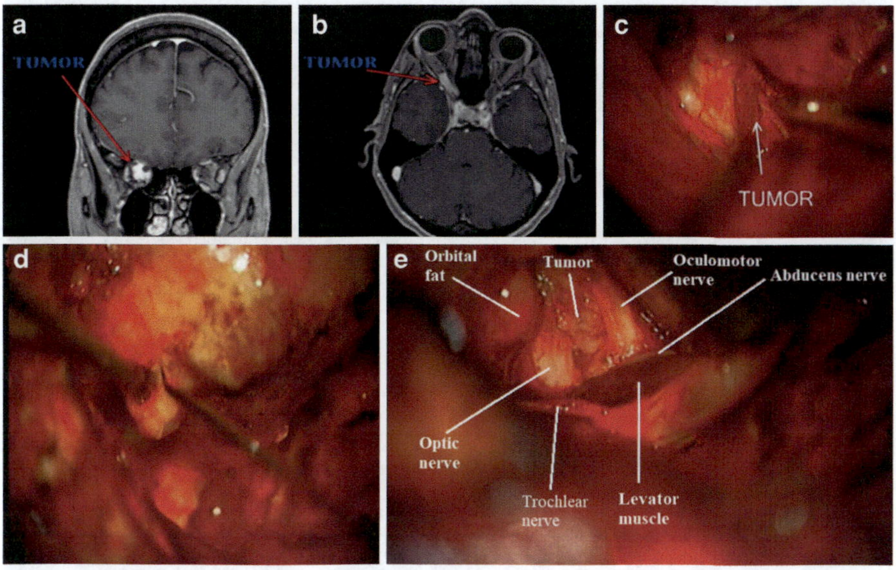

Fig. 19 Resection of an orbital apex meningioma using the focus fronto-orbito-zygomatic approach (FOCA). (**a**) Coronal T1-weighted MRI demonstrating a lesion with homogeneous contrast enhancement, in an intraconal situation, in the orbital apex region. (**b**) Axial T1-weighted MRI demonstrating a lesion with homogeneous contrast enhancement in the orbital apex region, with slight anterior displacement of the globe (proptosis). (**c**) Optic nerve in its intracanalicular segment distended by the tumor located below the nerve. (**d**) Periorbital opening with fat extrusion. (**e**) Intraconal tumor dissection with visualization of the intraorbital segment of the optic nerve, as well as the oculomotor, trochlear and abducens nerves

space. Malleable retractors were used to delineate the tumor. After identifying relevant anatomical structures in the region, the lesion is removed piecemeal (Fig. 19e—see Video 3).

At 2 years of follow-up, the patient regained 20/20 vision and had no diplopia, ptosis, or enophthalmos postoperatively.

7.2.2 Case # 2

A 44-year-old male with a 3-year history of lymphoma presented with bilateral retro-orbital pain.

MRI showed enlarging bilateral middle fossa meningioma. The left-sided tumor was compatible with a medial sphenoidal wing meningioma. The right-sided tumor was compatible with an anterior clinoid meningioma (Fig. 20a, b).

After extensive counseling on treatment options, the patient elected to undergo surgical resection.

The patient was operated on with bilateral extended eyebrow incision for FOCA, 7 months apart. On the right side, extradural clinoidectomy was carried out, and the

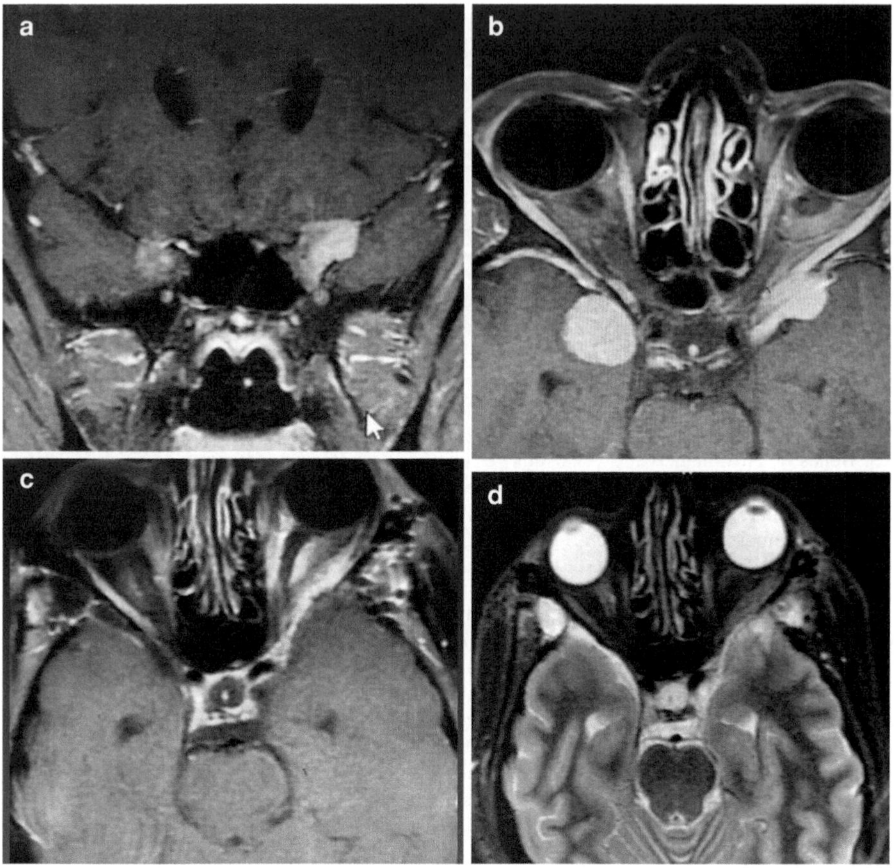

Fig. 20 Resection of bilateral meningiomas using the focus fronto-orbito-zygomatic approach—FOCA (in two different surgical times). (**a, b**) Coronal and axial T1-weighted MRI, respectively, demonstrating bilateral lesions with homogeneous contrast enhancement. The lesion on the left is located in the medial sphenoidal wing (orbital apex in an extraconal situation), and on the right, the lesion originates in the anterior clinoid process. (**c, d**) Axial T1- and T2-weighted MRI, respectively, demonstrating complete resection of these lesions, with preservation of adjacent structures. The brain parenchyma close to the approaches has no detectable signal alterations on MRI

tumor was followed into the CS and SOF, and on the left side, the orbital apex was approached (see Video 4). Bilateral gross total resection was achieved.

The patient has no extraocular muscle weakness or double vision at 4-year follow-up. There is no palsy of the frontalis branch of the facial nerve. The patient is satisfied with his cosmetic outcome.

Postoperative MRI shows no recurrent or residual tumor. There is no abnormal T2/FLAIR signal in the temporal or frontal lobes (Fig. 20c, d).

7.2.3 Case # 3

A 42-year-old male presented with 6-month history of left eye severe visual loss (finger counting) and loss of left temporal visual field. Brain MRI showed a 2.8 × 2.9 cm TS meningioma with significant left ON compression, OC tumor invasion, and optical chiasm displacement (Fig. 21a, b).

The patient was operated on through a left FOCA with an extended eyebrow incision. Anterior clinoidectomy was necessary for accessing the lateral aspect of the tumor under the ON and over the ICA. Gross total resection was achieved (see Video 5).

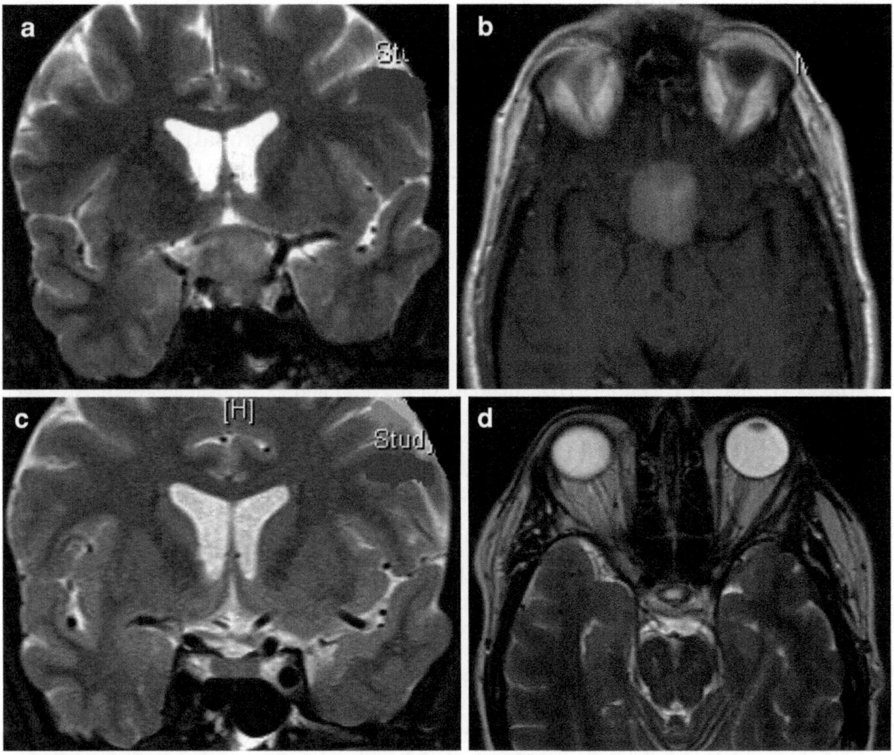

Fig. 21 Resection of a tuberculum sellae meningioma using the focus fronto-orbito-zygomatic approach (FOCA). (**a**) Coronal T2-weighted MRI demonstrating a sellar tumor with suprasellar extension, isointense in relation to the brain parenchyma, and with a close bilateral relationship with the internal carotid arteries. The optic chiasm is compressed in the upper portion of the lesion. (**b**) Axial T1-weighted MRI demonstrating a hyperintense lesion anterior to the third ventricle plane. (**c, d**) coronal and axial, respectively, T2-weighted MRI demonstrating complete resection of the lesion. The optic chiasm has returned to its usual position, and the pituitary has a preserved volume. There are no signs of surgical manipulation in the brain parenchyma

Postoperative exam 27 months after surgery shows left visual acuity 20/30 and the peripheral visual defect recovery in the left temporal field. The patient has no loss of olfaction or taste. The patient has had no seizure or cognitive changes. One year after surgery, a postoperative MRI shows no evidence of residual or recurrent tumor and no T2/flair abnormal signal in the frontal or temporal lobe (Fig. 21c, d).

8 Conclusions

Managing orbital lesions involve a multidisciplinary team with surgical experience in the skull base, including neurosurgeons, oculoplastic surgeons, and otolaryngologists. The type of approach to be used will depend on the disease etiology, surgical proposal, and the surgical team's experience and may include traditional direct transorbital approaches, primary transorbital endoscopy, endonasal endoscopic multiportal endoscopic approaches (endonasal + transorbital), or transcranial approaches. The topography of the lesion and its relationship to the surrounding neurovascular structures are also a key point in choosing the approach.

References

1. Hayek G, Mercier P, Fournier HD. Anatomy of the orbit and its surgical approach. Adv Tech Stand Neurosurg. 2006;31:35–71.
2. Lee RP, Khalafallah AM, Gami A, Mukherjee D. The lateral orbitotomy approach for intraorbital lesions. J Neurol Surg B Skull Base. 2020;81(4):435–41.
3. Hu S, Colley P. Surgical orbital anatomy. Semin Plast Surg. 2019;33(2):85–91.
4. Bejjani GK, Cockerham KP, Kennerdel JS, Maroon JC. A reappraisal of surgery for orbital tumors. Part I: Extraorbital approaches. Neurosurg Focus. 2001;10(5):E2.
5. Engin Ö, Adriaensen GFJPM, Hoefnagels FWA, Saeed P. A systematic review of the surgical anatomy of the orbital apex. Surg Radiol Anat. 2021;43(2):169–78.
6. Maroon JC, Kennerdell JS. Surgical approaches to the orbit. Indications and techniques. J Neurosurg. 1984;60(6):1226–35.
7. Shields JA, Shields CL, Scartozzi R. Survey of 1264 patients with orbital tumors and simulating lesions: the 2002 montgomery lecture, part 1. Ophthalmology. 2004;111(5):997–1008.
8. Abussuud Z, Ahmed S, Paluzzi A. Surgical approaches to the orbit: a neurosurgical perspective. J Neurol Surg B Skull Base. 2020;81(4):385–408.
9. Locatelli D, Dallan I, Castelnuovo P. Surgery around the orbit: how to select an approach. J Neurol Surg B Skull Base. 2020;81(4):409–21.
10. Abou-Al-Shaar H, Krisht KM, Cohen MA, Abunimer AM, Neil JA, Karsy M, et al. Cranio-orbital and Orbitocranial approaches to orbital and intracranial disease: eye-opening approaches for neurosurgeons. Front Surg. 2020;7:1.
11. Pai SB, Nagarjun MN. A neurosurgical perspective to approaches to the orbit: a cadaveric study. Neurol India. 2017;65(5):1094–101.
12. Kannan S, Hasegawa M, Yamada Y, Kawase T, Kato Y. Tumors of the orbit: case report and review of surgical corridors and current options. Asian J Neurosurg. 2019;14(3):678–85.
13. Shorr N, Baylis HI, Goldberg RA, Perry JD. Transcaruncular approach to the medial orbit and orbital apex. Ophthalmology. 2000;107(8):1459–63.

14. Bernardini FP, Nerad J, Fay A, Zambelli A, Cruz AAV. The revised direct Transconjunctival approach to the orbital floor. Ophthalmic Plast Reconstr Surg. 2017;33(2):93–100.
15. Harris GJ, Logani SC. Eyelid crease incision for lateral orbitotomy. Ophthalmic Plast Reconstr Surg. 1999;15(1):9–16; discussion 16–8.
16. Felippu A, Mora R, Guastini L, Peretti G. Transnasal approach to the orbital apex and cavernous sinus. Ann Otol Rhinol Laryngol. 2013;122(4):254–62.
17. Pant H, Bhatki AM, Snyderman CH, Vescan AD, Carrau RL, Gardner P, et al. Quality of life following endonasal skull base surgery. Skull Base. 2010;20(1):35–40.
18. Castelnuovo P, Dallan I, Locatelli D, Battaglia P, Farneti P, Tomazic PV, et al. Endoscopic transnasal intraorbital surgery: our experience with 16 cases. Eur Arch Otorhinolaryngol. 2012;269(8):1929–35.
19. Udhay P, Bhattacharjee K, Ananthnarayanan P, Sundar G. Computer-assisted navigation in orbitofacial surgery. Indian J Ophthalmol. 2019;67(7):995–1003.
20. Bleier BS, Healy DY, Chhabra N, Freitag S. Compartmental endoscopic surgical anatomy of the medial intraconal orbital space. Int Forum Allergy Rhinol. 2014;4(7):587–91.
21. Healy DY, Lee NG, Freitag SK, Bleier BS. Endoscopic bimanual approach to an intraconal cavernous hemangioma of the orbital apex with vascularized flap reconstruction. Ophthalmic Plast Reconstr Surg. 2014;30(4):e104–6.
22. Bleier BS, Castelnuovo P, Battaglia P, Turri-Zanoni M, Dallan I, Metson R, et al. Endoscopic endonasal orbital cavernous hemangioma resection: global experience in techniques and outcomes. Int Forum Allergy Rhinol. 2016;6(2):156–61.
23. Schmidt BL, Pogrel MA, Hakim-Faal Z. The course of the temporal branch of the facial nerve in the periorbital region. J Oral Maxillofac Surg. 2001;59(2):178–84.
24. Davidge KM, van Furth WR, Agur A, Cusimano M. Naming the soft tissue layers of the temporoparietal region: unifying anatomic terminology across surgical disciplines. Neurosurgery. 2010;67(3 Suppl Operative):ONS120–9; discussion ONS129–30.
25. Salas E, Ziyal IM, Bejjani GK, Sekhar LN. Anatomy of the frontotemporal branch of the facial nerve and indications for interfascial dissection. Neurosurgery. 1998;43(3):563–8; discussion 568–9.
26. Dolenc VV. Microsurgical anatomy and surgery of the central skull base. Vienna: Springer; 2003. http://public.ebookcentral.proquest.com/choice/publicfullrecord.aspx?p=3069327.
27. Vidal CHF, Nicácio JA, Hahn Y, Caldas Neto SS, Coimbra CJ. Tentorial peeling: surgical extradural navigation to protect the temporal lobe in the focused combined transpetrosal approach. Oper Neurosurg (Hagerstown). 2020;19(5):589–98.

Surgical Anatomy of the Middle Fossa

José Ernesto Chang, Thomas Meduneckas Tourinho, Sebastián Aníbal Alejandro, and Feres Chaddad-Neto

1 Introduction

The *skull* is a structure that serves as rest and protection for the encephalon, the most vital part of the central nervous system. It can be divided into, facial skeleton *(viscerocranium)* and cranium *(neurocranium)* [1], containing the encephalon, which can also be separated into two parts, a *voult* and a *base* by a transverse line that goes between the frontal eminence and the external occipital protuberance, and this line would be an oblique one with the horizontal plane. The cranial vault or *calvaria,* formed by four bones, the frontal anteriorly, two parietals, and the occipital posteriorly, cover and protect the brain. The base is made up of thicker bones that help protect a variety of neurovascular structures and the brain stem. It presents two surfaces, exocranial and endocranial, both communicated with each other and to the nasal cavity, the orbit, the infrapetrosal, and infratemporal fossae, thru a series of foramina and fissures. Classically, the cranial base is separated into anterior, middle, and posterior fossa using a series of bony landmarks. The frontal bone encloses the anterior fossa, and its floor is formed by the orbital processes of the frontal bone and the cribiform plate of the ethmoid bone [2]. The occipital bone surrounds the posterior fossa.

The *middle cranial fossa* or *central skull base* is a critical region containing many neurovascular and parenchymal structures, all subject of presenting a variety of pathologies, vascular, oncologic, infectious, among others, making it extremely

J. E. Chang · T. M. Tourinho · S. A. Alejandro
Department of Neurology and Neurosurgery, Universidade Federal de São Paulo—UNIFESP, São Paulo, SP, Brazil

F. Chaddad-Neto (✉)
Department of Neurology and Neurosurgery, Universidade Federal de São Paulo—UNIFESP, São Paulo, SP, Brazil

Hospital Beneficência Portuguesa de São Paulo, São Paulo, SP, Brazil

E. G. Figueiredo et al. (eds.), *Brain Anatomy and Neurosurgical Approaches*,
https://doi.org/10.1007/978-3-031-14820-0_23

important to the neurosurgeon the precise knowledge of anatomy and overall a good surgical correlation, how to extrapolate anatomical understanding in the planning of our surgical strategy and how this can guide us during the performance of a procedure. In addition to offering an anatomical review of the structures participating in this region, the objective of this chapter is to provide means for understanding the middle fossa as a surgical region in a three-dimensional manner, using osseous and neurovascular anatomical landmarks.

2 Osseous Anatomy of the Middle Fossa

The middle cranial fossa is formed mainly by the sphenoid and temporal bones, creating a space, where many critical structures convey. The classical limits of the middle fossa are as follows (Fig. 1):

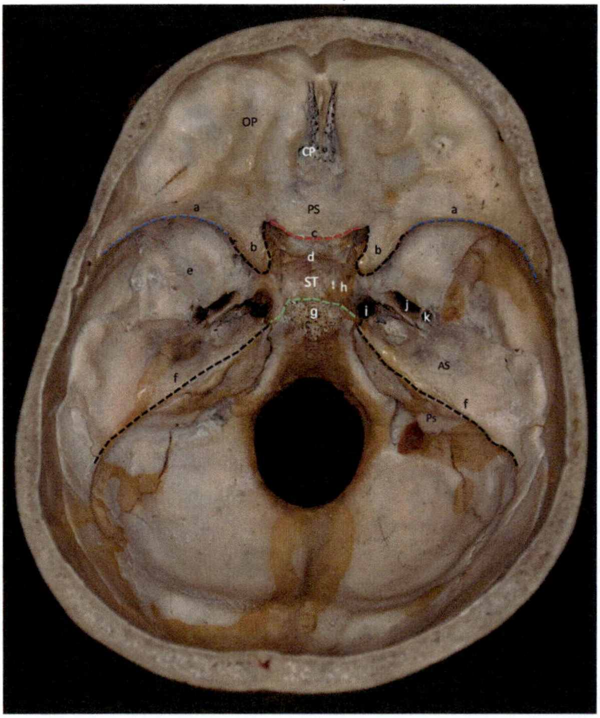

Fig. 1 Superior view of the endocranial surface of the skull base. The anterior and posterior limits of the middle cranial fossa are marked. The anterior border composed by (**a**) lesser sphenoid wing, (**b**) anterior clinoid process, and (**c**) limbus sphenoidale. The posterior border: (**f**) the superior margin of petrous bone and (**g**) dorsum sellae. Also can recognize: (OP) orbital process of the frontal bone; (CP) cribiform plate of the ethmoid bone, (PS) planum sphenoidale, (ST) sella turcica, (**d**) tuberculum sellae, (**e**) greater sphenoid wing, (**h**) carotid or cavernous canal, (**i**) foramen lacerum, (**j**) foramen ovale, and (**k**) foramen spinosum. Anterosuperior (AS) and posterosuperior surfaces of the petrous bone (Ps)

- **Anterolaterally:** lesser sphenoid wings and anterior clinoid process.
- **Anteromedially:** limbus sphenoidal.
- **Posterolaterally:** superior margin of the petrous bone.
- **Posteromedially:** dorsum sellae.
- **Lateral walls:** squamous part of the temporal bone and (the inferior aspect of a parietal bone).
- **Floor:** sphenoid greater wings, squamous and petrous part of the temporal bone.

It is important to recognize these and familiarize with its nomenclature for a better understanding of this region and the rest of ligamentous and neurovascular structures; that is why, we are going to go deeper into the study of these osseous components.

3 Sphenoid Bone

The **sphenoid** bone, whose name derives from the Greek *[−sphēn-]* meaning "wedge," since it is embedded between the rest of the bones that make up the skull [1]. It resembles the shape of a butterfly or a bat with extended wings and is composed of *(2) greater wings, (2) lesser wings, (2)* pterygoid processes, and *(1) body.*

The **body** has a cuboid shape with six surfaces (superior, inferior, lateral, medial, anterior, and posterior), and the anterior surface with an anterior crest in the midline, articulated with the perpendicular lamina of the ethmoid bone, also presents laterally in both sides, the *sphenoid concha*, containing an aperture that varies in size. The inferior surface presents the medially positioned crest that runs anteriorly until joined with the anterior crest forming a triangular shape spine or crest called *Rostrum sphenoidale,* which will articulate with a small depression in the Vomer bone *(Vomer wing/canal).* On each side of the inferior crest, just medial to the pterygoid process, there is a depression, the *pterygopalatine canal,* that when articulated with the *sphenoidal process* of the Palatine bone, more specifically, when this canal is combined with a notch located between the orbital and the sphenoidal process of the Palatine bone, called *sphenopalatine notch,* form together with the ***sphenopalatine canal***, through which the sphenopalatine artery and vein, the posterior superior lateral nasal nerve and the nasopalatine nerves pass.

The posterior surface with a quadrilateral surface is strongly articulated with the occipital bone. The lateral surface of the body serves as insertion area for the greater wings and is separated from the superior surface by the *carotid sulcus* on each side, this sulcus contains mainly the *cavernous sinus (CS)* and the cavernous segment of the *internal carotid artery (ICA)*, in its posterior origin, which is a continuation of the *carotid canal,* has a small posterolateral bony projection known as *lingula (of Meckel's),* and it continues anteriorly, making two opposite curves that cause an "*S*" shape until reaching the paraclinoid region. The superior surface is the most relevant for this chapter, and is completely endocranial; a series of important structures are described from anterior to posterior, one quadrilateral surface, *the jugum sphenoidale*, in a midline position, behind it, a bony ridge known as *Limbus sphenoidale*, determining a limit between the jugum sphenoidale and a transverse canal posteriorly known as *Prechiasmatic sulcus*, which is continued at its lateral ends with both optic canals; and the optic chiasm is located just above this sulcus. Posteriorly, it has an excavation that has been compared with a *saddle,* the *Sella turcica (Latin for Turkish saddle),* where the pituitary gland is found *(pituitary fossa)* (Fig. 1). The limits of the Sella turcica are anteriorly, a ridged bony process known as *Tubercullum sellae,* which separate it from the prechiasmatic sulcus [1, 3]; posteriorly, a quadrangular lamina that separates it from the occipital bone, the *Dorsum sellae*. We also can identify four angles limiting the sella known as *Clinoid processes,* two anterior that forms part of the medial aspect of the lesser sphenoid wings (and acts as the lateral wall of the optic canal) and two posteriors which are the free angles of the dorsum sellae; between them, in some specimens, a median eminence could be identified lateral to the pituitary fossa, normally reduced to a tuberculum, but sometimes, it can be larger even to the point of merging with either the anterior clinoid *(carotidoclinoid foramen)* or the posterior clinoid (Fig. 2) [1].

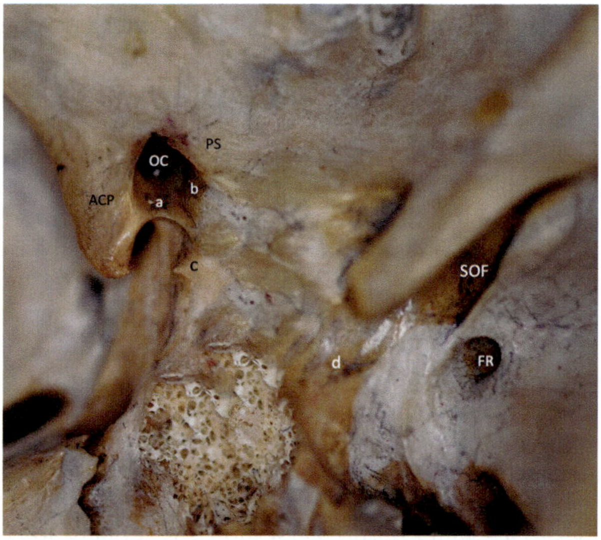

Fig. 2 Right posterolateral view of the optic canal region. The components of the optic canal (OC) can be identified: the roof or superomedial border by the planum sphenoidale (PS) and the superior aspect of the sphenoid body (**b**), the lateral wall or superolateral border by the anterior clinoid process (ACP) and the floor by the Optic strut (**a**). Also in this specimen we can identifiy a middle clinoid process, emerging from the lateral border of the sella turcica (**c**). Superior orbital fissure (SOF), foramen rotundum (FR), and carotid or cavernous canal (**d**)

The **lesser wings** *(Ingrassia apophyse, orbital wings, or alae parvae)* [1] are two elongated, triangular shape projections with a medial base and a lateral vertex, inserted on each side of the body of the sphenoid, in its most anterosuperior portion, flattened above and below; its posterior free edge, also with a triangle-shape and its vertex pointing backward, is known as the *anterior clinoid process* (klinē, in greek, that means bed, resembling a four-poster bed). The **greater sphenoid wings** are more extensive and wide projections from either side of the body, curving upward and laterally; being able to describe three surfaces, one *superior* which is concave, endocranial and contains the temporal lobe of the brain (Fig. 1); one anterior or orbital surface that forms part of the lateral wall of the orbit and one external with a lateral convexity, and with its anterior part more concave; and this face serves as a site of insertion of different muscles.

In the medial border of the superior surface, from anterior to posterior, we find four important openings or foramina. First, the *superior orbital fissure (SOF)* (Fig. 2), between this border and the medial aspect of the lesser sphenoid wing, wider medially, acts as the entrance to the orbit of the *oculomotor nerve (III CN), trochlear nerve (IV CN),* and *abducens nerve (VI CN)*, the *ophthalmic branch* of the *trigeminal nerve* (V_1) with its three branches, *nasociliary, lacrimal, and frontal nerves*, also the *superior ophthalmic vein*, orbital branches of the middle meningeal artery, sympathetic, and parasympathetic fibers [3]. Just posterior to the SOF, we identified a circle shape foramen, the *Foramen rotundum*, through which passes the maxillary nerve (V_2), then, posterior, and more lateral a bigger foramen with an oval shape, the *Foramen Ovale* is located, which transmits the *Mandibular nerve* (V_3), an accessory meningeal artery and the lesser superficial petrosal nerve (LSPN), and in some cases, a small orifice can be identified, *Canaliculus innominatus (of Arnold),* slightly medial-to-foramen spinosum transmitting the LSPN [1]. The fourth foramen is located behind and more lateral than the last one, the *Foramen spinosum,* through which the middle meningeal artery enters.

Another important foramina, the *optic canal*, through which the *optic nerve (II CN)* enters the orbit and is delimited superomedial *(roof)* by the *planum sphenoidale* and the body of the sphenoid, superolateral by the anterior clinoid process, and inferiorly *(floor)* by the *optic strut* (Fig. 2). These three structures must be sectioned or drilled when performing an intradural anterior clinoidectomy [4].

Finally, the **pterygoid processes** are two bony columns located on the lower surface of the sphenoid body, directed vertically from superior to inferior; its base is implanted by two roots, one from the inner edge of the greater wing and a smaller one from the lower face of the body; and it is also crossed by a rectilinear canal called the *vidian canal*, which gives way to the *Vidian nerve and artery.* Each process has two extensions, the medial and lateral pterygoid plates, which served as insertion areas for the muscles with the same name. In the internal aspect of the lateral plate, close to its root, there is a small bone lamina called *vaginal process,* which articulates with the sphenoid process of the palatine bone, forming the *pterygopalatine canal.*

4 Sphenoid Sinus

The sphenoid sinus are two large air cavities formed within the body of the sphenoid bone, normally located below the sella turcica and the chiasmatic sulcus, separated from each other in the midline by a thin lamina or septum of variable size, although smaller septa can be found. We believe that this structure deserves a separate mention, since it is part of the sphenoid bone and one of the most important surgical corridors to the middle fossa, with the continuous innovation of techniques thanks to the development of ever-better devices for endoscopic visualization as well as specialized instrumental. Its drainage toward the nostrils is carried out mainly through the superior meatus, and it also has a variable opening on its anterior face known as the sphenoid ostium, which is one of the main anatomical landmarks that serve as a safe entry point to the sinus cavity [1, 3, 5, 6].

Different variations have been described in the pneumatization pattern of the sphenoid sinus both in the sagittal plane *(Hammer and Radberg)* [7] and in the coronal plane *(Vaezi, A.* et al.*)* [8], as described below.

- **Sagittal plane:** (1) *conchal type,* very small sphenoid sinuses situated well laterally separated by a thick lamina of bone, (2) *pre-sellar,* when a medium-sized sinus cavity is located entirely anterior to the pituitary fossa, and (3) *sellar*, the anterior wall and frequently the floor of the sella appears as a protrusion into the sphenoidal cells. In addition, it has been described as a *post-sellar* type when it is located behind the pituitary fossa also a combination of two or more types to the complete absence.
- **Coronal plane:** *In type 1 (previdian),* the pneumatization extends from the midline to the medial edge of the vidian canal, in *type 2 (prerotundum)* extends to the lateral edge of foramen Rotundum, and in *type 3 (postrotundum),* the pneumatization extends laterally to the foramen rotundum.

To better understand the concept of the sphenoid sinus as a surgical corridor, it is merely necessary to recognize the anatomical relationships that it presents. The superior wall of the sinus is continued with the roof of the ethmoidal sinus in close contact with the olfactory nerves, the optic chiasma, and the pituitary gland (part of the floor of the anterior and middle fossa). The floor of the sinus forms part of the nasopharynx dome. The anterior aspect is connected to the vomer and the ethmoid bone. Finally, posterolateral is related to the sella turcica and according to the grade of pneumatization, a series of nervous and vascular structures create impressions in this wall. If the sinus has a lateral projection, it could identify the impressions of the maxillary nerve above and the vidian nerve below. The ICA impression in a more medial position [6, 9].

5 The Temporal Bone

The temporal bone is a paired bone that occupies the entire space between the occipital, the parietal, and the sphenoid on each side of the skull, forming part of both the vault and the cranium base. It is divided into s*quamous, petrous, mastoid, styloid*, and *tympanic* parts. It has an exocranial and an intracranial surface [1, 3].

The squamous part presents a large vertical portion with an asymmetrical circumference and a shape resembling a *fish scale,* from which its name derives. Its external aspect is concave and gives attachment to the temporalis muscle, forming part of the temporal fossa. We also can see on this surface that it is a series of vascular impressions. The most consistent impression is the *deep posterior temporal artery,* fusion with the greater sphenoid wing anteriorly *(sphenosquamous suture)* and with the parietal bone *(parietotemporal suture)* posteriorly. From the inferior part arises a large process, *the zygomatic process,* flattened from the inside out, it has a triangular base with two roots, the first *(transverse)* is the temporal condyle and the second *(longitudinal)*, extending posteriorly with the *supramastoid crest*. This longitudinal root presents in its upper aspect a concave surface, where posterior fibers of the temporalis muscle cross, while in its inferior aspect, it presents an excavation in the middle part that forms the external pole of the *glenoid cavity*. Anterior to this cavity, we find the *anterior zygomatic tubercle* and behind the *posterior zygomatic tubercle* (Fig. 3). The horizontal portion of the squama has a triangular shape with a superior surface covered by the *tegmen timpani* of the petrous bone, but we can recognize the impression of the middle meningeal artery. We can appreciate the *temporal condyle*, the *glenoid fossa on the inferior surface*, among other structures [1, 10].

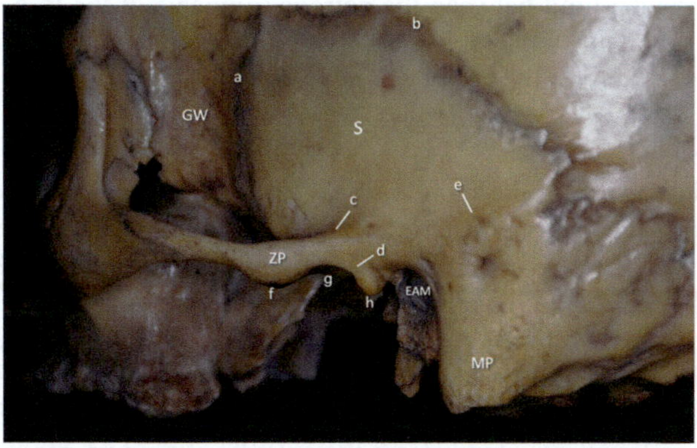

Fig. 3 Left posterolateral view of skull. Showing the exocranial surface of the greater sphenoid wing (GW), and the squamous part of the temporal bone (S). Sphenosquamous suture (**a**), spheno-parietal suture (**b**), zygomatic process (ZP) with its two roots, the horizontal (**c**) with its posterior extension, the supramastoid crest (**e**), and the transverse root (**d**). The glenoid cavity escavation (**g**) provoque two bony elevations, the anterior (**f**) and posterior (**h**) zygomatic tubercles. External auditive meatus (EAM), mastoid process (MP)

The petrous part has two surfaces endocranially, anterosuperior *(or cerebral surface)* and posterosuperior *(or cerebellar),* separated by *the superior edge of the petrous bone* (Fig. 1) and the apex, located at an angle between the greater sphenoid wing and the occipital bone. In the most inner portion, the anterosuperior surface has a depression or excavation, where the *Gasserian ganglion* rests, preceded by a less wide excavation for the trunk of the trigeminal nerve *(trigeminal impression).* At the junction of the external third with the middle third of this surface, close to the edge of the bone, and elevation can be identified, *the arcuate eminence*, classically related to the superior semicircular canal; nevertheless, there are studies using tomography reconstructions that reject this statement [11, 12]. We identified the *Fallopian hiatus* for the *greater petrosal nerve between the arcuate eminence and the trigeminal impression.*

The mastoid part is made up of trabeculated bone with a variable grade of pneumatization. The styloid part has a thin downward projection, oblique from posterior to anterior, serving as a site of attachment for different muscles. The tympanic part contributes to the tympanic cavity and the external acoustic meatus.

6 Dural Folds and Ligaments

In the endocranial surface of the middle fossa, there are projections of the dura mater; some miscalled "ligaments" that are folds of the dura. The *petrolingual ligament* is the only true ligament, which originates from the periosteum of the carotid canal and extends between the petrous apex and the sphenoidal lingula (from Latin [*−lingua*-], which means tongue) (Fig. 4). The petrolingual ligament is important, because it marks the limit between the petrous segments of the ICA, laterally, and the cavernous segment, medially. Its posterior portion is continued with the *Gruber's ligament* fibers, which form the superior border of the *Dorello's canal,* by which the abducent nerve enters the CS (Fig. 4). Also is a point of attachment of the lateral wall of the CS [13].

In the medial aspect of the middle fossa, laterally to the sella turcica, the dura mater that covers the osseous structures projects folds between the anterior clinoid process and the posterior clinoid process, the *interclinoid fold*; between the anterior clinoid process and the petrous apex, the *anterior petroclinoid fold*, and from the posterior clinoid process to the petrous apex, the *posterior petroclinoid fold*, which is occasionally calcified. These three folds form the oculomotor triangle, and through the center of this triangle, the oculomotor nerve enters the CS (Fig. 4).

The *falciform ligament* extends from the anterior clinoid process to the tuberculum sellae above the optic canal, and despite its name, it is a fold of dura mater that lies over the optic nerve and covers it in about average of 3 mm. The recognition of this structure is important when performing an anterior intradural clinoidectomy [4] or an optic nerve decompression at the moment of coagulation of the dura mater, which can lead to nerve injury (Fig. 4).

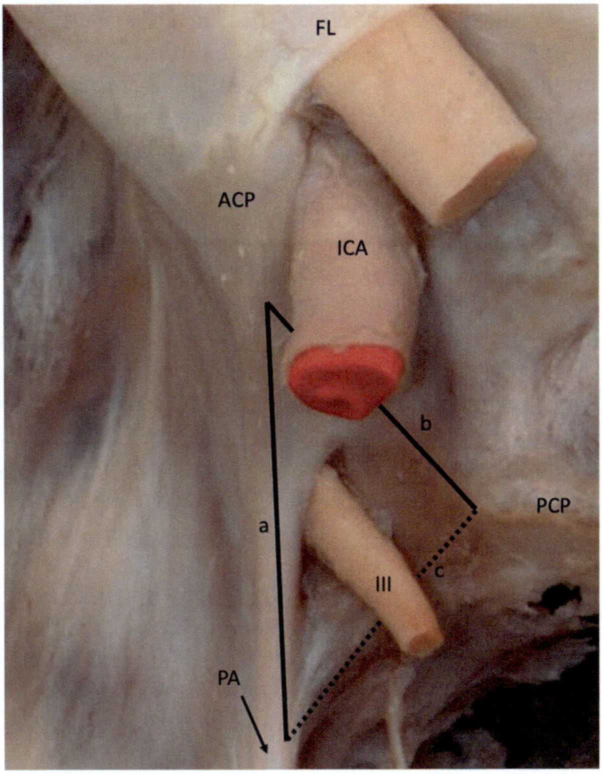

Fig. 4 Left superolateral view of the oculomotor triangle region. The borders of this triangle are the anterior petroclinoid fold (**a**), posterior petroclinoid fold (**c**) and the interclinoid dural fold (**b**). Also can recognize the anterior clinoid process (ACP), the falciform ligament (FL), the internal carotid artery (ICA), the left posterior clinoid process (PCP), and petrous apex (PA)

7 Sellar and Parasellar Regions

To facilitate the study of the middle fossa, it can be divided into *medial* and two *lateral* portions. The medial portion contains the pituitary fossa and its contents *(sellar region)* and the structures located on each side of it, such as the CS and its contents *(parasellar region)*. In addition, it can be added structures placed in the same axial plane but superior to the sellar region, the *suprasellar region*; however, in a more practical way and due to its surgical implications, we will understand as *parasellar region*, all the structures surrounding the sella [2, 3, 9].

The sellar region is located in the center of the cranial base. The pituitary gland comprises two parts with functional, anatomical, and embryological differences. The anterior lobe or *adenohypophysis* originated from the *Rathke's pouch,* and a smaller posterior lobe or *neurohypophysis* originated in the neuroectoderm. The first one has glandular tissue with endocrine function, producing and releasing a series of hormones. On the other hand, neurohypophysis is a projection of axons from the hypothalamus and releases oxytocin and vasopressine [3, 6, 14]. Its blood supply comes from the superior hypophyseal arteries (arising from the supraclinoid segment of the ICA), giving branches to the optic nerve, the infundibulum *(pituitary stalk)*, and the *diaphragma sellae,* and also by the inferior hypophyseal arteries, arising from the meningoypophyseal trunk (from ICA cavernous segment) which irrigates the posterior lobule. The *diaphragma sellae* is defined as a small circular or horizontal fold of dura mater, forming the roof for the sella turcica covering the pituitary gland, with a small central opening in its center transmitting the infundibulum. Anatomical studies have shown that the hypophysis is covered by two layers of the dura, an external or *periosteal* layer, in contact with the bone, and an inner or *meningeal* layer, both separated laterally by the CS, where the inner layer contributes to forming the medial wall [3] (Fig. 5).

Many important neurovascular structures surround the pituitary gland. Superiorly, the optic nerves, the optic chiasm, the anterior communicating artery complex, and the third ventricle. The distance between the optic chiasm and the *tuberculum sellae* varies, which has surgical implications, especially when approached by the transsphenoidal route. (Bergland et al.) described three types of disposition: (1) *normal*, when the chiasm is over the diaphragm, (2) *prefixed*, when it overlies the tuberculum sellae, and (3) *postfixed*, when overlies the dorsum sellae. The same restriction of a prefixed chiasma could occur when presenting protrusion of the tuberculum sellae over the chiasma [1, 15].

Laterally, the CS delimits the gland with all its content, especially the cavernous ICA, which has a medial disposition. Superolateral to the pituitary fossa, we identified the supraclinoid ICA with all its branches. Its anterior aspect has a close relation to the sphenoid sinus and posteriorly to the dorsum sellae and, therefore, with vascular structures of posterior circulation, such as the basilar artery and its branches.

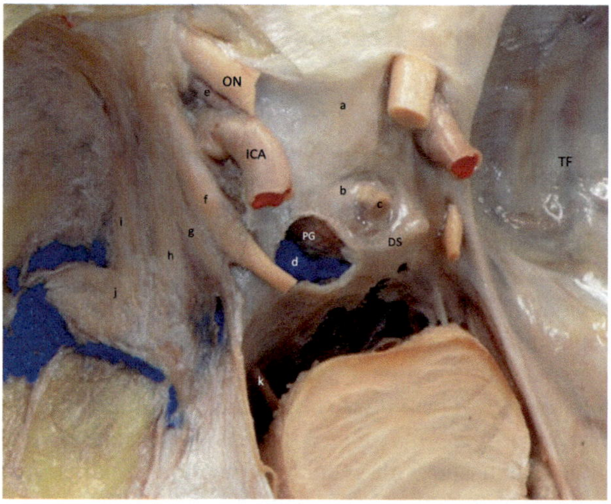

Fig. 5 Left superolateral view of the sellar and parasellar region with some of its components. Optic nerve (ON), internal carotid artery (ICA), pituitary gland (PG), dorsum sellae (DS), temporal fossa (TF), chiasmatic sulcus (**a**), diaphragma sellae (**b**), pituitary stalk (**c**), cavernous sinus (**d**), ophtalmic artery (**e**), oculomotor nerve (**f**), trochlear nerve (**g**), ophtalmic branch (**h**), mandibular branch (**i**) and maxillary branch (**j**) of the trigeminal nerve. Abducens nerve entering into Dorello's canal (**k**)

8 Cavernous Sinus

The CS is the historical definition of a *parasellar compartment* described by Winslow [16] in the eighteenth century. The name *cavernous* is a reference to the *corpus cavernosum* because of numerous filaments which gave the interior of the CS its plexiform appearance. The anatomic inexperience in this area meant that for some time, it was considered as a *no man's land,* at least for direct surgery intervention, until, in 1965, Parkinson [17, 18] described landmarks for safely approaching the cavernous portion of the carotid artery to treat a traumatic carotid–cavernous fistula. Understanding the normal microsurgical anatomy of the CS is essential for safe surgical approaches to the region. It is located nearly in the center of the head, deep-seated at the lateral sides of the body of the sphenoid and the sella. It is a large venous compartment enclosed by a dura mater, with many neurovascular structures closely related.

The CS looks like a jewelry box and has the shape of a hexahedron, with a larger *lateral wall*, than with the *medial wall; both* are in the sagittal plane. The *inferior wall* is oblique, the *anterior* and *posterior* wall straight in the coronal plane, and the *superior wall* is horizontal [19, 20] (Fig. 6).

As explained before, the anterior and medial walls are formed by the extension and separation of the two layers of the *diaphragma sellae*. The medial wall is composed only of a meningeal layer, and the inferior wall is just by the periosteal layer; all the rest of the walls are composed of both.

The *lateral wall* is formed by the extension of the dura mater of the middle fossa, while the *posterior wall* forms part of the dural covering of the *clivus.* The lateral wall has a smooth superficial layer formed by continuity with the periosteal layer of the middle cranial fossa and tentorium. A deep or *inner* layer (concerning the CS) is thinner and is formed by reticular tissue and epineurium of the cranial nerves that run inside it. The clinoidal and oculomotor triangle forms the roof, continuous with the *diaphragma sellae* [19, 21, 22] (Fig. 6).

The CS contains the ICA and its branches, the abducens nerve, and sympathetic nerves. The oculomotor and the trochlear nerve (superiorly) and V_1, V_2 divisions of the trigeminal nerve (inferiorly) are all located within the inner layer of the lateral wall of the CS [20] (Fig. 5).

The oculomotor nerve enters the CS through the center of the oculomotor triangle and is found immediately underneath the dura of the anterior half of the roof (clinoid triangle) (Fig. 4). Posteromedially, the compartment is continuous with the dura of the clivus, and inferolaterally, it extends into a funnel-shaped space around the ICA through the foramen lacerum. This is important, because incisions in this area should be done medially, avoiding the oculomotor nerve.

We will better explore the components and all the landmarks and triangles described for this important area.

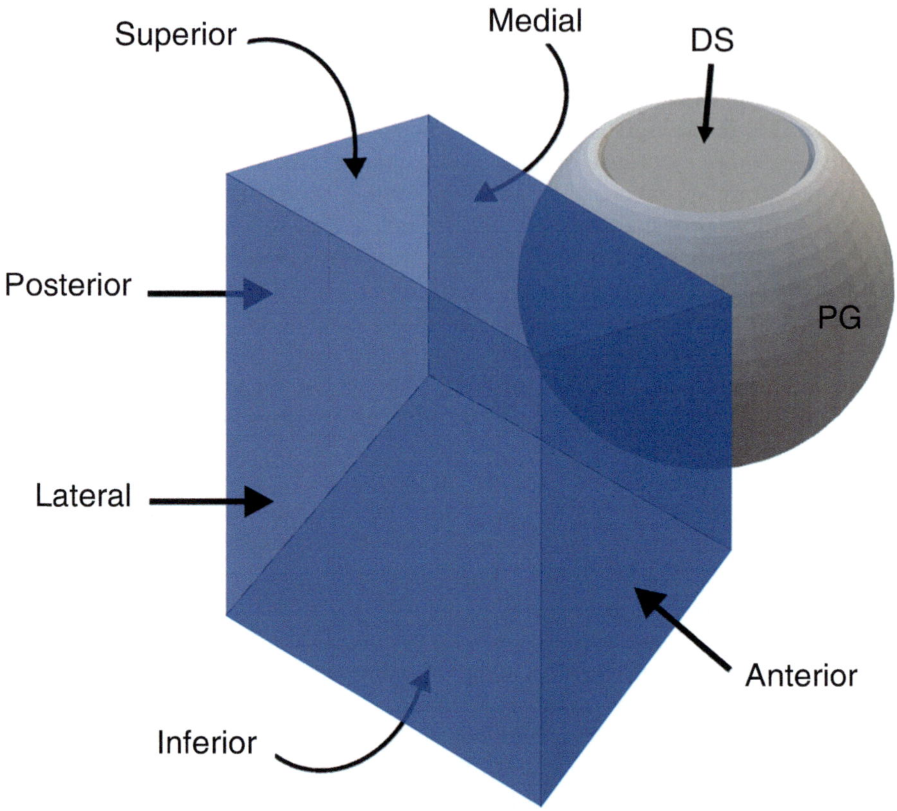

Fig. 6 Schematic figure of the cavernous sinus described by (Beom Sun Chung et al.) [19]. It can be seen that has a box shape with six walls, the lateral one is bigger than the rest and the inferior wall is oblique. The pituitary gland (PG) is in close relation with the medial wall, and the diaphragma sellae (DS) is in the same plane as its roof or superior wall, where the oculomotor nerve enters the CS. The posterior wall is in close relation with the clivus (Original illustration by José Chang MD & José Daboub.) based on the cited article

9 Oculomotor Nerve (III CN)

After leaving the interpeduncular fossa, the III CN runs in an anterolateral and slightly inferior direction in the anterior incisural space between the posterior cerebral artery and the superior cerebellar artery to cross the roof of the CS in the center of the oculomotor trigone and enters the orbit via the SOF, to innervates all the extrinsic muscles of the eye, except the lateral rectus and the superior oblique muscles (Figs. 4 and 5).

10 Trochlear Nerve (IV CN)

The trochlear nerve is the only one that originates dorsally to the brainstem. The IV CN passes by the dorsal surface of the midbrain just below the inferior colliculus at the level of the superior medullary velum. In the cerebellum–mesencephalic fissure, the nerve curves anteriorly around the lateral aspect of both the tectum and tegmentum in the quadrigeminal and ambiens cistern. The cisternal segment runs under the free edge of the tentorium, reaching the CS below the petroclinoid ligament. It runs into the lateral wall of the CS and enters the orbit via the SOF to innervate the superior oblique muscle (Figs. 5 and 7).

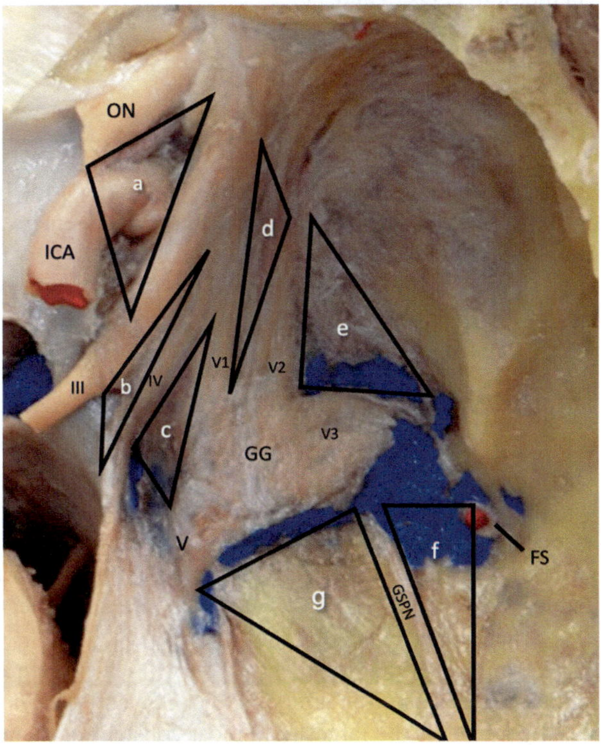

Fig. 7 Right superolateral view of the cavernous sinus region. The outer dural layer has been resected, opening the roof and lateral wall of the CS, exposing the content of the inner layer of the lateral wall of the CS. The cavernous sinus triangles has been marked. The anteromedial triangle (**a**), paramedial (**b**), infratochlear of Parkinson (**c**), anterolateral (**d**), lateral (**e**), posterolateral or Glasscock's (**f**), posteromedial or Kawase's (**g**). Internal carotid artery (ICA), optic nerve (ON), oculomotor nerve (III), trochlear nerve (IV), trigeminal nerve (V), Gasserian glanglia (GG), and foramen spinosum

11 Trigeminal Nerve (V CN)

After arising from the lateral pons, the motor and sensory roots of the trigeminal nerve run in an anterior and lateral direction through the middle incisural space of the posterior cranial fossa superior to the petrous apex to enter the subarachnoid and dural outpouching known as Meckel's cave. Both branches reach the semilunar ganglion (of Gasser), from which the nerve splits mediolaterally into three branches: the ophthalmic, maxillary, and the mandibular nerve (V_1, V_2, V_3). V_3 continues inferiorly, laterally, and slightly anteriorly to leave the middle cranial fossa via the foramen ovale. V_2 runs anteriorly and slightly inferiorly between the two dural layers to leave the middle cranial fossa through the foramen rotundum. V_1 enters the lateral wall of the CS running in the wall inferior to the trochlear nerve and into the orbit via the SOF (Figs. 5 and 7).

12 Abducens Nerve (VI CN)

The VI CN emerges from the lateroventral side of the brainstem, between the pons and medulla, medial to the facial nerve, and enters inside the CS through the Dorello's canal in the petrous apex. The abducens nerve innervates the lateral rectus muscle, making "abduction" (lateral movement) of the eye possible (Fig. 5).

13 Internal Carotid Artery

The ICA enters the cranial cavity through the carotid foramen accompanied by the sympathetic nervous plexus. The initial portion of the intracavernous carotid ascended through the bony carotid canal and over the foramen lacerum toward the posterior clinoid process. Then, it turned abruptly forward to its horizontal portion and terminated by passing upward on the medial aspect of the anterior clinoid process, where it perforated the roof of the CS.

The intracavernous segment of the ICA has three arterial branches: meningopituitary trunk, inferior artery of the CS, and *McConnell's artery*. The pituitary meningo trunk is the largest and most constant of them originating medially in the middle third of the medial curve of the ICA; it gives rise, in turn, to three other branches, which are the tentorial artery, the dorsal meningeal artery, and the inferior pituitary artery. The next artery originating from the intracavernous ICA is the inferior artery of the CS (or inferolateral trunk), in the inferolateral or lateral portion of the horizontal segment of the intracavernous carotid artery. An important finding is that although the ophthalmic artery usually arises distally to the distal dural ring on the medial half of the anterior wall of the ICA, in some cases can originate within the CS, from the clinoid segment of the ICA, or even from the middle meningeal artery.

14 Cavernous Sinus Triangles

This region was classically divided by Dolenc [6] into *three subregions* composed of 10 triangular spaces, four of which are in close relationship with the CS in the parasellar subregion, providing access routes to its interior, and six are adjacent to it:

- **Parasellar:** anteromedial triangle, paramedial triangle, oculomotor trigone, and Parkinson's triangle.
- **Middle cranial:** anterolateral triangle, lateral triangle, posterolateral (Glasscock's triangle), and posteromedial (Kawase's triangle).
- **Paraclival**: inferomedial triangle and inferolateral (trigeminal triangle).

15 Anteromedial Triangle

The complete exposure of the anteromedial triangle, also known as a clinoidal triangle or Dolenc's triangle [23], requires complete removal of the anterior clinoid process. The boundaries of this triangle are medially the lateral aspect of the II CN, laterally the medial aspect of the III CN, and a line between the III CN entry point in the CS and the II CN entry point into the optic canal. The contents of the anteromedial triangle include the clinoidal segment of the ICA, the venous trabecular channels of the anteromedial portion of the CS, and a thin layer of fibrous connective tissue continuous with that surrounding the anterior loop of the ICA, thus forming the proximal dural ring. Another membrane, the carotid oculomotor membrane, extends laterally from the carotid artery through the interval between the carotid artery and the oculomotor nerve and medially from the oculomotor nerve to the posterior clinoid process to attach superiorly to the interclinoid fold. This membrane extends inferiorly to the lateral part of the CS, where it is continuous with the inner reticular membrane of the lateral sinus wall. The anterior part of this membrane separates the CS from the clinoid segment of the ICA. The space between the two membranes is called the "clinoidal space." The clinoidal space is important, because it is extracavernous, therefore, bleeds only occur in this region only if the membrane is romped [23] (Fig. 7).

In surgery, the anteromedial triangle can be enlarged by dissecting the proximal dural ring from the III CN and incising the dura propria along the longitudinal axis of the optic nerve. This space can also be enlarged laterally with retraction of the trochlear and oculomotor nerves offering much better access to the horizontal portion of the ICA [24].

16 Paramedial Triangle

This space's medial and lateral boundaries are defined by the lateral aspect of the III CN and by the medial aspect of the IV CN, respectively. The apex of the paramedial triangle is formed by the point at which the trochlear nerve crosses the oculomotor nerve. The base of this triangle is the dural fold between dural entries of oculomotor and trochlear nerves. The content of this triangle is the horizontal segment of the intracavernous ICA, the meningohypophyseal trunk, and the proximal segment of the VI CN [25]. The paramedial triangle can be enlarged by retraction of the lateral wall. If the meningohypophyseal trunk is not visible through the paramedial triangle, it can be reached by further dissection from the paramedial into the inferomedial triangle. This procedure can be done by cutting the base of the paramedial triangle in the direction of the inferomedial triangle [23] (Fig. 7).

17 Oculomotor Triangle

The oculomotor triangle is between the anterior clinoid process, the posterior clinoid process, and the petrous apex. The projections of the dura mater that cover the surface of these structures and connect them form a geometric figure named the oculomotor triangle [25]. The oculomotor triangle boundaries are medially the interclinoid fold, laterally the anterior petroclinoid fold, and the base are formed by the posterior petroclinoid fold. The content of this triangle is the III CN, the proximal siphon, and the horizontal segment of the ICA. The oculomotor nerve enters the CS through the center of this triangle. Through this triangle, the posterior clinoid process can be exposed and resected in order to access the interpeduncular and prepontine cisterns, as well as the basilar artery when its bifurcation is below this process [26] (Fig. 4).

18 Parkinson's Triangle

The Parkinson's triangle, also known as the infratrochlear triangle, was first described by Parkinson in 1965 [18]. The limits are medially the lateral aspect of the IV CN, laterally the medial aspect of V_1, and the base is formed by the tentorial edge and clival dura between the entry points of the trochlear and trigeminal nerves. The content of this triangle is the horizontal segment of the cavernous ICA, the meningohypophyseal trunk, and the VI CN [25, 26]. This area has historical importance and was originally described as the main entry access to the CS. By dissecting the area, the entire intracavernous segment of the ICA and the meningohypophyseal trunk can be accessed. This triangle is also used in surgery for exposing the complete course of the sixth cranial nerve from its entry through Dorello's canal to its exit through the SOF. This triangular area is relatively narrow and can be enlarged by slight retraction of the IV CN medially and V_1 CN (Fig. 7).

19 Anterolateral Triangle

This triangle was also named the *anteromedial middle fossa triangle* by Rhoton A. [27] or *Mullan's triangle (1979)* [28]. Its limits are the lateral aspect of V_1 and the medial aspect of V_2, with an imaginary line between the foramen rotundum and SOF. The contents of this triangle are the inferolateral aspect of the distal horizontal portion of the cavernous ICA, the dura and the anterior bony floor of the middle cranial fossa, the venous trabecular channels of the inferolateral portion of the CS, the superior orbital vein, and the V_1 [23]. Removing bone in this triangle will expose the sphenoid sinus and the pterygopalatine fossa [29] (Fig. 7).

20 Lateral Triangle

The boundaries of the lateral triangle also named the *anterolateral middle fossa triangle* by Rhoton [27] are the maxillary and mandibular divisions of the trigeminal nerve imaginary line between the foramen rotundum and foramen ovale. The contents of this space are the anterior and lateral aspects of the lateral loop of the ICA, as well as the sympathetic fibers on the ICA [23]. Sometimes, the sphenoidal emissary foramen and vein can be found in this area, connecting the CS with the pterygoid venous plexus. The middle fossa bone can be drilled laterally in the surgery to gain additional space and greater lateral exposure. This procedure may approach masses that extend laterally in the CS and enter the infratemporal fossa [24] (Fig. 7).

21 Posterolateral Triangle (Glasscock)

The boundaries of this triangle are medially the greater superficial petrosal nerve, laterally by an imaginary line between the foramen spinosum and the arcuate eminence of the petrous bone. The base is formed by the posterior aspect of the mandibular division of the trigeminal nerve, the contents of this triangle is the posterior loop of the ICA, the proximal portion of the lateral loop of the ICA covered by the anterior aspect of the petrous apex forming the posteromedial floor of the middle cranial fossa, and the labyrinthine branch of the middle meningeal artery. In the surgery, the horizontal portion of the ICA can be exposed by carefully drilling this triangle from the spinous foramen and medially along the posterior margin of V_3 (Fig. 7).

22 Posteromedial Triangle (Kawase)

The posteromedial triangle has the following boundaries: the posterior border of the *Gasserian ganglion* anteriorly, the large petrosal nerve laterally, and the petrous border with the superior petrosal sinus medially. The content of this triangle is the petrous apex, posterior edge of the petrous ICA, and the cochlea. In the surgery, the bone from this area can be removed to do an anterior petrosectomy (Kawase approach) to connect the middle and posterior fossa. This procedure enables access to the petroclival area, the anterolateral brain stem, and the vertebrobasilar junction exposing the roots of the V, VII, and VII CNs. On the other hand, this can also be explored to drill the petrous apex. This procedure allows the visualization of the medial aspect of the posterior loop of the ICA and can be used for temporary clipping and grafting.

23 Inferomedial Paraclival Triangle

The boundaries of the inferomedial paraclival triangle are the dura between the posterior clinoid process and Dorello's canal, a line between Dorello's canal, and the entry point of the IV CN into the tentorium laterally, with the petrous apex at its base. The contents of this triangle are the porous abducens *(dural opening into Dorello's canal)*, the VI CN, the basilar venous plexus, the posterior petroclinoid fold, the lateral aspect of the ICA covered by the petroclinoid ligament, and the dorsal meningeal branch of themeningohypophyseal trunk (Fig. 8).

24 Inferolateral Paraclival Triangle (Trigeminal Triangle)

The boundaries of this triangle are the dura between the entry point of the IV and the VI CNs, the dura between Dorello's canal, and the entry point of the petrous vein into the superior petrosal sinus laterally, and the *petrous apex* as its base.

The inferolateral triangle can be subdivided into two more triangles by drawing a line between the *point of entry of the petrosal vein* into the superior petrosal sinus and to the *point of the exit of the trigeminal nerve* from the posterior fossa. The superior part is the *(tentorial triangle)*, where the petrous vein enters the superior petrosal vein, while the inferior part *(osseous triangle)* represents the posterior extension of Kawase's triangle (Fig. 8).

After reviewing the anatomical landmarks and boundaries of the different triangles of the middle fossa, it is important to recognize the true practical helpfulness. Many have been described for educational purposes; however, some have everyday use. For example, in our practice, the Parkinson's triangle is very important, since it is the one that offers the widest access lesions in the lateral aspect of the CS, being able to expose the cavernous ICA as well as the VI CN. The anterolateral and lateral triangle help us approach infiltrative lesions of the floor of the middle fossa, which affect extracranial cavities, since the SOF communicates the middle fossa with the *orbit,* the foramen rotundum with the *pterygopalatine fossa*, and the foramen ovale with the *infratemporal fossa* (Fig. 9). Some of these lesions will benefit from combined techniques (microsurgery and endoscopy), using different routes. The Kawase triangle gives us a reference to amplify our access to the anterolateral aspect of the pons via anterior petrosectomy.

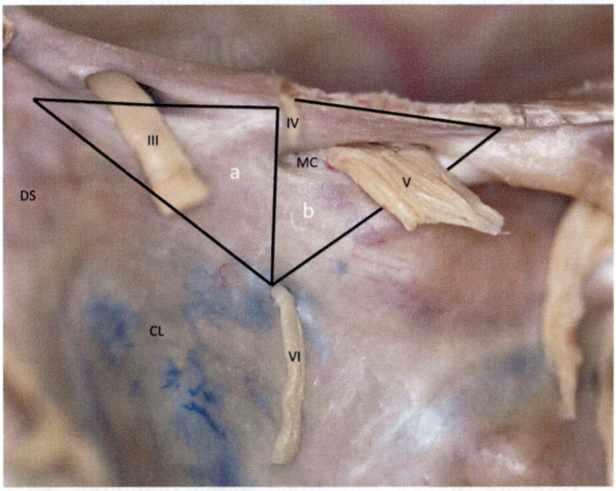

Fig. 8 Left posterolateral view of the clivus region. Dorsum sellae (DS), clivus (CL), oculomotor (III), troclhear nerve entering the free border of tentorium (IV), abducens (VI), Meckel's cave (MC), trigeminal nerve (V). Inferomedial paraclival triangle (**a**), inferolateral paraclival triangle (**b**)

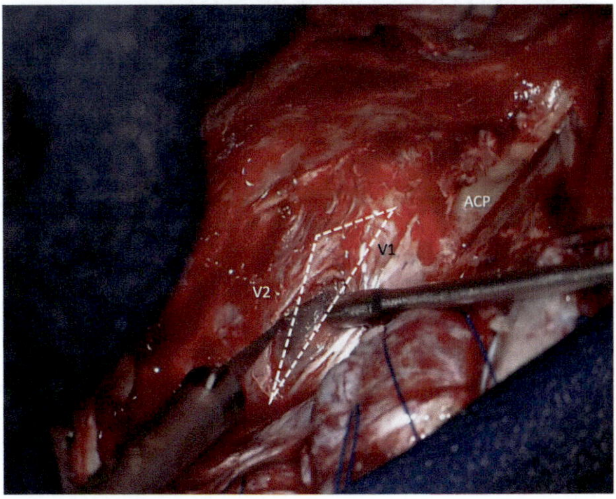

Fig. 9 Transoperative photograpy of a middle fossa pilling, opening the anterolateral triangle to access a trigeminal shwanoma. Ophtalmic branch (V1) and mandibullary branch (V2) of the trigeminal nerve. Anterior clinoid process (ACP)

25 Middle Cranial Fossa as a Surgical Region

We have reviewed the definition, limits, structure, and content of the middle fossa; however, how can we apply this anatomy to our clinical practice from a surgical point of view? Do we begin by asking what represents for us the middle fossa? We have been given this answer during this chapter; for us, it is *a* ***broader region*** than its bony limits. It is an area in which a series of venous, arterial, neural, and bony structures convey, in which a variety of pathologies can occur and, therefore, can share similar surgical approach strategies due to their close relation. Therefore, what structures will we face against or expose when treating a lesion located in this region? In most cases, they will be the same, many of which are not strictly within the classically defined area, so probably the limits of the middle fossa extend beyond. We will then describe some of these limits, which more than boundaries would act as landmarks that will help us surgically approach lesions in this region.

26 Neural Margins

- **Anteromedial:** optic nerves at their entrance to their respective optic canals.
- **Anterolateral:** temporal poles.
- **Lateral:** anterior aspect of the inferior temporal gyrus.
- **Posteromedial:** anterolateral aspect of the mesencephalon, anterior segment, and uncus apex.
- **Superior (roof):** optic chiasm, optic tracts, and the third ventricle floor *(hypothalamic part).*
- **Inferior (floor)**: trigeminal nerve including the *Gasserian ganglion* and its branches (V_1, V_2, V_3), and the GSPN.

27 Arterial Margins

- **Anterior:** supraclinoid ICA until its bifurcation and middle cerebral artery (M1 segment).
- **Posterior:** upper third of basilar artery, including the proximal segment of the posterior cerebral and superior cerebellar arteries.
- **Lateral:** middle cerebral artery (M4 branches) and distal cortical temporal branches of the posterior cerebral artery.
- **Superior (roof):** anterior communicating artery complex.
- **Inferior (floor):** ICA (petrous, lacerum, and cavernous segments) and middle meningeal artery.

28 Venous Margins

- **Anterior:** sphenoparietal sinus, anterior intercavernous, or *coronary* sinus.
- **Posterolateral:** superior petrosal sinus.
- **Posteromedial:** basilar sinus and mesencephalic veins.
- **Lateral:** superficial sylvian vein.
- **Superior:** anterior cerebral vein and olfactory vein.
- **Inferior:** CS and inferior petrosal sinus.

29 Cisternal Margins

- **Superior:** chiasmatic and lamina terminalis cistern.
- **Posterior:** interpeduncular cistern, the prepontine cistern bellow, and the crural cistern.
- **Inferior:** carotid cistern.
- **Lateral:** sylvian cistern.

As we can see, the surgical region of the middle fossa includes a wider area, which we can access virtually with the same approach and its extensions, either microsurgical or endoscopically, according to its characteristics and location. If we take these margins into account, we know our limits and our reach, familiarizing ourselves with access routes and different anatomical landmarks. Let us then briefly review our boarding options for this region.

30 Approaches to the Middle Fossa

The *Pretemporal* [30] and the *Orbitozygomatic* [31–33] craniotomies both are enlargement of the Pterional craniotomy popularized by Yasargil [34], and all of them allow, in their way, multiple possibilities to access to the middle cranial fossa. The transylvian approach and the lateral subfrontal approach are shared by the three, enabling the exposition of the basal cisterns and the multiple spaces (interoptic, opticocarotid, carotid–oculomotor, oculo-tentorial, and supracarotid).

The pretemporal craniotomy is a caudal modification of the pterional craniotomy, exposing the temporal lobe entirely, providing the temporo-polar approach [8] and the subtemporal approach [9, 10] while also expanding the transylvian corridor downward. The temporo-polar approach is achieved once the pole of the temporal lobe is retracted posteriorly, allowing the lateral view of the CS. An alternative is a subtemporal view, where the temporal lobe is retracted cranially, with a wide view of the middle fossa floor as a result.

The orbitozygomatic craniotomy combines the benefits of the pterional and pretemporal craniotomies, expanding the transylvian corridor inferiorly and superiorly, giving a wide view from caudal to cranial in the surgical field. If it is combined with a middle fossa peeling, it is possible to expose the triangles of the middle fossa and reach the lateral wall of the CS extradurally.

Once these approaches have been carried out, with a fine and careful dissection of the subarachnoid space, opening cisterns, expanding the natural corridors of the brain, we can verify the wide limits described as *the middle fossa region*, and we believe that having this three-dimensional notion of the relationships and structures involved, it can help us expand the limits of our approaches when planning our surgical strategies. The same approach is used for accessing a supraclinoid ICA aneurysm; for example, it can also be used for accessing an upper basilar tip aneurism by expanding its limits *(like we are expanding the described boundaries of middle cranial fossa)*, either by a small modification in the craniotomy or by opening different cisterns or just by mobilizing the brain in a particular way, it allows us to attack much more distal or profound lesions. At the same time, our work area continues to be considered the middle fossa (Figs. 10 and 11).

Finally, the surgeon must choose all these tools and surgical routes wisely, and more importantly, know when not to use them.

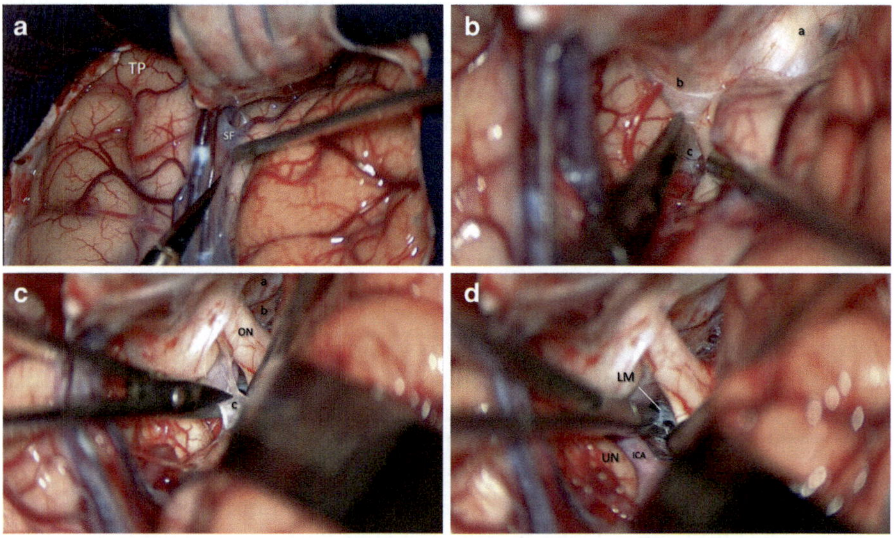

Fig. 10 Surgical secuence showing a Pretemporal approach as an extension of the Pterional approach and the progressive space and different surgical corridors available to access the middle fossa region as defined in this chapter. (**a**) Pretemporal craniotomy expose the temporal pole (TP) and the inferior temporal gyrus, both neural limits of the region, the surgery begin by opening the most lateral cisternal margin, the sylvian cistern and fissure (SF). (**b**) Progressing in the arachnoid dissection, we identified the lesser sphenoid wing (*b*) which is align with the fissure. Planun sphenoidale (*a*), carotid cistern (*c*). (**c**) Dissection continue by opening the carotid cistern and the chiasmatic cistern (*b*). Chiasmatic sulcus (*a*). (**d**) Opening the Liliquist membrane (LM), communitating all the basal cistern of the region upto the prepontine cistern. The anterior part of the uncus can be seen (UN), and the internal carotid artery (ICA)

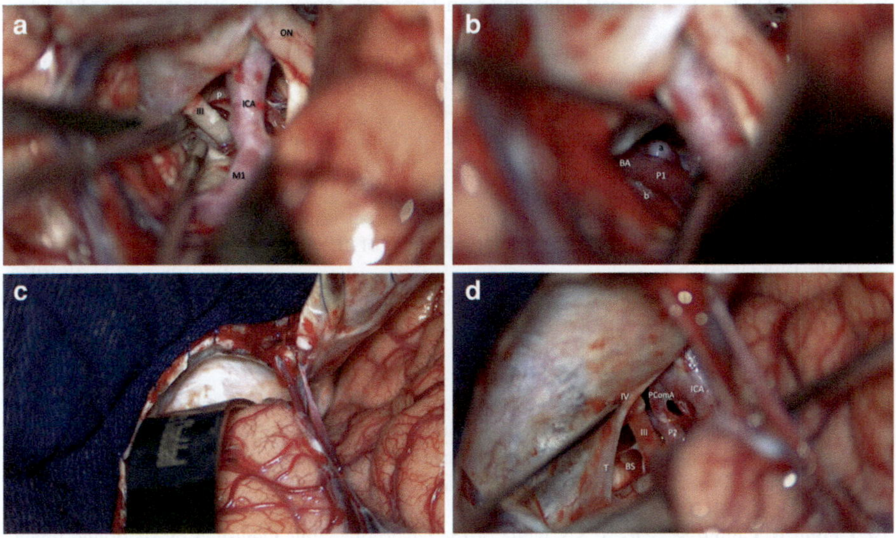

Fig. 11 (Cont.) Once all the cisterns were open, we can access to a great number of structures of the sellar and parasellar regions of the middle fosa. Optic nerve and chiasm (ON), Internal carotid artery (ICA), middle cerebral artery M1 segment (M1), oculomotor nerve (III). (**b**) As described before, with good anatomical knowledge, the middle fossa approach can permit us access to distal structures that we include in the region, here we can see the basilar artery (BA) at the level of its bifurcation with a tip aneurysm (**a**). Posterior cerebral artery (P1), perforating branches of the basilar trunk (**b**). (**c**) If necessary, by separating and gently retract the temporal pole, we can open the subtemporal corridor, always trying to respect the veins. (**d**) Final exposure thru a subtemporal corridor via Pretemporal craniotomy. Posterior communicating artery (PComA), Posterior cerebral artery P2 Segment (P2). Trochlear nerve (IV), tentorial edge (T), brainstem (BS)

References

1. Testut L, Latarjet A, Latarjet M. Trate D'Anatomie Humaine. 7th ed. Paris: G. Doin Ed; 1928.
2. Harsh GR IV, Vaz-Guimaraes F. Chordomas and chondrosarcomas of the skull base and spine. 2nd ed. London: Academic Press; 2017.
3. Rhoton AL Jr. The sellar region. Neurosurgery. 2002;51(4 Suppl):S335–74.
4. Alejandro SA, Carrasco-Hernández JP, da Costa MDS, Ferreira DS, Lima JVF, de Amorim BL, et al. Anterior clinoidectomy: intradural step-by-step en bloc removal technique. World Neurosurg. 2021;146:217–31.
5. Doubi A, Albathi A, Sukyte-Raube D, Castelnuovo P, Alfawwaz F, AlQahtani A. Location of the sphenoid sinus ostium in relation to adjacent anatomical landmarks. Ear Nose Throat J. 2020;100(10_Suppl):961S–8S.
6. Singh A, Wessell A, Anand V, Schwartz T. Surgical anatomy and physiology for the skull base surgeon. Oper Tech Otolaryngol Head Neck Surg. 2011;22:184–93.
7. Hammer G, Radberg C. The sphenoidal sinus. An anatomical and roentgenologic study with reference to transsphenoid hypophysectomy. Acta Radiol. 1961;56:401–22.
8. Vaezi A, Cardenas E, Pinheiro-Neto C, et al. Classification of sphenoid sinus pneumatization: relevance for endoscopic skull base surgery. Laryngoscope. 2015;125(3):577–81.
9. Patel C, Fernandez-Miranda J, Wang W, Wang E. Skull base anatomy. Otolaryngol Clin N Am. 2016;49:9–20.

10. Rhoton AL Jr. The temporal bone and transtemporal approaches. Neurosurgery. 2000;47(3 Suppl):S211–65.
11. Seo Y, Ito T, Sasaki T, Nakagawara J, Nakamura H. Assessment of the anatomical relationship between the arcuate eminence and superior semicircular canal by computed tomography. Neurol Med Chir (Tokyo). 2007;47(8):335–40.
12. Santos FP, Longo MG, May GG, Isolan GR. Computed tomography evaluation of the correspondence between the arcuate eminence and the superior semicircular Canal. World Neurosurg. 2018;111:e261–6.
13. Ziyal IM, Salas E, Wright DC, Sekhar LN. The petrolingual ligament: the anatomy and surgical exposure of the Posterolateral landmark of the cavernous sinus. Acta Neurochir. 1998;140(3):201–5.
14. Chin B, Orlandi R, Wiggins R. Evaluation of the sellar and parasellar regions. Magn Reson Imaging Clin N Am. 2022;20(2012):515–43.
15. Renn W, Rhoton AL Jr. Microsurgical anatomy of the sellar region. J Neurosurg. 1975;43:288–98.
16. Thakur JD, Sonig A, Khan IS, Connor DE Jr, Pait TG, Nanda A. Jacques Bénigne Winslow (1669–1760) and the misnomer cavernous sinus. World Neurosurg. 2014;81(1):191–7.
17. Parkinson D. Extradural neural axis compartment. J Neurosurg. 2000;92(4):585–8.
18. Parkinson D. A surgical approach to the cavernous portion of the carotid artery. J Neurosurg. 1965;23(5):474–83.
19. Chung B, Chung M, Park J. Six walls of the cavernous sinus identified by sectioned images and three-dimensional models: anatomic report. World Neurosurg. 2015;84(2):337–44.
20. Umansky F, Nathan H. The lateral wall of the cavernous sinus. J Neurosurg. 1982;56(2):228–34.
21. Yasuda A, Campero A, Martins C, Rhoton AL Jr, de Oliveira E, Ribas GC. Microsurgical anatomy and approaches to the cavernous sinus. Neurosurgery. 2008;62(6 Suppl 3):1240–63.
22. Bakan AA, Alkan A, Kurtcan S, et al. Cavernous sinus: a comprehensive review of its anatomy, pathologic conditions, and imaging features. Clin Neuroradiol. 2015;25(2):109–25.
23. Dolenc V (1989) Anatomy of the cavernous sinus. In: Anatomy and surgery of the cavernous sinus. Vienna: Springer; 1989. p. 3–137.
24. Inoue T, Rhoton AL Jr, Theele D, Barry M. Surgical approaches to the cavernous sinus. Neurosurgery. 1990;26(6):903–32.
25. Harris F, Rhoton AL Jr. Anatomy of the cavernous sinus. J Neurosurg. 1976;45:169–80.
26. van Loveren H, Keller J, El-Kalliny M, Scodary D, Tew J. The Dolenc technique for cavernous sinus exploration (cadaveric prosection). J Neurosurg. 1991;74:837–44.
27. Rhoton AL Jr. Cranial anatomy and surgical approaches. 2003. p. 363–438.
28. Mullan S. Treatment of carotid-cavernous fistulas by cavernous sinus occlusion. J Neurosurg. 1979;50:131–44. https://doi.org/10.3171/jns.1979.50.2.0131. PMID: 430123.
29. Watanabe A, Nagaseki Y, Ohkubo S, Ohhashi Y, Horikoshi T, Nishigaya K, Nukui H. Anatomical variations of the ten triangles around the cavernous sinus. Clin Anat. 2002;16:9–14.
30. Chaddad-Neto F, Dória-Netto HL, Campos Filho JM, Reghin Neto M, Oliveira E. Pretemporal craniotomy. Arq Neuropsiquiatr. 2014 Feb;72(2):145–51.
31. Hakuba A, Liu S, Shuro N. The orbitozygomatic infratemporal approach: a new surgical technique. Surg Neurol. 1986;26(3):271–6.
32. Hakuba A, Tanaka K, Suzuki T, Nishimura S. A combined orbitozygomatic infratemporal epidural and subdural approach for lesions involving the entire cavernous sinus. J Neurosurg. 1989;71(5):699–704.
33. Chaddad Neto F, Doria Netto HL, Campos Filho JM, Reghin Neto M, Silva-Costa MD, Oliveira E. Orbitozygomatic craniotomy in three pieces: tips and tricks. Arq Neuropsiquiatr. 2016;74(3):228–34.
34. Yasargil MG, Antic J, Laciga R, Jain KK, Hodosh RM, Smith RD. Microsurgical pterional approach to aneurysms of the basilar bifurcation. Surg Neurol. 1976;6(2):83–91.

Surgical Anatomy of the ParaSellar Region

Saniya S. Godil, Alexandre Todeschini, Benjamin McGahan, Douglas Hardesty, and Daniel Prevedello

1 Sellar Region

The sphenoid bone is a bat-shaped structure and consists of the sphenoid body, two lesser wings, two greater wings, and two pterygoid plates. The greater wings extend superiorly and anterolaterally and form part of the middle cranial fossa, the orbit's lateral wall, the inferior and lateral margins of the superior orbital fissure, and the roof of the infratemporal fossa. The lesser wing forms the superior margin of the superior orbital fissure and the posterior part of the roof of the orbit. The flat part of the lesser wing forms the planum sphenoidale [1]. The sphenoid body houses the pituitary gland within the sella and other surrounding structures, including the cavernous sinus.

The sphenoid bone is one of the largest components of the skull base. It has the most complex relationship with other neurovascular structures, including posterior frontal lobes, olfactory tracts, optic chiasm, temporal lobe, pons, and mesencephalon, as well as several foramina and fissures for the exit of cranial nerves (CN) II–VI. Understanding these anatomical relationships for sellar and parasellar lesions is important as the trans-sphenoidal approach is one of the most common routes to address lesions in this area [2].

The endoscopic anatomy of the trans-sphenoidal approach can be divided into the following areas, see Fig. 1:

- Sellar
- Suprasellar
- Parasellar
- Clival

S. S. Godil · A. Todeschini · B. McGahan · D. Hardesty · D. Prevedello (✉)
Department of Neurosurgery, Ohio State University, Wexner Medical Center, Columbus, OH, USA
e-mail: daniel.prevedello@osumc.edu

E. G. Figueiredo et al. (eds.), *Brain Anatomy and Neurosurgical Approaches*, https://doi.org/10.1007/978-3-031-14820-0_24

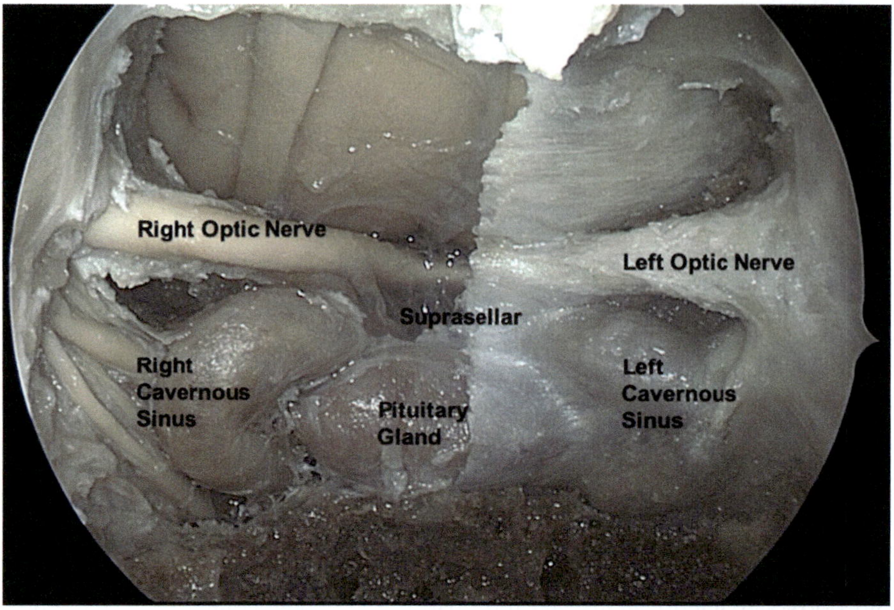

Fig. 1 Endoscopic cadaveric dissection and anatomy of sellar, suprasellar, and parasellar regions. The right-side dura has been removed, but the left side is still covered with dura

The sella is a saddle-shaped structure that houses the pituitary gland and does not have any bony covering superiorly and laterally. Tuberculum sella forms the anterior wall of the sella, whereas dorsum sella forms the posterior wall. The pituitary gland is surrounded laterally by the cavernous sinuses on each side, which are large venous plexuses between the inner and outer layer of the dura. The diaphragma sellae is formed by the inner reflection of the dura above the pituitary gland. The chiasmatic groove is the sulcus formed by the optic chiasm and is bounded anteriorly by planum sphenoidale and posteriorly by tuberculum sellae. The lesser wing forms the anterior clinoid process, and the dorsum sellae forms the posterior clinoid process. The dorsum sellae then continues inferiorly and forms the superior part of the clivus.

2 Parasellar Region

The parasellar region is a complex anatomical area containing several neurovascular structures in a small, confined space [1–4]. It comprises the cavernous sinuses and adjacent parts of the middle cranial fossa.

The paired cavernous sinuses are vascular structures located on either side of the sella limited by dural walls and containing essential neural and vascular structures, including sympathetic plexus, CN, and internal carotid artery (ICA). Due to the complex anatomy and proximity of important neurovascular structures, parasellar and cavernous sinus pathologies remain a challenge for surgical access and treatment. Therefore, it is important to understand the anatomy of the parasellar area and identify the surgical corridors that can be utilized to address different pathologies in this area [5, 6].

3 Anatomy of Cavernous Sinus

The cavernous sinus is a venous space surrounded by dural walls. Its bony limits include the greater and lesser wings of the sphenoid bone, anterior and posterior clinoid, and petrous apex. It is shaped like a boat extending from the superior orbital fissure anteriorly to the dorsum sellae and Meckel's cave posteriorly (Figs. 1 and 2). It contains CNs III, IV, V1, and VI as well as cavernous ICA [5–7].

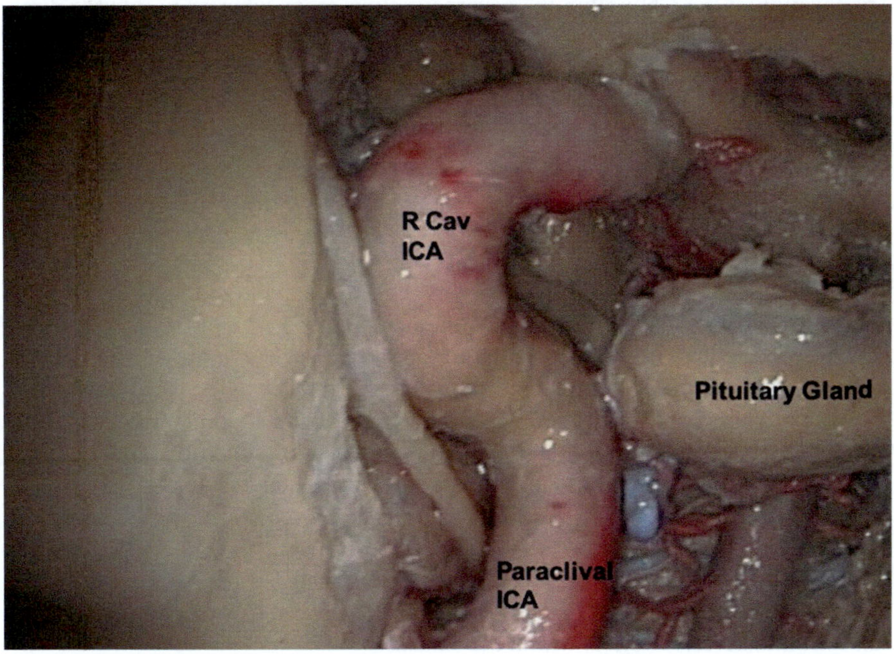

Fig. 2 Course and relationships of the right paraclival and right cavernous internal carotid artery

3.1 Walls of the Cavernous Sinus

The lateral wall of the cavernous sinus is formed by two layers of the dura, the periosteal and meningeal layers. The periosteal layer of the dura forms the floor and the medial wall of the cavernous sinus. It is the dura of the floor of the middle cranial fossa that extends medially over the trigeminal nerve to form the lateral wall of the cavernous sinus. The posterior part of the lateral wall of the cavernous sinus is also part of the medial wall of Meckel's cave. The medial wall of the cavernous sinus is very thin and compact, and there is a plane that separates the medial wall of the cavernous sinus from the pituitary gland [8]. Two layers of the dura form the roof of the cavernous sinus. It is attached to the anterior and posterior clinoid processes medially; it is continuous anteriorly with the diaphragma sella. It terminates at the superior orbital fissure and posteriorly at the tentorium. The posterior wall of the cavernous sinus is formed by the dual-layered dura of the clivus [6].

3.2 The Course of Cranial Nerves

The location of the CN within the cavernous sinus from superior to inferior direction includes oculomotor nerve followed by trochlear, ophthalmic, maxillary, and abducens nerve (Fig. 3). The oculomotor, trochlear, ophthalmic, and maxillary nerves travel between the two layers of the lateral wall of the cavernous sinus, whereas the abducens nerve and associated sympathetic plexus go through the cavernous sinus proper.

The oculomotor nerve enters the cavernous sinus through the roof, slightly anterior to the dorsum sella, and directly above the meningohypophyseal trunk from the cavernous ICA.

The trochlear nerve enters the cavernous sinus through the roof, posterolateral to the oculomotor nerve. It courses below the oculomotor nerve in the lateral wall of the cavernous sinus. Both the oculomotor and trochlear nerve are situated medial and beneath the free edge of the tentorium at their entrance at the cavernous sinus.

The ophthalmic nerve is the smallest division of the trigeminal nerve, and it forms the lower part of the lateral wall of the cavernous sinus. It forms lacrimal, frontal, and nasociliary nerves entering the superior orbital fissure.

The abducens nerve forms the lower part of the posterior wall of the cavernous sinus as it enters the Dorello's canal and crosses below the Gruber's ligament to enter the cavernous sinus. It is lateral to the cavernous ICA, but is the most medial nerve in the cavernous sinus. The nerve enters the cavernous sinus as a bundle, but can split into several rootlets within the sinus.

The bundles of sympathetic fibers run along the surface of the ICA and then briefly join the abducens nerve before joining the ophthalmic division of the trigeminal nerve. They eventually go via the superior orbital fissure to long and short ciliary nerves.

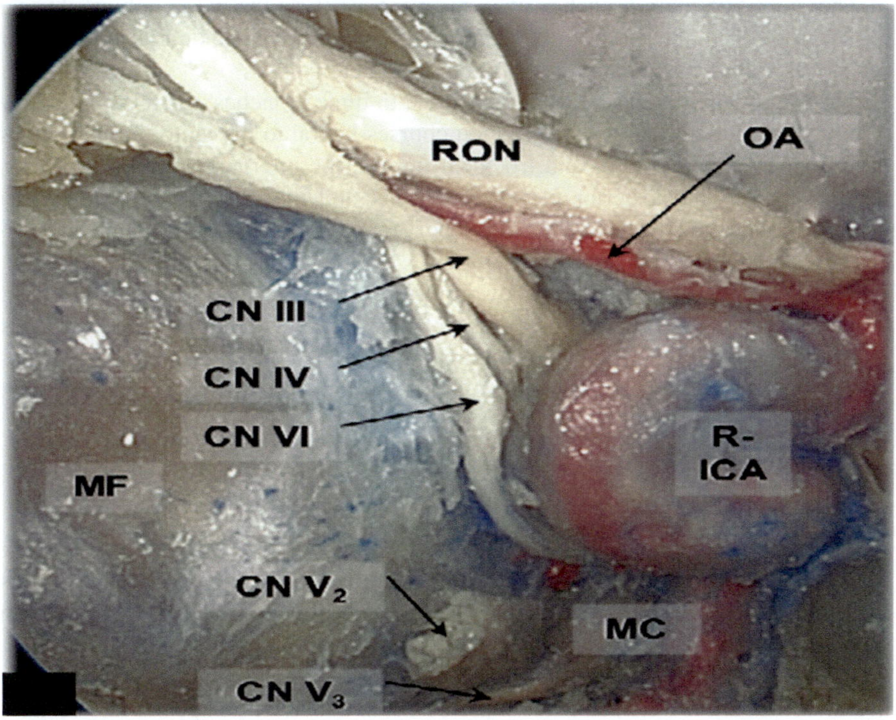

Fig. 3 Contents of the right cavernous sinus and the relationships of all the neurovascular structures within the cavernous sinus. OA, ophthalmic artery; RON, right optic nerve; MF, middle fossa

3.3 Cavernous Carotid Artery

The intracavernous segment of the ICA starts at the end of the carotid canal, above the foramen lacerum, and under the petrolingual ligament, where the petrous segment of the ICA enters the cavernous sinus. The cavernous ICA lies in the carotid sulcus on the lateral surface of the body of the sphenoid bone. With pneumatization of the sphenoid sinus and resorption of the sinus walls, the carotid sulcus makes a prominence along the lateral wall of the sella. The bone over the parasellar carotid arteries is much thinner than the bone over the sella. The inter-carotid distance is an important consideration for trans-sphenoidal pituitary surgery due to the risk of ICA injury.

The parasellar or intracavernous ICA is C-shaped and divided into five segments: the posterior vertical segment, the posterior bend, the horizontal segment, the anterior bend, and the anterior vertical segment (Fig. 2). The posterior vertical segment begins as the ICA emerges from the foramen lacerum and ends as the artery bends to form the posterior bend. Then, the cavernous ICA courses anteriorly as the horizontal segment and bends superomedially, forming the anterior bend. Finally, it takes a vertical course as the anterior vertical segment, medial to the anterior clinoid process, and exits through the roof of the cavernous sinus.

There are two main branches of intracavernous ICA: the meningohypophyseal trunk arising from the posterior bend and the inferolateral trunk or artery of inferior cavernous sinus arising from the horizontal segment.

The meningohypophyseal trunk gives rise to three branches: the tentorial artery (artery of Bernasconi–Cassinari), which courses lateral to the tentorium and gives off branches to oculomotor and trochlear nerves; the inferior hypophyseal artery, which supplies the posterior pituitary capsule; and the dorsal meningeal artery which supplies the clival dura, abducens nerve, and posterior portion of the cavernous sinus.

The inferolateral trunk passes above or below the abducens nerve toward the first trigeminal division to supply the dura of the inferior lateral wall of the cavernous sinus and the area of the foramen rotundum and foramen ovale. It gives off branches toward the foramen spinosum and the main blood supply to the trigeminal ganglion.

3.4 Venous Connections

The cavernous sinuses communicate with each other via intercavernous venous connections. The cavernous sinus receives venous drainage from the orbit, Sylvian fissure, anterior and middle cranial fossa via superior and inferior ophthalmic veins, the sphenoparietal sinus, the superficial middle cerebral vein, and the middle meningeal veins. The cavernous sinus has free connections with superior and inferior petrosal sinus and basilar sinus.

4 Parasellar Triangles

There are several triangles formed by traversing CN in the parasellar area. There are four triangles in the cavernous sinus, four in the middle fossa, and two in the paraclival area [9]. It is important to understand the anatomy of these triangles to plan surgical approaches to the parasellar area and cavernous sinus.

5 Cavernous Sinus Triangles

5.1 Clinoidal Triangle

It is the anteromedial triangle known as the Dolenc's triangle [9]. The borders are formed by the optic nerve medially, oculomotor nerve laterally, and tentorial edge at the base. It contains the clinoidal ICA and anterior clinoid process and can be exposed by drilling the anterior clinoid process either extradurally or intradurally. An anterior clinoid meningioma requires intradural resection via clinoidal triangle.

5.2 *Oculomotor Triangle*

The medial triangle is also known as the Hakuba's triangle [9]. The anterior petroclinoidal dural fold forms the borders (extending from anterior clinoid process to petrous apex), posterior petroclinoidal dural fold (extending from posterior clinoid process to petrous apex), and the interclinoidal dural fold (extending from anterior to posterior clinoid process). The oculomotor triangle contains the oculomotor nerve and horizontal segment of the ICA. This triangle is important for accessing vascular lesions in the horizontal segment of ICA and tumors in the medial cavernous sinus and interpeduncular lesions.

5.3 *Supratrochlear Triangle*

It is also called the paramedian triangle. It is a very narrow space bordered by the oculomotor nerve medially, trochlear nerve laterally, and the tentorial edge at the base. It contains the meningohypophyseal trunk and the medial loop of the intracavernous ICA [9].

5.4 *Infratrochlear Triangle*

It is the superolateral space and is also known as the Parkinson's triangle [9]. The borders are formed by the trochlear nerve medially, ophthalmic division of the trigeminal nerve laterally, and the dural edge formed by these nerves at the base. It contains the cavernous ICA, the origin of the meningohypophyseal trunk, and the abducens nerve. During surgical exposure of this triangle, the entire cavernous ICA can be explored from lateral to proximal ring and abducens nerve from Dorello's canal to superior orbital fissure. Thus, this area is valuable in the surgical management of vascular pathology and cavernous sinus/CN lesions. The posteroinferior part of the cavernous sinus can also be explored through this triangle.

6 Middle Fossa Triangles

6.1 *Anteromedial Triangle*

This triangle is located between the ophthalmic and maxillary nerve and the line connecting the entry point of the ophthalmic and maxillary nerve in the superior orbital fissure and foramen rotundum, respectively. Removing the bone in this triangle can open the sphenoid sinus.

6.2 Anterolateral Triangle

This triangle is between the maxillary and mandibular nerve and the line connecting the foramen rotundum and foramen ovale.

6.3 Posterolateral Triangle (Glasscock's Triangle)

The triangle is formed anteromedially by the mandibular nerve, posterolaterally by the greater petrosal nerve, and laterally by the floor of the middle cranial fossa. The middle meningeal artery passes through this triangle. Drilling the floor in this triangle gives access to the infratemporal fossa.

6.4 Posteromedial Triangle (Kawase's Triangle)

This triangle is formed by the greater petrosal nerve, lateral edge of the trigeminal nerve, arcuate eminence, and superior petrosal sinus. It consists of the petrous apex, ICA, and vertebrobasilar junction. Drilling the medial part of the triangle exposes the posterior fossa dura and gives access to the anterior clivus and anterolateral brainstem. Drilling the lateral part of the triangle exposes the cochlea and IAC.

7 Paraclival Triangles

7.1 Inferolateral Triangle

The triangle is formed by the line joining the dural entry of the trochlear and abducens nerve to the tentorium medially, the line joining the entry of the abducens nerve to the petrosal vein laterally, and the base is formed by the petrous apex [10].

7.2 Inferomedial Triangle

The triangle is formed by the dural entry of the trochlear and abducens nerve, the line joining the posterior clinoid, and the entry of the abducens nerve, and the base is formed by the petrous apex. It contains the Dorello's canal and Gruber's ligament [10].

8 Endoscopic Anatomy of Cavernous Sinus

The surgical routes and corridors for cavernous sinus have been studied extensively [7, 11, 12]. Above-mentioned cavernous sinus triangles have been utilized for microsurgical transcranial approaches to the cavernous sinus and usually access the cavernous sinus through the superior and lateral wall. However, these approaches generally cannot visualize the anteromedial surface of the cavernous sinus. Recent advances in endoscopic endonasal techniques over the last couple of decades have made it possible to approach cavernous sinus through medial and anterior walls in a medial to the lateral trajectory or anterior to the posterior trajectory [5, 11, 13, 14]. Thus, the cavernous sinus triangles and surgical anatomy described for transcranial microsurgical approaches can have limited value in understanding cavernous sinus anatomy via the endoscopic endonasal approach. Several authors have now developed a new classification system for different compartments of the cavernous sinus based on endoscopic anatomy and compared cavernous sinus anatomy from microscopic and endoscopic approaches (Figs. 1, 2 and 3) [7, 15].

In order to access the cavernous sinus from an endoscopic endonasal corridor, the approach can vary from a standard trans-sphenoidal approach to a more extended approach to involve transpterygoid approach, anterior and posterior ethmoidectomies, and dissection of the vidian canal, based on the pathology [16]. Lesions in the medial and postero-superior compartments can be accessed via a trans-sphenoidal transcavernous approach, but lesions in the lateral or antero-inferior compartments require an extended endoscopic transpterygoid approach [7, 12, 16–19].

The anteromedial triangle can be accessed via the endonasal approach by drilling the lateral wall of the sphenoid sinus and opening the periosteal layer of the dura. The anterolateral triangle can be accessed via the endonasal approach by drilling the pterygoid plates and the pterygoid process to find the foramen ovale. To access the Parkinson's triangle via endonasal approach, extensive mobilization of CN VI and parasellar ICA is required, and therefore, one must be cautious. It might not be possible to get complete exposure to this triangle through the endoscopic endonasal approach. The visualization of the quadrangular space can be limited via an endonasal approach based on the location of the ICA [20, 21]. If the ICA is more lateral, it can obscure the course of CN VI, but if the ICA is more medial, better exposure of quadrangular space is possible via the endonasal route [19–21].

To understand the endoscopic anatomy of the cavernous sinus, some authors have classified different compartments of the cavernous sinus based on the anatomy of the cavernous ICA [7, 15, 20, 21]. The *superior compartment* lies superior to the horizontal segment of the cavernous ICA. This compartment is bounded by the roof of the cavernous sinus, paraclinoidal ICA, and dura of the clinoidal triangle. The key structure of the compartment is CN III which runs in the lateral wall and the interclinoidal ligament. This compartment is generally accessed via the medial wall of the cavernous sinus, and the bony covering of paraclinoidal ICA must be removed to mobilize the ICA. The *posterior compartment* is located posterior to the vertical cavernous ICA. The key structures of this compartment include the meningohypophyseal trunk and CN VI. Access to this compartment requires extensive bony drilling to expose the anterior wall of the cavernous sinus and the entrance of ICA into the cavernous sinus. The *inferior compartment* is located inferior to the horizontal and anterior genu of the ICA. The key structures of this compartment include the sympathetic plexus and distal segment of CN VI. Access to this compartment requires drilling the bone over the anterior wall of the cavernous sinus all the way to the lateral dural fold marking the transition between the parasellar and middle fossa dura. The *lateral compartment* lies lateral to the anterior genu and horizontal ICA. The key structures of this compartment include CN III, CN IV, CN V1, and branches of the inferolateral trunk [19]. Access to this corridor is challenging and requires complete exposure of anterior genu and paraclinoidal ICA and anterior wall of the cavernous sinus to superior orbital fissure.

9 Parasellar Pathologies

Several pathologies can be encountered in the parasellar region [1, 3, 4]. Pituitary adenomas arising from the anterior lobe of the pituitary gland (both functioning and non-functioning tumors) can be invasive and invade the cavernous sinus. Craniopharyngiomas originate from epithelial remnants of Rathke's pouch and can invade sellar, suprasellar, and parasellar areas. Meningiomas originate from the dura and arachnoid. They can invade the cavernous sinus and present with ophthalmoplegia. Schwannomas in the parasellar region arise mostly from the trigeminal nerve and can rarely arise from CN III, IV, and VI. They can involve either the cisternal, cavernous, or extracranial components of the nerves. Skull base tumors in the parasellar region can include primary (chordoma, chondrosarcoma, and plasmacytoma), secondary (sinonasal tumors), and metastatic tumors. Vascular lesions including aneurysms and carotid–cavernous fistulas can also be found in this region. Several inflammatory (Tolosa–hunt syndrome) and infectious processes (bacterial and fungal infections, pituitary abscess) can also involve the parasellar region [4].

References

1. Zamora C, Castillo M. Sellar and parasellar imaging. Neurosurgery. 2017;80(1):17–38.
2. Patel CR, Fernandez-Miranda JC, Wang WH, Wang EW. Skull base anatomy. Otolaryngol Clin N Am. 2016;49(1):9–20.
3. Gatto F, Perez-Rivas LG, Olarescu NC, Khandeva P, Chachlaki K, Trivellin G, et al. Diagnosis and treatment of parasellar lesions. Neuroendocrinology. 2020;110(9–10):728–39.
4. Jipa A, Jain V. Imaging of the sellar and parasellar regions. Clin Imaging. 2021;77:254–75.

5. Chowdhury F, Haque M, Kawsar K, Ara S, Mohammod Q, Sarker M, et al. Transcranial microsurgical and endoscopic endonasal cavernous sinus (CS) anatomy: a cadaveric study. J Neurol Surg A Cent Eur Neurosurg. 2012;73(5):296–306.
6. Yasuda A, Campero A, Martins C, Rhoton AL Jr, de Oliveira E, Ribas GC. Microsurgical anatomy and approaches to the cavernous sinus. Neurosurgery. 2008;62(6 Suppl 3):1240–63.
7. Almeida JP, de Andrade E, Neto MR, Radovanovic I, Recinos PF, Kshettry VR. From above and below: the microsurgical anatomy of endoscopic endonasal and transcranial microsurgical approaches to the parasellar region. World Neurosurg. 2022;159:e139–60.
8. Truong HQ, Lieber S, Najera E, Alves-Belo JT, Gardner PA, Fernandez-Miranda JC. The medial wall of the cavernous sinus. Part 1: surgical anatomy, ligaments, and surgical technique for its mobilization and/or resection. J Neurosurg. 2018;131(1):122–30.
9. Drazin D, Wang JM, Alonso F, Patel DM, Granger A, Shoja MM, et al. Intracranial anatomical triangles: a comprehensive illustrated review. Cureus. 2017;9(10):e1741.
10. Wysiadecki G, Radek M, Tubbs RS, Iwanaga J, Walocha J, Brzezinski P, et al. Microsurgical anatomy of the inferomedial paraclival triangle: contents, topographical relationships and anatomical variations. Brain Sci. 2021;11(5):596.
11. Bassim MK, Senior BA. Endoscopic anatomy of the parasellar region. Am J Rhinol. 2007;21(1):27–31.
12. Erdogan U, Turhal G, Kaya I, Biceroglu H, Midilli R, Gode S, et al. Cavernous sinus and parasellar region: an endoscopic endonasal anatomic cadaver dissection. J Craniofac Surg. 2018;29(7):e667–e70.
13. Liu JF, Qu QY, Yang DZ, Han J, Zhang QH, Snyderman CH, et al. Transnasal endoscopic anatomy and approaches to the cavernous sinus. Zhonghua Er Bi Yan Hou Tou Jing Wai Ke Za Zhi. 2013;48(11):901–7.
14. Felippu A, Mora R, Guastini L, Peretti G. Transnasal approach to the orbital apex and cavernous sinus. Ann Otol Rhinol Laryngol. 2013;122(4):254–62.
15. Fernandez-Miranda JC, Zwagerman NT, Abhinav K, Lieber S, Wang EW, Snyderman CH, et al. Cavernous sinus compartments from the endoscopic endonasal approach: anatomical considerations and surgical relevance to adenoma surgery. J Neurosurg. 2018;129(2):430–41.
16. Kasemsiri P, Solares CA, Carrau RL, Prosser JD, Prevedello DM, Otto BA, et al. Endoscopic endonasal transpterygoid approaches: anatomical landmarks for planning the surgical corridor. Laryngoscope. 2013;123(4):811–5.
17. Abuzayed B, Tanriover N, Gazioglu N, Ozlen F, Cetin G, Akar Z. Endoscopic anatomy and approaches of the cavernous sinus: cadaver study. Surg Radiol Anat. 2010;32(5):499–508.
18. Alfieri A, Jho HD. Endoscopic endonasal cavernous sinus surgery: an anatomic study. Neurosurgery. 2001;48(4):827–36; discussion 36–7.
19. Amin SM, Fathy H, Hussein A, Kamel M, Hegazy A, Fathy M. Endoscopic endonasal approach to the lateral wall of the cavernous sinus: a cadaveric feasibility study. Ann Otol Rhinol Laryngol. 2018;127(12):903–11.
20. Dolci RL, Upadhyay S, Ditzel Filho LF, Fiore ME, Buohliqah L, Lazarini PR, et al. Endoscopic endonasal study of the cavernous sinus and quadrangular space: anatomic relationships. Head Neck. 2016;38(Suppl 1):E1680–7.
21. Dolci RL, Carrau RL, Buohliqah L, Zoli M, Mesquita Filho PM, Lazarini PR, et al. Endoscopic endonasal anatomical study of the cavernous sinus segment of the ophthalmic nerve. Laryngoscope. 2015;125(6):1284–90.

Surgical Anatomy of the Anterior Incisural Space

Nicollas Nunes Rabelo, Mateus Gonçalves de Sena Barbosa, Leonardo Luca Luciano, Leonardo Christiaan Welling, and Eberval Gadelha Figueiredo

1 Anatomy

The anterior incisural space (AIS) is positioned anteriorly to the midbrain, and it unfolds superiorly to the sellar diaphragm. Furthermore, around the optic chiasm to the area anterior to the *lamina terminalis* between the frontal lobes, the oculomotor, abducens and optic nerves, carotid arteries, olfactory tract, and infundibulum cross the AIS [1–7].

AIS is ventral to the brainstem; it has the posterior communicating artery, anterior choroidal artery, basilar bifurcation, and the supraclinoidal part of the internal carotid artery. Furthermore, it unfolds laterally in the portion of the Sylvian fissure, located inferiorly to the anterior perforated substance. This space expands inferiorly between the brainstem and clivus and obliquely forward and upward around the optic chiasm until reaching the subcallosal area. The portion of the AIS located inferiorly to the optic chiasm is restricted by the anterior third of the uncus, the cerebral peduncles, and the pons [1–7].

N. N. Rabelo (✉)
Neurosurgery Department, Atens University Center, Passos, MG, Brazil

Neurosurgery Department, University of São Paulo, São Paulo, SP, Brazil

M. G. de Sena Barbosa · E. G. Figueiredo
Neurosurgery Department, Atens University Center, Passos, MG, Brazil

L. L. Luciano
Alfenas University, Alfenas, MG, Brazil
e-mail: leonardo.luciano@sou.unifal-mg.edu.br

L. C. Welling
Neurosurgery Department, State University of Ponta Grossa, Ponta Grossa, PR, Brazil

E. G. Figueiredo et al. (eds.), *Brain Anatomy and Neurosurgical Approaches*,
https://doi.org/10.1007/978-3-031-14820-0_25

The AIS roof is composed of the optic nerves, chiasma and tracts, the posterior and anterior perforated substances, tuber cinereum, olfactory tracts, and mammillary bodies. Moreover, this space is crossed by the oculomotor and optic nerves and by the infundibulum as well [1–7].

The uncinate gyrus, the Giacomini band, and the interlimbic gyrus are gyral prominences on the medial surface of the uncus. However, only the uncinate gyrus is associated with the AIS, since the others are related to the middle incisural space [1–7].

2 Cisternal Relations

In the posterior region of the AIS between the cerebral peduncles and the dorsum of the sella, the interpeduncular cistern is located; this is laterally associated with the Sylvian cistern inferiorly to the anterior perforated substance and anteriorly with the chiasmatic cistern, which is located inferiorly to the optical chiasm. Liliequist's membrane separates the interpeduncular and chiasmatic cisterns, this membrane being an arachnoid lamina, extending from the dorsum of the sella to the anterior edge of the mammillary bodies. The cisterna of the laminae terminalis is located anteriorly to the lamina terminalis, and this cisterna contains communication around the optic chiasm [1, 4–6, 8, 9] (Fig. 1).

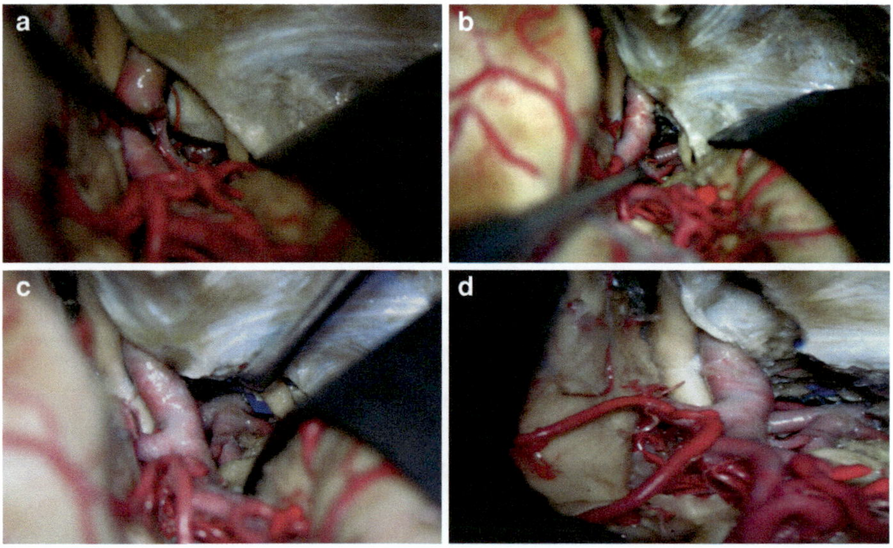

Fig. 1 Some of the steps of the transcavernous approach are illustrated in this cadaverous dissection. Cadaveric exposure of the right internal carotid artery after dividing the Sylvian fissure (**a**). Removal of the posterior clinoid process extends surgical exposure and illumination (**b**) to provide the most inferior surgical view, making it adequate for lower lying basilar apex aneurysms and lesions in retrosellar topography (**c, d**)

3 Ventricular Relations

In the medial plane, the AIS receives the projection of the anterior region of the third ventricle, being fragmented into infra and suprachiasmatic areas. Superior to the AIS are the frontal horns of the lateral ventricles. The amygdaloid nucleus separates the tip of the temporal horn from the uncal surface, thus constructing the posterolateral wall of the AIS. The anterior wall of the AIS ventrally receives the cerebral aqueduct [1, 4–6].

4 Cranial Nerve Relationships

The oculomotor and optic nerves and the posterior portion of the olfactory tract travel through the AIS. The entire olfactory tract proceeds posteriorly, and when it reaches the clinoid process, it fragments to generate medial and lateral olfactory streaks, which cross the anterior margin of the anteriorly perforated substance. The AIS is crossed by the optic nerves, chiasm, and the anterior part of the optic tracts. From the optic canal arise the optic nerves, being in the medial portion the stabilization of the free edge to the anterior clinoid processes; such nerves are conducted superiorly, posteriorly, and medially toward the optic chiasm [1, 2].

Typically, the location where the optic chiasm is installed is superior to the saddle diaphragm, and, less frequently, it is pre-installed in the saddle tubercle or post-installed and located in the saddle dorsum. Starting at the chiasm, the optic tract advances in a posterolateral direction around the cerebral peduncle to penetrate the mid incisural space. The oculomotor nerve arises from the midbrain on the medial surface of the cerebral peduncle. This nerve crosses the AIS between the superior cerebellar and posterior cerebral arteries and continues from the inferomedial to the unco to infiltrate the cavernous sinus roof through the trigone oculomotor. The abducens nerve rises deeply in the infratentorial region of the AIS. The pontomedullary sulcus rises in the pre-pontine cistern to transfix the dura mater that covers the clivus and pursues inferiorly to the petroclinoid ligament to invade the cavernous sinus [1, 2, 4–6, 10, 11] (Fig. 2).

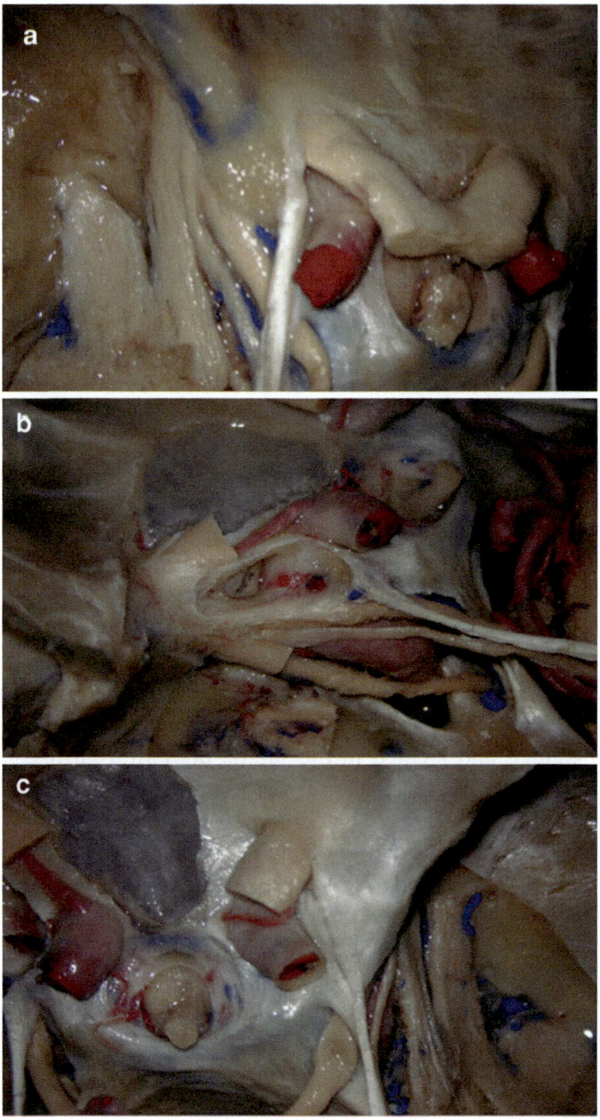

Fig. 2 Arterial and nerve structures of the anterior incisural space. (**a**) Superior oblique view showing the internal carotid artery and its relationship with the optic nerve and the lateral wall of the cavernous sinus. After removing the dural fold, neural structures V1, V2, III, IV, and VI cranial nerves are evident. (**b**) Removal of the anterior clinoid and part of the orbital roof, exposing the ophthalmic artery. In this dissection, the two dural folds before the vertical segment of the intracavernous carotid artery are also demonstrated, as well as the exposure of its horizontal segment and its relationship with the III, IV, and VI nerves. (**c**) Superior view with the dural cover and relationship of the tentorium edge with the entry point of the III nerve into the cavernous sinus as well as the internal carotid artery, optic nerve, and anterior clinoid process

5 Arterial Relations

The AIS has all components of the circle of Willis and the bifurcation of the internal carotid and basilar arteries, which is why the arterial associations of this space are considered difficult [1, 4–6, 12–14].

The internal carotid artery enters the AIS, running through the medial portion of the anterior clinoid process, and continues superiorly, posteriorly, and laterally toward its bifurcation inferiorly to the anterior perforated substance [1, 4–6, 12–14].

The posterior communicating artery arises from the posteromedial aspect of the carotid artery and continues superomedial to the oculomotor nerve to unite with the posterior cerebral artery in the AIS [1, 4–6, 12–14].

The anterior choroidal artery arises from the posterior surface of the carotid artery, about 0.1–3.0 mm distally to the point of appearance of the posterior communicating artery, and proceeds under the optic tract before crossing between the uncus and the cerebral peduncle to penetrate the middle incisural space. The proximal portion of the anterior cerebral artery passes into the AIS, and it originates below the anterior perforated substance and proceeds anteromedially and above the optic chiasm. It joins its counterpart on the opposite side by the communicating artery previous at this location. Afterward, there is an upward continuation in front of the terminal lamina, after which it detaches from the orbitofrontal artery before reaching the corpus callosum. In the case of the middle cerebral artery, it persists laterally from its starting point inferiorly to the anterior perforated substance. Lateral portion of the AIS is located in the most important bifurcation of the middle cerebral artery and the origins of its orbitofrontal, prefrontal, anterior temporal, and temporopolar branches. In the posterior portion of the AIS, there is an elevation of the basilar artery, generating the posterior cerebral and superior cerebellar arteries between the clivus and the posterior perforated substance. The basilar tip and bifurcation location change from the caudal to 1.3 mm inferiorly to the pontomesencephalic sulcus until rostral to the mammillary bodies. Superior to the oculomotor nerve, there is the path of the posterior cerebral artery around the cerebral peduncle. It goes beyond the anterior portion and penetrates the middle incisural space, crossing between the cerebral peduncle and the uncus [1, 4].

The superior cerebellar artery arises from the AIS below the posterior cerebral artery and proceeds inferolateral to the oculomotor nerve. Typically, the starting point is rostral to the freeboard point. To reach the superior surface of the cerebellum at the point of unification of the middle and AIS, it crosses under the tentorium. Furthermore, the organization of the AIS receives perforating ramifications from several different arteries, thus justifying its anatomical complexity. The perforating branches of the carotid, posterior communicating, anterior and middle cerebral, anterior choroid, and recurrent arteries invade the perforated substance [1, 4–6, 12–14].

In the AIS, the carotid arteries generate the communicating posterior, anterior, and middle choroidal, and anterior cerebral arteries and the arteries of the basilar artery generate the superior cerebellar and posterior cerebral arteries. Furthermore, the medial posterior choroidal artery comes from the proximal portion of the posterior cerebral artery [1, 4–6, 12–14].

6 Venous Relations

The most relevant venous trunk to the AIS is the basal vein. It crosses the anterior, middle, and posterior incisural spaces to reach the vein of Galen [1, 4–6, 15, 16].

In the superficial part of the brainstem, some veins make up the posterior wall of the AIS, which are fragmented into transverse or vertical groups. The transverse veins are the peduncular vein, which runs horizontally around the anterior surface of the cerebral peduncle and ends in the basal vein connecting the middle and AIS, and the transverse pontine veins, which run transversely around the anterior surface of the pons. They generate bridging veins, which enter the tentorial or superior petrous sinus, and the vein of the pontomesencephalic groove, which runs under the peduncular vein in the pontomesencephalic groove. The vertically shaped veins on the posterior wall of the AIS are: the lateral anterior pontomesencephalic veins, which run along the anterolateral surface of the cerebral peduncle and pons and join the basal vein superiorly, and the vein of the pontomesencephalic groove or a transverse pontine vein inferiorly, and the median anterior pontomesencephalic vein, which follows the midline and joins the peduncular veins superiorly to the inferior transverse pontine veins [1, 4–6, 15, 16].

In the AIS, the veins that flow to the limit of the basal vein in this space are: the deep middle cerebral, fronto-orbital, olfactory, uncal, anterior cerebral, paraterminal, anterior pericalosus, and inferior striatum. The unification of the veins that drain the walls of the AIS inferiorly to the anterior perforated substance generates the basal vein [1, 4–6, 15, 16].

7 Aneurysms in the Anterior Incisural Space

Of all saccular arterial aneurysms, about 95% of them appear in the AIS. Aneurysms located in the circle of Willis, which are anterior to the position of Liliequist's membrane, middle cerebral artery, and internal carotid artery, are accessed by a frontotemporal craniotomy. In the case of middle cerebral artery aneurysms positioned in the lateral region of the AIS, the superior temporal gyrus should be opened to display the aneurysm in this location. Furthermore, in the case of aneurysms of the anterior communicating artery, the posterior part of the straight gyrus is removed. Frontotemporal or subtemporal craniotomy is recommended for presenting aneurysms positioned posteriorly to Liliequist's membrane at the basilar apex in the interpeduncular fossa if they are superior to the dorsum. Furthermore, subtemporal craniotomy with incision or retraction of the tentorium is indicated for aneurysms located in the prepontine cistern or inferiorly to the dorsum. For basilar tip aneurysm surgeries, it is interesting to consider the delicate retraction of the cerebral peduncle, the optic tract, and the nipple body, since several studies show results without neurological complications applying this technique [1, 4–6, 17–22].

To gain access to the lesions surrounding the notch, it is essential, in most cases, to make an incision and retraction of the territory. In the section of the tentorium, there are venous sinuses and arteries, so it can be used to reduce the pressure of the brainstem due to giant aneurysms and incisural tumors, which cannot be removed. Furthermore, in the AIS, in most cases, the risk is much higher in the case of perforating arteries of the brainstem, since they are stretched around tumors and aneurysms in this region [1, 4–6, 17–22].

The superior temporal gyrus can be moved away to expose aneurysms of the cerebral artery in the lateral portion of the AIS [1, 4–6, 17–22] (Fig. 3).

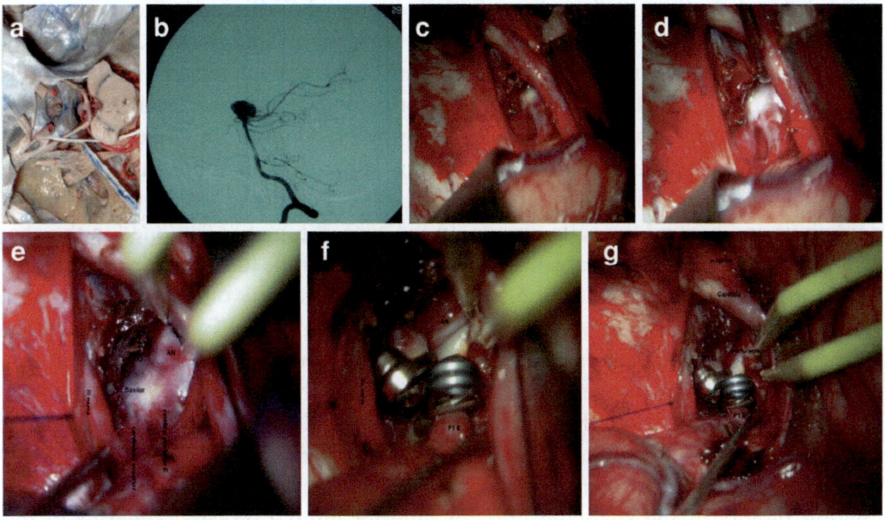

Fig. 3 Basilar aneurysm case. (**a**) Anatomical dissection showing all anterior incisural space structures with brain removal. (**b**) Basilar aneurysm in angiography. (**c**) Pre-temporal approach for a basilar aneurysm. (**d**) More exposure of the aneurysm. (**e**) All structures without posterior clinoid. It is possible to see the superior cerebellar artery, left cerebral posterior, and the third nerve. (**f**) Aneurysm clipped. (**g**) All the approaches show P1's (left and right), the perforators, optic nerve, and internal carotid artery

8 Arteriovenous Malformations in the Anterior Incisural Space

Arteriovenous malformations (AVMs) can affect the vessels in the anterior portion and other incisural spaces. Such malformations are not built from functional nervous tissue, so they can be removed without causing sequelae, even if they are profound. The AVMs in the AIS can be treated by cutting the anterior perforated substance and occluding the distribution arteries [1, 4–6, 23, 24].

9 Tumors in the Anterior Incisural Space

The types of tumors located in the AIS , whether originating in the same space or by extension, are pituitary adenomas, meningiomas arising from the tubercle of the sella, clivus, and medial portion of the sphenoid crest, craniopharyngiomas, gliomas of the optic nerve and hypothalamus, clival chordomas, oculomotor nerve neuromas, certain dermoid cysts, and teratomas [1, 4–6, 25, 26].

In the AIS, tumors may appear that cannot be completely resected, as some tumors involve delicate components, such as cranial nerves, optic pathways, carotid, basilar artery and its branches, polygon of Willis, among other structures [1, 4–6, 25, 26].

The main access routes to tumors in this space are frontotemporal, subtemporal, bifrontal, subfrontal, frontal–interhemispheric, and trans-sphenoidal. Normally, tumors located anteriorly to Liliequist's membrane between the optic chiasm and the sella diaphragm are accessed via the trans-sphenoidal or subfrontal route. The trans-sphenoidal route is more recommended if it is superior to a pneumatized sphenoid sinus and beyond the sella turcica. In the case of tumors positioned in the chiasmatic cistern, the subfrontal approach is suggested. If the tumor expands ventrally into both anterior children's fossae and there is no possibility of reaching the tumor via a unilateral subfrontal view, then a bifrontal craniotomy is recommended. In tumors limited to the portion of the anterior inter-hemispheric space, where it lies inferiorly to the rostrum, the inter-hemispheric pathway is indicated [25, 26].

For tumors that occupy the AIS and invade the mid incisural space, it is recommended to use the association of the frontotemporal route with a temporal lobectomy. Moreover, the frontotemporal approach without association is indicated in treating tumors of the sphenoid crest or anterior clinoid process, superior to the diaphragm and with a long extension through the sphenoid crest or in the middle cranial fossa [1, 4–6, 25, 26].

10 Anterior Incisural Width as a Preoperative Indicator for Intradural Space Assessment

Two anatomical triangles are used to access the interpeduncular region: the opticocarotid triangle (OCT) and the carotico-oculomotor triangle. Some studies suggest that the anterior incisural width (LIA) can serve as a powerful preoperative indicator for analyzing the intradural space, as it is the size of the OCT. LIA smaller than 26 mm is defined as "thin," demonstrating a disadvantageous condition for performing a trans-sylvian approach in lesions in the interpeduncular cistern. In this condition, a transcavernous approach may be a more favorable option. Thus, among several components for surgeries in this region, LIA is one of these [27–33].

11 A Possible New Approach to Treating Posterior Circulation Brain Aneurysms

A study describes several patients treated by a modified technique that uses a subtemporal transtentorial approach extended to the AIS and superior clival region, which obtained positive results, as it improves visibility, maneuverability, and anatomical direction, so that it increases the anterolateral and rostrocaudal exposure, without causing a permanent deficit of the postoperative trochlear nerve [33–38].

12 Conclusions

The AIS is positioned anterior to the midbrain and superior to the sellar diaphragm. Furthermore, it is considered around the optic chiasm anteriorly to the terminal lamina between the frontal lobes, the oculomotor, abducens and optic nerves, carotid arteries, olfactory tract, and infundibulum crossing the AIS. Due to the numerous structures, his anatomical and surgical knowledge can be considered one of the most important for neurosurgical practice. In parallel to the fact that it is home to numerous pathologies, from pituitary tumors, AVMs, aneurysms, hamartomas, and glial neoplasms, the AIS can serve as a "route" to address diseases located in the posterior fossa and adjacent areas.

References

1. Ono M, Ono M, Rhoton AL Jr, Barry M. Microsurgical anatomy of the region of the tentorial notch. J Neurosurg. 1984;60(2):365–99. https://doi.org/10.3171/jns.1984.60.2.0365.
2. Harris FS, Rhoton AL Jr. Anatomy of the cavernous sinus. A microsurgical study. J Neurosurg. 1976;45:169–80.
3. Bakan AA, Alkan A, Kurtcan S, Aralaşmak A, Tokdemir S, Mehdi E, Özdemir H. Cavernous sinus: a comprehensive review of its anatomy, pathologic conditions, and imaging features. Clin Neuroradiol. 2015;25(2):109–25. https://doi.org/10.1007/s00062-014-0360-0.
4. Rhoton AL Jr. Tentorial notch. Neurosurgery. 2000;47(3 Suppl):S131–53. https://doi.org/10.1097/00006123-200009001-00015.
5. Uğur Türe MD. Book review: Rhoton's cranial anatomy and surgical approaches. Oper Neurosurg. 2020;19(2):E218–9.
6. Rhoton AL, Surgeons CN. Rhoton's cranial anatomy and surgical approaches. Oxford University Press; 2019. https://books.google.com.br/books?id=HIK0DwAAQBAJ
7. Azab WA, Almanabri M, Yosef W. Endoscopic treatment of middle fossa arachnoid cysts. Acta Neurochir. 2017;159(12):2313–7. https://doi.org/10.1007/s00701-017-3320-z. Epub 2017 Sep 13
8. Liliequist B. The subarachnoid cisterns. An anatomic and roentgenologic study. Aeta Radiol Suppl. 1959;185:1–108.
9. Yasargil MG, Kasdaglis K, Jain KK, et al. Anatomical observations of the subarachnoid cisterns of the brain during surgery. J Neurosurg. 1976;44:298–302.
10. Ausman JI, Diaz FG, de los Reyes RA, et al. Posterior circulation revascularization. Superficial temporal artery to superior cerebellar artery anastomosis. J Neurosurg. 1982;56:766–76.
11. Renn WH, Rhoton AL Jr. Microsurgical anatomy of the sellar region. J Neurosurg. 1975;43:288–98.
12. Saeki N, Rhoton AL Jr. Microsurgical anatomy of the upper basilar artery and the posterior circle of Willis. J Neurosurg. 1977;46:563–78.
13. Perlmutter D, Rhoton AL Jr. Microsurgical anatomy of the anterior cerebral-anterior communicating-recurrent artery complex. J Neurosurg. 1976;45:259–72.
14. Gibo H, Lenkey C, Rhoton AL Jr. Microsurgical anatomy of the supraclinoid portion of the internal carotid artery. J Neurosurg. 1981;55:560–74.
15. Matsushima T, Rhoton AL Jr, Oliveria E, et al. Microsurgical anatomy of the veins of the posterior fossa. J Neurosurg. 1983;59:63–105.
16. Matsushima T, Rhoton AL Jr, Lenkey C. Microsurgery of the fourth ventricle: part 1. Neurosurgery. 1982;11:631–67.

17. Rhoton AL Jr. Anatomy of saccular aneurysms. Surg Neurol. 1980;14:59–66.
18. Rhoton AL Jr, Fujii K, Saeki N, et al. Microsurgical anatomy of aneurysms, part 1. In: Hopkins LN, Long DM, editors. Clinical management of intracranial aneurysms. New York: Raven; 1982. p. 201–43.
19. Samson DS, Hodosh RM, Clark WK. Microsurgical evaluation of the pterional approach to aneurysms of the distal basilar circulation. Neurosurgery. 1978;3:135–41.
20. Sano K. Aneurysms of the vertebral and basilar arteries. Techniques of direct operation. In Shinkei Geka. 1973;1:193–9.
21. Zeal AA, Rhoton AL Jr. Microsurgical anatomy of the posterior cerebral artery. J Neurosurg. 1978;48:534–59.
22. Drake CG. The treatment of aneurysms of the posterior circulation. Clin Neurosurg. 1979;26:96–144.
23. Drake CG. Cerebral arteriovenous malformations: considerations for and experience with surgical treatment in 166 cases. Clin Neurosurg. 1979;26:145–208.
24. Viale GL, Turtas S, Pau A. Surgical removal of striate arteriovenous malformations. Surg Neurol. 1980;14:321–4.
25. Rhoton AL Jr, Hardy DG, Chambers SM. Microsurgical anatomy and dissection of the sphenoid bone, cavernous sinus and sellar region. Surg Neurol. 1979;12:63–104.
26. Rhoton AL Jr, Yamamoto I, Peace DA. Microsurgery of the third ventricle: part 2. Operative approaches. Neurosurgery. 1981;8:357–73.
27. Zhao X, Labib M, Ramanathan D, Eastin TM, Song M, Little AS, et al. The previous incisural width as a preoperative indicator for intradural space evaluation: an anatomical investigation. Surg Neurol Int. 2020;11:207.
28. Adler DE, Milhorat TH. The tentorial notch: anatomical variation, morphometric analysis, and classification into 100 human autopsy cases. J Neurosurg. 2002;96:1103–12.
29. Benok BR, Getch CC, Parkinson R, O'Shaughnessy BA, Batjer HH. Extended lateral transsylvian approach for basilar bifurcation aneurysms. Neurosurgery. 2004;55:174–8.
30. Dolenc VV, Škrap M, Šušteršič J, Skrbec M, Morina A. A transcavernous-transsellar approach to the basilar tip aneurysms. Br J Neurosurg. 1987;1:251–9.
31. Figueiredo EG, Zabramski JM, Deshmukh P, Crawford NR, Preul MC, Spetzler RF. Anatomical and quantitative description of the transcavernous approach to interpeduncular and prepontine cisterns. J Neurosurg. 2006;104:957–64.
32. Hsu FP, Clatterbuck RE, Spetzler RF. Orbitozygomatic approach to basilar apex aneurysms. Neurosurgery. 2005;56(Suppl 1):172–7.
33. Krisht AF, Kadri PA. Surgical clipping of complex basilar apex aneurysms: a strategy for successful outcome using the pretemporal transzygomatic transcavernous approach. Neurosurgery. 2005;56(Suppl 2):261–73.
34. McLaughlin N, Martin NA. Extended subtemporal transtentorial approach to the anterior incisural space and upper clival region: experience with posterior circulation aneurysms. Oper Neurosurg. 2014;10(1):15–24. https://doi.org/10.1227/NEU.0000000000000175.
35. MacDonald JD, Antonelli P, Day AL. The anterior subtemporal, medial transpetrosal approach to the upper basilar artery and punto-mesencephalic junction. Neurosurgery. 1998;43(1):84–9.
36. Rice BJ, Peerless SJ, Drake CG. Surgical treatment of unruptured aneurysms of the posterior circulation. J Neurosurg. 1990;73(2):165–73.
37. Ammirati M, Musumeci A, Bernardo A, Bricolo A. The microsurgical anatomy of the cisternal segment of the trochlear nerve, as seen through different neurosurgical operative windows. Acta Neurochir. 2002;144(12):1323–7.
38. Nagashima H, Kobayashi S, Tanaka Y, Hongo K. Endovascular therapy versus surgical clipping for basilar artery bifurcation aneurysm: retrospective analysis of 117 cases. J Clin Neurosci. 2004;11(5):475–9.

Surgical Anatomy of the Cavernous Sinus

Marc Sindou and Timothée Jacquesson

Abbreviations

ACP	Anterior clinoid process
APCF	Anterior petroclinoid fold
CN	Cranial nerves
CS	Cavernous sinus
FL	Falciform ligament
ICA	Internal carotid artery
ICL	Interclinoidal ligament
LS	Limbus sphenoidale
LWSph	Lesser wing of the sphenoid bone
MCF	Middle cranial fossa
PCP	Posterior clinoid process
PG	Pituitary gland
PPCF	Posterior petroclinoid fold
PPL	Petrolingual ligament
SOF	Superior orbital fissure
SSPL	Superior sphenopetrosal ligament

M. Sindou
University of Lyon 1, Lyon, France

T. Jacquesson (✉)
University of Lyon 1, Lyon, France

Department of Anatomy, Lyon, France

Skull Base Unit, Department of Neurosurgery B, Neurological Hospital Pierre Wertheimer, Hospices Civils de Lyon, Lyon, France
e-mail: timothee.jacquesson@neurochirurgie.fr

E. G. Figueiredo et al. (eds.), *Brain Anatomy and Neurosurgical Approaches*,
https://doi.org/10.1007/978-3-031-14820-0_26

1 Introduction

The specific space on each side of the sella, known since Winslow [1] as a "cavernous sinus" and further by neurosurgeons as a "no man's land," has long been and still is a matter of controversy. However, since the past mid-century, important names have brought insights for a better understanding of the true nature of this space, among them: Taptas [2], Bonnet [3], Parkinson [4], and Krivosic [5].

Whereas some defended the classical conception of an unbroken trabeculated canal, i.e., a sinus [6, 7], Taptas was the first to consider the so-called cavernous sinus an extradural—better said, an interperiosteodural—space, whose content is not venous sinus but plexus [8].

Because its comprehensive anatomy has surgical implications, namely, challenging the dogma that the cavernous sinus is a surgical no man's land, revisiting its development and architecture is the matter of this chapter.

2 Histo-Embryological Background

The cavernous sinus, which should be better named the parasellar lodge, is a crossroad that receives from one side the neural structures coming from the encephalon (cranial nerves III, IV, and VI, and the trigeminal ophthalmic branch) and the other side the internal carotid artery going to the brain. This interperiosteodural space contains venous plexuses that collect and redistribute the venous blood of the neighboring structures (Fig. 1a, b).

The architecture of this lodge is remarkable by its important variations: absence of similarity between individuals, even between sides in the same individual. Krivosic has carefully studied this on 600 specimens [5]. The architecture of this parasellar lodge is the result of continuous histo-architectural remodeling. Among the changing factors, various hydrodynamic pressures are likely to play an

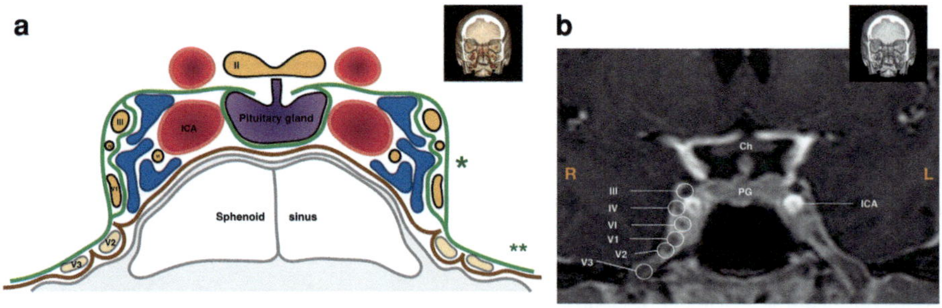

Fig. 1 Interperiosteodural parasellar lodge. Coronal schematic view (**a**) and cerebral MRI coronal slice (**b**). Sphenoid periosteum in brown, dura mater in green of the lateral wall of the cavernous sinus*, and dura mater of the temporal fossa**. The cavernous sinus is an interperiosteodural space between the sellar lodge medially and the temporal fossa dura laterally. It is crossed by cranial nerves (III, IV, V1, and VI) and the internal carotid artery

important role. The weakening of the internal carotid artery elastic layer leads to progressive dilatation of the artery and age. The cerebrospinal fluid pressure allows the extension of the arachnoid between the lodge and the venous cavities. Furthermore, even more important, veins—because of their walls limited by a simple endothelium more sensitive to pressure "à-coups" (a sudden increase of blood flow)—anastomose through fenestration mechanism. These confluences lead to the formation of cavities that give the aspect of a dural venous sinus in the elderly.

Inside the lodge, the veins, internal carotid artery, and intracavernous branches are surrounded by connective tissue from mesenchymal origin containing adipose cells, fibrocytes, and collagenous fibers. This connective tissue forms trabeculae and reticular structures. In places, it presents reinforcements as cords or rings, the latter particularly concerning the internal carotid artery. This has operative implications for the surgical exposure of the internal carotid artery.

The mesenchymal cells can form islands of arachnoid tissue inside the cavernous sinus; this may explain the possibility of genuine intracavernous meningiomas and tumor extension to the contralateral cavernous sinus [5].

Since nerves cross it with their sheaths and vessels, those structures can give specific tumors: schwannomas, neurofibromas, hemangiomas, among others. Also, the lodge can be the site of dysembryoplasias (epidermoid, dermoid cysts), lymphomas, and inflammatory pseudotumors (frequently extending to the retro-orbital region).

As the cavernous sinus is an extradural space, it may be invaded by tumors originating from neighboring structures. The most frequent ones are chondrosarcomas and chordomas, locoregional carcinomas and, metastases of all types. Basically, the cavernous sinus is often the site of invasive pituitary adenomas, more rarely mucoceles from the sphenoid sinus. All these pathologies justify that the surgical anatomy of the cavernous sinus surrounding structures should be known as well.

3 Surgical Anatomy of the Cavernous Sinus

3.1 The Parasellar Lodge

The denomination of cavernous sinus is not appropriate as this anatomical entity is not a venous dural sinus, and its structure is not trabecular, as is the cavernous type. The so-called cavernous sinus is an extradural space-delimited medially by the periosteum of the sphenoid bone and laterally by the dura of the middle cranial fossa attached superiorly to the sella turcica dura. This interperiosteodural space extends from the petrous apex posteriorly to anteriorly the orbit with which it directly communicates through the superior orbital fissure (Figs. 1a, b and 2a).

Superiorly, the cavernous sinus is covered by dura mater forming the oculomotor triangle with the entry point of the oculomotor nerve. Ligaments and dura folds hence limit the roof of the cavernous sinus: the interclinoidal ligament between the anterior and the posterior clinoid processes—medially—the anterior petroclinoid fold between the anterior clinoid process and the petrous apex—laterally—and the

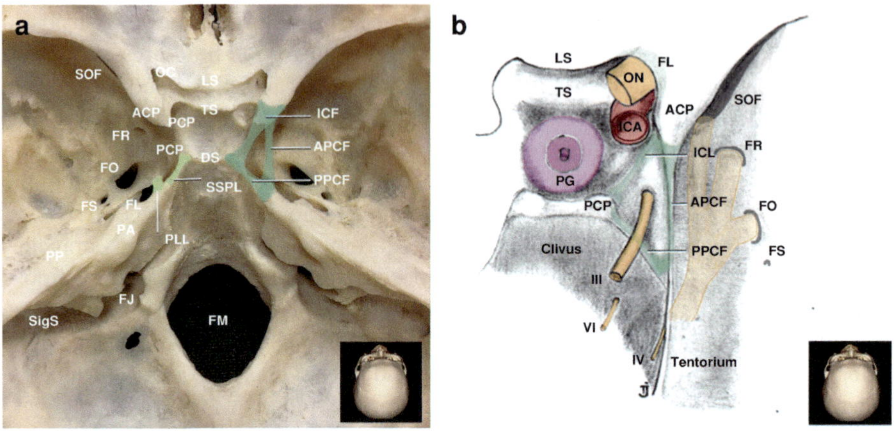

Fig. 2 Cavernous sinus area: superior view of bone, dura mater, and ligaments. Photography (**a**) and drawing (**b**) of skull base from a superior view. The roof of the cavernous sinus is limited by the anterior clinoid process (ACP) anteriorly, the posterior clinoid process (PCP), and the petrous apex (PA) posteriorly. Medially is the sellar lodge between the tuberculum sellae (TS) and the dorsum sellae (DS). The skull base foramina are successively: the foramen lacerum (FL), the foramen rotundum (FR), the foramen ovale (FO), the foramen spinosum (FS), and the superior orbital fissure (SOF). Anteriorly, the limbus sphenoidal (LS) draws the chiasmatic groove till the optic canals (OC). Posteriorly, the clivus dives to the foramen magnum (FM) and the sigmoid sinus (SigS) joins the foramen jugularis (FJ). The petrolingual ligament (PLL) arises from the PA to join the anterior margin of the carotid canal. The superior sphenopetrosal (Gruber's) ligament connects PA to PCP. The cavernous sinus roof is formed by: the interclinoidal ligament (ICL) between ACP and PCP, the anterior petroclinoidal fold (APCF) between ACP and PA, and the posterior petroclinoidal fold (PPCF) between PCP and AP. The abducens nerve enters the cavernous sinus through the Dorello's canal

posterior petroclinoid fold between the posterior clinoid process and the petrous apex—posteriorly. Also, a superior sphenopetrosal ligament strengthens the dura over the trigeminal root entering the cavernous sinus. These ligaments are important landmarks when accessing the cavernous sinus from above and lateral (Fig. 2b).

3.2 *The Venous Plexuses of the Parasellar Lodge*

The parasellar lodge is occupied by venous plexuses with proper endothelial walls, which is part of the venous network of the endocranial skull base.

The main **affluents** are the ophthalmic veins coming from the orbit through the superior orbital fissure. They are classically two: the inferior and the superior ophthalmic veins. Their obliteration can have harmful consequences on the ocular visual function; this is especially true when acute occlusion occurs, for instance, during anterior clinoidectomy.

The sphenoparietal sinus of Breschet is also an important affluent coming from the anterior and deep Sylvian territory and running along the lesser sphenoid wing. It can be of large caliber when draining a major part of the Sylvian venous

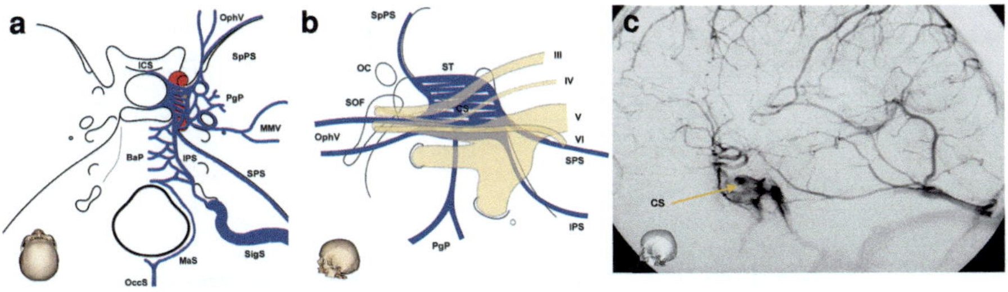

Fig. 3 Cavernous sinus venous supply and drainages: the venous plexus. Superior (**a**) and lateral (**b**) schematic views and lateral view of digital subtraction angiography (DSA) in the venous phase (**c**). These images depict the cavernous sinus receiving the ophthalmic veins and the sphenoparietal sinus of Breschet (draining the deep Sylvian veins) and draining into the superior and inferior petrosal sinuses that join the transverse sinus in the foramen jugularis. Another draining path could be to the pterygoid plexus through the skull base foramina

system—if so—its interruption on pterional approaches could lead to a hemorrhagic infarct of the cerebral hemisphere. This would also be the case if interrupting the Labbé vein on subtemporal approaches to the parasellar region.

There are also intracavernous affluents from the pituitary lodge with no clear systematization.

Veins or venous plexus can be opened during a surgical approach leading to high-flow bleeding. This has to be controlled using the classical microsurgical techniques: pledges, packing, and hemostatic agents.

Main **drainages** are toward the lateral sinus, upward via the superior petrosal sinus coursing along the upper edge of the petrous bone to the transverse-sigmoid sinus junction, and downward via the (generally small) inferior petrosal sinus that reaches the sigmoid sinus before entering the foramen jugularis. Accessory drainages are toward the pterygoid plexus, located below the skull base, via veins passing through the several foramina of the base. The cavernous sinus also drains toward the basilar system, which for some authors is a plexual system, and for others, a true dural sinus contained within the dura of the dorsum sellae and the clivus, and that anastomoses with the venous network of the posterior fossa and foramen magnum. Effraction of those can be difficult to control during surgery. The right and left venous plexus have rich anastomoses between both sides and the other components of the skull base venous system (Fig. 3a–c).

3.3 *The Motor Ocular Nerves*

The three motor ocular nerves (IIIrd, IVth, and VIth cranial nerves) cross the cavernous sinus from their cisternal intradural location to reach the orbit through the superior orbital fissure. The trigeminal nerve is adjacent to the cavernous sinus, running apart inside the trigeminal cave of Meckel (Figs. 4a and 5).

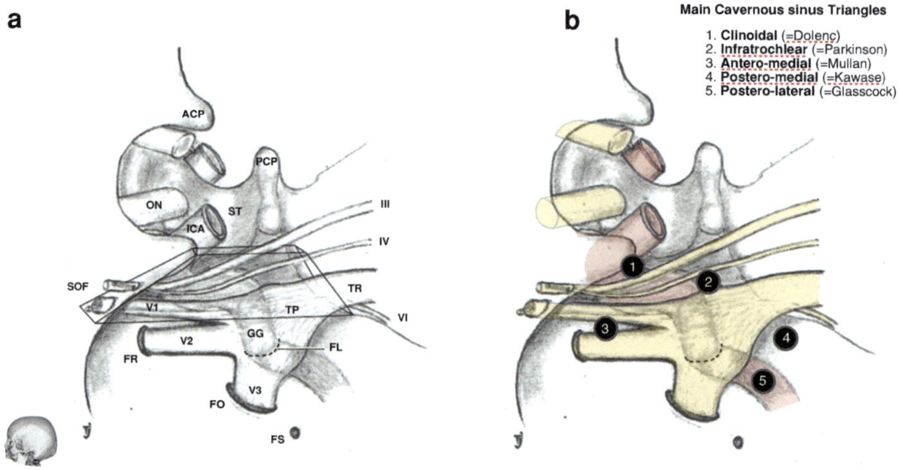

Fig. 4 Lateral wall of the cavernous sinus. Artistic drawing from the lateral view of cranial nerves (**a**), main triangles for surgical access (**b**). Cranial nerves on the lateral view, artistic drawing (**a**) and main surgical triangles (**b**): (1) superior through the roof of the cavernous lodge, along the III and IV (Dolenc); (2) Lateral superior between IV and V1 (Parkinson); (3) lateral inferior between V1 and V2 (Mullan); (4) posterior medial through the petrous apex using an anterior petrosectomy (Kawase); (5) posterior along the horizontal intrapetrous internal carotid artery segment (Glasscock)

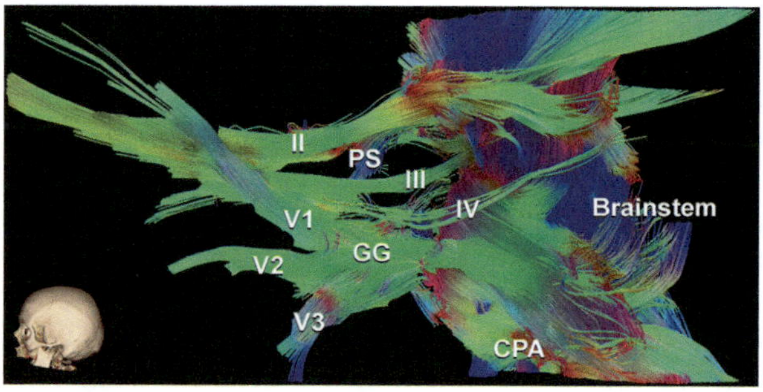

Fig. 5 MR diffusion tractography: lateral view of the cranial nerves of the cavernous sinus. The cavernous sinus is a transit region for cranial nerves III, IV, V, and VI. They emerge from the brainstem in the intradural space to reach the orbit through the superior orbital fissure. The trigeminal cave of Meckel is aside from the cavernous sinus and contains the V2 and V3 branches of the trigeminal nerve. The specific shape drawn by cranial nerves within the cavernous sinus is well-documented using three-dimensional tractography reconstruction

3.3.1 The Oculomotor Nerve

The oculomotor nerve (IIIrd cranial nerve) emerges from the interpeduncular fossa and penetrates the roof of the cavernous sinus after an approximately 20 mm long cisternal trajectory. Its entry point—forming a porus at the center of the oculomotor triangle—is at 8 ± 4 mm posterior to the anterior clinoid process and 7 ± 3 mm anterior to the posterior clinoid process [9]. The oculomotor porus is medial to the dura fold of the tentorial incisure attached to the anterior clinoid process [10, 11]. There are frequent arachnoid membranes adherent to the cisternal segment of the IIIrd nerve before its entry point into the cavernous sinus; consequently, the retraction of the temporal lobe may stretch and damage the oculomotor nerve if not freed before.

3.3.2 The Trochlear Nerve

The trochlear nerve (IVth cranial nerve) courses within the crural cistern, then along the tentorial incisura, stuck to it, medial and 2 mm below its free edge, so it is hidden when performing a pterional subtemporal approach of the incisura. Its penetration into the roof of the cavernous sinus is 10 mm posterior to the IIIrd nerve entry point, masked by the fold of the tentorial incisura [9, 11]. There, the nerve is vulnerable to manipulation of the tentorial incisura if the tentorium must be incised, as in the subtemporal transtentorial approach, to access the upper part of the cerebellopontine angle, the cut must be done as posteriorly as possible in order not to damage the trochlear nerve.

3.3.3 The Abducens Nerve

The abducens nerve (VIth cranial nerve) emerges from the anterior aspect of the brainstem. After a short cisternal trajectory, it penetrates the posterior wall of the cavernous sinus through an osteo-ligamentous canal named Dorello's canal, lateral to the clivus. The Dorello's canal is delimited by the petrous apex, the petroclinoidal ligament, and the dorsum sellae [9]. At this point, the nerve passes underneath the superior petrosal sinus. After a trajectory within the lodge lateral to the C4 internal carotid artery segment, the nerve joins the superior orbital fissure to enter the orbit. Blocked in the Dorello's canal, the abducens nerve is particularly sensitive to compression by growing tumor, as observed in chordomas arising from the adjacent petroclival suture. The abducens nerve is also early affected by the tumorous or vascular space-occupying lesions that develop within the cavernous sinus.

All the three motor ocular nerves penetrate independently into the cavernous sinus through an individual porus that corresponds to an invagination of the double sheet formed by dura and arachnoid layers which form a small cave inside the lodge. These two layers merge and become the epineurium at the level of the superior orbital fissure. This architecture is even more pronounced at the level of the trigeminal system, where the cave is a real cavity containing cerebrospinal fluid and corresponds to a cistern—the trigeminal cave of Meckel.

While the abducens nerve courses inside the parasellar lodge laterally to the internal carotid artery, surrounded by venous plexuses, the oculomotor and trochlear nerves are located in between the two layers of the lateral dural wall of the cavernous sinus, the surgical implication is that the IIIrd and IVth cranial nerves can be dissected free by incision of the superficial layer of the dural wall, from their porus to their penetration into the superior orbital fissure. Such an incision can be done without opening the lodge's venous compartment, which is of great importance for tumor resection and nerve decompression in this region.

3.4 The Trigeminal Cave and the Trigeminal Nerve

The trigeminal cave of Meckel is formed by the dura and the arachnoid of the posterior cranial fossa that goes anteriorly to enter the cavernous sinus. The trigeminal cave is different anatomical space from the cavernous sinus.

The trigeminal cave contains cerebrospinal fluid and forms the trigeminal cistern identifiable on a high-resolution T2 MRI. The cave harbors the trigeminal ganglion of Gasser—in its anterior part—and the triangular plexus of the trigeminal sensory root—in its posterior part. The trigeminal cistern surrounds both at their superior part [12].

After its superior medial course along the trigeminal sensory root in the cerebellopontine angle, known as the *pars minor*, the trigeminal motor root runs inferiorly to the triangular plexus and the trigeminal ganglion, to exit with the V3 mandibular sensory branch through the foramen ovale, to innervate the masticatory muscles.

The trigeminal ganglion has a semi-lunar shape and presents a convex anterior edge where the three peripheral branches terminate and a concave posterior margin from which the sensory root emerges. Between their emergence of the trigeminal ganglion and the upper petrous ridge, rootlets form a triangle, named the triangular plexus, because of its nervous anastomoses [12]. At the triangular plexus level, sensory fibers still have a somatotopic organization before merging into the posterior root. Fibers of V3 are inferior lateral, fibers of V1 are superior medial, fibers of V2 course in between [13].

The trigeminal ganglion and the triangular plexus are identifiable and dissectible inside the trigeminal cave after its opening either by extradural or intradural subtemporal approaches. This is particularly useful for surgery of trigeminal schwannomas or soft cavernous sinus meningiomas enclosed in the trigeminal cave.

Extensions of some petroclival meningiomas within the trigeminal cave can be extracted through the porus itself. When so performed, care has to be taken not to damage the superior petrosal vein (the Dandy's vein) and the superior petrosal sinus that crosses over the porus. Endoscopic assistance may help in that condition.

While V2 (maxillary) and V3 (mandibular) branches do not belong to the cavernous sinus region per se, the V1 (ophthalmic) branch penetrates in between the two layers of the lateral dural wall of the cavernous sinus to reach the superior orbital fissure. The triangular space delimited by the IIIrd and the IVth superiorly and the V1 inferiorly has been demonstrated by Parkinson as an appropriate window to access the vascular compartment of the parasellar lodge [4].

3.5 The Internal Carotid Artery and Its Intracavernous Segments

The internal carotid artery (ICA)—5 mm of average in diameter—emerges from the horizontal portion of the petrous canal (**C6 segment** of the classical classification) at the foramen lacerum, where a fibrous ring fixes it without piercing any floor [14]. Then it goes upward and enters the parasellar lodge (Fig. 6a–c).

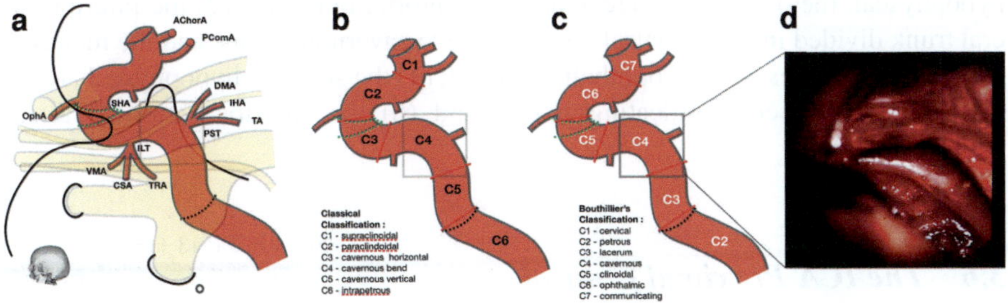

Fig. 6 ICA segmentation. Lateral schematic views of ICA segments (**a**). ICA classical classification (**b**) and ICA Bouthillier classification (**c**). Microsurgical view of the left internal carotid artery C4 segment through the Parkinson's triangle, i.e., between the IVth cranial nerve superiorly and the V1 ophthalmic branch inferiorly (**d**)

In the parasellar lodge, the **C5 ascending segment** is in contact medially with the periosteum of the sphenoid bone, where it draws a groove. Laterally, the internal carotid artery is adjacent to the medial aspect of the trigeminal ganglion, separated from it by a thin bone layer or, more frequently, a simple dural membrane. There, the internal carotid artery may be in danger during any surgery performed within the trigeminal cave.

Then, the internal carotid artery makes a right angle coursing anteriorly. This **C4 segment** is in contact with the dura of the sella and may bulge medially into the pituitary fossa; therefore, it could be in danger during pituitary intrasellar surgery. This segment—of about 20 mm in length—can be accessed through the triangle described by Parkinson, between the IVth and the IIIrd cranial nerve superiorly and the V1 branch inferiorly [4]. The C4 portion is in close contact with the intracavernous trajectory of the VIth cranial nerve that courses just lateral to it. The C4 segment is almost surrounded by the venous plexuses, which fulfill the true vascular compartment of the cavernous sinus.

Then the internal carotid artery curves upward to reach the medial aspect of the anterior clinoid process. This **C3 vertical segment** pierces the more anterior part of the roof of the cavernous sinus through a strong fibrous ridge. This dural ring can be cut after anterior clinoidectomy to access this C3 segment; Dolenc has well described this to perform the direct clipping of intracavernous arterial aneurysms [15].

The **C2 segment** is above the dural ring of the cavernous sinus roof, and this segment gives the ophthalmic artery that courses in the optic canal along the inferior aspect of the optic nerve. This C2 paraclinoid segment can be completely exposed after an anterior clinoidectomy, as the anterior clinoid process masks over 10 mm of length the superior lateral aspect of this C2 segment.

The **C1 segment** is superior to the anterior clinoid process, and it corresponds to the supraclinoidal segment of the internal carotid artery where the posterior communicating and the anterior choroidal arteries arise.

The **main branches** of the intracavernous portion of the internal carotid artery are proximal to distal: (1) the superior posterior trunk divided into the inferior hypophyseal, the dorsal meningeal, and the tentorial branches; (2) the inferior lateral trunk divided into the ventral meningeal, the cavernous sinus, and the recurrent trigeminal branches; (3) the ophthalmic artery; (4) the superior hypophyseal artery; (5) the posterior communicating artery; and (6) the anterior choroidal artery (Fig. 6a).

3.6 The ICA Proximal Control

The proximal control of the intracavernous **C5** portion of the internal carotid artery might be wise for temporary clipping in the exceptional situation of accidental downstream vascular damage (fissuration). This approach could also be useful in cases of highly vascularized tumors, as many meningiomas, to suppress their arterial supply.

The extradural subtemporal approach includes coagulation of the middle meningeal artery at the foramen spinosum and external carotid artery branches that cross the skull base foramina—the superior orbital fissure, foramen rotundum, foramen ovale—along with the cranial nerves. The proximal control of the intracavernous segment of the internal carotid artery can be obtained by reaching the posterior margin of the foramen lacerum at the posterior edge of the V3 dural sheath before it enters into the foramen ovale (Fig. 7a). Then, the internal carotid artery is unroofed in its intrapetrous carotid canal—anteriorly to posteriorly—following the landmarks described by Glasscock [16] (Fig. 7b). Care has to be taken not to injure the Eustachian tube, the tensor tympani muscle, and above all, the cochlea posteriorly (Fig. 7c and d).To expose the Eustachian canal is not necessary to access the internal carotid artery; a Kerrison rongeur is less aggressive than extensive drilling, as observed in our 150-case experience (Fig. 7e).

Fig. 7 Proximal control of ICA. Operative microsurgical views of the steps of the proximal control of the internal carotid artery (ICA) with their orientation CT-scan axial slices. Extradural exposure of the floor of the right middle cranial fossa from the superior orbital fissure (SOF) to the foramen spinosum (FS), using a pterional subtemporal craniotomy (**TOP**). Then from top to bottom: steps of proximal control of ICA. After coagulation and cutting of the middle meningeal artery at the FS, the dura is elevated to expose the foramen ovale (FO) and the great superficial petrosal nerves (GSPN) arising from the *fallopian hiatus(***)*. The axis of both the hiatus and GSPN gives the trajectory of the internal carotid artery. The V3 dural sheath is strongly retracted anteriorly before the exit to FO (arrow on CT-scan) to expose the internal carotid artery (ICA) at the posterior edge of the foramen lacerum (FL) (**a**). The carotid canal is carefully unroofed from FL, anterior to posterior, using a drill or a Kerrison rongeur (**b**). Having exposed the entire horizontal segment of the ICA, lateral drilling skeletonizes the Eustachian tube (ET), and a thin bony wall between this latter and the internal carotid artery is identified (**c**). Also, the tensor tympani (TT) muscle is shown with its insertion lodge (**d**). These previous steps allow for proximal control of the C6 internal carotid artery segment by temporary clipping if required (**e**)

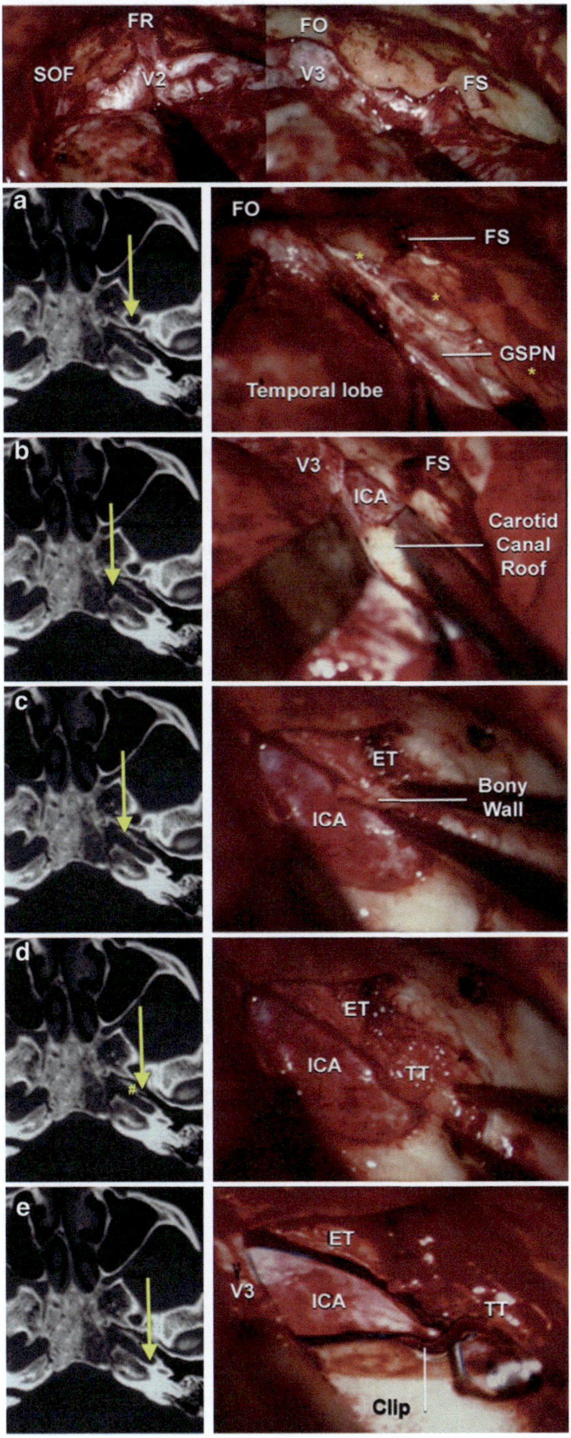
FR
SOF
V2
FO
V3
FS
a
FO
FS
GSPN
Temporal lobe
b
V3
FS
ICA
Carotid Canal Roof
c
ET
Bony Wall
ICA
d
ET
ICA
TT
#
e
ET
V3
ICA
TT
Clip

3.7 *The ICA Distal Control*

Access to the distal, i.e., paraclinoid **C2** portion of the internal carotid artery, can be useful to control the artery and expose the optic nerve in its canal. This requires anterior clinoidectomy, preferably performed extradurally according to our experience.

In cases with vascularized tumors—especially meningiomas of the cavernous sinus, the anterior clinoid process, or the inner third of the sphenoid wing—anterior clinoidectomy is an efficient way to reduce the arterial supply from the C2/C3 carotid segments and the ophthalmic artery, and so make the surgical resection easier.

The procedure starts by opening the superior edge of the superior orbital fissure, then unroofing the optic canal, and breaking the inferior pilar of the clinoid process from the sphenoid bone—the optic strut. According to our experience, a tiny Kerrison rongeur is preferred to skeletonize the dural sheath of the optic nerve rather than a drill. The anterior clinoid process can be extracted in one piece since drilling risks damaging the optic nerve function. The removal of the anterior clinoid process is often associated with venous bleeding from the ophthalmic veins at the anterior pole of the cavernous sinus; this can easily be controlled by packing a small pledge of absorbable tissue like Surgicel (Fig. 8a–c).

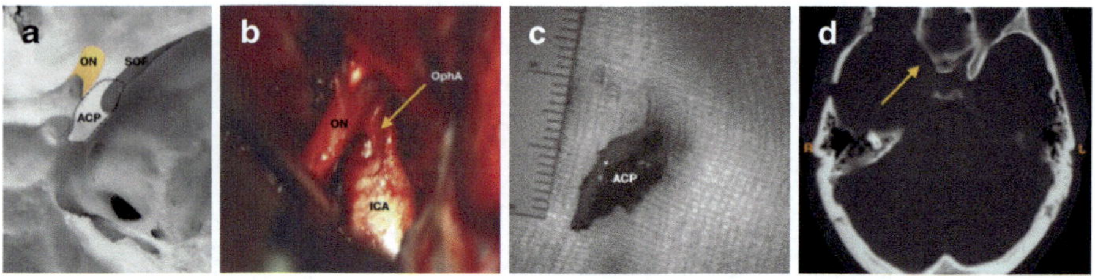

Fig. 8 Distal control of right ICA after anterior clinoidectomy. Right anterior clinoidectomy: skull base superior view (**a**); operative microsurgical view of the optic nerve (ON) and the paraclinoid portion of the internal carotid artery (ICA) (**b**); photography (**c**); and CT-scan axial slice (**d**). The anterior clinoidectomy involves the opening of the supraorbital fissure (SOF), the unroofing of the optic canal (OC) to skeletonize the ON, and the resection of the anterior clinoid process (ACP) by drilling or extraction (**a**). This approach exposes ICA and the ophthalmic artery (OphA) that reaches the inferior aspect of ON (**b**). An "en bloc" extraction of ACP avoids extensive drilling that would jeopardize ON and ICA (**c**). The postoperative control imaging displays the bone resection that includes ACP, lateral wall of the OC, and medial part of the SOF (CT-scan axial slice, **d**)

4 Surgical Anatomy of the Cavernous Sinus Approaches

4.1 Endocranial Microsurgical Approaches

The supratentorial access to the cavernous sinus needs a frontal pterional temporal craniotomy. Additional osteotomy—of the orbital rim, or the zygomatic bone, or the whole orbito-zygomatic arch depending on the chosen "working cones" and related corridors—may be performed. This can be particularly useful if enlarged access is required because of superior tumor extension to the base of the brain. Such osteotomies increase the surgical exposure up to 70%, avoiding excessive brain retraction or large opening of the Sylvian fissure [17]. A semicircular dural incision is centered at the pterion, and the dural flap is reflected anteriorly. On demand, a subtemporal dural incision can be done, not too extensively, to prevent damage to the inferior anastomotic vein of Labbé.

According to the concept of "cone of approach," large access to the cisterns offers various trajectories to attack the tumor, according to the concept of "cone of approach" [18]. Such an approach exposes the optic nerve and the chiasma anteriorly; the internal carotid, the posterior communicating, and the anterior choroidal arteries medially; and the oculomotor and trochlear nerves posteriorly. Then, the superior and lateral walls of the cavernous sinus are accessed through the suprasellar basal cisterns and their corridors, then through the various triangles of the cavernous sinus walls (Fig. 4b).

Several triangles have been described between the anatomical structures within the cavernous sinus (Fig. 4b). We choose here those with a true surgical role, from superior to inferior: (1) the clinoidal triangle above the CN III described by Dolenc [19] and Hakuba [20]; (2) the infratrochlear triangle between CN IV and V1 described by Parkinson [4] (Fig. 6d); (3) the anterior medial triangle between CN V2 and V3 described by Mullan [21]; (4) the posterior lateral triangle along the internal carotid artery horizontal segment described by Glasscock [16, 22]; and the posterior medial triangle through the petrous apex described by Kawase [23] and Velut [24]. Sekhar reported an inferior lateral approach to the cavernous sinus using a pre-auricular infratemporal approach [25].

In cases with tumor extension into the trigeminal cave, its roof is opened to free the trigeminal ganglion and the triangular plexus. If necessary, after identifying the trochlear nerve course posteriorly, the tentorium can be cut to enlarge the exposure to reach the cerebellopontine angle through a transtentorial route. Thus, the trigeminal root can be freed, and the petroclival region visualized down to the Dorello's canal and the basilar arterial system.

In some circumstances, as tumor extension to the petroclival region, the surgical approach can be retrosigmoid and presigmoid-retrolabyrinthine. A craniotomy is hence suboccipital retromastoid; it can be enlarged by tailored bone flap extension above the transverse sinus. To add a presigmoid-retrolabyrinthine approach, the sigmoid sinus is skeletonized with cautious drilling of the mastoid and the superior petrosal sinus, the antrum, and the labyrinth bloc. Care should be taken to respect the fallopian canal of the facial nerve, the endolymphatic duct, and the posterior semicircular canal. Drilling can be conducted anteriorly to the internal auditory meatus and superiorly to the petrous ridge to reach the trigeminal cave of Meckel [26, 27]. During this extradural step, the sigmoid sinus should not be retracted too much not to create avulsion at the foramen jugularis. Then the opening of the dura is in a retro- and presigmoid location to access the petroclival region [28].

4.2 Expanded Endonasal Transsphenoidal Approach

An endoscopic transsphenoidal route provides another anterior door to the cavernous sinus [29–31]. After a middle turbinectomy, a large sphenoidotomy allows identifying intrasphenoidal anatomical landmarks: sella turcica, optic canal, lateral and medial opticocarotid recesses, clival recess, paraclival and paraclinoid segments of the internal carotid artery. The thin bone of the lateral recess of the sphenoid sinus can be carefully drilled to skeletonize the internal carotid artery. Laterally, the medial wall of the cavernous sinus can be exposed together with cranial nerves IV and V1. The intracavernous portion of the IIIrd cranial nerve is covered by the anterior bend of the internal carotid artery. Entering the cavernous sinus involves venous plexus bleeding, and this approach is used when cavernous sinus tumors open the route. For making such an anterior approach, knowledge of its specific surgical anatomy and instrumentation is mandatory. From an anterior perspective, four compartments of the cavernous have been described: superior, posterior, inferior, and lateral [32, 33] (Fig. 9a–c).

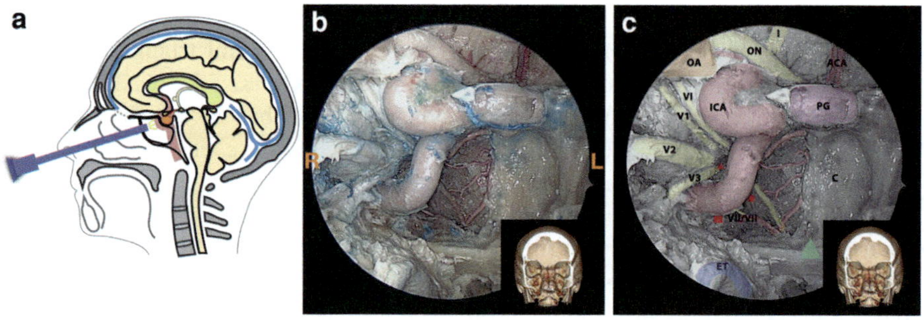

Fig. 9 Endoscopic endonasal approach to the Cavernous Sinus. Schematic drawing (**a**), anterior endoscopic views without (**b**), and with legends (**c**). After an expanded endoscopic endonasal approach through the two nostrils using a 0° lens endoscope, the cavernous sinus is fully opened, and its content is exposed, including the internal carotid artery and cranial nerves (CN) III to VI. Note that only V1 belongs to the cavernous sinus while V2 and V3 are part of the trigeminal cave of Meckel. CN III, IV, V1 travel within the lateral dural wall of the cavernous sinus until the superior orbital fissure. CN VI courses within the interdural Dorello's canal at the middle clivus then cross the cavernous sinus anterior and close to the vertical paraclival segment of the internal carotid artery

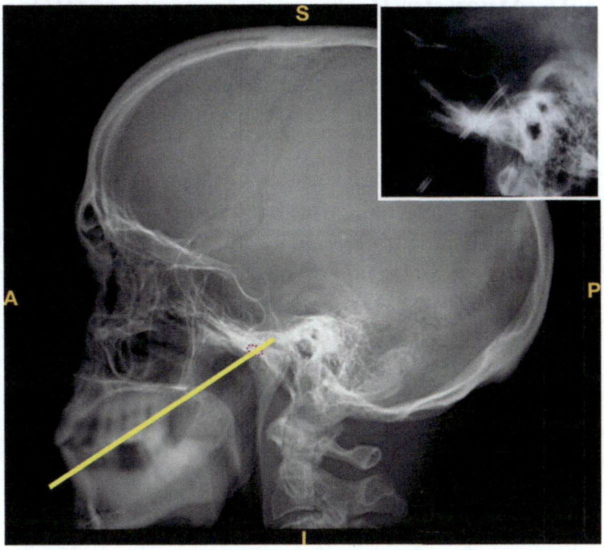

Fig. 10 Percutaneous Cavernous Sinus access using Hartel's approach through the *foramen ovale*. A percutaneous trajectory through the *foramen ovale* has been described by Hartel, and it allows to perform biopsies or a thermic lesioning of the trigeminal nerve in cases with trigeminal neuralgias

4.3 Percutaneous Access to the Cavernous Sinus

When imaging is insufficient to ascertain the histopathological nature of cavernous sinus tumors, it may be useful to get a tumor sample to adjust the strategy. The biopsy can be performed by a percutaneous approach, using the Hartel trajectory, under fluoroscopy, or CT guidance [34, 35] (Fig. 10). With local anesthesia and short neuroleptanalgesia, a dedicated needle crosses the pterygomaxillary fossa and the foramen ovale (Fig. 10). For doing so, the surgical anatomy of the Hartel and cavernous sinus approaches should be known. Tumors located inside the trigeminal cave of Meckel, in the posterior part of the cavernous sinus, and at the petroclival region are particularly amenable to this procedure. Soft tumors, lymphoma, or inflammatory pseudotumors are more propitious to biopsy than hard tissular neoplasms. The method offers chances to decide the most appropriate treatment to be applied between removal surgery (meningiomas, schwannomas), radiosurgery (meningiomas enclosed within the parasellar lodge), chemo-radiotherapy (malignant tumors, metastases), or steroid medications (inflammatory pseudotumors).

5 Anatomical Imaging of the Cavernous Sinus

Surgical strategy begins with a thorough imaging assessment, particularly for tumor management.

5.1 Magnetic Resonance Imaging (MRI)

MRI is crucial for defining the tumor features and their growth consequences on the skull base and the brain, thus designing the best surgical approach with fewer operative risks.

High-resolution MRI sequences visualize the tumor anatomical environment, including the internal carotid artery and its branches, the optic nerve, the motor ocular nerves, the trigeminal system, and the pituitary gland. It shows the presence or absence of a cleavage plane from the brain. Additional angiographic sequence informs of any displacement, compression, or invasion of the internal carotid artery. Images in the venous phase depict the main venous drainage of the brain to the skull base [36].

5.2 Computed Tomography (CT)

The CT-scan with bone window is a useful complement to MRI, especially to delineate the lesion relationships with the neighboring bone structures: the sella turcica, anterior and posterior clinoid processes, sphenoid wing, middle cranial fossa floor, orbital apex, petrous apex, petrous pyramid. Furthermore, the CT-scan can add important data about tumor extension to the sphenoid sinus, subtemporal fossa, petrous pyramid, orbit, skull base foramina—superior orbital fissure, *foramen rotundum, foramen ovale*, among others. The morphology of the intrapetrous carotid canal, Eustachian tube, otologic intrapetrous structures can also be precised to prepare the surgical approaches. Variations of sphenoid sinus septations, pneumatization of the anterior clinoid process, hypertrophy, or lysis of the bone around the cavernous sinus should be checked, as they could influence the surgical approach.

5.3 Digital Subtraction Angiography (DSA)

Selective four arteries angiography via the femoral artery can reveal the relationships between cavernous sinus tumors and the internal carotid artery, its segments, and branches. Angiography remains the best imaging to precise the arterial feeders of skull base meningiomas. Studying the external and internal carotid artery independently is of great interest since these feeders may arise from the various segments of the internal carotid artery or maybe branches of the external carotid artery. The caliber of each feeder can be assessed and leads at the same time to embolization. Moreover, angiography explores the suppliance capacity of the Willis circle arteries by the anterior and posterior communicating arteries, using neck compression or even balloon occlusion of the carotid arteries. The late venous phase is also important to identify the main draining veins—the anterior Sylvian vein and the inferior anastomotic vein of Labbé—whose sacrifice might cause a cerebral venous infarction [36].

Detailed imaging is crucial to provide surgeons maximal anatomical information that can be converted into practical decisions for optimal surgical strategy. This is especially true when dealing with the extracavernous extensions of cavernous sinus meningiomas for designing the basal cistern corridors to access the cavernous sinus walls (Fig. 11a and b).

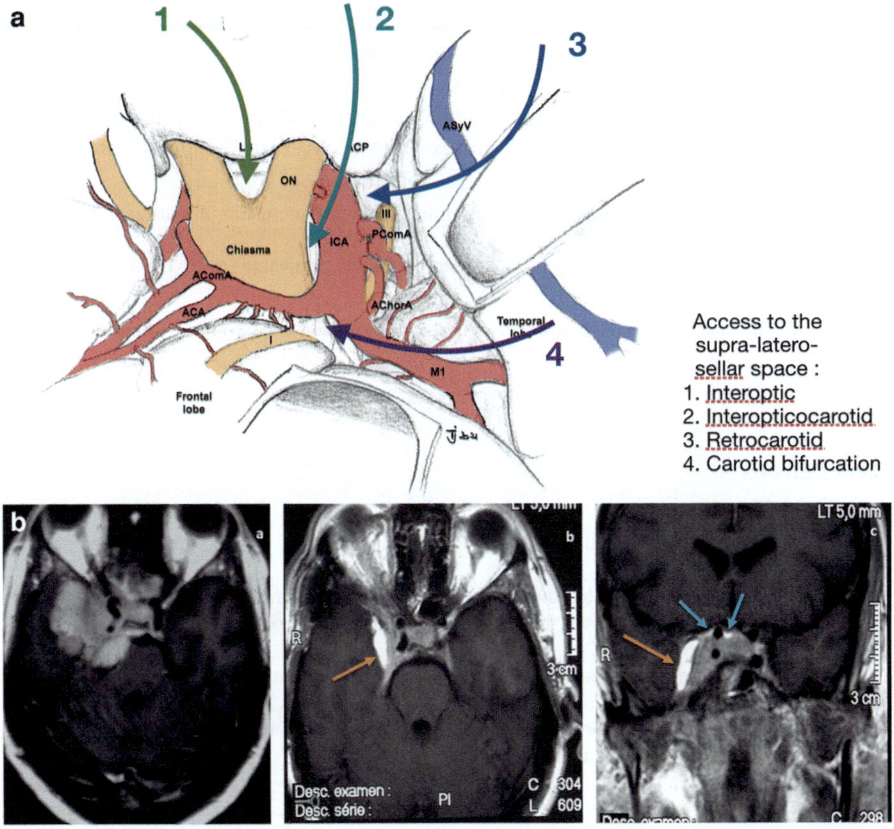

Fig. 11 Artistic drawing of a microsurgical view of the supra- and latero-sellar approaches of the cavernous sinus (**a**) and imaging of a cavernous sinus meningioma with superior and lateral extensions (**b**). On drawing (**a**), the various corridors to the cavernous sinus through the basal cisterns are the following: interoptic (1), interopticocarotid (2), retrocarotid (3), carotid bifurcation (4). Through a pterional craniotomy, the greater and lesser wings of the sphenoid bone are followed to reach the anterior clinoid process and the surrounding structures: the optic nerves; the oculomotor nerve; the internal carotid artery and its branches: the posterior communicating, the anterior choroidal, the middle cerebral, the anterior cerebral, and the perforating arteries. All of them are exposed after a careful opening of the arachnoid membranes. Access to the superior and lateral sellar regions can be obtained through those various corridors. On the MRI imaging (**b**), all the extra cavernous extensions have been surgically removed; note the remnant within the cavernous sinus and the pieces of fat tissue that have been interposed between the residue and the basal structures (arrows)

6 Conclusion

Surgical anatomy and its counterpart imaging are determinants to provide surgeons with all useful information that can be converted into practical decisions to set up the optimal treatment strategy. In this chapter, the authors aimed to critically revise the anatomy of the cavernous sinus, providing a practical insight that would help any surgery across this region. The cavernous sinus cannot be therefore considered as a no-man's land surgical region. However, the cavernous sinus contains vulnerable neurovascular structures jeopardized by surgery; in-depth knowledge of its anatomy surely helps decide reasonable surgery and perform an appropriate approach according to the specific features of each cavernous sinus disease.

Disclosure of Funding No personal or institutional financial interest in any drugs, materials, or devices.

We hereby confirm that this work has not been published elsewhere in any form.

Conflict of Interest We hereby confirm that the authors have no conflict of interest in this manuscript.

References

1. Winsløw JB. (médecin) A du texte. Exposition anatomique de la structure du corps humain, vol. 2. London; 1734.
2. Taptas JN. The parasellar osteo-dural chamber and the vascular and neural elements that traverse it. An anatomical concept that would replace the cavernous sinus of classical anatomy. Neurochirurgie. 1990;36(4):201–8.
3. Bonnet P. Cavernous sinus region syndromes. Arch Ophtalmol Rev Gen Ophtalmol. 1955;15(4):357–72.
4. Parkinson D. A surgical approach to the cavernous portion of the carotid artery. Anatomical studies and case report. J Neurosurg. 1965;23(5):474–83.
5. Krivosic I. Histoarchitecture of the cavernous sinus. In: Dolenc VV, editor. The cavernous sinus: a multidisciplinary approach to vascular and tumorous lesions [Internet]. Vienna: Springer; 1987. [cited 2021 Aug 29]. p. 117–29. https://doi.org/10.1007/978-3-7091-6982-7_8.
6. Bedford MA. The "cavernous" sinus. Br J Ophthalmol. 1966;50(1):41–6.
7. Harris FS, Rhoton AL. Anatomy of the cavernous sinus. A microsurgical study. J Neurosurg. 1976;45(2):169–80.
8. Taptas JN. The so-called cavernous sinus: a review of the controversy and its implications for neurosurgeons. Neurosurgery. 1982;11(5):712–7.
9. Lang J. Topographical anatomy of the cranial nerves. In: Samii M, Jannetta PJ, editors. The cranial nerves: anatomy pathology pathophysiology diagnosis treatment [Internet]. Berlin, Heidelberg: Springer; 1981. [cited 2021 Aug 30]. p. 6–15. https://doi.org/10.1007/978-3-642-67980-3_2.
10. Guyotat J. L'incisure tentorielle. "Les tumeurs de la loge caverneuse / Cavenous sinus tumors". Sindou, Neurochirurgie, Masson, Paris. 1995;41(3):163–5.
11. Guyotat J, Bret P, Rémond J, Fischer G. Meningioma of the apex of the tentorial incisura. Diagnostic and surgical aspects apropos of a series of 7 cases. Neurochirurgie. 1991;37(1):12–7.

12. Bernard F, Mercier P, Sindou M. Morphological and functional anatomy of the trigeminal triangular plexus as an anatomical entity: a systematic review. Surg Radiol Anat SRA. 2019;41(6):625–37.
13. Sindou M, Brinzeu A. Topography of the pain in classical trigeminal neuralgia: insights into somatotopic organization. Brain J Neurol. 2020;143(2):531–40.
14. Taptas J. Must we still call cavernous sinus the parasellar vascular and nervous crossroads? The necessity of a definite topographical description of the region. 1987.
15. Dolenc V. Direct microsurgical repair of intracavernous vascular lesions. J Neurosurg. 1983;58(6):824–31.
16. Glasscock ME, Miller GW, Drake FD, Kanok MM. Surgery of the skull base. Laryngoscope. 1978;88(6):905–23.
17. Alaywan M, Sindou M. Fronto-temporal approach with orbito-zygomatic removal. Surg Anatomy Acta Neurochir (Wien). 1990;104(3–4):79–83.
18. Sindou MP. Working area and angle of attack in three cranial base approaches: pterional, orbitozygomatic, and maxillary extension of the orbitozygomatic approach. Neurosurgery. 2002;51(6). http://journals.lww.com/neurosurgery/Fulltext/2002/12000/Working_Area_and_Angle_of_Attack_in_Three_Cranial.28.aspx
19. Dolenc VV. The Cavernous Sinus - A multidisciplinary approach to vascular and tumorous lesions | V.V. Dolenc | Springer [Internet]. [cited 2021 Aug 23]. https://www.springer.com/gp/book/9783709174609
20. Hakuba A, Nishimura S, Shirakata S, Tsukamoto M. Surgical approaches to the cavernous sinus. Neurol Med Chir (Tokyo). 1982;22(4):295–308.
21. Mullan S. Treatment of carotid-cavernous fistulas by cavernous sinus occlusion. J Neurosurg. 1979;50(2):131–44.
22. Glasscock ME. A surgical approach to the cavernous portion of the carotid artery. Anatomical studies and case report - PubMed [Internet]. [cited 2021 Aug 23]. https://pubmed.ncbi.nlm.nih.gov/5858438/
23. Kawase T, Toya S, Shiobara R, Mine T. Transpetrosal approach for aneurysms of the lower basilar artery. J Neurosurg. 1985;63(6):857–61.
24. Velut S, Jan M. Petrectomy of the point during approach to the clivus: technic, values and limitations. Apropos of a case of meningioma. Neurochirurgie. 1988:17–25.
25. Sekhar LN, Sen CN, Jho HD, Janecka IP. Surgical treatment of intracavernous neoplasms: a four-year experience. Neurosurgery. 1989;24(1):18–30.
26. Jacquesson T, Berhouma M, Tringali S, Simon E, Jouanneau E. Which routes for petroclival tumors? A comparison between the anterior expanded endoscopic endonasal approach and lateral or posterior routes. World Neurosurg. 2015;83(6):929–36.
27. Jacquesson T, Simon E, Berhouma M, Jouanneau E. Anatomic comparison of anterior petrosectomy versus the expanded endoscopic endonasal approach: interest in petroclival tumors surgery. Surg Radiol Anat SRA. 2015;37(10):1199–207.
28. Samii M, Tatagiba M. Experience with 36 surgical cases of petroclival meningiomas. Acta Neurochir. 1992;118(1–2):27–32.
29. Zada G, Agarwalla PK, Mukundan S, Dunn I, Golby AJ, Laws ER. The neurosurgical anatomy of the sphenoid sinus and sellar floor in endoscopic transsphenoidal surgery. J Neurosurg. 2011;114(5):1319–30.
30. Kassam AB, Prevedello DM, Carrau RL, Snyderman CH, Gardner P, Osawa S, et al. The front door to meckel's cave: an anteromedial corridor via expanded endoscopic endonasal approach-technical considerations and clinical series. Neurosurgery. 2009;64(3 Suppl):ons71–82; discussion ons82–83.
31. Kassam AB, Gardner P, Snyderman C, Mintz A, Carrau R. Expanded endonasal approach: fully endoscopic, completely transnasal approach to the middle third of the clivus, petrous bone, middle cranial fossa, and infratemporal fossa. Neurosurg Focus. 2005;19(1):E6.

32. Truong HQ, Lieber S, Najera E, Alves-Belo JT, Gardner PA, Fernandez-Miranda JC. The medial wall of the cavernous sinus. Part 1: surgical anatomy, ligaments, and surgical technique for its mobilization and/or resection. J Neurosurg. 2018;131(1):122–30.
33. Fernandez-Miranda JC, Zwagerman NT, Abhinav K, Lieber S, Wang EW, Snyderman CH, et al. Cavernous sinus compartments from the endoscopic endonasal approach: anatomical considerations and surgical relevance to adenoma surgery. J Neurosurg. 2018;129(2):430–41.
34. Sindou M, Messerer M, Alvernia J, Saint-Pierre G. Percutaneous biopsy through the foramen ovale for parasellar lesions: surgical anatomy, method, and indications. Adv Tech Stand Neurosurg. 2012;38:57–73.
35. Sindou M, Chavez JM, Saint Pierre G, Jouvet A. Percutaneous biopsy of cavernous sinus tumors through the foramen ovale. Neurosurgery. 1997;40(1):106–10; discussion 110–111.
36. Sindou M, Dumot C. Planning of endocranial supratentorial basal cistern and skull base approaches depending on venous patterns using a topogram. World Neurosurg. 2020;134:365–71.

Surgical Anatomy of the Petroclival Region

Kaan Yağmurlu, Hasan Barış Ilgaz, and Feres Chaddad-Neto

The petroclival region is anatomically divided along the petroclival line into a superior space, which is related with the midbrain, the contents of the interpeduncular cistern, sellar and parasellar region; a middle space that is related with pons and prepontine and cerebellopontine angle; and inferior space that is related to the medulla and structures around the foramen magnum [1] (Figs. 1a–c and 2a–c). The superior petroclival region also corresponds to the anterior part of the tentorial incisura, which divides the superior space into supra- and infratentorial compartments. It extends anteriorly and laterally to the sellar and parasellar regions. Its inferior limit is the origin of the trigeminal nerve. It contains the intradural parts of the oculomotor and trochlear nerves, the basilar artery, posterior cerebral and superior cerebellar arteries, cavernous segment of the ICA. The superior limit of the middle petroclival space is at the pontomesencephalic sulcus, and the inferior limit is at the pontomedullary sulcus. Its lateral limits are the level of the cerebellopontine angle, including the CN V. Clinically, the true petroclival region corresponds to the superior and middle spaces between the level of the CN V laterally since the tumors in the inferior space are specifically called the foramen magnum lesion or jugular foramen lesion [2].

K. Yağmurlu (✉)
Department Neurosurgery, University of Tennessee Health Science Center, Memphis, TN, USA
e-mail: kyagmurl@uthsc.edu

H. B. Ilgaz
Department of Neurological Surgery, University of Virginia School of Medicine, Charlottesville, VA, USA

F. Chaddad-Neto
Department of Neurology and Neurosurgery, Universidade Federal de São Paulo, São Paulo, SP, Brazil

E. G. Figueiredo et al. (eds.), *Brain Anatomy and Neurosurgical Approaches*, https://doi.org/10.1007/978-3-031-14820-0_27

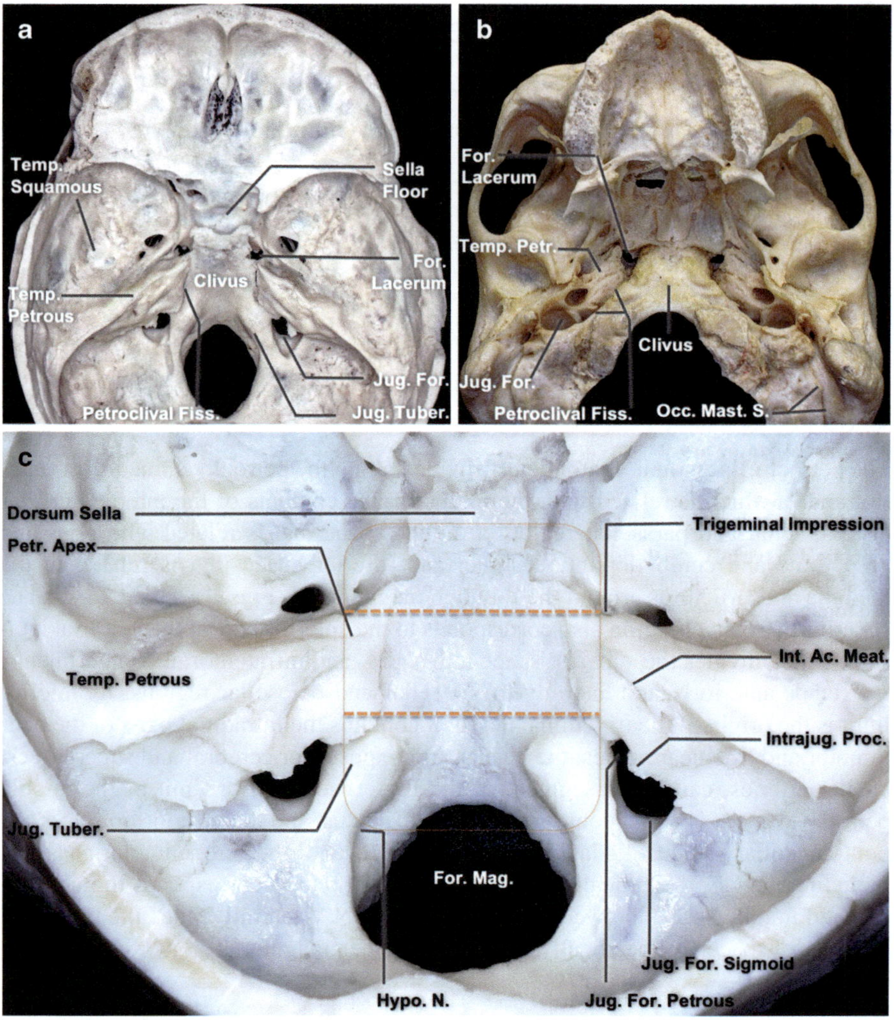

Fig. 1 Bony anatomy of the petroclival region. (**a**), the endocranial surface of the skull base. The petroclival fissure runs between the petrous part of the temporal bone and the clivus of the occipital bone. (**b**), the extracranial surface of the skull base. The relationship of the petroclival fissure with foramen lacerum and jugular foramen was exposed. (**c**), the true petroclival region formed by superior and middle petroclival spaces extends from the top of the petroclival fissure to the upper edge of the jugular tubercle. Its lateral border is the level of the medial edge of the trigeminal impression on both sides, including clivus and petrous apex. Abbreviations: *Ac* acoustic, *Fiss* fissure, *For* foramen, *Hypo* hypoglossal, *Int* internal, *Intra* Intrajug, intrajugular, *Jug* jugular, *Mag* magnum, *Mast* mastoid, *N* nerve, *Occ* occipital, *Petr* petrous, *Proc* process, *S* suture, *Temp* temporal, *Tuber* tubercle

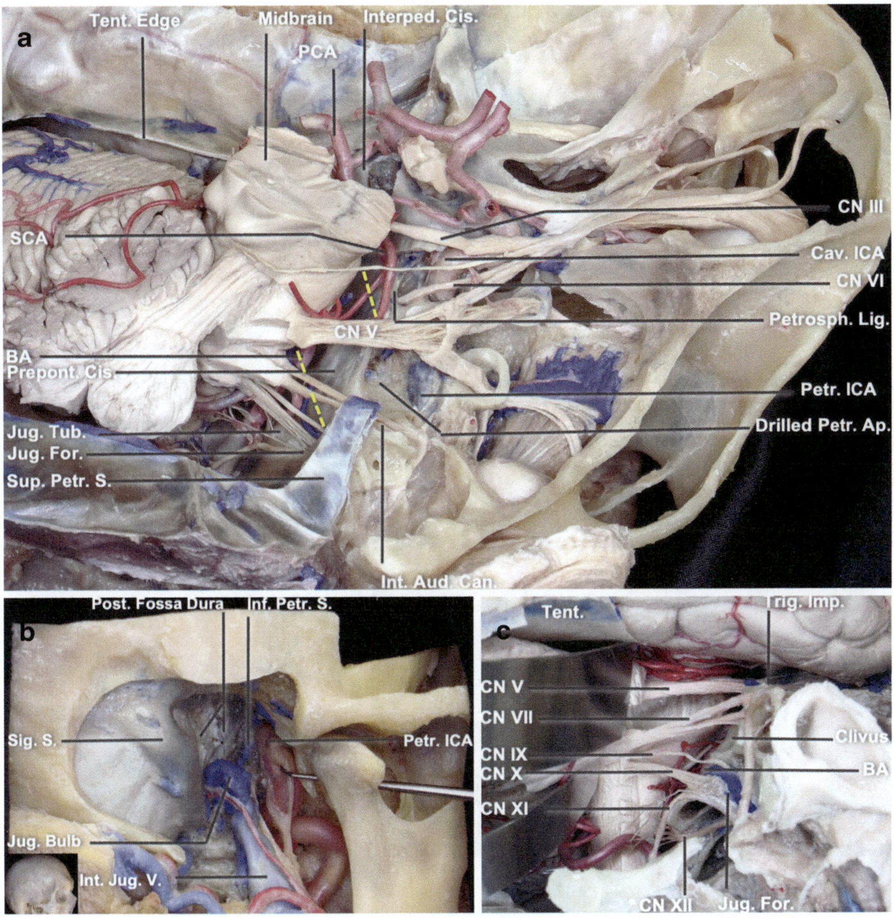

Fig. 2 Neurocritical structures around the petroclival region. (**a**) lateral view of the petroclival region (yellow lines are the borders of the superior, middle, and inferior petroclival spaces). The petroclival region contains the prepontine cistern, basilar artery, and brainstem. The CN VI runs at the level of the petroclival fissure in the prepontine cistern to enter Dorello's canal. The trigeminal impression is the depression under the trigeminal nerve on the petrous apex. The superior petrosal sinus was removed to drill the petrous apex. (**b**) after total petrosectomy, the inferior petrosal sinus was exposed, which runs in the petroclival fissure. (**c**) the relationship of the petroclival region and brainstem. (**d**) neurocritical structures in the petrous part of the temporal bone and petroclival region. (**e**) ventral view of the skull base, ICA segments (yellow circle indicates the external orifice of the carotid canal). (**f**) ventral surface of the dry skull same angle as 2C. Abbreviations: *Ac* acoustic, *Ap* apex, *Aud* auditory, *BA* basilar artery, *Can* canal, *Car* carotid, *Cav* cavernous, *Cis* cistern, *CN* cranial nerve, *Dist* distal, *Fiss* fissure, *For* foramen, *Hypo* hypoglossal, *ICA* internal carotid artery, *Imp* impression, *Inf* inferior, *Int* internal, *Interped* interpeduncular, *Jug* jugular, *Lig* ligament, *Paraclin* paraclinoid, *Paracl paraclival, PCA* posterior cerebral artery, *Parap* parapharyngeal, *Petr* petrous, petrosal, *Petrosph* petrosphenoidal, *Prepont* prepontine, *Post* posterior, *Prox* proximal, *S* sinus, *Sig* sigmoid, *Sup* superior, *Tent* tentorium, *Tub* tubercle, *Trig* trigeminal, *V* vein

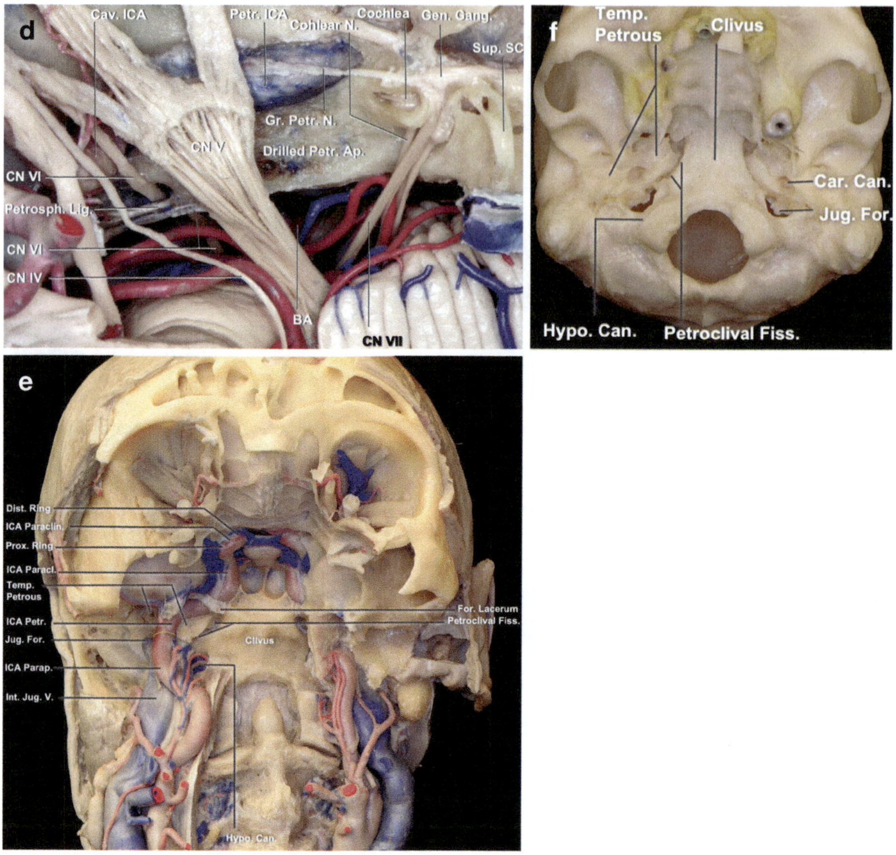

Fig. 2 (continued)

The petrous bone is a highly complicated structure with middle ear and vascular structures within or around it, such as the petrous ICA and superior and inferior petrosal sinuses (Fig. 2d).

Greater Superficial Petrosal Nerve is the branch of the facial nerve that joins the geniculate ganglion laterally and deep petrosal nerve medially to form the vidian nerve (Fig. 2d). ***The arcuate eminence*** corresponds to the position of the superior semicircular canal and is situated just lateral to the ***meatal depression,*** which roofs the internal auditory canal. ***The internal auditory canal*** is located nearly the middle portion of the angle between the arcuate eminence and the greater petrosal nerve, approximately 120° degrees. The tegmen tympani that roofs the middle ear and auditory ossicles is lateral to the arcuate eminence. ***The cochlea*** is located at the apex of the angle between the greater petrosal nerve and the labyrinthine segment of the facial nerve. The cochlea is preserved if a hearing is present. ***The petrous apex*** is the medial one-third of the petrous bone that faces the middle and posterior fossa and locates the medial to the internal acoustic meatus. The petrous apex is bordered by the superior petrosal sinus superiorly, the inferior petrosal sinus inferiorly, and an imaginary line passing along the meatal impression laterally. ***The superior petrosal sinus*** most commonly drains into the junction of the transverse and sigmoid sinuses. In approximately one-third of cases, some temporal bridging veins drain the superior petrosal sinus [3, 4]. In rare cases, the vein of Labbe drains directly into the superior petrosal sinus [5].

The superior petrosal sinus is sectioned at its most anterior part in these circumstances. In a total (anterior plus posterior petrosectomy) transpetrosal approach, the superior petrosal sinus may be dissected away to spare instead of its scarify [6]. ***The trochlear nerve*** originates the dorsal surface of the midbrain and runs in the basal cistern before entering the tentorium. In the transpetrosal approaches, the tentorium is cut behind the point in which the trochlear nerve reaches the free edge of the tentorium. The trochlear nerve is hidden from the surgeon's view, and the safest area to avoid the trochlear nerve injury was at least 15 mm behind the oculomotor nerve contacts with tentorium [7]. ***The abducens nerve*** has either a lateral or a medial location in the Dorello's canal, located at the uppermost edge of the petroclival fissure, and the nerve passes through it [8]. For its lateral location, the abducens nerve has a greater angle and is closer to the petrous apex, increasing the risk of nerve injury. The abducens nerve is the neurocritical structure in the petroclival region, the medial limit for anterior petrosectomy, and the lateral limit for the endoscopic transclival approach [8, 9]. ***The inferior petrosal sinus*** communicates the cavernous sinus and basilar venous plexus and enters the petrosal part of the jugular foramen [1] (Fig. 2b). It forms a plexiform confluence with the venous plexus of the hypoglossal canal, inferior petroclival vein, and tributaries from the vertebral venous plexus and posterior condylar emissary vein. The inferior petroclival vein runs on the extracranial surface of the petroclival fissure and corresponds to the inferior petrosal sinus on the endocranial surface. ***The jugular tubercle*** is a bony protrusion on either side of the occipital bone. The lateral edge of the jugular tubercle relates to the petroclival fissure superiorly. It descends to form the medial border of the jugular foramen and then continues with the posterior edge of the sigmoid sinus, which corresponds to the occipitomastoid suture on the exocranial surface (Fig. 1c). The medial edge of the jugular tubercle usually originates the junction of the middle and posterior thirds of the clivus to form the roof of the hypoglossal canal and then joins the foramen magnum. The jugular tubercle is more obstacle than condyle, especially for reaching the ventral part of the foramen magnum and petroclival region [10–12]. The most dramatic increase in petroclival exposure is obtained by removing the jugular tubercle [13]. **The jugular foramen** is formed by the occipital bone and the petrous part of the temporal bone [1] (Fig. 1c). The jugular foramen is divided into three compartments: two venous, which are sigmoid and petrous, and a neural or intrajugular compartment, which is the glossopharyngeal, vagus, and accessory nerves. The petrous venous part receives drainage from the inferior petrosal sinus, and the sigmoid venous part receives the flow from the sigmoid sinus. The junction of the sigmoid and petrosal parts is the intrajugular processes of the temporal and occipital bones (Fig. 1c).

The interdural space of the petroclival region has access to the cavernous sinus. The interdural space, which includes the abducens nerve, petrosphenoidal ligament, and clival arteries, forms a venous junction of the superolateral part of the basilar plexus, anterior part of the superior petrosal sinus, the superior part of the inferior petrosal sinus, and the posterior part of the cavernous sinus around the petrous apex [8, 14, 15](Fig. 2a).

The ICA is subdivided into parapharyngeal, petrous, paraclival, parasellar, paraclinoid, and intradural segments [16] (Fig. 2e and f). The parapharyngeal segment begins at the common carotid artery bifurcation level and ends at the carotid canal of the petrous bone. The parapharyngeal ICA runs medial to the internal jugular vein, which exits jugular foramen together with CN IX–XI. The petrous ICA extends from the external orifice of the carotid canal to the caudal surface of the foramen lacerum. The paraclival ICA extends from the foramen lacerum to the upper edge of the petroclival fissure, which is inferior to the level of the sellar floor. From the transcranial standpoint, the paraclival ICA corresponds to the petrous ICA, which extends from the carotid canal to the level of the petrolingual ligament and some part of the cavernous ICA that starts at the level of the petrolingual ligament to the proximal dural ring [17–19]. The parasellar ICA extends from the upper edge of the petroclival fissure up to the level of the proximal dural ring of the ICA. The paraclinoid ICA is located between the proximal and distal dural rings.

1 Surgical Considerations

The description of the petroclival region tumors varies in the literature. The petroclival region is considered a tumor originating from the petroclival junction located along the clivus between the CNs V–XI. One study includes only the upper two-thirds of the clivus, named the true petroclival region [2, 20] (Fig. 3a). On the other

Fig. 3 Surgical approaches to the petroclival region and possible directions of the growing petroclival tumors. (**a**) superior view of the skull base. The potential extensions of the petroclival region tumors. The surgical route possibilities to reach the petroclival region. *1. Endoscopic endonasal, 2. Orbitozygomatic, 3. Anterior transpetrosal, 4. Posterior transpetrosal, 5. Presigmoid, 6. Retrosigmoid approaches.* (**b**) Endoscopic endonasal route, (**c**) the anterior, middle and posterior (clivus) midline skull base. (**d**) paraclival ICA is the main obstacle for accessing the petroclival region through the nose. (**e**) the right modified orbitozygomatic craniotomy. (**f**) the exposure of the upper one-third of the petroclival region and cavernous sinus. (**g**) the presigmoid retrolabyrinthine approach on the right side. (**h**) the mastoid segment of the facial nerve was exposed. (**i**) after opening the posterior fossa dura. (**j**) the presigmoid translabyrinthine approach. (**k**) the retrosigmoid approach. (**l**) after drilling the internal auditory canal. (**m**) exposure after retrosigmoid approach. Abbreviations: *Ac* acoustic, *Aud* auditory, *BA* basilar artery, *Can* canal, *Cav* cavernous, *Crib* cribriform, *CN* cranial nerve, *Endo* endolymphatic, *Ext* external; *For* foramen, *ICA* internal carotid artery, *Int* internal, *Jug* jugular, *Lab* labyrinthine, *Mast* mastoid, *Meat* meatus, *Paracl* paraclival, *PCA* posterior cerebral artery, *Petr* petrosal, *Pl* planum, *Post* posterior, *S* sinus, *SC* semicircular canal, *SCA* superior cerebellar artery, *Sig* sigmoid, *Sup* superior, *Temp* temporal

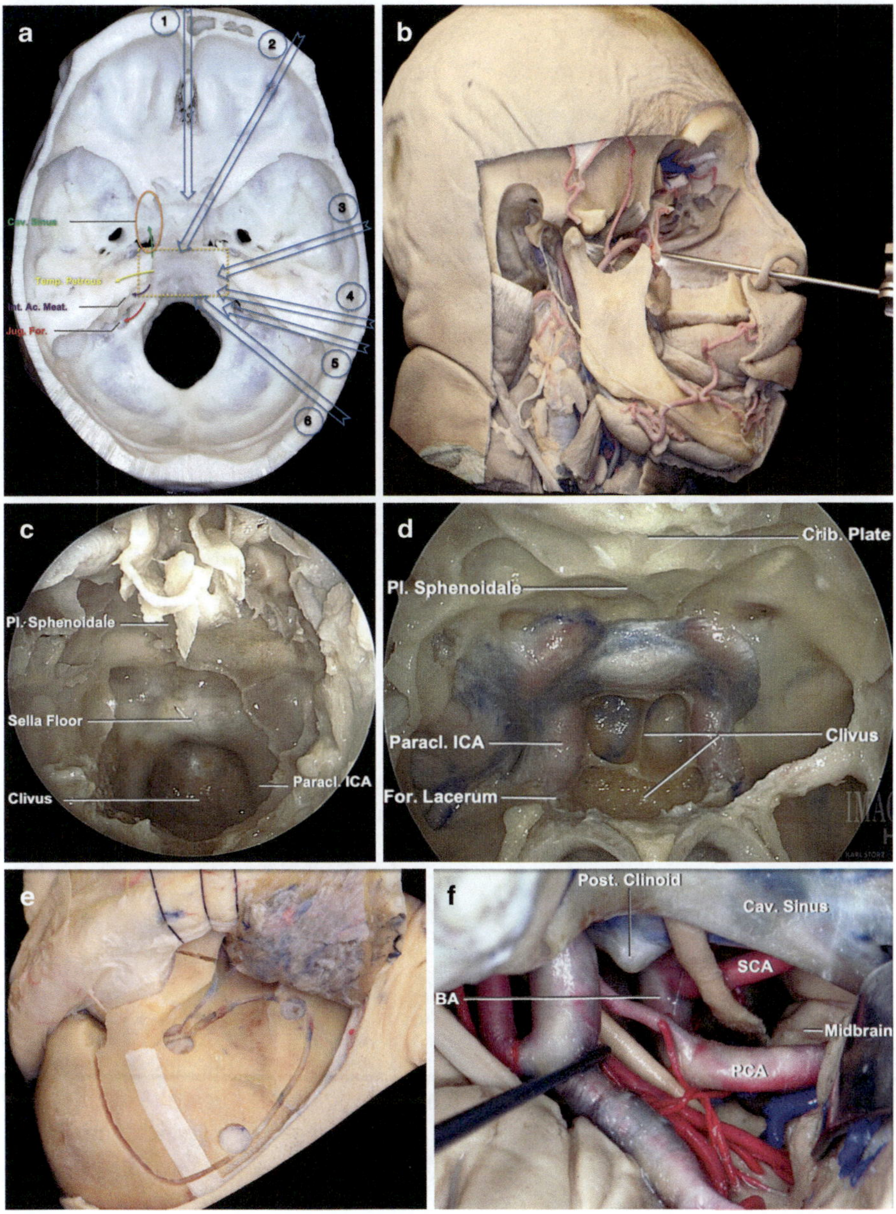
a
1
2
3
4
5
6
Cav. Sinus
Temp. Petrous
Int. Ac. Meat.
Jug. For.
b
c
Pl. Sphenoidale
Sella Floor
Clivus
Paracl. ICA
d
Crib. Plate
Pl. Sphenoidale
Paracl. ICA
Clivus
For. Lacerum
e
f
Post. Clinoid
Cav. Sinus
SCA
BA
Midbrain
PCA

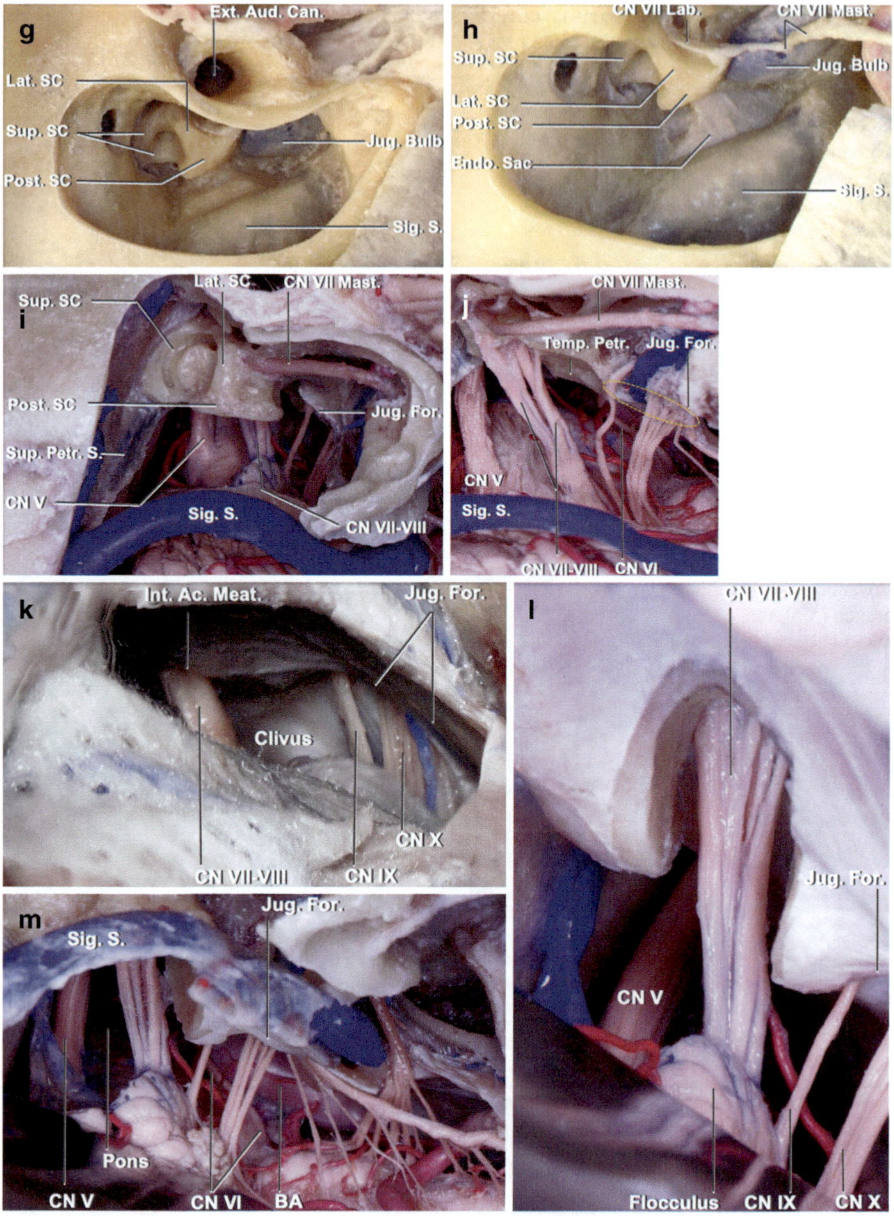

Fig. 3 (continued)

hand, some studies involve the cerebellopontine angle additional to the clivus to describe the petroclival tumor [21–23]. The petroclival tumors may encroach the foramen magnum, jugular foramen, cerebellopontine angle, petrous apex, clivus, tentorial incisura, middle fossa, and cavernous sinus [24]. Meningiomas may displace the basilar artery and brainstem posteriorly and toward the contralateral side; might involve CNs III, IV, V, VI, and VII, and might have extension into the internal auditory canal, jugular foramen, Meckel's cave, Dorello's canal, or ipsilateral cavernous sinus [2]. The petroclival lesions in the middle third of the clivus tend to displace the trigeminal nerve laterally and abducens nerve inferiorly.

The petroclival tumors differentiate from sphenopetroclival, midclival, and posterior petrosal tumors [25]. The sphenopetroclival tumors extend through cavernous sinuses, with en plaque extension along the clival dura and tentorium, frequently encroaching the clivus, petrous apex, or sphenoid sinus [2]. The midclival tumors grow broad in the dural base of the clivus and may infringe the cranial nerves on both sides and either push the basilar artery posteriorly or encase it centrally. The posterior petrosal tumors arise from lateral to the level of the internal acoustic meatus and push the trigeminal and abducens nerves medially.

Access to the petroclival region is one of the most formidable challenges due to its proximity to the brainstem, cranial nerves, basilar artery, perforating arteries of the brainstem, superior and inferior petrosal sinuses. The optimal surgical approach is based on the patient's symptoms like CNs paralysis, hearing loss, tumor extensions, and venous anatomy [2]. The lateral skull base approaches are more suitable if the tumor has an extension lateral to the level of the CN V. The lateral skull base approaches are subdivided into an anterolateral (anterior petrosal and transcavernous), which passes in front of the cochlea, and a posterolateral either through or behind the cochlea (posterior petrosal, presigmoid, and retrosigmoid) (Fig. 3). Transcranial surgery or endonasal endoscopic surgery, or both, are used to remove the petroclival region. The main obstacle for the endoscopic endonasal approach is the paraclival ICA and abducens nerve (Figs. 2d, e, and 3b–d). The endoscopic endonasal approach is superior to the anterior transpetrosal approach for accessing medial or caudal lesions to the abducens nerve [26].

In contrast, the anterior transpetrosal approach is superior to lesions located posterior or lateral to the paraclival ICA and lesions with extension to the middle fossa or infratemporal fossa [9].

The anterior transpetrosal approach is indicated for the petroclival tumors, which extend into the middle fossa and Meckel's cave, such as trigeminal neurinomas or prepontine lesions, meningiomas chordomas, or epidermoids. The maximum exposure of the anterior transpetrosal approach is the foramen ovale anteriorly, oculomotor nerve superiorly, mid-clivus inferiorly, internal auditory canal posteriorly, and contralateral abducens nerve medially. For small tumors that are superior to the internal acoustic meatus, the anterior petrosal approach can be used; for large tumors located below the internal acoustic meatus, the posterior transpetrosal approach or total (anterior plus posterior) transpetrosal approach can be used. For large tumors with crossing midline or large tumors with extension into the cavernous sinus, the total petrosal approach can be used. If venous anatomy is unfavorable, the retrosigmoid approach is chosen instead of the transpetrosal approach. Another surgical option for petroclival tumors with extension to the middle fossa, the combination of the orbitozygomatic and retrosigmoid approaches, can be used (Fig. 3e and f) [27]. One group uses the OZ approach for the petroclival tumors located medial to the internal acoustic meatus without extensive involvement in the posterior fossa [28].

In contrast, the posterior transpetrosal approach is used for the tumor situated lateral to the internal acoustic meatus with deep extension into the posterior fossa [27]. The posterior transpetrosal approach has three options: retrolabyrinthine, translabyrinthine, and transcochlear [29–31] (Fig. 3g–j). The retrolabyrinthine approach avoids entering the labyrinth and skeletonizing the facial nerve, preserving hearing and facial nerve function. In the translabyrinthine approach, all semicircular canals are removed, and the lateral segment of the IAC is entirely skeletonized. This approach is preferred if the patient has little or no hearing. The transcochlear approach requires the resection of the cochlea, cutting off the greater petrosal nerve that provides maximum exposure of the petroclival region; it also involves the sacrifice of hearing and risks of facial nerve function. Thus, the transcochlear approach is used in a patient with preoperative facial weakness and no hearing function.

The anterior and posterior transpetrosal approaches can be combined, called the total transpetrosal approach or combined transpetrosal approach, and can be used in some cases to preserve hearing by performing the partial labyrinthectomy and partial removal of the petrous apex [32–34]. Although the total transpetrosal approach with partial labyrinthectomy, which has a risk of hearing loss, is considered a greater exposure, some studies obtain that it provides a minor benefit in the exposure of the clivus alone compared with the retrolabyrinthine approach [35, 36]. Thus, for patients with fair hearing and large-sized petroclival tumors, the total transpetrosal approach, including the retrolabyrinthine approach and anterior transpetrosal approach, provides maximum exposure to the petroclival region with less risk of cranial nerves injury and no chance of hearing loss.

The retrosigmoid route, which relatively less aggressive technique, gained popularity over time [37]. The retrosigmoid approach has been used in 20–70% of petroclival meningioma cases with extensions into the cerebellopontine angle, middle fossa, or cavernous sinus extensions rather than pure centrally located lesions, named the true petroclival tumor [2, 37–41] (Fig. 3k–m). Based on the involvement of the middle fossa, the retrosigmoid approach might be used alone or combined with other approaches, such as the orbitozygomatic approach, middle fossa approach, presigmoid approach, or other combinations (Figs. 3e–m, 4, and 5).

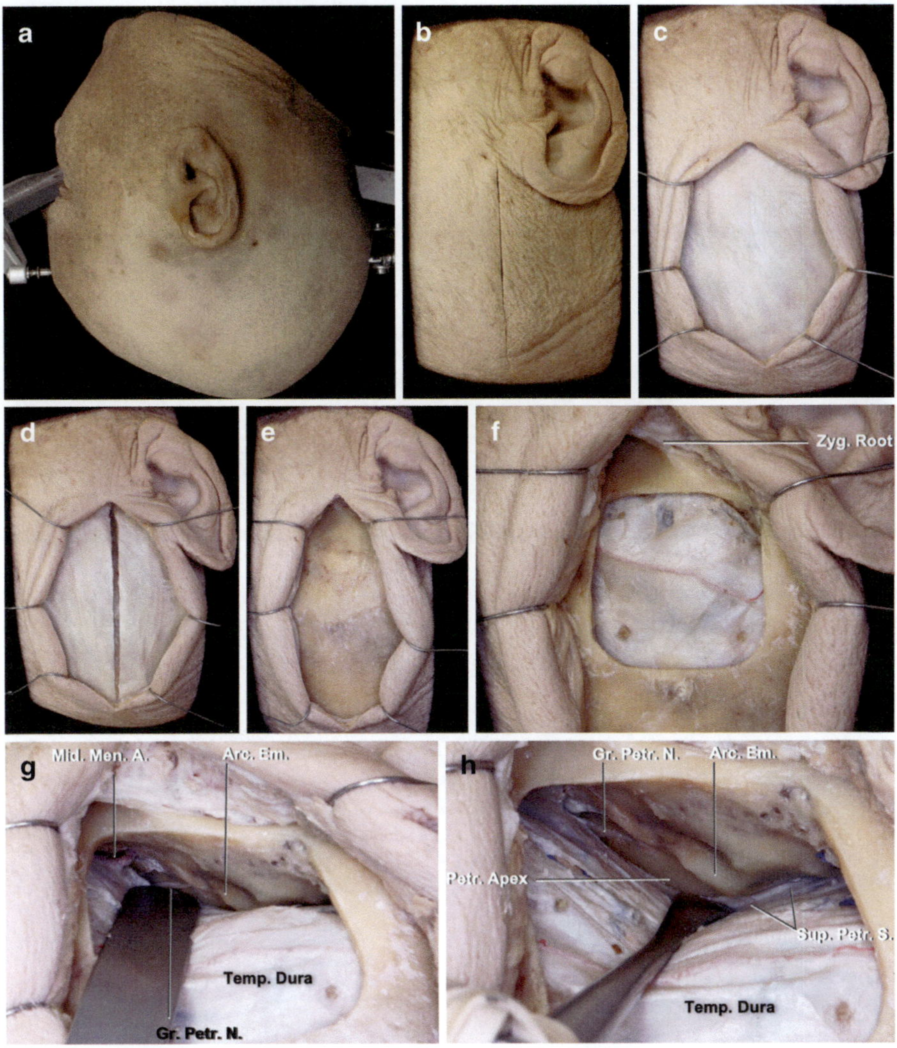

Fig. 4 Anterior transpetrosal approach in a stepwise manner. (**a**) the head positioning. (**b**) preauricular linear incision. (**c**) scalp retraction. (**d, e**) the cutting and retraction of the temporalis muscle. (**e, f**) roughly a 4x4 craniotomy placed middle fossa. (**g**) after peeling the dura off the middle fossa to expose the landmarks. (**h**) after cutting the middle meningeal artery, proceeding the dural dissection. (**i**) another specimen demonstrating the landmarks on the petrous ridge and middle fossa. (**j**) drilling the internal auditory canal. (**k**) further drilling to expose the petrous apex, semicircular canals, and the meatal, cisternal, geniculate, and labyrinthine segments of the facial nerve. (**l**) after drilling the petrous apex. (**m**) the sacrification and coagulation of the superior petrosal sinus, and cutting of the tentorium. (**n**) the final exposure of the petroclival region after drilling the petrous apex. Abbreviations: *A* artery, *Ac* acoustic, *Ap* apex, *Arc* arcuate, *BA* basilar artery, *Can* canal, *Cav* cavernous, *CN* cranial nerve, *Depr* depression, *Em* eminence, *Gang* ganglion, *Gen* geniculate, *Gr* greater, *ICA* internal carotid artery, *Int* internal, *Lig* ligament, *Meat* meatal, *Men* meningeal, *Mid* middle, *N* nerve, *Petr* petrous, petrosal, *Petrosph* petrosphenoidal, *Post* posterior, *Prom* prominence, *S* sinus, *SC* semicircular canal, *SCA* superior cerebellar artery, *Sup* superior, *Temp* temporal, *Tent* tentorium, *Trig* trigeminal, *Zyg* zygomatic

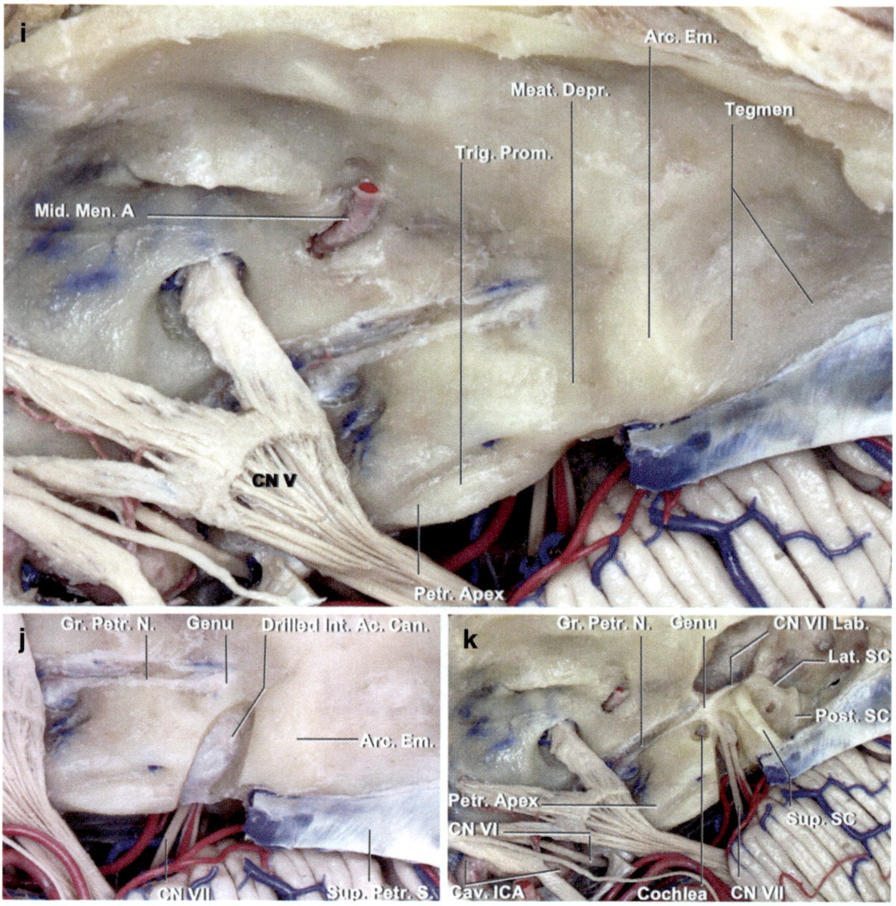

Fig. 4 (continued)

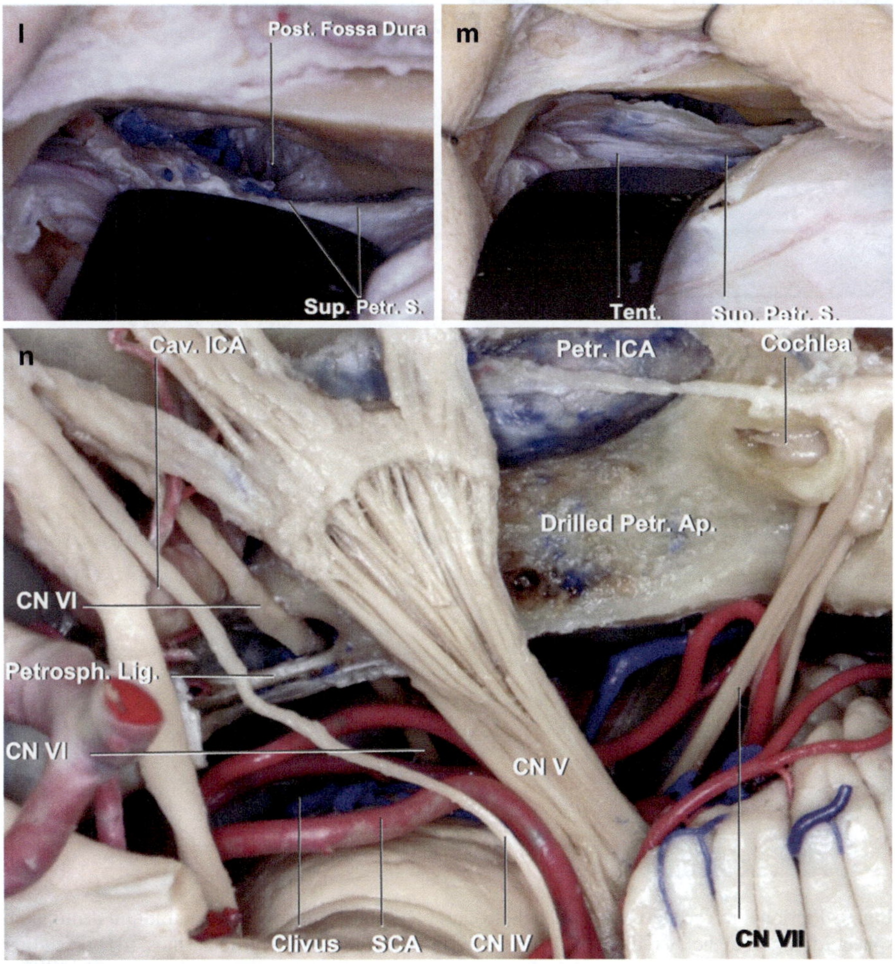

Fig. 4 (continued)

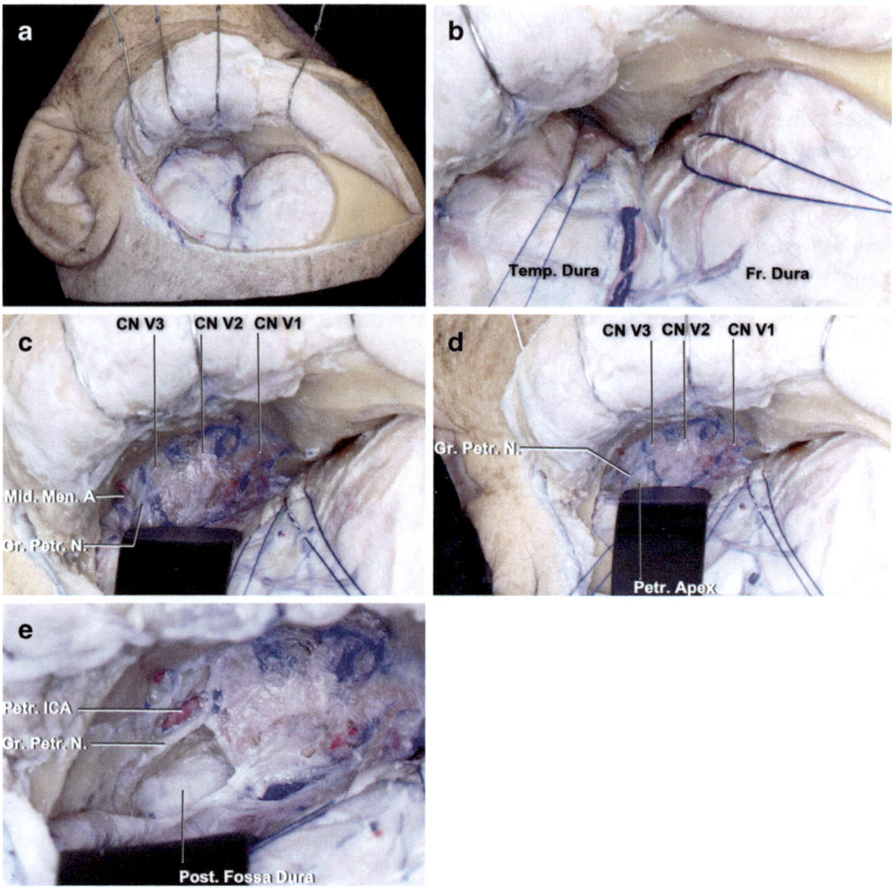

Fig. 5 Pterional transcavernous approach. (**a**) left pterional craniotomy. (**b**) extradural drilling the sphenoid ridge. (**c**) the extradural exposure of the middle fossa and cavernous sinus. (**d**) proceeding dural dissection away from the middle fossa and the exposure of the petrous apex. (**e**) after drilling the petrous apex, disclosure of the dura of the posterior fossa. Abbreviations: *A* artery, *CN* cranial nerve, *Fr* frontal, *Gr* greater, *Men* meningeal, *Mid* middle, *N* nerve, *Petr* petrous, *Post* posterior

2 Surgical Approaches to the Petroclival Region

Anterior Transpetrosal Approach. The patient is placed in the supine position, and the head is rotated 60° degrees to the contralateral side of the tumor (Fig. 4a). A preauricular linear skin incision is made starting from the level of the zygomatic arch to the parietal suture (Fig. 4b). Temporal muscle is cut in the same manner as the skin incision and retracted laterally (Fig. 4c and e). A 3 × 4 cm temporal craniotomy is performed as its lowest border is at the same level of the middle fossa (Fig. 4f). A small dural incision is made to release CSF, making the temporal lobe's retraction easier. The dura is dissected away from the middle fossa from back to front, and the middle meningeal artery is exposed first in the foramen spinosum (Fig. 4g). This artery may be obliterated and divided to continue the dural peeling. It is essential to expose the greater superficial petrosal nerve and geniculate ganglion of the facial nerve. The greater superficial petrosal nerve is either preserved or cut for avoiding the facial paralysis caused by the nerve traction and injury.

Further elevation of the dura exposes the arcuate eminence and meatal depression on the petrous ridge (Fig. 4h). If the hearing is preserved, the cochlea is maintained in the angle between the meatal depression and greater superficial petrosal nerve and about 5 mm anteromedial from the center of the geniculate ganglion. Proceeding medial to the meatal depression, the trigeminal prominence and trigeminal nerve are exposed (Fig. 4i). The third segment of the trigeminal nerve is displaced anteriorly for better exposure to the petrous apex. The borders of the petrous apex, the greater superficial petrosal nerve, trigeminal nerve, and arcuate eminence are exposed. The meatal depression may be drilled from back to front and medial (trigeminal prominence) to lateral to expose the facial and vestibulocochlear nerves in the auditory canal (Fig. 4j). The superior petrosal sinus on the petrous ridge is obliterated and divided in the region lateral to the trigeminal nerve, and the dural incision is extended across the tentorium (Figs. 4k–m). The drilling of the petrous apex is carried on until encountering the inferior petrosal sinus by giving attention to the abducens nerve (Fig. 4n).

Pterional Transcavernous Approach. The patient is placed in a supine position with 45° degrees contralateral head rotation and vertex down. As noted elsewhere, a standard pterional craniotomy like a curvilinear skin incision, one-layer skin flap, and frontotemporal craniotomy is performed [42]. The sphenoid ridge is drilled flat downward extradurally, and the meningo-orbital band is cut (Fig. 5a and b). Whether or not the anterior clinoidectomy is performed based on size and extension of the tumor. The dura of the anterolateral cavernous sinus is peeled off starting from the superior orbital fissure, foramen rotundum, and toward the foramen ovale to expose the oculomotor trochlear and all three branches of the trigeminal nerve (Fig. 5c and d). Proceeding dissection posteriorly reveals the middle meningeal artery, arcuate eminence, and petrous ridge (Fig. 5e). The third segment of the trigeminal nerve is translocated anteriorly to expose the underlying bone part of the petrous apex. The petrous apex is drilled until the inferior petrosal sinus exposes the petroclival region.

References

1. Rhoton AL Jr. Jugular foramen. Neurosurgery. 2000;47(suppl_3):S267–85.
2. Almefty R, Dunn IF, Pravdenkova S, Abolfotoh M, Al-Mefty O. True petroclival meningiomas: results of surgical management. J Neurosurg. 2014;120(1):40–51.
3. Hafez A, Nader R, Al-Mefty O. Preservation of the superior petrosal sinus during the petrosal approach. J Neurosurg. 2011;114(5):1294–8.

4. Sakata K, Al-Mefty O, Yamamoto I. Venous consideration in petrosal approach: microsurgical anatomy of the temporal bridging vein. Neurosurgery. 2000;47:153–61.
5. LabbéC:Anomaliedes sinus deladure-mère. Dévelopement de ces sinus. Considérations sur la suppléance réciproque de ces canaux veineux dans les cas d'absence de l'un d'eux. Description de quelques sinus peu connus. Arch Physio Norm Pathol. 1883;3:1–27.
6. Almefty KK, Al-Mefty O. Petrosal approach with preservation of the superior petrosal sinus (the graceful petrosal) for resection of Giant trigeminal schwannoma: 2-dimensional operative video. Operat Neurosurg. 2021;20(5):E342–3.
7. Gupta T, Gupta SK, Sahni D. Anatomy of the tentorial segment of the trochlear nerve in reference to its preservation during surgery for skull base lesions. Surg Radiol Anat. 2014;36(10):967–71.
8. Ozveren MF, Uchida K, Aiso S, Kawase T. Meningovenous structures of the petroclival region: clinical importance for surgery and intravascular surgery. Neurosurgery. 2002;50(4):829–37.
9. Kawase T, Shiobara R, Toya S. Middle fossa transpetrosal- transtentorial approaches for petroclival meningiomas: selective pyramid resection and radicality. Acta Neurochir. 1994;129:113–20.
10. Matsushima T, Natori Y, Katsuta T, Ikezaki K, Fukui M, Rhoton AL. Microsurgical anatomy for lateral approaches to the foramen magnum with special reference to transcondylar fossa (supracondylar transjugular tubercle) approach. Skull Base Surg. 1998;8:119–25.
11. Nanda A, Vincent DA, Vannemreddy PS, Baskaya MK, Chanda A. Far-lateral approach to intradural lesions of the foramen magnum without resection of the occipital condyle. J Neurosurg. 2002;96:302–9.
12. Matsushima T, Matsukado K, Natori Y, Inamura T, Hitotsumatsu T, Fukui M. Surgery on a saccular vertebral artery-posterior inferior cerebellar artery aneurysm via the transcondylar fossa (supracondylar transjugular tubercle) approach or the transcondylar approach: surgical results and indications for using two different lateral skull base approaches. J Neurosurg. 2001;95:268–74.
13. Spektor S, Anderson GJ, McMenomey SO, Horgan MA, Kellogg JX, Delashaw JB. Quantitative description of the far-lateral transcondylar transtubercular approach to the foramen magnum and clivus. J Neurosurg. 2000;92(5):824–31.
14. Halbach VV, Higashida RT, Hieshima GB, Hardin CW, Yang PJ. Transvenous embolization of direct carotid cavernous fistulas. AJNR Am J Neuroradiol. 1988;9:741–7.
15. Oishi H, Arai H, Sato K, Iizuka Y. Complications with transvenous embolization of cavernous dural arteriovenous fistula. Acta Neurochir. 1999;141:1265–71.
16. Labib MA, Prevedello DM, Carrau R, Kerr EE, Naudy C, Abou Al-Shaar H, Corsten M, Kassam A. A road map to the internal carotid artery in expanded endoscopic endonasal approaches to the ventral cranial base. Operat Neurosurg. 2014;10(3):448–71.
17. Rhoton AL Jr. The supratentorial arteries. Neurosurgery. 2002;51(suppl 4):S53–S120.
18. Osawa S, Rhoton AL Jr, Tanriover N, Shimizu S, Fujii K. Microsurgical anatomy and surgical exposure of the petrous segment of the internal carotid artery. Operative. Neurosurgery. 2008;63(suppl_4):ONS210–39.
19. Bouthillier A, van Loveren HR, Keller JT. Segments of the internal carotid artery: a new classification. Neurosurgery. 1996;38(3):425–32; discussion 432–433.
20. Couldwell WT, Fukushima T, Giannotta SL, Weiss MH. Petro- clival meningiomas: surgical experience in 109 cases. J Neurosurg. 1996;84:20–8.
21. Spetzler RF, Daspit CP, Pappas CT. The combined supra- and infratentorial approach for lesions of the petrous and clival regions: experience with 46 cases. J Neurosurg. 1992;76:588–99.
22. Spetzler RF, Hamilton MG, Daspit CP. Petroclival lesions. Clin Neurosurg. 1994;41:62–82.
23. Yasargil MG, Mortara RW, Curcic M. Meningiomas of basal posterior cranial fossa. Adv Technol Stand Neurosurg. 1980;7:3–118.
24. Hedeschi H, Rhoton AL Jr. Lateral approaches to the petroclival region. Surg Neurol. 1994;41(3):180–216.
25. Al-Mefty O. Operative atlas of meningiomas. New York: Lippincott Williams & Wilkins; 1998.

26. Muto J, Prevedello DM, Ditzel Filho LF, Tang P, Oyama K, Kerr EE, Otto BA, Kawase T, Yoshida K, Carrau RL. Comparative analysis of the anterior transpetrosal approach with the endoscopic endonasal approach to the petroclival region. J Neurosurg. 2016;125(5):1171–86.
27. Bambakidis NC, Kakarla UK, Kim LJ, Nakaji P, Porter RW, Daspit CP, Spetzler RF. Evolution of surgical approaches in the treatment of petroclival meningiomas: a retrospective review. Operative. Neurosurgery. 2007;61(suppl_5):ONS202–11.
28. Erkmen K, Pravdenkova S, Al-Mefty O. Surgical management of petroclival meningiomas: factors determining the choice of approach. Neurosurg Focus. 2005;19:E7.
29. House WF, Hitselberger WE. The transcochlear approach to the skull base. Arch Otolaryngol. 1976;102:334–42.
30. Siwanuwatn R, Deshmukh P, Figueiredo EG, Crawford NR, Spetzler RF, Preul MC. Quantitative analysis of the working area and angle of attack for the retrosigmoid, combined petro Sal, and transcochlear approaches to the petroclival region. J Neurosurg. 2006;104:137–42.
31. Xu F, Karampelas I, Megerian CA, Selman WR, Bambakidis NC. Petroclival meningiomas: an update on surgical approaches, decision making, and treatment results. Neurosurg Focus. 2013;35(6):E11.
32. Horgan MA, Anderson GJ, Kellogg JX, Schwartz MS, Spektor S, McMenomey SO, Delashaw JB. Classification and quantification of the petrosal approach to the petroclival region. J Neurosurg. 2000;93(1):108–12.
33. Horgan MA, Delashaw JB, Schwartz MS, Kellogg JX, Spek tor S, McMenomey SO: Transcrusal approach to the petro- clival region with hearing preservation. Technical note and illustrative cases. J Neurosurg 94:660–666, 2001.
34. Sekhar LN, Schessel DA, Bucur SD, Raso JL, Wright DC. Partial labyrinthectomy petrous apicectomy approach to neoplastic and vascular lesions of the petroclival area. Neurosurgery. 1999;44:537–52.
35. Miller CG, van Loveren HR, Keller JT, Pensak M, el-Kalliny M, Tew JM Jr. Transpetrosal approach: surgical anatomy and technique. Neurosurgery. 1993;33:461–9.
36. Ramina R, Neto MC, Fernandes YB, Silva EB, Mattei TA, Aguiar PH. Surgical removal of small petroclival meningiomas. Acta Neurochir. 2008;150:431–9.
37. Safavi-Abbasi S, Zabramski JM, Deshmukh P, Reis CV, Bambakidis NC, Theodore N, Crawford NR, Spetzler RF, Preul MC. Moving toward the petroclival region: a model for quantitative and anatomical analysis of tumor shift. J Neurosurg. 2007;107(4):797–804.
38. Bricolo A, Turazzi S. Petroclival meningiomas. In: Schmidek HM, Sweet WH, editors. *Operative neurosurgical techniques*. Philadelphia, PA: W.B. Saunders; 2000. p. 933–55.
39. Bricolo AP, Turazzi S, Talacchi A, Cristofori L. Microsurgical removal of petroclival meningiomas: a report of 33 patients. Neurosurgery. 1992;31:813–28.
40. Samii M, Ammirati M, Mahran A, Bini W, Sepehrnia A. Surgery of petroclival meningiomas: report of 24 cases. Neurosurgery. 1989;24:12–7.
41. Sekhar LN, Jannetta PJ, Burkhart LE, Janosky JE. Meningiomas involving the clivus: a six-year experience with 41 patients. Neurosurgery. 1990;27:764–81.
42. Chaddad-Neto F, Campos Filho JM, Dória-Netto HL, Faria MH, Ribas GC, Oliveira E. The pterional craniotomy: tips and tricks. Arq Neuropsiquiatr. 2012;70:727–32.

Surgical Anatomy of the Far Lateral Approach and Jugular Foramen

Arnau Benet, Lea Scherschinski, and Michael T. Lawton

Abbreviations

CN	Cranial nerve
PICA	Posterior inferior cerebellar artery

1 Introduction

The far lateral approach is one of the most versatile skull base approaches, providing maximum exposure to the posterior fossa at the craniocervical junction [1]. The posterior fossa comprises two anatomical compartments—the ventromedial compartment, located anterior to the lower cranial nerves, and the posterolateral compartment, located posterior to the lower cranial nerve [2]. Both compartments may be safely accessed to target complex vascular, neoplastic, and other pathologies [1–5]. Alternative surgical approaches to this area are the retrosigmoid and the endoscopic endonasal approach [2–4, 6]. Both approaches may also be combined with the far lateral approach for an extended exposure [7]. The type and location of the lesion dictate the definite approach and its extension [1]. We describe the surgical anatomy of the far lateral approach, its variations, and its clinical implications. A thorough understanding of the far lateral approach and its surrounding anatomical structures helps neurosurgeons design the safest and most efficient approach to target lesions in the posterior cranial fossa.

A. Benet · L. Scherschinski · M. T. Lawton (✉)
Department of Neurosurgery, Barrow Neurological Institute, St. Joseph's Hospital and Medical Center, Phoenix, AZ, USA
e-mail: Neuropub@barrowneuro.org

E. G. Figueiredo et al. (eds.), *Brain Anatomy and Neurosurgical Approaches*,
https://doi.org/10.1007/978-3-031-14820-0_28

2 Osseous Anatomy

The far lateral approach uses a posterolateral trajectory to access the posterior fossa [8]. Fundamental knowledge of the anatomy of the occipital bone and C1 vertebra is essential to execute the far lateral approach safely. Anatomy-based osteotomies extend the microsurgical field of view because the angle of exposure is critical for accessing the various lesions. When a transcondylar, supracondylar, or paracondylar extension of the far lateral approach is planned, the anatomy of the C1 vertebra and the petrous part of the temporal bone also becomes relevant.

2.1 Occipital Bone

The occipital bone comprises three structures: the squamous, basilar, and condylar parts. The squamous part forms the posterior wall of the posterior fossa, and the basilar part forms the lower anterior wall of the posterior fossa. Superiorly, the basilar part fuses with the dorsum sellae. The condylar part is attached to the petrous part of the temporal bone, forming the lateral border of the posterior fossa. This chapter focuses on the squamous and condylar parts of the occipital bone as they provide anatomical landmarks for the far lateral approach.

2.1.1 Squamous Part of the Occipital Bone

The squamous part of the occipital bone has an internal and external surface. The internal surface forms a concavity that accommodates the occipital lobes, cerebellar hemispheres, and transverse and superior sagittal sinuses (Fig. 1). The external surface has several bony landmarks that assist in designing the craniotomy. The inion is a midline osseous protuberance that projects from the center of the squamous part of the occipital bone. This external occipital protuberance serves as a reliable landmark to identify the position of the torcular, which is located 1 cm superior to the inion on the endocranial surface (Fig. 2).

From the inion, two transverse osseous crests project laterally. The most superior is the supreme nuchal line, and the most inferior is the superior nuchal line. The superior nuchal line serves as the origin for the nuchal muscles, including the sternocleidomastoid, trapezius, splenius capitis, and semispinalis capitis muscles. Laterally, the superior nuchal line meets the asterion, which constitutes the confluence of the lambdoid, parieto-mastoid, and occipitomastoid sutures. The supreme nuchal line is less prominent and is the insertion of the occipitofrontal muscle. Both the superior and the supreme nuchal lines estimate the tentorium's location when designing the craniotomy. The inferior nuchal line resides between the superior nuchal line and the foramen magnum. This ridge extends laterally to both sides of the external occipital crest as it slightly curves upward and receives the insertions of

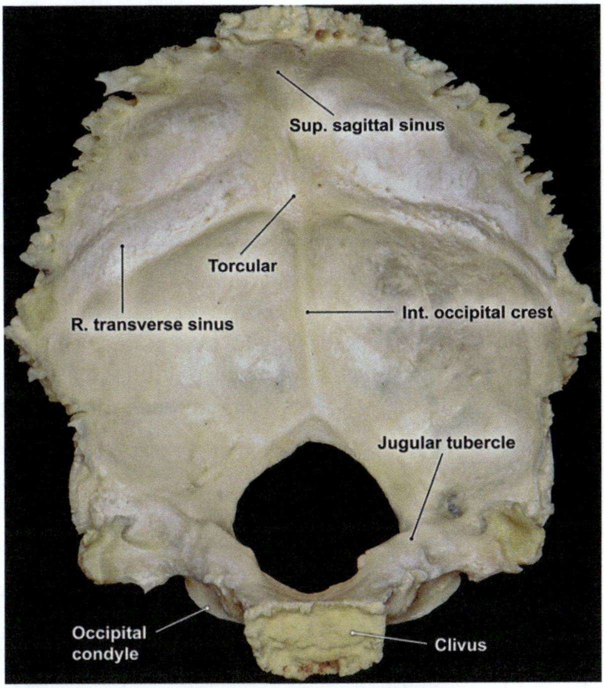

Fig. 1 Cadaveric view of the endocranial surface of the occipital bone. The superior sagittal and transverse sinuses form bone impressions on the inner surface of the occipital bone. They provide landmarks for the falx cerebri and the tentorium. The bilateral condylar parts of the occipital bone transition into the basilar part inferiorly and the squamous part superiorly. *Int* internal, *R* right, *Sup* superior. *Image provided by Arnau Benet. Used with permission of the University of California San Francisco's Skull Base & Cerebrovascular Laboratory*

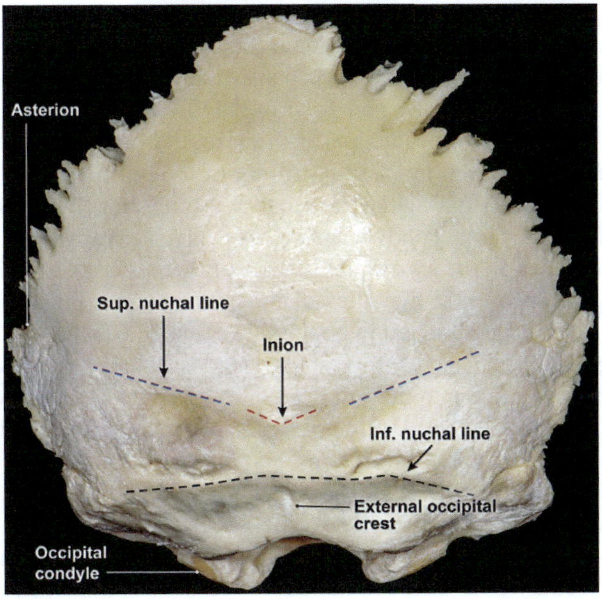

Fig. 2 Cadaveric view of the exocranial surface of the occipital bone. The squamous part of the occipital bone harbors the superior and inferior nuchal lines. The inion serves as a landmark for the confluence of sinuses (torcular Herophili). The asterion is formed by the lambdoid, occipitomastoid, and parieto-mastoid sutures and marks the transition between the transverse sinus and the sigmoid sinus. *Ext* external, *Inf* inferior, *Sup* superior. *Image provided by Arnau Benet. Used with permission of the University of California San Francisco's Skull Base & Cerebrovascular Laboratory*

the suboccipital muscles. Identifying the suboccipital muscles is of utmost importance during the far lateral approach, given their proximity to the vertebral artery [9].

The asterion is a craniometric point on the exocranial surface that marks the transition between the transverse sinus and the sigmoid sinus. Inferior to the asterion, the occipital artery ascends in a furrow parallel to the occipitomastoid suture. Anterior to the occipitomastoid suture, the posterior belly of the digastric muscle inserts in the digastric groove [8]. The digastric groove serves as a landmark of the mastoid and the sigmoid sinus in the supracondylar extension and the descending portion of the facial nerve in the stylomastoid foramen in the paracondylar extension [8].

Along the midline of the squamous part of the occipital bone runs a bony projection called the occipital crest or median nuchal line. This ridge extends from the inion to the opisthion, marking the posterior midline border of the foramen magnum. The external occipital crest serves as a reference for the falx cerebelli and the midline.

2.1.2 Condylar Part of the Occipital Bone

The condylar part of the occipital bone encompasses the condylar fossa, the occipital condyle, and the jugular tubercle. The condylar fossa is located between the squamous part and the occipital condyle in the exocranial surface of the occipital bone. This osseous depression is particularly important because it may contain a rich venous channel known as the posterior condylar emissary vein [10]. The posterior condylar emissary vein connects the vertebral venous sinus and the sigmoid sinus [10]. Avulsion of the posterior condylar emissary vein may result in profuse bleeding [10].

The occipital condyle is an oval-shaped structure at the posterior skull base that articulates with the superior articular facet of the C1 vertebra, forming the atlanto-occipital joint. A thorough evaluation of the size and orientation of the occipital condyle should be followed by careful drilling of the occipital condyle to prevent craniocervical instability during the transcondylar extension. The hypoglossal canal lies superior to the occipital condyle and crosses anterolaterally at a 45° angle from the foramen magnum to the exocranial surface of the posterior skull base. It serves as a traversing channel for the hypoglossal nerve (cranial nerve [CN] XII), the hypoglossal artery (if present), and the hypoglossal venous plexus.

The hypoglossal canal and its traversing hypoglossal nerve separate the condylar part of the occipital bone into the inferior condylar compartment and the superior jugular tubercle compartment. Drilling the medial aspect of the condylar compartment exposes the lower third of the clivus, and drilling of the jugular tubercle compartment exposes the jugular foramen and the lateral and anterior aspects of the medullary cavity. The jugular tubercle overlies the hypoglossal canal and also serves as the floor of the jugular foramen. The jugular foramen is a large aperture at the skull base formed by both the occipital bone inferiorly and the petrous part of the temporal bone superiorly.

Although the jugular foramen is described as a single orifice, it is a complex structure comprising three compartments that transmit their neurovascular structures (Figs. 3 and 4). The three compartments of the jugular foramen are the sigmoid compartment posteriorly, the neural compartment intermediately, and the petrous compartment anteriorly. The following anatomical landmarks help divide these compartments. The superior intrajugular process, a central bony spur that runs

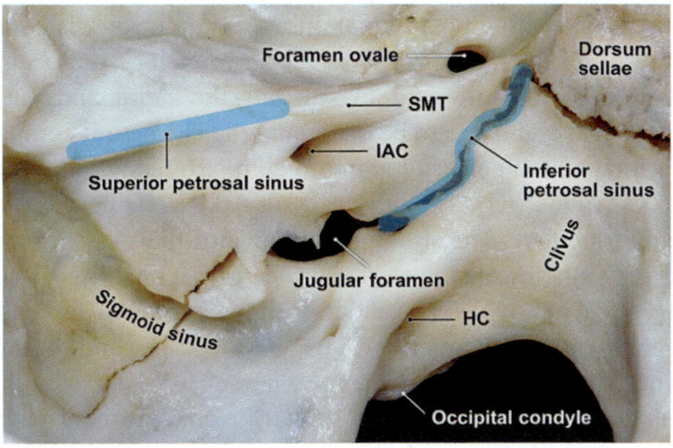

Fig. 3 Cadaveric endocranial view of the middle and posterior cranial fossae. The jugular foramen is formed by the petrous part of the temporal and the corresponding margin of the occipital bone. It is located in the center of the lateral wall of the posterior fossa and transmits the inferior petrosal and sigmoid sinus, the glossopharyngeal, vagus, and accessory nerves (CN IX, X, XI). The blue lines mark the locations of the internal petrosal sinus and the superior petrosal sinus. *HC* hypoglossal canal, *IAC* internal auditory canal, *SMT* suprameatal tubercle. *Image provided by Arnau Benet. Used with permission of the University of California San Francisco's Skull Base & Cerebrovascular Laboratory*

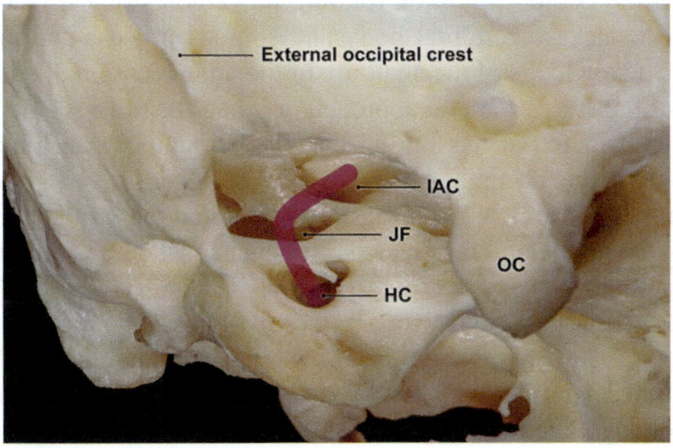

Fig. 4 Cadaveric view of the posterior fossa through the foramen magnum. The lower cranial nerves divide the posterior fossa into the ventromedial and the posterolateral compartment (*red line*). The hypoglossal canal transmits the hypoglossal nerve (CN XII), the hypoglossal artery, and the hypoglossal venous plexus. *IAC* internal auditory canal, *JF* jugular foramen, *HC* hypoglossal canal, *OC* occipital condyle. *Image provided by Arnau Benet. Used with permission of the University of California San Francisco's Skull Base & Cerebrovascular Laboratory*

across the jugular surface of the temporal bone, is connected to the jugular tubercle by a dural fold that attaches to the inferior intrajugular process, a promontory in the jugular tubercle. The superior and inferior intrajugular processes and the dural fold delimit the posterior sigmoid compartment of the jugular foramen. The posterior sigmoid compartment transmits the sigmoid sinus, the jugular bulb, and the meningeal branch of the ascending pharyngeal artery. Sometimes this compartment may also contain the posterior meningeal artery or a branch of the occipital artery. As drilling of the jugular tubercle continues during the supracondylar extension, the sigmoid compartment is the first to be encountered. The neural compartment, which is located anteriorly to the posterior sigmoid compartment, contains the glossopharyngeal (CN IX), vagus (CN X), and accessory (CN XI) nerves. A dural septum separates the intermediate neural compartment into the glossopharyngeal meatus anteriorly and the vagal meatus posteriorly. The glossopharyngeal meatus contains the glossopharyngeal nerve (CN IX) and its tympanic branch, also called Jacobson's nerve. The vagal meatus contains the vagus nerve (CN X) and its auricular branch, also called Arnold's nerve, and the accessory nerve (CN XI).

The petrous compartment is the most anterior compartment within the jugular foramen and is confined posteriorly by the neural compartment and anteriorly by the petroclival synchondrosis. The petrous compartment transmits the inferior petrosal sinus.

2.2 Atlas

The atlas is the first cervical vertebra (C1). It articulates with the dens of the axis (C2) and with the occiput to allow rotating movements of the head. The atlas is unique in that it lacks a vertebral body and spinous process and instead has two lateral masses. Knowledge of the anatomy of the atlas is essential when executing a caudal extension of the far lateral approach or when the third segment of the vertebral artery is considered at risk during the suboccipital approach because it loops superiorly toward the occipital bone [1, 9]. The atlas is a ring-shaped vertebra formed by an anterior arch and a posterior arch united by two lateral masses. Only the posterior arch and lateral masses are of relevance when performing an extended far lateral approach.

The lateral masses of the atlas are a pair of ovoid osseous protuberances on each side of the atlas. They support the head's weight and enable mobility in flexion, extension, rotation, and lateral bending because they transmit motion forces from the occipital condyles to the axis (C2 vertebra). Each lateral mass has two articular surfaces, one on the inferior and one on the superior side. The oval inferior articular facet of the lateral mass articulates with the underlying superior articular facet of the axis (C2). On the superior side, the classically kidney-shaped, concave superior articular facets of the lateral masses articulate with the inferior facets of the occipital condyle, forming the atlanto-occipital joint. Identifying and preserving the atlanto-occipital joint during the transcondylar extension of the far lateral approach is critical to avoid neck instability and ensure a proper anteromedial trajectory.

Lateral to the lateral masses lie the transverse foramen and the transverse process. The transverse foramen of C1 serves as a traversing channel for the vertebral artery before it takes a medial course along the superior aspect of the posterior arch of the atlas [11]. The serpentine course of the vertebral artery in the craniovertebral region makes it susceptible to injury during the far lateral approach [9]. Therefore, an opening in the transverse foramen can mobilize the vertebral artery laterally away from the surgical field [9, 11, 12]. The transverse process confines the transverse foramen anteriorly, laterally, and posteriorly and provides attachments for several suboccipital and cervical muscles.

The posterior arch is the most important structure of the atlas when performing the far lateral approach. This elongated, semicircular bone carries the vertebral artery groove on each side as the third segment of the vertebral artery runs alongside the superior aspect of the posterior arch [11]. The C1 cervical nerve is embedded in the vertebral venous plexus and crosses the superior surface of the posterior arch of the atlas [10]. Because the course of the C1 nerve is unprotected after it exits the spinal cord, neurosurgeons are advised to proceed carefully to prevent accidental injury to it during vertebral artery manipulation [9]. To prevent injury to the C1 nerve, the neurosurgeon may completely or partially remove the posterior arch of the atlas to gain access to the spinal cord and upper cervical nerves during the far lateral approach.

3 Sequential Steps of the Far Lateral Approach

3.1 Positioning

Skull base procedures require well-planned patient positioning to ensure a head position that offers the best working corridor while maintaining the patient's safety [1, 13]. To fulfill both of these objectives, the neurosurgeon positions the head to account for the effect of gravity on the brain after opening the dura. The park bench and three-quarter prone positions are suitable operative positions that allow for a natural retraction of the cerebellar hemisphere away from the surgical corridor [1, 13–16]. In the park bench position, the patient is partially lifted off the operating table at a 45° angle from a prone position, with the patient's shoulder and arm reaching beyond the operating table and resting on an armrest [13, 15, 16]. This position allows for extended head rotation movements when adjusting for the lateral craniocervical access site [1, 16]. Then, the head is rotated 45° away from the lesion and flexed laterally toward the floor [5, 13, 16, 17]. The suitability of patient positioning may be assessed by palpating the mastoid tip, the inion, and the spinous processes of the cervical vertebrae from a standing position (Fig. 5).

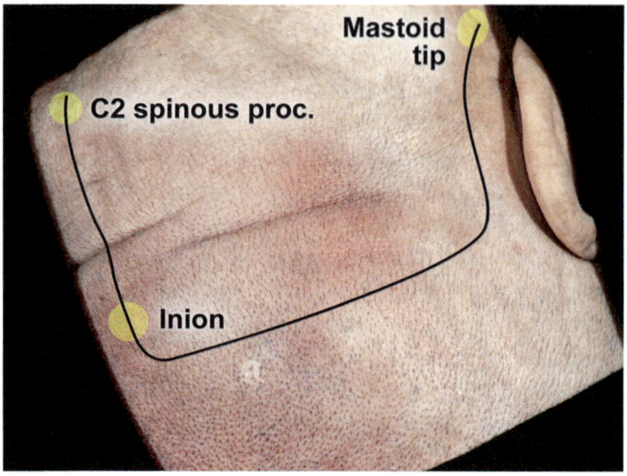

Fig. 5 Outline of the inverted hockey-stick incision during the far lateral approach. The cadaver is positioned in the park bench position. The inverted hockey-stick incision starts at the mastoid tip and continues over the inion to the spinous process of the second cervical vertebra along the midline. *proc* process. *Image provided by Arnau Benet. Used with permission of the University of California San Francisco's Skull Base & Cerebrovascular Laboratory*

3.2 Skin Incision

The type and location of the lesion dictate the surgical exposure and the design of the skin incision [1]. The far lateral approach uses two types of skin incisions: the inverted hockey stick and the lazy S.[1, 13, 15] For the inverted hockey-stick incision, the skin is incised 2 cm inferior to the mastoid process, where the incision is continued cranially until above the superior nuchal line (Fig. 5) [1, 13]. Here, the incision is turned to a medial course toward the midline, extending inferiorly to the C3 or C4 vertebral level [13]. The advantages of the inverted hockey-stick incision are a wide exposure of the suboccipital musculature, access to the lateral aspects of the occipital condyle, access to the transverse process of the C1 vertebra, and use of an avascular plane along the midline over the suboccipital space and spinous processes in the neck. The drawbacks of the inverted hockey-stick incision are that it is associated with increased surgery duration, potentially increased blood loss, and less satisfying aesthetic outcomes postoperatively.

The lazy-S incision uses a flat S-shaped incision. This incision begins at the asterion and continues to the foramen magnum in a diagonal trajectory, where the incision then curves medially toward the spinous process of the axis (C2). This incision provides sufficient exposure during the standard far lateral approach and the transcondylar and supracondylar variants of the far lateral approach. However, the lazy-S incision may not be suitable for the paracondylar variant because it requires the complete exposure of the transverse process of the atlas.

3.3 Muscular Dissection

The preparation of muscular flaps aims to preserve the muscular architecture as much as possible while exposing the bone sufficiently [1]. We do not recommend multiple-layer dissection because it may lead to dehiscence, loss of function, or ischemic atrophy [1, 12]. However, the far lateral approach is unique in that preservation of the vertebral artery needs to be ensured. Therefore, a two-step muscular dissection is used, including a nuchal and suboccipital dissection phase [9, 14, 18]. The insertions of the nuchal muscles are encased between the superior and inferior nuchal lines. The nuchal muscular flap is dissected from approximately 1 cm inferior to the superior nuchal line for muscle attachment during closure [18]. The nuchal muscular flap comprises the sternocleidomastoideus, trapezius, longissimus capitis, splenius capitis, and semispinalis capitis muscles (Fig. 6) [15, 18]. Extension of the nuchal muscular flap inferiorly and laterally exposes the suboccipital triangle (Fig. 7) [15, 18], formed by three triangularly aligned nuchal muscles. The superior oblique muscle, which originates from the transverse process of the atlas and attaches to the occipital bone, forms the lateral border of the occipital triangle [18]. The inferior oblique muscle, which originates from the spinous process of the axis and attaches to the transverse process of the atlas, forms the inferior border of the

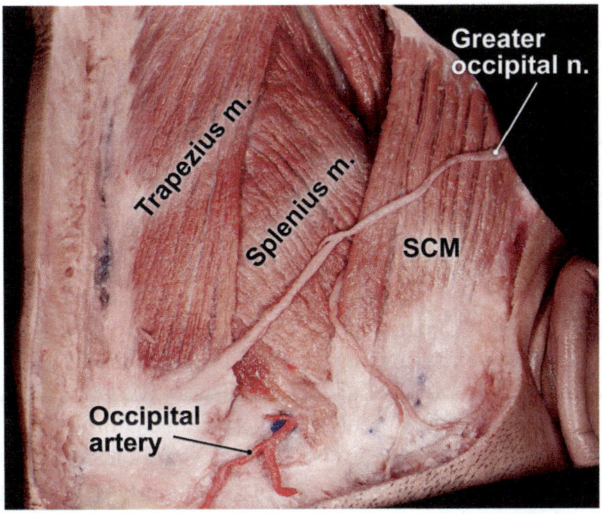

Fig. 6 Cadaveric view of the superficial nuchal muscular dissection through the far lateral approach. The sternocleidomastoid, trapezius, and splenius capitis muscles are exposed. The greater occipital nerve originates from the medial branch of the dorsal ramus of C2 and pierces the trapezius muscle before ascending to the occiput. *m* muscle, *n* nerve, *SCM* sternocleidomastoid muscle. *Image provided by Arnau Benet. Used with permission of the University of California San Francisco's Skull Base & Cerebrovascular Laboratory*

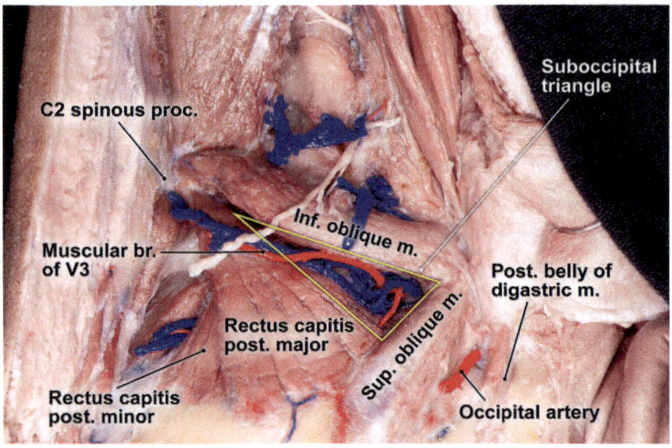

Fig. 7 Cadaveric view of the deep nuchal muscular dissection through the far lateral approach. The suboccipital triangle (*yellow triangle*) is formed by the superior oblique, rectus capitis posterior major, and inferior oblique muscles. It is located at the center of dissection following reflection of these muscles. The suboccipital triangle harbors the V3 segment of the vertebral artery, the dorsal ramus of the C1 cervical nerve (suboccipital nerve), and the suboccipital venous plexus. *br* branch, *Inf* inferior, *m* muscle, *post* posterior, *proc* process, *Sup* superior. *Image provided by Arnau Benet. Used with permission of the University of California San Francisco's Skull Base & Cerebrovascular Laboratory*

suboccipital triangle [18]. Lastly, the rectus capitis posterior major muscle, which originates from the spinous process of the axis and attaches to the lateral aspect of the inferior nuchal line, closes the triangle at its inferomedial apex [18]. Three important neurovascular structures traverse the suboccipital triangle and can be encountered during dissection: the dorsal ramus of the C1 cervical nerve (the suboccipital nerve), the third segment of the vertebral artery, and the suboccipital venous plexus [10, 18]. Identifying these structures within the suboccipital triangle is critical to prevent inadvertent neural injury or venous or arterial bleeding [9, 10].

The suboccipital vertebral venous plexus is a tangle of densely anastomosed, valveless veins. This venous plexus resides in the suboccipital triangle and drains into the sigmoid sinus and jugular bulb via the posterior condylar emissary vein [10]. Enclosed in this venous plexus are the muscular branches of the vertebral artery and the dorsal roots of the C1 cervical nerves [10, 18]. Particular caution is required when dissecting the venous plexus due to anatomical variations in the course of the third segment of the vertebral artery because it sometimes passes more superiorly [10, 18]. Although rare, the posterior spinal artery and the posterior inferior cerebellar artery can branch off this vertebral artery segment and must not be mistaken for muscular branches. Upon exposure of the suboccipital triangle, three deep-seated muscles are encountered that serve as orienting landmarks during the far lateral approach. In line with the vertical trajectory of the more superficial superior oblique muscle runs the rectus capitis lateralis muscle, which spans from the superior surface of the transverse process of the atlas to the inferior surface of the jugular process at the posterior aspect of the jugular tubercle. Despite its short length, the rectus capitis lateralis muscle is extraordinarily valuable in guiding the paracondylar dissection toward the jugular foramen. Inferior to the rectus capitis lateralis muscle, the levator scapulae muscle extends from the posterior tubercles of the transverse processes of the C1 to C4 vertebrae to the vertebral margin of the scapula between the superior angle and the root of the spine. Early identification of the levator scapulae muscle during cervical dissection is critical because it serves as the lateral landmark for the V2 segment of the vertebral artery and as the medial landmark for the carotid compartment of the parapharyngeal space. Medially, the rectus capitis posterior minor muscle becomes apparent as it extends from the tubercle on the posterior arch of the atlas to the medial aspect of the inferior nuchal line of the occipital bone. Medial reflection of this muscle allows complete exposure of the posterior atlanto-occipital membrane (Fig. 8).

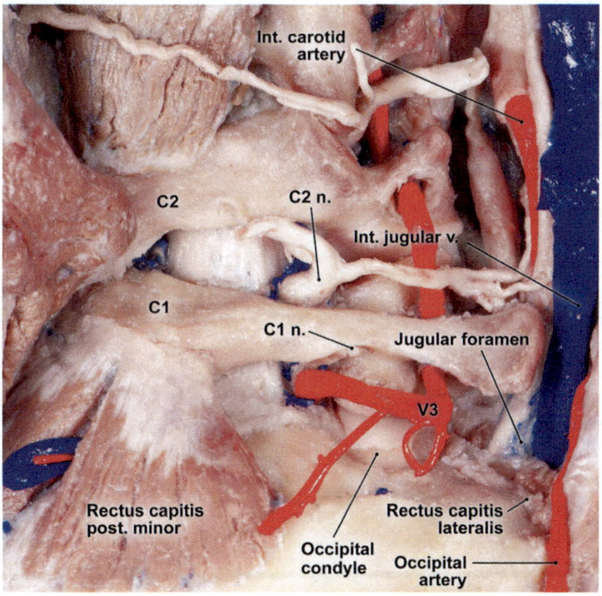

Fig. 8 Cadaveric view of deep dissection through the suboccipital triangle during the far lateral approach. The V3 segment of the vertebral artery courses vertically through the transverse foramina of the atlas and axis vertebrae. The rectus capitis lateralis is a small muscle that originates from the transverse process of the C1 vertebrae and attaches to the inferior surface of the jugular process at the squamous part of the occipital bone. *Int* internal, *n* nerve, *post* posterior, *v* vein. *Image provided by Arnau Benet. Used with permission of the University of California San Francisco's Skull Base & Cerebrovascular Laboratory*

3.4 Craniotomy

The craniotomy is designed to maximize exposure to the lesion [2]. Depending on the type and location of the lesion, a combined craniotomy and craniectomy may be performed [19]. The far lateral approach uses a lateral suboccipital craniotomy that may be augmented using a transcondylar, supracondylar, or paracondylar extension. The suboccipital craniotomy provides access to the cerebellar hemisphere, cerebellar vermis, lateral sinus, and the confluence of the sinuses, foramen magnum, and mastoid process. Furthermore, this craniotomy allows access to the inferior vermian and hemispheric veins that drain the ipsilateral cerebellum toward the torcular and the transverse and tentorial sinuses.

For the lateral suboccipital craniotomy, a square-shaped bone flap is drilled into the squamous part of the occipital bone. The bone window is confined superiorly by the transverse sinus and laterally by the sigmoid sinus and the occipital condyle (Fig. 9). The suboccipital craniotomy includes drilling of the condylar fossa. Extreme caution is needed where the posterior condylar vein bridges from the jugular bulb to the suboccipital venous plexus. Several landmarks are used to assist the

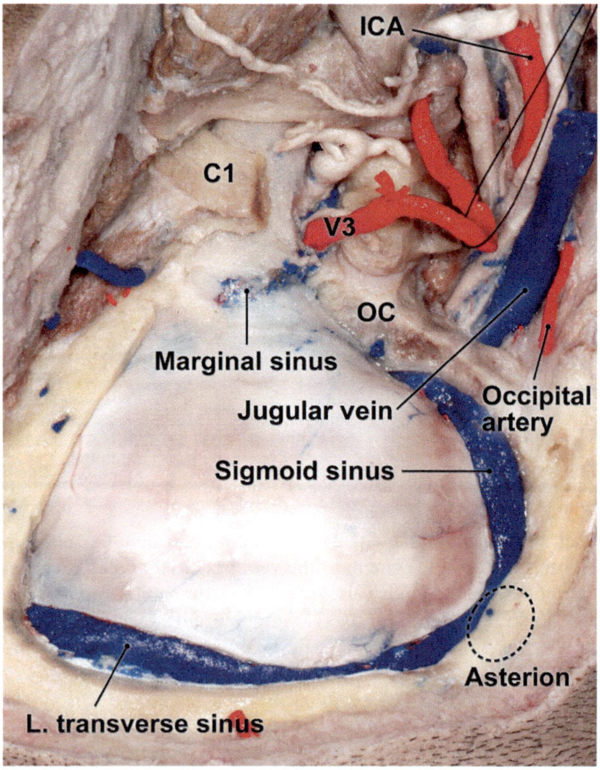

Fig. 9 Cadaveric view of the suboccipital craniotomy during the far lateral approach. The asterion marks the transition from the transverse into the sigmoid sinus. The marginal sinus runs along the inner margin of the foramen magnum and communicates with the basilar venous plexus anteriorly, the occipital sinus posteriorly, the vertebral venous plexus inferiorly, and the sigmoid sinus laterally. *ICA* internal carotid artery, *L* left, *OC* occipital condyle. *Image provided by Arnau Benet. Used with permission of the University of California San Francisco's Skull Base & Cerebrovascular Laboratory*

neurosurgeon in planning the craniotomy alongside these significant vascular structures. The asterion indicates the location of the sigmoid sinus laterally, and the superior nuchal line indicates the most cranial level of extension for protection of the transverse sinus and the torcular. Medially, the lateral suboccipital craniotomy is guided by the external occipital crest. Because the medial border of the suboccipital craniotomy has no limiting vascular structure, it can be extended beyond the occipital crest if the lesion occupies the majority of the cisterna magna. In most cases, a medial extension up to the occipital crest provides sufficient space for cerebellar distention.

The standard suboccipital craniotomy often includes drilling of the posterior arch of the atlas. The first step when planning to remove the posterior arch of the atlas is a meticulous dissection of the vertebral artery from its connective sheath. This step should be followed by identifying the root of the C2 cervical nerve medial to the atlantoaxial joint. En bloc removal of the posterior arch of the atlas is achieved by placing two cuts; the first is in the midline medial to the rectus capitis posterior minor muscle, and the second is on the lateral aspect of the posterior arch of the atlas. Attention is directed laterally toward the transverse foramen of the atlas, whose roof is then drilled carefully. This exposure is adequate for transposing the distal V2 segment of the vertebral artery laterally, and it enables safer access to the posterior fossa and the occipital condyle (Fig. 10).

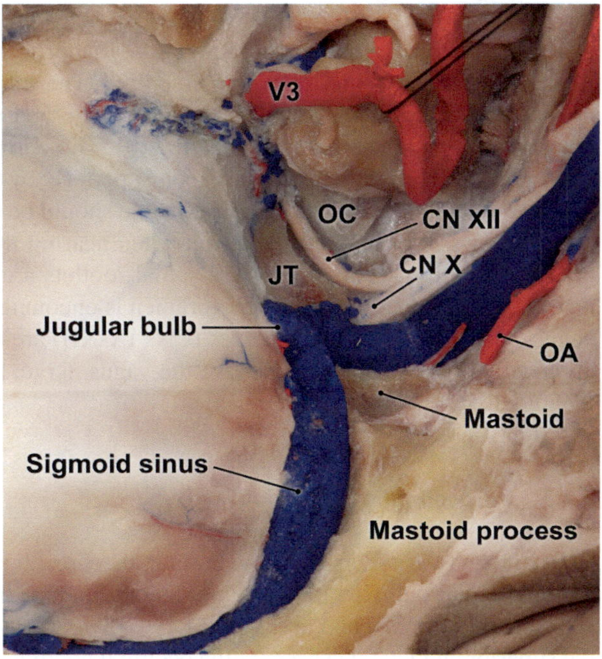

Fig. 10 Cadaveric view of a maximized exposure during the extended far lateral approach. Drilling of the occipital condyle provides access to the jugular tubercle and the jugular bulb. The jugular tubercle is located between the hypoglossal canal and the jugular foramen. *CN* cranial nerve, *JT* jugular tubercle, *OA* occipital artery, *OC* occipital condyle. *Image provided by Arnau Benet. Used with permission of the University of California San Francisco's Skull Base & Cerebrovascular Laboratory*

3.4.1 Transcondylar Extension

Of the three craniectomy variants that can accompany the suboccipital craniotomy, the transcondylar extension is used most frequently. This extension's trajectory is directed above the atlanto-occipital joint through the occipital condyle and below the hypoglossal canal [20]. The course of the hypoglossal canal is critically important during the transcondylar extension because it is the first structure to be encountered in the medial aspect of the exposure when drilling the occipital condyle. The trajectory of the hypoglossal canal is followed laterally as the drilling continues anteriorly. Removal of the posterior third of the occipital condyle is continued until a change in bone consistency and color is noticed as the cancellous bone of the condyle becomes the inner cortical posterior wall of the hypoglossal canal. The hypoglossal canal contains a venous plexus that causes the wall of the hypoglossal canal to appear dark blue. Preserving the hypoglossal canal is advised as long as it does not directly involve the lesion. However, the condylectomy may be extended inferiorly and medially to the hypoglossal canal to better access the lower clivus. The transcondylar approach also gives access to the intracranial segment of the vertebral artery and the anterior medulla. A complete condylectomy may be needed if the pathology invades the condyle or its vicinity.

3.4.2 Supracondylar Extension

The supracondylar extension of the far lateral approach is used when access to the jugular foramen or upper medulla is required. The trajectory of the supracondylar approach is directed above the occipital condyle and includes the removal of the posterior aspect of the jugular tubercle [20]. The narrow surgical corridor is confined inferiorly by the hypoglossal canal, superiorly by the sigmoid sinus, and laterally by the jugular bulb. The jugular tubercle should be drilled using a diamond bur, which allows for maximum precision when nearing the jugular foramen's medial border where the accessory nerve's spinal rootlets are contained in the dura. The neural compartment of the jugular foramen is encountered next, and the accessory, vagus, and glossopharyngeal nerves (CNs IX-XII) are exposed. The supracondylar approach allows maneuvering above the vagus nerve, the petroclival junction, and the midclivus.

3.4.3 Paracondylar Extension

The paracondylar extension of the far lateral approach is feasible for lesions involving the jugular bulb, the lower sigmoid sinus, and the meatal segments of the glossopharyngeal, vagus, and accessory nerves. This variant's trajectory is directed to the superior and lateral aspects of the occipital condyle, the exocranial aspect of the jugular foramen, and the posterior aspect of the mastoid process (Fig. 10) [20]. A more anterolateral muscular flap must be raised and can include the posterior belly of the digastric muscle. The digastric muscle inserts at the digastric groove of the mastoid bone and provides a landmark for the stylomastoid foramen on the inferior surface of the petrous temporal bone [8]. The facial nerve exits the mastoid process through the stylomastoid foramen. A diamond bur is used when drilling around the posterolateral aspect of the jugular foramen, which yields to allow removal of the jugular process. This drilling results in the exposure of several neurovascular structures, including the lower part of the sigmoid sinus, the jugular bulb, the jugular vein, the neural compartment of the jugular foramen, and the pharyngeal segment of the internal carotid artery.

3.5 Dural Opening

After the dura has been exposed via the suboccipital craniotomy, the neurosurgeon must be aware of the underlying structures that border the exposure, which in this case represent a significant source of potential hemorrhage [21]. The craniotomy window is confined by the transverse sinus superiorly, the sigmoid sinus laterally, and the marginal sinus inferiorly. Additionally, the posterior meningeal artery sometimes originates from the V4 segment of the vertebral artery and crosses the surgical field. In addition, the posterior spinal artery may branch off the V4 segment of the vertebral artery at its dural cuff. Finally, the meningeal branch of the ascending pharyngeal artery may be encountered in the inferolateral aspect of the dural opening. Usually, the patient's anatomy dictates the design of the dural flap. However, most dural flaps share an incision alongside the midline and the superior border of the craniotomy for a lateral reflection of the flap. The dural opening is designed to easily access the lesion based on its size and location, allowing for a surgical corridor that accounts for all anticipated trajectories [21]. A wide dural flap approach allows for more cerebellar retraction and better access to the superior spinal region [21]. If the cerebellopontine angle is involved, opening the dura near the transition of the transverse sinus into the sigmoid sinus is advisable while leaving a margin for safe closure. Upon dural opening, an incision is made to release the cerebrospinal fluid from the cisterna magna (also known as cisterna cerebellomedullaris), allowing atraumatic retraction of the cerebellar hemispheres and maximizing the effect of gravity [5].

3.6 Intradural Exposure

The intradural exposure of the far lateral approach offers extensive access to the dorsolateral compartment of the posterior cranial fossa. However, this exposure provides only limited access to the ventromedial compartment, the petroclival region, and the lateral aspect of the middle and lower thirds of the clivus (Fig. 4). During the transcondylar, supracondylar, and paracondylar extensions of the far lateral

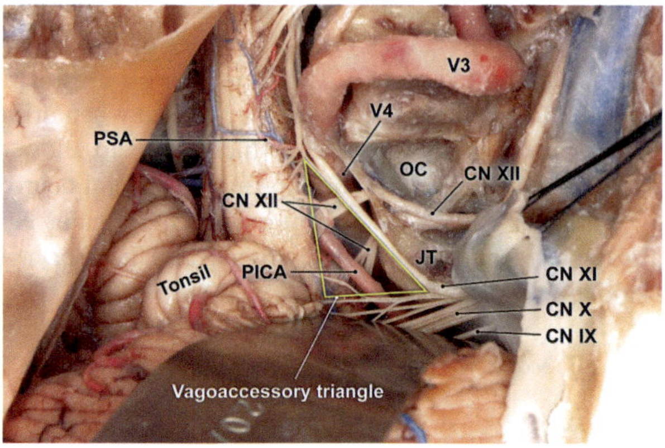

Fig. 11 Cadaveric view of the intradural anatomy during a left far lateral approach. The vagoaccessory triangle (*yellow triangle*) is formed by the vagus and accessory nerves (CNs X and XI). The hypoglossal nerve divides the vagoaccessory triangle into the infrahypoglossal and suprahypoglossal triangles. Both triangles provide access to the vertebral artery and the posterior inferior cerebellar artery. *CN* cranial nerve, *JT* jugular tubercle, *OC* occipital condyle, *PICA* posterior inferior cerebellar artery, *PSA* posterior spinal artery. *Image provided by Arnau Benet. Used with permission of the University of California San Francisco's Skull Base & Cerebrovascular Laboratory*

approach, the extent of structures being exposed increases as bone removal progresses. The conventional suboccipital craniotomy provides access to the cisterna magna and the ipsilateral cerebellar hemisphere (Fig. 11). Within the cisterna magna reside the inferomedial aspect of the cerebellar hemisphere, the cerebellar tonsil, and the inferior part of the medulla. The obex, which represents the inferior angle of the fourth ventricle before it continues as the central canal of the spinal cord, is visualized through dissection of the arachnoid. It serves to help identify the foramen of Magendie. Opening the cisterna magna leads to exposure of the intradural vertebral artery. The vertebral artery enters the dura at the inferomedial aspect of the occipital condyle. It pierces the lateral aspect of the posterior atlanto-occipital membrane and is stabilized by the topmost denticulate ligament at the craniocervical junction [20].

The V4 segment of the vertebral artery passes through the cisterna magna and the premedullary cisterns toward the clivus in an anteromedial trajectory. It then joins the contralateral vertebral artery at the vertebrobasilar junction to form the basilar artery. The posterior inferior cerebellar artery (PICA) is the largest branch of the vertebral artery. It usually branches from the vertebral artery at the premedullary cistern near the inferior olive; this first segment of the PICA is called the anterior medullary segment. The next segment, the lateral medullary segment, crosses the rootlets of the vagus and accessory nerves (CNs X and XI) before taking a looping course in the cisterna magna below the cerebellar hemisphere [7]. The looping course of the PICA can be followed medially and superiorly around the cerebellar tonsil. This segment of the PICA is called the tonsillomedullary segment [7]. The final segment of the PICA, the telovelomedullary segment, passes between the cerebellum and the posterior wall of the fourth ventricle, progressively moving away from the surgical field exposed through the far lateral approach [7]. Finally, the PICA takes a superficial turn (cortical segment) to supply the posterior aspect of the ipsilateral cerebellar hemisphere [7]. In addition, the posterior spinal artery may arise from the PICA or the V3 or V4 segment of the vertebral artery. Early recognition and protection of the posterior spinal artery are vital to prevent inadvertent bleeding when dissecting the arachnoid space of the cisterna magna [15].

Within the surgical field of the far lateral approach, the accessory nerve (CN XI), which innervates the ipsilateral sternocleidomastoid and trapezius muscles, emerges at the lateral aspect of the cisterna magna. Its long cisternal segment receives nerve rootlets from the cervical spine and the lower aspect of the posterolateral medullary sulcus. The accessory nerve follows a superolateral trajectory from its spinal origin toward the jugular foramen and enters the skull base. Notably, it is the only nerve to enter and exit the skull base. Upon exiting the skull base through the jugular foramen, the accessory nerve joins the vagus nerve. Together, they define the vagoaccessory triangle. The vagoaccessory triangle serves as the main corridor to access the ventromedial compartment of the posterior fossa (Fig. 11). The triangle is bordered superiorly by the vagus nerve and medullary rootlets of the accessory nerve, laterally by the main trunk of the accessory nerve, and medially by the medulla oblongata [5, 12]. The vagoaccessory triangle is further subdivided into the suprahypoglossal and infrahypoglossal triangles based on the intercrossing course of the hypoglossal nerve (CN XII) [5]. Within the cranium, the accessory nerve courses alongside the denticulate ligament, which is located anteriorly to the accessory nerve rootlets. As both structures ascend, their fibers cross each other as the denticulate ligament anchors to the dural cuff of the vertebral artery. The accessory nerve then runs in an anterosuperior trajectory toward the posterior aspect of the jugular foramen (in the neural compartment). The denticulate ligament courses between the accessory nerve and the vertebral artery and is microscopically attached to the vertebral adventitia in 20% of patients [22]. Knowledge of the relationship of the three structures is of the utmost importance when operating at the craniocervical junction [22].

The transcondylar extension of the far lateral approach allows access to the cisternal, canalicular, and cervical segments of the hypoglossal nerve (CN IX). The extended working window provides a maximized angle of exposure to both the intradural part of the vertebral artery and the inferior aspects of the cisterna magna and premedullary cisterns. Access to the cisterna magna enables manipulation of the medullary rootlets of the glossopharyngeal, vagus, accessory, and hypoglossal nerves (CNs IX-XII). It also provides access to the lateral aspects of the medulla and the lateral medullary segment of the PICA. In addition, the transcondylar craniectomy increases the infrahypoglossal working window of the vagoaccessory triangle.

The supracondylar extension of the far lateral approach allows greater access to the superior aspect of the premedullary and cerebellomedullary cisterns. A combined transcondylar and supracondylar extension is used when even greater access to these cisterns is needed. The supracondylar approach is the variant of choice to gain maximal access to the vagoaccessory triangle and beyond. Specifically, the vertebral artery, vertebrobasilar junction, and lower basilar artery may be accessed through the working corridors of the suprahypoglossal and infrahypoglossal triangles. In addition, the petroclival junction and medial portion of the clivus are exposed as the drilling of the jugular tubercle is continued anteriorly. The extent of accessibility of neurovascular structures through the far lateral approach depends greatly on the working angle of the microscope and the placement of the retractors. Drilling the jugular tubercle and retraction of the midportion of the cerebellar hemisphere increase visualization of the hypoglossal canal, jugular foramen, internal acoustic meatus, and inferior surface of the tentorium while allowing the neurosurgeon to explore a wide variety of anatomical landmarks and surgical lesions in a single approach. For instance, when the neurosurgeon directs the microscope toward the petrous part of the temporal bone, the distal anterior inferior cerebellar artery, the trigeminal, facial, and vestibulocochlear nerves (CNs V, VII, and VIII), and the foramen of Luschka appear in the surgical field. Three working corridors may be created from this perspective: (1) the space between the vestibulocochlear and glossopharyngeal nerves (CNs VIII and IX); (2) the space between the trigeminal nerve (CN V) and the facial-vestibulocochlear nerve bundle (CNs VII-VIII); and (3) the space between the trigeminal nerve (CN V) and the tentorium (Fig. 12). Finally, directing the microscope toward the brainstem exposes the pons at the level of the cerebellopontine cistern, the lateral aspect of the pontomedullary sulcus, and the medulla at the level of the premedullary and cerebellomedullary cisterns.

Using an endoscope during the far lateral approach allows the neurosurgeon to gain greater access to the ventromedial compartment of the posterior cranial fossa. The major advances of the endoscope, especially when used in combination with

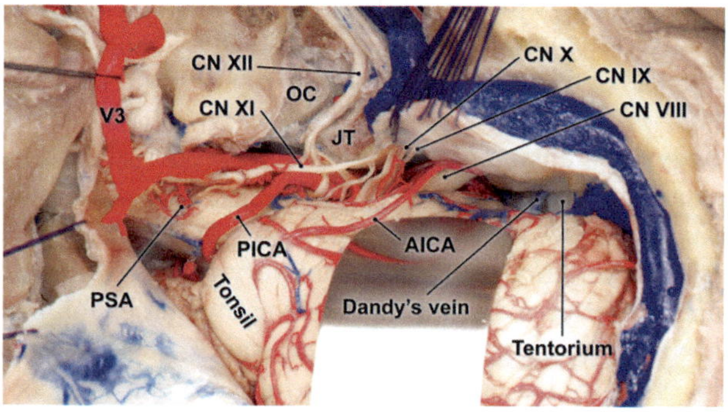

Fig. 12 Cadaveric superior view of the intradural anatomy during the left far lateral approach. Viewed through a lateral approach to the posterior fossa, the surgical corridors include the supratrigeminal, trigeminal-cochlear, cochlear-glossopharyngeal, and vagoaccessory spaces. *AICA* anterior inferior cerebellar artery, *CN* cranial nerve, *JT* jugular tubercle, *OC* occipital condyle, *PICA* posterior inferior cerebellar artery, *PSA* posterior spinal artery. *Image provided by Arnau Benet. Used with permission of the University of California San Francisco's Skull Base & Cerebrovascular Laboratory.*

the microscope, are a view and lighting that reach deeper than that of the microscope alone. This advantage becomes even more significant with the use of angled endoscopes. For example, a 30° angled endoscope passed through the vagoaccessory triangle enables the neurosurgeon to explore the medial aspect of the medulla, the abducens nerve (CN VI) at the pontomedullary sulcus medially, and the medial aspect of the jugular foramen laterally. Likewise, when the angled endoscope is used in the surgical corridor between the vestibulocochlear and glossopharyngeal nerves (CNs VIII and XII), it exposes the origin of the trigeminal nerve (CN V), the prepontine cistern, the proximal segment of the anterior inferior cerebellar artery, and the cisternal segment of the abducens nerve (CN VI) on the medial aspect of the surgical field, as well as the internal acoustic meatus in the cerebellopontine angle in the lateral aspect of the surgical field. Despite these advantages, the endoscope has some limitations, including a limited capacity for passing additional instruments through the already narrow surgical corridor without limiting the view. Furthermore, the bypassed tissue may be at risk of retraction damage as the endoscope is advanced through the long surgical corridors without visual control on its external surface.

4 Case Illustration

A woman in her 70s presented to the emergency department with progressive right leg weakness and headaches. Contrast-enhanced magnetic resonance imaging revealed a large dolichoectatic left vertebral artery aneurysm compressing the medulla (Fig. 13a). The patient underwent diagnostic catheter angiography revealing an aneurysm of the V4 segment of the vertebral artery. The aneurysm was located close to the branching point of the posterior inferior cerebellar artery (Fig. 13b). Several endovascular treatment options, including stent placement with coiling and flow diversion, were discussed with the patient. The patient elected open vascular surgery with clipping of the aneurysm and decompression of the medulla. The posterior cranial fossa was accessed via the far lateral approach and a standard suboccipital craniotomy. The cisterna magna was entered to access the posterolateral compartment of the posterior cranial fossa (Fig. 13c). Following exposure of the V4 segment of the vertebral artery through the vagoaccessory triangle (Fig. 13d), the aneurysm was softened by placing a temporary clip proximally.

Next, the aneurysm was obliterated using two crosswise 45° angled fenestrated clips that overlapped at their interface, allowing complete closure at the aneurysm neck [23, 24] (Fig. 13e). Intraoperative indocyanine green video angiography confirmed complete occlusion of the aneurysm (Fig. 13f). Incision of the aneurysm deflated the aneurysm and led to decompression of the medulla. Intraoperative video angiography with Yellow 560 was used to confirm the patency of the medullary perforating arteries (Fig. 13g). Postoperative angiography demonstrated complete occlusion of the aneurysm, showing excellent positioning of the clip construction (Fig. 13h). The patient tolerated the surgery well. She had no new focal deficits, and her weakness had significantly improved on the 3-month follow-up.

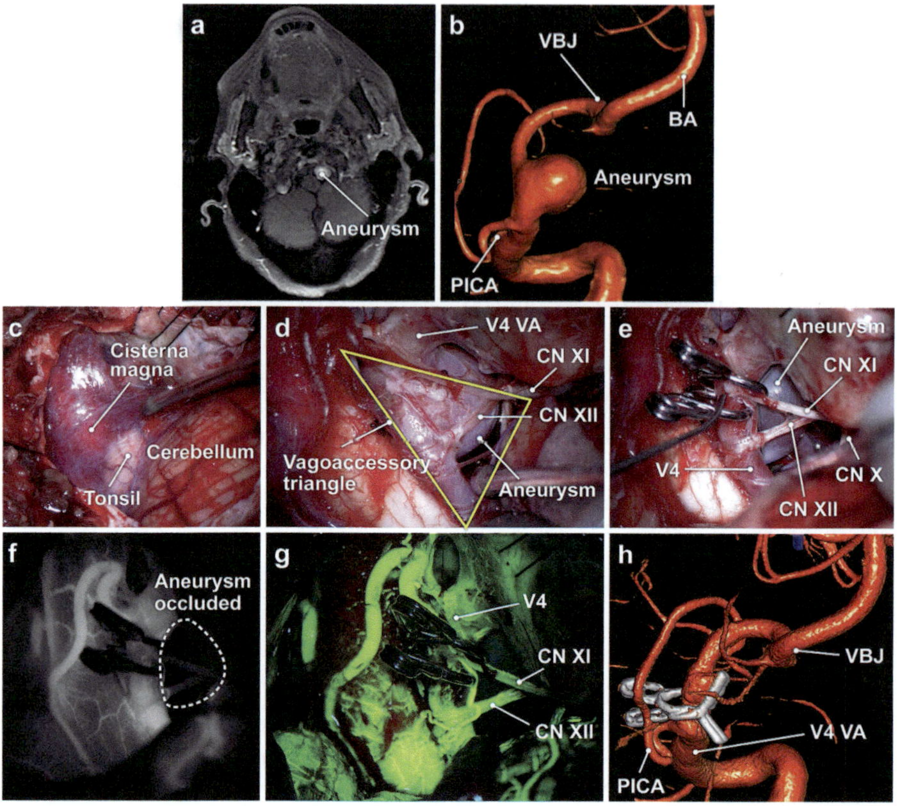

Fig. 13 Imaging studies and intraoperative photographs of a patient with a wide-necked aneurysm of the V4 segment of the vertebral artery accessed via the far lateral approach. (**a**) Axial T1-weighted contrast magnetic resonance imaging of the posterior fossa showing a left-sided vertebral aneurysm with medullar compression. (**b**) Three-dimensional (3D) reconstruction of a digital subtraction angiogram revealing a wide-necked left vertebral aneurysm located distally to the posterior inferior cerebellar artery. (**c**) Intraoperative photograph of the intradural exposure of the cisterna magna and the left inferior cerebellum following a suboccipital craniotomy. (**d**) Intraoperative photograph of the intradural anatomy exposing the vagoaccessory triangle (*yellow triangle*). The V4 aneurysm compressed the accessory and hypoglossal nerves and the medulla oblongata. (**e**) Intraoperative photograph showing the tandem crosswise clip occlusion of the V4 aneurysm accessed through the vagoaccessory triangle. Two 45° angled fenestrated aneurysm clips were placed in a crosswise direction, and their overlapping interface resulted in complete occlusion of the aneurysm. (**f**) Intraoperative indocyanine green videoangiography showing complete occlusion of the aneurysm. (**g**) Intraoperative Yellow 560 videoangiography showing patency of the perforating arteries adjacent to the positioning of the aneurysm clips. (**h**) Postoperative 3D reconstruction of a digital subtraction angiogram demonstrating the final clip configuration with no remnant and confirming patency of adjacent perforating arteries. *BA* basilar artery, *CN* cranial nerve, *PICA* posterior inferior cerebellar artery, *VA* vertebral artery, *VBJ* vertebrobasilar junction. *Used with permission from Barrow Neurological Institute, Phoenix, Arizona*

This illustrative case highlights the significant aspects of the far lateral approach, including wide exposure of the vertebral artery, allowing the neurosurgeon to follow its course from outside to inside the cranium [1, 5, 19]. Furthermore, the location of the aneurysm required entering the ventromedial compartment of the posterior cranial fossa, which was followed by advancing through the vagoaccessory triangle to achieve proximal and distal control of the aneurysm [5]. Finally, retraction of the cerebellar hemisphere allowed the neurosurgeon to perform a complex tandem counter-clipping technique through a wide working angle [24].

Acknowledgments We thank the staff of Neuroscience Publications at Barrow Neurological Institute for assistance with manuscript preparation.

Disclosures The authors have no personal, financial, or institutional interest in any drugs, materials, or devices described in this manuscript.

Financial Support None provided for this chapter.

References

1. Lanzino G, Paolini S, Spetzler RF. Far-lateral approach to the craniocervical junction. Neurosurgery. 2005;57(4 Suppl):367–71; discussion-71. https://doi.org/10.1227/01.neu.0000176848.05925.80.
2. Benet A, Prevedello DM, Carrau RL, Rincon-Torroella J, Fernandez-Miranda JC, Prats-Galino A, et al. Comparative analysis of the transcranial "far lateral" and endoscopic endonasal "far medial" approaches: surgical anatomy and clinical illustration. World Neurosurg. 2014;81(2):385–96. https://doi.org/10.1016/j.wneu.2013.01.091.
3. Jahangiri A, Chin AT, Wagner JR, Kunwar S, Ames C, Chou D, et al. Factors predicting recurrence after resection of clival chordoma using variable surgical approaches and radiation modalities. Neurosurgery. 2015;76(2):179–85; discussion 85-6. https://doi.org/10.1227/NEU.0000000000000611.
4. Chan AK, Benet A, Ohya J, Zhang X, Vogel TD, Flis DW, et al. The endoscopic transoral approach to the craniovertebral junction: an anatomical study with a clinical example. Neurosurg Focus. 2016;40(2):E11. https://doi.org/10.3171/2015.11.FOCUS15498.
5. Abla AA, Benet A, Lawton MT. The far lateral transpontomedullary sulcus approach to pontine cavernous malformations: technical report and surgical results. Neurosurgery. 2014;10(Suppl 3):472–80. https://doi.org/10.1227/NEU.0000000000000389.
6. de Notaris M, Cavallo LM, Prats-Galino A, Esposito I, Benet A, Poblete J, et al. Endoscopic endonasal transclival approach and retrosigmoid approach to the clival and petroclival regions. Neurosurgery. 2009;65(6 Suppl):42–50; discussion-2. https://doi.org/10.1227/01.NEU.0000347001.62158.57.
7. Nisson PL, Ding X, Tayebi Meybodi A, Palsma R, Benet A, Lawton MT. Revascularization of the posterior inferior cerebellar artery using the occipital artery: a cadaveric study comparing the p3 and p1 recipient sites. Oper Neurosurg (Hagerstown). 2020;19(2):E122–E9. https://doi.org/10.1093/ons/opaa023.

8. Tayebi Meybodi A, Lawton MT, Mokhtari P, Kola O, El-Sayed IH, Benet A. Exposure of the external carotid artery through the posterior triangle of the neck: a novel approach to facilitate bypass procedures to the posterior cerebral circulation. Oper Neurosurg (Hagerstown). 2017;13(3):374–81. https://doi.org/10.1093/ons/opw024.
9. Osorio JA, Benet A, Hess CP, McDermott MW, Abla AA. Primary vertebral artery reanastomosis during retrosigmoid skull base approach following iatrogenic near-transection with monopolar electrocautery. Neurosurgery. 2014;10(Suppl 4):631–9. https://doi.org/10.1227/NEU.0000000000000526.
10. Tabani H, Tayebi Meybodi A, Benet A. Venous anatomy of the supratentorial compartment. Handb Clin Neurol. 2020;169:55–71. https://doi.org/10.1016/B978-0-12-804280-9.00003-2.
11. Tayebi Meybodi A, Lawton MT, Benet A. Sequential extradural release of the V3 vertebral artery to facilitate intradural V4 vertebral artery reanastomosis: feasibility of a novel revascularization technique. Oper Neurosurg (Hagerstown). 2017;13(3):345–51. https://doi.org/10.1093/ons/opw015.
12. Tabani H, Yousef S, Burkhardt JK, Gandhi S, Benet A, Lawton MT. Macrovascular decompression of facial nerve with anteromedial transposition of a dolichoectatic vertebral artery: 3-dimensional operative video. Oper Neurosurg (Hagerstown). 2019;16(1):E4. https://doi.org/10.1093/ons/opy117.
13. Baldwin HZ, Miller CG, van Loveren HR, Keller JT, Daspit CP, Spetzler RF. The far lateral/combined supra- and infratentorial approach. A human cadaveric prosection model for routes of access to the petroclival region and ventral brain stem. J Neurosurg. 1994;81(1):60–8. https://doi.org/10.3171/jns.1994.81.1.0060.
14. Benet A, Tabani H, Ding X, Burkhardt JK, Rodriguez Rubio R, Tayebi Meybodi A, et al. The transperiosteal "inside-out" occipital artery harvesting technique. J Neurosurg. 2018;130(1):207–12. https://doi.org/10.3171/2017.6.JNS17518.
15. Tayebi Meybodi A, Lawton MT, Benet A. Dual origin of extradural posterior inferior cerebellar artery from vertebral and occipital arteries: anatomic case report. Oper Neurosurg (Hagerstown). 2015;11(4):564–8. https://doi.org/10.1227/NEU.0000000000000923.
16. Raabe A, Meyer B, Schaller K, Vajkoczy P, Winkler PA. The craniotomy atlas. Thieme. 2019:11–2.
17. Tayebi Meybodi A, Lawton MT, Tabani H, Benet A. Tonsillobiventral fissure approach to the lateral recess of the fourth ventricle. J Neurosurg. 2017;127(4):768–74. https://doi.org/10.3171/2016.8.JNS16855.
18. Tayebi Meybodi A, Benet A, Lawton MT. The V3 segment of the vertebral artery as a robust donor for intracranial-to-intracranial interpositional bypasses: technique and application in 5 patients. J Neurosurg. 2018;129(3):691–701. https://doi.org/10.3171/2017.4.JNS163195.
19. Lawton MT, Abla AA, Rutledge WC, Benet A, Zador Z, Rayz VL, et al. Bypass surgery for the treatment of dolichoectatic basilar trunk aneurysms: a work in progress. Neurosurgery. 2016;79(1):83–99. https://doi.org/10.1227/NEU.0000000000001175.
20. Rhoton AL Jr. The far-lateral approach and its transcondylar, supracondylar, and paracondylar extensions. Neurosurgery. 2000;47(3 Suppl):S195–209. https://doi.org/10.1097/00006123-200009001-00020.
21. Tayebi Meybodi A, Tabani H, Benet A. Arachnoid and dural reflections. Handb Clin Neurol. 2020;169:17–54. https://doi.org/10.1016/B978-0-12-804280-9.00002-0.
22. Tubbs RS, Mortazavi MM, Loukas M, Shoja MM, Cohen-Gadol AA. The intracranial denticulate ligament: anatomical study with neurosurgical significance. J Neurosurg. 2011;114(2):454–7. https://doi.org/10.3171/2010.9.JNS10883.
23. Benet A, Tabani H, Griswold D, Yousef S, Meybodi AT, Lawton MT. Clip reconstruction of large posterior inferior cerebellar artery aneurysm: 3-dimensional operative video. Oper Neurosurg (Hagerstown). 2018;14(5):590. https://doi.org/10.1093/ons/opx182.
24. Benet A, Griswold D, Tabani H, Rubio RR, Yousef S, Meybodi AT, et al. Cross-wise counter clipping of a dolichoectatic left vertebral artery aneurysm: 3-dimensional operative video. Oper Neurosurg (Hagerstown). 2018;14(2):204–5. https://doi.org/10.1093/ons/opx083.

Surgical Anatomy of the Foramen Magnum

Kaan Yağmurlu, Musa Çırak, and James Liu

1 Bone Anatomy

The foramen magnum is an oval or circular-shaped foramen surrounded by the occipital bone. The occipital bone has a basal (clivus), squamous, and condylar (Fig. 1a) [1]. The clivus forms the anterior part of the foramen magnum, and the squamous part forms the posterior part. At the same time, occipital condyles are situated anterior half of the foramen on both sides. The clivus articulates with the petrous part of the temporal bone laterally and sphenoid bone superiorly. The petroclival fissure is the junction of the petrous bone of the temporal bone and clivus that the inferior petrosal sinus runs along the endocranial surface of the fissure while the inferior petrosal vein runs on the exocranial surface. On the exocranial surface of the clivus, the pharyngeal tubercle is an important landmark for transoral or endonasal routes to the foramen magnum (Fig. 1b). The squamous part of the occipital bone articulates with parietal bone by the lambdoid suture superiorly, mastoid part of the temporal bone by the occipitomastoid suture laterally (Fig. 1c). The squamous part has a vertical crest on the extracranial surface, and the external occipital

K. Yağmurlu (✉)
Department Neurosurgery, University of Tennessee Health Science Center, Memphis, TN, USA
e-mail: kyagmurl@uthsc.edu

M. Çırak
Department of Neurosurgery, University of Health Sciences, Bakirkoy Dr. Sadi Konuk Training and Research Hospital, Istanbul, Turkey

J. Liu
Department of Neurosurgery, Neurological Surgery, Center for Skull Base and Pituitary Surgery, Rutgers New Jersey Medical School/Saint Barnabas Medical Center, Livingston, NJ, USA

E. G. Figueiredo et al. (eds.), *Brain Anatomy and Neurosurgical Approaches*,
https://doi.org/10.1007/978-3-031-14820-0_29

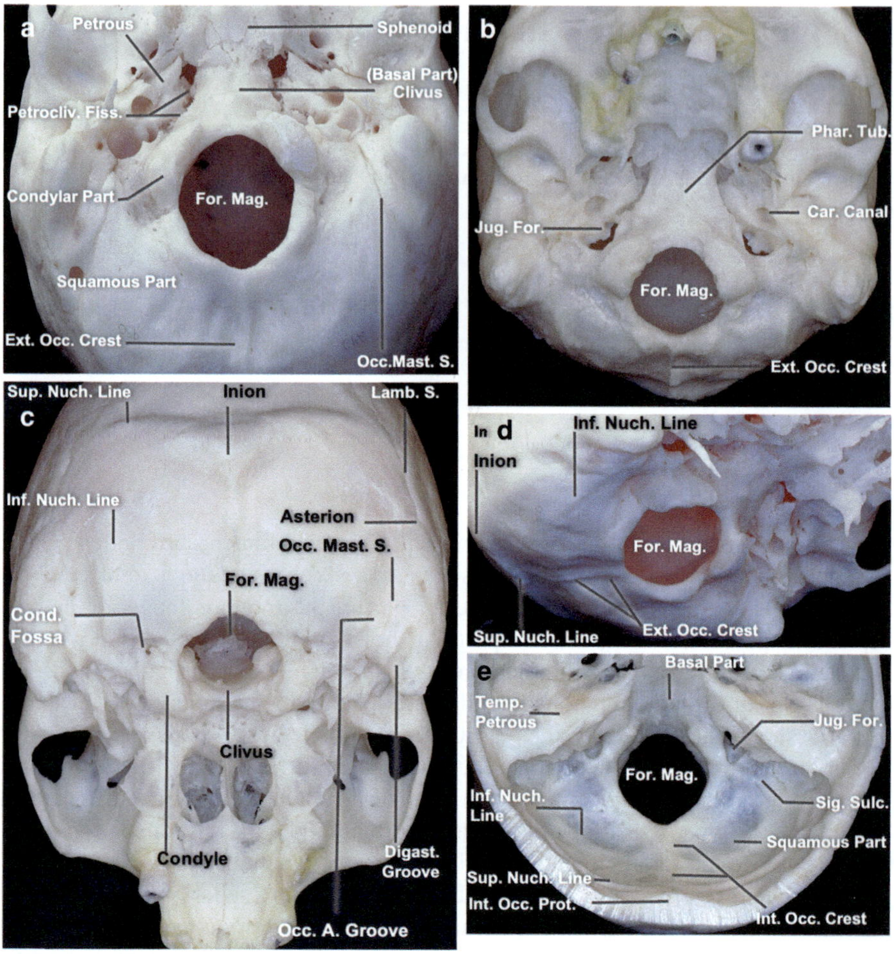

Fig. 1 Bone anatomy of the posterior fossa. (**a**) inferior view of skull exposes the exocranial surface of the occipital bone. Occipital bone has basilar part (clivus), squamous part, and condylar part. (**b**) anterior view of the exocranial surface of the occipital bone. (**c**) posteroinferior view of the exocranial surface of the skull base. The landmarks for suboccipital midline and lateral approaches are the mastoid groove, asterion, and nuchal lines are exposed. (**d**) a lateral view of the exocranial surface of the occipital bone, superior and inferior nuchal lines are attached by the suboccipital muscles. The external occipital crest extends from the superior nuchal line to the foramen magnum in the midline. (**e**) the endocranial surface of the posterior fossa, all parts of the occipital bone are exposed. (**f**) the endocranial surface of the half of the posterior fossa. (**g**) another angle of the endocranial surface of the posterior fossa, the jugular tubercle that roofs the hypoglossal canal, is revealed. (**h**) another lateral view angle, the condylar fossa is situated behind the condyle used as a surgical trajectory. The hypoglossal canal passes through the condyle, and jugular tubercle roofs the hypoglossal canal. (**i**) an anterior view of the atlanto-occipital junction. The anterior arch of the C1 and muscles were removed to expose the odontoid process of C2. (**j**) 3D printing of the cervical vertebrae. Abbreviations: *A* artery, *Can* canal, *Car* carotid, *Ext* external, *For* foramen, *Hypo* hypoglossal, *Inf* inferior, *Int* internal, *Jug* jugular, *Mag* magnum, *Mast* mastoid, *Nuch* nuchal, *Occ* occipital, occipito, *Phar* pharyngeal, *Post* posterior, *S* suture, *Sig* sigmoid, *Sup* superior, *Temp* temporal, *Tub* tubercle

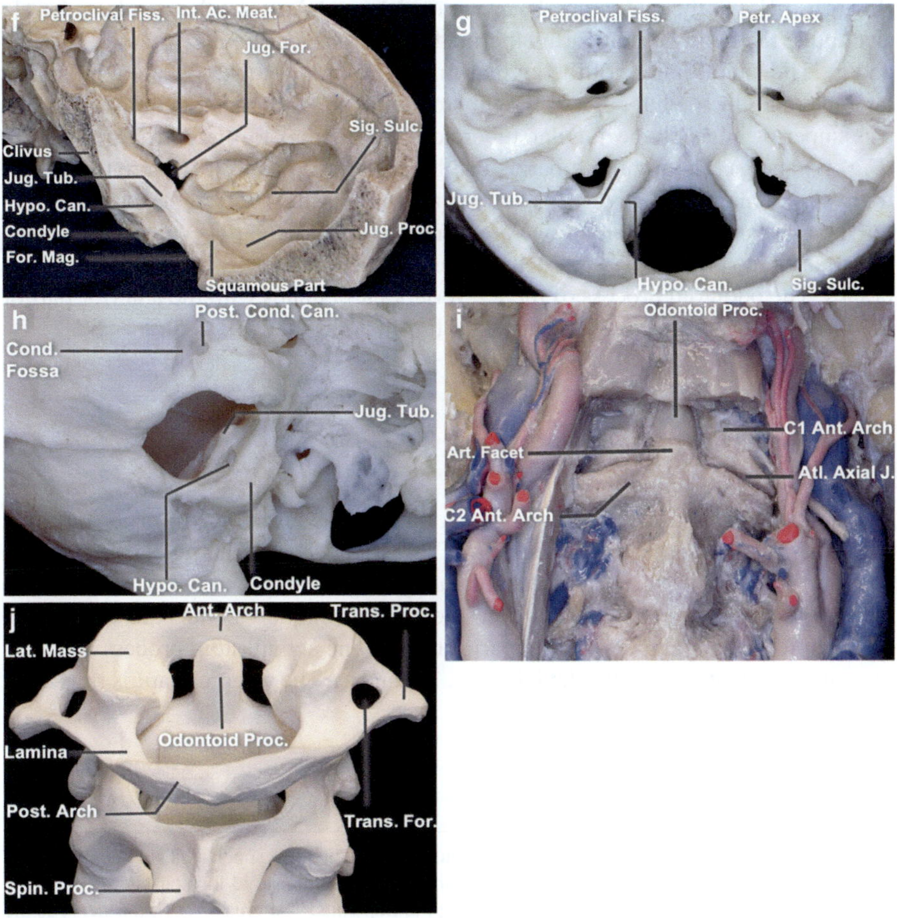

Fig. 1 (continued)

crest runs between the inion and foramen magnum in the midline (Fig. 1d). The internal occipital crest corresponds to the external crest on the endocranial surface (Fig. 1e). The jugular process, the bridge between squamous and condylar parts, extends laterally from the posterior portion of the condyle to form the occipital part of the jugular foramen (Fig. 1f). The sigmoid sulcus, where the sigmoid sinus sits, is located lateral to the jugular foramen and is a portion of the squamous part of the occipital bone.

The condyles are located at the anterior half of the foramen magnum on both sides and articulate with C1 facets (Fig. 1f and g). The hypoglossal canal, which the hypoglossal nerve transmits, passes through the superior portion of the condyle directed forward and laterally with 45° angle (Fig. 1h). On the exocranial surface, the condylar fossa, a depression behind the condyle, is perforated by the posterior condylar canal, which passes above the hypoglossal canal and from the condylar fossa to provide the venous drainage from vertebral venous plexus to the sigmoid sinus. One or both condylar foramina may be absent or incompletely perforated

(Fig. 1h) [2]. On the endocranial surface of the condylar part, the jugular tubercle sits above the condyle that forms the medial edge of the jugular foramen (Fig. 1f and g). The lateral border of the jugular tubercle continues with the sigmoid sulcus laterally, and the medial edge of the tubercle joins the foramen magnum inferiorly. The superior edge of the jugular tubercle starts medial to the petroclival fissure (Fig. 1f) and g). The jugular tubercle can be drilled for greater exposure to the petroclival region, middle clivus, or foramen magnum lesions with superior extension.

The atlas and the occipital bone are joined by the articular capsules encompassing atlanto-occipital joints and by the anterior and posterior atlanto-occipital membranes. The odontoid process of the C2 vertebra articulates with the anterior part of the foramen magnum, called the articular facet (Fig. 1i and j)).

2 Suboccipital Muscles

The suboccipital muscles surround the foramen magnum (Fig. 2). The trapezius runs the back of the head and neck (Fig. 2a). It attaches to the medial half of the superior nuchal line, inion, and the spinous processes of the cervical and thoracic vertebrae. The sternocleidomastoid muscle travels obliquely at the side of the neck to attach to the lateral half of the superior nuchal line and mastoid process and sternum and adjacent part of the clavicle (Fig. 2b) and c). The splenius capitis runs deep

Fig. 2 Stepwise dissection of the suboccipital muscles. (**a**) The most superficial suboccipital muscle is the trapezius located next to the midline, and sternocleidomastoid located laterally. (**b**) the splenius capitis muscle is exposed after removing the trapezius. The greater occipital nerve is the posterior primary ramus of C2 and a sensory nerve. (**c**) the splenius capitis was removed to expose the semispinalis capitis, longissimus capitis, and levator scapula muscles. (**d**) removing the semispinalis capitis reveals the muscles forming the suboccipital triangle. (**e**) superior oblique and inferior oblique muscles, and rectus capitis posterior major muscles form the suboccipital triangle. The muscles attaching between the inferior nuchal line and the edge of the foramen magnum are the rectus capitis posterior major and minor and the superior oblique muscle. (**f**) the third segment of the vertebral artery and venous complex are seen in the suboccipital triangle. (**g**) the second part of the vertebral artery located in the transverse foramina becomes the third segment after exiting the transverse foramen of C2. (**h**) the posterior condylar vein enters the posterior condylar canal in the condylar fossa. (**i**) the posterior arch of the C1 and C2 were removed to expose the vertebral artery. The suboccipital nerve is the posterior primary ramus of C1 and innervates the suboccipital triangle muscles and the rectus capitis posterior minor. The removal of the rectus capitis lateralis, which extends from the jugular process of the occipital bone to the transverse process of C1, exposes the internal jugular vein. (**j**) opening the foramen magnum and drilling the condyle. Abbreviations: *A* artery, *Atl* atlanto, *C* cervical, *Cap* capitis, *Cerv* cervicalis, *Cond* condylar, *For* foramen, *Gr* greater, *Hypo* hypoglossal, *Inf* inferior, *Int* internal, *Jug* jugular, *Lat* lateral, *Lev* levator, *M* muscle, *Mag* magnum, *Maj* major, *Min* minor, *N* nerve, *Obl* oblique, *Occ* occipital, *Post* posterior, *Proc* process, *Rec* rectus, *Scap* scapula, *SCM* sternocleidomastoid muscle, *Semispin* semispinalis, *Splen* splenius, *Spin* spinous, *Subocc* suboccipital, *Sup* superior, *V* vein, *Vert* vertebral

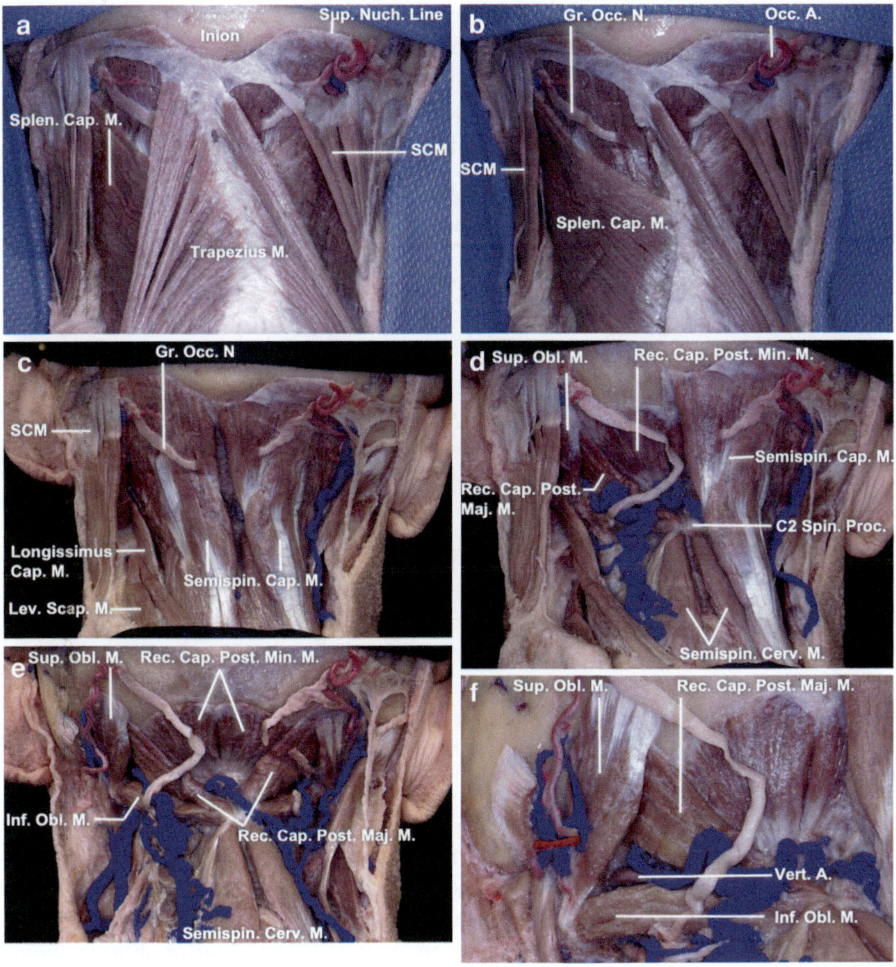
a
Sup. Nuch. Line
Inion
Splen. Cap. M.
SCM
Trapezius M.
b
Gr. Occ. N.
Occ. A.
SCM
Splen. Cap. M.
c
Gr. Occ. N
SCM
Longissimus Cap. M.
Semispin. Cap. M.
Lev. Scap. M.
d
Sup. Obl. M.
Rec. Cap. Post. Min. M.
Semispin. Cap. M.
Rec. Cap. Post. Maj. M.
C2 Spin. Proc.
Semispin. Cerv. M.
e
Sup. Obl. M.
Rec. Cap. Post. Min. M.
Inf. Obl. M.
Rec. Cap. Post. Maj. M.
Semispin. Cerv. M.
f
Sup. Obl. M.
Rec. Cap. Post. Maj. M.
Vert. A.
Inf. Obl. M.

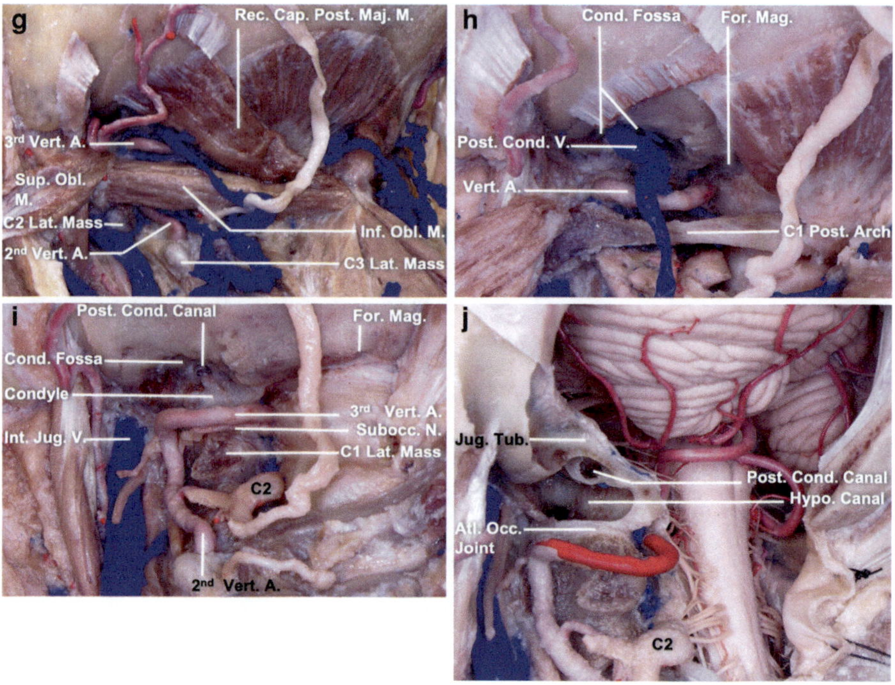

Fig. 2 (continued)

to and partially covered by the trapezius and sternocleidomastoid, running from the lateral third of the superior nuchal line to the spinous processes of the lower cervical and upper thoracic vertebrae. Deep to the splenius capitis, the semispinalis capitis attaches the area between the superior and inferior nuchal lines and extends laterally to the occipitomastoid junction (Fig. 2c). Proceeding deeply, the splenius, semispinalis, and longissimus capitis muscles are encountered. The superior oblique, which extends from the superior and inferior nuchal lines to the transverse process of the C1 vertebra; the inferior oblique extends from the spinous process and lamina of the axis to the transverse process of the C1; the rectus capitis posterior major extends from inferior nuchal line to the spine of the C2; the rectus capitis posterior minor is located medial to the rectus capitis posterior major to rise from the inferior nuchal line to the tubercle on the posterior arch of the C1 (Fig. 2d–f). The suboccipital triangle is formed by the superior and inferior oblique muscles and rectus capitis posterior major muscle. It is covered by the semispinalis capitis medially and the splenius capitis laterally. The structures in the triangle are the V3 segment of the vertebral artery, venous plexus, and C1 nerve (Fig. 2g–j).

3 Surgical Considerations

The foramen magnum tumors might be extra- or intra-axial. To reach the foramen magnum lesions can be reached with an anterior trajectory through the nose or mouth such as the anterior transoral approach, endoscopic endonasal approach; a posterior trajectory such as the suboccipital midline approach or posterolateral trajectories such as lateral transcervical approach, far lateral approach, or extreme lateral approach [3–8]. Each approach has advantages and disadvantages. Because of high mortality and morbidity rates of anterior approaches through oral or nasal cavities, posterolateral approaches have been modified [9, 10]. The far-lateral approach includes the removal of the lateral edge of the foramen magnum, and the transcondylar approach consists of the resection of some part of or all the condyle additional to the classic far lateral approach. The removal of the posterior one-third or two-thirds or complete condyle is called as transcondylar approach. Drilling out the superior part of the condyle above the hypoglossal canal is called the supracondylar approach; the drilling jugular process located lateral to the condyle is called the paracondylar the drilling through the condylar fossa located behind the condyle is called the transcondylar fossa approach [11, 12]. Drilling out the jugular tubercle is called the transtubercular approach and might be combined with the transcondylar approach, supracondylar approach, or transcondylar fossa approach. Notable, drilling out the condyle might be necessary for tumors only ventrally located and do not have a large lateral extension, or the large meningiomas with bone destruction or vertebral artery encasement, giant vertebral artery aneurysms [13].

The meningiomas are the most seen lesion in the foramen magnum and account for 1.8–3.2% of all meningiomas [14–16]. Foramen magnum meningiomas arise at the craniocervical junction defined from the lower third of the clivus to the upper margin of the C2 body and laterally from jugular tubercle to upper margin of the C2 lamina. The far lateral transcondylar approach provides greater exposure for intra- and extradural tumors. The disadvantages include atlanto-occipital instability, postoperative pain, lengthening of surgical procedure, increased injury risk of the hypoglossal nerve, lower cranial nerves, vertebral artery, and jugular bulb. The condylar resection is unnecessary since the sizeable foramen magnum tumors create a surgical route by pushing the brain tissue, cranial nerves, and vascular structures away [9, 17–19].

The superior, inferior, or lateral extensions of the tumor the following are the key in decision making the surgery, the relationship of the tumor with vertebral artery like vessel encasement, prior surgery or radiotherapy, the patient's preoperative deficits are the critical considerations for surgical decision-making [20].

4 Far Lateral Approach

The patient is placed in a park bench position, and the thorax is elevated 15°. The head is in a neutral position or is turned 30° to the contralateral side of the lesion. A linear incision extends from the level of the external acoustic meatus to the level of the C4 vertebra by passing just behind the posterior border of the

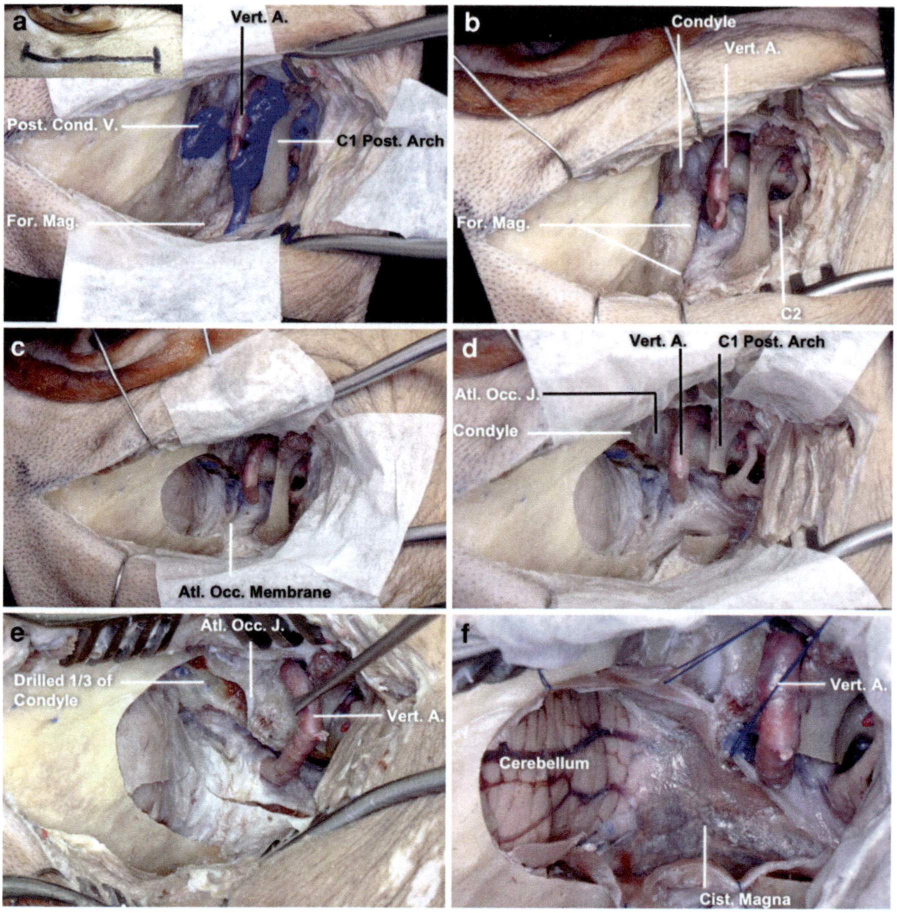

Fig. 3 The far lateral approach in a stepwise manner. (**a**) The far lateral approach on the right side, the small box indicates the linear incision. After linear skin and muscles cuts, the exposure of the edge of the foramen magnum, vertebral artery, and venous plexus. (**b**) The removal of the venous plexus exposes the condyle. (**c**) After removing the edge of the foramen magnum for craniotomy or craniectomy. (**d**) C1 posterior arch can be removed. (**e**) the posteromedial one-third of the condyle can be removed for greater exposure to the ventral part of the brainstem. (**f**) the letter "Y" shaped dural opening exposes the cisterna magna. (**g**) the intradural disclosure of the classic far lateral craniotomy, the opening of cisterna magna. Intradural vertebral artery and lower cranial nerves are exposed. (**h**) the removal of the posteromedial one-third of the condyle, called the transcondylar approach. (**i**) the exposure after total removal of the condyle. (**j**) final exposure of the medulla, pons, and neurocritical structures. (**k**) watertight dural closure. (**l**) the replacement of the bone flap, (**m**) fascia, and muscle closure. Abbreviations: *A* artery, *Acc* accessory, *Atl* atlanto, *C* cervical, *Can* canal, *Cist* cisternal, *Cond* condylar, *For* foramen, *Hypo* hypoglossal, *J* joint, *Jug* jugular, *Mag* magnum, *N* nerve, *Occ* occipital, *Post* posterior, *PICA* posterior inferior cerebellar artery, *Tub* tubercle, *V* vein, *Vert* vertebral

Fig. 3 (continued)

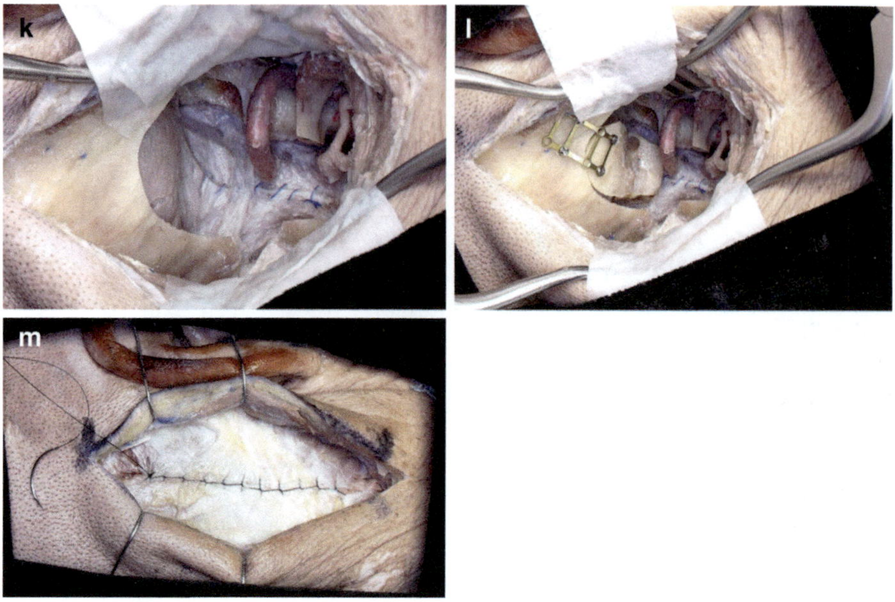

Fig. 3 (continued)

sternocleidomastoid muscle (Fig. 3a). After skin incision, fascia and neck muscles are cut linearly. The vertebral artery is exposed in the suboccipital triangle. Enriched fat tissue and venous oozing or bleeding can be seen when approaching the suboccipital triangle (Fig. 3b). After exposure of the vertebral artery, the subperiosteal dissection is performed laterally to expose the condyle and lateral edge of the foramen magnum. The lateral suboccipital craniotomy is performed (Fig. 3c). The posterior arch of the C1 is removed for more exposure (Fig. 3d). The dural cut is made in the letter of "Y" shape (Fig. 3e). After that, the dura is opened and tacked up; the cisterna magna is opened for CSF release (Fig. 3f). The extradural (V3 segment) and intradural (V4) segments of the vertebral artery, cerebellar tonsil, C1 nerve rootlets, spinal root of the accessory nerve, and dentate ligament are exposed (Fig. 3g–j). The far lateral approach has variations based on the removal of the condyle (Fig. 3g–i). Depending on how much ventral exposure, such as the premedullary area, ventral part of the foramen magnum is needed, the posterior one-third or more of the condyle can be removed, which is called the transcondylar approach. The removal of the area above the hypoglossal canal or drilling through the posterior condylar fossa are called the supracondylar approach and supracondylar fossa approach, respectively. Another variation is the paracondylar approach, which includes drilling the jugular process of the occipital bone located lateral to the condyle. During the closure, the dura is sutured in a watertight fashion. The bone flap can be replaced and fixed to the bone edge by using a titanium plate or not be replaced, called a craniectomy. The suboccipital muscle, fascia, and subcutaneous layers are tightly sutured (Fig. 3k–m).

5 Midline Suboccipital Approach

The patient is placed prone position at the operation table. The table is elevated 20–30° in a reverse-Trendelenburg position to keep the head above the heart level. The head is 45° anteflexed in concord position. A straight midline skin incision is extended from 2 cm above the inion to the level of the C4 spinous process (Fig. 4a). The skin is retracted laterally, and the nuchal ligament is incised until reaching the occipital bone and spinous processes in the midline. The suboccipital muscles are stripped off from bones and retracted laterally to expose the posterior edge of the foramen magnum (Fig. 4b). The paraspinal muscles are released off the posterior arches of the C1 to C4 spines to retract laterally. Two paramedian burr holes are placed next to the midline (Fig. 4c). The 3 × 3 cm craniotomy or craniectomy is performed with the posterior edge of the foramen magnum as the craniotomy's inferior edge. Attention is given to the inferior border of the craniotomy due to the tight attachment of the atlanto-occipital membrane to the posterior edge of the foramen magnum (Fig. 4d). The dura is opened in the letter of "Y" shape, starting the level of the foramen magnum to the upper corner of the craniotomy. The free edge of the dura is sutured and tacked up, and cisterna magna is opened to release CSF (Fig. 4e). The posterior arch of the C1 can be removed for greater exposure (Fig. 4f and g). The suboccipital muscles and fascia are tightly closed using running locked sutures, and subcutaneous and skin layers are closed (Fig. 4h–j).

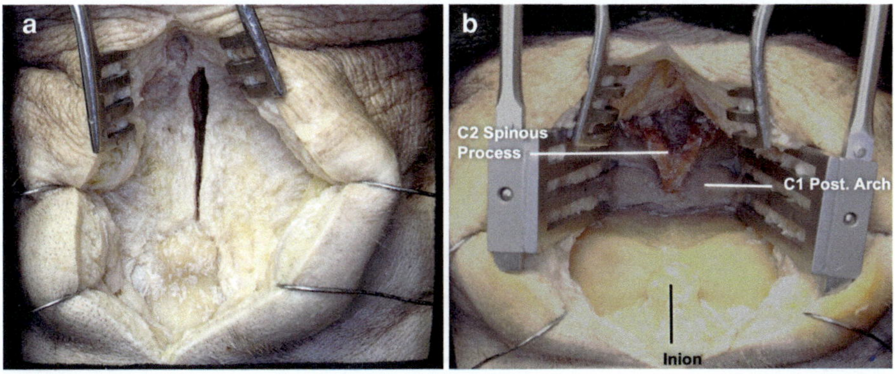

Fig. 4 Suboccipital midline approach step by step. (**a**) A linear incision of the skin and nuchal line in the midline. (**b**) Exposure of the posterior arch of the C1 and C2. (**c**) a craniotomy or craniectomy flap. (**d**) The exposure of the posterior fossa dura and atlanto-occipital membrane. (**e**), a "Y" shaped dural opening exposes the cisterna magna. (**f**) the posterior arches of C1–C2 were removed. The medulla, upper cervical cord, and lower cranial nerves. (**g**) complete exposure after suboccipital midline approach. The occipital condyles were drilled to expose the hypoglossal canal. (**h**) replacement of the bone flap that is an alternative to the craniectomy. (**i**) tightly closure of the suboccipital muscles and fascia. (**j**) subcutaneous and skin closure. Abbreviations: *A* artery, *Acc* accessory, *Atl* atlanto, *Can* canal, *Cist* cisterna, *Dent* dentate, *Hypo* hypoglossal, *Jug* jugular, *Lig* ligament, *Men* meningeal, *N* nerve, *Occ* occipital, *PICA* posterior inferior cerebellar artery, *Post* posterior, *Tub* tubercle, *Vert* vertebral

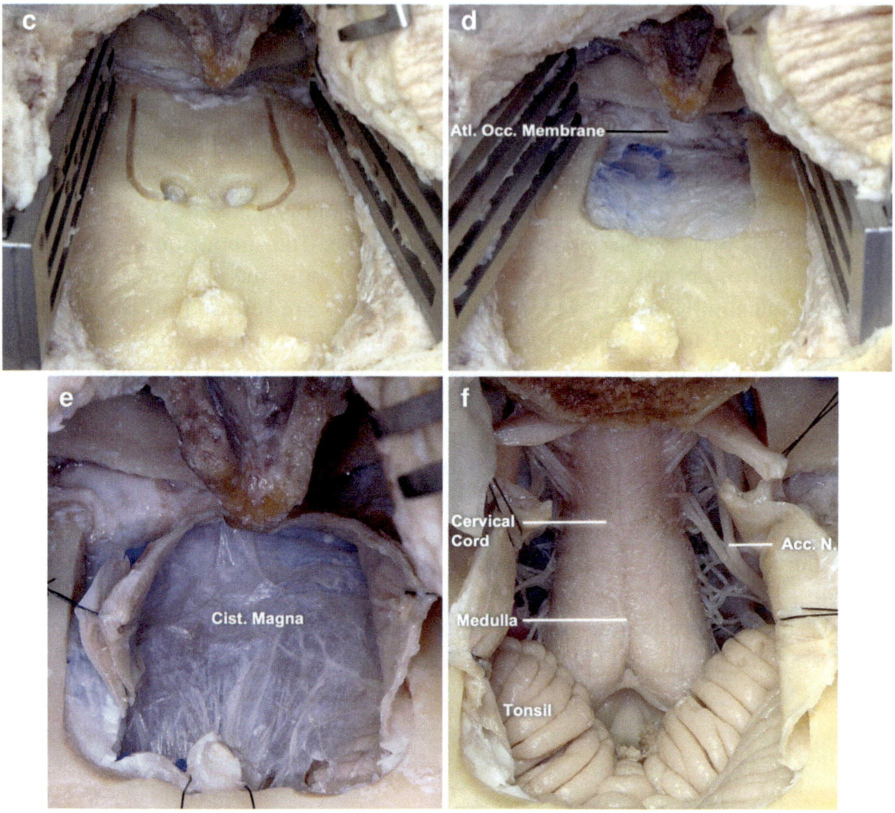

Fig. 4 (continued)

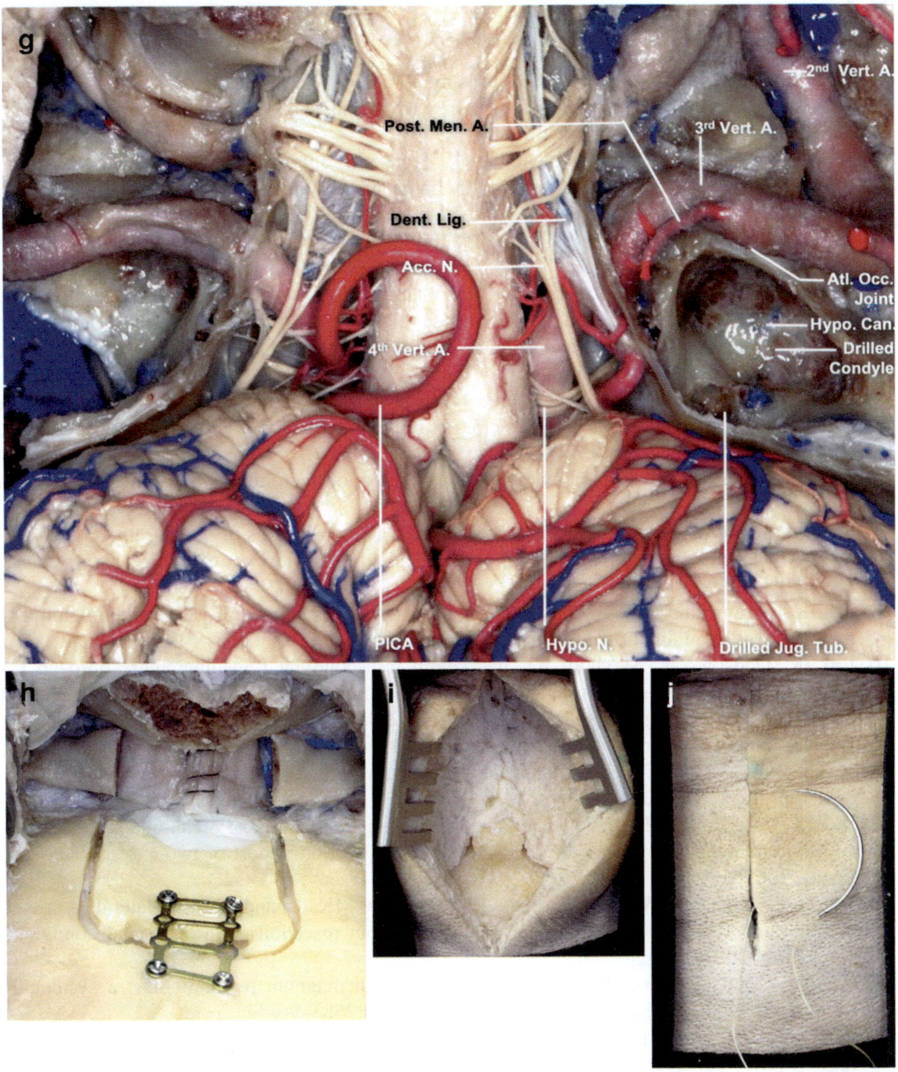

Fig. 4 (continued)

References

1. Rhoton AL. The foramen magnum. Neurosurgery. 2000;47(3):S155–94.
2. Haas LL. The posterior condylar fossa, foramen and canal, and the jugular foramen. Radiology. 1957;69(4):546–52.
3. Liu JK, Couldwell WT, Apfelbaum RI. Transoral approach and extended modifications for lesions of the ventral foramen magnum and craniovertebral junction. Skull Base. 2008;18(03):151–66.
4. Miller E, Crockard AH. Transoral transclival removal of anteriorly placed meningiomas at the foramen magnum. Neurosurgery. 1987;20(6):966–8.
5. Khattar N, Koutourousiou M, Chabot JD, Wang EW, Cohen-Gadol AA, Snyderman CH, Fernandez-Miranda JC, Gardner PA. Endoscopic endonasal and transcranial surgery for microsurgical resection of ventral foramen magnum meningiomas: a preliminary experience. Oper Neurosurg. 2018;14(5):503–14.
6. Stevenson GC, Stoney RJ, Perkins RK, Adams JE. A transcervical transclival approach to the ventral surface of the brain stem for removal of a clivus chordoma. J Neurosurg. 1966;24(2):544–51.
7. Guidetti B, Spallone A. Benign extramedullary tumors of the foramen magnum. Surg Neurol. 1980;13(1):9–17.
8. Salas E, Sekhar LN, Ziyal IM, Caputy AJ, Wright DC. Variations of the extreme-lateral craniocervical approach: anatomical study and clinical analysis of 69 patients. J Neurosurg Spine. 1999;90(1):206–19.
9. Liu JK. Extreme lateral transcondylar approach for resection of ventrally based meningioma of the craniovertebral junction and upper cervical spine. Neurosurg Focus. 2012;33(Suppl 1):1.
10. Liu JK, Dodson VN, Meybodi AT. Far lateral transcondylar transtubercular approach for microsurgical resection of foramen magnum meningioma: operative video and technical nuances. J Neurol Surg Part B Skull Base. 2021;82(S 01):S19–21.
11. Liu JK, Rao G, Schmidt MH, Couldwell WT. Far lateral transcondylar transtubercular approach to lesions of the ventral foramen magnum and craniovertebral junction. Contemp Neurosurg. 2007;29(10):1–7.
12. Liu JK, Couldwell WT. Far-lateral transcondylar approach: surgical technique and its application in neurenteric cysts of the cervicomedullary junction: report of two cases. Neurosurg Focus. 2005;9(2):1–7.
13. Nanda A, Vincent DA, Vannemreddy PS, Baskaya MK, Chanda A. Far-lateral approach to intradural lesions of the foramen magnum without resection of the occipital condyle. J Neurosurg. 2002;96(2):302–9.
14. Paun L, Gondar R, Borrelli P, Meling TR. Foramen magnum meningiomas: a systematic review and meta-analysis. Neurosurg Rev. 2021;44(5):2583–96.
15. Akalan N, Seckin H, Kilic C, Ozgen T. Benign extramedullary tumors in the foramen magnum region. Clin Neurol Neurosurg. 1994;96:284–9.
16. Arnautovic KI, Al-Mefty O, Husain M. Ventral foramen magnum meninigiomas. J Neurosurg. 2000;92:71–80.
17. Heros RC. Lateral suboccipital approach for vertebral and vertebrobasilar artery lesions. J Neurosurg. 1986;64:559–62.
18. Koos WT, Spetzler RF. Color atlas of microneurosurgery. New York: Thieme; 1985. p. 125–34.
19. Samii M, Klekamp J, Carvalho G. Surgical results for meningiomas of the craniocervical junction. Neurosurgery. 1996;39:1086–95.
20. Sekhar LN, Shenoy VS, Qazi Z. Surgical management of benign tumors at the foramen magnum region. Int J Neurooncol. 2021;4(3):65.

Surgical Anatomy of the Approaches to the Brainstem

Luciano Furlanetti, Matheus Fernando Manzolli Ballestero, and Ricardo Santos de Oliveira

1 Introduction

The brainstem is the portion of the neural axis between the diencephalon and the cervical spinal cord [1–3], comprising structures and vital functions necessary for sustaining life [4]. Due to the high density of neural pathways and nuclei, even small intra- or extra-axial lesions determining some degree of mass-effect may lead to clinically relevant neurological impairment [5]. The brainstem may be affected by various neoplastic and non-neoplastic, intra- and extra-axial lesions. The most common intra-axial neoplastic tumors encompass brainstem gliomas (BSGs), lymphoma, and metastatic disease [2, 4]. On the other hand, cavernous malformations are the most common type of clinically relevant non-neoplastic lesions of the brainstem, accounting for approximately 20% of all intracranial cavernomas [6]. Extra-axial tumors, such as meningiomas, schwannoma, and epidermoid cysts, can also arise from various locations adjacent to the brainstem causing mass effect and clinical manifestation [5].

BSGs are the most common primary tumor arising from the brainstem, accounting for around 10% to 15% of primary pediatric intracranial tumors and ca. 1–2% of all brain tumors in adults [7, 8]. Until about two decades ago, tumors within the

---

L. Furlanetti (✉)
Department of Basic and Clinical Neuroscience, Institute of Psychiatry, Psychology, and Neuroscience, King's College London, London, UK
e-mail: Luciano.furlanetti@kcl.ac.uk

M. F. M. Ballestero
Department of Medicine, Federal University of São Carlos, São Carlos, SP, Brazil

R. S. de Oliveira
Division of Neurosurgery, Department of Surgery and Anatomy, University of São Paulo, Ribeirao Preto, SP, Brazil

E. G. Figueiredo et al. (eds.), *Brain Anatomy and Neurosurgical Approaches*,
https://doi.org/10.1007/978-3-031-14820-0_30

brainstem were considered inoperable due to the high morbidity and mortality associated with surgery in this anatomical location [1, 4, 8]. The fact that nuclei of the cranial nerves, neural networks, and other vital structures lay compactly arranged in such a tiny space represents a great challenge for the neurosurgeon to approach intrinsic brainstem lesions [8–10]. Nevertheless, experience over the last several decades has demonstrated that brainstem gliomas comprise a heterogeneous group of tumors, some of which with favorable long-term survival [8, 11, 12].

Pool et al. 1968 reported three patients diagnosed with small focal or cystic brainstem tumors successfully treated surgically and associated with adjuvant postoperative "x-ray therapy," achieving encouraging results [3]. The improved neuroimaging with the introduction of computed tomography in the 1970s and magnetic resonance later led to a remarkable advance in the diagnosis and surgical planning of these tumors [1, 4, 8, 10]. From the 1980s, several authors started publishing case series, some with expressive results, initiating a new era in brainstem tumor surgery. It became clear that BSGs encompass a group of heterogeneous lesions, so the treatment strategy needed to be tailored according to their anatomic localization and clinical and radiological presentation [1, 4]. In line with this, further refinement of the neuroanatomical knowledge, development of neuroimaging allied to the improvement of surgical tools, such as magnification and visualization of the surgical field with more powerful surgical microscopes and sources of light, followed by the introduction of the ultrasonic aspirator, advancement of anesthesia and intraoperative neurophysiological monitoring provided the neurosurgeon with conditions for a much safer surgical management of these tumors in carefully selected cases [5, 8–10]. Importantly, postoperative care has evolved since then, mainly preventing respiratory complications and swallowing disorders [4]. Continuous technological advances, mainly in neuroradiology and MRI techniques, such as diffusion tensor imaging tractography, have led to a better understanding of the relationship of the tumor with brainstem structures and surroundings, allowing improved surgical planning and less transgression of adjacent normal neural structures. Furthermore, minimally invasive surgical techniques, such as stereotactic navigated brain biopsy or even stereotactic guided brachytherapy, consist of safer and less aggressive histological diagnosis and targeted treatment in carefully selected cases [13].

In the following sessions, the authors review the current literature regarding the relevant neuroanatomy of the brainstem, the famous "safe entry zones," and possible surgical approaches that may be applied in the treatment of neoplastic and non-neoplastic brainstem lesions.

2 Surgical Technique and Approaches

Optimal patient positioning in neurosurgery is often debated [5, 10, 14]. Nevertheless, there is a consensus that adequate positioning is fundamental for obtaining optimal results and avoiding perioperative complications [5, 14, 15]. Positioning the patient for prolonged and complex procedures must be a team effort, requiring good interaction between theater staff, anesthesiology, and surgical team. Attention should be directed to preventing uneven and potentially excessive pressure distribution, preventing excessive stretching or compression, allowing chest expansion for proper ventilation, absorbing compressive forces, minimizing exposure of the patient's skin, and controlling the ambient theater temperature. Besides that, adequate surgical positioning and preoperative planning often allow the surgeon to take advantage of gravity to reduce or even eliminate the need for brain retraction. The supine position offers good exposure to the anterior and middle fossae of the cranium and access to the posterior fossa and anterior/anterolateral surface of the brainstem, as detailed below. Access to the lateral skull base is often preferably achieved with the patient in a supine position with the head slightly extended, turned to the contralateral side, and fixed on a three-point headrest stabilizing device. The posterior fossa is usually exposed with the patient in the prone position, lateral/park-bench, or semi-sitting position. Although the semi-sitting position confers a wider upper angle of visualization and cleaner operating field, it has been associated with an increased risk of pulmonary air embolism and the necessity of preoperative exclusion of a patent foramen ovale.

2.1 *Orbitozygomatic*

The patient is placed supine with the head above the level of the heart, slightly extended, and rotated 15–30° contralaterally. A frontotemporal flap with interfascial dissection of the temporal muscle is performed. The zygomatic arch is exposed, and a pterional craniotomy is performed. The orbitozygomatic osteotomy is usually made as a second step [16].

In carefully selected cases, lesions on the anterior surface of the midbrain, pontomesencephalic junction, and upper pons can be accessed via dissection of the Sylvian fissure and carotid-oculomotor triangle. Nevertheless, one must bear in mind the complex anatomy of this region, the motor tracts traveling ventrally in the brainstem, and, therefore, the high risks involved when approaching this region [10].

2.2 *Subtemporal/Transtentorial*

The patient is positioned in lateral decubitus, with the sagittal suture parallel to the floor [5]. Alternatively, a supine position with padding under the ipsilateral shoulder and rotation of the head contralaterally can be used. A straight vertical incision anterior to the tragus is placed. Next, a temporal craniotomy just above the zygomatic arch is made, and the dura opens in a U-shaped fashion. The dissection is performed below the base of the temporal lobe toward the tentorial edge and crural, ambient, and interpeduncular cisterns, as needed. This approach yields access to the anterior mesencephalic zone and the lateral mesencephalic sulcus. The transtentorial extension allows access to the pontomesencephalic junction and the anterolateral upper pons [10].

2.3 Anterior Petrosectomy

The Kawase approach can be an extension of the subtemporal approach described above, which gives better access to the anterolateral pons and the trigeminal nerve [5]. The anterior petrosectomy has a lateral limit at the greater superficial petrosal nerve and the V3 segment of the trigeminal nerve anteriorly [10].

2.4 Retrosigmoid

The retrosigmoid approach is probably one of the most used to access posterior fossa lesions, including brainstem tumors [5, 10, 17]. It can be performed with the patient positioned in a lateral position with the head slightly flexed and rotated ipsilaterally or in a dorsal position with padding under the ipsilateral shoulder, and the head rotated toward the contralateral side. Some centers prefer the advantages of the semi-sitting position, which allows better visualization of the tentorium and upper posterior fossa, as discussed above [14]. A straight or semi-curved skin incision is placed about 3 cm behind the earlobe with subperiosteal dissection of the myofascial layer. A bur hole is placed just above the asterion on the parietomastoid suture. The upper and lateral limits of the craniotomy are the transverse and sigmoid sinus, respectively. This approach gives access to the cerebellopontine angle, allowing a wide exposure of the middle cerebellar peduncle and lateral pons [18].

The extreme lateral supracerebellar infratentorial approach is a variation of the retrosigmoid approach. Nevertheless, once the dura is open, the dissection above the cerebellum, along its tentorial surface, toward the cerebellomesencephalic fissure. This approach gives access to the ambient cistern, trochlear nerve, and superior cerebellar artery laterally and the quadrigeminal cistern medially. The tentorium can be opened if needed, providing extended access to the medial temporal lobe and thalamus [5, 8, 10].

2.5 Far Lateral

The far lateral approach is performed with the patient in a lateral/park-bench or semi-setting position [5, 8, 10]. A hockey-stick skin incision is placed about 3 cm behind the earlobe and extended toward the posterolateral cervical region. It is crucial to identify the muscular layers to avoid damaging vital structures such as the vertebral artery. The first muscular layer is composed of the sternocleidomastoid and trapezius muscles. Secondly, the longissimus capitis (laterally) and the semispinalis capitis are dissected and elevated, giving access to the suboccipital triangle. The V3 segment of the vertebra artery is found within the triangle formed by the superior oblique, inferior oblique, and rectus capitis posterior major muscles. The rectus capitis posterior minor and major are retracted, exposing the posterior arch of C1 and the vertebral artery. Next, the ipsilateral half of the posterior arch of C1 is removed, and a lateral suboccipital craniotomy is performed. The posterior third of the occipital condyle can be drilled, preserving the hypoglossal canal if an improved angle of view toward the anterolateral medulla is needed. Microsurgical dissection of the cisterna magna allows a wide view of the posterolateral cervicomedullary surface, whereas ventral dissection of the premedullary cistern gives access to the anterolateral sulcus and olivary region [5, 8, 10].

2.6 Transpetrosal Retrolabyrinthine Presigmoid

Apart from the anterior petrosal approach described above, petrosal approaches can be presigmoid retrolabyrinthine, presigmoid translabyrinthine (removal of the semicircular canals), or total petrosectomy [19]. The patient is positioned in dorsal decubitus with padding under the ipsilateral shoulder, and the head is rotated contralaterally. A curved skin incision is placed in the temporal region 4 cm above the zygomatic arch, passing 3 cm behind the earlobe and extending 2 cm behind the mastoid tip. To avoid CSF leak, it is advisable to carefully dissect and preserve the myofascial layers during the approach for posterior skull-base reconstruction. The temporalis muscle fascia is dissection downward with the mastoid periosteum, craniocervical fascial, and sternocleidomastoid muscle to form a large vascularized myofascial flap, which can be later used to cover the surgical field. Next, the cortical bone of the mastoid is drilled out as previously described. The labyrinth block, the fascial nerve canal, the superior petrosal, and the sigmoid sinus are identified and preserved. A craniotomy is performed, exposing the middle and posterior fossa above and below the sigmoid sinus. The dura anterior to the sigmoid sinus is exposed, and the superior petrosal sinus coagulated and ligated. The dura is opened, and the tentorium is incised up to the incisura. Caution should be taken with the trochlear nerve running at the edge of the tentorium. These approaches provide access to the posterior, middle fossa, clivus region, and the supratrigeminal, peritrigeminal, and lateral pontine safe zones [5, 10, 19].

2.7 Suboccipital Telovelar

The telovelar approach is performed based on a suboccipital craniotomy, usually associated with a laminoplasty of the posterior C1 arch, which allows a wide vertical angle to visualize the superior aspect of the superior rhomboid fossa [8, 10, 20].

The patient is positioned in a prone, park-bench, or semi-sitting position, according to the preference of the surgical team. Y-shaped dura opening, gentle lateral retraction of the cerebellar tonsils and the posterior inferior cerebellar arteries, followed by the microsurgical dissection of the tela choroidea bilaterally, gives access to both the supra- and infracollicular zones, as well as to the lateral recesses [8, 10].

2.8 *Median Supracerebellar Infratentorial*

This approach is initially quite like the one above described, apart from the fact that the suboccipital craniotomy should have its superior edge at the level of the transverse sinus [10]. Other than that, it is usually not necessary to take the posterior arch of C1 in this case. The infratentorial supracerebellar microsurgical dissection gives access to the quadrigeminal cistern and its contents, such as the internal cerebral veins, basal veins of Rosenthal, and vein of the cerebellomesencephalic fissure, quadrigeminal plate, and pulvinar of the thalamus laterally [21, 22].

2.9 *Endoscopic/Endoscopic-Assisted Approaches*

In recent years, the endoscope has become an important tool in the neurosurgical armamentarium and gained attention to treat numerous intracranial skull-base tumors successfully and intraventricular and intrinsic neoplastic lesions, including those affecting the brainstem. Several publications highlight the importance of the hybrid microscopic and endoscopic-assisted technique that allows the surgeon to "look around the corner" and ensure that a complete tumor removal has been achieved [23–26]. Furthermore, the endoscopic approach may be an efficient option in managing brainstem tumor-associated hydrocephalus, or for instance, in the context of minimally invasive endoscopic tumor biopsy of intraventricular or exophytic lesions.

3 Safe Entry Zones to the Brainstem

Awareness of the main safe entry zones on the brainstem is crucial to reducing surgical morbidity. Improvement of neuroanatomical and neurophysiological knowledge has led to the development of different skull-base approaches and increased safety for surgical management of brainstem lesions that do not emerge from the pial or ependymal surface. Below, the main access to the midbrain, pons, and medulla have been detailed as follows:

3.1 *Midbrain*

3.1.1 Anterior Mesencephalic Zone (AMZ)

Lesions located in the anterior portion of the midbrain can be approached through a narrow corridor between the oculomotor nerve (medially) and the corticospinal tract (laterally), the posterior cerebral artery (superiorly) and the superior cerebellar artery (inferiorly).

Possible Approaches: orbitozygomatic, subtemporal, subtemporal transtentorial, anterior petrosectomy

3.1.2 Lateral Mesencephalic Sulcus (LMS)

The LMS runs from the medial geniculate body to the pontomesencephalic sulcus with an average length of less than 1 cm. The region is limited by the posterior portion of P2 (superiorly), the medial posterior choroidal artery, the cerebellomesencephalic superior cerebellar artery, trochlear never, and tentorium (inferiorly), the substantia nigra anterolaterally, the medial lemniscus (posteriorly) and the oculomotor fibers (anteriomedially).

Possible Approaches: median supracerebellar infratentorial, extreme lateral supracerebellar infratentorial, retrosigmoid

3.1.3 Intercollicular Region

Small space in the midline, between the right and left superior and inferior colliculi.

Possible Approaches: median supracerebellar infratentorial, extreme lateral supracerebellar infratentorial

3.2 *Pons*

3.2.1 Peritrigeminal Zone

Entry to the anterolateral pons, anterior to the trigeminal nerve, lateral to the corticospinal tract, and anterior to the motor and sensory nuclei of the trigeminal nerve.

Possible Approaches: anterior petrosectomy, retrosigmoid, retrolabyrinthine

3.2.2 Supratrigeminal Zone

Also used to approach lesions in the anterior/anterolateral pons, with entry right above the emergence of the trigeminal root entry zone within the middle cerebellar peduncle.

Possible Approaches: subtemporal transtentorial, anterior petrosectomy, retrolabyrinthine

3.2.3 Lateral Pontine Zone

The corridor between the middle cerebellar peduncle and the pons and between the entry zones of the V and VII-VIII complex.

Possible Approaches: retrosigmoid, retrolabyrinthine

3.2.4 Rhomboid Fossa

On the floor of the IV ventricle, the supra- and infracollicular zones and the median sulcus of the fourth ventricle have been described as potential safe entries to the pons. The supracollicular triangle is delimited by the facial nerve (caudally), the cerebellar peduncles (laterally), and by the medial longitudinal fascicle (medially). The infrafacial triangle has the striae medullaris (inferiorly), the facial nerve (laterally), and the medial longitudinal fascicle (medially) as its borders. Dissection of the median sulcus can damage the medial longitudinal fascicle and cause extraocular movement issues.

Possible Approaches: suboccipital telovelar

3.3 Medulla

3.3.1 Anterolateral Sulcus of the Medulla

The area between the rootlets of the XII and C1 nerve is lateral to the decussation of the corticospinal tract at the pyramid.

Possible approaches: retrosigmoid, retrolabyrinthine, far lateral

3.3.2 Posterior Median Sulci of the Medulla

The area on the midline, below the obex, and having the gracile nucleus laterally on each side, the median sulcus can be carefully dissected.

Possible approaches: suboccipital

3.3.3 Olivary Zone

The vertical area is limited by the anterolateral sulcus (anteriorly) and the posterolateral sulcus. Tectospinal and spinothalamic tracts (posteriorly), the hypoglossal nerve fibers and medial lemniscus (medially), and the XII nerve fibers and corticospinal tract in depth.

Possible Approaches: far lateral, retrosigmoid

3.3.4 Lateral Medullary Zone/Inferior Cerebellar Peduncle

It has been described for the approach to dorsolateral lesions of the medulla through a small vertical incision in the inferior cerebellar peduncle, inferior to the cochlear nuclei, and posterior to the IX and X entry zone.

Possible Approaches: retrosigmoid, far lateral, retrolabyrinthine

4 Clinical Presentation

The average age of onset of BSG ranges from 5 to 10 years, with an equal incidence among boys and girls, comprising 10 to 20% of all central nervous system tumors in children. There are no recognized premalignant lesions; however, several familial cancer syndromes have been associated with it, including neurofibromatosis type 1, Li-Fraumeni syndrome, and tuberous sclerosis, among others [4, 11]. Brainstem tumors may develop as focal or diffuse lesions, therefore being classified as diffuse, focal, dorsal exophytic, and cervicomedullary tumors, according to their presentation [4].

The specific signs and symptoms of brainstem tumors are dependent on their anatomical location [4, 7]. Almost 75% of brainstem tumors are diffuse gliomas with an unfavorable prognosis, whereas local growing and exophytic tumors may be amenable to surgical resection with a better prognosis [2, 11, 27, 28]. Patients with diffuse intrinsic pontine glioma (DIPG) typically have a rapidly progressive course, usually with dysfunction of cranial nerves and pyramidal and cerebellar pathological signs. In contrast, the presentation of focal brainstem gliomas is more insidious with localizing signs such as isolated cranial nerve deficits and contralateral hemiparesis spanning months to years [4, 7, 27]. Tumors of the cervicomedullary junction commonly present with lower cranial nerve palsy, pyramidal tract signs, ataxia, spinal cord dysfunction, and nystagmus. Signs and symptoms of hydrocephalus usually manifest later along with the progression of the disease, exception for tectal tumors, which may present with obstructive hydrocephalus due to its close relationship with the cerebral aqueduct [8, 29].

5 Neuroradiology

The most utilized classification system segregates brainstem tumors based on their radiological appearance into three major categories: (1) diffuse, (2) focal, or (3) exophytic [4, 7, 8]. The focal and exophytic types are usually low-grade gliomas and carry a much better prognosis than diffuse high-grade glioma.

The radiological investigation of brainstem tumors has advanced a lot using computed tomography (CT) and mainly magnetic resonance imaging (MRI). Tractography (DTI) has facilitated planning the best surgical approach for these lesions, showing that the fibers are often deviated and not infiltrated by the tumor (Fig. 1).

Epstein and McCleary et al. classified these lesions into three distinct groups: diffuse, focal, and cervicobulbar. At that time, the authors recommended radical surgical resection only for cervicobulbar tumors [30]. Later, the classification was revised based on radiological and anatomopathological findings into three main types: (1) diffuse growth tumors in the brainstem with no precise delimitation (DIPG); (2) tumors with well-defined limits concerning neural structures; (3) infiltrative tumors with a well-defined macroscopic appearance [1, 2, 5].

Choux et al. 1999 revisited this classification, subdividing brainstem tumors into four different types based on radiological findings: type I—intrinsic and diffuse tumor; type II—intrinsic and focal tumor (which may be solid or cystic and located in one of the three anatomical subdivisions of the brainstem; type III—exophytic tumor, which may be lateral or dorsal; and type IV – cervicomedullary tumors (Fig. 2). This classification started being used as a reference guide for the neuro-oncology team to plan the most appropriate treatment strategy for each case [2, 5].

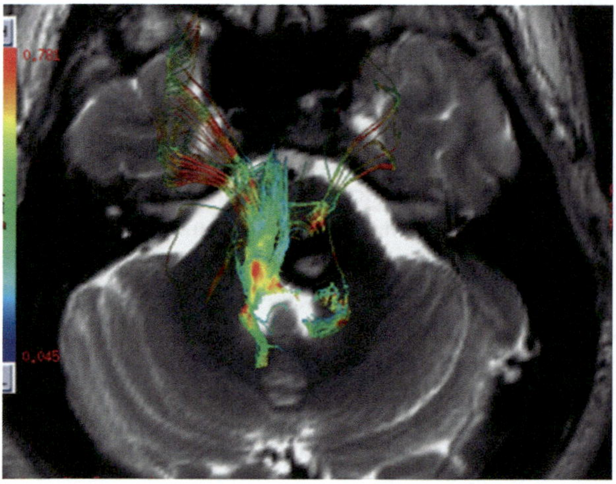

Fig. 1 Diffusion tensor imaging (DTI) tractography image showing a pontine lesion deflecting fibers from the long tracts

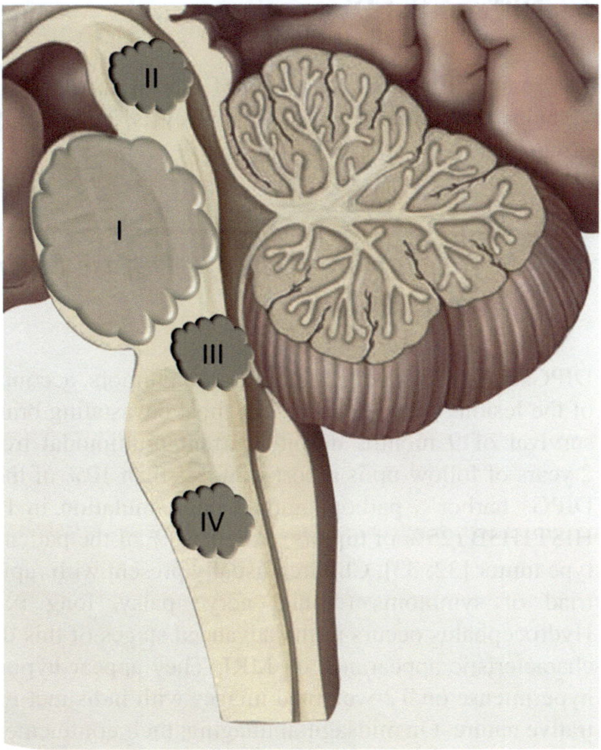

Fig. 2 Classification for brainstem tumors (I) Diffuse intrinsic, (II) Focal tumor, (III) Exophytic, (IV) cervicomedullary transition

Low-grade gliomas are typically focal or exophytic in appearance, with well-circumscribed margins, demonstrating hypointense to isointense signal on T1- and hyperintense signal on T2-weighted imaging relative to the adjacent parenchyma. Focal tumors may show mild or no post-contrast enhancement, predominantly involving the exophytic component. On the other hand, adult low-grade gliomas may have focal or diffuse infiltrative behavior and demonstrate post-contrast enhancement. Pediatric diffuse low-grade glioma may demonstrate a similar T1/T2 signal but no contrast enhancement, restricted diffusion on DWI/ACD, or necrosis [11]. DTI tractography has been proven a useful tool, providing invaluable preoperative information to the surgical team regarding the distorted brainstem anatomy [8, 10, 11, 31].

On the other hand, diffuse midline gliomas, formerly known as DIPG, have been revised as a new distinct tumor under the 2016 World Health Organization classification of central nervous systems tumors [32]. The *H3 K27M-mutant* is high-grade midline glioma with predominant astrocytic differentiation and a *K27M* mutation in the *H3F3A* or *HIST1H3B/C* genes. Most of the tumors previously diagnosed as DIPG belong to this category, occurring mainly in children, and determining poor prognosis [32].

6 Differential Diagnosis

The main differential diagnosis of brainstem tumors are metastases to the CNS, vascular malformations, ependymal cysts, parasitic diseases, and demyelinating diseases [13].

6.1 Diffuse tumors (Type I)—Diffuse Intrinsic Pontine Glioma (DIPG)

DIPGs are the most common brainstem tumors, accounting for around 60% to 80% of the lesions [9]. DIPGs are the most devastating brainstem tumor with a median survival of 9 months despite current multimodal treatment. Overall survival at 2 years of follow-up is reported in less than 10% of the patients. Up to 90% of the DIPGs harbor a pathognomonic point mutation in H3F3A (65% of tumors) or HIST1H3B (25% of tumors). About 10% of the patients have a histone three wild-type tumor [32, 33]. Children usually present with rapid onset and progression of a triad of symptoms (cranial nerve palsy, long tract, and cerebellar signs). Hydrocephalus occurs in the advanced stages of this disease. These lesions have a characteristic appearance on MRI. They appear hypointense on T1-weighted and hyperintense on T2-weighted images with indistinct margins, reflecting their infiltrative nature. On midsagittal imaging, their confinement to the pons is better delineated. Gadolinium enhancement is variable and provides no additional prognostic information. Magnetic resonance spectroscopy (MRS) is an evolving imaging modality in brainstem tumors. DIPGs have increased metabolic ratios of choline to creatine and *N*-acetylaspartate, which may be useful in delineating DIPG from demyelination, encephalitis, and radionecrosis [14, 15] (Fig. 3).

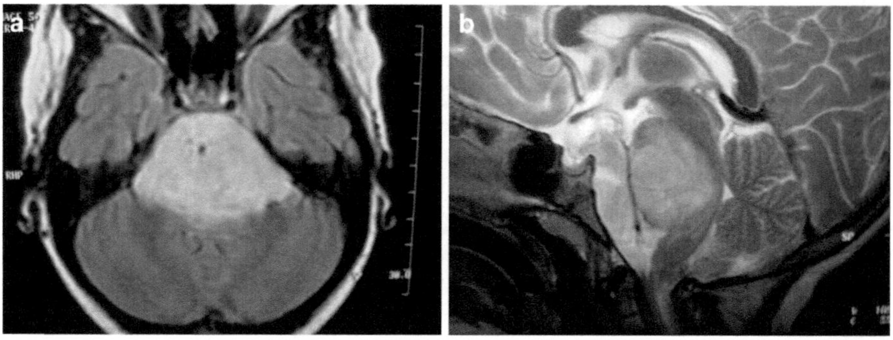

Fig. 3 Diffuse tumor (Type I) affecting the pons. (**a**) Axial FLAIR and (**b**) Sagittal T2-weighted MRI scans

6.2 *Focal Brainstem Tumors (Type II)*

Tectal tumors are intrinsic lesions in the mesencephalon region, representing 5% of brainstem lesions, and are typically WHO grade I or II. Although most tectal gliomas exhibit an indolent benign course (>85%), a subtype of this tumor may behave more aggressively. MRI may help distinguish between them, as the more aggressive subtype usually has a larger volume at presentation, may invade the adjacent tegmentum, thalamus, and pons, and present post-gadolinium enhancement. Given their proximity to the cerebral aqueduct, tectal tumors may cause hydrocephalus, rapid deterioration, and death even when small-sized.

The most common benign intrinsic tectal tumors generally follow an indolent course. These are typically well-circumscribed non-enhancing low-grade astrocytomas, presenting clinically with signs and symptoms of intracranial hypertension due to obstructive hydrocephalus [15]. Other signs include gait disturbance, ataxia, Parinaud syndrome, and strabismus. There is still a debate regarding the best treatment strategy for nonsuspicious tectal lesions due to the high morbidity in this region's surgical management. [8, 10] Therefore, the treatment of underlying obstructive hydrocephalus followed by watchful waiting may be the safest approach to these lesions. Nevertheless, some centers favor a more aggressive surgical approach, given that 18% to 30% of tectal tumors eventually progress [16].

Hydrocephalus can be treated with a ventriculoperitoneal shunt or via endoscopic third ventriculostomy (ETV), which has also been demonstrated to be safe and effective. It is crucial to follow up with these patients closely because of the risk of tumor progression and eventual ETV failure, leading to serious complications [7, 24, 34] (Fig. 4).

The more aggressive subtype of tectal tumors is typically larger on presentation than the indolent variety. It can be associated with neurofibromatosis, and sometimes surgical resection followed by radiation therapy could be indicated [2, 28].

6.3 *Dorsally Exophytic Brainstem Tumors (Exophytic tumors) – (Type III)*

The dorsally exophytic subtype accounts for approximately 10–20% of brainstem tumors. Children usually present with an insidious history of headaches, vomiting, ataxia, and cranial nerve dysfunction. Papilledema, torticollis, and long tract signs can be found on neurological examination. These tumors protrude into the fourth ventricle, but occasionally they can be dorsolaterally exophytic, projecting into the cerebellopontine angle (CPA). The hypointense signal on T1 and the hyperintense signal on T2 will generally display consistent tumor edges reflecting its less infiltrative nature. Contrast enhancement is typical. These tumors are predominantly pilocytic astrocytomas, occasionally diffuse astrocytoma (local growth), or gangliogliomas, generally carrying a good prognosis. Dorsally exophytic tumors

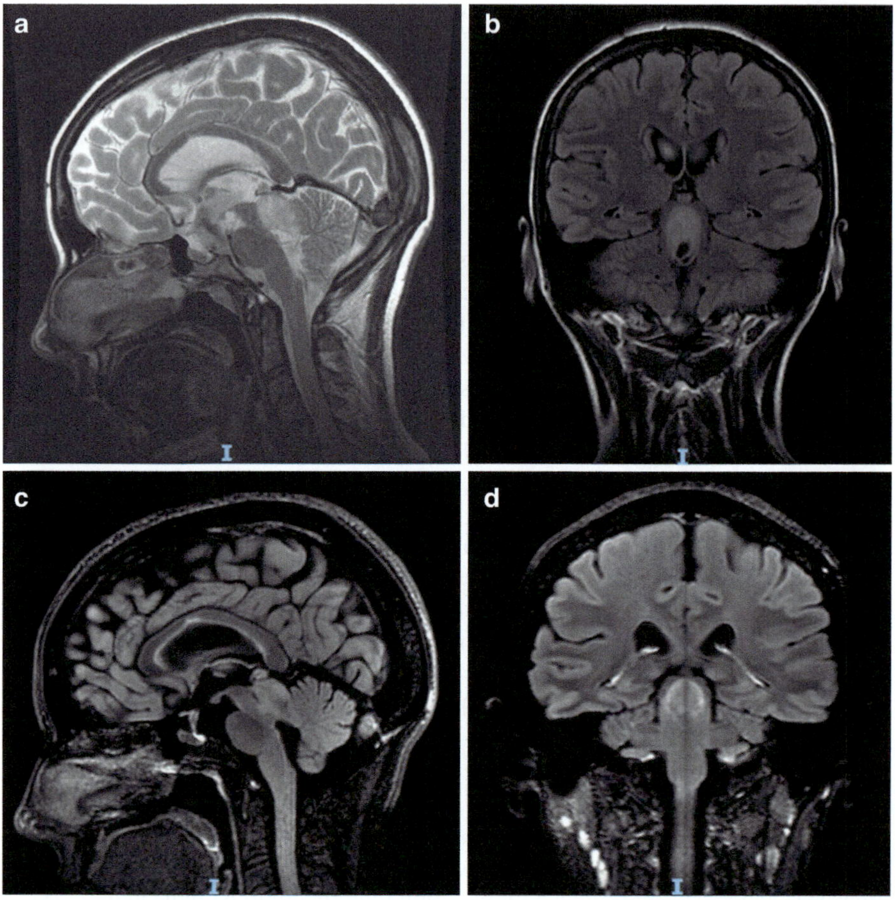

Fig. 4 Focal tumor (Type II) of the tectal plate, leading to hydrocephalus. (**a** and **b**) scans at diagnosis; (**c** and **d**) postoperative scans after ETV, at 9-years follow-up

are the most surgically accessible of all brainstem gliomas. Like pilocytic astrocytomas elsewhere in the neural axis, the treatment of choice is microsurgical resection [2, 4, 28] (Fig. 5).

6.4 Cervicomedullary Tumors (Type IV)

Children with cervicomedullary tumors commonly present with slowly progressive lower cranial nerve palsy, pyramidal tract signs, ataxia, spinal cord dysfunction, or nystagmus. Lower cranial nerve deficits may include dysphagia, nasal speech,

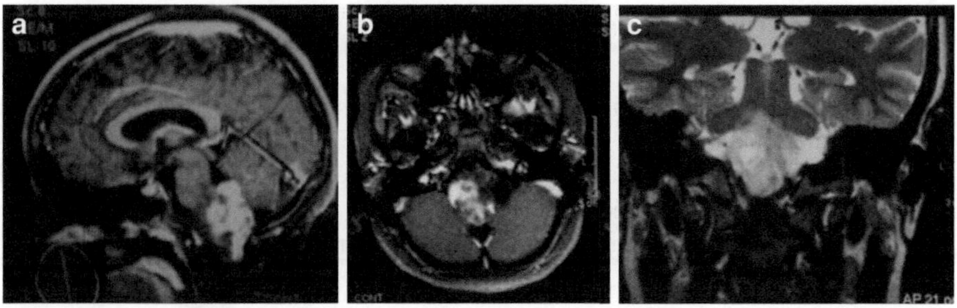

Fig. 5 Exophytic bulbar ganglioglioma (Type III) following partial resection and stable follow-up at 17 years. (**a**) sagittal contrast-enhanced T1-weighted; (**b**) axial enhanced T1-weighted; (**c**) coronal T2-weighted scan

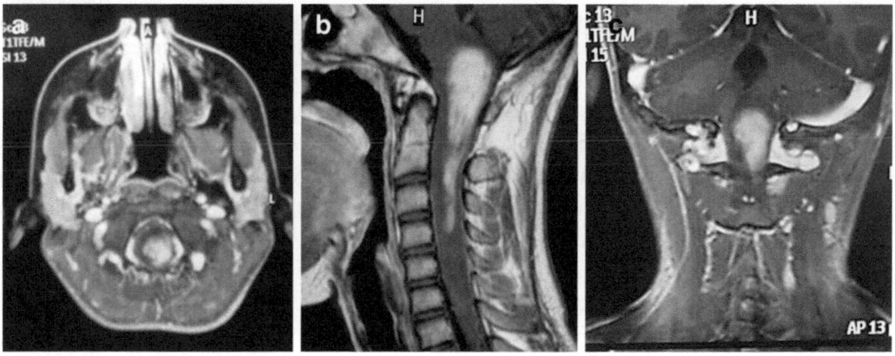

Fig. 6 Pilocytic astrocytoma in the cervicomedullary transition (type IV) in a patient with the diagnosis of type 1 neurofibromatosis. T1-weighted axial (**a**) sagittal; (**b**) and coronal; (**c**) post-gadolinium MRI scans

nausea, vomiting, palate deviation, facial nerve palsy, head tilt, apnea, or irregular breathing patterns. The gradual onset of these symptoms reflects the slowly growing, relatively benign histological picture that is mostly found, namely, the pilocytic astrocytoma. These tumors typically extend from the caudal two-thirds of the medulla to the rostral portion of the cervical spinal cord. The prognosis is generally favorable following neurosurgical resection, given the low-grade nature of most of these lesions [2, 4, 8, 11]. An aggressive surgical resection is usually undertaken, given that these tumors are more likely to have a defined surgical plane than other brainstem tumors. However, surgery in this location carries significant risks, including quadriparesis, sleep apnea, cranial nerve palsy, proprioceptive deficits, and spasticity. Intraoperative neuromonitoring and an experienced surgical-anesthesiology team are mandatory to improve outcomes. Postoperative radiation therapy may be indicated although most reference centers would opt for watchful waiting until there is no evidence of recurrence [11, 35] (Fig. 6).

7 Surgical Indications and Treatment

The decision-making regarding the best treatment strategy for brainstem tumors should be undertaken case by case, within a multidisciplinary team, and shared with the patient's carers. Several factors may influence the choice approach, such as the tumor location, risks involved, surgical routes, and adjuvant treatments available [36, 37]. Here, the main treatment strategies are summarized:

7.1 Diffuse Intrinsic Pontine Glioma (DIPG)

Because the natural history and malignant course of DIPGs are well established, there is currently little clinical role for a diagnostic biopsy [28]. A biopsy is reserved for indeterminate lesions on MRI, unusual presentations, or in a clinical study setting. Pincus et al. carrying out a retrospective study of 182 cases of stereotaxic biopsy, collected in 13 series in the literature for DIPG, found that the diagnosis of the tumor was verified in 75–100% of the cases. In 87% of the cases, the lesions were gliomas, whereas in the remaining 13%, primitive neuroectodermal tumors, neurocytomas, ependymomas, and demyelinating lesions. Morbidity ranged from zero to 16%, and mortality reached 5% [38].

On the other hand, many authors have shown that biopsies can be accomplished with reasonable safety, sufficient tissue for a histological diagnosis, and complication rates ranging from 10% to 30% [39]. Roujeau et al. published the results of stereotactic biopsy in 24 pediatric patients [40]. There was no surgical mortality in that series. The authors' arguments for performing a stereotaxic biopsy were the need for a sample to define treatment strategy and the need for material for molecular biology studies. Albright [22] argued that the biopsy should only be performed when there is a great doubt regarding the radiological aspect or when the lesion extends beyond the limits of the pons.

The rationale is that nearly all DIPGs are high-grade astrocytomas, outcomes are poor regardless of pathological grade, and biopsy may add additional morbidity. Nevertheless, in the era of targeted therapies, stereotactic/navigated biopsy may become useful, considering that new insights into these tumors' biological and genetic behavior could influence the treatment strategy and impact on outcome [41]. Recent molecular biology studies indicate promising therapeutic strategies, including epigenetic modifying agents such as histone deacetylase inhibitors, bromodomain, extra-terminal motif protein, and CDK7 inhibitors [41].

7.2 Focal Brainstem Tumors

Brainstem tumor surgery requires experience from the neurosurgical team, detailed knowledge of neuroanatomy and neurophysiology, and the availability of the appropriate surgical armamentarium and neurophysiological monitoring.

Surgery may be indicated for focal and exophytic tumors, with some exceptions, for instance, those located within the quadrigeminal plate. The overall survival for focal pylocytic astrocytomas ranges from around 80% in 5 years and 70% in 10 years. Therefore, surgical treatment may be of advantage. Total surgical resection should be aimed at maximum safety and preservation of function whenever possible. Thus, the neurosurgeon must be prepared to interrupt surgery and plan a second intervention when appropriate.

Intrinsic medullary tumors tend to have good survival rates and stable progression after surgery; however, the perioperative must be considered when counseling. Patients with cervicomedullary tumors can achieve favorable results with surgery, with 4- and 5-year survival rates of 72–100%. In cervicomedullary tumors, the degree of preoperative morbidity predicts postoperative function.

The selection of an appropriate surgical approach and route to the brainstem must be carefully evaluated preoperatively, as previously discussed (Fig. 7). The surgical position may vary according to the surgeon's preferences and experience. For instance, Lesions presenting extension towards the cerebellopontine angle but

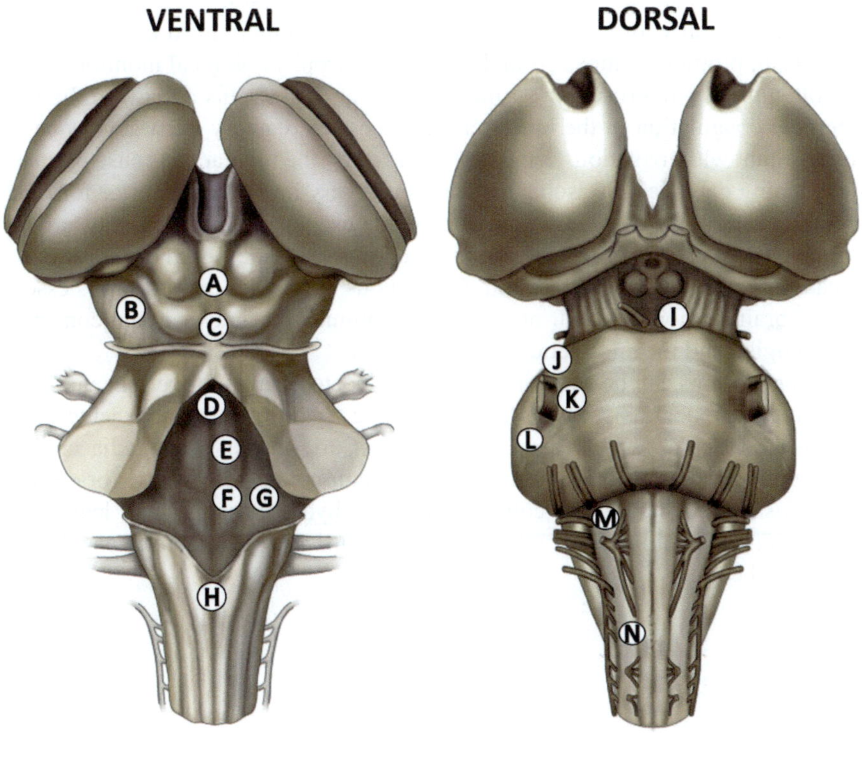

Fig. 7 Entry zones for surgical access to the brainstem. (**a**) supracollicular zone; (**b**) lateral mesencephalic sulcus; (**c**) infracollicular zone; (**d**) median sulcus of fourth ventricle; (**e**) supracollicular paramedian; (**f**) infracollicular paramedian; (**g**) acoustic area; (**h**) dorsal sulcus of the medulla; (**i**) anterior mesencephalic zone; (**j**) supratrigeminal zone; (**k**) peritrigeminal zone; (**l**) lateral pontine zone; (**m**) olivary zone; (**n**) anterolateral sulcus

arising within the brainstem can be approached through a classic suboccipital/telovelar approach through the fourth ventricle or directly via a retrosigmoid approach, as discussed above.

Tumors located in the midbrain require a different route. For instance, an inter-fornicial transcallosal approach or even an endoscopic/microscopic approach through the foramen of Monro may be chosen for exophytic tumors projecting themselves anteriorly inside the third ventricle [28]. Lesions located in the posterior portion of the midbrain or even in the quadrigeminal plate, reaching the third ventricle, can be removed through a supracerebellar infratentorial route. In some situations, the combined pre- and subtemporal supratentorial approaches can be employed to manage mesencephalic lesions emerging on the anterolateral surface of the midbrain [8, 10].

Regarding tumors within the rhomboid fossa, they can be approached above the facial colliculus or below the medullary striae [9]. However, when the tumor dislocates anatomical structures, it may be challenging to identify the midline accurately. In this situation, intraoperative monitoring guidance is quintessential. If any autonomic changes occur, such as tachycardia, high blood pressure, and bradycardia, surgery must be stopped until the improvement of physiological signs to normal ranges. Irrigation with warm saline solution is also important to keep the surgical field clean, promote hemostasis and facilitate neurophysiological monitoring.

Following the microsurgery principles, preserve the veins and avoid the use of electrical coagulation in the brainstem. If necessary, cautery must always be used concomitant with irrigation, as the heat released by the bipolar can cause permanent surrounding damage. Ultrasonic aspiration is a very important resource for removing brainstem tumors, avoiding traction and, consequently, autonomic disorders. It is important to use the ultrasonic aspirator at low power settings and gently so as not to cause bleeding during dissection. Continuous anesthetic monitoring is essential during brain tumor removal, and good communication between surgeon, neurophysiologist, and the anesthetic team is required.

Complications can be acute/intraoperative, postoperative, or permanent, leading to neurological deficits or major life-threatening conditions. Possible acute complications are vascular lesions (deep draining veins, arteries), edema of adjacent brainstem/cerebellar tissue, and intraoperative autonomic disorders (bradycardia, hypertension). Common late complications are hydrocephalus, CSF leak, cranial nerve injuries, hemiparesis or other long tract injuries, mutism, balance and coordination, and coma.

8 Conclusions

Despite improvements in the understanding of biology and advances in treatment strategies, brainstem tumors continue to present a therapeutic challenge to neurosurgeons and neuro-oncology teams. Surgical advancements in techniques, neurophysiological monitoring, and neuroimaging have improved long-term survival in cases of focal and exophytic brainstem tumors amenable to surgical treatment. Diffuse intrinsic pontine gliomas still confer a poor prognosis. The use of neuronavigation, neurophysiological monitoring, and technological advancements in terms of neurosurgical armamentarium are indispensable for maximizing the safety and extent of resection.

The general principles for surgical debulking of BSGs include (1) a deep understanding of the neuroanatomy; (2) identifying the most direct route to the tumor, either by exploring tumor cysts or locating pial surfaces with discoloration or tumor bulge; and (3) debulking the center of the lesion prior to dissecting the tumor margins (inside-out technique), under continuous neurophysiological monitoring. Finally, on the one hand, neuronavigation can be extremely helpful. On the other hand, it can also be seriously misleading due to brain shifts, demanding extra attention from the surgical team during the entire procedure.

Financial Support This research did not receive specific grants from funding agencies in public, commercial, or not-for-profit sectors.

Conflict of Interest Statement Authors have no conflict of interest related to this subject.

References

1. Morcos JJ, Haines SJ. History of brain stem surgery. Neurosurg Clin N Am. 1993;4(3):357–65.
2. Recinos PF, Sciubba DM, Jallo GI. Brainstem tumors: where are we today? Pediatr Neurosurg. 2007;43(3):192–201.
3. Pool JL. Gliomas in the region of the brain stem. J Neurosurg. 1968;29(2):164–7.
4. Grimm SA, Chamberlain MC. Brainstem glioma: a review. Curr Neurol Neurosci Rep. 2013;13(5):346.
5. Youmans JR, Winn HR. Youmans and Winn neurological surgery, 4-Volume Set, 7e. 7th edição. Philadelphia, PA: Elsevier; 2016.
6. Gross BA, Batjer HH, Awad IA, Bendok BR. Brainstem cavernous malformations. Neurosurgery. 2009;64(5):E805–18. discussion E818.
7. Eisele SC, Reardon DA. Adult brainstem gliomas. Cancer. 2016;122(18):2799–809.
8. Cavalheiro S, Yagmurlu K, da Costa MDS, Nicácio JM, Rodrigues TP, Chaddad-Neto F, et al. Surgical approaches for brainstem tumors in pediatric patients. Childs Nerv Syst. 2015;31(10):1815–40.
9. Párraga RG, Possatti LL, Alves RV, Ribas GC, Türe U, de Oliveira E. Microsurgical anatomy and internal architecture of the brainstem in 3D images: surgical considerations. J Neurosurg. 2016;124(5):1377–95.
10. Cavalcanti DD, Preul MC, Kalani MYS, Spetzler RF. Microsurgical anatomy of safe entry zones to the brainstem. J Neurosurg. 2016;124(5):1359–76.
11. Puget S, Boddaert N, Veillard AS, Garnett M, Miquel C, Andreiuolo F, et al. Neuropathological and neuroradiological spectrum of pediatric malignant gliomas: correlation with outcome. Neurosurgery. 2011;69(1):215–24.
12. Puget S, Crimmins DW, Garnett MR, Grill J, Oliveira R, Boddaert N, et al. Thalamic tumors in children: a reappraisal. J Neurosurg. 2007;106(5 Suppl):354–62.
13. Rajshekhar V, Moorthy RK. Status of stereotactic biopsy in children with brain stem masses: insights from a series of 106 patients. Stereotact Funct Neurosurg. 2010;88(6):360–6.

14. Schackert G, Ralle S, Martin KD, Reiss G, Kowalski M, Sobottka SB, et al. Vestibular schwannoma surgery: outcome and complications in lateral decubitus position versus semi-sitting position-a personal learning curve in a series of 544 cases over 3 decades. World Neurosurg. 2021;148:e182–91.
15. St-Arnaud D, Paquin MJ. Safe positioning for neurosurgical patients. AORN J. 2008;87(6):1156–68. quiz 1169–72
16. Zabramski JM, Kiriş T, Sankhla SK, Cabiol J, Spetzler RF. Orbitozygomatic craniotomy. Technical note. J Neurosurg. 1998;89(2):336–41.
17. Volpon Santos M, Furlanetti LL, Jeanne Bezerra MD, Santos de Oliveira R. Epidermoid cyst mimicking an intrinsic brainstem tumor. Neurocirugia (Astur). 2013;24(3):135–8.
18. Tedeschi H, Rhoton AL. Lateral approaches to the petroclival region. Surg Neurol. 1994;41(3):180–216.
19. Janjua MB, Caruso JP, Greenfield JP, Souweidane MM, Schwartz TH. The combined transpetrosal approach: anatomic study and literature review. J Clin Neurosci. 2017;41:36–40.
20. Cohen-Gadol A. Telovelar approach: a practical guide for its expansion to the fourth ventricle. World Neurosurg. 2021;148:239–50.
21. Rhoton AL. The posterior fossa cisterns. Neurosurgery. 2000;47(3 Suppl):S287–97.
22. Rhoton AL. The posterior fossa veins. Neurosurgery. 2000;47(3 Suppl):S69–92.
23. de Divitiis O, Conti A, Somma T, Angileri F, Cappabianca P. Ventral brainstem anatomy: an endoscopic transoral perspective. Acta Neurochir Suppl. 2019;125:45–50.
24. Furlanetti LL, Santos MV, de Oliveira RS. The success of endoscopic third ventriculostomy in children: analysis of prognostic factors. Pediatr Neurosurg. 2012;48(6):352–9.
25. Topczewski TE, Di Somma A, Culebras D, Reyes L, Torales J, Tercero A, et al. Endoscopic endonasal surgery to treat intrinsic brainstem lesions: correlation between anatomy and surgery. Rhinology. 2021;59(2):191–204.
26. Essayed WI, Singh H, Lapadula G, Almodovar-Mercado GJ, Anand VK, Schwartz TH. Endoscopic endonasal approach to the ventral brainstem: anatomical feasibility and surgical limitations. J Neurosurg. 2017;127(5):1139–46.
27. Vitanza NA, Monje M. Diffuse intrinsic pontine glioma: from diagnosis to next-generation clinical trials. Curr Treat Options Neurol. 2019;21(8):37.
28. Albright AL. Diffuse brainstem tumors: when is a biopsy necessary? Pediatr Neurosurg. 1996;24(5):252–5.
29. Maher CO, Raffel C. Neurosurgical treatment of brain tumors in children. Pediatr Clin N Am. 2004;51(2):327–57.
30. Epstein F, McCleary EL. Intrinsic brainstem tumors of childhood: surgical indications. J Neurosurg. 1986;64(1):11–5.
31. Prabhu SP, Ng S, Vajapeyam S, Kieran MW, Pollack IF, Geyer R, et al. DTI assessment of the brainstem white matter tracts in pediatric BSG before and after therapy: a report from the pediatric brain tumor consortium. Childs Nerv Syst. 2011;27(1):11–8.
32. Wesseling P, Capper D. WHO 2016 classification of gliomas. Neuropathol Appl Neurobiol. 2018;44(2):139–50.
33. Castel D, Philippe C, Calmon R, Le Dret L, Truffaux N, Boddaert N, et al. Histone H3F3A and HIST1H3B K27M mutations define two subgroups of diffuse intrinsic pontine gliomas with different prognosis and phenotypes. Acta Neuropathol. 2015;130(6):815–27.
34. Furlanetti LL, Santos MV, de Oliveira RS. Neuroendoscopic surgery in children: an analysis of 200 consecutive procedures. Arq Neuropsiquiatr. 2013;71(3):165–70.
35. Hoffman LM, Veldhuijzen van Zanten SEM, Colditz N, Baugh J, Chaney B, Hoffmann M, et al. Clinical, radiologic, pathologic, and molecular characteristics of long-term survivors of diffuse intrinsic pontine glioma (DIPG): a collaborative report from the international and European Society for Pediatric Oncology DIPG registries. J Clin Oncol. 2018;36(19):1963–72.
36. Faulkner H, Arnaout O, Hoshide R, Young IM, Yeung JT, Sughrue ME, et al. The surgical resection of brainstem glioma: outcomes and prognostic factors. World Neurosurg. 2021;146:e639–50.

37. Green AL, Kieran MW. Pediatric brainstem gliomas: new understanding leads to potential new treatments for two very different tumors. Curr Oncol Rep. 2015;17(3):436.
38. Pincus DW, Richter EO, Yachnis AT, Bennett J, Bhatti MT, Smith A. Brainstem stereotactic biopsy sampling in children. J Neurosurg. 2006;104(2 Suppl):108–14.
39. Morais BA, Solla DJF, Matushita H, Teixeira MJ, Monaco BA. Pediatric intrinsic brainstem lesions: clinical, imaging, histological characterization, and predictors of survival. Childs Nerv Syst. 2020;36(5):933–9.
40. Roujeau T, Machado G, Garnett MR, Miquel C, Puget S, Geoerger B, et al. Stereotactic biopsy of diffuse pontine lesions in children. J Neurosurg. 2007;107(1 Suppl):1–4.
41. Aziz-Bose R, Monje M. Diffuse intrinsic pontine glioma: molecular landscape and emerging therapeutic targets. Curr Opin Oncol. 2019;31(6):522–30.

Index

E. G. Figueiredo et al. (eds.), *Brain Anatomy and Neurosurgical Approaches*,
https://doi.org/10.1007/978-3-031-14820-0

D

R

T

GPSR Compliance

The European Union's (EU) General Product Safety Regulation (GPSR) is a set of rules that requires consumer products to be safe and our obligations to ensure this.

If you have any concerns about our products, you can contact us on ProductSafety@springernature.com

In case Publisher is established outside the EU, the EU authorized representative is:

Springer Nature Customer Service Center GmbH
Europaplatz 3
69115 Heidelberg, Germany

Batch number: 10726131

Printed by Printforce, the Netherlands